Security and Privacy in Blockchains and the IoT II

Security and Privacy in Blockchains and the IoT II

Editors

Christoph Stach
Clémentine Gritti

MDPI

Basel • Beijing • Wuhan • Barcelona • Belgrade • Novi Sad • Cluj • Manchester

Editors
Christoph Stach
Department of Applications of Parallel and Distributed Systems/Department of Data Engineering, University of Stuttgart
Stuttgart, Germany

Clémentine Gritti
Department of Computer Science and Software Engineering, University of Canterbury
Christchurch, New Zealand

Editorial Office
MDPI
St. Alban-Anlage 66
4052 Basel, Switzerland

This is a reprint of articles from the Special Issue published online in the open access journal *Future Internet* (ISSN 1999-5903) (available at: https://www.mdpi.com/journal/futureinternet/special_issues/SP_BI_II).

For citation purposes, cite each article independently as indicated on the article page online and as indicated below:

Lastname, A.A.; Lastname, B.B. Article Title. *Journal Name* **Year**, *Volume Number*, Page Range.

ISBN 978-3-0365-8772-1 (Hbk)
ISBN 978-3-0365-8773-8 (PDF)
doi.org/10.3390/books978-3-0365-8773-8

Cover image courtesy of Pexels

Contents

About the Editors

Christoph Stach

Dr. rer. nat. Christoph Stach is a postdoctoral researcher at the Applications of Parallel and Distributed Systems Department of the University of Stuttgart. He completed his studies in computer science at the University of Stuttgart in 2009. In 2017, he received his Ph.D. in computer science from the University of Stuttgart for his research in information security and data privacy in mobile applications. Following his successful doctorate, he was appointed Academic Councilor at the Institute for Parallel and Distributed Systems of the University of Stuttgart. From June 2020 to September 2021, he held the deputy professorship in Data Engineering at the University of Stuttgart. Today, he is head of the working area of Information Systems and Applications at the Applications of Parallel and Distributed Systems department of the University of Stuttgart. His current research focuses on the concepts and tools required to enable trustworthy and demand-oriented data provisioning for users, such as data scientists and analysts. To this end, his research addresses research questions regarding data acquisition, management, security, and protection. In this regard, he has managed and participated in various research projects covering a wide range of application areas, including Smart Health, Industry 4.0, and Connected Cars. He has published more than 60 peer-reviewed papers about his research and presented the results at international conferences. For his work, he has received four awards. He also shares his knowledge and experience by giving lectures, such as Introduction to Data Science and Applied Data Science using Python, as well as holding seminars on these topics.

Clémentine Gritti

Clémentine Gritti, Ph.D., received an M.Sc. degree in computer science from Grenoble Alpes University, France, in 2012, and a Ph.D. in computer science from the University of Wollongong, Australia, in 2017. She has been a Lecturer at the Computer Science and Software Engineering Department at the University of Canterbury since 2020. At the end of 2023, she will join the digital security team at Eurecom, France, as a research fellow. She previously worked on several research projects dealing with information security and privacy for electronic health and electronic voting. Her current research interests include designing and evaluating public-key cryptographic protocols for security and privacy in various environments, such as cloud computing, the Internet of Things, and blockchain.

Preface

In the rapidly evolving digitalized landscape of the 21st century, where data have emerged as the quintessential driving force behind modern technologies, it is imperative to delve into the intricate interplay between security, privacy, and innovation. The advent of the Internet of Things (IoT) and blockchain technologies has ushered in a new era of possibilities, introducing an unprecedented range of capabilities and conveniences across various domains, including smart homes, smart healthcare, smart manufacturing, and more. This book addresses the topic of "Security and Privacy in Blockchains and the Internet of Things" by meticulously exploring the synergistic relationship between these two transformative forces and their impact on our digital society.

The motivation behind this scientific endeavor stems from the profound impact these technologies have on our lives. As data-driven applications and smart services become integral to our daily routines, it is increasingly important to overcome fundamental security and privacy challenges in data handling. This book provides answers to the pertinent questions that arise in this context and aims to shed light on the complex interrelation between technological advances, the proliferation of data, the protection of user interests, and compliance with privacy laws such as the EU General Data Protection Regulation (GRPR).

The scope of this book is wide-ranging, covering a variety of topics inherently related to security and privacy in blockchains and the IoT. It addresses the multi-layered aspects of the tasks, from data collection and transmission to data management, and explores how blockchain technologies and the IoT converge. By covering novel use cases of smart services and cutting-edge technologies, as well as research approaches addressing the complexities of protection goals, such as confidentiality, data integrity, and availability, this book provides a comprehensive overview of both challenges and solutions to ensuring the security and usability of modern data ecosystems.

This book is aimed at a broad audience, including practitioners and researchers at the forefront of technological innovation. Developers looking to harness the potential of blockchain technologies and IoT systems find valuable insights into secure and privacy-compliant methodologies for implementation. Researchers exploring the frontiers of data security can find a wealth of recent research articles and literature reviews in this book that highlight key trends, challenges, and future research directions. In addition, this book is also intended for practitioners and educators who want to deepen their understanding of the complicated relationship between such data-driven technologies and users' security and privacy needs.

This Special Issue of Future Internet assembles a wide range of contributors, each bringing their unique expertise and dedication to realizing this seminal work. A total number of seventeen papers (fourteen research articles and three literature reviews) contribute to providing a holistic view of this rapidly evolving landscape. These contributions stem from sixty-four authors from thirty-three distinct research institutions in sixteen countries across all six continents.

Our heartfelt gratitude goes to all who have played an instrumental role in making this second volume possible. First and foremost, we want to thank the esteemed authors who are the bedrock upon which this Special Issue stands. It is only through their exemplary research and innovative thinking that such a compilation of highly informative papers becomes possible in the first place. Pursuing high scientific standards would not have been possible without our diligent reviewers' careful review and critical evaluation. We express our appreciation to these experts for dedicating their time and expertise to ensure the quality and scientific integrity of the manuscripts published in this Special Issue. Behind the scenes, a proficient editing team has redacted and orchestrated the manuscripts into a coherent ensemble. We cordially thank the editing team for their commitment the fact that smart things are ubiquitous in our everyday lives. As a result, more and more sensitive data are being collected and processed, providing comprehensive insights into the personal lives of the data subjects.

In this Special Issue, fourteen research articles and three literature reviews therefore address research questions in the thematic area of security and privacy in blockchains and the IoT, while Volume I [1] addresses general security and privacy aspects in the context of blockchain technologies and the IoT, the focus of Volume II is rather on a holistic understanding that, for instance, takes into account mutual effects between individual protection goals and end-to-end security measures from acquisition to provisioning. The seventeen papers in Volume II are outlined in the following.

Articles. In the opening research article, Stach [2] highlights the similarities and differences between physical goods, such as oil, and virtual goods, such as data, in terms of handling and processing. Here, it becomes apparent that raw data, just like crude oil, needs to be refined in order to be useful. Therefore, an end-to-end data platform called REFINERY Platform is designed to enable effective and efficient data handling. The work focuses on the special characteristics of data and their implications for modern data management, covering challenges in data acquisition, refinement, storage, and delivery. In data processing, specific attention is paid to protection goals such as confidentiality, integrity, availability, and authenticity, as well as compliance with data protection laws such as the European General Data Protection Regulation (GDPR). In this way, the REFINERY Platform represents an information retrieval and provision platform that enables data to be handled in a trustworthy manner. Any data-heavy application benefits from such a platform.

One example of such a data-heavy application is presented by Fragkou et al. [3] in the form of the Internet of Battlefield Things (IoBT). This relatively novel application domain is about a concept in which various devices on the battlefield—both smart military equipment that makes autonomous decisions and conventional human-controlled military devices—are interconnected. The work focuses on a fundamental problem in IoBT networks, namely the routing of information and the calculation of a backbone network. To overcome these issues, the authors model the IoBT network as a multilayer network and adopt the concept of dominance to select a set of nodes as a backbone for the IoBT network. However, the domination concept of single-layer networks cannot be transferred straightforwardly to the more complex IoBT networks. To this end, the authors introduce the Cross-layer Connected Dominating Set (CCDS) algorithm, which is specifically designed for IoBT networks. Compared to state-of-the-art approaches, CCDS achieves more compact network spanners, while still ensuring network-wide flow of information.

It is evident that confidential communication between devices is mandatory in a setting such as the IoBT. Virtual Private Network (VPN) technology allows us to establish private and secure communication channels over the Internet. Gentile et al. [4] therefore analyze in their paper the performance of different open-source firmware and operating systems for deploying VPNs in outdoor locations with poor network connectivity. For this purpose, they compare the performance of OpenWrt 21.x, Debian 11 x64, and Mikrotik 7.x as server-side operating systems, and various client-side operating systems, including Windows 10/11, iOS 15, and Android 11. As VPNs establish secure connections between client devices and remote endpoints by encrypting traffic, the applied VPN protocol also has an effect on the achievable level of security and performance parameters such as latency and throughput. The main goals of the research are therefore to identify algorithms that ensure efficient data transmission and encryption, achieve compatibility with existing VPN infrastructure, and enable the use of open firmware on constrained routers.

In addition to secure communications, it is also essential to monitor the IoT devices in terms of the software running on them. In this regard, the early detection of cyberattacks plays a key role. Therefore, Lightbody et al. [5] study whether malicious behavior of IoT devices can be detected based on their power consumption behavior. To this end, they identify unique power usage patterns for typical operations performed on an IoT device. A Raspberry Pi 3 model B and a DragonBoard 410c, two of the most common

IoT systems today, are used for this purpose. If the energy consumption of an operation deviates noticeably from the expected pattern, this is an indication of an attack. In particular, the authors consider reconnaissance, brute force, and denial of service attacks. The findings in the paper demonstrate that this non-intrusive side-channel method is a useful addition to existing protection measures. Despite the focus on low-power IoT devices, the findings can be applied to any type of computing system, up to high-performance computers.

In the IoT context, not only the data transmission and the IoT devices pose potential threats but also the gathered data themselves. As more and more personal data are involved, the processing of such data inevitably raises privacy concerns. Therefore, Stach et al. [6] investigate which privacy-enhancing technologies (PETs) can be used to systematically conceal certain information patterns contained in the data without compromising the overall data utility. However, in a highly heterogeneous landscape like the IoT, there is no compelling one-size-fits-all solution to this end. Accordingly, the authors identify the required data processing tasks and the resulting privacy or confidentiality concerns for seven representative IoT use cases and propose suitable PETs for all types of data involved. A SWOT-like analysis (strengths, weaknesses, opportunities, and threats) reveals open challenges, e.g., how to configure and deploy these PETs or how to prevent the misuse of such PETs. The paper draws conclusions on how to overcome these challenges.

Besides IoT, i.e., mainly data acquisition and interconnection topics, this Special Issue also deals with blockchain technologies. These were originally intended for the management of digital currencies, i.e., rather homogeneous data and simple data structures. However, the inherent distributed, tamper-resistant, and immutable nature as a data store makes blockchain a key technology whenever data need to be shared securely within trustless environments. Five papers therefore discuss other emerging use cases for blockchain technologies. For instance, Ahmed et al. [7] introduce the concept of Self-Sovereign Banking Identity (SSBI), a blockchain-based self-sovereign identity system, to address challenges in open finance and open banking. In open finance, financial service providers are not only able to retrieve transactional details from bank customers but also have write access to such information. In such a scenario, privacy leakage and misuse obviously need to be prevented. The SSBI prototype therefore relies on industry-standard bank cards that allow users to grant fine-grained permission for data sharing with financial service providers to ensure that no detailed personal identity or financial information is disclosed. The prototype implementation, based on Veramo SDK and Ethereum, demonstrates that SSBI can overcome the limitation of signing curves on current state-of-the-art Java Card approaches.

Santos et al. [8] look at another application case for blockchain technologies. In their studies, they have identified the limitations of loyalty platforms, i.e., programs designed to increase customer loyalty and encourage repeat purchases on behalf of specific brands by offering some type of reward for each purchase. Their main problems arise from rewards, which are seen as insufficient incentives by many customers and still cause high costs for brands. Furthermore, there is no option for customers to manage their earned loyalty points across different platforms. The authors therefore propose a blockchain-based approach to manage customer data from multiple loyalty programs on a single meta platform. The accurate and secure handling of transactions and loyalty points is ensured by means of smart contracts. The application of these blockchain technologies not only enables a more seamless, efficient, and user-friendly experience for customers but also creates advantages for brands in terms of reduced costs and increased customer loyalty.

Blockchain technologies can also be applied in the agricultural business. In their paper, Rocha et al. [9] analyze the benefits, disadvantages, challenges, and opportunities associated with the use of blockchain-based approaches in this sector. For this purpose, they conducted interviews with ten agribusiness professionals working with blockchain technologies to gain insights into their practical experiences. To this end, they applied a two-phase survey approach, in which the same participants were consulted again two years later, in order to reflect shifts in their assessment. In general, blockchain technologies play an increasingly important role in this application area, especially with regard to governance and information flow within supply chains. However, the findings of the study also clearly reveal that the transition to blockchain-based solutions also entails drawbacks, e.g., high implementation costs. These drawbacks are also critically assessed by the authors. Overall, they come to the conclusion that the benefits and opportunities associated with blockchain application in agribusiness outweigh the challenges and disadvantages.

Another application for blockchain technologies is the management of certificates in public key infrastructures (PKIs). In this regard, Honecker et al. [10] discuss how DLTs can address the challenges of modern PKIs, namely their heavy focus on a central certificate authority (CA) that manages the root certificates. Since security in PKIs is based on certificate chains, tampering with or blocking access to the root certificate compromises the entire chain and thus renders the PKI useless or inoperable. To this end, the authors therefore propose using a DLT-based data storage. Due to the decentralized nature of DLTs, attackers would only be able to compromise the CA if they had access to all nodes. Furthermore, DLTs establish trust in the authenticity of the certificates provided. The authors additionally adapt the Near-Field Communication Key Exchange scheme, which enables lightweight handling of keys. Evaluation results show that by using blockchain technologies that support smart contracts, the security of the certificates can be increased, but at the expense of performance, e.g., in terms of execution times.

The entertainment industry has also realized the potential of blockchain technologies. Digital assets are increasingly distributed as non-fungible tokens (NFTs), which can be verifiably linked to a specific user, e.g., unique player items in a video game, while this can be solved relatively easily from a technical point of view via blockchain entries certifying ownership and authenticity, player perception of such techniques is still largely unresearched. Paajala et al. [11] therefore study the opinion of gamers towards the inclusion of such blockchain-based features in games. Based on a game called IkuneRacer, they investigate how players perceive transparency, trust, and user-generated content in blockchain-based games. Their overall research goal is to evaluate whether the implementation of blockchain technologies in games influences player engagement and retention. In this context, it is observed that a combination of asset ownership and user-generated content has a positive effect on the willingness to devote time to a game.

With such applications for blockchain technologies, the volume of stored data is increasing, and the underlying data structures are becoming more and more complex. Accordingly, the demand to carry out comprehensive data analytics directly in the blockchain is on the rise as well. In DLTs, however, data access is usually realized via low-level block data instead of high-level data assets, which makes such analytics significantly more difficult. Current blockchain analytics tools address this challenge by using an internal middleware that maps high-level user queries to low-level data interfaces. To improve the analytics capabilities of such tools, Vinceslas et al. [12] introduce an abstraction layer that provides blockchain data in aggregated and pre-processed form to block explorers and other analytics dashboards. The work aims to improve the auditability and intuitiveness of DLTs by providing users with lightweight data interfaces such as dashboards. The benefits of the proposed abstraction layer architecture are illustrated by means of an industrial case study.

Similar to IoT-based applications, the extensive accumulation of data from a wide range of domains raises confidentiality and privacy concerns in the context of blockchain as well. For this purpose, Xu et al. [13] consider the data collected using 3D sensing technology. Such sensors create point clouds, i.e., annotated geometric information which is arranged in a 3D space. This technology is used, for instance, in autonomous driving. In such a use case, the data are always related to a driver in one way or another, so special attention must be paid to data security, e.g., in terms of access, processing, and sharing. To this end, the authors present SAUSA, an approach that combines blockchain technologies with concepts of software-defined networking (SDN). SAUSA monitors and controls any kind of interaction with the point cloud data. Using a hybrid on-chain/off-chain storage strategy, less sensitive metadata can be stored in a distributed data store, whereas sensitive payload data can reside on a private storage. This way, SAUSA enables efficient and resilient point cloud applications while still giving users full control over their data.

Such an approach is suitable whenever it is possible to share no or very little data with certain parties. In the context of elections, however, this is not possible as the election results have to be verifiable, i.e., it must be ensured that no votes are cast illegitimately and that every legitimate vote is counted. Notwithstanding, privacy has a high priority with regard to elections as well since it must be ensured that a vote can be cast confidentially and that it cannot be traced back to the voter. Sallal et al. [14] address this paradox by leveraging blockchain technologies. In their proposed PVPBC e-voting system, a permissioned distributed ledger and smart contracts are applied, achieving effective authorization while preserving revocable anonymity properties. The adopted Selene voting scheme ensures that voters can easily verify that their votes are accurately captured by the system, without having to deal with the underlying cryptographic complexity. This way, not only the most crucial security requirements for e-voting systems are addressed—e.g., fairness, verifiability, and privacy—but experimental results also demonstrate that PVPBC is highly scalable.

In the final research article, the two thematic blocks of this Special Issue—namely, IoT and blockchain technology—are brought together. In this paper, Xevgenis et al. [15] discuss next-generation networks, which aim to provide features such as ultra-low latency, high availability, and wide service coverage to users. This is achieved by the integration of SDN and network function virtualization. To improve elasticity and dynamics and reduce the necessity for human interaction, the Zero-touch Service Management (ZSM) framework can achieve complete end-to-end automation of network service management, even across different domains. However, this raises security concerns. The authors argue that the use of blockchain technology in a trustless environment such as the ZSM can create inherent security in terms of data integrity and transaction validity. For this reason, they introduce an architecture for multi-domain network infrastructures that are managed via ZSM, yet the management functions are implemented as smart contracts. The authors discuss different blockchain systems that are suitable for this purpose and provide implementation guidelines for their proposed architecture.

Reviews. This Special Issue is complemented by three interesting literature reviews. The first review is focused on how a network can be secured using blockchain technology. Ismail et al. [16] specifically consider wireless sensor networks (WSNs), which are the enabler technology for the IoT. Since the devices deployed in these networks are usually rather rudimentary, they often lack sufficient security measures, which is why they are prone to cyberattacks. In their work, the authors therefore address the question of which protection goals are particularly relevant in WSNs and which unique characteristics of WSNs impede security measures. Furthermore, they investigate which cyberattacks are particularly prevalent and effective in the WSN context and how they affect the WSN. As existing security measures are insufficient against these attacks, the paper proposes the inclusion of blockchain technologies and machine learning techniques in WSNs. It also identifies challenges in this regard and discusses different implementation strategies to ensure reliable cyberattack detection and cyberattack prevention.

The second review covers a specialized application domain involving WSN technologies. Sayeed et al. [17] examine the Internet of Robotic Things (IoRT), which supplements the IoT with concepts of cloud computing and artificial intelligence to enable cyber-physical systems to make autonomous decisions and carry out appropriate actions. The IoRT can be leveraged in various fields, including manufacturing, healthcare, security, and transportation. In their review, the authors outline the techniques used in IoRT, the relevant architectures in the IoRT landscape, and the capabilities of cyber-physical systems in this context. The gained insights provide the foundation for taxonomies that reflect the different classes into which IoRT environments can be categorized. The review also addresses security aspects that are characteristic for the IoRT. It is noticeable that there are still major limitations related to data security, for which the authors also suggest the use of blockchain technologies. The paper concludes with open research questions, including ethical issues, trust issues, and data quality issues.

The final review brings the blockchain topic to the forefront. Blockchain technology offers a high level of data protection, as all data are stored in an immutable manner and tampering with the data is impossible. However, this very characteristic is also an inherent problem of this technology when it comes to data privacy. Data protection laws such as the GDPR grant data subjects a right to rectification (Art. 16 GDPR) as well as a right to erasure (Art. 17 GDPR). That is, whenever personal data are stored in a blockchain-based data repository, there have to be redacting capabilities. Abd Ali et al. [18] therefore investigate which concepts towards a mutable blockchain, which allows controlled and supervised amendments to specific content, are discussed in research. In this respect, they also identify the particular implementation challenges that a redactable blockchain has to overcome and address security criteria that must still apply despite the redacting capabilities. Based on their findings, the authors outline future research directions and open issues in the field of effective redaction mechanisms for blockchain technologies.

The seventeen excellent papers included in this Special Issue provide a well-rounded overview of the current state of the art and current state of research in the field of security and privacy issues in blockchain technologies and the IoT. In this respect, it illustrates how versatile and complex this topic area is. On the one hand, the literature reviews highlight the potential of these technologies and how they can significantly facilitate our everyday lives in all kinds of domains, but also which open research questions still need to be solved in the future when it comes to security and privacy. On the other hand, the research articles present highly innovative approaches to overcome such problems without having to sacrifice the conveniences that blockchain technologies and the IoT have to offer. This Special Issue is therefore aimed at developers and researchers seeking to implement effective and efficient blockchain-based and/or IoT-based solutions as well as users of such technologies who want to learn more about their capabilities but also their limitations and about future trends in this field of research.

As guest editors, we would like to take this opportunity to thank all authors for submitting their interesting and informative manuscripts to Volume II of this Special Issue. We would further like to acknowledge all the reviewers, whose thorough and substantive reviews have helped to improve the quality of the manuscripts and without whom this Special Issue would not have been possible. Last but not least, we would like to express our special thanks to the MDPI editorial team, who have strongly supported us in the work on this Special Issue.

Author Contributions: All authors have read and agreed to the published version of the manuscript.

Conflicts of Interest: The authors declare no conflict of interest.

References

1. Stach, C. (Ed.) *Security and Privacy in Blockchains and the IoT*; MDPI: Basel, Switzerland, 2023. [CrossRef]
2. Stach, C. Data Is the New Oil—Sort of: A View on Why This Comparison Is Misleading and Its Implications for Modern Data Administration. *Future Internet* **2023**, *15*, 71. [CrossRef]
3. Fragkou, E.; Papakostas, D.; Kasidakis, T.; Katsaros, D. Multilayer Backbones for Internet of Battlefield Things. *Future Internet* **2022**, *14*, 186. [CrossRef]
4. Gentile, A.F.; Macrì, D.; De Rango, F.; Tropea, M.; Greco, E. A VPN Performances Analysis of Constrained Hardware Open Source Infrastructure Deploy in IoT Environment. *Future Internet* **2022**, *14*, 264. [CrossRef]
5. Lightbody, D.; Ngo, D.M.; Temko, A.; Murphy, C.C.; Popovici, E. Attacks on IoT: Side-Channel Power Acquisition Framework for Intrusion Detection. *Future Internet* **2023**, *15*, 187. [CrossRef]
6. Stach, C.; Gritti, C.; Bräcker, J.; Behringer, M.; Mitschang, B. Protecting Sensitive Data in the Information Age: State of the Art and Future Prospects. *Future Internet* **2022**, *14*, 302. [CrossRef]
7. Ahmed, K.A.M.; Saraya, S.F.; Wanis, J.F.; Ali-Eldin, A.M.T. A Blockchain Self-Sovereign Identity for Open Banking Secured by the Customer's Banking Cards. *Future Internet* **2023**, *15*, 208. [CrossRef]
8. Santos, A.F.; Marinho, J.; Bernardino, J. Blockchain-Based Loyalty Management System. *Future Internet* **2023**, *15*, 161. [CrossRef]

9. Rocha, G.d.S.R.; Mühl, D.D.; Chingamba, H.A.; de Oliveira, L.; Talamini, E. Blockchain, Quo Vadis? Recent Changes in Perspectives on the Application of Technology in Agribusiness. *Future Internet* **2023**, *15*, 38. [CrossRef]
10. Honecker, F.; Dreyer, J.; Tönjes, R. Comparison of Distributed Tamper-Proof Storage Methods for Public Key Infrastructures. *Future Internet* **2022**, *14*, 336. [CrossRef]
11. Paajala, I.; Nyyssölä, J.; Mattila, J.; Karppinen, P. Users' Perceptions of Key Blockchain Features in Games. *Future Internet* **2022**, *14*, 321. [CrossRef]
12. Vinceslas, L.; Dogan, S.; Sundareshwar, S.; Kondoz, A.M. Abstracting Data in Distributed Ledger Systems for Higher Level Analytics and Visualizations. *Future Internet* **2023**, *15*, 33. [CrossRef]
13. Xu, R.; Chen, Y.; Chen, G.; Blasch, E. SAUSA: Securing Access, Usage, and Storage of 3D Point CloudData by a Blockchain-Based Authentication Network. *Future Internet* **2022**, *14*, 354. [CrossRef]
14. Sallal, M.; de Fréin, R.; Malik, A. PVPBC: Privacy and Verifiability Preserving E-Voting Based on Permissioned Blockchain. *Future Internet* **2023**, *15*, 121. [CrossRef]
15. Xevgenis, M.; Kogias, D.G.; Karkazis, P.A.; Leligou, H.C. Addressing ZSM Security Issues with Blockchain Technology. *Future Internet* **2023**, *15*, 129. [CrossRef]
16. Ismail, S.; Dawoud, D.W.; Reza, H. Securing Wireless Sensor Networks Using Machine Learning and Blockchain: A Review. *Future Internet* **2023**, *15*, 200. [CrossRef]
17. Sayeed, A.; Verma, C.; Kumar, N.; Koul, N.; Illés, Z. Approaches and Challenges in Internet of Robotic Things. *Future Internet* **2022**, *14*, 265. [CrossRef]
18. Abd Ali, S.M.; Yusoff, M.N.; Hasan, H.F. Redactable Blockchain: Comprehensive Review, Mechanisms, Challenges, Open Issues and Future Research Directions. *Future Internet* **2023**, *15*, 35. [CrossRef]

Short Biography of Authors

Dr. rer. nat. Christoph Stach is a postdoctoral researcher at the Applications of Parallel and Distributed Systems department of the University of Stuttgart. He completed their studies in computer science at the University of Stuttgart in 2009. In 2017, he received their Ph.D. in computer science from the University of Stuttgart for their research in the area of information security and data privacy in mobile applications. Following their successful doctorate, he was appointed Academic Councilor at the Institute for Parallel and Distributed Systems of the University of Stuttgart. From June 2020 to September 2021, he held the deputy professorship in Data Engineering at the University of Stuttgart. Today, he is head of the working area of Information Systems and Applications at the Applications of Parallel and Distributed Systems department of the University of Stuttgart. His current research focuses on concepts and tools required to enable trustworthy and demand-oriented data provisioning for users, such as data scientists and data analysts. To this end, their research addresses research questions regarding data acquisition, data management, data security, and data protection.

Clémentine Gritti received the M.Sc. degree in computer science from Grenoble Alpes University, France, in 2012, and the Ph.D. degree in computer science from the University of Wollongong, Australia, in 2017. She has been a Lecturer at the Computer Science and Software Engineering Department, University of Canterbury since 2020. At the end of 2023, she will join the digital security team at Eurecom, France, as a research fellow. She previously worked on several research projects dealing with information security and privacy for electronic health and electronic voting. Her current research interests include design and evaluation of public-key cryptographic protocols for security and privacy in various environments, such as cloud computing, the Internet of Things, and blockchain.

While the allegory of data and oil primarily invokes associations with industrial application domains—often coined as Industry 4.0 [8]—IoT-based data capturing also enables a general improvement of everyone's daily life. Such data-driven services are often referred to as Smart Living [9]. This umbrella summarizes, e.g., services in the area of Smart Mobility [10], Smart Health [11], and Smart Homes [12]. This also shares some similarities with oil, as the extensive use of oil initially transformed the industry but gradually found its way into the private sector in the form of new types of products and finally changed society as a whole [13]. Data also possess this potential—it is not a coincidence that IoT technologies are seen as a booster for the digital age [14].

Besides their strategic and economic importance, data and oil share another common feature: like oil, data initially have to be refined before they can be used profitably [15] Peter Sondergaard, Senior Vice President of Gartner Research, addresses this fact with his statement that "information is the oil of the 21st century, and analytics is the combustion engine". So, in order to extract interpretable information from raw data, they must first be structured and put into context [16]. This requires processes to systematically transform the data [17] as well as tools and techniques to support such information management [18]. Only when the data have been properly refined can they reach their full potential [19].

While it therefore seems that the metaphorical comparison is sound and that data are just another commodity to be managed and processed like any other asset, a closer look reveals significant differences between the digital commodity 'data' and physical commodities such as oil [20]. These differences are so fundamental that it is necessary to rethink the way data are handled in order to be effective and efficient in the process [21].

In this paper, we therefore study and discuss which characteristics are unique to the intangible commodity 'data' and the resulting implications for modern data administration. To this end, we make the following three contributions:

(a) We discuss the key differences between data and oil. For that purpose, we identify ten specific characteristics of data that need to be taken into account in data administration. In the context of this work, we focus on inherent challenges that arise due to technical characteristics of big data, often referred to as Big Vs [22]. Ethical social aspects, e.g., data liberation and fair distribution [23], or green processing, e.g., energy-efficient data acquisition and storage [24], are out of scope since such factors must also be considered for any other commodity.

(b) For each identified special characteristic, we derive the resulting implications for data administration. In the context of this work, we take an end-to-end look at the data management process, i.e., we deal with data acquisition, data refinement, data storage, and data provision.

(c) We present a concept for a novel reliable information retrieval and delivery platform, called REFINERY Platform, which addresses the data administration challenges, we have identified. In this context, 'reliable' refers to both the data producers—in those terms, it is ensured that sensitive data are handled in a trustworthy manner—and the data consumers—in those terms, it is ensured that data have the promised quality.

The remainder of this paper is structured as follows: In Section 2, we elaborate on the characteristics of data that inherently distinguish them from tangible commodities and address specific and novel challenges that arise when handling such intangible assets. We provide an overview of the state of the art in modern data administration in Section 3 and discuss how it responds to these challenges. Based on these findings, we introduce our concept of the REFINERY Platform in Section 4. This holistic end-to-end concept not only takes into account the unique characteristics of data but also addresses the weaknesses of current approaches. We then assess our approach in Section 5. To this end, we present a security and privacy assessment, a feature discussion, a case study, and a performance evaluation. Finally, the lessons learned are summarized in Section 6.

2. Characteristics of Data and Resulting Consequences for Data Administration

As outlined in the introduction, besides the metaphorical level, there are also many similarities between oil and data when thinking about the handling of these two commodities. For instance, both have first to be discovered and then extracted. The extracted product, i.e., the crude oil or the raw data, must then be refined to transform it into a usable resource (i.e., value-added products such as kerosene or information). For this purpose, the commodity has to be cleansed and preprocessed. The resources refined in this way must then be stored and delivered to customers [25].

While the handling of both commodities involves the same steps, there are considerable differences in the actual implementation. These are due to special characteristics of data that distinguish them significantly from oil. From our point of view, ten characteristics must be taken into account to this end. In the following, we present these characteristics and discuss their consequences for data administration.

I. *Data are nonconsumable:* When tangible commodities are transformed into value-added products, they are consumed in the process. This is completely different for the intangible commodity 'data'. Even after data have been fully analyzed and a value-added data product has been generated (e.g., information or knowledge), the raw data are still available. They do not lose their meaning in the process, since they can be processed again, in another way, in order to derive new information or knowledge from it. Therefore, the volume of the data to be administered increases constantly, since processed data are neither consumed nor become worthless. For this reason, data management systems are needed that store these volumes of raw data in a resource-efficient manner as well as concepts that enable efficient access to the data. Without such concepts, data can no longer be retrieved as needed, resulting in an economic loss of value.

II. *Data can be duplicated losslessly:* The supply of tangible commodities is finite. For instance, every single drop of oil is unique and can only be consumed once. Tangible commodities can be thinned down to a certain degree, but this reduces their quality and thus their value. Data, on the other hand, can be duplicated indefinitely and even without any loss. This initially sounds promising to data producers and data providers since it means that their product can never run out. However, this fact also means that the value of data is measured differently than that of oil. While the price of oil is determined primarily by supply and demand, the value of data is determined by how original and unique their content is. If previously unknown correlations can be determined with them, they represent a clear competitive advantage and thus a particularly high value. Whereas the more the data are reproduced—i.e., their content becomes common knowledge—the lower their value becomes. Concepts are therefore needed to ensure that the contents of the data remain as confidential as possible and that only certain data consumers gain insight.

III. *Data are generated at high velocity:* Tangible commodities are available wherever they arise naturally. For instance, crude oil remains in the Earth's crust until it is extracted. This can be done based on demand and free capacities in the refineries. Data, on the other hand, are generated at any point in time. A processable data object is obtained only if they are captured at exactly this point in time. If they are not captured, they are lost. However, since many IoT devices have limited memory resources, they cannot store the captured data indefinitely but rely on a stream-based processing concept, i.e., they process the data on the fly. Therefore, while oil requires a pull model (i.e., it is acquired from the source when needed), data require a push model (i.e., the source transmits the data when captured). Since data currently accumulate at a high velocity, data storage systems must either have large input buffers to temporarily store new data until screening and further processing or have the necessary capacities to handle voluminous data streams.

IV. *Data are volatile:* Oil has no expiration date, which is why processing is not time critical. Yet, data are volatile. Although a data object can be stored indefinitely, its content is sometimes only relevant for a very short time. For instance, if a sensor in a driverless car detects that there is an obstacle in the lane, this information must be processed immediately, since it is only relevant until the collision occurs. In other cases, data also become invalid. For instance, if the driverless car detects that a traffic light is green, this information is rendered invalid as soon as the light changes to red. Data storage systems must therefore be able to cope with this limited lifespan and have the capability to process data in (near) real time. While some tangible commodities also have a limited shelf life, the volatility of data dynamically differs from data object to data object, and it is often not possible to specify its expiration date in advance, i.e., how quickly the data must be processed.

V. *Data are heterogeneous:* Tangible commodities are usually homogeneous. Although each drop of oil is unique (i.e., it exists only once), all drops from one source are identical in terms of properties such as purity and quality. Therefore, all extracted oil can be stored in a common container and refined similarly. Data, meanwhile, are heterogeneous. For instance, they can have different data formats or schemata and have different levels of completeness and accuracy. Furthermore, the contents of the data differ. Therefore, data cannot be stored in a common storage. Either all data must initially be transformed into a common structure or data stores that support heterogeneous structures are required. Metadata management is also required to track the properties of the data so that they can be handled appropriately.

VI. *Data refinement has to be in accordance with the data source and intended use:* There are established refinement processes for tangible commodities in order to convert them into certain value-added products. Even though these processes may be adapted over time due to new findings or technical innovations, they can be seen as static processes. With data, this is completely different. On the one hand, new and improved cleansing and preparation techniques are constantly developed, which require adjustments to the data refinement process. On the other hand, due to the heterogeneity of the data, a variety of processing steps geared to the raw data are required. This is aggravated by the fact that there is no one-size-fits-all data refinement process. Rather, adjustments must be made to the steps of the data refinement process depending on the intended use of the data. Only if the process is tailored to both the raw data and the intended use can an optimal result can be achieved. Therefore, data refinement requires flexible adjustments to dynamically respond to changes in sources (i.e., the raw data) and sinks (i.e., the intended use).

VII. *The economic value of data is uncertain:* The value of tangible commodities is generally known. There are some fluctuations due to supply and demand, and over time, commodities can gain or lose value. However, these fluctuations tend to be rather small, while substantial changes are extremely rare. With data, this is completely different. Here, the economic value is initially completely unknown. Data that appear to be worthless today may prove to be needle-movers tomorrow. The reason for this is on the one hand that the derivable knowledge cannot be identified in advance but only when the data have been processed. On the other hand, in such a highly dynamic environment, new use cases for data are constantly emerging, which subsequently define the need and thus the value of the data. Since it is almost impossible to anticipate this need in advance, data administration must be able to manage and process data as cost-effectively as possible, since it is not feasible to distinguish between worthless and valuable data.

VIII. *Data can be manipulated indiscernibly:* Tangible commodities are usually relatively resilient to manipulation. For instance, crude oil could be deliberately contaminated, but this can be detected and subsequently purified. In the worst case, sources can be corrupted to such an extent that they become unusable. However, this problem is far worse in the case of intangible commodities and, in particular, data. Data can be manipulated indiscernibly and, above all, in a targeted manner. Malicious parties can falsify data either in their favor or to harm the data consumers, blend fake data with real data, or withhold data. This can happen both when transferring data from the sources and while storing the data. Since the manipulation generally goes unnoticed, it is also almost impossible to undo the contamination. To make matters worse, besides third parties, data producers themselves may have an interest in falsifying the data they provide. Measures must therefore be taken to verify the authenticity and genuineness of data and to prevent subsequent manipulation.

IX. *Data may be subject to special restrictions:* Tangible commodities such as oil are primarily subject to rights related to ownership. Whoever owns the oil well may extract, refine, and sell the oil. With regard to the last two issues, there may be further restrictions, e.g., regarding the environmental friendliness of the refining process or regarding sanctions that affect exports to certain customers. However, these restrictions always relate to the product as a whole. With data, the situation is much more complex. In particular, when it comes to personal data, the data subject (which is not necessarily the data producer) has far-reaching rights when it comes to data processing. For instance, the consent of the data subject is required for the processing of such data. This consent can be withdrawn at any time. However, even with the consent of the data subject, there are further restrictions to be observed when processing personal data, such as a purpose limitation or data minimization. Furthermore, the data subject has the right to request that all data about him or her be erased. Yet, this applies not only to the raw data themselves, but also to all data products in which the raw data in question have been incorporated. Data administration must therefore take measures to implement such privacy rights. These include, e.g., the use of privacy filters that either anonymize data or reduce the amount of contained information to a required minimum, or provenance mechanisms that make it possible to trace which raw data has been incorporated into which data products.

X. *Data require new trading concepts and infrastructures:* When trading tangible commodities, the main problem is to build distribution infrastructures that bring the goods to international trading partners in time. This is not the case with data. Thanks to the Internet, data can be made available in an instant anywhere in the world. Explicit distribution channels therefore do not need to be established. With data, however, three novel trade problems arise: First, due to the large amount of constantly emerging data, there is an elevated risk of losing track of the available data. However, potential customers must be able to find data that are relevant to them. Second, customers must be able to rely on the provided data. This means that they must be able to use the data for their purposes and that there are no conflicting confidentiality or privacy restrictions. For instance, if privacy filters have to be applied to the data in advance, this contaminates the data and reduces the quality of the data. Data administration must therefore ensure that a customer can rely on the authenticity and quality of the data despite the use of such privacy techniques. Third, the privacy requirements of data subjects as well as the quality requirements of data consumers change dynamically. Therefore, it is not possible to offer static data products, but the data refinement must be constantly adapted to offer tailored data products. A trading platform for data must therefore establish concepts to cope with these three problems.

In summary, these ten inherent differences we identified between oil and data also lead to a significant difference regarding the handling of these commodities. The differences are related to three pillars in particular: Novel concepts and techniques must be developed regarding the administration of data so that this can be done efficiently. That is, the large volumes of data that are generated at high velocity must be handled, tailored cleansing and transformation measures must be applied, and access structures must be established for facilitating the retrieval of the data products. Due to their economic value today, data must be protected against illegal access, manipulation, and unauthorized erasures. This requires end-to-end measures, from the authentication of data sources and the secure storage of data to an appropriate access control system for ready-to-use data products. Finally, data protection laws now make it essential to implement privacy by design and by default concepts whenever personal data are processed. In the following section, we therefore look at how related work addresses these challenges.

3. Related Work

In this section, we review the state of the art in data handling. In the context of our work, three research directions are of particular interest, namely data administration [26], data security [27], and data privacy [28]. We discuss these three research areas in Section 3.1 to Section 3.3 and identify research focuses within these areas. The resulting hierarchical classification of related work is shown in Figure 1. We summarize in Section 3.4 the findings regarding the state of research and discuss what open questions remain to be addressed.

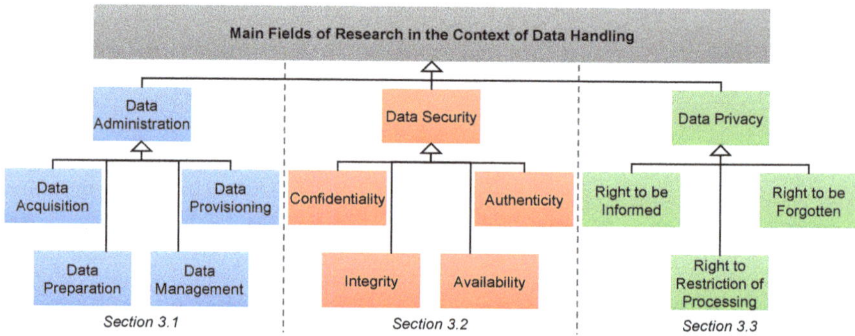

Figure 1. Hierarchical Classification of Research in the Context of Data Handling.

3.1. Data Administration

Data administration comprises all data science tasks in the context of data refinement, i.e., all steps necessary to gain knowledge from raw data [29]. We divide them into the core tasks of a data refinement process, namely selection and extraction of raw data (i.e., data acquisition), data cleansing and data transformation (i.e., data preparation), data storage and data curation (i.e., data management), and distribute refined data (i.e., data provisioning) [30]. Research approaches to assist with these tasks are discussed next.

Data Acquisition. In the context of this work, data acquisition refers to the process of selecting relevant data from a wide variety of sources and then gathering them in a central data management architecture [31]. This represents the first step in the big data value chain and thus enables data to become an asset in the first place [32]. Due to the prevailing heterogeneity among data sources and schemas in which the raw data are available, a systematic approach is required for data acquisition. The so-called ETL process (ETL stands for extraction, transformation, loading) represents a well-established three-step process in which adapter technologies are used to first gather the selected data from the sources, then sanitize them, and finally store them in a structured form [33]. However, this process assumes that there is a central data management architecture with a uniform data schema and that the data in the sources are at rest, i.e., can be retrieved at any point in time [34]. However, due to the IoT, such a conception is outdated. Here, data have a high variety of features even within a single source. Therefore, a rigid target schema is not practicable, since information would be lost in the merger [35]. Moreover, the data are in motion, i.e., they are sent out as a data stream immediately after they are captured by the source, and they accrue in a large volume and at a high velocity, which requires adjustments to the ETL process [36]. Thus, modern-day data acquisition approaches must offer a combination of near real-time processing for streaming data (e.g., via Kafka [37]) and traditional batch-based processing for data at rest [38]. To store the collected raw data, modern data management systems such as Apache Hive (see https://hive.apache.org/; accessed on 6 February 2023) are suitable, as they are not only able to store large volumes of data efficiently, but also cope with heterogeneity within the data [39]. This way, an acquisition infrastructure for big data can be implemented based on the lambda architecture [40]. In the lambda architecture principle, a batch processing layer and a stream processing layer acquire and preprocess data independently of each other and then make them available via a merging layer [41]. A fundamental problem with this architecture is that two separate implementations of the same preprocessing logic need to be maintained for the batch processing layer and the stream processing layer, respectively. In the kappa architecture, all data are therefore gathered and preprocessed as micro-batches by a single stream processing system and made available in a schemaless mass storage system [42]. However, the composition of the micro-batches results in either latency (if the batches are too big which results in long waiting times until sufficient data are available) or a high overhead (if the batches are too small and therefore a lot of batches have to be processed). So, there are sacrifices to be made with both approaches. Therefore, approaches such as the delta architecture aim to combine these two architectures to achieve real-time processing with the ability to handle bulk data efficiently [43]. Nevertheless, more comprehensive preprocessing operations should be performed detached from data acquisition as part of subsequent data preparation [44].

Data Preparation. Once data acquisition has been completed, the collected raw data must be converted into a machine-processable form via data cleansing, transforming, adding metadata, and harmonizing the schemas. These activities are collectively referred to as data preparation [45]. To carry out data preparation effectively, both data knowledge and domain knowledge are urgently needed [46]. The term 'human in the loop' encompasses approaches that empower experts without IT knowledge to actively participate in the data preparation process and contribute their expertise [47]. Research approaches therefore aim to cluster thematically related data sources and thereby represent the sources as a knowledge network so that users can easily identify further relevant sources [48]. Alternative approaches aim to group sources based on the features of their data, as they may need similar data preparation steps [49] or to suggest which data transforming operations are appropriate for such data [50]. Furthermore, the knowledge of the experts can be persisted in the form of a knowledge base that users can leverage in data preparation [51]. Sampling approaches aim directly at facilitating the work of experts by presenting them with only a representative sample of the complete base data. On this manageable sample, the expert defines cleansing and transforming steps, which are subsequently applied to the entire dataset [52]. Here, efficiency and effectiveness can be significantly increased if the data are initially divided into semantically related blocks, which are then taken into account for sampling [53]. Such defined cleansing and transforming steps can be converted into data preparation rules, which can be applied semi-automatically also to new datasets in the future [54]. These rules describe how to obtain processable data from raw data. For reasons of transparency, however, the backward direction must also be provided in order to be able to disclose later on which base data a result was obtained [55]. Why- and how-provenance can be used for this purpose [56]. In addition to human-in-the-loop approaches, however, there is also the countertrend, namely AI-assisted fully automated data preparation [57]. However, this results in a chicken-and-egg problem, since good and reliable training data are required to train the AI—this training data, however, also requires sound data preparation [58].

Data Management. For the management of the processed data, data warehouses were state-of-the-art technology for a long time. Here, the data from multiple data sources are organized in a unified structure that is optimized for tailorable but predefined analysis purposes [59]. However, due to the IoT, the heterogeneity of data sources as well as the amount of semistructured or outright unstructured data increased drastically. Moreover, as data became an essential asset, there is a need for comprehensive and flexible data analysis. The rigid structure of data warehouses is not designed for either [60]. Although there are approaches to describe the semantics of inherently unstructured data to make them processable in a data warehouse [61], they are always limited to specific types of data. Since a rigid data structure is an inherent property of a data warehouse, such approaches do not solve the fundamental issues when dealing with IoT data. Data lakes are intended to overcome these challenges. The basic idea is that all raw data are stored (almost) untouched and data preparation takes place dynamically depending on the respective use case [62]. Thus, a data lake pursues a schema-on-read philosophy, i.e., only when data are processed, a schema that is appropriate for the data and the intended usage is defined and applied [63]. To reduce the resulting overhead that occurs with every data access and to facilitate data governance in general, a zone architecture for the data lake is highly recommended. Each zone provides data at a certain processing stage [64]. However, data lakes are rather concepts than clearly specified architectures [65]. Research approaches therefore attempt to create a reference model for a data lake in which, besides the raw data, harmonized data (i.e., data with a consolidated schema) and distilled data (i.e., aggregated data) are also kept in dedicated zones. In a sandbox zone, data scientists can play around with the data at will, in order to enable exploratory data analytics and provide the full flexibility of a data lake [66]. While this unlimited freedom initially sounds appealing, this might flood the data lake with too much irrelevant data—the data lake increasingly degenerates into a data swamp in which no useful data can be found. This can be prevented on the one hand by a systematic metadata management to keep an overview of all collected data [67] and on the other hand by sanitizing the raw data in terms of detecting integrity violations in the data and dealing with them to maintain the quality [68]. In practice, however, such monolithic data stores are prone to be exceedingly difficult to manage in terms of governance and operation. Research therefore aims to develop a data mesh in which the central data lake is split into distributed, independently managed data silos [69]. Other approaches focus on achieving an optimal trade-off between the flexibility of data lakes (i.e., support for all current and future use cases) and the structured organization of a data warehouses (i.e., efficient data processing and effective information retrieval) [70]. To this end, the data lakehouse architecture supports both classic BI and exploratory data analytics by means of a transaction layer that provides a logical ETL process on top of a data lake [71].

Data Provisioning. In recent years, self-service BI has become increasingly relevant, i.e., users should be able to conduct customized analytics autonomously for their individual use cases [72]. However, a basic requirement to this end is that there is simple access to the relevant data of the required quality [73]. Due to the large amount of data required for today's analyses, the data available internally in a corporation is often not sufficient. Therefore, data from external providers are required as well. As a consequence, data is not only a commodity but has also become a tradable good. To address this new strategic role of data, infrastructures for data marketplaces are being developed to allow customers to find and obtain data products [74]. However, a data marketplace is not a traditional warehouse, but because of the intangible nature of data, rather, a storefront for the available data products. Customers can select the data they want from a data catalog and the marketplace then acts as an interface to the respective data store [75]. From a data provider perspective, one of the most important functionalities that a data marketplace has to offer for this purpose is a comprehensive metadata management system that allows them to describe their data. This includes descriptive information about the data themselves (e.g., their data model or content descriptions) as well as information about the conditions under which they are permitted to be accessed (e.g., their price or their permitted usage) [76]. Since a marketplace usually represents the storefront for multiple third-party data providers, the metadata of all these providers must be merged to assemble a holistic data catalog [77]. From a customer perspective, the data marketplace must facilitate data retrieval. To this end, two main functionalities must be supported: On the one hand, it must be possible to find relevant data from all available sources (e.g., content- or quality-wise), and on the other hand, data acquisition has to be simple [78]. Comprehensive metadata management is required to this end as well [79]. One of the most important aspects of a data marketplace for both sides, however, is trust. Only if data providers can assume that confidentiality and privacy are guaranteed with regard to their data and customers can rely on the authenticity and quality of the offered data, they will use a data marketplace [80]. Therefore, data security and data privacy are central issues in the context of data provisioning.

3.2. Data Security

In modern data administration, four protection goals in particular have to be addressed, namely confidentiality, integrity, availability, and authenticity [81]. Next, we discuss research approaches designed to fulfill these protection goals. We look at the protection goals separately. In the practical application, however, there are correlations with other protection goals. For instance, effective authenticity is a cornerstone of access control and thus a prerequisite for all the other protection goals. The protection goals are also contradictory to some extent, e.g., the highest level of confidentiality can be achieved if no one has access to the data. Yet, this conflicts with availability [82]. Such mutual effects have to be taken into account when adopting solutions to ensure the protection goals.

In our discussion, we refer to the definitions of these four protection goals given in the ISO/IEC 27000-series [83], which is the internationally recognized standard for information security management systems. Confidentiality means that information is not disclosed to unauthorized third parties. Integrity ensures that the information made available is complete and uncorrupted. Availability means that authorized parties have access to the information at all times. Finally, authenticity ensures that both the origin of the information and the identity of the parties interacting with the information are verified.

Confidentiality. To protect against the disclosure of sensitive information, cryptography approaches are typically used. That is, data are available in encrypted form and can only be decrypted (and thus read) with the appropriate key. Both symmetric encryption—in which the same key is used for encryption and decryption—and asymmetric encryption—in which a key pair with different keys for encryption and decryption is used—can be applied to this end. While symmetric encryption approaches generally require less encryption time, asymmetric encryption approaches facilitate key distribution, since the private key always remains with the key owner, while the corresponding public key is shared with anybody without compromising confidentiality [84]. Combinations of these two techniques are also found, particularly in IoT environments, in order to reconcile simple key management with reduced hardware requirements [85]. To reduce the overall decryption effort required for each data access, homomorphic encryption can be used. Here, encrypted data are also unreadable without a corresponding key, but certain predefined operators can still be applied to them, such as aggregation functions, for statistical surveys [86] or search queries [87]. That is, the data can be preprocessed and even analyzed without being fully exposed [88]. An access control policy can be used to specify who has access to which data and for what purpose [89]. However, since the IoT is a dynamic environment, this must also be reflected in an access control system [90]. Thus, policy rules also have to consider the current context in which data are accessed (e.g., a spatiotemporal context in which the access takes place or a role-based context of the accessor) [91]. As a result, such a policy becomes highly complex, which is why access control systems in practice have to solve problems regarding conflicting policy rules [92] and scalable access management [93].

Integrity. Currently, data integrity is often ensured by the use of blockchain technologies. A blockchain can be regarded as an immutable and tamper-resistant data store. By organizing the data in blocks that are inseparably linked to each other via cryptographic hashing, it can be ensured that neither individual data items within a block nor entire blocks can be manipulated. As this chain of blocks is managed in a distributed manner, i.e., multiple parties manage an equivalent copy of the chain, manipulations can be easily detected and reversed [94]. In addition to providing a secure storage for IoT data [95], however, blockchain technologies also facilitate the trustworthy sharing of sensitive data in inherently semi-trusted or unreliable environments [96]. Yet, inherent problems of blockchain-based data stores are their low throughput due to their serial operating principle in terms of query processing [97] and their limited data access support which results in minimalistic query capabilities [98]. Therefore, there are a variety of research approaches to improve the query performance as well as the query capabilities of blockchain-based data stores. For instance, SQL-based query languages are being developed for blockchain systems to improve usability [99]. In addition, there are query extensions for specific application domains tha exceed the SQL standard, such as spatiotemporal queries [100] or top-k queries [101]. Othe approaches aim to create schemata for the data in the blocks [102] or cross-block index structures, in order to improve the performance of query processing [103]. However, as blockchain systems not only have low throughput but also high storage costs [104], it is necessary to keep the volume of stored data as low as possible. Therefore, an off-chain strategy is often applied. In this case, the actual payload data are stored in an external data store. The blockchain itself only stores references to the payload data and digital finger prints in the form of hash codes that can be used to verify the integrity of the data [105]. This way, even traditional relational databases can be extended by the integrity properties of blockchain storages by means of a lightweight blockchain-based verification layer on top of the database [106]. While such an approach can ensure the integrity of the payload data the same cannot be said for the queries and the query results [107]. For this reason, there are also approaches aimed at deeper integration of blockchain technologies in a relational database system to enable more holistic integrity assurances [108].

Availability. IoT devices generally do not have the capability to permanently store the vast amounts of data they collect, let alone the computing power to adequately process them As a result, IoT applications rely on cloud providers to store, manage, and analyze their data [109]. Despite the undeniable advantages that cloud services offer in this context, the nontransparent nature of cloud computing requires blind trust on the part of the data owner in the cloud provider [110]. A key concern for data owners is that they have to hand over control of their data to the provider. This also includes where the provider stores the data and whether there are enough replicas of the data to ensure permanent availability [111]. In general, a semihonest provider is assumed in the cloud environment, i.e., a basic level of trust is appropriate, but a provider will always act to maximize its own benefit [112]. For instance, a provider could keep significantly fewer replicas of the data than promised in order to cut storage costs. Initially, there is no noticeable disadvantage for the data owner, but protection against data loss deteriorates considerably as a result [113]. Data owners therefore need tools to enable them to verify whether a cloud provider is storing their data reliably. So-called Proofs of Ownership and Retrievability (PoOR) are one option for this purpose [114]. Here, digital fingerprints of the data managed in the cloud are stored in the form of homomorphic verifiable tags [115] in a Merkle tree [116]. A data owner can pose challenges to the cloud provider, which the provider can only solve if it is in possession of the data. The user can verify the provider's answers using the homomorphic verifiable tags. If this is successful, proof is provided that the data are available without having to download the full data. However, this does not ensure that there is also the promised number of replicas available. Proof of Retrievability and Reliability (PoRR) approaches can be applied to verify this as well [117]. Here, a verifiable delay function (VDF) is applied to the data, which is slow to compute but easy to verify [118]. Therefore, if a data owner poses challenges to instances of the cloud provider that are supposed to hold the data in question that relate to this function, the provider can only answer them if the data are actually at rest here. If the response takes too long, this is proof that there is no replica on the instance and the cloud provider needs to compute the VFD on the fly. In addition to unreliable cloud providers, a central server always represents a bottleneck and thus an inherent weak point with regard to the availability of data in traditional client-server structures. If such a server is flooded with requests, e.g., due to a distributed denial of service attack (DDoS), and thus becomes unavailable, all data managed by it are also no longer available to the clients. IoT environments in particular are severely vulnerable to such attacks [119]. In order to ensure data availability, it is therefore necessary to replace such centralized structures with highly distributed models that store the data in multiple replicas on different nodes to be resilient in the event of a single node failure [120]. Consequently, the use of blockchain technologies is also suitable for ensuring data availability as the blockchain is based on the distributed ledger technology [121]. This refers to technologies that enable data to be stored and shared over distributed computer networks. In simplified terms, a distributed ledger is a data storage system that manages data on multiple computer nodes with equal rights [122]. Due to the decentralized nature, no central authority has control and interpretational sovereignty over the data. Moreover, the collective of nodes can decide which data should be available and thus keep false or harmful data out of the data store [123]. As blockchain-based database systems typically require each node to manage the entire blockchain, this incurs excessive storage costs. Therefore, there are approaches in which only a few server nodes need to store the entire blockchain, while clients can still verify the authenticity of the data by means of authenticated data structures (ADS) [124]. Since this constrains the data distribution, which can endanger availability if the number of expected malicious or compromised nodes is remarkably high, other approaches rely on data partitioning. In sharding, the complete data stock of a blockchain is divided into several parts and distributed to the available nodes according to well-defined rules [125].

Authenticity. In order to identify users, i.e., to verify their authenticity, passwords are commonly used. These can either be real words or PIN codes or lock patterns that have to be entered for authentication [126]. The IoT also offers other authentication options based on biometrics features such as voice, fingerprints, or facial expressions [127]. All of these methods have in common, however, that they can be easily exploited by shoulder surfing during input [128] or replay attacks [129]. To reduce the number of authentications required and thus mitigate some of these threats, there are OAuth-based approaches for the IoT. Here, an authentication service issues a token that authorizes the use of devices or services for a certain period of time [130]. However, this only shifts the problem of illegally acquired authentication data to the OAuth service. To address this, the ownership model relies on using a technical device for authentication that has a unique hardwired fingerprint [131], e.g., by means of physical unclonable functions (PUF) [132]. Yet, the loss of such a device inevitably enables another person to gain possession of the authentication data. In the IoT, this issue is further exacerbated as devices are not linked to specific users but are used by several people. Moreover, IoT devices have limited input capabilities, which means that users cannot enter their credentials like on a traditional computer [133]. For these reasons, there are trends away from 'what you know' (e.g., password-based approaches) or 'what you have' (token-based approaches) authentication toward 'what you are' authentication [134]. In attribute-based approaches, an entity is authenticated based on certain properties it has in the current situation. Especially in dynamic and rapidly changing environments, such a context-based description is advantageous [135]. Due to the high flexibility of attribute-based approaches, they are particularly suitable for IoT applications [136] or cloud-based applications [137]. In addition to users, however, it must also be ensured that data are authentic, in terms of, they come from the specified sources and have not been falsified [138]. For digital media, such as image, sound, or video data, digital watermarking can be used for this purpose. That is, an identification tag is inseparably burned into the carrier medium. If it can be ensured that no unauthorized third parties have access to the identification tag, authenticity can be verified by the presence of the digital watermark [139]. Similar techniques can also be applied to data in relational databases. Here, individual marker bits are inserted into the payload data [140]. While the origin of the data can be proven in this way, watermarking approaches inevitably contaminate the actual data with the inserted identification tags. Digital signatures represent a noise-free approach to ensure the authenticity of data [141]. Here, methods of asymmetric cryptography are used. While asymmetric cryptography uses the public key of a recipient for encryption—thereby ensuring that the message can only be decrypted with the corresponding private key, i.e., only by the recipient—the sender uses his or her own private key for signing. Since the sender's public key is freely available, anyone can decrypt the message. However, this verifies that the sender has used the corresponding private key, which proves the origin of the data beyond doubt [142]. In the IoT, attribute-based signatures are suitable. Here, the attributes of a data source are stored in the signature (e.g., technical specifications of a sensor) and the receiver can verify whether these attributes are sufficient for the data to be authentic (e.g., does the sender have the capabilities to capture the data in the required quality) [143]. Yet, attribute-based signatures are computationally expensive, which is a particular problem in the context of lightweight IoT devices. Thus, research approaches aim to outsource part of the heavy workload to the cloud [144]. Other approaches deal with the problem that the attributes in the signature might contain sensitive information. To this end, a trusted intermediary is installed that preverifies the signatures and removes the sensitive attributes from it [145]. For areas like social media, where such an authentication of sources is not possible, the authenticity of data can be verified based on their content [146]. Then, fake news can either be blocked [147] or overridden by authentic data [148].

3.3. Data Privacy

Due to the Quantified Self movement, i.e., the continuous self-tracking with IoT technology, the amount of personal data is growing exponentially [149]. The analysis of such data is also of high economic value for the industry, as it reveals a lot about potential customers [150]. Therefore, one aspect of data security is becoming increasingly relevant in the context of data processing and management, namely data privacy. The idea of data privacy originates from philosophical theories that predate the information age. These theories reflect the basic need of humans to keep certain information about themselves secret. Historical examples include the Hippocratic Oath [151], the seal of confession [152], and the secrecy of correspondence [153]. In these examples, the owner of the secret trusts that the person keeping the secret will adhere to his duty of professional secrecy [154]. In a broader sense, however, data privacy is motivated by the fundamental human desire not to be at the mercy of external control and to have autonomy over one's personal data [155]. A pioneering work that attempts to define this philosophical idea describes privacy as follows: "Privacy is the claim of individuals, groups, or institutions to determine for themselves when, how, and to what extent information about them is communicated to others" [156].

To this end, two aspects are particularly noteworthy: On the one hand, it is evident that privacy, unlike the protection goals discussed in Section 3.2, is an individual right—that is, each data subject must be able to decide individually what information s/he wants to shares with society. On the other hand, this definition implies a necessity to provide controlling and regulating measures for the sharing of personal information [157]. Therefore, in our digitalized world, it is particularly important that laws comprehensively preserve the privacy of all individuals. This includes a plethora of organizational, technical, and legal measures necessary to enable data subjects to enforce these two aspects in the information society of today [158].

Data protection laws such as the General Data Protection Regulation (GDPR) [159] therefore demand technical measures that give data subjects full control over their data. However, the so-called privacy paradox arises here—although data subjects have fundamental privacy concerns, they do not want to give up the comfort and respective benefit they experience from the analysis of their data. In the context of the IoT and its smart services, this problem is even more aggravated [160]. For effective data protection, it is therefore not only important to protect private data, but also to maintain the perceived quality of service. Otherwise, the privacy measures would be rejected by users. Technical privacy measures can be divided into three categories: measures that inform data subjects about potential privacy risks, measures that restrict the sharing and processing of personal data, and measures that erase data permanently [161]. Only with such technical measures it is possible to comply with the requirement for data protection by design and default [162]. Besides these technical measures, data protection laws also deal with organizational measures (e.g., the appointment of a dedicated contact person for data subjects or the obligation that data must not leave a predefined territorial scope) [163]. In the context of this work, however, we are focusing on technical challenges, only.

Right to be Informed. Data protection regulations give data subjects the right to be informed about the processing of their data. Three main research directions can be identified in this context. First, techniques are being developed to explain what knowledge can be derived from certain types of data. For instance, IoT data sources are described with the help of ontologies. Thus, it can be disclosed which data is collected by a sensor (including, e.g., the accuracy and frequency) and which information is derived from the respective data [164]. Based on such ontologies, reasoning can be done about what knowledge can be derived from this information [165]. These ontologies can be regarded as persisted knowledge of technical experts and domain experts. By involving privacy experts, the ontology can be extended in order to identify potential privacy threats and thereby make data subjects aware of critical issues. From this information, abnormal and questionable data usage patterns can be detected [166]. Second, in addition to informing about correlations between data and knowledge, research approaches also aim to describe the actual data processing. For this purpose, a holistic view of data-driven applications and processes is provided in order to identify inherent privacy threats. In the field of system safety, the STAMP framework (STAMPS stands for System-Theoretic Accident Model and Processes) provides such a top-down approach. Here, causal relationships between hazards and actions are initially modeled. Based on this model, the System-Theoretic Process Analysis (STPA) can be performed to identify problems in the design of the system that lead to or facilitate these hazardous actions [167]. This approach can be adapted to the field of privacy threats. Here, potential privacy threats posed by a data-processing application have to be identified initially, e.g., using the ontologies described earlier. The components of the application are then analyzed to determine whether they are sufficiently protected against these privacy threats [168]. Such a top-down analysis is ideal for modeling and analyzing complex data-driven systems and informing data subjects about inherent privacy-related vulnerabilities with respect to the data sources involved [169]. Third, data subjects also have a right to be informed about the data products for which their data was used as input. Machine learning models represent a prime example. Although these models can be processed by machines easily, they are usually a black box for humans. When applying such a model, it is generally nontransparent why it came to a certain result (e.g., a prediction or a suggestion). Since the models may be flawed or unfair due to an insufficient amount of appropriate training data, it is crucial that the results (and thus the models themselves) are comprehensible to data subjects [170]. To this end, there are three research areas: In the field of interpretable models, methods are developed which generate explainable models. In the field of model induction, models, which represent a black box, are transformed into an explainable model. Yet, both approaches have limitations when it comes to deep learning, as models in this context are far more complex. In deep explanation, deep learning algorithms are therefore adapted in such a way that not the model but at least the relevance of individual input factors are identified across the layers of the model, in order to determine what eventually led to a decision [171]. Yet, there are still many open questions to be solved, especially in the area of deep learning, before full transparency is achieved [172].

Right to Restriction of Processing. However, all this information is of little use to a data subject in exercising his or her digital self-determination if s/he cannot also object to the data processing and enforce this technically. To this end, however, a binary consent system is far too restrictive. Here, a rejection leads to a massive reduction in service quality, which tempts data subjects to agree to all requests. Therefore, this is not an actual informed consent [173]. Instead, privacy-enhancing technologies (PET) should be used to minimize the data—or rather the amount of information they contain—in a target-oriented manner in accordance with the privacy requirements of the data subjects. Three fundamentally different types of PET can be identified: obfuscation techniques for users, statistical disclosure control for mass data providers, and distribution of data across multiple trusted third parties [174]. The obfuscation techniques for users apply privacy filters to the data. Very simple filters allow horizontal filtering [175]—which corresponds to the selection operator in relational algebra, i.e., filtering out specific data items from a dataset—or vertical filtering [176]—which corresponds to the projection operator in relational algebra, i.e., filtering out specific features of a data item. Also, an aggregation, i.e., condensing many data items to a single representative data item (e.g., the mean), can be used as a privacy filter [177]. However, such generic privacy filters are rather coarse grained and therefore severely impair the data quality. Privacy filters that are tailored to a specific type of data are therefore more appropriate, as they are fine-grained and thus less invasive [178]. For instance, there are privacy filters that are tailored to location data and can optionally obfuscate individual locations or entire trajectories [179], privacy filters for health data that enable certain types of examinations only, while rendering the data unusable for all other analyses [180], or privacy filters that mask the voice or remove revealing background noise in speech data [181]. Other approaches focus on complex events that represent a specific sequence of data items instead of individual data items. Privacy-critical events are concealed, e.g., by reordering the data items or dropping or inserting data items. This preserves privacy without reducing the data quality of the data themselves [182]. Statistical disclosure controls for mass data providers include methods in which individual data subjects are hidden in the faceless masses formed by all users [183]. For instance, k-anonymity approaches ensure that a data item can only be mapped to a group of k users. With a sufficiently large k, no private information can be derived from the data about each of the k individuals [184]. Differential privacy is intended to protect an individual even more extensively. It ensures, e.g., by adding noise, that statistical analyses cannot determine whether the data of an individual data subject contributed (significantly) to the outcome [185]. While this sounds tempting in theory, differential privacy often turns out to be complex to implement and very costly to compute in practice [186]. If parameterized inappropriately, it even offers little protection and therefore leads to a false sense of security [187]. Federated learning can be regarded as an approach in the field of distribution of data across multiple trusted third parties. Here, the data with which a machine learning model is to be trained is distributed among several parties. Each party calculates its own local model, which is then incorporated into a global model. By partitioning the data appropriately, it can be ensured that no party gains a comprehensive insight into the data [188]. By using privacy filters when creating the local models, it can also be ensured that no unintended conclusions can be drawn from the global model [189].

Right to be Forgotten. The right to be informed and the right to restriction of processing are, however, not sufficient to provide all-round protection. As data can become outdated, the privacy requirements of data subjects can change, or data may have been unlawfully transferred to data providers in the first place, there is also a need for a right to have all personal data erased. This also includes all related metadata as well as all data products in which these data have been integrated. In addition, personal data also have an inherent lifespan, after which they must also be completely deleted, as they may not be stored longer than required for the intended purpose. In terms of today's data protection regulations, data erasure has to ensure that the data cannot be restored in any way or form [190]. From a technical point of view, therefore, measures are needed to enable secure deletion of all the data concerned. In the context of providers of big data, the large volume of data from a vast number of users means that, from an economic point of view, it is not feasible to apply deletion methods that destroy the data carriers. To this end, there are nondestructive approaches for both hard disk drives and solid-state drives. While overwriting-based approaches are effective for hard disk drives, where the sectors containing the data in question are overwritten several times with other data, erasure-based approaches for solid-state drives ensure that the flash block that provides access to the flash page containing the data is erased. As a result, the data can no longer be accessed and is finally dumped by the garbage collection [191]. So, while in theory, there are approaches to enable secure deletion for those physical data carriers, these approaches have limitations in terms of logical data storage. For instance, in the case of databases, such an approach is not possible as abstraction layers prevent a direct mapping of data to sectors or flash pages [192]. Blockchain-based data stores represent another example where this type of secure erasure is unsuccessful, as here the data are immutable, i.e., a deletion renders the blockchain invalid [193]. Cloud-based data stores also require other approaches, since only the cloud provider has control over the physical storages and the knowledge of how many replicas are held on which nodes [194]. In these cases, encryption-based approaches can be used. Here, the data are stored fully encrypted. This way, it is sufficient to delete the decryption key, which renders the encrypted data unreadable in the process. Hierarchical key management enables fine-grained deletion of individual data items, clusters of related data, or all data of a data subject at once [195]. Provenance analyses can be used to identify all data products that have been created based on a specific data item [196]. These data products must also be erased if a data subject requests to have their base data (or any part of it) deleted. It is evident that such a provenance analysis also generates a lot of additional data that potentially disclose private information [197]. To this end, when answering this kind of provenance queries, it is therefore necessary to rely on special privacy-aware approaches, such as intensional provenance answers [198].

3.4. Lessons Learned

As illustrated in Section 2, there are ten key differences between data and tangible commodities such as oil that make them special assets. These special characteristics must also be taken into account when handling and managing data in order to use this commodity effectively and efficiently. Our review of the state of the art shows that there are many research approaches to this end. They can be divided into three categories: In the area of data administration, methods and techniques are being explored that enable or facilitate data acquisition, data preparation, data management, and data provisioning. Thereby, the special challenges in today's big data context (namely, volume, variety, and velocity) are addressed. In the area of data security, approaches are explored that ensure the confidentiality, integrity, and authenticity of the data in order to verify their veracity and thus preserve their value. Furthermore, methods and techniques are developed that guarantee high availability of the data and ensure that only authorized entities have access to them. Data privacy is a branch of data security that plays an important role currently due to the increasing importance of personal data. In the context of this work, we focus on technical issues, whereas legal, ethical, and organizational research is out of scope. In this regard, there are approaches that provide data subjects with more information regarding the processing of their data as well as identifying potential privacy threats. Other approaches address how data processing can be restricted in a fine-grained manner. PET ensure that no sensitive knowledge can be gained from the data without compromising the overall data quality. Finally, methods are being developed to erase data permanently to ensure that they can no longer be processed. In addition to the actual data, this also includes all replicas and metadata as well as all data products that have been created based on those data.

However, all of these research approaches are island solutions to individual data administration, security, and privacy problems. Yet, these three aspects are highly interrelated. For instance, a data marketplace cannot provide accountability for data if their origin cannot be traced through provenance and authentication of the sources, while privacy can only be guaranteed if confidentiality is ensured as well. Moreover, data administration can only be effective if availability is ensured. Thus, these aspects must not be considered independently of each other. Even within the individual research areas, the isolated approach represents a disadvantage. For instance, obfuscation techniques for users are immensely powerful because they give data subjects full control over their data. With statistical disclosure control for mass data providers, they lose this control. However, due to the holistic view on all available data, data providers are able to apply privacy techniques in a much more target-oriented manner. Therefore, the optimal solution would be a mutually coordinated mix that first provides data subjects with a prefiltering option and then allows data providers to make comprehensive readjustments from their side. Island solutions cannot achieve synergy effects and, even worse, some of them are mutually exclusive. For instance, privacy can easily be achieved by completely randomizing all data. This, however, minimizes data quality and thus renders the data worthless. Another example is the use of blockchain-based data stores to ensure integrity and availability. Yet, immutable data storages inevitably prevent data subjects from exercising their right to be forgotten.

Although these individual solutions are efficient for the respective application context, a holistic end-to-end view from data acquisition to data delivery is required to enable trustworthy and demand-driven data provisioning. 'Trustworthy' refers to enabling data producers to make their data available without losing control over them and without having to fully disclose their assets. 'Demand-driven' refers to ensuring that data consumers are provided with authentic data of the required quality. To this end, it is required to implement and apply appropriate security and privacy mechanisms in all data administration processing steps. Furthermore, such an approach must be both generic and flexible to cope with the great heterogeneity in today's big data landscape.

For this purpose, we have developed a set of solutions for all data processing steps. The novelty of our research is its holistic approach. That is, all our solutions can be integrated seamlessly into an end-to-end platform. Therefore, by combining these concepts appropriately, the result is significantly more than the sum of the individual parts. In the following section, we present this integrated concept called REFINERY Platform as a whole for the first time and describe how its individual components interact with each other.

4. The REFINERY Platform

In our REFINERY Platform, we aim to provide comprehensive support for data administration, starting from any data source and ending with tailored data products. Despite this holistic view of the data administration process, the REFINERY Platform consists of a wide range of individual solutions for each process step, e.g., data acquisition from smart devices and databases, customizable data preparation rules, secure data management, and making the data products visible to data consumers. The main focus here is on ensuring that all concepts are geared to the specific characteristics of data (see Section 2). This includes in particular that the commodity 'data' is handled reliably. On the one hand, this means that it must be feasible for data producers or data subjects to regulate data processing in accordance with their data privacy and confidentiality requirements. On the other hand, the customers of the data products, i.e., the data consumers, must be able to fully trust the authenticity of the data and be reassured that the data are of the promised quality. A high-level overview of the general concept of the REFINERY Platform is shown in Figure 2.

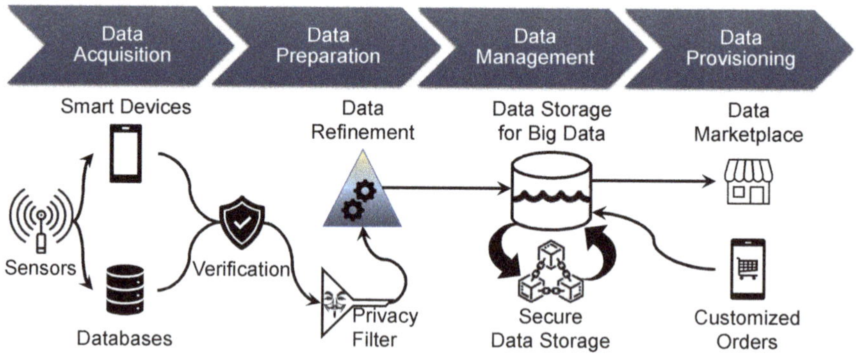

Figure 2. High-Level Overview of the General Concept of the REFINERY Platform.

Due to the IoT, two different types of data sources need to be supported by a data administration platform currently. On the one hand, there is a multitude of smart devices that are able to collect a plethora of different data via connected sensors and share them due to their connectivity. On the other hand, there are mass storage devices in the form of databases, both relational databases and NoSQL data stores, which provide data at rest. In the REFINERY Platform, both types of data sources can be integrated. During acquisition, the data are verified by the REFINERY Platform in terms of whether it possesses the specified properties, e.g., regarding data quality. In addition, the data sources can assert privacy requirements, which are then enforced in the REFINERY Platform in the form of PET, e.g., privacy filters. All three types of PET are supported, i.e., obfuscation techniques for users, statistical disclosure control for mass data providers, and distribution of data across multiple trusted third parties.

The received data prepared in this way are subsequently processed semiautomatically in the REFINERY Platform and data products are generated. For this purpose, domain experts specify rules for the data cleansing and data transformation steps to be applied to the data. These rules can then be automatically applied to the acquired raw data. The semi-automatic approach is necessary to generate customized and high-quality data products, which would not be possible without the involvement of human experts. However, the resulting rule base increasingly reduces the experts' workload in this context, as they can draw on already specified preparation rules (either in parts or as a whole).

Then, both the raw data and the data products have to be managed. The REFINERY Platform uses a big data storage system that enables efficient management of the raw data and the processed data, as well as demand-oriented access to the actual data products. Due to the high economic value of these intangible commodities and digital products, we apply blockchain technologies to secure them against malicious tampering and deletion.

Finally, it must be possible for customers to find the products they need from the vast amount of available data products. For this purpose, the REFINERY Platform has data management structures that support an electronic storefront, which can be used to search the product range efficiently. The metadata gathered and generated by the REFINERY Platform enables straightforward information retrieval. In addition, customers can use the metadata to inform themselves about existing raw data and place orders for customized data products. That is, they describe their own data refinement steps, which are then carried out by the REFINERY Platform to provide data products tailored to their needs.

The concepts used in the REFINERY Platform to this end are described in more detail in the following. The structure of the section reflects the data administration tasks as defined in Section 3.1 since these tasks represent the value generation in terms of data refinement. First, the concepts of data acquisition are outlined in Section 4.1. Subsequently, Section 4.2 describes the data preparation in the REFINERY Platform. The management concepts are covered in Section 4.3, while data provisioning is addressed in Section 4.4. The protection goals regarding data security and privacy constitute value-added services, only. Yet, these matters are taken into account in every component of the REFINERY Platform.

4.1. Data Acquisition

In the area of data acquisition, it is important to give data producers extensive options to control which of their data they want to share with third parties, i.e., which information they want to disclose and which knowledge they want to reveal. Without such options, they would be faced with the binary choice of releasing all or none of their data. However, this would inevitably mean that the REFINERY Platform would have to exclude many potential data sources upfront.

To this end, we have developed a fine-grained permission system that allows data producers to specify which information content they want to share with the REFINERY Platform at all, and which knowledge third parties are allowed to derive from it for a specific purpose. As observed in Section 3.3, there are two fundamentally different approaches to regulating the information content of data, namely the application of PET either on the user side (e.g., obfuscation techniques for users or a distribution of data across multiple trusted users, respectively) or on the side of the mass data providers (e.g., statistical disclosure control). Both approaches have their intrinsic strengths and weaknesses. For instance, if data regulation takes place in the user-controlled area, the data producer has full control over his or her data. In contrast, privacy measures applied by mass data providers are often much more effective as here all available data sources are known. Moreover, a mass data provider can select PET which are adjusted to the subsequent data processing steps.

As a result, PET can be applied to the data in a target-oriented manner, which reduces the impact on data quality.

In the REFINERY Platform, we therefore combine these two approaches to obtain the best of both worlds. Our solution is shown in Figure 3. First, the data producer specifies his or her privacy requirements (more on this in Section 4.4). These privacy requirements are then split using a tailorable metric based on a data quality and privacy budget into policies that have to be deployed in the user-controlled area and policies that have to be deployed at the mass data provider. That is, it is determined how much data can be disclosed unmodified to the mass data provider in order to achieve a higher data quality and which data are so confidential that they have to be distorted already by the data producers themselves. As a mass data provider usually receives data from more than one data producer, this also achieves data distribution, since some raw data never leave the sphere of influence of the respective data producer in unprocessed form [199]. How the privacy policies are applied in the REFINERY Platform at the mass data provider is described in Sections 4.2 and 4.3. In this subsection, we discuss how the policies are applied to the data during acquisition, i.e., in the user-controlled area.

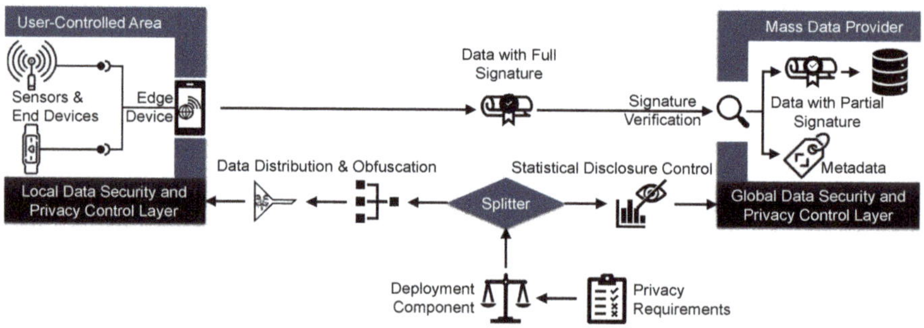

Figure 3. Deployment Concept of PET as Part of the Data Acquisition in the REFINERY Platform.

For this purpose, we have developed the Privacy Management Platform (PMP) for Android-based IoT edge devices (see https://www.android.com/; accessed on 6 February 2023). In this context, an IoT edge device is a device that is capable of running third-party applications (i.e., not just a fixed set of applications hardcoded into it by the manufacturer) and has sufficient computing power to process a reasonable amount of data. In addition, an edge device has the ability to connect to the Internet or cloud services. Thus, in our use case, it represents the interface to the mass data provider for all connected sensors and IoT end devices of a single user.

Our PMP is able to enforce privacy policies that describe which data consumer (or which application, respectively) is allowed to have access to which data. In the most basic case, this is implemented by means of horizontal or vertical filtering [200]. However, since such an approach is restrictive and severely compromises either the data volume or the data quality, more fine-grained privacy filters can also be applied. For this purpose, we have studied privacy filters that can remove different information contents from time-series data (i.e., the predominant type of data in the IoT domain). For instance, individual data items can be concealed while preserving the overall temporal progression, or resilient noise can be added to all data to reduce the accuracy of all values (and thus the information content). These filters are rather generic, so they can be applied to data from any sensor [201]. In addition to these generic privacy filters, specialized filters can also be applied, which are adjusted to specific types of data. For instance, in the case of location data, the accuracy can be decreased so that only the approximate location can be determined. Our approach is also extensible in the sense that additional specialized filters for other types of data can be added to the PMP retroactively [202].

From a logical perspective, the PMP therefore represents an intermediate layer that isolates the data-consuming applications from the data-producing operating system (OS). In other words, any data flow has to be handled by the PMP, which means that the PMP has full data sovereignty. In order to assure this, two properties must apply to the PMP: First, the PMP has to be able to provide all kinds of data that can be requested by applications. Second, the PMP has to be able to prevent applications from bypassing it, i.e., there must be no data flow that is not controlled by the PMP.

The former is programmatically realized by means of an adapter concept. There is an adapter in the PMP for each data source that is accessible via the underlying OS. In this context, it is irrelevant whether streaming data (e.g., data originating from a sensor) or data at rest (e.g., data stored in a database or a file system) are involved. Each adapter essentially replicates the interfaces of its underlying data source. This way, any data from the data source can be simply passed through. In addition, however, privacy filters are implemented in the adapters, which are applicable to the respective type of data. This allows the adapter also to forward modified versions of the data instead of the original data to meet the privacy requirements of the data subject. The interactions between these components are shown in Figure 4 as a sequence diagram.

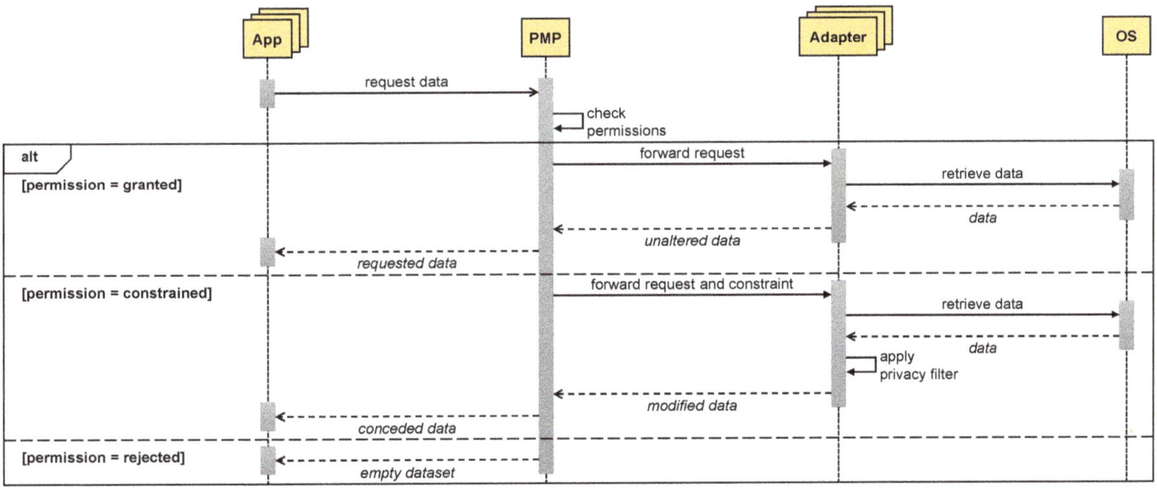

Figure 4. Sequential Processing of a Data Request Including All Interactions Between Applications, the PMP, Adapters, and the OS.

When an application (referred to as 'app' in the figure) requests data from the PMP, the PMP checks whether the data subject has permitted this access. Data subjects can assign one of three permissions: a request can be granted, constrained, or rejected. The latter is the default if no permission is specified for the requesting application. If the request is granted, the PMP forwards it to the corresponding adapter, which fetches the data from the OS, and the unaltered data are provided to the requesting application. If constraints are specified for the data access, these constraints are forwarded to the adapter along with the request. In this case, the adapter also retrieves the corresponding data from the OS but applies the appropriate privacy filters to the data. The modified data are then provided to the application. Whereas if the request is rejected, the PMP returns an empty dataset to the application. This way, the application cannot tell whether the data access has been denied or not. Otherwise, the application could penalize the data subject (e.g., by means of a reduced functionality) to force him or her to disclose the data.

In order to ensure that an application cannot bypass this data request process, the PMP is also deeply integrated into the Android installation process. During installation, an application in Android is assigned all the permissions required to interact with the OS, e.g., access a data source. The PMP revokes all requested Android permissions from an application. As a result, even if an application tries to bypass the PMP, it cannot request data directly from the OS due to the lack of permissions [203].

In addition to applying privacy filters to data, the extensible adapter concept of the PMP also facilitates the integration of complementary external data sources. That is, adapters can be used to connect external IoT components to an edge device in a hub-and-spoke architecture. These external components include, among others, sensors or IoT end devices such as fitness wristbands, which themselves do not have a direct connection to the Internet. Instead, they transfer their data to a more powerful IoT device, e.g., a smartphone, via technologies such as Bluetooth. Our adapters ensure that such external data providers are fully integrated into the edge device on a logical level. This eliminates the need to distinguish between data sources that are embedded in the edge device and those that originate from external IoT components. In both cases, the edge device acts as a gateway for the data to the outside world. It also allows the privacy filters to operate on the data of the external IoT components directly. Otherwise, this would not be possible since such components neither have the necessary computing power nor offer the possibility of running third-party software, such as the PMP with its privacy filters [204].

While stream-based data processing is commonly used in the IoT as the lightweight IoT devices do not have sufficient memory to persistently store the captured data, in the context of a data delivery platform such as the REFINERY Platform, a different approach is required. To counteract the volatility of the IoT data, they are buffered on the edge device and made available via the PMP. To this end, we have developed a secure data store for IoT edge devices. This data store is secured against illegal access and manipulation from the outside by the fact that the data it contains are fully encrypted. Only the PMP has the key to decrypt the data. This ensures that data can only be accessed via the PMP, i.e., in a fully controlled and regulated manner. Internally, our data store manages the data in a relational database [205] or a NoSQL data store (e.g., a key-value store or a document store) [206], depending on the type of data. This way, it is able to handle both structured and unstructured data efficiently. However, the PMP completely abstracts from this internal data management, as the secure data store on a logical level is just another of its data sources. Furthermore, via the PMP, we enable a systematic synchronization mechanism for our data store. This allows data from multiple IoT edge devices to be synchronized with a mass data provider, i.e., as soon as an edge device has connectivity, it propagates all changes in the secure data store to its mass data provider [207].

For reliable information delivery, as we are targeting in the REFINERY Platform, two factors are crucial in data acquisition: On the one hand, the data quality must be transparent to data consumers. That is, if the quality has deteriorated, e.g., due to the application of privacy filters, this must be communicated in a transparent manner. On the other hand, confidentiality must also be maintained with respect to the data producers. In particular, this entails that data consumers must not know, e.g., what data have been withheld, as this could disclose what information has been concealed. This also means that data consumers do not gain complete insight into how the data have been tampered with. In addition, it must be ensured that the mass data provider meets its obligations with regard to privacy policies. To master this balancing act, the REFINERY Platform applies a two-stage attribute-based signature procedure.

To this effect, data are digitally signed on the edge devices. The full signature used here contains both the information on how the data were captured and which privacy filters were applied to them, as well as which privacy requirements still have to be addressed by the mass data provider. This means that the information about the data (e.g., data quality or accuracy) and the privacy policies that still need to be applied are inseparably linked to the payload data. The mass data provider checks and verifies the signature of the incoming data and uses the information contained in the signature for metadata management, as information on data quality is mandatory for subsequent processing and provision to data consumers. However, not all of this information is intended for the data consumer, as it can be used to draw conclusions about the data. If, e.g., it is evident that a privacy filter was applied to spoof certain location data, then assumptions can be made as to why these locations are considered to be compromising by the data subject. After the signature has been verified, all of this information is therefore removed from the signature. On a technical level, this is realized by means of a second key pair, the delegated keys. The resulting partial signature only contains the privacy policies that have to be considered during data preparation (see Section 4.2) and applied during data management (see Section 4.3). Yet, the remaining information is not lost as it is still available as part of the metadata in the REFINERY Platform. It was only detached from the payload data [208].

4.2. Data Preparation

Data preparation encompasses all activities required to cleanse raw data, transform them into a processable form, and finally turn them into a marketable data product. In the REFINERY Platform, we split data preparation into two separate steps. In the first step, we transform data into information. In contrast to raw data, information is organized and has a structure. General data impurities (e.g., missing values or outliers) are also addressed in this first step. In a second step, we transform information into knowledge. Unlike the rather generic information, knowledge is geared to a specific use case. This means that data products on the knowledge level can be applied by data consumers according to their purposes, e.g., as base data for their analyzes or as training data for their machine learning models. For both data preparation steps, we have developed techniques for the REFINERY Platform that are tailored to the specific characteristics of the commodity 'data'.

In order to bring raw data to the information level, the first step is to improve the data quality. For this purpose, missing data or attributes must be filled in, outliers or integrity violations must be identified as well as treated, and the data must be harmonized (e.g., by the unification of the value units). Although some of this can be done automatically (e.g., identification of null values), human participation is essential for successful data preparation, so that they can contribute their data knowledge to resolve the data impurities. For this purpose, human experts, often referred to as data stewards, need extensive insight into the base data in order to be able to identify the problems and apply appropriate countermeasures. However, it is necessary to comply with the privacy requirements of the data producers in the process.

Therefore, we have developed a sample-based approach for the REFINERY Platform, in which the data steward only has access to a representative sample of the data and works on this sample [209]. In synthesizing this sample, statistical disclosure techniques are applied. In a first step, a data sample is automatically generated from the stock of raw data. According to the privacy requirements that are still available in the partial signature of the data (i.e., the privacy requirements that have not yet been applied by the PMP in the user-controlled area), it is assessed whether the sample meets these requirements. Thresholds for various metrics (e.g., regarding the uncertainty, information gain, or accuracy of the sample) can be used to this end. Only if the sample meets these requirements, it is forwarded to the data steward. If it does not, a new sample must be selected. In addition to privacy metrics, fairness metrics can also be specified to ensure that the sample is not biased, i.e., that it is truly representative for the base data.

The data steward then cleanses the sample. Based on the applied techniques, data cleansing rules are derived, which are then applied to the complete base data. The data steward has access to metrics about the base data and can request and cleanse additional samples until s/he is fully satisfied with the overall data quality. User studies show that this dynamic sample-based approach even helps to fix more data issues compared to a data steward working on the entire base data. Since the volume of base data is large, the data steward cannot inspect all of the data. In contrast, the representative (and significantly smaller) sample can be analyzed in much greater detail, enabling the steward to identify and correct any errors it contains. That is, our approach is not only privacy-friendly but also leads to better overall data quality. With this approach, the data steward can also transform raw data into a uniform (and thus processable) structure.

In order to bring the data from the information level to the knowledge level, individual processing tailored to the respective intended use cases is required. This necessitates extensive domain knowledge. However, domain experts generally do not have the necessary IT knowledge to implement the processing steps themselves. For the REFINERY Platform, we have therefore developed an ontology approach that allows domain experts to specify in a simple way how the data needs to be processed for a given use case. In doing so, we use RDF/XML (see https://www.w3.org/TR/rdf-syntax-grammar/; accessed on 6 February 2023) for the internal representation of the ontology.

We have implemented a processing engine for our ontology that automatically applies the specified data processing rules to the data in question [210]. Figure 5 shows a simplified excerpt from such an ontology.

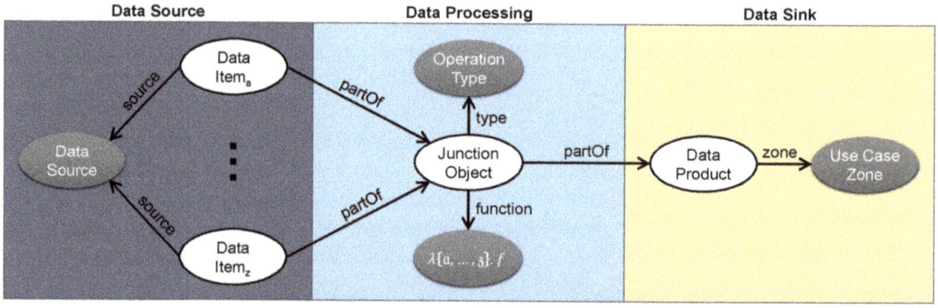

Figure 5. A Data Processing Rule Specified in the Ontology Provided by the REFINERY Platform.

A data processing rule of the ontology always consists of three segments. These three segments correspond to the three questions 'What?', 'How?', and 'To what end?'. First, data items must be selected to which the processing rules should be applied. Basically, all cleansed data on the information level can be selected. For instance, all temperature data should be converted into a data product. With this selection, however, the domain expert gains no insight into actual data items but only into which types of data are available and which attributes these items have. Therefore, this does not pose a threat to the privacy requirements of individual data producers.

After the source data have been selected, the core part of the data processing rule must describe which kind of processing has to be performed. To this end, we support the three key operators from functional programming, namely the map operator, the filter operator, and the reduce operator. These operators are each applied to a data sequence (e.g., to all data items from temperature sensors). In the following, D and E denote two arbitrary data types and $\mathbb{B}$ stands for Boolean data whereas a_i and b_i are instances of these data types.

$$\text{map operator} : (D \to E) \times (a_0, \ldots, a_n) \to (b_0, \ldots, b_n)$$

The map operator applies a unary function to all n elements of the sequence. This results in a new sequence consisting also of n elements—namely the n elements from the original sequence after they have been processed. The data type of the elements may change during processing. In our example, the domain expert could use a map operator to change the unit of temperature data from Celsius to Fahrenheit.

$$\text{filter operator} : (D \to \mathbb{B}) \times \underbrace{(a_0, \ldots, a_n)}_{n} \to \underbrace{(a_0, \ldots, a_n)}_{m} \mid m \leq n$$

The filter operator validates all n elements of a sequence using a unary predicate logical expression. The result is a sequence with the m elements from the original sequence for which the expression is evaluated to true ($0 \leq m \leq n$). However, the elements themselves are not changed in the process. In our example, the domain expert could use a filter operator to filter out the temperature data that are below a certain threshold.

$$\text{reduce operator} : (E \times D \rightarrow E) \times (a_0, \ldots, a_n) \times E \rightarrow E$$

The reduce operator aggregates the n elements of a sequence to a single result value. Unlike the map and filter, reduce applies a binary function. With this function, the reduce operator initially combines the first element of the sequence with an initial value. The result is then combined with the second element of the sequence using the same function. This is repeated until all elements of the sequence have been condensed to a single value. In our example, a reduce operator could calculate the average temperature of the n data items.

We apply the functional programming paradigm since functional programming is very well suited for composing such functions at runtime and applying them to arbitrary data streams. The actual program logic is defined via lambda expressions. As lambda expressions are essentially a concise mathematical description of functions and require a simplified syntax only, even non-IT experts can handle them.

In addition, our ontology approach also supports user-defined procedures, if the expression power of these three operators is not sufficient. To this end, arbitrary program code can be specified in the ontology, which is then applied to the data.

Finally, a data sink has to be selected, i.e., it has to be specified which data product was generated with this processing rule and for which use case it is intended. Intermediate data products can also be specified in our ontology, i.e., a data sink can be the data source for another data processing rule.

These data products can be further refined, e.g., they can be used to train machine learning and AI models. Such models also represent data products in their own right. In contrast to the data products addressed by the REFINERY Platform, however, such models have a more complex life cycle [211]. Maintaining these complex data products therefore requires additional measures, e.g., monitoring the validity of the models or providing the models in different data formats [212]. Yet, this is not in the scope of our work. Rather, the REFINERY Platform is the precursor to such a model management platform, providing it with the data needed to train these models. Furthermore, with our privacy filters, it is also possible to provide the data foundation for several variants of a model in which certain privacy-relevant aspects are not included [213]. For more information on such a model management platform, please refer to the work of Weber and Reimann [214].

However, even without considering such complex data products, a large number of (intermediate) data products need to be managed. Section 4.3 describes how this is solved in the REFINERY Platform. In addition, data consumers can retrieve available data products and also tailor them to their needs, which is outlined in Section 4.4.

4.3. Data Management

Data lake architectures are well suited for managing big data. They can hold not only any heterogeneous raw data but also processed variants of these data that have been refined for specific purposes. However, it is extremely important that the stored data are organized appropriately, because otherwise, the lake can easily degenerate into a data swamp, i.e., although the data are basically available, users are unable to retrieve them.

For the REFINERY Platform, we have therefore used the concepts of a data lake architecture and extended its basic zone concept. We apply pass-through zones, in which data are only temporarily buffered until they are processed, and persistent zones, in which data are permanently stored. Figure 6 illustrates the zone architecture we designed for the REFINERY Platform. Pass-through zones are shown in light blue and persistent zones are shown in white. In addition to these zones, there is a Data Security and Privacy Control Layer that manages the privacy requirements and access policy for the data lake as a whole. More details on the access control are given in Section 4.4.

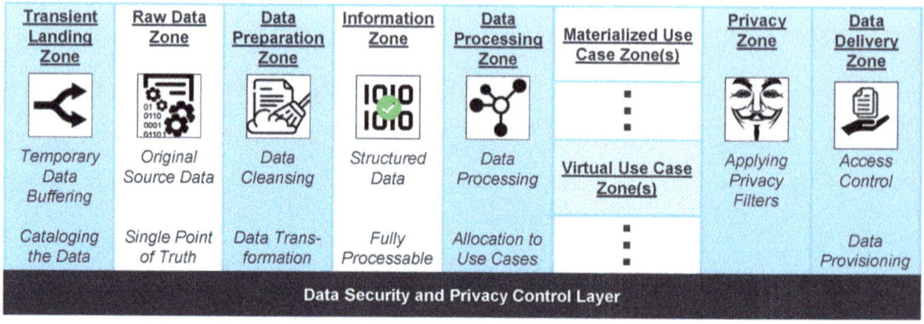

Figure 6. Data Lake Zone Architecture Applied in the REFINERY Platform.

The Transient Landing Zone represents the entry point of the data lake. Any incoming data is initially buffered here. This is also where the verification of the full signature takes place. If successful, the metadata contained in the signature is used to catalog the data so that they can be retrieved later. Then, the partial signature is added, to retain the privacy requirements to be observed in further processing. Via the event streaming platform Kafka (see https://kafka.apache.org/; accessed on 6 February 2023), the Transient Landing Zone then forwards the raw data to the Raw Data Zone for persistent storage.

In the first persistent zone, the incoming data are stored in an appropriate storage system. Since these raw data are heterogeneous and partly unstructured, a distributed file system like HDFS (see https://hadoop.apache.org/; accessed on 6 February 2023) is suitable for this purpose. To live up to the promise of being a reliable information retrieval and delivery platform, the original raw data must be protected as they represent the single point of truth. On the one hand, it must be ensured that they cannot be deleted, and on the other hand, it must be possible to prove to data consumers that the data products are based on authentic facts which have not been tampered with. To ensure both, the Raw Data Zone is made immutable and tamper-resistant by means of blockchain technologies. Whether the data are stored completely on-chain or only a digital fingerprint is stored in the blockchain depends on the respective data (e.g., their volume or their required level of protection) [215]. In case the data are stored on-chain, we have developed privacy-by-design concepts for blockchain systems [216], as well as concepts to improve the query capabilities of blockchain systems [217].

These base data are subsequently considered ground truth for the entire REFINERY Platform. The Raw Data Zone is also the basis for the Data Preparation Zone, in which our sample-based concepts are applied to cleanse and transform the data (see Section 4.2). The result of this preparation (i.e., the data at the information level) is then persisted in the Information Zone. As information is structured (i.e., a predefined data schema exists), relational databases such as PostgreSQL (see https://www.postgresql.org/; accessed on 6 February 2023) are suitable for storage. The schema provided by these databases facilitates the handling of the data in the following zones. Special protection measures such as blockchain technologies are not required here—the contents of this zone can be restored at any time based on the raw data and the defined cleansing and transformation rules.

In the Data Processing Zone, the structured data are converted into a data product using our ontology with the processing rules (see Section 4.2). Since some of the data products might be tailored to rather uncommon use cases, our concept allows them to be stored in Virtual Use Case Zones in addition to Materialized Use Case Zones. Materialized Use Case Zones store the data products in a fully persistent manner. The choice of storage technology depends on the data product, e.g., document stores such as MongoDB (see https://www.mongodb.com/; accessed on 6 February 2023). In contrast, data products in Virtual Use Case Zones are maintained only temporarily, e.g., in a pure in-memory database such as Redis (see https://redis.io/; accessed on 6 February 2023), or via Spark Structured Streaming (see https://spark.apache.org/streaming/; accessed on 6 February 2023) as a data stream that is generated live on demand. To enable such a mix of batch processing and stream processing, we have developed a hybrid processing model [218].

As described in Section 4.2, statistical disclosure techniques are applied during data preparation in accordance with the privacy requirements. However, due to data refinement activities (e.g., by combining data from various sources), data products can still violate privacy requirements. Therefore, there is a dedicated Privacy Zone in our REFINERY Platform before any data product is made available. In this Privacy Zone, privacy filters are applied to the data products if necessary. Since mass data providers have extensive processing capabilities, computationally intensive data obfuscation techniques can also be applied here to conceal highly specific information content [219]. This zone represents the final audit, to ensure that all privacy requirements of the data producers have been met. If this is the case, the partial signature is removed, and the data products are made available to data consumers via the Data Delivery Zone. We address data provisioning in Section 4.4.

4.4. Data Provisioning

There are two main tasks to be fulfilled in data provisioning: There are two main tasks to be fulfilled in data provision: On the one hand, the available data products must be retrievable for data consumers, and on the other hand, a control mechanism must ensure that only authorized entities (in accordance with the privacy requirements of the data producers) are granted access to the products. Both tasks are facilitated by our ontology of processing rules.

The ontology maps the complete lineage of a data product. This lineage includes where the raw data came from, what was done with them (i.e., what processing steps were taken during data preparation, but also what PET were applied to them), and where the data products are stored (i.e., in which Use Case Zone they can be found). With the help of this data lineage, data consumers can retrieve the data products they are looking for. As the ontology is machine-processable and can be analyzed, e.g., it can be used to feed a recommender system that can suggest related data products to data consumers that are also suitable for their purposes. In addition, data consumers can also design their own tailored data products. Since the ontology is extensible, data consumers can define their own data preparation processes if the existing data products do not meet their requirements. Either existing processing rules can be customized, or entirely new ones can be specified. Existing partial solutions can be used for this purpose, i.e., any subset of the processing rule in the ontology as well as any data product can be reused as a starting point for the new rules.

An access control system regulates which data products a data consumer has access to. We have developed a purposed-based access policy for the REFINERY Platform. The primary focus here is that it is comprehensible for data producers. That is, it must be transparent what knowledge a certain data product might expose about them. The model of our access control policy is shown in Figure 7 in a notation based on UML (see https://www.omg.org/spec/UML/; accessed on 6 February 2023).

From a technical perspective, access control is about defining who—in terms of which smart service—is allowed to access which data sources. Yet, humans cannot grasp what these data expose about them, as they are often abstract and only reveal meaningful knowledge after processing and analysis. That is why our access policy focuses on precisely that wisdom that can be disclosed due to the processing of data. To quantify this, we have studied a privacy risk elicitation procedure based on STPA, which allows privacy experts to systematically audit data processes. In this way, they can identify potential exposure threats in terms of disclosed wisdom [220]. The data processes are defined by our ontology as the sum of all processing rules. Therefore, the threats identified by the privacy experts can be mapped to one or more knowledge patterns. For this purpose, domain experts have to specify for which purposes the data products can be used, e.g., which analyses can be performed with them. These knowledge patterns are composed of the information prepared by the data stewards who process the raw data provided by the data producers. In the IoT, these data generally originate from sensors. Therefore, sensor experts are needed to describe from which sensors—or more generally, from which data sources—this kind of data can originate. This hierarchical top-down approach enables a comprehensible mapping of disclosed wisdom to data sources. That is, data subjects define their privacy requirements at a level of abstraction they can understand, whereas the rules are mapped to a technical level and applied to the respective components [221].

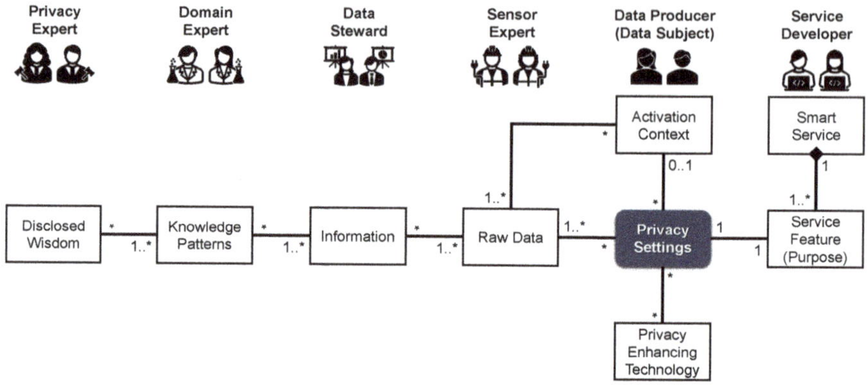

Figure 7. Model of the Purposed-Based Access Control Policy Designed for the REFINERY Platform.

On the opposite side, service developers in their role as data consumers must define which data their services need to access. Our policy allows to break down a smart service into individual service features. These service features each correspond to a purpose as the GDPR stipulates that data access is restricted to a specific purpose. With the help of our policy model, a data subject can thus clearly identify the purpose for which s/he exposes what kind of knowledge. Finally, PET can be attached to each policy rule. To this end, we evaluated a variety of privacy filters for specific types of data (e.g., location data, health data, or audio data) [222]. These reflect the privacy requirements expressed by data producers. The PET are applied in the Privacy Zone before a data product is made available to the smart service in question.

As static access control rules are too restrictive in a dynamic environment like the IoT, each rule in our model can be subject to an activation context. This could be, e.g., a temporal context (e.g., data collected during free time is subject to stricter access rules than data collected during working hours) or a spatial context (e.g., data collected at home is subject to stricter access rules than data collected at the workplace). Any type of data that is available and evaluable can be used to define this activation context [223]. This way, demand-driven data provisioning is made possible in the REFINERY Platform.

5. Assessment

After introducing the REFINERY Platform as our end-to-end approach toward reliable information retrieval and delivery and outlining how it carries out data administration tasks while complying with data security and data privacy, we now critically review our work. For this purpose, we will first perform a security and privacy assessment for the individual stages of the REFINERY Platform in Section 5.1. Subsequently, we provide a feature discussion in Section 5.2, in which we assess whether the REFINERY Platform has the required functionality to address the special data characteristic, i.e., whether it is able to effectively handle the commodity 'data'. Then, in Section 5.3, we discuss the practicality of the two key components of the REFINERY Platform with which a user primarily interacts, namely its privacy control measures and the selection and specification of data products, based on two case studies. Since the best functionality is of little use if it cannot be provided efficiently, we also perform a performance evaluation for the REFINERY Platform in Section 5.4. These four assessments reflect whether the REFINERY Platform is able to manage data in a secure, effective, practical, and efficient manner, i.e., whether it is a useful tool for modern data administration.

5.1. Security and Privacy Assessment

As described in Section 4, the REFINERY Platform fulfills all functional data administration tasks that arise in the context of data refinement, namely data acquisition, data preparation, data management, and data provisioning. In addition, data security and data privacy precautions are intrinsically integrated into each of these process steps, as required to ensure a reliable handling of the valuable commodity 'data'. Since these precautions are coordinated and seamlessly intertwine due to the holistic approach, end-to-end data protection is guaranteed. Since the application of these precautions is coordinated and all individual measures are closely intertwined, end-to-end data protection is achieved. In the following, we assess whether the protection goals (see Section 3.2) and privacy requirements (see Section 3.3) in this context are addressed in the REFINERY Platform.

The confidentiality of the data is ensured on the side of the data producer since they are stored solely in encrypted form in our secure data store on the edge devices. That is, they are protected against illegal access right after they have been captured. The data producer defines via the privacy requirements which of these data are forwarded by the PMP to the REFINERY Platform. This further promotes confidentiality, since only a portion of the data leaves the data producer's sphere of influence and can thus be leaked in the first place. For transmission and during data refinement, the data are signed. In general, a digital signature does not protect against unwanted access, since the data are encrypted with the private key of the data owner and thus anyone who has the public key can verify the signature, i.e., decrypt the data. However, in our case, only the REFINERY Platform has the necessary public keys. That is, the digital signature automatically guarantees confidentiality during transmission and storage in the REFINERY Platform as well. Since for the data preparation only samples that meet the privacy requirements can be accessed by the data stewards, confidentiality is also maintained in this process step. Even the data processing rules are specified without granting third parties deep insights into the data. External third parties (e.g., data consumers) can only access the data products via restricted interfaces, namely the Data Delivery Zone. Here, an access policy set by the data producers regulates who may access which data products and for what purpose.

Two main techniques are used to verify the integrity of the data: On the one hand, the REFINERY Platform can use the digital attribute-based signature to check how the data were captured, as it contains, e.g., information about the sensor used for this purpose and its accuracy. Furthermore, the privacy requirements are included in the signature as well. That is, the REFINERY Platform knows which distortions have been made to the data on the part of the data producer and which constraints have to be respected during data refinement. This transparency makes it possible to determine whether the quality of the raw data meets the quality requirements of a data consumer. Furthermore, manipulations by any third parties (e.g., during transmission) can be detected due to the signatures. While this does not prevent such manipulations, it can be ensured that no corrupted data can make its way into the data store of the REFINERY Platform. On the other hand, the integrity of the raw data after they have been transmitted is ensured by blockchain technologies. By means of the information stored in the blockchain (either the data items themselves or their digital fingerprints), it is possible to verify that the data in the Raw Data Zone have not been tampered with. Since the ontology with the processing rules provides a complete lineage of each data product, this also ensures the integrity of these products. If there is any doubt about the integrity of a data product, the raw data can first be verified and then the data product can be reproduced using the ontology.

No custom-made approaches for availability and authenticity are introduced in the REFINERY Platform. Rather, established techniques are also used for this purpose. For instance, the raw data are managed using HDFS. This distributed file system ensures high availability, as the data are redundantly distributed on different nodes. The relational databases in the Information Zone also have a high fault tolerance and allow recovering the stored data in the unlikely event of a failure. Thus, this also applies to the data products in the Use Case Zones, as they can be rebuilt based on the data from the Raw Data Zone and the Information Zone via the ontology. For IoT devices (i.e., the data sources), permanent availability cannot be achieved. However, our synchronization mechanism for the secured data store ensures that in case of connection failure, all new data are transmitted to the REFINERY Platform as soon as the connection is re-established. The digital signatures ensure the authenticity of the data as they certify the origin of the data. We have not addressed issues related to the authenticity of data consumers. For us, it is only important that the REFINERY platform is able to identify them. There are numerous mature approaches such as attribute-based credentials that can be used for this purpose. Authentication of data consumers is not within the scope of our work, as it has no implications for the REFINERY Platform.

Our access policy approach describes the insights that can be gained from certain data by mapping the rather abstract raw data to human-comprehensible knowledge patterns. This illustrates to data producers what can be extracted from their data. Furthermore, the ontology reflects the full data processing workflow. In this way, we enable data subjects to obtain detailed information about the processing of their data in a way that even non-IT experts can understand. Thus, our REFINERY Platform inherently implements the right to be informed. In addition, due to the PET which are geared to the privacy requirements, e.g., privacy filters or statistical disclosure control, we also offer technical support to implement the right to restriction of processing. This is particularly effective in our approach for two reasons: On the one hand, we use a variety of dedicated privacy filters and techniques. Thus, certain information contents in the data can be specifically concealed without degrading the overall data quality. On the other hand, we deploy the privacy filters in both the user-controlled area (i.e., on the edge devices in the form of the PMP) and in the REFINERY Platform mass data storage (i.e., in the Data Preparation Zone and in the Privacy Zone). This distribution allows us to apply the privacy filter in a much more targeted manner. It also leverages the strengths of both approaches. As a result, the usability of the data can be maintained without having to make any sacrifices in terms of privacy. Our access policy model also contributes to demand-driven data provisioning, as data accesses are mapped to actual purposes instead of a smart service as a whole. The introduction of activation contexts enables an even more fine-grained permission management. This reduces the privacy paradox since users are no longer faced with a more or less binary choice between privacy and service quality. We have not taken any explicit measures to enforce the right to be forgotten. Our concept is based on the assumption that the REFINERY Platform is operated by a trusted party. If a data subject therefore makes use of the right to be forgotten, we provide technical support for this, e.g., our ontology to easily identify and subsequently delete all data products related to the raw data in question. Explicit means for data subjects to verify that the REFINERY Platform actually deletes all data in question are not necessary given our basic assumption.

Table 1 summarizes the key data security and data privacy concepts integrated into our REFINERY Platform for the four data administration tasks.

5.2. Feature Discussion

As this initial investigation demonstrated that the REFINERY Platform provides comprehensive solutions for all protection goals regarding data security and privacy, we now assess to what extent the ten data characteristics, which we have identified and discussed in Section 2, are addressed in our approach.

The fact that data are not consumed during processing, i.e., the data volume grows continuously (Characteristic I), is addressed by the deployment of a distributed file system, namely HDFS, for the management of raw data. Since HDFS distributes the data across a multitude of nodes, it is suitable for the efficient handling of big data. In addition to the raw data, the number of data products to be managed in the REFINERY Platform is also growing steadily. With the introduction of Virtual Use Case Zones, i.e., a way to store data products only temporarily in volatile storages, data products that are tailored to rather uncommon use cases and therefore rarely in demand are automatically removed after usage. However, the knowledge gained during production is not lost, since the processing steps required to manufacture these data products are still available in the ontology. That is, if required, the data products can be restored from the raw data at any time.

Table 1. Summary of the Key Contributions of the REFINERY Platform with Regard to Data Security and Data Privacy.

Data Administration Task	Key Contributions to Data Security	Key Contributions to Data Privacy
Data Acquisition	On edge devices, data are fully encrypted. The data producers stipulate which data are transmitted. Digital signatures secure the transmission.	Data subjects specify privacy requirements that are applied by means of PET (e.g., privacy filters) on edge devices and in the REFINERY Platform.
Data Preparation	The applied sample-based data preparation ensures that no unauthorized insights into the data can be gained.	Data subjects specify a privacy threshold that is respected when selecting the data samples for the data stewards.
Data Management	Blockchain technologies ensure the integrity of the transmitted data via digital fingerprints.	Additional PET can be applied to a data product before it is distributed.
Data Provisioning	Our access control model enables to define who is allowed to access which data. Data access may be subject to additional privacy restrictions.	Since our model maps retrieved data to revealed knowledge patterns, this approach allows for truly informed consent.

Since data can be losslessly duplicated, which inevitably leads to a loss of value (Characteristic II), data consumers are not given direct access to the raw data or the processable information retrieved from them. Only the data products can be accessed by data consumers via restricted interfaces. This preserves the value of the raw data. The REFINERY Platform does not provide any special protection against the duplication of data products after they have been delivered to data consumers. However, this is not necessary from our point of view, as it is in the interest of data consumers that the value of the data products acquired by them is not diminished, e.g., by means of unregulated reproduction. The assets of the REFINERY Platform (i.e., the raw data, the information, and the production rules) remain unaffected by such a reproduction in any case.

In the REFINERY Platform, we first address the fact that data are generated at high velocity and are partially volatile (Characteristics III & IV) by developing a secure data store for edge devices that serves as a buffer until the data can be transmitted to the mass data provider. Our synchronization mechanism for this data store ensures that changes are always transmitted in a timely manner, as soon as connectivity is available. The Virtual Use Case Zones take into account the fact that raw data—and thus the data products derived from them—are volatile. The lifespan of these zones is limited by design, which acts as a kind of garbage collection in our data storage.

The fact that data are heterogeneous (Characteristic V) is addressed in the REFINERY Platform on the one hand by using HDFS in the Raw Data Zone. That is, even unstructured data can be managed without having to transform them first. In addition, metadata about the acquired data are provided by the digital signatures, which facilitate further processing of the data. For the data preparation, we rely on a human-in-the-loop approach. A data steward defines the necessary preparation steps. In contrast to a fully automated approach, our approach therefore does not require a predefined data schema.

Our approach to data preparation also addresses the fact that data refinement has to be in accordance with the data source and intended use (Characteristic VI). While the data preparation by the data steward is rather generic (whereby s/he can also take special characteristics of data sources into account), the subsequent data processing is fully geared to the intended use. The processing rules in our ontology describe how the processable information is turned into a data product. Data consumers can add further processing rules or adapt existing ones if no available data product meets their needs. Our PET include specialized privacy filters that are tailored to certain types of data. These filters can be used to conceal specific information contents without rendering the data unusable for an intended purpose.

HDFS is used for the implementation of the Raw Data Zone. With this file system, it is possible to add further nodes at any time in order to increase capacity. Due to decreasing prices for hard disk space, it is therefore possible to store all available raw data, even if their economic value is initially uncertain (Characteristic VII).

We address the fact that data can be manipulated indiscriminately (Characteristic VIII) by means of our data integrity measures. Blockchain technologies are used to verify that the stored raw data, which represent the ground truth for all data products, have not been manipulated. Digital signatures ensure that they have not been secretly falsified by third parties during transmission. These signatures also describe which PET have already been applied to the data in the user-controlled area. Therefore, it is not only possible to prevent all illegitimate data manipulations but also to communicate these justified distortions by the data producers (in accordance with their privacy requirements) to the data consumers in a transparent manner. In addition, the selection of PET is matched to the intended use case, i.e., their impact on the data attributes relevant for the data consumer is as low as possible.

In addition to the PET, our purposed-based access policy model ensures that special restrictions regarding the handling of certain data can be complied with (Characteristic IX). This access policy is used in the Data Delivery Zone and maps for which purpose which data consumer has access to which data products. Access can be further restricted by means of activation contexts that can be attached to each access policy rule.

All data products are described by means of our machine-processable ontology. Recommender systems can therefore use the ontology to point data consumers to similar data products that might also be relevant to them. Data consumers can also extend the ontology to design customized data products for their particular use cases. This forms the foundation for a data marketplace. That is, our REFINERY Platform provides the required concepts and infrastructures to trade the commodity 'data' (Characteristic X). To this end, only an implementation of an electronic storefront as an interface to the data marketplace is missing and has to be addressed in future work.

This feature discussion demonstrates that our REFINERY Platform is effective in handling the commodity 'data'. The concepts responsible for this are recapped in Table 2.

5.3. Case Study

After this feature discussion, showing that our REFINERY Platform provides all required functionalities to enable an appropriate modern data administration, we now focus on the practicality of our solution by means of two case studies.

This study is divided into two parts, as we aim to evaluate the two user interfaces. On the one hand, this involves the privacy control capabilities enabled by the PMP, which provide the foundation for demand-driven data acquisition (see Section 4.1). On the other hand, this also concerns the specification of tailored data products in the course of data preparation (see Section 4.2). These two case studies also reflect the two key user groups of the REFINERY Platform, namely data producers and data consumers.

Privacy Control Capabilities. In our first case study, we focus on data acquisition of health applications for smartphones. Since it is evident that such applications are particularly beneficial, there is a mobile health application for literally all aspects of life [224]. One of the main drivers for these applications is the fact that the built-in sensors in standard smartphones can capture many health-related factors without a great deal of user input. For instance, the stress level [225] or the mood [226] can be determined passively (i.e., without explicit user interaction) by recording and analyzing the user's voice via the microphone. Or image analysis techniques can be used to analyze pictures of a food product on a smartphone in order to determine its ingredients, such as bread units [227]. Furthermore, there is a large variety of IoT-enabled medical metering devices that can be connected to a smartphone and thus provide these applications with more specific health data [228].

Table 2. Summary of the Key Concepts Applied in the REFINERY Platform to Address the Special Characteristics of Data.

Data Characteristic	Concept in the REFINERY Platform to Handle this Characteristic
I. Data are nonconsumable.	All incoming data are stored in the Raw Data Zone in an expandable big data storage. Data products that are not permanently demanded can be made available temporarily via a Virtual Use Case Zone and reproduced when needed.
II. Data can be duplicated losslessly.	Third parties such as data consumers only have access to data products, not the underlying raw data or the refined information.
III. Data are generated at high velocity.	Data are buffered on edge devices in our secure data store and updates are synchronized with the mass data storage automatically.
IV. Data are volatile.	Data products that are based on volatile data can be stored in Virtual Use Case Zones to ensure automatic sanitization of the provided products.
V. Data are heterogeneous.	The data storage in the Raw Data Zone is schemaless. Data preparation is based on a human-in-the-loop approach, i.e., no strict data schema is required here either.
VI. Data refinement has to be in accordance with the data source and intended use.	Data products are generated by means of an ontology, which can be extended if needed. Besides generic privacy filters, specialized filters tailored to specific types of data can be applied in order to preserve the data quality for the intended usage.
VII. The economic value of data is uncertain.	Virtually unlimited amounts of data can be stored in the Raw Data Zone at almost no cost. Therefore, their potential value does not need to be known in advance.
VIII. Data can be manipulated indiscernibly.	Digital signatures assure the integrity of the data in transit and the data in use while blockchain technologies enable to verify the integrity of the data at rest.
IX. Data may be subject to special restrictions.	The access policy model enables data subjects to define who can access which data and for what purpose. Access can be further constrained, e.g., via privacy filters.
X. Data require new trading concepts and infrastructures.	The information available in the Delivery Zone (e.g., metadata on raw data, specification of data products, and access policy) provide the foundation for a data marketplace.

While this kind of data is an obvious choice in the context of a health application, smartphones provide another very relevant piece of information that is often overlooked. It can be observed that the location of a user is also relevant to health applications. This spatiotemporal aspect is important for the interpretation of health data, because, e.g., a higher stress level in a noisy and hectic environment has to be evaluated differently than if it occurs in a peaceful place [229]. A smartphone can usually determine the current location accurately and therefore put every medical reading into its context.

By combining this data collection with gamification aspects, it is possible to address children in particular and encourage them to regularly capture and document their health values, which are otherwise often perceived as a chore [230]. With Candy Castle [231], we have therefore developed such a game aimed at young children suffering from diabetes. The main functionality of this application is shown in Figure 8.

Listing 1. Ontology Excerpt to Specify a Data Preparation Process in the Food Chemistry Domain.

```
<rdf:RDF xmlns:rdf="http://www.w3.org/1999/02/22-rdf-syntax-ns#"
 ↪ xmlns:dl="http://barents.dl/">
  <rdf:Description rdf:about="http://barents.dl/walnut">
    <dl:layer>Raw Data Zone</dl:layer>
    <dl:source>chocolate</dl:source>
    <dl:partOf rdf:resource="http://barents.dl/peptides"/>
  </rdf:Description>
  <rdf:Description rdf:about="http://barents.dl/hazelnut">
    <dl:layer>Raw Data Zone</dl:layer>
    <dl:source>chocolate</dl:source>
    <dl:partOf rdf:resource="http://barents.dl/peptides"/>
  </rdf:Description>
  <rdf:Description rdf:about="http://barents.dl/peptides">
    <dl:layer>Information Zone</dl:layer>
    <dl:function>lambda x : x.hazelnut or x.walnut</dl:function>
    <dl:type>filter</dl:type>
    <dl:partOf rdf:resource="http://barents.dl/results"/>
  </rdf:Description>
  <rdf:Description rdf:about="http://barents.dl/results">
    <dl:layer>Use Case Zone</dl:layer>
    <dl:zone>allergens</dl:zone>
  </rdf:Description>
</rdf:RDF>
```

Such tasks represent worst-case scenarios for our processing engine since the accesses to the Information Zone account for the majority of the processing costs compared to the actual data processing. In this process, the data must not only be read from the relational database used in the Information Zone but must also be converted into the data structure on which our processing engine operates, namely a data frame. As soon as the data are contained in this data structure, we can make use of indexes, which enables the execution environment to perform computations efficiently. This overhead caused by reading the data is inherent to any type of processing. Therefore, in general, it can be assumed that for more complex processing tasks, the overall overhead is lower, since in these cases, those access costs are negligible compared to the actual computation costs. The latter, however, also accrues without the use of the REFINERY Platform when manufacturing the respective data product.

For our performance evaluation, the Information Zone, which contains the base data, is implemented using SQLite DB in version 3.39.4 (see https://www.sqlite.org/; accessed on 6 February 2023). The data products are stored in a Virtual Use Case Zone, which is implemented using TinyDB in version 4.7.0 (see https://tinydb.readthedocs.io/; accessed on 6 February 2023). However, the choice of these two databases does not affect the evaluation results. They can be replaced by any other relational database or NoSQL data store without loss of generality since the actual processing is fully decoupled from the input and output technologies. Our prototype of the processing engine is implemented in Python 3.10.9 (see https://www.python.org/; accessed on 6 February 2023) and uses pandas 1.5.2 (see https://pandas.pydata.org/; accessed on 6 February 2023) for data processing. To this end, the data from the Information Zone are initially loaded into a pandas DataFrame, a tablelike data structure. All operators specified in the ontology are then applied to this DataFrame and the result is forwarded to the Virtual Use Case Zone. Pandas is well suited for this purpose since in addition to import and export functions that support a variety of data sources and data sinks, it also provides index structures that allow efficient computations on a DataFrame. For this reason, pandas is a de facto standard in the field of data science for these kinds of tasks [237].

However, when it comes to processing large amounts of data, there are a few decisive limitations in pandas. On the one hand, pandas is by design only able to use a single threat on a single CPU for processing the data. Yet, especially with large amounts of data, significant performance improvements can be achieved by splitting the data into smaller chunks that are processed in parallel by multiple cores. On the other hand, pandas holds the entire DataFrame as well as all intermediate processing artifacts in main memory. Therefore, this represents another bottleneck in terms of an upper limit for the maximum amount of data that can be processed. Regardless of the available main memory, pandas is not designed to handle more than 100 GB of data [238]. Modin addresses these scalability issues by providing its own distributed DataFrame that can be processed by multiple cores. The Modin DataFrame is almost fully compatible with the pandas API, which makes it easy to parallelize pandas applications [239]. To this end, a Modin DataFrame can be partitioned either horizontally (i.e., by data item) or vertically (i.e., by attribute) [240].

Modin uses either Dask [241] or Ray [242] as its execution engine. Apart from some minor differences, the main distinctive feature is their respective scheduling strategies. While Dask uses a centralized scheduler that distributes the tasks to the workers and monitors the progress, Ray applies a distributed bottom-up scheduling strategy. Here, local schedulers distribute the tasks independently to the workers assigned to them. Workers can exchange data with each other via a shared-memory object store. Local schedulers can also forward tasks to a global scheduler, which can assign them to another local scheduler to achieve load balancing between the local schedulers. In effect, Dask is particularly suited for general-purpose data science tasks such as standard data analytics and data wrangling, while Ray shows its strengths in complex machine learning and AI-related tasks [243].

Therefore, we also implemented our prototype of the processing engine in a parallelized version using Modin 0.18.0 (see https://modin.readthedocs.io/; accessed on 6 February 2023) with Dask 2022.12.1 (see https://www.dask.org/; accessed on 6 February 2023) and Ray 2.2.0 (see https://www.ray.io/; accessed on 6 February 2023) as execution engines. As a baseline, we implemented the three data processing tasks as SQL commands that are executed directly by the SQLite DB, i.e., in a highly optimized manner that is beyond the implementation knowledge of the domain experts or the data consumers. This baseline represents the minimum processing cost. The more our processing machine approximates this baseline, the better its efficiency. Keep in mind that a certain overhead is inevitable. This is the price to pay for supporting tailorable processing rules and a wide range of data sources and sinks.

For the evaluation, we first adopted a deployment scenario in which a data producer runs an instance of the REFINERY Platform on his or her hardware for his or her own data only. That is, it is a limited amount of data and a significant limitation regarding computational power. To reflect this, we generated stepwise from 500 to 5000k data items, i.e., a data volume between 1 MB and 10 GB. In each step, we increased the amount of data by a factor of ten. We applied the three data processing tasks to these synthetic base data, using a desktop computer with an Intel Core i7-1165G7 with four cores and 16 GB DDR4-3200 of main memory. For each data volume and processing task, we measured the time it takes to process all data items. We carried out these measurements ten times each and after each run, we fully rolled back the SQLite DB and the TinyDB to exclude distortions due to warm caches. The medians of these runtime measurements are shown in Figure 9. Due to the use of medians, outliers (e.g., due to influences of concurrent background processes) do not skew the results.

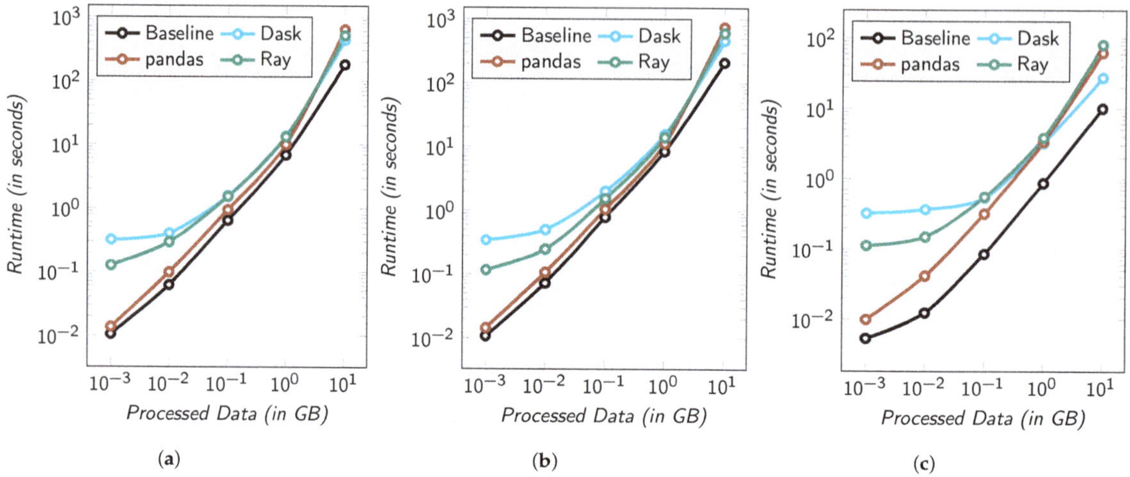

Figure 9. Performance Evaluation Results Regarding the Runtime Overhead of the REFINERY Platform on a Desktop Computer. (**a**) Selection Task; (**b**) Projection Task; (**c**) Aggregation Task.

Two things can be observed across all three tasks: On the one hand, Modin causes a basic overhead with both execution engines due to the initial partitioning and distribution as well as the merging of the results. Especially for small data volumes, this overhead is excessive, since the data processing tasks are not time-consuming. On the other hand, the costs caused by pandas increase significantly for larger data volumes. Since the complete DataFrame must be kept permanently in main memory in this case, a lot of memory paging is required once the remaining free main memory runs out. Especially with the highly parallelizable aggregation task—here, the mean value for each group can be computed independently—one can see that for large data volumes, Modin has the advantage. Apparently, for such rather simple tasks, the centralized scheduling strategy of Dask is more advantageous. For more complex tasks, however, in particular in the area of machine learning, the Ray-based execution engine should clearly outperform the other implementation approaches. Apart from the small data volumes, both parallelized approaches are even for these simple processing tasks in $\mathcal{O}$(baseline), i.e., the asymptotic behavior of the runtime costs incurred by the REFINERY Platform are identical to the one of the highly optimized baseline.

In a second deployment scenario, we assumed a large mass data provider running the REFINERY Platform to refine data from many data producers on a high-performance server cluster. To this end, we incrementally generated from 1250 k to 20,000 k data items, i.e., a data volume between 5 GB and 80 GB. In each step, we increased the data volume by a factor of two. We applied the three data processing tasks to these synthetic base data, using a server cluster with 188 CPUs and 3 TB of main memory, organized as one master and ten worker nodes. Again, for each data volume and processing task, we measured the time it takes to process all data items. Figure 10 also presents the median of ten consecutive runs for each runtime measurement.

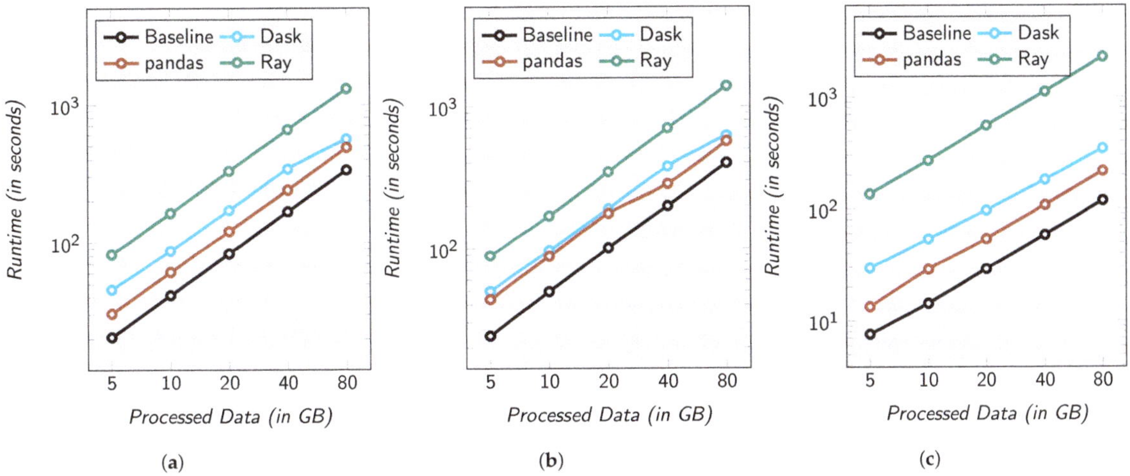

Figure 10. Performance Evaluation Results Regarding the Runtime Overhead of the REFINERY Platform on a Server Cluster. (**a**) Selection Task; (**b**) Projection Task; (**c**) Aggregation Task.

Here, the result is much more uniform for all three tasks. All three implementation variants as well as the baseline show a linear growth in processing costs across the board. Since all datasets easily fit into main memory, the runtime behavior of pandas does not deteriorate, even for the large data volumes. The processing tasks are so simple that the organizational overhead of splitting the dataset and distributing the chunks in parallel does not pay off. For more complex tasks, however, this would be the case. In addition, the pandas DataFrame already reaches nearly maximum capacity with the largest set of base data. Thus, a partitioning strategy is needed for larger volumes anyway. In any case, the runtime behavior of the REFINERY Platform is also in $\mathcal{O}$(baseline). As it therefore only causes a constant overhead in relation to the baseline in both deployment scenarios, this can be considered a great success. In return for this overhead, our approach offers the possibility to model data products in our ontology without requiring IT knowledge. Furthermore, our approach offers maximum flexibility in terms of the involved data sources and sinks.

If the latter is not required, the implementation of the Information Zone could be limited to relational databases. In this case, it is possible to use a tool such as Grizzly (see https://github.com/dbis-ilm/grizzly; accessed on 6 February 2023) for the implementation of our processing engine in order to further reduce the processing costs [244]. Grizzly also operates on pandas-like DataFrames. However, those DataFrames are not realized as actual data objects in memory that contain the data. Instead, all operations are translated into SQL commands that are executed directly by the data source. Lazy evaluation ensures that processing only occurs when a result has to be output, which can further reduce processing costs [245]. Since all operations used by our processing engine are fully supported by Grizzly, a migration to this execution engine is also possible if flexibility regarding the supported data sources is not required.

This performance evaluation demonstrates that our REFINERY Platform is efficient in handling the commodity 'data'. As it is therefore effective, practically applicable, and efficient in data administration and, due to its security and privacy features, respects both the interests of the data producers (i.e., protection of their sensitive data) and the interests of the data consumers (i.e., compliance with the promised data quality), it can be concluded that we have achieved our research goal to develop a reliable information retrieval and delivery platform.

6. Conclusions

Currently, data are boldly pithily referred to as the oil of the 21st century. On a metaphorical level, this statement is quite accurate, as the IoT and the resulting large-scale systematic collection of data not only enabled the fourth industrial revolution but also marked a major evolutionary step in the information age as it led to the digitization of society. Data-driven services significantly shape our everyday lives.

Even on a less figurative level, there are similarities between oil and data when it comes to their handling. Both commodities first have to be discovered and extracted, then refined, and finally delivered to consumers. A closer look, however, reveals inherent differences between the intangible commodity 'data' and a tangible commodity such as oil, which must be addressed when processing them.

Therefore, the goal of this work was to elaborate a modern data administration strategy that takes into account the strategic and economic importance of this special commodity. To this end, we made three contributions:

(a) Our investigation of the commodity 'data' revealed that ten unique characteristics have to be taken into account when handling this intangible resource. For instance, data are not only nonconsumable but can also be duplicated losslessly, which means that their volume is constantly growing. Furthermore, data accumulate at high velocity and have to be processed quickly, as they are partially volatile. Their heterogeneous nature and the need to apply individual refinement techniques to the data further complicate this endeavor. Since the economic value of data cannot be estimated in advance, and indiscernibly data manipulations can impair the quality of the data, it is essential to avoid unreasonably high handling costs. Finally, data products can be subject to special restrictions in terms of processing and provisioning. Therefore, there is a fundamental need for new trading concepts and infrastructures for the commodity 'data'.

(b) Based on this knowledge base, our review of state-of-the-art techniques related to data administration indicated that there are four aspects in particular where these characteristics need to be taken into account in order to enable effective and efficient data handling. First, data have to be acquired appropriately (in terms of, e.g., quality and quantity) from heterogeneous sources. These data must then be cleansed and made processable by means of data preparation and transformed into custom-made data products. The data products, together with all the high-volume data artifacts generated during manufacturing, must be managed and made retrievable. Only then can they be offered to data consumers in a digital storefront as part of data provisioning. In addition to these data administration tasks, security and privacy aspects also have to be taken into account in each of these work steps.

(c) Our review of related work revealed that there are many island solutions to individual aspects of these data administration problems. However, there is no holistic end-to-end solution addressing all data characteristics at once. This is necessary in order to achieve synergy effects and thus exploit the full potential of the commodity 'data'. To this end, we presented our own concept toward a reliable information retrieval and delivery platform called REFINERY Platform. Our REFINERY Platform not only addresses all the challenges we identified in the area of data administration but also provides both data producers and data consumers with assertions regarding data security and privacy on the one hand and data quality on the other hand. An in-depth assessment confirms that our approach is effective (in terms of provided functionality), practicable (in terms of operability), and efficient (in terms of data throughput) in this respect.

Despite its undeniable advantages in terms of modern data administration—namely its capability to deal with the challenges of big data while satisfying both the privacy requirements of data subjects and the data quality demands of data consumers—our presented REFINERY Platform also has some limitations. In this regard, it is important to keep in mind that we are presenting a concept. That is, although the various components of the REFINERY Platform have been implemented and their isolated application shows good results in terms of practicality and performance, it is an open task to implement and comprehensively evaluate a full-fledged prototype of the REFINERY Platform. However, since the more than promising results presented in this paper demonstrate the soundness of our approach and the effectiveness, practicality, and efficiency of its key components, we are confident about this future work.

Funding: This research received no external funding.

Institutional Review Board Statement: Not applicable.

Informed Consent Statement: Not applicable.

Data Availability Statement: Not applicable.

Conflicts of Interest: The author declares no conflict of interest.

Abbreviations

The following abbreviations are used in this paper:

ADS	Authenticated Data Structures
AI	Artificial Intelligence
App	(Mobile) Application
BI	Business Intelligence
CPU	Central Processing Unit
DB	Database
DDoS	Distributed Denial of Service Attack
DDR	Double Data Rate
ETL	Extraction, Transformation, Loading
GB	Gigabyte
GDPR	General Data Protection Regulation
IoT	Internet of Things
IT	Information Technology
MB	Megabyte
NoSQL	Not only SQL
OAuth	Open Authorization
OS	Operating System
PET	Privacy-Enhancing Technologies
PIN	Personal Identification Number
PMP	Privacy Management Platform
PoOR	Proofs of Ownership and Retrievability
PoRR	Proofs of Retrievability and Reliability
PUF	Physical Unclonable Function
RDF	Resource Description Framework
REFINERY Platform	Reliable Information Retrieval and Delivery Platform
SQL	Structured Query Language
STAMP	System-Theoretic Accident Model and Processes
STPA	System-Theoretic Process Analysis

TB	Terabyte
VDF	Verifiable Delay Function
XML	Extensible Markup Language

References

1. Schwab, K.; Marcus, A.; Oyola, J.R.; Hoffman, W.; Luzi, M. Personal Data: The Emergence of a New Asset Class. An Initiative of the World Economic Forum. 2011. pp. 1–40. Available online: https://www.weforum.org/reports/personal-data-emergence-new-asset-class/ (accessed on 6 February 2023).
2. Javornik, M.; Nadoh, N.; Lange, D. Data Is the New Oil. In *Towards User-Centric Transport in Europe: Challenges, Solutions and Collaborations*; Müller, B., Meyer, G., Eds.; Springer: Cham, Switzerland, 2019; pp. 295–308.
3. Klingenberg, C.O.; Borges, M.A.V.; Antunes, J.A.V., Jr. Industry 4.0 as a data-driven paradigm: A systematic literature review on technologies. *J. Manuf. Technol. Manag.* **2021**, *32*, 570–592. [CrossRef]
4. Sisinni, E.; Saifullah, A.; Han, S.; Jennehag, U.; Gidlund, M. Industrial Internet of Things: Challenges, Opportunities, and Directions. *IEEE Trans. Ind. Inform.* **2018**, *14*, 4724–4734. [CrossRef]
5. Singh, M.; Fuenmayor, E.; Hinchy, E.P.; Qiao, Y.; Murray, N.; Devine, D. Digital Twin: Origin to Future. *Appl. Syst. Innov.* **2021**, *4*, 36. [CrossRef]
6. Philbeck, T.; Davis, N. The Fourth Industrial Revolution: Shaping a New Era. *J. Int. Aff.* **2018**, *72*, 17–22.
7. Schwab, K. (Ed.) *The Fourth Industrial Revolution*, illustrated ed.; Crown Business: New York, NY, USA, 2017.
8. Lasi, H.; Fettke, P.; Kemper, H.G.; Feld, T.; Hoffmann, M. Industry 4.0. *Bus. Inf. Syst. Eng.* **2014**, *6*, 239–242. [CrossRef]
9. Leelaarporn, P.; Wachiraphan, P.; Kaewlee, T.; Udsa, T.; Chaisaen, R.; Choksatchawathi, T.; Laosirirat, R.; Lakhan, P.; Natnithikarat, P.; Thanontip, K.; et al. Sensor-Driven Achieving of Smart Living: A Review. *IEEE Sens. J.* **2021**, *21*, 10369–10391. [CrossRef]
10. Paiva, S.; Ahad, M.A.; Tripathi, G.; Feroz, N.; Casalino, G. Enabling Technologies for Urban Smart Mobility: Recent Trends, Opportunities and Challenges. *Sensors* **2021**, *21*, 2143. [CrossRef]
11. Al-rawashdeh, M.; Keikhosrokiani, P.; Belaton, B.; Alawida, M.; Zwiri, A. IoT Adoption and Application for Smart Healthcare: A Systematic Review. *Sensors* **2022**, *22*, 5377. [CrossRef] [PubMed]
12. Yar, H.; Imran, A.S.; Khan, Z.A.; Sajjad, M.; Kastrati, Z. Towards Smart Home Automation Using IoT-Enabled Edge-Computing Paradigm. *Sensors* **2021**, *21*, 4932. [CrossRef]
13. Taffel, S. Data and oil: Metaphor, materiality and metabolic rifts. *New Media Soc.* **2021**. [CrossRef]
14. Urbach, N.; Ahlemann, F. *IT Management in the Digital Age: A Roadmap for the IT Department of the Future*; Springer: Cham, Switzerland, 2019.
15. Possler, D.; Bruns, S.; Niemann-Lenz, J. Data Is the New Oil–But How Do We Drill It? Pathways to Access and Acquire Large Data Sets in Communication Science. *Int. J. Commun.* **2019**, *13*, 3894–3911.
16. Liew, A. Understanding Data, Information, Knowledge And Their Inter-Relationships. *J. Knowl. Manag. Pract.* **2007**, *8*, 1–10.
17. Sarker, I.H. Data Science and Analytics: An Overview from Data-Driven Smart Computing, Decision-Making and Applications Perspective. *SN Comput. Sci.* **2021**, *2*, 377. [CrossRef]
18. Arfat, Y.; Usman, S.; Mehmood, R.; Katib, I. Big Data Tools, Technologies, and Applications: A Survey. In *Smart Infrastructure and Applications: Foundations for Smarter Cities and Societies*; Mehmood, R., See, S., Katib, I., Chlamtac, I., Eds.; Springer: Cham, Switzerland, 2020; Chapter 19, pp. 453–490.
19. Rowley, J. The wisdom hierarchy: Representations of the DIKW hierarchy. *J. Inf. Sci.* **2007**, *33*, 163–180. [CrossRef]
20. Mandel, M. *The Economic Impact of Data: Why Data Is Not Like Oil*; ppi Radically Pragmatic: London, UK, 2017; pp. 1–20. Available online: https://www.progressivepolicy.org/publication/economic-impact-data-data-not-like-oil/ (accessed on 6 February 2023).
21. Nolin, J.M. Data as oil, infrastructure or asset? Three metaphors of data as economic value. *J. Inf. Commun. Ethics Soc.* **2020**, *18*, 28–43. [CrossRef]
22. Katal, A.; Wazid, M.; Goudar, R.H. Big data: Issues, challenges, tools and Good practices. In Proceedings of the 2013 Sixth International Conference on Contemporary Computing (IC3), Noida, India, 8–10 August 2013; IEEE: Piscataway, NJ, USA, 2013; pp. 404–409.
23. Mladenović, M.N. Data is not the new oil, but could be water or sunlight? From ethical to moral pathways for urban data management. In Proceedings of the 17th International Conference on Computational Urban Planning and Urban Management (CUPUM), Espoo, Finland, 9–11 June 2021; pp. 9–11.
24. Hirsch, D.D. The Glass House Effect: Big Data, the New Oil, and the Power of Analogy. *Maine Law Rev.* **2014**, *66*, 373–395.
25. van der Aalst, W.M.P. Data Scientist: The Engineer of the Future. In Proceedings of the 7th International Conference on Interoperability for Enterprises Systems and Applications (I-ESA), Albi, France, 24–28 March 2014; Springer: Cham, Switzerland, 2014; pp. 13–26.
26. Siddiqa, A.; Hashem, I.A.T.; Yaqoob, I.; Marjani, M.; Shamshirband, S.; Gani, A.; Nasaruddin, F. A survey of big data management: Taxonomy and state-of-the-art. *J. Netw. Comput. Appl.* **2016**, *71*, 151–166. [CrossRef]
27. Moreno, J.; Serrano, M.A.; Fernández-Medina, E. Main Issues in Big Data Security. *Future Internet* **2016**, *8*, 44. [CrossRef]

28. Binjubeir, M.; Ahmed, A.A.; Ismail, M.A.B.; Sadiq, A.S.; Khurram Khan, M. Comprehensive Survey on Big Data Privacy Protection. *IEEE Access* **2020**, *8*, 20067–20079. [CrossRef]
29. Löcklin, A.; Vietz, H.; White, D.; Ruppert, T.; Jazdi, N.; Weyrich, M. Data administration shell for data-science-driven development. *Procedia CIRP* **2021**, *100*, 115–120. [CrossRef]
30. Jeyaprakash, T.; Padmaveni, K. Introduction to Data Science—An Overview. *Int. J. Sci. Manag. Stud.* **2021**, *4*, 407–410. [CrossRef]
31. Lyko, K.; Nitzschke, M.; Ngonga Ngomo, A.C. Big Data Acquisition. In *New Horizons for a Data-Driven Economy: A Roadmap for Usage and Exploitation of Big Data in Europe*; Cavanillas, J.M., Curry, E., Wahlster, W., Eds.; Springer: Cham, Switzerland, 2016; Chapter 4, pp. 39–61.
32. Curry, E. The Big Data Value Chain: Definitions, Concepts, and Theoretical Approaches. In *New Horizons for a Data-Driven Economy: A Roadmap for Usage and Exploitation of Big Data in Europe*; Cavanillas, J.M., Curry, E., Wahlster, W., Eds.; Springer: Cham, Switzerland, 2016; Chapter 3, pp. 29–37.
33. Vassiliadis, P.; Simitsis, A.; Skiadopoulos, S. Conceptual Modeling for ETL Processes. In Proceedings of the 5th ACM International Workshop on Data Warehousing and OLAP (DOLAP), McLean, VA, USA, 8 November 2002; ACM: New York, NY, USA, 2002; pp. 14–21.
34. Simitsis, A. Modeling and managing ETL processes. In Proceedings of the VLDB 2003 PhD Workshop co-located with the 29th International Conference on Very Large Databases (VLDB), Berlin, Germany, 9–12 September 2003; CEUR-WS.org: Aachen, Germany, 2003; pp. 1–5.
35. Lau, B.P.L.; Marakkalage, S.H.; Zhou, Y.; Hassan, N.U.; Yuen, C.; Zhang, M.; Tan, U.X. A survey of data fusion in smart city applications. *Inf. Fusion* **2019**, *52*, 357–374. [CrossRef]
36. Diouf, P.S.; Boly, A.; Ndiaye, S. Variety of data in the ETL processes in the cloud: State of the art. In Proceedings of the 2018 IEEE International Conference on Innovative Research and Development (ICIRD), Bangkok, Thailand, 11–12 May 2018; IEEE: Piscataway, NJ, USA, 2018; pp. 1–5.
37. D'silva, G.M.; Khan, A.; Gaurav.; Bari, S. Real-time processing of IoT events with historic data using Apache Kafka and Apache Spark with dashing framework. In Proceedings of the 2017 2nd IEEE International Conference on Recent Trends in Electronics, Information & Communication Technology (RTEICT), Bangalore, India, 19–20 May 2017; IEEE: Piscataway, NJ, USA, 2017; pp. 1804–1809.
38. Geng, D.; Zhang, C.; Xia, C.; Xia, X.; Liu, Q.; Fu, X. Big Data-Based Improved Data Acquisition and Storage System for Designing Industrial Data Platform. *IEEE Access* **2019**, *7*, 44574–44582. [CrossRef]
39. Huai, Y.; Chauhan, A.; Gates, A.; Hagleitner, G.; Hanson, E.N.; O'Malley, O.; Pandey, J.; Yuan, Y.; Lee, R.; Zhang, X. Major Technical Advancements in Apache Hive. In Proceedings of the 2014 ACM SIGMOD International Conference on Management of Data (SIGMOD), Snowbird, UT, USA, 22–27 June 2014; ACM: New York, NY, USA, 2014; pp. 1235–1246.
40. Lee, G.; Lin, J.; Liu, C.; Lorek, A.; Ryaboy, D. The Unified Logging Infrastructure for Data Analytics at Twitter. *Proc. VLDB Endow.* **2012**, *5*, 1771–1780. [CrossRef]
41. Marz, N. How to Beat the CAP Theorem. Thoughts from the Red Planet. 2011. Available online: http://nathanmarz.com/blog/how-to-beat-the-cap-theorem.html (accessed on 6 February 2023).
42. Kreps, J. Questioning the Lambda Architecture. *O'Reilly*, 2 July 2014. Available online: https://www.oreilly.com/radar/questioning-the-lambda-architecture/ (accessed on 6 February 2023).
43. Kraetz, D.; Morawski, M. Architecture Patterns—Batch and Real-Time Capabilities. In *The Digital Journey of Banking and Insurance, Volume III: Data Storage, Data Processing and Data Analysis*; Liermann, V., Stegmann, C., Eds.; Palgrave Macmillan: Cham, Switzerland, 2021; pp. 89–104.
44. Lin, J. The Lambda and the Kappa. *IEEE Internet Comput.* **2017**, *21*, 60–66. [CrossRef]
45. Terrizzano, I.; Schwarz, P.; Roth, M.; Colino, J.E. Data Wrangling: The Challenging Journey from the Wild to the Lake. In Proceedings of the 7th Biennial Conference on Innovative Data Systems Research (CIDR), Asilomar, CA, USA, 4–7 January 2015; pp. 1–9.
46. Ding, X.; Wang, H.; Su, J.; Li, Z.; Li, J.; Gao, H. Cleanits: A Data Cleaning System for Industrial Time Series. *Proc. VLDB Endow.* **2019**, *12*, 1786–1789. [CrossRef]
47. Behringer, M.; Hirmer, P.; Mitschang, B. A Human-Centered Approach for Interactive Data Processing and Analytics. In Proceedings of the 19th International Conference on Enterprise Information Systems (ICEIS), Porto, Portugal, 26–29 April 2017; Springer: Cham, Switzerland, 2018; pp. 498–514.
48. Diamantini, C.; Lo Giudice, P.; Potena, D.; Storti, E.; Ursino, D. An Approach to Extracting Topic-guided Views from the Sources of a Data Lake. *Inf. Syst. Front.* **2021**, *23*, 243–262. [CrossRef]
49. Bogatu, A.; Fernandes, A.A.A.; Paton, N.W.; Konstantinou, N. Dataset Discovery in Data Lakes. In Proceedings of the 2020 IEEE 36th International Conference on Data Engineering (ICDE), Dallas, TX, USA, 20–24 April 2020; IEEE: Piscataway, NJ, USA, 2020; pp. 709–720.
50. Megdiche, I.; Ravat, F.; Zhao, Y. Metadata Management on Data Processing in Data Lakes. In Proceedings of the 47th International Conference on Current Trends in Theory and Practice of Computer Science (SOFSEM), Bolzano-Bozen, Italy, 25–29 January 2021; Springer: Cham, Switzerland, 2021; pp. 553–562.

51. Castro Fernandez, R.; Abedjan, Z.; Koko, F.; Yuan, G.; Madden, S.; Stonebraker, M. Aurum: A Data Discovery System. In Proceedings of the 2018 IEEE 34th International Conference on Data Engineering (ICDE), Paris, France, 16–19 April 2018; IEEE: Piscataway, NJ, USA, 2018; pp. 1001–1012.
52. Behringer, M.; Hirmer, P.; Fritz, M.; Mitschang, B. Empowering Domain Experts to Preprocess Massive Distributed Datasets. In Proceedings of the 23rd International Conference on Business Information Systems (BIS), Colorado Springs, CO, USA, 8–10 June 2020; Springer: Cham, Switzerland, 2020; pp. 61–75.
53. Behringer, M.; Fritz, M.; Schwarz, H.; Mitschang, B. DATA-IMP: An Interactive Approach to Specify Data Imputation Transformations on Large Datasets. In Proceedings of the 27th International Conference on Cooperative Information Systems (CoopIS), Bozen-Bolzano, Italy, 4–7 October 2022; Springer: Cham, Switzerland, 2022; pp. 55–74.
54. Mahdavi, M.; Abedjan, Z. Semi-Supervised Data Cleaning with Raha and Baran. In Proceedings of the 11th Annual Conference on Innovative Data Systems Research (CIDR), Chaminade, CA, USA, 11–15 January 2021; pp. 1–7.
55. Wulf, A.J.; Seizov, O. "Please understand we cannot provide further information": Evaluating content and transparency of GDPR-mandated AI disclosures. *AI & Soc.* **2022**, 1–22. [CrossRef]
56. Auge, T.; Heuer, A. ProSA—Using the CHASE for Provenance Management. In Proceedings of the 23rd European Conference on Advances in Databases and Information Systems (ADBIS), Bled, Slovenia, 8–11 September 2019; Springer: Cham, Switzerland, 2019; pp. 357–372.
57. Lam, H.T.; Buesser, B.; Min, H.; Minh, T.N.; Wistuba, M.; Khurana, U.; Bramble, G.; Salonidis, T.; Wang, D.; Samulowitz, H. Automated Data Science for Relational Data. In Proceedings of the 2021 IEEE 37th International Conference on Data Engineering (ICDE), Chania, Greece, 19–22 April 2021; IEEE: Piscataway, NJ, USA, 2021; pp. 2689–2692.
58. Ilyas, I.F.; Rekatsinas, T. Machine Learning and Data Cleaning: Which Serves the Other? *J. Data Inf. Qual.* **2022**, *14*, 1–11. [CrossRef]
59. Devlin, B.; Cote, L.D. *Data Warehouse: From Architecture to Implementation*; Addison-Wesley Longman Publishing Co., Inc.: Boston, MA, USA, 1996.
60. Aftab, U.; Siddiqui, G.F. Big Data Augmentation with Data Warehouse: A Survey. In Proceedings of the 2018 IEEE International Conference on Big Data (Big Data), Seattle, WA, USA, 10–13 December 2018; IEEE: Piscataway, NJ, USA, 2018; pp. 2785–2794.
61. Wongthongtham, P.; Abu-Salih, B. Ontology and trust based data warehouse in new generation of business intelligence: State-of-the-art, challenges, and opportunities. In Proceedings of the 2015 IEEE 13th International Conference on Industrial Informatics (INDIN), Cambridge, UK, 22–24 July 2015; IEEE: Piscataway, NJ, USA, 2015; pp. 476–483.
62. Mathis, C. Data Lakes. *Datenbank-Spektrum* **2017**, *17*, 289–293. [CrossRef]
63. Taniar, D.; Rahayu, W. Data Lake Architecture. In Proceedings of the 9th International Conference on Emerging Internet, Data & Web Technologies (EIDWT), Chiang Mai, Thailand, 25–27 February 2021; Springer: Cham, Switzerland, 2021; pp. 344–357.
64. Ravat, F.; Zhao, Y. Data Lakes: Trends and Perspectives. In Proceedings of the 30th International Conference on Database and Expert Systems Applications (DEXA), Linz, Austria, 26–29 August 2019; Springer: Cham, Switzerland, 2019; pp. 304–313.
65. Giebler, C.; Gröger, C.; Hoos, E.; Schwarz, H.; Mitschang, B. Leveraging the Data Lake: Current State and Challenges. In Proceedings of the 21st International Conference on Big Data Analytics and Knowledge Discovery (DaWaK), Linz, Austria, 26–29 August 2019; Springer: Cham, Switzerland, 2019; pp. 179–188.
66. Giebler, C.; Gröger, C.; Hoos, E.; Schwarz, H.; Mitschang, B. A Zone Reference Model for Enterprise-Grade Data Lake Management. In Proceedings of the 2020 IEEE 24th International Enterprise Distributed Object Computing Conference (EDOC), Eindhoven, The Netherlands, 5–8 October 2020; IEEE: Piscataway, NJ, USA, 2020; pp. 57–66.
67. Hai, R.; Geisler, S.; Quix, C. Constance: An Intelligent Data Lake System. In Proceedings of the 2016 International Conference on Management of Data (SIGMOD), San Francisco, CA, USA, 26 June–1 July 2016; ACM: New York, NY, USA, 2016; pp. 2097–2100.
68. Farid, M.; Roatis, A.; Ilyas, I.F.; Hoffmann, H.F.; Chu, X. CLAMS: Bringing Quality to Data Lakes. In Proceedings of the 2016 International Conference on Management of Data (SIGMOD), San Francisco, CA, USA, 26 June–1 July 2016; ACM: New York, NY, USA, 2016; pp. 2089–2092.
69. Machado, I.A.; Costa, C.; Santos, M.Y. Data Mesh: Concepts and Principles of a Paradigm Shift in Data Architectures. *Procedia Comput. Sci.* **2022**, *196*, 263–271. [CrossRef]
70. Oreščanin, D.; Hlupić, T. Data Lakehouse—A Novel Step in Analytics Architecture. In Proceedings of the 2021 44th International Convention on Information, Communication and Electronic Technology (MIPRO), Opatija, Croatia, 9–11 June 2021; IEEE: Piscataway, NJ, USA, 2021; pp. 1242–1246.
71. Armbrust, M.; Ghodsi1, A.; Xin, R.; Zaharia, M. Lakehouse: A New Generation of Open Platforms that Unify Data Warehousing and Advanced Analytics. In Proceedings of the 11th Annual Conference on Innovative Data Systems Research (CIDR), Chaminade, CA, USA, 11–15 January 2021; pp. 1–8.
72. Alpar, P.; Schulz, M. Self-Service Business Intelligence. *Bus. Inf. Syst. Eng.* **2016**, *58*, 151–155. [CrossRef]
73. Lennerholt, C.; van Laere, J. Data access and data quality challenges of self-service business intelligence. In Proceedings of the 27th European Conference on Information Systems (ECIS), Stockholm and Uppsala, Sweden, 8–14 June 2019; AIS: Atlanta, GA, USA, 2019; pp. 1–13.
74. Huang, L.; Dou, Y.; Liu, Y.; Wang, J.; Chen, G.; Zhang, X.; Wang, R. Toward a research framework to conceptualize data as a factor of production: The data marketplace perspective. *Fundam. Res.* **2021**, *1*, 586–594. [CrossRef]
75. Gröger, C. There is No AI without Data. *Commun. ACM* **2021**, *64*, 98–108. [CrossRef]

76. Eichler, R.; Gröger, C.; Hoos, E.; Schwarz, H.; Mitschang, B. From Data Asset to Data Product – The Role of the Data Provider in the Enterprise Data Marketplace. In Proceedings of the 17th Symposium and Summer School On Service-Oriented Computing (SummerSOC), Heraklion, Greece, 3–9 July 2022; Springer: Cham, Switzerland, 2022; pp. 119–138.
77. Eichler, R.; Giebler, C.; Gröger, C.; Hoos, E.; Schwarz, H.; Mitschang, B. Enterprise-Wide Metadata Management: An Industry Case on the Current State and Challenges. In Proceedings of the 24th International Conference on Business Information Systems (BIS), Hannover, Germany, 14–17 June 2021; pp. 269–279.
78. Eichler, R.; Gröger, C.; Hoos, E.; Schwarz, H.; Mitschang, B. Data Shopping—How an Enterprise Data Marketplace Supports Data Democratization in Companies. In Proceedings of the 34th International Conference on Advanced Information Systems Engineering (CAiSE), Leuven, Belgium, 6–10 June 2022; Springer: Cham, Switzerland, 2022; pp. 19–26.
79. Eichler, R.; Giebler, C.; Gröger, C.; Schwarz, H.; Mitschang, B. Modeling metadata in data lakes—A generic model. *Data Knowl. Eng.* **2021**, *136*, 101931. [CrossRef]
80. Driessen, S.W.; Monsieur, G.; Van Den Heuvel, W.J. Data Market Design: A Systematic Literature Review. *IEEE Access* **2022**, *10*, 33123–33153. [CrossRef]
81. Chanal, P.M.; Kakkasageri, M.S. Security and Privacy in IoT: A Survey. *Wirel. Pers. Commun.* **2020**, *115*, 1667–1693. [CrossRef]
82. Samonas, S.; Coss, D. The CIA Strikes Back: Redefining Confidentiality, Integrity and Availability in Security. *J. Inf. Syst. Secur.* **2014**, *10*, 21–45.
83. *ISO/IEC 27000:2018(en)*; Information Technology—Security Techniques—Information Security Management Systems—Overview and Vocabulary. International Organization for Standardization: Geneva, Switzerland, 2018.
84. Maqsood, F.; Ali, M.M.; Ahmed, M.; Shah, M.A. Cryptography: A Comparative Analysis for Modern Techniques. *Int. J. Adv. Comput. Sci. Appl.* **2017**, *8*, 442–448. [CrossRef]
85. Henriques, M.S.; Vernekar, N.K. Using symmetric and asymmetric cryptography to secure communication between devices in IoT. In Proceedings of the 2017 International Conference on IoT and Application (ICIOT), Nagapattinam, India, 19–20 May 2017; IEEE: Piscataway, NJ, USA, 2017; pp. 1–4.
86. Shafagh, H.; Hithnawi, A.; Burkhalter, L.; Fischli, P.; Duquennoy, S. Secure Sharing of Partially Homomorphic Encrypted IoT Data. In Proceedings of the 15th ACM Conference on Embedded Network Sensor Systems (SenSys), Delft, The Netherlands, 6–8 November 2017; ACM: New York, NY, USA, 2017; pp. 1–14.
87. Do, H.G.; Ng, W.K. Blockchain-Based System for Secure Data Storage with Private Keyword Search. In Proceedings of the 2017 IEEE World Congress on Services (SERVICES), Honolulu, HI, USA, 25–30 June 2017; IEEE: Piscataway, NJ, USA, 2017; pp. 90–93.
88. Sun, X.; Zhang, P.; Liu, J.K.; Yu, J.; Xie, W. Private Machine Learning Classification Based on Fully Homomorphic Encryption. *IEEE Trans. Emerg. Top. Comput.* **2020**, *8*, 352–364. [CrossRef]
89. Ouaddah, A.; Mousannif, H.; Abou Elkalam, A.; Ait Ouahman, A. Access control in the Internet of Things: Big challenges and new opportunities. *Comput. Netw.* **2017**, *112*, 237–262. [CrossRef]
90. Qiu, J.; Tian, Z.; Du, C.; Zuo, Q.; Su, S.; Fang, B. A Survey on Access Control in the Age of Internet of Things. *IEEE Internet Things J.* **2020**, *7*, 4682–4696. [CrossRef]
91. Alagar, V.; Alsaig, A.; Ormandjiva, O.; Wan, K. Context-Based Security and Privacy for Healthcare IoT. In Proceedings of the 2018 IEEE International Conference on Smart Internet of Things (SmartIoT), Xi'an, China, 17–19 August 2018; IEEE: Piscataway, NJ, USA, 2018; pp. 122–128.
92. Alkhresheh, A.; Elgazzar, K.; Hassanein, H.S. Context-aware Automatic Access Policy Specification for IoT Environments. In Proceedings of the 2018 14th International Wireless Communications & Mobile Computing Conference (IWCMC), Limassol, Cyprus, 25–29 June 2018; IEEE: Piscataway, NJ, USA, 2018; pp. 793–799.
93. Novo, O. Blockchain Meets IoT: An Architecture for Scalable Access Management in IoT. *IEEE Internet Things J.* **2018**, *5*, 1184–1195. [CrossRef]
94. Raikwar, M.; Gligoroski, D.; Velinov, G. Trends in Development of Databases and Blockchain. In Proceedings of the 2020 Seventh International Conference on Software Defined Systems (SDS), Paris, France, 20–23 April 2020; IEEE: Piscataway, NJ, USA, 2020; pp. 177–182.
95. Li, R.; Song, T.; Mei, B.; Li, H.; Cheng, X.; Sun, L. Blockchain for Large-Scale Internet of Things Data Storage and Protection. *IEEE Trans. Serv. Comput.* **2019**, *12*, 762–771. [CrossRef]
96. Chowdhury, M.J.M.; Colman, A.; Kabir, M.A.; Han, J.; Sarda, P. Blockchain as a Notarization Service for Data Sharing with Personal Data Store. In Proceedings of the 2018 17th IEEE International Conference on Trust, Security And Privacy In Computing and Communications/12th IEEE International Conference on Big Data Science and Engineering (TrustCom/BigDataSE), New York, NY, USA, 1–3 August 2018; IEEE: Piscataway, NJ, USA, 2018; pp. 1330–1335.
97. Gupta, S.; Hellings, J.; Rahnama, S.; Sadoghi, M. Building High Throughput Permissioned Blockchain Fabrics: Challenges and Opportunities. *Proc. VLDB Endow.* **2020**, *13*, 3441–3444. [CrossRef]
98. Li, Y.; Zheng, K.; Yan, Y.; Liu, Q.; Zhou, X. EtherQL: A Query Layer for Blockchain System. In Proceedings of the 22nd International Conference on Database Systems for Advanced Applications (DASFAA), Suzhou, China, 27–30 March 2017; Springer: Cham, Switzerland, 2017; pp. 556–567.
99. Bragagnolo, S.; Rocha, H.; Denker, M.; Ducasse, S. Ethereum Query Language. In Proceedings of the 1st International Workshop on Emerging Trends in Software Engineering for Blockchain (WETSEB), Gothenburg, Sweden, 27 May 2018; ACM: New York, NY, USA, 2018; pp. 1–8.

100. Qu, Q.; Nurgaliev, I.; Muzammal, M.; Jensen, C.S.; Fan, J. On spatio-temporal blockchain query processing. *Future Gener. Comput. Syst.* **2019**, *98*, 208–218. [CrossRef]
101. Hao, K.; Xin, J.; Wang, Z.; Yao, Z.; Wang, G. On efficient top-k transaction path query processing in blockchain database. *Data Knowl. Eng.* **2022**, *141*, 102079. [CrossRef]
102. Han, J.; Kim, H.; Eom, H.; Coignard, J.; Wu, K.; Son, Y. Enabling SQL-Query Processing for Ethereum-Based Blockchain Systems. In Proceedings of the 9th International Conference on Web Intelligence, Mining and Semantics (WIMS), Seoul, Republic of Korea, 26–28 June 2019; ACM: New York, NY, USA, 2019; pp. 1–7.
103. Przytarski, D. Using Triples as the Data Model for Blockchain Systems. In Proceedings of the Blockchain enabled Semantic Web Workshop and Contextualized Knowledge Graphs Workshop Co-Located with the 18th International Semantic Web Conference (BlockSW/CKG@ISWC), Auckland, New Zealand, 26–30 October 2019; pp. 1–2.
104. Kurt Peker, Y.; Rodriguez, X.; Ericsson, J.; Lee, S.J.; Perez, A.J. A Cost Analysis of Internet of Things Sensor Data Storage on Blockchain via Smart Contracts. *Electronics* **2020**, *9*, 244. [CrossRef]
105. Hepp, T.; Sharinghousen, M.; Ehret, P.; Schoenhals, A.; Gipp, B. On-chain vs. off-chain storage for supply- and blockchain integration. *IT Inf. Technol.* **2018**, *60*, 283–291. [CrossRef]
106. Schuhknecht, F.; Sharma, A.; Dittrich, J.; Agrawal, D. chainifyDB: How to get rid of your Blockchain and use your DBMS instead. In Proceedings of the 11th Annual Conference on Innovative Data Systems Research (CIDR), Chaminade, CA, USA, 11–15 January 2021; pp. 1–10.
107. Wang, H.; Xu, C.; Zhang, C.; Xu, J. vChain: A Blockchain System Ensuring Query Integrity. In Proceedings of the 2020 ACM SIGMOD International Conference on Management of Data (SIGMOD), Portland, OR, USA, 14–19 June 2020; ACM: New York, NY, USA, 2020; pp. 2693–2696.
108. Nathan, S.; Govindarajan, C.; Saraf, A.; Sethi, M.; Jayachandran, P. Blockchain Meets Database: Design and Implementation of a Blockchain Relational Database. *Proc. VLDB Endow.* **2019**, *12*, 1539–1552. [CrossRef]
109. Cai, H.; Xu, B.; Jiang, L.; Vasilakos, A.V. IoT-Based Big Data Storage Systems in Cloud Computing: Perspectives and Challenges. *IEEE Internet Things J.* **2017**, *4*, 75–87. [CrossRef]
110. Habib, S.M.; Hauke, S.; Ries, S.; Mühlhäuser, M. Trust as a facilitator in cloud computing: A survey. *J. Cloud Comput. Adv. Syst. Appl.* **2012**, *1*, 19. [CrossRef]
111. Khan, K.M.; Malluhi, Q. Establishing Trust in Cloud Computing. *IT Prof.* **2010**, *12*, 20–27. [CrossRef]
112. Gilad-Bachrach, R.; Laine, K.; Lauter, K.; Rindal, P.; Rosulek, M. Secure Data Exchange: A Marketplace in the Cloud. In Proceedings of the 2019 ACM SIGSAC Conference on Cloud Computing Security Workshop (CCSW), London, UK, 11 November 2019; ACM: New York, NY, USA, 2019; pp. 117–128.
113. Gritti, C.; Chen, R.; Susilo, W.; Plantard, T. Dynamic Provable Data Possession Protocols with Public Verifiability and Data Privacy. In Proceedings of the 13th International Conference on Information Security Practice and Experience (ISPEC), Melbourne, Australia, 13–15 December 2017; Springer: Cham, Switzerland, 2017; pp. 485–505.
114. Du, R.; Deng, L.; Chen, J.; He, K.; Zheng, M. Proofs of Ownership and Retrievability in Cloud Storage. In Proceedings of the 2014 IEEE 13th International Conference on Trust, Security and Privacy in Computing and Communications (TrustCom), Beijing, China, 24–26 September 2014; IEEE: Piscataway, NJ, USA, 2014; pp. 328–335.
115. Erway, C.C.; Küpçü, A.; Papamanthou, C.; Tamassia, R. Dynamic Provable Data Possession. *ACM Trans. Inf. Syst. Secur.* **2015**, *17*, 1–29. [CrossRef]
116. Merkle, R.C. A Digital Signature Based on a Conventional Encryption Function. In Proceedings of the Conference on the Theory and Applications of Cryptographic Techniques (CRYPTO), Santa Barbara, CA, USA, 16–20 August 1987; Springer: Berlin/Heidelberg, Germany, 1988; pp. 369–378.
117. Gritti, C.; Li, H. Efficient Publicly Verifiable Proofs of Data Replication and Retrievability Applicable for Cloud Storage. *Adv. Sci. Technol. Eng. Syst. J.* **2022**, *7*, 107–124. [CrossRef]
118. Boneh, D.; Bonneau, J.; Bünz, B.; Fisch, B. Verifiable Delay Functions. In Proceedings of the 38th International Cryptology Conference (Crypto), Santa Barbara, CA, USA, 17–19 August 2018; Springer: Cham, Switzerland, 2018; pp. 757–788.
119. Salim, M.M.; Rathore, S.; Park, J.H. Distributed denial of service attacks and its defenses in IoT: A survey. *J. Supercomput.* **2020**, *76*, 5320–5363. [CrossRef]
120. Zhang, H.; Wen, Y.; Xie, H.; Yu, N. *Distributed Hash Table: Theory, Platforms and Applications*; Springer: New York, NY, USA, 2013.
121. Singh, R.; Tanwar, S.; Sharma, T.P. Utilization of blockchain for mitigating the distributed denial of service attacks. *Secur. Priv.* **2020**, *3*, e96. [CrossRef]
122. Liu, X.; Farahani, B.; Firouzi, F. Distributed Ledger Technology. In *Intelligent Internet of Things: From Device to Fog and Cloud*; Firouzi, F., Chakrabarty, K., Nassif, S., Eds.; Springer: Cham, Switzerland, 2020; Chapter 8, pp. 393–431.
123. Zhu, Q.; Loke, S.W.; Trujillo-Rasua, R.; Jiang, F.; Xiang, Y. Applications of Distributed Ledger Technologies to the Internet of Things: A Survey. *ACM Comput. Surv.* **2019**, *52*, 1–34. [CrossRef]
124. Peng, Y.; Du, M.; Li, F.; Cheng, R.; Song, D. FalconDB: Blockchain-Based Collaborative Database. In Proceedings of the 2020 ACM SIGMOD International Conference on Management of Data (SIGMOD), Portland, OR, USA, 14–19 June 2020; ACM: New York, NY, USA, 2020; pp. 637–652.
125. El-Hindi, M.; Binnig, C.; Arasu, A.; Kossmann, D.; Ramamurthy, R. BlockchainDB: A Shared Database on Blockchains. *Proc. VLDB Endow.* **2019**, *12*, 1597–1609. [CrossRef]

126. Barkadehi, M.H.; Nilashi, M.; Ibrahim, O.; Zakeri Fardi, A.; Samad, S. Authentication systems: A literature review and classification. *Telemat. Inform.* **2018**, *35*, 1491–1511. [CrossRef]
127. Ferrag, M.A.; Maglaras, L.; Derhab, A. Authentication and Authorization for Mobile IoT Devices Using Biofeatures: Recent Advances and Future Trends. *Secur. Commun. Netw.* **2019**, *2019*, 5452870. [CrossRef]
128. Cheong, S.N.; Ling, H.C.; Teh, P.L. Secure Encrypted Steganography Graphical Password scheme for Near Field Communication smartphone access control system. *Expert Syst. Appl.* **2014**, *41*, 3561–3568. [CrossRef]
129. Baig, A.F.; Eskeland, S. Security, Privacy, and Usability in Continuous Authentication: A Survey. *Sensors* **2021**, *35*, 5967. [CrossRef]
130. Sciancalepore, S.; Piro, G.; Caldarola, D.; Boggia, G.; Bianchi, G. OAuth-IoT: An access control framework for the Internet of Things based on open standards. In Proceedings of the 2017 IEEE Symposium on Computers and Communications (ISCC), Heraklion, Crete, Greece, 3–6 July 2017; IEEE: Piscataway, NJ, USA, 2017; pp. 676–681.
131. Kulseng, L.; Yu, Z.; Wei, Y.; Guan, Y. Lightweight Mutual Authentication and Ownership Transfer for RFID Systems. In Proceedings of the 2010 30th IEEE International Conference on Computer Communications (INFOCOM), San Diego, CA, USA, 14–19 March 2010; IEEE: Piscataway, NJ, USA, 2010; pp. 1–5.
132. Maes, R. *Physically Unclonable Functions*; Springer: Berlin/Heidelberg, Germany, 2013.
133. He, W.; Golla, M.; Padhi, R.; Ofek, J.; Dürmuth, M.; Fernandes, E.; Ur, B. Rethinking Access Control and Authentication for the Home Internet of Things (IoT). In Proceedings of the 27th USENIX Security Symposium (USENIX Security), Baltimore, MD, USA, 15–17 August 2018; USENIX Association: Berkeley, CA, USA, 2018; pp. 255–272.
134. Almuairfi, S.; Veeraraghavan, P.; Chilamkurti, N. A novel image-based implicit password authentication system (IPAS) for mobile and non-mobile devices. *Math. Comput. Model.* **2013**, *58*, 108–116. [CrossRef]
135. Hu, V.C.; Kuhn, D.R.; Ferraiolo, D.F.; Voas, J. Attribute-Based Access Control. *Computer* **2015**, *48*, 85–88. [CrossRef]
136. Hemdi, M.; Deters, R. Using REST based protocol to enable ABAC within IoT systems. In Proceedings of the 2016 IEEE 7th Annual Information Technology, Electronics and Mobile Communication Conference (IEMCON), Vancouver, BC, Canada, 13–15 October 2016; IEEE: Piscataway, NJ, USA, 2016; pp. 1–7.
137. Hüffmeyer, M.; Schreier, U. Formal Comparison of an Attribute Based Access Control Language for RESTful Services with XACML. In Proceedings of the 21st ACM on Symposium on Access Control Models and Technologies (SACMAT), Shanghai, China, 6–8 June 2016; ACM: New York, NY, USA, 2016; pp. 171–178.
138. Liu, S.; You, S.; Yin, H.; Lin, Z.; Liu, Y.; Yao, W.; Sundaresh, L. Model-Free Data Authentication for Cyber Security in Power Systems. *IEEE Trans. Smart Grid* **2020**, *11*, 4565–4568. [CrossRef]
139. Arnold, M.; Schmucker, M.; Wolthusen, S.D. *Techniques and Applications of Digital Watermarking and Content Protection*; Artech House, Inc.: Norwood, MA, USA, 2003.
140. Agrawal, R.; Haas, P.J.; Kiernan, J. Watermarking relational data: Framework, algorithms and analysis. *VLDB J.* **2003**, *12*, 157–169.
141. Subramanya, S.; Yi, B.K. Digital signatures. *IEEE Potentials* **2006**, *25*, 5–8. [CrossRef]
142. Kaur, R.; Kaur, A. Digital Signature. In Proceedings of the 2012 International Conference on Computing Sciences (ICCS), Phagwara, India, 14–15 September 2012; IEEE: Piscataway, NJ, USA, 2012; pp. 295–301.
143. Gritti, C.; Molva, R.; Önen, M. Lightweight Secure Bootstrap and Message Attestation in the Internet of Things. In Proceedings of the 33rd Annual ACM Symposium on Applied Computing (SAC), Pau, France, 9–13 April 2018; ACM: New York, NY, USA, 2018; pp. 775–782.
144. Gritti, C.; Önen, M.; Molva, R. CHARIOT: Cloud-Assisted Access Control for the Internet of Things. In Proceedings of the 2018 16th Annual Conference on Privacy, Security and Trust (PST), Belfast, Ireland, 28–30 August 2018; IEEE: Piscataway, NJ, USA, 2018; pp. 1–6.
145. Gritti, C.; Önen, M.; Molva, R. Privacy-Preserving Delegable Authentication in the Internet of Things. In Proceedings of the 34th ACM/SIGAPP Symposium on Applied Computing (SAC), Limassol, Cyprus, 8–12 April 2019; ACM: New York, NY, USA, 2019; pp. 861–869.
146. Antoniadis, S.; Litou, I.; Kalogeraki, V. A Model for Identifying Misinformation in Online Social Networks. In Proceedings of the 2015 Confederated International Conferences on the Move to Meaningful Internet Systems: CoopIS, ODBASE, and C&TC (OTM), Rhodes, Greece, 26–30 October 2015; Springer: Cham, Switzerland, 2015; pp. 473–482.
147. Litou, I.; Kalogeraki, V.; Katakis, I.; Gunopulos, D. Efficient and timely misinformation blocking under varying cost constraints. *Online Soc. Netw. Media* **2017**, *2*, 19–31. [CrossRef]
148. Litou, I.; Kalogeraki, V. Influence Maximization in Evolving Multi-Campaign Environments. In Proceedings of the 2018 IEEE International Conference on Big Data (Big Data), Seattle, WA, USA, 10–13 December 2018; IEEE: Piscataway, NJ, USA, 2018; pp. 448–457.
149. Lupton, D. *The Quantified Self*; Polity: Cambridge, UK; Malden, MA, USA, 2016.
150. Jiang, B.; Li, J.; Yue, G.; Song, H. Differential Privacy for Industrial Internet of Things: Opportunities, Applications, and Challenges. *IEEE Internet Things J.* **2021**, *8*, 10430–10451. [CrossRef]
151. Perr, I.N. Privilege, confidentiality, and patient privacy: Status 1980. *J. Forensic Sci.* **1981**, *26*, 109–115. [CrossRef]
152. Read, G. The Seal of Confession. *Law Justice Christ. Law Rev.* **2022**, *188*, 28–37.
153. Roba, R.M. The legal protection of the secrecy of correspondence. *Curentul Jurid. Jurid. Curr. Le Courant Juridique* **2009**, *36*, 135–154.
154. Bok, S. The Limits of Confidentiality. *Hastings Cent. Rep.* **1983**, *13*, 24–31. [CrossRef] [PubMed]
155. Hirshleifer, J. Privacy: Its Origin, Function, and Future. *J. Leg. Stud.* **1980**, *9*, 649–664. [CrossRef]

156. Westin, A.F. *Privacy and Freedom*; Atheneum Books: New York City, NY, USA, 1967.
157. Margulis, S.T. On the Status and Contribution of Westin's and Altman's Theories of Privacy. *J. Soc. Issues* **2003**, *59*, 411–429. [CrossRef]
158. Diamantopoulou, V.; Lambrinoudakis, C.; King, J.; Gritzalis, S. EU GDPR: Toward a Regulatory Initiative for Deploying a Private Digital Era. In *Modern Socio-Technical Perspectives on Privacy*; Knijnenburg, B.P., Page, X., Wisniewski, P., Lipford, H.R., Proferes, N., Romano, J., Eds.; Springer: Cham, Switzerland, 2022; Chapter 18, pp. 427–448.
159. European Parliament and Council of the European Union. Regulation on the protection of natural persons with regard to the processing of personal data and on the free movement of such data, and repealing Directive 95/46/EC (Data Protection Directive). Legislative Acts L119. *Off. J. Eur. Union* **2016**. Available online: https://eur-lex.europa.eu/eli/reg/2016/679/oj (accessed on 6 February 2023).
160. Williams, M.; Nurse, J.R.C.; Creese, S. The Perfect Storm: The Privacy Paradox and the Internet-of-Things. In Proceedings of the 2016 11th International Conference on Availability, Reliability and Security (ARES), Salzburg, Austria, 31 August–2 September 2016; IEEE: Piscataway, NJ, USA, 2016; pp. 644–652.
161. Jung, F.; von Holdt, K.; Krüger, R.; Meyer, J.; Heuten, W. I Do. Do I?—Understanding User Perspectives on the Privacy Paradox. In Proceedings of the 25th International Academic Mindtrek Conference (Academic Mindtrek), Tampere, Finland, 16–18 November 2022; ACM: New York, NY, USA, 2022; pp. 268–277.
162. Rubinstein, I.S.; Good, N. The trouble with Article 25 (and how to fix it): The future of data protection by design and default. *Int Data Priv. Law* **2020**, *10*, 37–56. [CrossRef]
163. Georgiopoulou, Z.; Makri, E.L.; Lambrinoudakis, C. GDPR compliance: Proposed technical and organizational measures for cloud provider. *Inf. Comput. Secur.* **2020**, *28*, 665–680. [CrossRef]
164. Gyrard, A.; Zimmermann, A.; Sheth, A. Building IoT-Based Applications for Smart Cities: How Can Ontology Catalogs Help? *IEEE Internet Things J.* **2018**, *5*, 3978–3990. [CrossRef]
165. Chen, G.; Jiang, T.; Wang, M.; Tang, X.; Ji, W. Modeling and reasoning of IoT architecture in semantic ontology dimension *Comput. Commun.* **2020**, *153*, 580–594. [CrossRef]
166. Gheisari, M.; Najafabadi, H.E.; Alzubi, J.A.; Gao, J.; Wang, G.; Abbasi, A.A.; Castiglione, A. OBPP: An ontology-based framework for privacy-preserving in IoT-based smart city. *Future Gener. Comput. Syst.* **2021**, *123*, 1–13. [CrossRef]
167. Leveson, N.G. *Engineering a Safer World: Systems Thinking Applied to Safety*; MIT Press: Cambridge, MA, USA, 2011.
168. Shapiro, S.S. Privacy Risk Analysis Based on System Control Structures: Adapting System-Theoretic Process Analysis for Privacy Engineering. In Proceedings of the 2016 IEEE Security and Privacy Workshops (SPW), San Jose, CA, USA, 22–26 May 2016; IEEE: Piscataway, NJ, USA, 2016; pp. 17–24.
169. Renganathan, V.; Yurtsever, E.; Ahmed, Q.; Yener, A. Valet attack on privacy: A cybersecurity threat in automotive Bluetooth infotainment systems. *Cybersecurity* **2022**, *5*, 30. [CrossRef]
170. Angerschmid, A.; Zhou, J.; Theuermann, K.; Chen, F.; Holzinger, A. Fairness and Explanation in AI-Informed Decision Making. *Mach. Learn. Knowl. Extr.* **2022**, *4*, 556–579. [CrossRef]
171. Hagras, H. Toward Human-Understandable, Explainable AI. *Computer* **2018**, *51*, 28–36. [CrossRef]
172. Holzinger, A.; Saranti, A.; Molnar, C.; Biecek, P.; Samek, W. Explainable AI Methods—A Brief Overview. In Proceedings of the International Workshop on Extending Explainable AI Beyond Deep Models and Classifiers Held in Conjunction with ICML 2020 (xxAI), Vienna, Austria, 18 July 2020; Springer: Cham, Switzerland, 2022; pp. 13–38.
173. Utz, C.; Degeling, M.; Fahl, S.; Schaub, F.; Holz, T. (Un)Informed Consent: Studying GDPR Consent Notices in the Field. In Proceedings of the 2019 ACM SIGSAC Conference on Computer and Communications Security (CCS), London, UK, 11–15 November 2019; ACM: New York, NY, USA, 2019; pp. 973–990.
174. Kaaniche, N.; Laurent, M.; Belguith, S. Privacy enhancing technologies for solving the privacy-personalization paradox: Taxonomy and survey. *J. Netw. Comput. Appl.* **2020**, *171*, 102807. [CrossRef]
175. Li, N.; Qardaji, W.; Su, D. On Sampling, Anonymization, and Differential Privacy or, k-Anonymization Meets Differential Privacy. In Proceedings of the 7th ACM Symposium on Information, Computer and Communications Security (ASIACCS), Seoul, Republic of Korea, 2–4 May 2012; ACM: New York, NY, USA, 2012; pp. 32–33.
176. Pattuk, E.; Kantarcioglu, M.; Ulusoy, H.; Malin, B. Privacy-aware dynamic feature selection. In Proceedings of the 2015 IEEE 31st International Conference on Data Engineering (ICDE), Seoul, Republic of Korea, 13–17 April 2015; IEEE: Piscataway, NJ, USA, 2015; pp. 78–88.
177. Dou, H.; Chen, Y.; Yang, Y.; Long, Y. A secure and efficient privacy-preserving data aggregation algorithm. *J. Ambient. Intell. Humaniz. Comput.* **2022**, *13*, 1495–1503. [CrossRef]
178. Alpers, S.; Oberweis, A.; Pieper, M.; Betz, S.; Fritsch, A.; Schiefer, G.; Wagner, M. PRIVACY-AVARE: An approach to manage and distribute privacy settings. In Proceedings of the 2017 3rd IEEE International Conference on Computer and Communications (ICCC), Chengdu, China, 13–16 December 2017; IEEE: Piscataway, NJ, USA, 2017; pp. 1460–1468.
179. Jiang, H.; Li, J.; Zhao, P.; Zeng, F.; Xiao, Z.; Iyengar, A. Location Privacy-Preserving Mechanisms in Location-Based Services: A Comprehensive Survey. *ACM Comput. Surv.* **2021**, *54*, 4. [CrossRef]
180. Wang, Z.; Wang, B.; Srivastava, M. Protecting User Data Privacy with Adversarial Perturbations: Poster Abstract. In Proceedings of the 20th International Conference on Information Processing in Sensor Networks Co-Located with CPS-IoT Week 2021 (IPSN), Nashville, TN, USA, 18–21 May 2021; ACM: New York, NY, USA, 2021; pp. 386–387.

181. Hernández Acosta, L.; Reinhardt, D. A Survey on Privacy Issues and Solutions for Voice-Controlled Digital Assistants. *Pervasive Mob. Comput.* **2022**, *80*, 101523. [CrossRef]
182. Palanisamy, S.M.; Dürr, F.; Tariq, M.A.; Rothermel, K. Preserving Privacy and Quality of Service in Complex Event Processing through Event Reordering. In Proceedings of the 12th ACM International Conference on Distributed and Event-Based Systems (DEBS), Hamilton, New Zealand, 25–29 June 2018; ACM: New York, NY, USA, 2018; pp. 40–51.
183. Kwecka, Z.; Buchanan, W.; Schafer, B.; Rauhofer, J. "I am Spartacus": Privacy enhancing technologies, collaborative obfuscation and privacy as a public good. *Artif. Intell. Law* **2014**, *22*, 113–139. [CrossRef]
184. Slijepčević, D.; Henzl, M.; Klausner, L.D.; Dam, T.; Kieseberg, P.; Zeppelzauer, M. k-Anonymity in practice: How generalisation and suppression affect machine learning classifiers. *Comput. Secur.* **2021**, *111*, 102488. [CrossRef]
185. Dwork, C.; McSherry, F.; Nissim, K.; Smith, A. Calibrating Noise to Sensitivity in Private Data Analysis. *J. Priv. Confidentiality* **2017**, *7*, 17–51. [CrossRef]
186. Machanavajjhala, A.; He, X.; Hay, M. Differential Privacy in the Wild: A Tutorial on Current Practices & Open Challenges. In Proceedings of the 2017 ACM International Conference on Management of Data (SIGMOD), Chicago, IL, USA, 14–19 May 2017; ACM: New York, NY, USA, 2017; pp. 1727–1730.
187. Dwork, C.; Kohli, N.; Mulligan, D. Differential Privacy in Practice: Expose your Epsilons! *J. Privacy Confid.* **2019**, *9*, 1–22. [CrossRef]
188. Yang, Q.; Liu, Y.; Cheng, Y.; Kang, Y.; Chen, T.; Yu, H. *Federated Learning*; Morgan & Claypool: San Rafael, CA, USA, 2019.
189. Wu, X.; Zhang, Y.; Shi, M.; Li, P.; Li, R.; Xiong, N.N. An adaptive federated learning scheme with differential privacy preserving. *Future Gener. Comput. Syst.* **2022**, *127*, 362–372. [CrossRef]
190. István, Z.; Ponnapalli, S.; Chidambaram, V. Software-Defined Data Protection: Low Overhead Policy Compliance at the Storage Layer is within Reach! *Proc. VLDB Endow.* **2021**, *14*, 1167–1174. [CrossRef]
191. Wang, W.C.; Ho, C.C.; Chang, Y.M.; Chang, Y.H. Challenges and Designs for Secure Deletion in Storage Systems. In Proceedings of the 2020 Indo—Taiwan 2nd International Conference on Computing, Analytics and Networks (Indo-Taiwan ICAN), Rajpura, India, 7–15 February 2020; IEEE: Piscataway, NJ, USA, 2020; pp. 181–189.
192. Zhang, Q.; Jia, S.; Chang, B.; Chen, B. Ensuring data confidentiality via plausibly deniable encryption and secure deletion—A survey. *Cybersecurity* **2018**, *1*, 1. [CrossRef]
193. Politou, E.; Alepis, E.; Virvou, M.; Patsakis, C. Privacy in Blockchain. In *Privacy and Data Protection Challenges in the Distributed Era*; Springer: Cham, Switzerland, 2022; Chapter 7, pp. 133–149.
194. Meng, W.; Ge, J.; Jiang, T. Secure Data Deduplication with Reliable Data Deletion in Cloud. *Int. J. Found. Comput. Sci.* **2019**, *30*, 551–570. [CrossRef]
195. Waizenegger, T.; Wagner, F.; Mega, C. SDOS: Using Trusted Platform Modules for Secure Cryptographic Deletion in the Swift Object Store. In Proceedings of the 20th International Conference on Extending Database Technology (EDBT), Venice, Italy, 21–24 March 2017; OpenProceedings.org: Konstanz, Germany, 2017; pp. 550–553.
196. Auge, T. Extended Provenance Management for Data Science Applications. In Proceedings of the VLDB 2020 PhD Workshop co-located with the 46th International Conference on Very Large Databases (VLDB), Tokyo, Japan, 31 August–4 September 2020; CEUR-WS.org: Aachen, Germany, 2020; pp. 1–4.
197. Davidson, S.B.; Khanna, S.; Roy, S.; Stoyanovich, J.; Tannen, V.; Chen, Y. On Provenance and Privacy. In Proceedings of the 14th International Conference on Database Theory (ICDT), Uppsala, Sweden, 21–24 March 2011; ACM: New York, NY, USA, 2011; pp. 3–10.
198. Auge, T.; Scharlau, N.; Heuer, A. Privacy Aspects of Provenance Queries. In Proceedings of the 8th and 9th International Provenance and Annotation Workshop (IPAW), Virtual Event, 19–22 July 2021; Springer: Cham, Switzerland, 2021; pp. 218–221.
199. Stach, C.; Alpers, S.; Betz, S.; Dürr, F.; Fritsch, A.; Mindermann, K.; Palanisamy, S.M.; Schiefer, G.; Wagner, M.; Mitschang, B.; et al. The AVARE PATRON—A Holistic Privacy Approach for the Internet of Things. In Proceedings of the 15th International Joint Conference on e-Business and Telecommunications (SECRYPT), Porto, Portugal, 26–28 July 2018; SciTePress: Setúbal, Portugal, 2018; pp. 372–379.
200. Stach, C. How to Deal with Third Party Apps in a Privacy System—The PMP Gatekeeper. In Proceedings of the 2015 16th IEEE International Conference on Mobile Data Management (MDM), Pittsburgh, PA, USA, 15–18 June 2015; IEEE: Piscataway, NJ, USA, 2015; pp. 167–172.
201. Stach, C. VAULT: A Privacy Approach towards High-Utility Time Series Data. In Proceedings of the Thirteenth International Conference on Emerging Security Information, Systems and Technologies (SECURWARE), Nice, France, 27–31 October 2019; IARIA: Wilmington, DE, USA, 2019; pp. 41–46.
202. Stach, C.; Mitschang, B. Privacy Management for Mobile Platforms—A Review of Concepts and Approaches. In Proceedings of the 2013 IEEE 14th International Conference on Mobile Data Management (MDM), Milan, Italy, 3–6 June 2013; IEEE: Piscataway, NJ, USA, 2013; pp. 305–313.
203. Stach, C.; Mitschang, B. Design and Implementation of the Privacy Management Platform. In Proceedings of the 2014 IEEE 15th International Conference on Mobile Data Management (MDM), Brisbane, Australia, 14–18 July 2014; IEEE: Piscataway, NJ, USA, 2014; pp. 69–72.

future internet

MDPI

Article

Multilayer Backbones for Internet of Battlefield Things †

Evangelia Fragkou, Dimitrios Papakostas, Theodoros Kasidakis and Dimitrios Katsaros *

Department of Electrical and Computer Engineering, University of Thessaly, 38334 Volos, Greece; efragkou@uth.gr (E.F.); papdimit@inf.uth.gr (D.P.); tkasidakis@uth.gr (T.K.)
* Correspondence: dkatsar@inf.uth.gr

† This paper is an extended version of our paper "Backbones for Internet of Battlefield Things" published in the Proceedings of the IEEE/IFIP Annual Conference on Wireless On-Demand Network Systems and Services, Klosters, Switzerland, 9–11 March 2021.

Abstract: The Internet of Battlefield Things is a newly born cyberphysical system and, even though it shares a lot with the Internet of Things and with ad hoc networking, substantial research is required to cope with its scale and peculiarities. This article examines a fundamental problem pertaining to the routing of information, i.e., the calculation of a backbone network. We model an IoBT network as a network with multiple layers and employ the concept of *domination* for multilayer networks. This is a significant departure from earlier works, and in spite of the huge literature on the topic during the past twenty years, the problem in IoBT networks is different since these networks are multilayer networks, thus making inappropriate all the past, related literature because it deals with single layer (flat) networks. We establish the computational complexity of our problem, and design a distributed algorithm for computing connected dominating sets with small cardinality. We analyze the performance of the proposed algorithm on generated topologies, and compare it against two—the only existing—competitors. The proposed algorithm establishes itself as the clear winner in all experiments concerning the dominating set from a size-wise and an energy-wise perspective achieving a performance gain of about 15%.

Keywords: dominating sets; multilayer networks; Internet of Battlefield Things; ad hoc networking

Citation: Fragkou, E.; Papakostas, D.; Kasidakis, T.; Katsaros, D. Multilayer Backbones for Internet of Battlefield Things. *Future Internet* **2022**, *14*, 186. https://doi.org/10.3390/fi14060186

Academic Editors: Christoph Stach and Clémentine Gritti

Received: 24 May 2022
Accepted: 10 June 2022
Published: 15 June 2022

1. Introduction

The progress in IoT inevitably impacted upon the modern battlefield, which is populated by thousands of "things", such as humans, sensors, vehicles, unmanned aerial vehicles (UAVs), aircrafts, etc. carrying out various tasks including environmental sensing, communicating, acting in isolation and/or in cooperation [1,2]. Therefore, the Internet of Battle(field) Things [1] or the Internet of Military Things (IoMT) was born, which interconnects "devices" aiming to meet multiple and diverse missions, to operate in a (semi)autonomic mode, and to execute battlefield operations supporting end-to-end control and command; its ultimate goal is to carry out a commander's intent in a safe, responsive and resilient manner. Thus, IoBT presents at the same time all of the following significant and very challenging characteristics [3] which distinguish it from traditional IoT:

- *Diversity in tasks and aims.* There will be many networks operating simultaneously to achieve a particular goal, e.g., tracking, surveillance, attack.
- *Operation in dynamically changing and resource starving environments.* Some "devices" of the IoBT network might be energy-starving (sensors, drones), others might not have energy issues (armored vehicles), others might be obliged to travel in non-chartered territories (e.g., planes).
- *Extreme device heterogeneity.* IoBT networks are expected to include from tiny little sensors to some as big as armored fighting vehicles.
- *High variance in network's density and size.* An IoBT network might be comprised e.g., by the cluttered network of a swarm of drones, or by the "union" of the ad hoc network of a battalion's soldiers and the ad hoc network of a tank platoon.

The so-called *assured synthesis* is one of the major challenges identified for IoBT, and in particular the *recruitment* and *network composition* [3] tasks. The former task is about the discovery of cyberphysical assets, the human assets and their particularities, and also about the resilience to adversaries. The latter task is about the issue of dictating the nodes that must be considered in order for the requirements and constraints of the planned mission to be satisfied.

1.1. Related Literature

The Internet of Battlefield Things concept emerged some five years ago [1,4], and since then research is conducted in designing backbones [5,6], in designing reconfigurable and secure IoBT networks [7], in developing solutions for enemy localization [8], in controling/monitoring communication links, in combating attacks at nodes [9], in detecting malware and fake news [10,11], and in protecting human assets [12,13].

The past literature on designing backbones for routing support in wireless ad hoc networks comprises the most closely related work to the present article. This literature is literally enormous during the past two decades [14–16]. However, the same issue in the realm of military/IoBT networks is different; these networks are actually multilayer networks [5,7] (see Figure 1), and therefore this past literature is in principle inappropriate, because it concentrates only on single layer networks. Moreover, we showed in ([5] Theorem 1), that these algorithms usually produce suboptimal solutions in the sense that, instead of announcing as a dominator a node with a few interlayer links, they tend to announce someone with many intralayer links. Finally, the work in [17] deals with domination in multiplex networks which is a far more restricted version of multilayer networks, since interlayer links exist only between *clones* of the same node in different layers; moreover, that work does not establish the computational complexity of their problem.

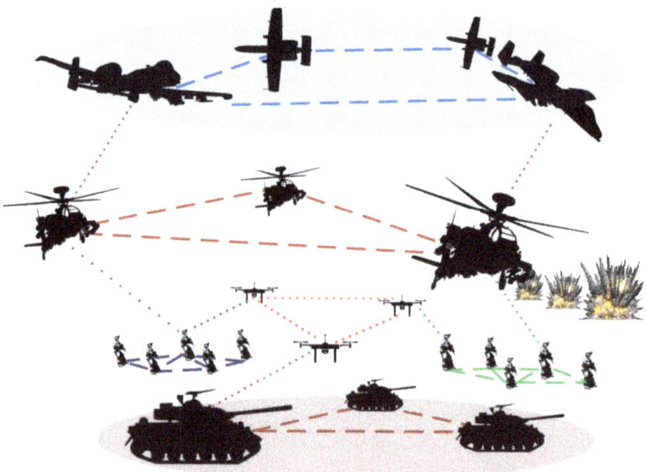

Figure 1. A sample multilayer IoBT ad hoc network.

The very first algorithm for calculating a backbone for multilayer military networks is described in [5], where a distributed backbone establishment algorithm, namely *clPCI* was presented. It is based on the concept of connected dominating sets (CDS). The concept of a CDS was selected for the following reasons: (a) only/mainly distributed algorithms are appropriate in the battlefield, and despite the fact that finding a Minimum CDS is a NP-complete problem even in the centralized setting, there exist some very efficient distributed algorithms for the problem; (b) the battlefield requires resilient solutions, and fortunately the scientific knowledge on how to build a multi-connected, multi-dominating

CDS is quite mature [18]; and finally (c) an IoBT is composed by many energy-starving devices, and therefore energy-aware solutions (e.g., sleep scheduling) is almost mandatory; again, the theory on how to create multiple dominating sets, i.e., domatic partitions [19] is quite rich. Until before the introduction of *clPCI* for constructing a connected dominating set in a distributed fashion for multilayer networks, there was no prior work on the topic. This complete lack may be attributed to the mistaken assumption that a multilayer network is equivalent (from the perspective of computing a CDS) to a single layer network after ignoring layer information.

Multilayer networks have also been used in biological sciences in order to model interacting networks. In [20], *FAST-MDSM* is developed to calculate a dominating sets in multilayer networks. However, this algorithm is *centralized*, and it results in an *unconnected* dominating set, incurring also a high computation cost because it runs as an integer programming problem (Calculating a minimum (connected) dominating set after formulating it as an integer programming model is a mature technique [21]). In conclusion, all aforementioned algorithms consider only single layer networks, with the exception of [5], and of the centralized *FAST-MDSM* which comprises a departure from this literature.

1.2. Motivation and Work's Contributions

In [5] we dealt with the heterogeneity of an IoBT network and modelled it as a *multilayer network*. We proved analytically that if we treat it as a plain union of independent subnetworks, then we do not reap efficient solutions for facilitating fast information routing. So, despite the past, very rich literature on the domination concept in ad hoc networks, the problem in IoBT networks is different, since these are multilayer networks and considering them as a single (flat) network or considering each layer in isolation and calculating dominating set produces either suboptimal or bad solutions. Therefore, the whole past literature is in principle inappropriate. Later, we incorporated in this first work energy issues [6], and we devised algorithms with fact-finding methods to address aspects of social sensing, e.g., characterize human assets/sources. We note here, that in [5] we developed distributed, communication-efficient algorithms based on the concept of node domination for choosing a set of nodes to be a backbone for a multilayer network, i.e., the IoBT network. In the biological sciences area, the work described in [20] developed centralized algorithms for computing dominating sets in multilayer networks.

Nevertheless, neither the algorithms developed in [5] (and of course in [6]) nor those developed in [20] can adequately serve the purposes of an IoBT backbone network. The main issue with the former algorithms is that they do not produce very compact network spanners, so there is plenty of room for improvement with regard to the produced connected dominating set. This is very significant if we wish to design solutions scalable to enormous IoBT networks. The two main disadvantages of the latter algorithm is its centralized nature, and thus it can not work in IoBT, and additionally because it produces unconnected spanners, and therefore it can not guarantee network-wide flow of information. Thus, our article contributes the following:

- It establishes the computation complexity (NP-completeness) of the problem of calculating a minimum connected dominating set for multilayer networks.
- It presents a new distributed algorithm, namely the *Cross layer Connected Dominating Set* (*CCDS*) for calculating connected dominating sets for IoBT networks, by applying an efficient mechanism to reduce the size of the dominating set.
- It enhances *FAST-MDSM* [20] so as to produce connected dominating sets thus getting a new algorithm called *FAST-CMDSM*, and compares our proposed algorithm against the only existing competitors, namely the algorithm in [5] and *FAST-CMDSM*.

The rest of this article is organized as follows: Section 2 formulates the problem and proves its complexity, and Section 3 develops distributed solutions for it. Section 4 evaluates the competing algorithms, and finally, Section 5 summarizes the article.

2. Formulation of the Problem

A dominating set (DS) [22] of a graph (i.e., the set of *dominators*) is defined as a subset of the nodes with the property that the rest of the nodes are adjacent (i.e., 1-hop neighbors) from some dominator(s). If the network induced by the DS is additionally connected, then the DS is called connected DS (CDS). In the IoBT setting, we seek for minimum CDS (MCDS), i.e., CDS with minimum cardinality. We treat an IoBT network as a multilayer network, i.e., as a multilayer graph [5,7]. We assume that there exist only bidirectional links (In principle the paper ideas can be applied to networks with unidirectional links as well). A multilayer network comprised of n layers is a pair (G^{ML}, E^{ML}), where $G^{ML} = \{G^i, i = 1, \ldots, n\}$ is a set of networks (G_i, E_i) (i.e., with G_i nodes and E_i links), and a set of interlayer links $E^{ML} = \{E_{i,j} \subseteq G_i \times G_j; i, j \in \{1, \ldots, n\}, i \neq j\}$. Figure 1 illustrates such a network where we can see a *layer* of soldiers, a layer comprised by helicopters, the *intralayer* links connecting "nodes" of the same layer, and *interlayer* links connecting "nodes" belonging to different layers.

Apparently, the requirement of calculating a CDS with minimum cardinality stems from scalability issues [3]. Additionally, from the discussion in Section 1 and in ([3] Section III.B) where "... resilience and latency requirements for synthesizing a near-optimal network" [3] are emphasized, we conclude that the more nodes with many interlayer links are included in our IoBT backbone the better it is. The inclusion of many nodes with "a lot" of interlayer links supports low-latency communication among layers; we can consider them as the hubs encountered in the literature on complex networks that reduce the "degrees of separation". Moreover, the existence within the backbone of many nodes with "a lot" of interlayer links, reduces the danger of partitioning among layers. Therefore, we provide Definition 1 for the problem of calculating a MCDS for IoBT networks, and establish its computational complexity (Proposition 1) in the centralized setting, closing a gap in the literature which is open since [5] and it was not dealt with in [23].

Definition 1 (Multilayer MCDS problem for IoBT). *Solve the MCDS for a multilayer graph in a distributed manner, i.e., calculate the set $MCDS^{ML}$ comprised by the minimum number of nodes with the following properties: (a) their induced subgraph is connected (with intra and/or inter-layer links) and the rest of the nodes are adjacent to one node (or more) belonging to $MCDS^{ML}$, (b) maximize the number of dominators with many interlayer links, (c) knowing only the k-hop neighborhood of each node; we work with $k = 2$ here. (Working with broader neighborhoods (i.e., $k > 2$) would cause a severe broadcast storm problem [24] in order to acquire the topology deploying successive rounds of "beacon" messages).*

In the rest of this section, we will investigate and establish the computational complexity of the centralized version of our problem for connected dominating sets. In particular, we will define the decision and optimization version of the examined problem, and then establish their complexities. The validity of these results for the case of connected dominating sets calculated in a distributed fashion is straightforward.

MULTILAYER CONNECTED DOMINATING SET PROBLEM
INSTANCE: A positive integer K, and a multilayer network consisting of n layers, i.e., a set of n pairs (G_i, E_i), where G_i is the node set of a usual network and E_i is a set of edges, for $1 \leq i \leq n$, and a set of interlayer links $E^{ML} = \{E_{i,j} \subseteq G_i \times G_j; i, j \in \{1, \ldots, n\}, i \neq j\}$.
QUESTION: Is there a dominating set of size K for (G^{ML}, E^{ML}), i.e., a subset $V' \subseteq (\cup V_i)$ with $|V'| = K$ such that for all $u \in (\cup V_i) - V'$ there is some $v_j \in (\cup V_i)$ for which $(u, v_j) \in ((\cup E_i) \cup E^{ML})$, and moreover the induced subgraph defined by V' is connected?

MULTILAYER MINIMUM CONNECTED DOMINATING SET PROBLEM
INSTANCE: A multilayer network consisting of n layers, i.e., a set of n pairs (G_i, E_i), where G_i is the node set of a usual network and E_i is a set of edges, for $1 \leq i \leq n$, and a set of interlayer links $E^{ML} = \{E_{i,j} \subseteq G_i \times G_j; i, j \in \{1, \ldots, n\}, i \neq j\}$.

QUESTION: What is the minimum cardinality dominating set for (G^{ML}, E^{ML}), i.e., what is the minimum cardinality subset $V' \subseteq (\cup V_i)$ such that for all $u \in (\cup V_i) - V'$ there is some $v_j \in (\cup V_i)$ for which $(u, v_j) \in ((\cup E_i) \cup E^{ML})$, and moreover the induced subgraph defined by V' is connected?

We can now proceed to establish the computational complexity of the second problem. Our proof method will also establish the complexity of the first problem as well. The result is described in Proposition 1, but firstly we remind some background on proving the computational complexity of problems. So, a problem is NP-complete if the following two conditions hold:

(a) we can prove that the problem belongs to the class of NP problems, *and*
(b) we can
   (b1) *either* transform in polynomial-time a known NP-complete problem to the problem at hand ([25] p. 45), *or*
   (b2) prove that the problem at hand contains a known NP-complete problem as a special case ([25] p. 63, Section 3.2.1); this is called the "restriction" approach to proving NP-completeness.

We will now construct the proof that the problem of finding a minimum connected dominating set for (V, E_i) is NP-complete. Recall that our problem is an *optimization* problem, i.e., "…finding the *minimum* dominating set".

Proposition 1. *The Multilayer MDS problem is NP-complete.*

Proof. Our proof consists of three steps: (a) Transform the problem into a decision version, (b) Prove that it belongs to the class of NP problems. (c) Establish its NP-completeness complexity by following the "restriction" approach, i.e., by showing that our problem contains a known NP-complete problem as a special case.
[STEP 1. Transform our optimization problem into the decision version of the same problem.]
Without violating the validity of the proof, we will work with the *decision* version of the problem:
Given an integer $k > 0$, does our multilayer network contain a connected dominating set of cardinality equal to k?

Apparently, the smallest k for which the answer to the above problem is 'yes' is the size of the minimum connected dominating set of our multilayer network.

From the perspective of computational complexity, the two problems are equivalent (see item (b1) above): With the use of binary search, we need to solve the decision version for $O(log(|\cup_i G^i|))$ different values of k, i.e., which is a polynomial time of searches.
[STEP 2. Proving that the decision problem belongs to the NP class.]
The MULTILAYER CONNECTED DOMINATING SET PROBLEM is clearly in NP, as given a graph (G^{ML}, E^{ML}), a set $S \subseteq \cup_i G^i$ of nodes, and a number k, we can test if S is a connected dominating set of (G^{ML}, E^{ML}) of size k or less by first checking if its cardinality is less than or equal to k and then checking if every node in (G^{ML}, E^{ML}) is either in S or adjacent to a node in S. This process clearly takes polynomial time, i.e., $O(|\cup_i G^i| + |\cup_i E^i| + |E^{ML}|)$. Should we wish to check whether S is actually connected, we can resort to any polynomial-time algorithm, e.g., breadth/depth first search. Therefore, the MULTILAYER CONNECTED DOMINATING SET PROBLEM is indeed in NP.
[STEP 3. Proving that the decision problem contains a known NP-complete problem as a special case.]
The third step of our proof involves proving that the MULTILAYER CONNECTED DOMINATING SET PROBLEM contains as a special case the problem of finding a connected dominating set for a single layer graph $G(V, E)$—a known NP-complete problem ([25] p. 190, Problem GT2).

Algorithm 2: FAST-CMDSM

precondition : All nodes are designated as dominators
postcondition : Completed MCDS election process
remarks : mlNetwork G = (V, E) where V and E are vertex & edge set, $M(u)$: (T)rue/(F)alse indicator for node u to being a DS node, Vm : DS node set, MDS_V : Minimum DS node set of V, CDS_V : Connected DS node set of V.

repeat
  if $\exists\ u_{j\ (1\leq\ j\ \leq n)}\ \in V\ |$
  $d(u_j) = 1\ \&\ u_{i\ (1\leq\ i\ \leq n,\ i\neq\ j)}\ \in\ N(u_j)$ **then**
    $M(u_{i\ (1\leq i\leq n)})$ = T;
    if $u_{i\ (1\leq i\leq n)} \notin Vm$ **then**
      Add node $u_{i\ (1\leq i\leq n)}$ to Vm;
    end
  end
until *all nodes at every layer have been examined*
repeat
  if $M(u_{j\ (1\leq j\leq n)}) = T$ **then**
    if $\exists\ u_{i\ (1\leq i\leq n)} \in N(u_j)\ \&\ M(u_{i\ (1\leq i\leq n)}) = T$ **then**
      $M(u_{j\ (1\leq j\leq n)})$ = F;
      continue;
    else
      if $u_{j\ (1\leq j\leq n)} \notin Vm$ **then**
        Add node $u_{j\ (1\leq j\leq n)}$ to Vm;
      end
    end
  end
until *no more nodes are added to* Vm
MDS_V = **Minimize** $\sum_{i=1}^{n} x_i$
Subject to $x_i + \sum_{j\ (u_j,u_i)\in E_k}^{n} x_j \geq 1\ \forall\ u_i \in V_{k\ (1\leq k\leq n)}$
repeat
  Add to CDS_V the least possible nodes from $N(u)$ that are needed to cover $N^2(u)$ neighborhood
until *any node* $u \in MDS_V$ *has been examined*
$CDS_V = Vm \bigcup (CDS_V - (CDS_V \bigcap Vm))$
Use Pruning in order to decrease the size of the CDS_V

3.3. *Computation and Communication Complexities*

3.3.1. Complexities of $CCDS$

$CCDS$ requires 7 rounds of communication among nodes to complete. In each round, at most one packet is sent over the wireless channel.

- The 2-hop neighborhood information used by the relay node set election process is collected via two rounds of information exchanges.
- In round 1, each node advertises its ID and builds its 1-hop neighbor set based on the advertisement of its neighbors.
- In round 2, each node advertises its 1-hop neighbor set and identifies links among 1-hop neighbors.
- In round 3, each node calculates its clPCI index value and advertises it together with its 2-hop neighbor set. Then, it identifies links among 2-hop neighbors.
- In round 4, each node calculates and advertises its own relay node set and updates 1-hop neighbor status.
- In round 5, the restricted Rule-*k* is applied to each relay node and each one of them advertises its status.
- In round 6, each node advertises its updated relay node set
- In round 7, the composition of the updated relay node set is advertised (if needed).

The computation complexity of its comprising parts is: $O(\Delta^2)$ for the $clPCI$ calculation, and $O(\Delta^3)$ for the relay node set election process and for the pruning phase, where Δ is the maximum node degree found in the network.

3.3.2. Complexities of *FAST-CMDSM*

FAST-CMDSM is a centralized algorithm and so its communication complexity is not an issue for investigation.

FAST-CMDSM' s computation complexity is exponential, similar to all branch-and-cut integer programming solvers.

4. Performance Evaluation

Competing Algorithms. We compare the proposed algorithms, namely *CCDS* and *FAST-CMDSM*, but in order to explain the usefulness of their pruning procedures, we develop and include in our comparison their annotated with an asterisk version of them, namely *CCDS** and *FAST-CMDSM**; these versions are simply the same algorithms but without activating the pruning procedure. Solely for illustrative purposes we include the experimentation, and *FAST-MDSM* algorithm which constructs minimum but *unconnected DS*; we proved in [5] that any (unconnected) *DS* of size $\|DS\|$ can be turned into a *CDS* by adding $2 \times \|DS\|$ additional nodes in the *DS* in the worst case. Table 1 summarizes the competing algorithms.

Table 1. Competitor characteristics.

Competitor	CDS Calculation	Pruning Heuristic	Complexities CPU [a] / Comm [b]
CCDS	Distributed	✓	Δ^3 / 7
*CCDS**	Distributed	-	Δ^3 / 7
FAST-CMDSM	Centralized	✓	*exp* / [c]
*FAST-CMDSM**	Centralized	-	*exp* / [c]
FAST-MDSM	Centralized	-	*exp* / [c]

[a] Δ is the maximum node degree; [b] Number of transmitted messages per node; [c] Not applicable due to its centralized nature.

Simulation testbed. Due to the lack of available, real military networks, and the inability (The requirement of modern battlefields is to able to operate ad hoc networks consisting of an order of magnitude more nodes; for instance a battalion would need a thousand nodes. e.g., https://www.darpa.mil/news-events/2013-04-30 (accessed on 15 May 2022)). of wireless testbeds and emulation environments for ad hoc networks to deal with several hundreds of nodes, we developed a generator for multilayer networks in MATLAB. The details of our generator can be found in [6,23], and here we present its basic features. The construction of a multilayer network is driven by the average node degree, by the nodes' number per layer (i.e., the so-called layer size), and the number of layers. The interconnection of the layers is driven by: (a) the number of a node's links towards nodes in different layers, and (b) the distribution of the interconnections to the nodes within a particular layer. We apply the *Zipfian* distribution to our connectivity generator. Skewness is managed by parameter $s \in (0,1)$. We make use of four distinct *Zipfian* distributions, one per parameter of interest:

- s_{degree}: to generate the frequency of appearance of highly interconnected nodes,
- s_{layer}: to choose how frequently a specific layer is selected,
- s_{node}: to choose how frequently a specific node is selected in a specific layer.
- s_{weight}: to choose how uniformly the energy is distributed in the nodes.

We call these parameters as the *topology skewness*, and represent it as a sequence of four floats, meaning that $s_{degree} = 0.5$, $s_{layer} = 0.5$, $s_{node} = 0.5$ and $s_{weight} = 0.5$ (which are the default settings we used to create the datasets).

Performance measures. The competitors are compared in terms of the cardinality of the *CDS*; apparently an algorithm is more efficient than another, if it generates a *CDS* having smaller cardinality [16]. Moreover, an algorithm that establishes a (per node) relay set with larger residual energy is naturally considered to be more efficient in terms of energy than another algorithm whose per node relay set has less residual energy.

Datasets. We created networks which vary with respect to the topology density, the network diameter, the number of network layers and their size. The topology density's impact on the performance is evaluated with 4-layer networks. Each layer consists of 50 nodes, and the mean node degree is 3, or 6, or 10, or 16, or 20. The network diameter's impact on the performance is evaluated with 4-layer networks as well. Each layer consists of 50 nodes, and the mean node degree is 6. The diameter of each layer is 3, or 5, or 8, or 12, or 17. The number of layers' impact is examined in networks with 2, or, 3, or 4, or 5, or 7 layers. Each layer consists of 50 nodes, and the mean node degree is 6. The increase in the layer size's impact is evaluated with 4-layer networks. The "base" layer consists of 50 nodes, and each next layer is larger than the previous by 10%, or 20%, or 30%, or 50%, or 70%. The mean node degree in each layer is 6. Table 2 records all the independent parameters.

Table 2. Experimentation parameters values.

Parameter	Range	Default
avg. node degree (D)	3, 6, 10, 16, 20	6
network diameter (H)	3, 5, 8, 12, 17	5
number of network layers (L)	2, 3, 4, 5, 7	4
size of a layer relative to its adjacent layers	10%, 20%, 30%, 50%, 70%	-

4.1. Experimental Evaluation

Each experiment was repeated 5 times. The variation around the mean was so negligible that the error bars are hardly recognizable in the plots. We conduct experiments with s_{degree}, s_{layer}, s_{node}, s_{weight} parameters into $0.5-0.5-0.5-0.5$ (Medium skewness), and $0.1-0.1-0.1-0.1$ (Low skewness), and $0.9-0.9-0.9-0.9$ (High skewness) setting, respectively.

4.1.1. Impact of Topology Density

Firstly, we consider the impact of topology density on the performance of each competitor. In Figure 2 we evaluate, for medium skewness, the *per layer* size of the *CDS*. The main conclusion is that the size of the CDS is practically a decreasing function of the node density. This happens because the higher the network density is, the greater the coverage capability of the network nodes becomes, and therefore the size of the *CDS* gets smaller. In the case we succeed medium skewness, there is no clear winner between *CCDS* and *FAST-CMDSM* as the topology becomes denser (degree ≥ 6) and both competitors present similar performance (<10% variance). The explanation is that in such topologies, there exist multiple redundant paths towards nodes, and thus both pruning mechanisms work equally well. On the contrary, in sparse topologies (degree $= 3$) *FAST-CMDSM* is up to 15% more efficient. This is due to the fact that during the pruning process, the redundant paths are less in sparse topologies, and their discovery requires knowledge of the topology further away than 2-hops, which is beyond the capabilities of localized algorithms deployed in wireless ad hoc networks. Apparently, the centralized nature of *FAST-CMDSM* provides a clear advantage, since its pruning task has a broader overview of the topology. If we exclude the pruning task, then *CCDS** and *FAST-CMDSM** present similar performance (less than 10% variance) when degree $\leq$10, and the performance of the latter is up to 15% better to the former when degree >10. However, these results are not good news, because both algorithms do not perform well in terms of the—*per layer*—CDS size; i.e., the—*per layer*—*CDS* size is up to 98% of that of the total number of nodes in that layer. This is considered natural in multilayer networks when the traditional methods are used (2-hop neighborhood coverage).

DS redundancy justifies the use of the pruning mechanism (up to 88% and 85% *CDS* size reduction for *CCDS* and *FAST-CMDSM*, respectively, in this particular case).

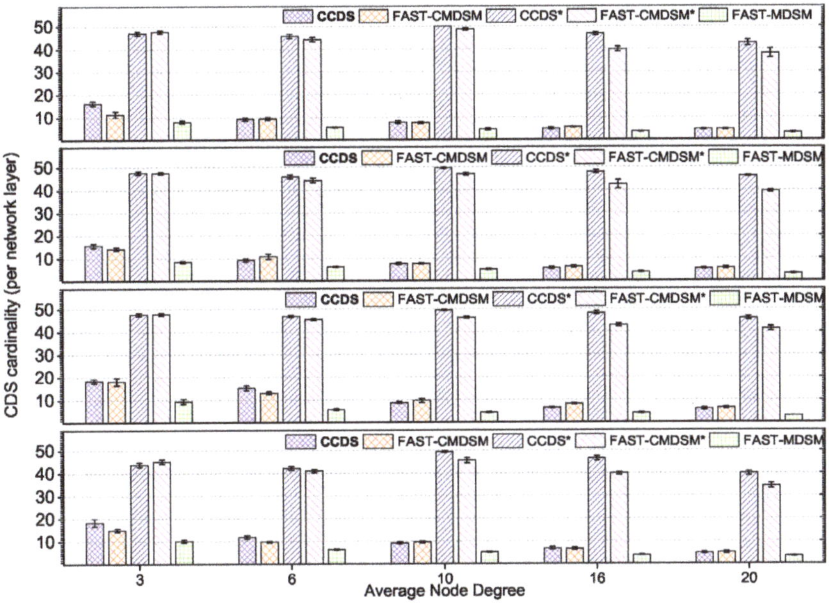

Figure 2. Impact of topology density (Medium skewness).

We expanded our experiments in case of low skewness. The results, depicted in Figure 3 show that there are no important differences regarding the efficiency, between *CCDS* and *FAST-CMDSM*, when topology becomes denser (degree $\geq$ 6). As mentioned above, this happens because pruning mechanisms work well, when nodes have multiple ways to connect with one another. When degree = 3 (sparse topologies), *CCDS* is about 10% less efficient than *FAST-CMDSM*, something which is justified due to the centralized control of the *FAST-CMDSM*. In this case, the reduction size of *CDS* is about 90% for *CCDS* algorithm and up to 88%.

In case of high skewness (see Figure 4), when degree $\geq$ 10, both the proposed algorithms behaves in a quite similar way (less than 10% variance), too, regarding the size of *CDS*. Here, *FAST-CMDSM* outperforms *CCDS*, due to the centralized control of the first one which makes the procedure of finding surplus paths easier inside the Multilayer network. Examining the case when topology is sparse (degree = 3), we see that *CCDS* is less efficient than the other pruning proposed algorithm, bearing in mind that due to the lack of enough connections, it is more difficult for a distributed algorithm to find minimum *DS*. Also, when degree = 6, *CCDS* continues to outperform *FAST-CMDSM*, but both algorithms create *DS* with smaller size than the above implementations do. So, in this case the efficiency of the proposed algorithms are up to 86% and up to 82% for both *CCDS* and *FAST-CMDSM*, respectively.

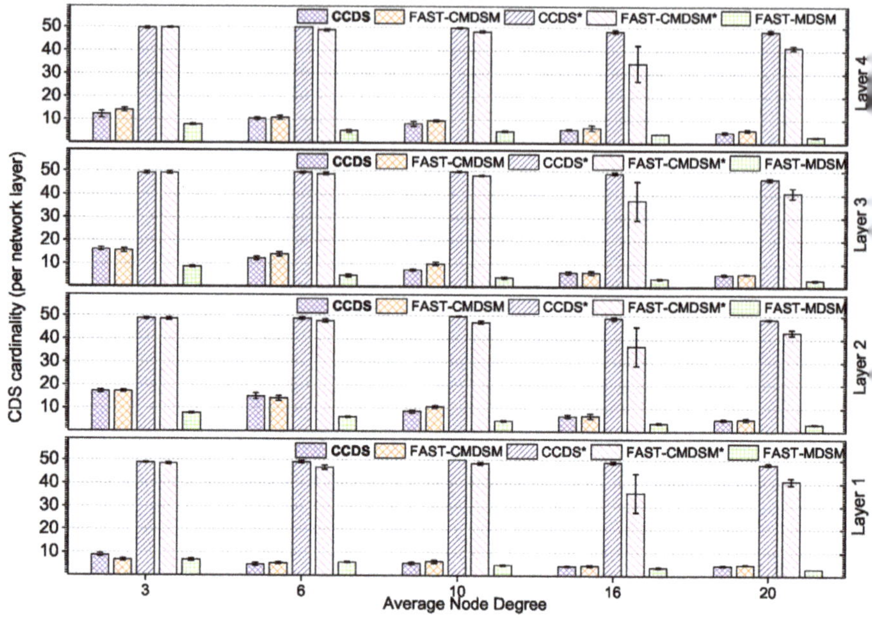

Figure 3. Impact of topology density (Low skewness).

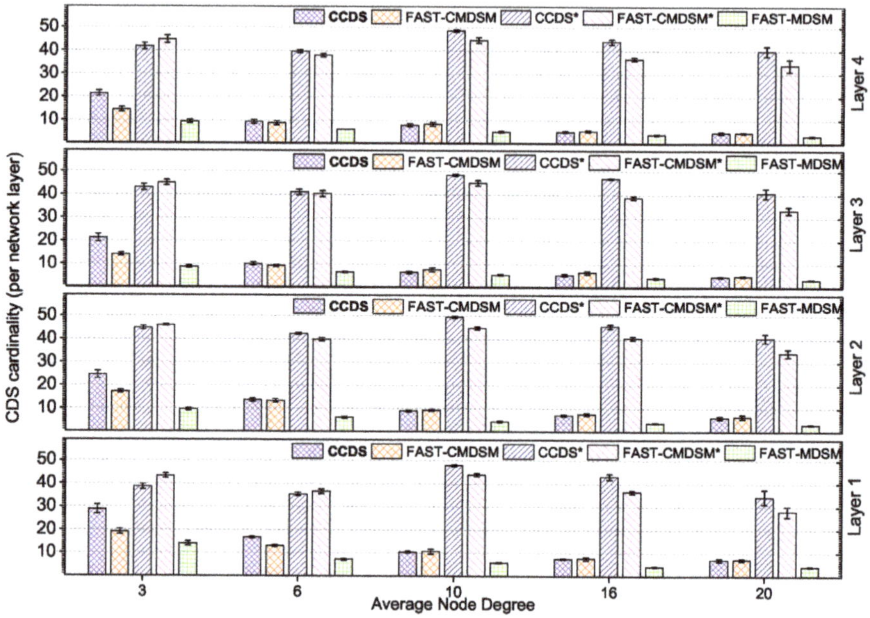

Figure 4. Impact of topology density (High skewness).

So, as it is also shown in Figure 5 which shows CDS sizes aggregated over all layers, for all the parameters tested, the results are quite similar regarding both *CCDS* and *FAST-CMDSM* algorithm, with the corresponding results with parameters $0.1 - 0.1 - 0.1 - 0.1$ to be a little more efficient, probably due to the more uniform distribution this case implements.

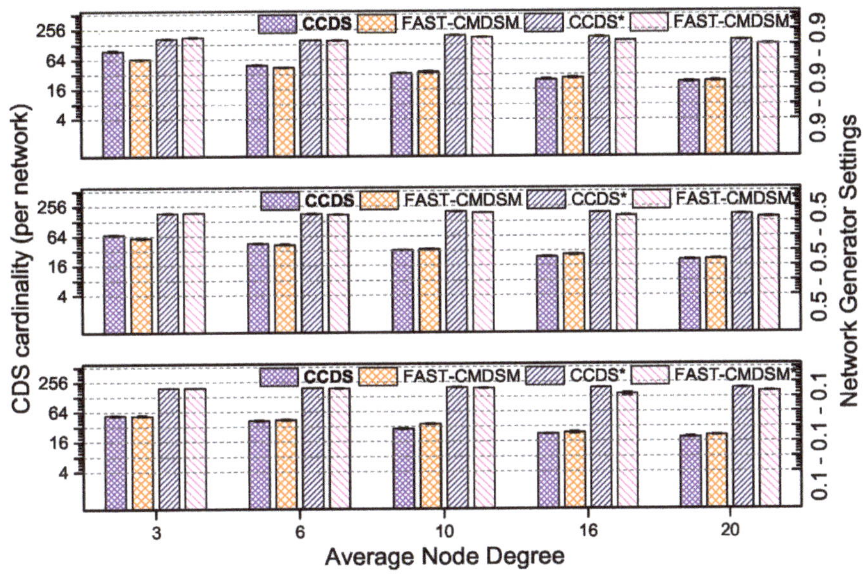

Figure 5. Impact of topology density (CDS size aggregated over all layers).

4.1.2. Impact of Network Diameter

In Figures 6–8, we evaluate the effect of the network diameter in the size of the *CDS*. The main conclusion is that as the network diameter increases the size of the resulting *CDS* increases, and this observations is valid for all competitors. In case of medium skewness (Figure 6), the diameter increases and also the topology gets sparser. So, we have fewer, longer (in hops), and less redundant paths. In that way, the selection of 1-hop neighbor nodes that cover the N^2 neighborhood of a particular node requires more 1-hop neighbors. Now, in terms of the competitors' performance, we see that in bushy topologies (diameter ≤ 5) *CCDS* presents up to 18% smaller *CDS* compared to *FAST-CMDSM*. This gain is attributed to the employment of the *clPCI* by *CCDS* and which helps in getting a better pruning. On the other hand, in bushy topologies an unfortunate erase from the *CDS* of a strategically located *DS* node will probably result in keeping a lot of (practically "useless") *DS* nodes just to ensure connectivity. When diameter $= 8$ or diameter $= 12$ the competitors have similar performance (less than 10% variance). As expected, in long and skinny topologies (diameter $= 17$) *FAST-CMDSM* outperforms *CCDS* by 16%, due to its centralized nature. Examining the pruning-free versions of the competitors, we see that both of them practically exhibit the worst-case behaviour, i.e., almost all nodes are selected as *DS* nodes; the performance of the pruning mechanism for *CCDS* and *FAST-CMDSM* regarding the *CDS* size reduction is up to 83% and 79%, respectively.

In case of low skewness as it is illustrated in Figure 7, we can also say that when diameter ≤ 5, *CCDS* algorithm outperforms FAST-CMDSM and when diameter $= 8$ or diameter $= 12$, both proposed algorithms have similar performance (less than 10% variance). Finally, we see that for diameter = 17, *FAST-CMDSM* has a better performance than *CCDS*, due to the benefits of its centralized form, in long paths. So, in this case, *CDS* reduction is up to 85% for *CCDS* and 81% for *FAST-CMDSM*.

Finally, when the interconnectivity generator parameters are $0.9 - 0.9 - 0.9 - 0.9$ (high skewness), as it is depicted in Figure 8, we observe a general increase in *CDS* size, due to the fact that the degrees of the nodes have a great variance in this case and as a result, more nodes needed to join the *DS*. When diameter $= 3$, *CCDS* has better performance than the other proposed algorithm, due to the fact that we have smaller paths to check and as a result the distributed algorithm can discover easier the redundant paths (2-hop coverage). When diameter $= 5$, or diameter $= 8$, *FAST-CMDSM* starts to have a better performance, as it creates *CDS* with smaller size. This is expected, since the centralized control of this algorithm contributes efficiently in finding the redundant paths. The longer the diameter is, the better performance *FAST-CMDSM* has as it is shown in Figure 8. To sum up, in this case, *CCDS* can achieve up to 81% reduction in the size of total *CDS*, while *FAST-CMDSM* can achieve 79%. As a summary, we provide Figure 9 to show average performance of the competitors.

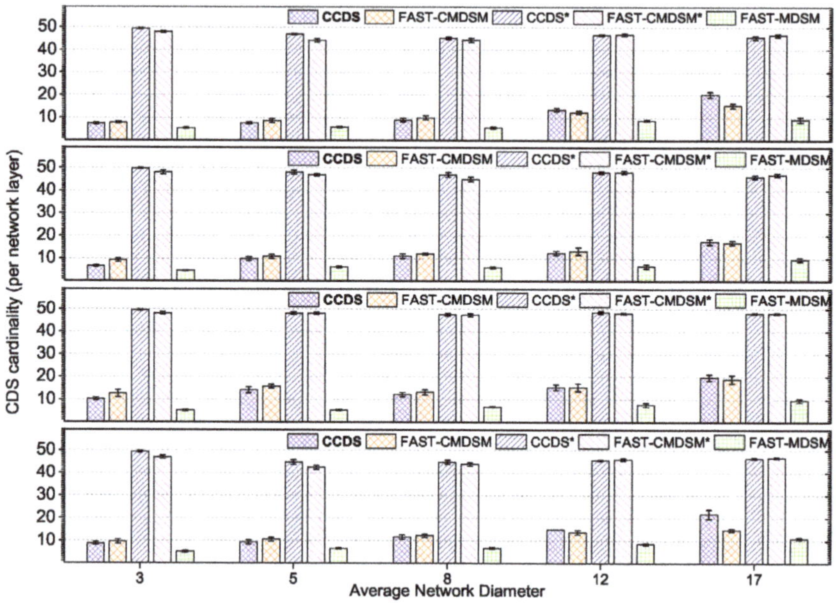

Figure 6. Impact of network diameter (Medium skewness).

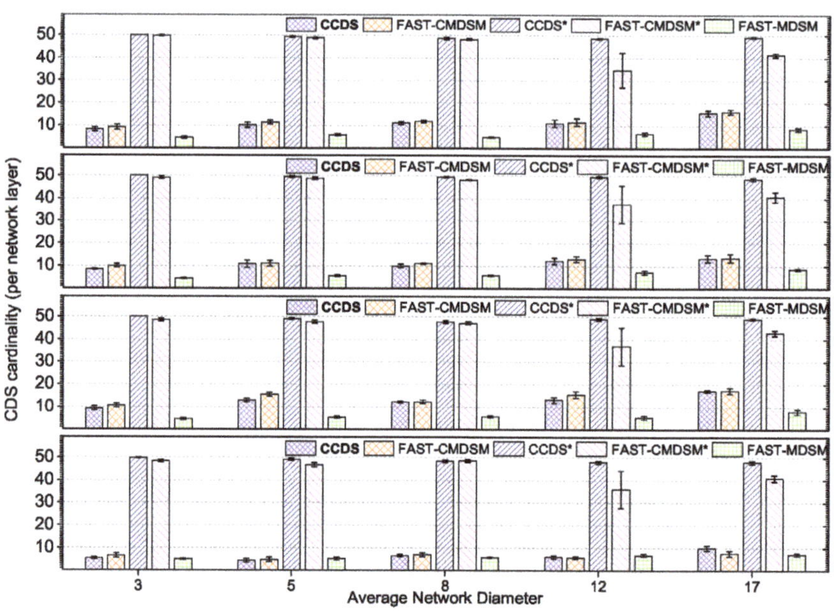

Figure 7. Impact of network diameter (Low skewness).

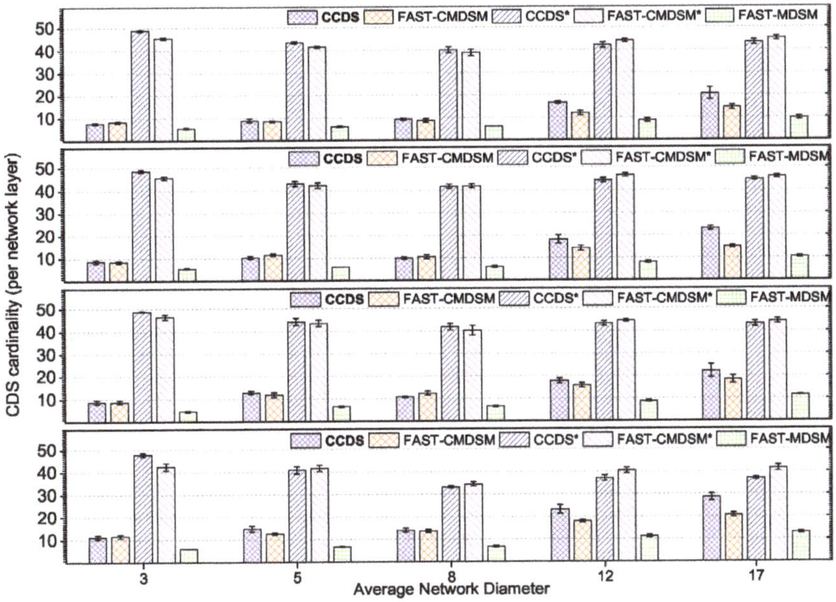

Figure 8. Impact of network diameter (High skewness).

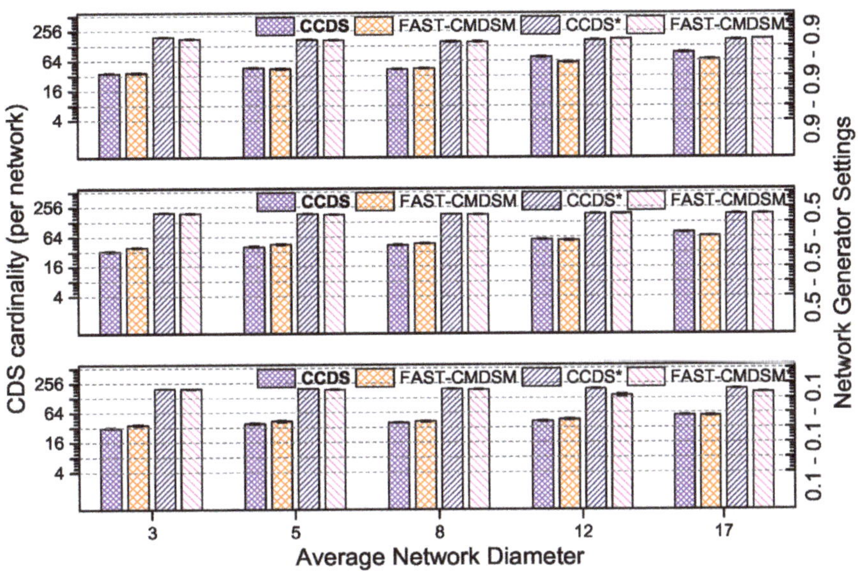

Figure 9. Impact of network diameter (CDS size aggregated over all layers).

4.1.3. Impact of the Number of Layers

We investigated the impact of the network layers' number on the competitors. In Figure 10 we show the *per layer* CDS cardinality for the competing algorithms (for medium skewness). The first observation we make is somewhat counter-intuitive: we see that the *per layer* CDS cardinality is independent on the number of layers(!) even though we would expect to be a decreasing function of the number of layers, because when a multilayer network has more layers, then it will (most probably) have more connections among layers (i.e., interlayer links), and thus the network will become more dense. So, even though we expected an increase in the coverage capability of the nodes, this does not happen, and it is attributed to the generic topology of the network. Turning now our focus on the competitors, we observe that with 5 or less layers both *CCDS* and *FAST-CMDSM* perform very good (10% or less variance). However, *CCDS* is the best of the two presenting a 14% better performance. Overall, the obtained results are consistent with our earlier which state that both competitors perform very good in dense networks. In general, the main reason for *CCDS'* superior performance with respect to *FAST-CMDSM'* s performance is the very effective pruning mechanism. When looking at the version of these algorithms without the pruning mechanism, we see that as expected they perform poor; in particular the performance of the pruning mechanism for *CCDS* and *FAST-CMDSM* regarding the *CDS* size reduction is up to 79% and 75%, respectively.

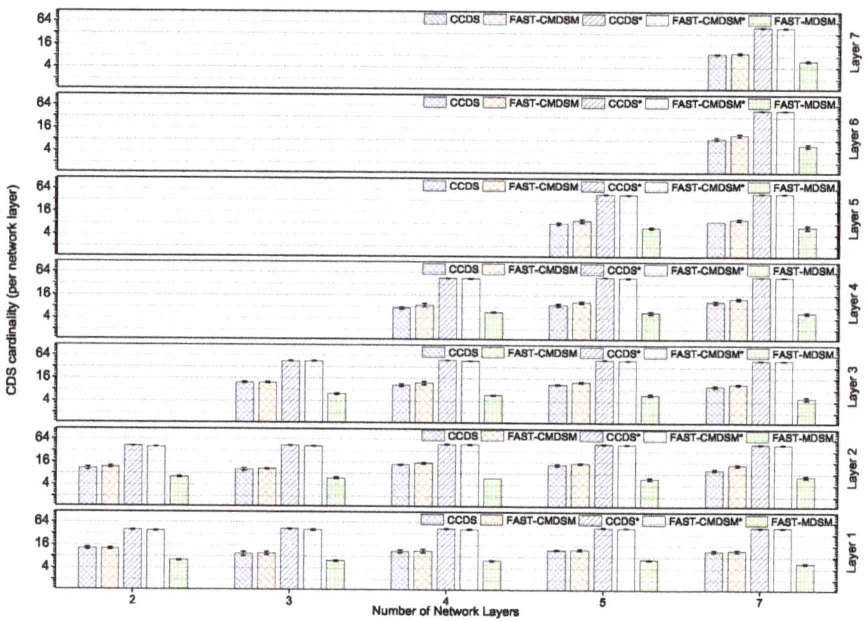

Figure 10. Impact of number of layers (Medium skewness).

Next, we present the case where the parameters of the interconnectivity generator are: $0.1 - 0.1 - 0.1 - 0.1$ (low skewness). We see in Figure 11, that both the proposed algorithms achieve similar performance, with the *CCDS* algorithm to be a little bit more efficient (less than 10% variance). The total reduction in *CDS* size is up to 80% for *CCDS* and up to 78% for *FAST-CMDSM*. The last experiment in this section is conducted by setting the respective interconnectivity generator variables to: $0.9 - 0.9 - 0.9 - 0.9$ (high skewness). In this case, when number of layers is less than 4, *FAST-CMDSM* outperforms *CCDS*, while when the number of layers is greater than 4, then the two proposed algorithms are barely equally efficient, as it is shown in Figure 12. Specifically, the total reduction provided by this experiment is up to 74% for CCDS and up to 73% for *FAST-CMDSM* (1% variance). In Figure 13 we show the competitors average performance.

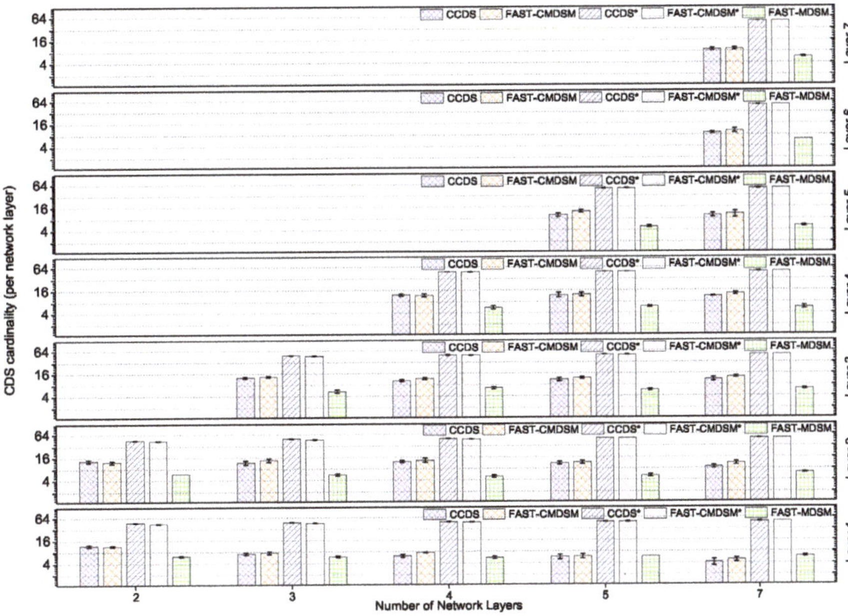

Figure 11. Impact of number of layers (Low skewness).

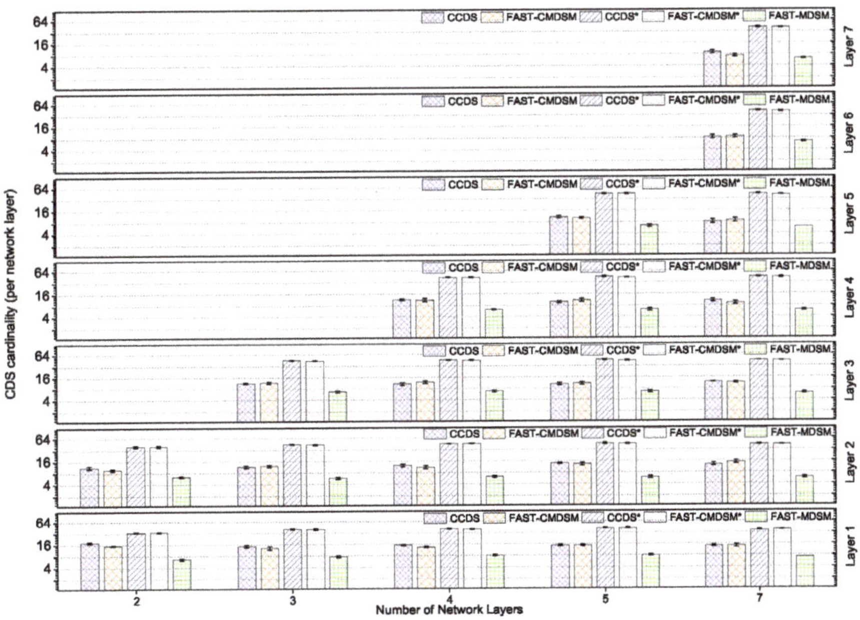

Figure 12. Impact of number of layers (High skewness).

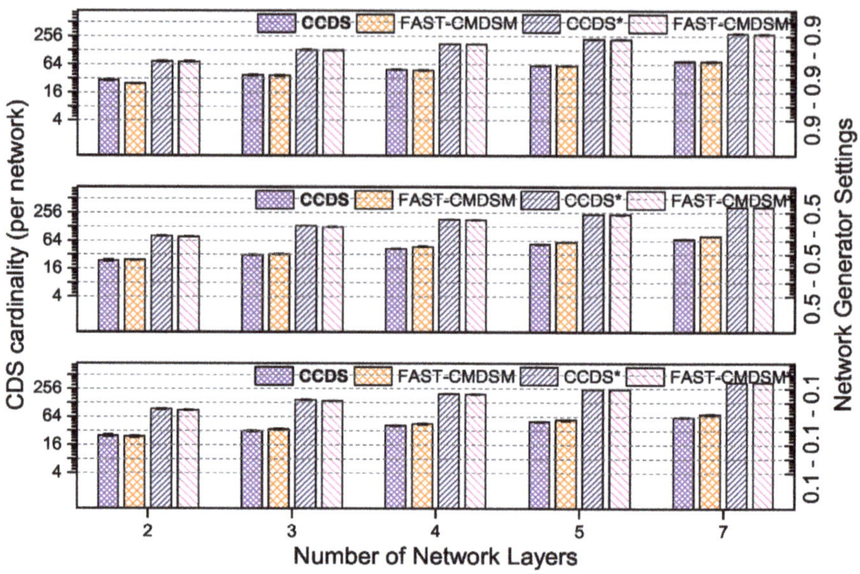

Figure 13. Impact of number of layers (CDS size aggregated over all layers).

4.1.4. Impact of Layer's Size Increase

Here we consider the impact of layer's size increase on the competitors' performance; we evaluate the *per layer* size of the *CDS* for medium skewness, and depict the results in Figure 14. A first, generic observation is that the cardinality of CDS increases with increasing layer size. This is easily explained by the fact that the as the size of each layer increases, so does the need for more nodes to act as connectors, and consequently we get a larger CDS. Looking at the performance of the competitors, we note that an increase in the size of each subsequent layer by 30% or less results in having *CCDS* to outperform *FAST-CMDSM* with a margin from 11% up to 16%. The basic reason behind that is the dense topology, with the consequence that many redundant paths remain within the vicinity of *CCDS* (i.e., 2-hop), and therefore the pruning process eliminates many redundant dominators. On the contrary, an increase in the size of each subsequent layer by 50% or more results in having the centralized *FAST-CMDSM* outperform *CCDS* from 18% up to 21%. This is expected, since the large difference in the cardinality of the layers implies a large number of interlayer links, and therefore the calculation of the redundant paths by the pruning process requires a broader/global view of the topology, which is only available to the centralized *FAST-CMDSM* algorithm and not the distributed *CCDS* algorithm. As a last comment to this experiment, we see that the versions without pruning of the algorithms select almost all nodes as dominators; in particular the performance of the pruning mechanism for *CCDS* and *FAST-CMDSM* regarding the *CDS* size reduction is up to 80% and 83%, respectively.

Conducting the experiment by changing the set of variables in the interconnectivity generator to $0.1 - 0.1 - 0.1 - 0.1$ (almost uniform distribution), we observe that *CCDS* is the champion algorithm, because of the dense topologies formed, as it is shown in Figure 15. In this case, *CCDS* achieves 82% reduction in the total *CDS* size, while *FAST-CMDSM* achieves 80%. The last experiment in this section is conducted by setting the respective interconnectivity generator parameters to: $0.9 - 0.9 - 0.9 - 0.9$ and the results are presented in Figure 16. In this case, we have a barely arbitrarily-formed distribution (high skewness). We see that while increasing the number of nodes, set in every layer, the *CCDS* algorithm remains our best option, as it is justified above. In this case, *CCDS* achieves 76% reduction in the total *CDS* size, while *FAST-CMDSM* achieves 75%.

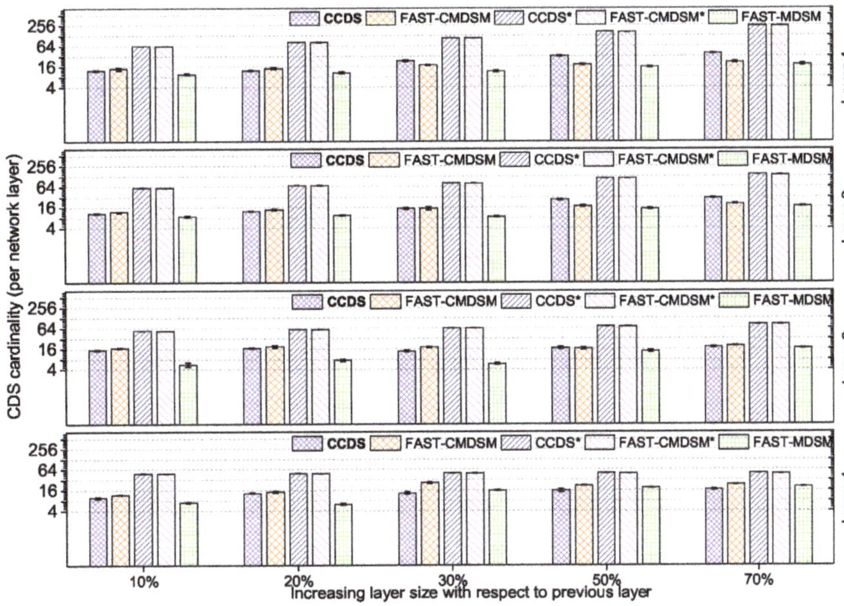

Figure 14. Impact of increasing layer cardinality (Medium skewness).

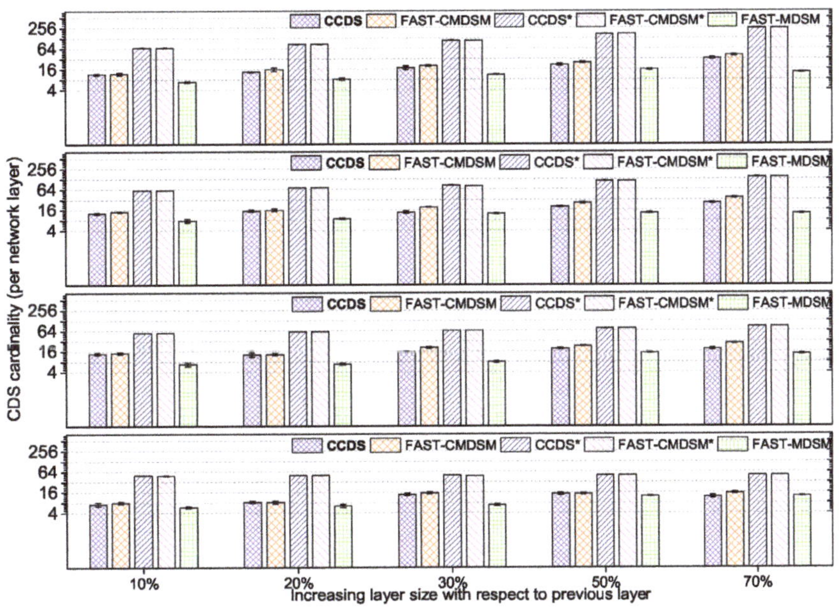

Figure 15. Impact of increasing layer cardinality (Low skewness).

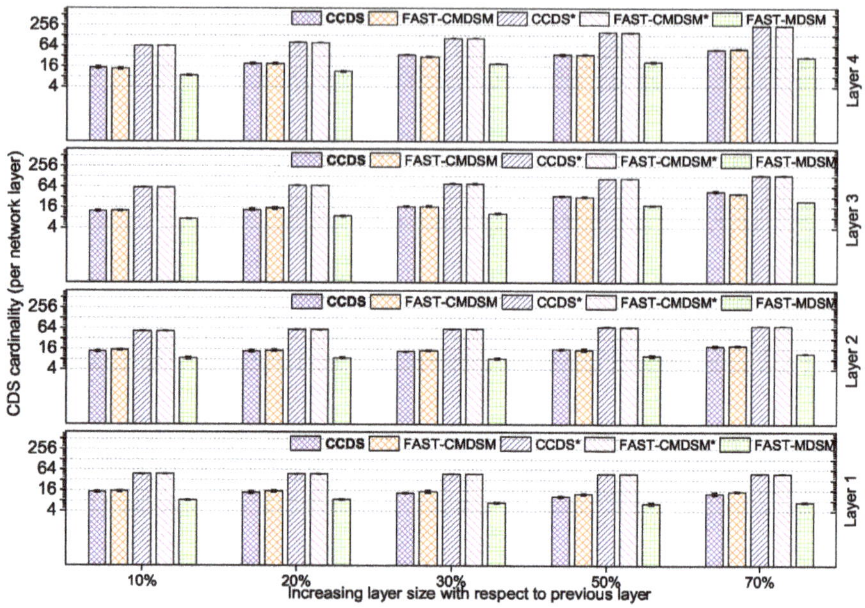

Figure 16. Impact of increasing layer cardinality (High skewness).

Comparing the experiments, done, by changing the values in variables of skewness, we conclude that having a more uniform distribution (Low Skewness) is the best scenario, in which, our algorithms achieves their highest efficiency. This happens because this type of distribution, normalizes well the degree of every node in every layer, which means less nodes are important for the coverage of the network and this results in smaller *CDS* size. On the other hand, when we have high skewness, nodes are arbitrarily connected which means, more nodes in total *DS* are needed in order to cover the whole network properly. Furthermore, in every case, *CCDS* is the more efficient algorithm to use, bearing in mind that in every section, it provides better reduction from *FAST-CMDSM*, except for the one in which we examine low skewness and our variables are set to 0.5 in which *FAST-CMDSM* outperforms *CCDS* (3% variance).

4.1.5. Energy Awareness of the Competing Algorithms

We repeated all experiments taking now into account the residual energy of each node, and we show here the obtained results which concern the aggregated over all network layers performance of the competitors. In Figure 17, we report the results for medium skewness and we see that *CCDS* selects the most energy efficient *CDS* in, almost, any case, followed by *FAST-CMDSM*. We end up in the same conclusion about *CCDS* algorithm's energy consumption regarding both the experiments of low and high skewness, as it is depicted in both Figures 18 and 19, respectively.

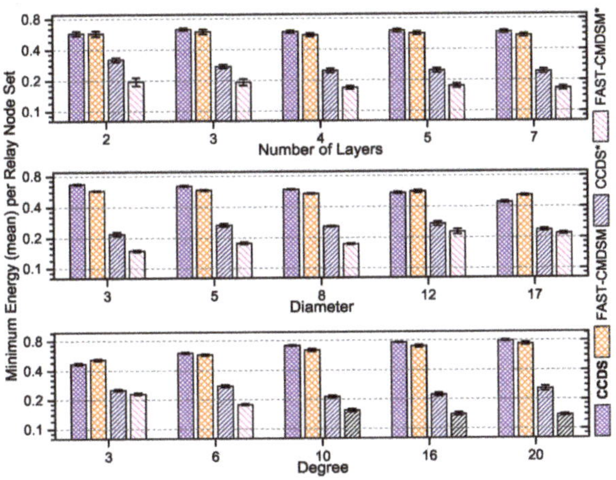

Figure 17. Energy awareness of the competitors (Medium skewness).

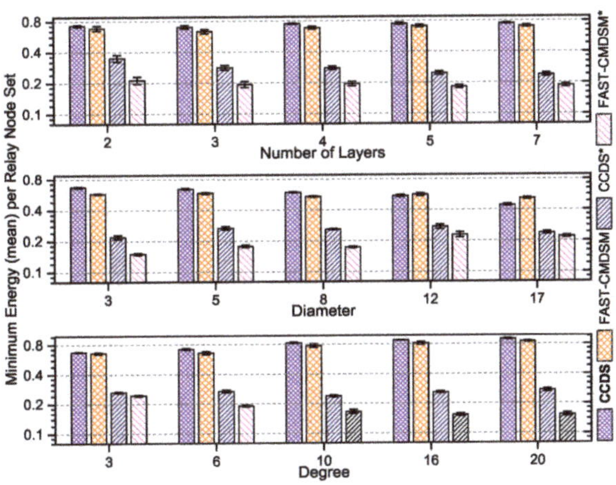

Figure 18. Energy awareness of the competitors (Low skewness).

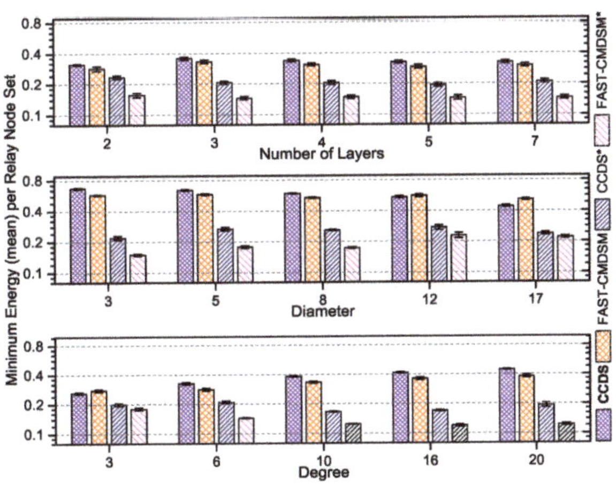

Figure 19. Energy awareness of the competitors (High skewness).

4.1.6. Results' Summary

In Table 3 we provide a summary of *CCDS* average performance gain (percentage-wise) over the second best performing distributed algorithm across the independent parameter space.

Table 3. Results' summary.

Parameter	Avg Performance Gain of *CCDS*
topology density	≈12%
network diameter	≈8.5%
number of layers	≈14%
layers' heterogeneity	≈18%

5. Conclusions

The Internet of Battlefield Things is a new cyberphysical system originating from the Internet of Things, but at a much larger scale, and with stringent robustness and latency requirements. Its most significant and challenging goal is to carry out commander's intent in a safe, responsive and resilient manner. This article investigated the issue of building small and resilient backbones for IoBT networks. We abstracted the topology of an IoBT network as a multilayer graph, and we resorted to the concept of dominating sets to achieve our goal. Then, we presented a distributed algorithm for calculating connected dominating sets with small-cardinality in this multilayer network context. We implemented and contrasted our proposed algorithm to two state-of-the-art algorithms for computing connected dominating sets for multilayer networks using generated topologies. Our algorithm showed constantly better performance against these competitors across a range of topologies with various representative features. Our future research involves algorithmic issues of detecting multiple MCDS for energy conservation via sleep-scheduling.

Author Contributions: Conceptualization, D.K.; methodology, E.F., D.P. and D.K.; software, D.P. and T.K.; validation, E.F., D.P. and D.K.; formal analysis, D.K.; investigation, E.F., D.P., T.K. and D.K.; resources, D.K.; data curation, D.P. and T.K.; writing original draft preparation, E.F. and D.P.; writing review and editing, D.K.; visualization, D.P.; supervision, D.K.; project administration, D.K.; funding acquisition, D.K. All authors have read and agreed to the published version of the manuscript.

Funding: E.Fragkou's research work is supported by the Hellenic Foundation for Research and Innovation (HFRI) under the 3rd Call for HFRI Ph.D. Fellowships (Fellowship Number: 5631).

Data Availability Statement: Not applicable.

Conflicts of Interest: The authors declare no conflict of interest.

References

1. Kott, A.; Swami, A.; West, B.J. The Internet of Battle Things. *IEEE Comput. Mag.* **2016**, *49*, 70–75. [CrossRef]
2. Russel, S.; Abdelzaher, T. The Internet of Battlefield Things: The next generation of command, control, communications and intelligence (C3I) decision-making. In Proceedings of the 2018 IEEE Military Communications Conference (MILCOM 2018), Los Angeles, CA, USA, 29–31 October 2018; pp. 737–742.
3. Abdelzaher, T.; Ayanian, N.; Basar, T.; Diggavi, S.; Diesner, J.; Ganesan, D.; Govindan, R.; Jha, S.; Lepoint, T.; Marlin, B. Will distributed computing revolutionize peace? The emergence of Battlefield IoT. In Proceedings of the IEEE International Conference on Distributed Computing Systems, Vienna, Austria, 2–6 July 2018; pp. 1129–1138.
4. Abdelzaher, T.; Ayanian, N.; Basar, T.; Diggavi, S.; Diesner, J.; Ganesan, D.; Govindan, R.; Jha, S.; Lepoint, T.; Marlin, B.; et al. Toward and Internet of Battle Things: A resilience perspective. *IEEE Comput. Mag.* **2018**, *51*, 24–36. [CrossRef]
5. Papakostas, D.; Basaras, P.; Katsaros, D.; Tassiulas, L. Backbone formation in military multi-layer ad hoc networks using complex network concepts. In Proceedings of the IEEE Military Communications Conference, Baltimore, MD, USA, 1–3 November 2016; pp. 842–848.
6. Papakostas, D.; Eshghi, S.; Katsaros, D.; Tassiulas, L. Energy-aware backbone formation in military multilayer ad hoc networks. *Ad Hoc Netw.* **2018**, *81*, 17–44. [CrossRef]
7. Farooq, M.J.; Zhu, Q. On the secure and reconfigurable multi-layer network design for critical information dissemination in the Internet of Battlefield Things (IoBT). *IEEE Trans. Wirel. Commun.* **2018**, *17*, 2618–2632. [CrossRef]

8. Gaikwad, N.B.; Ugale, H.; Keskar, A.; Shivaprakash, N.C. The Internet-of-Battlefield-Things (IoBT)-based enemy localization using soldiers location and gunshot direction. *IEEE Internet Things J.* **2020**, *7*, 11725–11734. [CrossRef]
9. Abuzainab, N.; Saad, W. Dynamic connectivity game for adversarial Internet of Battlefield Things Systems. *IEEE Internet Things J.* **2018**, *5*, 378–390. [CrossRef]
10. Abuzainab, N.; Saad, W. A multiclass mean-field game for thwarting misinformation spread in the Internet of Battlefield Things. *IEEE Trans. Commun.* **2018**, *66*, 6643–6658. [CrossRef]
11. Azmoodeh, A.; Dehghantanha, A.; Choo, K.K.R. Robust malware detection for Internet of (Battlefield) Things devices using deep eigenspace learning. *IEEE Trans. Sustain. Comput.* **2019**, *4*, 88–95. [CrossRef]
12. Castiglione, A.; Choo, K.K.R.; Nappi, M.; Ricciardi, S. Context aware ubiquitous biometrics in edge of military things. *IEEE Cloud Comput. Mag.* **2017**, *4*, 16–20. [CrossRef]
13. Nasim, I.; Kim, S. Human EMF exposure in wearable networks for Internet of Battlefield Things. In Proceedings of the IEEE Military Communications Conference, Norfolk, VA, USA, 12–14 November 2019.
14. Stojmenovic, I.; Seddigh, M.; Zunic, J. Dominating sets and neighbor elimination-based broadcasting algorithms in wireless networks. *IEEE Trans. Parallel Distrib. Syst.* **2002**, *13*, 14–25. [CrossRef]
15. Wu, J.; Li, H. On calculating connected dominating set for efficient routing in ad hoc wireless networks. In Proceedings of the Workshop on Discrete Algorithms and Methods for Mobile Computing and Communications, Seattle, WA, USA, 20 August 1999; pp. 7–14.
16. Yu, J.; Wang, N.; Wang, G.; Yu, D. Connected dominating sets in wireless ad hoc and sensor networks: A comprehensive survey. *Comput. Commun.* **2013**, *24*, 121–134. [CrossRef]
17. Zhao, D.; Xiao, G.; Wang, Z.; Wang, L.; Xu, L. Minimum dominating set of multiplex networks: Definition, application, and identification. *IEEE Trans. Syst. Man Cybern. Syst.* **2021**, *51*, 7823–7837. [CrossRef]
18. Wu, Y.; Li, Y. Construction algorithms for *k*-connected *m*-dominating sets in wireless sensor networks. In Proceedings of the ACM International Symposium on Mobile Ad Hoc Networking and Computing, Hong Kong, China, 26–30 May 2008; pp. 83–90.
19. Pemmaraju, S.V.; Pirwani, I.A. Energy conservation via domatic partitions. In Proceedings of the ACM International Symposium on Mobile Ad Hoc Networking and Computing, Florence, Italy, 22–25 May 2006; pp. 143–154.
20. Nacher, J.C.; Ishitsuka, M.; Miyazaki, S.; Akutsu, T. Finding and analysing the minimum set of driver nodes required to control multilayer networks. *Nat. Sci. Rep.* **2019**, *9*, 1–12. [CrossRef] [PubMed]
21. Fan, N.; Watson, J.P. Solving the connected dominating set problem and power dominating set problem by integer programming. In Proceedings of the International Conference on Combinatorial Optimization and Applications, Banff, AB, Canada, 5–9 August 2012; pp. 371–383.
22. Haynes, T.W.; Hedetniemi, S.; Slater, P. *Fundamentals of Domination in Graphs*; Chapman & Hall/CRC Pure and Applied Mathematics; CRC Press: Boca Raton, FL, USA, 1998.
23. Papakostas, D.; Kasidakis, T.; Fragkou, F.; Katsaros, D. Backbones for Internet of Battlefield Things. In Proceedings of the IEEE/IFIP Annual Conference on Wireless On-Demand Network Systems and Services, Klosters, Switzerland, 9–11 March 2021; pp. 116–123.
24. Tseng, Y.C.; Ni, S.Y.; Chen, Y.S.; Sheu, J.P. The broadcast storm problem in a mobile ad hoc network. *Wirel. Netw.* **2002**, *8*, 153–167. [CrossRef]
25. Garey, M.R.; Johnson, D.S. *Computers and Intractability: A Guide to the Theory of NP-Completeness*; W. H. Freenan and Company: New York, NY, USA, 1979.

future internet

MDPI

Article

A VPN Performances Analysis of Constrained Hardware Open Source Infrastructure Deploy in IoT Environment

Antonio Francesco Gentile [1], Davide Macrì [1], Floriano De Rango [2], Mauro Tropea [2,*] and Emilio Greco [1]

1 Institute for High-Performance Computing and Networking (ICAR), National Research Council of Italy (CNR), Via P. Bucci, 8/9C, 87036 Rende, Italy
2 DIMES Department, University of Calabria, Via P. Bucci 39/C, 87036 Rende, Italy
* Correspondence: m.tropea@dimes.unical.it; Tel.: +39-0984-494786

Abstract: Virtual private network (VPN) represents an HW/SW infrastructure that implements private and confidential communication channels that usually travel through the Internet. VPN is currently one of the most reliable technologies to achieve this goal, also because being a consolidated technology, it is possible to apply appropriate patches to remedy any security holes. In this paper we analyze the performances of open source firmware OpenWrt 21.x compared with a server-side operating system (Debian 11 x64) and Mikrotik 7.x, also virtualized, and different types of clients (Windows 10/11, iOS 15, Android 11, OpenWrt 21.x, Debian 11 x64 and Mikrotik 7.x), observing the performance of the network according to the current implementation of the various protocols and algorithms of VPN tunnel examined on what are the most recent HW and SW for deployment in outdoor locations with poor network connectivity. Specifically, operating systems provide different performance metric values for various combinations of configuration variables. The first pursued goal is to find the algorithms to guarantee a data transmission/encryption ratio as efficiently as possible. The second goal is to research the algorithms capable of guaranteeing the widest spectrum of compatibility with the current infrastructures that support VPN technology, to obtain a connection system secure for geographically scattered IoT networks spread over difficult-to-manage areas such as suburban or rural environments. The third goal is to be able to use open firmware on constrained routers that provide compatibility with different VPN protocols.

Keywords: VPN; IPsec; SSTP; OpenVPN; OpenConnect; Accel-PPP; Mikrotik; Linux; Windows 10/11; iOS; Android; Mac OS X; OpenWrt; IoT; Libreswan; Strongswan IKE; TLS; SSL

Citation: Gentile, A.F.; Macrì, D.; De Rango, F.; Tropea, M.; Greco, E. A VPN Performances Analysis of Constrained Hardware Open Source Infrastructure Deploy in IoT Environment. *Future Internet* **2022**, *14*, 264. https://doi.org/10.3390/fi14090264

Academic Editors: Christoph Stach and Clémentine Gritti

Received: 31 July 2022
Accepted: 9 September 2022
Published: 13 September 2022

1. Introduction

Modern IT infrastructures can cover large geographic areas, and therefore, secure and reliable IT infrastructures are needed, while also guaranteeing low-cost factors, both in terms of space and time. The virtual private network (VPN) is one of the most reliable technology to satisfy this type of need, passing both through the "old" (PSTN) and through the most modern 4G/5G architectures [1–3].

VPNs create a secure connection between the user client and the remote endpoint. Traffic is encrypted to protect it from others. All this takes place via the VPN servers, to which the user's internet traffic is forwarded before reaching its destination [4]. This allows to use public Wi-Fi connections more quietly. Such connections can be easily intercepted, but thanks to a VPN you can avoid nasty intrusions while connected from a hotel, a shopping center or at the airport waiting for your next flight, especially accessing sensitive services, such as your home banking. In this case, it is the encryption offered by the VPN that makes the difference. A tunnel is built between the VPN client and the server on the public network to protect the connection and guaranteeing security and privacy for the users. The connection is initiated by the client that communicates with the server and, once the connection is created, the tunnel is established and the messages are encrypted between the two authenticated parties of the VPN link.

Many types of VPN protocols have been proposed offering different security levels and features. So they present also different performance in terms of latency and throughput output parameters. The choice of the correct VPN solution is fundamental for the performance of the system. Moreover the compatibility with the hardware is important for the user experience. A clear separation between the protocols used on IPsec [5,6] or SSL [7,8] can be defined as follows. In this case, the different typologies of IPsec considered in our work are listed :

- IKEv1-XAUTH, IKEv1-L2TP, IKEv1 RSA
- IKEv2 RSA
- IKEv1 SITE TO SITE PSK, SITE TO SITE RSASIG
- IKEv2 SITE TO SITE RSASIG, IKEv2 SITE TO SITE PSK
- IKEv2 XFRMI ROUTE BASED

For the considered VPN SSL based [8,9]:

- OpenVPN: ROAD WARRIORS, SITE TO SITE, SITE TO MULTI SITE
- Accel-PPP: SSTP ROAD WARRIORS, SSTP SITE TO SITE, SSTP SITE TO MULTI SITE
- Wireguard: SITE TO SITE, SITE TO MULTI-SITE, ROAD WARRIORS
- Ocserv : SITE TO SITE, ROAD WARRIORS

The main contribution of this work focuses on demonstrating the power of OpenWRT Open Source firmware to implement different types of VPN infrastructures having a native Linux operating system "under the hood" with the possibility of installing it on various boards, from enterprise ones (Motherboard APU) to those TP-LINK/D-LINK for constrained environments and with poor connectivity problems. The main goal is to build an Open Source infrastructure for Smart Devices in IoT Environment for deployment in outdoor locations with poor network connectivity. The OpenWRT firmware allows to manage IPsec VPN initiator/responder (Xauth, L2TP, X509 IKEV1, and IKEV2), VPN client/server for OpenVPN, OpenConnect (aka Cisco AnyConnect) SSTP (Microsoft replaced the old and now insecure PPTP) and Wireguard. It also allows using dynamic routing management systems through software such as Quagga ("Quagga Routing Suite", https://www.quagga.net/ (accessed on 15 June 2022)/Babeld ("Babel Routing Suite", https://www.irif.fr/~jch/software/babel/ (accessed on 15 June 2022))/FRR ("FRRouting Routing Suite", https://frrouting.org/ (accessed on 15 June 2022)). These software packages support protocols such as OSPF ("OSPF Routing Protocol", https://www.cisco.com/c/en/us/td/docs/ios-xml/ios/iproute_bgp/configuration/xe-16/irg-xe-16-book/configuring-a-basic-bgp-network.html (accessed on 15 June 2022)) and BGP ("BGP Routing Protocol", https://www.cisco.com/c/en/us/td/docs/ios-xml/ios/iproute_bgp/configuration/xe-16/irg-xe-16-book/configuring-a-basic-bgp-network.html (accessed on 15 June 2022)), thus enabling the system to create mesh networks realized through point-to-point and point-to-multipoint VPN. Once the infrastructure has been built, the Mosquitto software ("Mosquitto Service", https://mosquitto.org/ (accessed on 15 June 2022)) can be installed to manage the IoT part. Mosquitto is a very light service and can be configured both as a client and as a broker, and therefore has a general-purpose functionality concentrator available in a single device.

This paper has the following structure: Section 2 shows related works about VPN implementations and analyses present in the literature; Section 3 presents the components of a VPN architecture; Section 4 provides the deployed scenarios used for the testbeds; Section 5 provides the experimental results in the implemented scenarios; lastly, Section 6 summarizes the paper.

2. Related Works

Many works exist that propose security systems at different protocol stack layers and in different network typologies, showing cryptography performance analysis such as in [10] where authors provide a MAC layer security study on Wireless Sensor Networks (WSNs) also with the use of new technology as Elliptic Curve Cryptography (ECC) [11] proposing different solution at different network attacks such as denial of services attack [12,13]. Many studies are proposed that deal with the VPN issues showing performance evaluation in terms of throughput and delay in different testbed environments. In this section, we show a brief state-of-the-art summary of the VPN literature in order to present a brief overview about this important topic.

In [7] a general VPN study approach is provided. The work focuses on the concept that VPN technology can only be meaningfully analyzed by referring to the values and motivations of the people who make up companies. A key finding reports the observed differences in "how" the terms "VPN" "security," and "privacy" are perceived and understood in the corporate employee population.

In [14], network layer-based VPN is evaluated. VPN provides secure encrypted communications between remote LANs around the world using Internet Protocol (IP) tunnels and a shared medium such as the Internet. The document focuses in particular on showing the strength of each VPN, through several studies on state-of-the-art and then analyzing the protocol Wireguard. A series of comparisons between deployments based on IPsec and GRE [15] and Wireguard is discussed in [15]. SSL VPN works at secure sockets layer. Again in this work [16], it is highlighted how an SSL-VPN is used extensively to guarantee authentication and confidentiality of data between the client (Web browser) and the server (SSL Server).

In [17] an approach to using VPN technologies for secure encryption of traffic in untrusted networks, such as bars, lounges, conferences, free hotspots, free wifi at the airport, and generically during any trip are discussed.

In [18,19] VPN technology is discussed in depth. In particular, this work focuses on the weaknesses and poor performance in the presence of a high load of a web browser-based SSL VPN, which among other things, only supports Windows [16] operating systems.

In paper [20] the choice of the IPSEC-IKEv2 suite is justified against other (SSL/TLS VPN, SSH Tunnel) possible implementations to ensure the security of the IP communication layer. The motivation lies in the ownership of "transparency" for the layers higher than the IP of IPSEC-IKEv2, as well as allowing a quick update of the encryption in case of data channel compromise. In this work [21] we evaluate the use of IPSEC to create VPNs suitable for "e-business" needs. A router Linksys, with the functionality of both an access point and a VPN concentrator and OpenSWAN as a system daemon, is used. This router supports the IKEv1-L2TP tunnel type (authentication/traffic via L2TP and encryption guaranteed by IPSEC).

The IPSec stack consists of three subcomponents: Encapsulating Security Payload (ESP), Authentication Header (AH), and Internet Key Exchange Protocol (IKE). The AH protocol is also discussed in [22] the security and quality of service aspects of VPNs are considered. The IKE protocol allows the various VPN endpoints to produce session keys for secure communications through a series of message exchanges. IKE currently consists of two versions, IKEv1 [23,24] and IKEv2 [16]. IKEv1 is very flexible and allows different configuration options, but it has a great architectural complexity that makes it increasingly difficult to deploy in modern networks. IKEv2 collects the legacy of the previous version and overcomes its limits, also providing more modern and performing encryption systems for network traffic. An important feature of IKEv2 is native support on current platforms (OS X 10.11+, iOS 9.1+, Linux 3.x + and Windows 8+). On the Mobile front, IKEv2 is supported both by native clients and with third-party apps on iOS, Android, Blackberry, and Windows devices.

QUIC is a UDP-based transport protocol, particularly suitable for IoT devices with limited resources [25,26] and general TRANSPORT SERVICES (TAPS) [18] implementation guidance can be found in [20]. Ref. [25] discusses Layer 3 VPN Tunnel Traffic Leakages and discusses possible mitigations.

In the work [19] we deal with the issue of the parameters to be provided to authenticate to a VPN server. Mainly, in many cases, the classic coupled username/password valid is enough to access, while for other implementations, it is necessary to provide additional parameters, as in the case of an IKEv1-XAUTH deployment.

The Secure Socket Tunneling Protocol (SSTP) [27,28] proposed by Microsoft allows channeling the traffic of a PPP device through an encrypted VPN tunnel via HTTP using SSL/TLS transport. This component, in particular, deals with negotiating keys, encrypting the channel, and ensuring the integrity of the traffic.

OpenConnect, discussed in [26], is an software designed to create point-to-point and point-to-multi-point secure architectures. Born as an open-source implementation of Cisco's AnyConnect SSL VPN client, it guarantees a good level of protection and data exchange at the application level [29]. This result allows to securely manage communications between applications and services running on IoT nodes and on cloud/edge infrastructures.

With the term "VPNaaS," we mean a simple way to parameterize network security centralizing the necessary configurations in a single product to meet numerous corporate security requirements and access to virtualized networks on the cloud. In particular, this work [30] proposes the implementation of such a service using WireGuard with a WAN backbone based on 5G transport.

In [22], experimental analysis was performed on Debian environment Linux by implementing the IPsec tunneling protocol with different encryption algorithms. The paper concludes that IPSec AES-sha1 provides fair and reasonable terms performance comparable to IPSec 3DES-sha1. It is also underlined how the encryption/decryption of the UDP VPN (User Datagram Protocol traffic) requires a large amount of CPU and memory that contributes to performance degradation.

The document [31] compares the VPN technologies "L2TP", "PPTP", "OpenVPN", "Ethernet over IP" (EoIP) and "MPLS". This analysis stemmed from having to choose the technology that best-suited business needs. For each VPN technology examined, accurate analysis of both performance and packets in transit was made.

3. Components of VPN Architecture

This section introduces the two currently most popular VPN protocols / implementations (IPsec, VPN based on SSL/TLS). They represent both the state-of-the-art and a starting point for new research in the sector and are supported both in commercial (e.g., Mikrotik) and open source products (e.g., OpenWrt, Linux Distros). In the following we present the used protocols for VPN implementation, the Linux OS for implementing the VPN and the considered hardware platforms, proprietary and free.

3.1. IPsec

IPsec is used to manage encrypted VPN tunnels at OSI Layer 3 [5,6]. IPsec is part of a series of protocols and its architecture has been proposed as a standard by the IETF, an organization that is responsible for continuing the technical development of the Internet. It was created for the new version of the network protocol (IPv6) and later also for IPv4. It can be configured according to three functional implementations:

- Transfer protocols : Encapsulating Security Payload (ESP) and Authentication Header (AH)
- Encryption Process Management: Internet Key Exchange (IKE) and Internet Security Association and Key Management Protocol (ISAKMP)
- Database: Security Policy Database (SPD) and Security Association Database (SAD)

Using the AH and ESP protocols, IPsec makes it possible to guarantee the integrity and authenticity of the data sent. The AH protocol, via the Packet Accelerator Extension, authenticates the data source and protects against modification of packets during transmission. Finally, it adds to the header a sequence number for preventing packets from being sent.

The ESP protocol guarantees, in addition to verifying the identity and integrity of the data, also the encryption of the pending data. It should be noted that ESP authentication does not consider the most external IP header and therefore it is not complete, therefore an additional encapsulation is required because the ESP contents are delivered correctly, traversing networks connected via Network Address Translation (NAT), as is generally the case in private xDSL networks. IPsec is configured to work in two modes: Tunnel and Transport.

By using "transport mode", the corresponding transfer protocol is added between the IP packet header, which remains unchanged, and the data area. The protection starts on the outgoing computer and comes up until it reaches the target computer. After the package is received, the original data are decompressed and made available. Transport mode has a very low elaboration time, and guarantees data security only to the detriment of that source and destination addresses. This is used for "Host 2 Host" or "Host 2 Router" connections.

By using "tunnel mode", the data packets receive a fresh complete IP header, then original address, destination address, and data are hidden. In addition to this the header of the respective transfer protocol is generated, creating the so-called "encapsulated" mode. The new outermost IP header defines the endpoints, encrypted and differentiated from the communication endpoints fixed in the actual IP header. Only when the packet has been unpacked on the encrypted endpoints it will be forwarded to the recipients. Data transmission in tunnel mode occurs at "Site 2 Site", "Host 2 Site", and "Host 2 Host".

3.2. SSL/TLS Based VPN

SSL/TLS VPN (Secure Sockets Layer VPN) provides a standard VPN solution based on a Web browser in Transport Layer or developed using specific client/server applications [7,8]. Sockets are used for data transfer between sender and receiver. There are two general deployments for SSL/TLS VPN implementation.

SSL/TLS Portal VPN - SSL/TLS Portal VPN: In this scenario, secure service access is achieved with a single standard SSL/TLS connection to the target web server. The client can access the SSL/TLS VPN gateway via a standard web browser, providing the necessary parameters to negotiate authentication [9].

SSL/TLS Tunnel VPN: In this scenario the VPN client can access multiple network services/hosts (this is the case with OpenVPN, OpenConnect, and SSTP). In the classic SSL/TLS communication, two keys are used to encrypt the data, the public one, shared among all, and the private one, typical of each endpoint. To further increase the level of security, two-factor authentication can be used by far or also OTP. In IPsec communication, once the client is authenticated to the VPN, it has full access to the private network, while in SSL/TLS VPN you have control of the more granular access, which allows you to create specific tunnels for applications via sockets rather than for the entire network, also creating specific "access roles" (access profile with specific rights for different users).

3.3. Linux Daemons

A VPN based on IPsec on Linux systems consists of the implementation via SW of the IKE, AH, and ESP protocols, using appropriate modules made available by the kernel.

IKE: As the name indicates, the purpose of the IKE protocol is to authenticate (using a pre-shared key, typing a public key, freeradius) the VPN peers, dynamically generate the keys, and share them with VPN peers. Keys are also used for the second phase of IPsec from IKE. Libreswan implements the IKE protocol using the project's bar program. ESP: The ESP protocol is the actual specification of the policy agreed by colleagues which is implemented in the Linux kernel's IPsec stack (NETEY/XFRM).

3.3.1. Libreswan/Strongswan

LibreSwan [32] and StrongSwan [33] are open source implementations of the IPsec protocol, powered by OpenSwan and based on the FreeSwan project, available as a ready-to-use package on Linux distributions on RedHat. However, detailed instructions are provided in the project's code for compiling on non-packaged Linux platforms. After the installation process, after the proper configuration, one will have an IPsec VPN gateway capable of protecting data in transit between the members of the network. Below, Table 1 compares LibreSwan and Strongswan features.

Table 1. Strongswan vs. Libreswan comparison Table.

Feature	LibreSwan	Strongswan
Pre-shared key authentication	Yes	Yes
Public-key authentication	Yes	Yes
IKEv1 key exchange	Yes	Yes
IKEv2 key exchange	Yes	Yes
AH support	Yes	Yes
NSS cryptographic library	Yes	No
Xauth and DNSSec	Yes	Yes
Network Manager	Yes	Yes
Virtual IP Addresses	Yes	Yes
MOBIKE	Yes	Yes
NAT Traversal	Yes	Yes
Route-based VPN	Yes	Yes
Policy-based VPN	Yes	Yes
Policy-based VPN and Route-based VPN simultaneous	Yes	No
Enable weak ciphersuites for backwards compatibility	No	Yes
High Availability	Yes	Yes

Libreswan configuration files are not compatible with strongSwan ones, although the format of ipsec.conf is identical. Different options have different meanings, others are mutually absent due to the supported architectures. For example, in Libreswan it is not possible to enable the L2TP IPsec [25] support of Android, because it is fixed in the client with the DH2 version (MODP1024), considered too weak to be supported, and it is disabled in the compilation phase. Strongswan and Libreswan both support both Policy-Based and Route-Based VPN, but in the former case they can be used in a mutually exclusive way unless you switch from the "classic" configuration (ipsec.conf) to the new one (swanctl.conf), while in the second, there are special flags, both policies are supported by default.

3.3.2. Accel-PPP

Accel-PPP [34] is a VPN concentrator designed to have high performance for Linux systems and allows the user to manage standard VPN technologies with a single application. Many open source projects provide VPN services but specialize in a specific technology. With Accel-PPTP, one has an all-in-one system with centralized configuration, management, and monitoring. Accel-PPP allows managing protocols: PPTP PPPoE L2TPv2 SSTP IPoE. The accounting is configurable via file or the use of Radius services. Authentication is managed through the mechanisms: PAP, CHAP (md5), MSCHAP-v1, and MSCHAP-v2 extensions, while EAP is not supported. All PPPoE, PPTP, and L2TP tunnels use special kernel modules to optimize performances.

3.4. Operative Systems and Hardware Platforms

In the current panorama, many operating systems and HW/SW platforms are dedicated to networking and secure communications. We have: MikroTik develops MikroTik RouterOS, the operating system of RouterBOARD boards; OpenWrt, developed by OpenWrt Project, an embedded Linux operating system, used on embedded devices for routing network traffic; PfSense, based on FreeBSD, an open source router and firewall with features that allow to manage unified threat, multi WAN and load balancing; OPNsense, another FreeBSD-based open source firewall that guarantees high security, Intrusion Prevention, Traffic Shaping and Captive Portal services; IPFire, based on Linux, a distribution designed to use the machine as an internal or perimeter firewall; VyOS, based on open Linux and distributed by the Sentrium company. It is open source and designed to protect the network and corporate data with high performance; Gargoyle, a firmware designed to use the machine as an internal or perimeter firewall; LibreMesh, an Open Source Sofware for Geek-free Mesh Community Networks. In this article, we will focus on the OpenWRT firmware and the Mikrotik platform for comparison.

3.4.1. OpenWRT

OpenWRT [35] is a distribution originally intended for use on wireless routers, to extend their functionality over manufacturer-supplied firmware. The operating system guarantees a filesystem with write permissions by the user, allowing among other things the installation of third-party software and therefore the possibility of extending its functionality.

This allows you to make use of the most recent routing software, and guarantees greater security and fewer bugs than pre-installed stock manufacturer software, especially in older devices no longer supported.

It can also be installed on custom HW, such as specific boards, and in the case of x86-64 platforms, it also supports virtualization, allowing small networks of containers for ad-hoc services, as well as providing network access to all devices in the LAN.

3.4.2. MikroTik

MikroTik [36] is a Latvian company based in Riga and produces equipment for networking and internet connectivity, in particular routers and wireless broadband equipment for Wireless ISPs. It is present in almost all countries of the world.

MikroTik's experience in creating hardware and routing systems highly compatible with the most widespread industry-standard systems led to the creation in 1997 of the RouterOS software, with high control, and flexibility for all types of routers and interfaces, developed on the Linux kernel. Thanks to RouterOS any PC or MikroTik RouterBOARD can become a dedicated router. Being proprietary devices, the flexibility in installing packages for additional features, however respectable (supports IoT, Lora, and 4G/5G), is lower than that of OpenWRT. Figure 1: highlights the ability of a Mikrotik router to act as a "Publisher" to an MQTT Broker (deployed on an OpenWRT router).

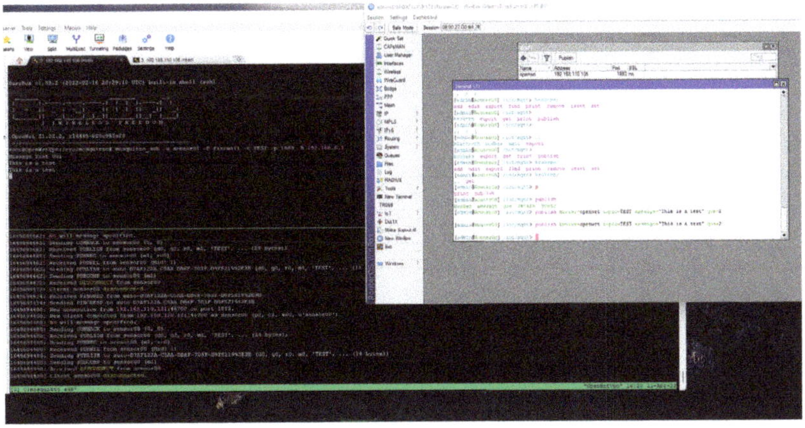

Figure 1. A publish/subscribe scenario from Mikrotik to OpenWRT MQTT Broker.

4. Description of Different Deployed VPN Solutions

To create a VPN, there are different solutions, both HW and SW (e.g., use Mikrotik RouterBoards, ("Mikrotik ", https://mikrotik.com/ (accessed on 15 June 2022)) or implement the appropriate stack on your own server, in this case with Debian OS ("DEBIAN OS " https://www.debian.org/index.it.html) (accessed on 15 June 2022)), with peculiar properties depending on the chosen implementation. In particular, we will focus on three macro application scenarios:

- A Site to Site scenario: (where two remote offices communicate securely), shown in Figure 2
- A Site to Multi-Site scenario: (where multiple remote offices communicate securely), shown in Figure 3
- A Road Warriors scenario: (in which multiple users from remote offices communicate securely with a particular local office), showed in Figure 4.

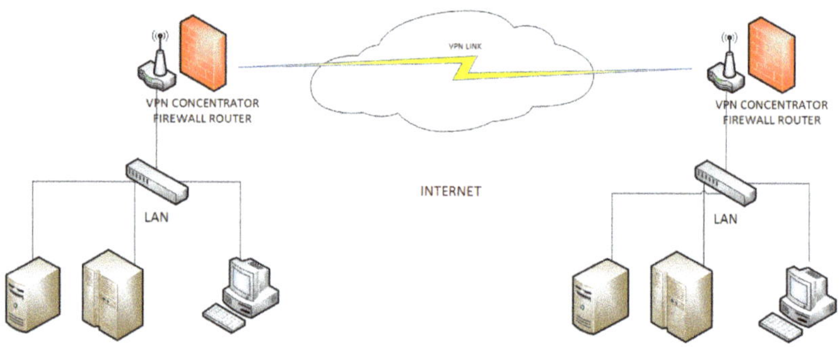

Figure 2. Site to Site VPN Scenario.

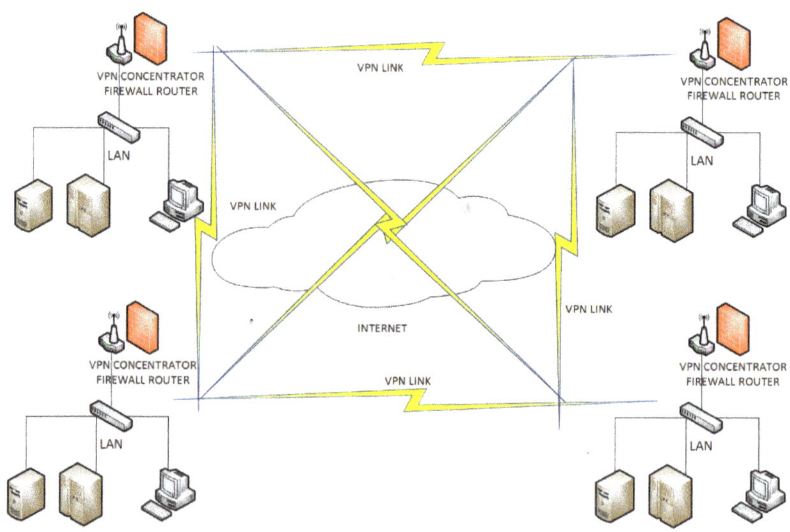

Figure 3. Site to Multi Site VPN Scenario.

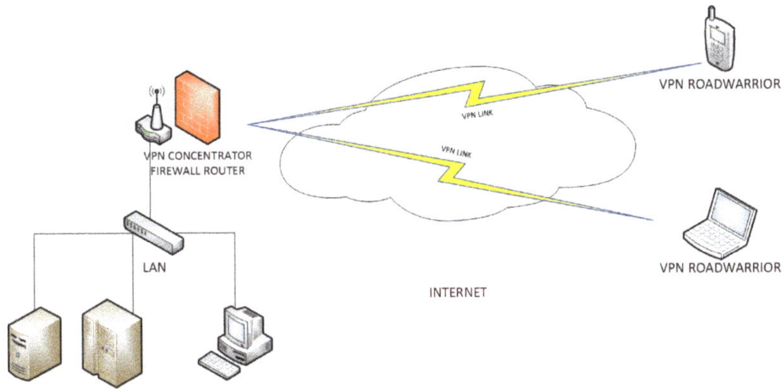

Figure 4. Site to Road Warriors VPN Scenario.

These solutions provide encryption and integrity of data in transit by relying on encryption algorithms ranging from Blowfish to AES (128–512) up to implementations that use elliptic curves (e.g., ecp192). Depending on the type of device/network to be interconnected, various alternatives are available, especially in the case of Road Warriors (RW), passing from authentication via PSK + XAUTH (e.g., in IPsec) to that through certificates, or even to two-step solutions, through one-time password (OTP) or supplementary credential files (case RADIUS + VPN IPsec/OpenVPN/OpenConnect). In this work we wanted to make a "field" assessment of the current landscape both at the level of communication infrastructures (WIFI, CABLE 4G/5G), and at the level of devices that it is possible to connect (NUC, mini PC, Raspberry Pi, Smartphones). Involved variables are many, ranging from guaranteeing the widest support to the greatest number of devices and technologies, trying to guarantee the best performance with the same safety.

Table 2 provides a summary of the implemented testbeds: from a description of the network infrastructure to the software used.

Many of the architectures proposed were built in the projects mentioned above and are still in production. OpenVPN and Wireguard are part of the Cogito project to interconnect the cloud and local offices. One used IPSec in the Domus project to manage data exchanges and resource access policies. Finally, one used OpenConnect and OpenVPN in the RES-NOVAE project for the same purpose.

Moreover, given the exponential increase of technologies such as the IoT [37], and the consequent development of communication paradigms such as "Fog Computing" or "Cloud oriented approaches" (e.g., AMAZON AWS), or multisite clusters (e.g., realized through Kubernetes [38]) both on fiber/PSTN carriers and on 4G/5G carriers, this ever-growing volume of data, also in compliance with the recent "General data protection regulation" (GDPR) regulations, must be managed properly. Another great advantage, deriving from the use of the OpenWRT ("OpenWRT Project ", https://openwrt.org/ (accessed on 15 June 2022)) operating system, allows the implementation of these technologies on all models of compatible routers and devices. The wide range of supported "bridges" allows, especially in the case of IoT networks, to interface with third-party protocols, to be used as a perimeter gateway, such as LoRa and Zigbee, as well as the classic ones supported by TCP/IP stack. If you consider that the x86-64 versions also easily support virtualization, you can create Fog networks of micro-services and IoT multi-protocol with negligible figures and connect them securely. An example could be given by the installation of outdoor sensor networks on areas covered in WAN by 4G/5G networks and locally configurable and processable with an "Edge Computing" approach.

Table 2. A summary of the implemented testbeds.

VPN Software Managed	VPN Deploy	VPN Implementers	Operative Systems	Platforms
IPsec LIBRESWAN 3.27 STRONGSWAN 5.9.1	ROAD WARRIORS	IKEv1-XAUTH IKEv1-L2TP IKEv2 RSA	LINUX DEBIAN 11 WINDOWS 10 (client) WINDOWS 11 (client) DEBIAN 11/10 (client) ANDROID 11 (client) iOS 15 (client) MAC OS X 13 (client) RASPBERRY Pi 2/3/4 MIKROTIK 7.X OpenWRT 21.x	armv7 ×86 ×86-64 ARM64 ARM MIPSBE MMIPS SMIPS PPC
IPsec LIBRESWAN 3.27 STRONGSWAN 5.9.1	SITE TO SITE	IKEv1 SITE TO SITE PSK IKEv2 SITE TO SITE RSASIG	Same as above	Same as above
IPsec LIBRESWAN 3.27 STRONGSWAN 5.9.1	SITE TO MULTI SITE	IKEv2 XFRMI ROUTE BASED	Same as above	Same as above
OpenVPN 2.5.1	ROAD WARRIORS SITE TO SITE SITE TO MULTI SITE		Same as above	armv7 ×86 ×86-64
Accel-PPP Latest github stable realease	SITE TO SITE SITE TO MULTI SITE	SSTP SITE TO SITE SSTP SITE TO MULTI SITE	Same as above	Same as above
Accel-PPP Latest github stable realease	ROAD WARRIORS	SSTP ROADWARRIORS	Same as above	Same as above
Wireguard 1.0.2	SITE TO SITE SITE TO MULTI SITE ROAD WARRIORS		Same as above	armv7 ×86 ×86-64 ARM64 ARM MIPSBE MMIPS SMIPS PPC
Ocserv 1.1.2	SITE TO MULTI SITE SITE TO SITE ROAD WARRIORS		Same as above	armv7 ×86 ×86-64

5. Experimental Results in Implemented Scenarios

In order to evaluate the performance of different VPN protocols on different operating systems and hardware as represented in Table 2, different topologies were configured as in Figures 2–4. Both Iperf software and Atop are used to record the activities of the server's operating system during the various VPN sessions. Network topologies nodes are shown in Table 3.

Table 3. Network topologies HW components.

Hardware	Quantity
Workstations with Intel® i7 @ 1.80 GHz processor, 8GB RAM	2
VMWare Mikrotik RouterOS x86-64 virtualized	1
TP-Link Archer C7 v5 Qualcomm Atheros QCA956X v1 128 MB RAM	1
TP-LINK TL-WR841N (RO) v11 Qualcomm Atheros QCA9533 32 MB RAM	1
TP-Link TL-SG105E Switch 10/100/1000 Mbps	1

Table 4 shows a comparison of Authentication and Encryption algorithms for deployed scenarios and proposed by default client-side.

Table 4. Comparison table of Authentication and Encryption used in the experiments and proposed by default client-side.

VPN Type	Encryption	Authentication
SSTP	AES-256	Username/Password, Certificates
IKEv1-L2TP	AES-256-GCM128	Username/Password/PSK
OPENCONNECT	AES-256	Username/Password, Certificates, OTP
OPENVPN	AES-256, CHACHA20	Username/Password, Certificates, OTP
IKEv2 X509	AES-256	Certificates
IKEv2 PEAP	AES-256	Username/Password, Certificates
IKEv1-XAUTH	AES-128	Username/Password, PSK
WIREGUARD	CHACHA20, CURVE25519	Certificates/Keys

The nodes are connected to a 10/100 Ethernet switch with 100 Mbps UTP links. The complete topology consists of a subnet representing the WAN (with public IP assigned, for testing from 4G/5G networks), two private subnets for each workstation, one subnet for Mikrotik and one subnet for OpenWRT, as shown in Figure 5. In the "Site to Site" topology, two VPN servers act as software routers and endpoints of the VPN tunnel and they are connected to local workstations where traffic is generated. In the "Site to Multisite Site" topology, the servers become three (the number of nodes can be increased), they maintain the function of the software router and endpoint of the VPN tunnels and are connected to local workstations on which traffic is generated. In the "Road Warriors" topology, a VPN server acts as a software router and endpoint of the VPN tunnel and it is connected to local workstations on which traffic is generated, this time directed towards mixed clients (smartphones or laptops connected via wireless). To generate the network traffic, the "Iperf" tool was used, in versions 2 and 3 (both UDP and TCP tests). This tool measured productivity and round-trip time (RTT) while several counters of the same operating system measured the CPU usage.

5.1. Considered VPN Deployed Topology

Some deploys described in this paper were also used in the context of the "DOMUS" ("Progetto DOMUS ", https://www.gruppotim.it/it/archivio-stampa/mercato/2016/TIM-Distretto-Domus-Cosenza-14Dicembre2016.html (accessed on 10 June 2022)), "COGITO" ("Progetto COGITO ", https://www.icar.cnr.it/progetti/cogito-sistema-dinamico-e-cognitivo-per-consentire-agli-edifici-di-apprendere-ed-adattarsi/ (accessed on 10 June 2022)) and "RES-NOVAE" ("Progetto RES-NOVAE ", https://www.cueim.org/progetti/res-novae-reti-edifici-strade-nuovi-obiettivi-virtuosi-per-lambiente-e-lenergia-smart-city/ (accessed on 10 June 2022)) projects and can be replied in any environment. The reason for the choice of the HW considered is given by the need to measure the performance of components already put into operation in the three research projects referred to above and also to ensure maximum backward compatibility.

The basic network topology with example IP address ranges is depicted in Figure 5. From time to time, the right endpoint is replaced in the site-to-site scenario, and the connections whose data are shown in the figures are established. For Wi-Fi-only devices, acting as a VPN client, the connection is established with the left endpoint. The Wi-Fi network these devices are originally connected to is the dummy WAN.

The dummy WAN is a simple network with a router (IP 192.168.110.254) on the segment 192.168.110.0/24, which simulates a public network and allows communication between VPNs endpoints (connected wired / wireless clients to the router or switch). In particular, two private IP of the fictitious WAN simulate the public IP of the gateways of two LANs located at a lower level and used to test site-to-site topology, respectively, with "internal" networks 192.168.111.0/24 (site one) and 192.168.112.0/24 (site two). The Wi-Fi clients are connected directly to the access point of the primary router. All the RTT tests were carried out by reaching the local network IP of the OpenWRT router, which acts as a VPN centralizer from the client on duty, as shown in Figure 5.

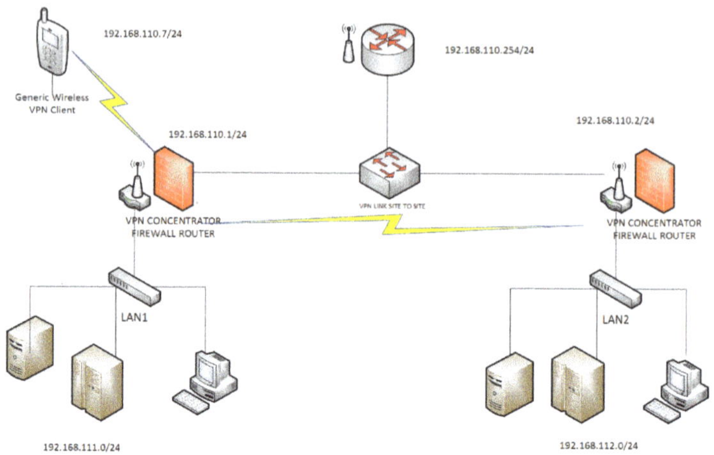

Figure 5. Generic VPN Deploy Topology.

Both tools, Iperf2 and Iperf3, were used to get throughput results, and the "ping" tool was used to get RTT data. The graphics were generated using Iperf3 because there were not significant differences with Iperf2.

5.2. *System Tools for Performance Analysis: Iperf and Atop*

Iperf allows to measure the bandwidth available for IP networks. It is available in two versions, Iperf2 and Iperf3, released as open source and thanks to the support for all the most popular operating systems (including mobile ones such as Android and iOS) and the ability to adjust numerous parameters in the analysis of network performance for various protocols, is undoubtedly one of the reference tools for network diagnostics.

Atop is conceived to be a system performance monitor, which works through CLI (Command Line Interface) and can record and report the activity of all operating system processes and its HW/SW components. It allows to analyze the server at runtime and automatically in the long term. It shows the resource usage of all processes and devices and provides the ability to monitor threads within processes and highlights critical lines using colors (red). Moreover, it allows to analyze disk I/O and network usage. Thanks to the kernel module netatop it can acquire TCP, UDP data, and network bandwidth. For each established VPN connection, an instance of atop was launched to collect CPU and network statistics data. These data was passed to *atopsar* to generate readable and archival reports via CSV. With special scripts, these CSVs were tested to create successive graphics using regex and Gnuplot. Subsequently, the aggregated graphs were produced using Excel. All the figures relating to the following topologies have a double scale and display respectively the average throughput (left axis) and the standard deviation (right axis), for wired and wireless endpoints (left figure and right figure).

Without using VPN protocols, the average throughput of the network is 94.1 Mbps, in the fast ethernet sections, with an RTT of 1417 ms. In the Gigabit sections, the throughput is 760 Mbps, with an average RTT of 0.730 ms.

5.3. *IPsec/L2TP Road Warriors Scenario*

The IPsec/L2TP scenario is one of the most common deployments today. It is used for its excellent compatibility, but being outdated, it should be avoided whenever possible. IPsec in transport mode and NAT with multiple connected clients often suffer from performance and stability issues. Many L2TP clients are configured using aggressive/PSK mode, which is problematic from a security point of view. The encapsulation of PPP over IPsec can cause MTU problems, and to remedy this, you force the MTU of clients into a range of 1200–1500 KB, depending on the type of WAN connection. For IPsec/L2TP servers behind

NAT, the registry settings on Windows to allow clients to connect have to be changed. The supported clients are many, all Apple iPhone, iPad, Mac OS X, Android, Linux with a command line, and Microsoft Windows. The server has three components to configure: libreswan for IPsec, xl2tpd for L2TP, and pppd for PPP. Figure 6 shows the comparison of the average throughput and standard deviation in IKEv1 L2TP Road Warriors scenario, wired (on the left) and wireless (on the right).

As seen from the figure, OpenWRT as an Ikev1-L2TP client turns out to be the best-performing, closely followed by Windows 10, where, however, it is necessary to work at the registry level to enable Diffie–Hellman group 14, 2048 bit. As for the wireless clients, the iPhones are slightly better performing than Android.

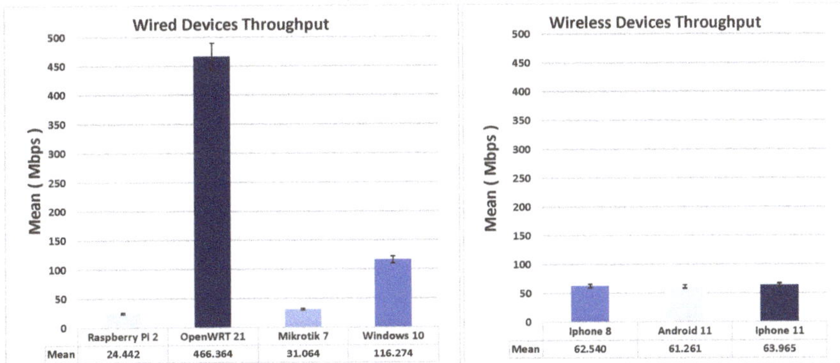

Figure 6. Comparison of the average Throughput in IKEv1-L2TP Road Warriors scenario.

5.4. *IPsec Xauth Road Warriors Scenario*

Implementing a VPN service for remote clients using IKEv1/XAUTH with PSK is the least secure way to run IKE/IPsec. The reason is that everyone in the "group" has to know the PSK (also called secret). Even further authentication is required, such as a username and password, someone that knows the PSK can launch a man-in-the-middle attack pretending to be the VPN server. If the client connects to the rogue server, it will tell the attacker their username and password. On Android, this mode is called PSK XAUTH. On iOS/OS X this mode is confusingly called type "IPSec". Some other vendors call it "Group PSK". Supported clients are: all Apple iphones, ipads, Mac OS X, Android, Linux with NetworkManager or commandline and Windows Shrew client, or Cisco routers/firewalls. It is based on draft [19]. XAUTH also requires a username and password. The password can also contain a OTP such as Google Authenticator. Figure 7 shows the comparison of the average throughput and standard deviation in IKEv1 XAUTH Road Warriors scenario, wired (on the left) and wireless (on the right). As seen from the figure, OpenWRT as an Ikev1-XAUTH client turns out to be the best performing, closely followed by Windows 10, where, however, it is necessary to use a third-party application to allow the connection. As for the Wireless clients, the iPhones are the only compatible smartphones.

5.5. *IPsec IKEv2 EAP Road Warriors Scenario*

The following EAP connection definition allows multiple Windows clients to connect to the strongSwan VPN gateway via any EAP method over IKEv2. Cypher suites aes256-sha256-modp2048 for IKE and aes256-sha1-modp2048 for ESP are the strongest proposals; the Windows client is able to offer unless PowerShell is used. These proposals are not explicitly configured here to accept stronger algorithms proposed by other clients via strongSwan's default proposals.

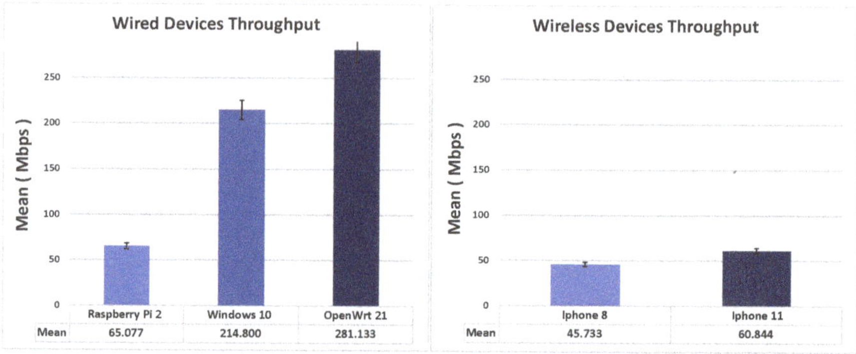

Figure 7. Comparison of the average Throughput in IPsec XAUTH Road Warriors scenario.

Figure 8 shows comparison of the average throughput and standard deviation in IKEv2-EAP Road Warriors scenario, wired. As seen from the figure, Windows 10 as an Ikev2-EAP client turns out to be the best performing, while Mikrotik and Raspberry Pi are almost equivalent. The virtual machine running the demo version of Mikrotik is locked to Ethernet 100 Mbps). Using IKEv2 with certificates, the performance of OpenWRT (client) and the Raspberry Pi 2 are noticeably close. As for the wireless clients, the Android, this time, turns out to be slightly better performing than iPhone 11 and iPhone 8.

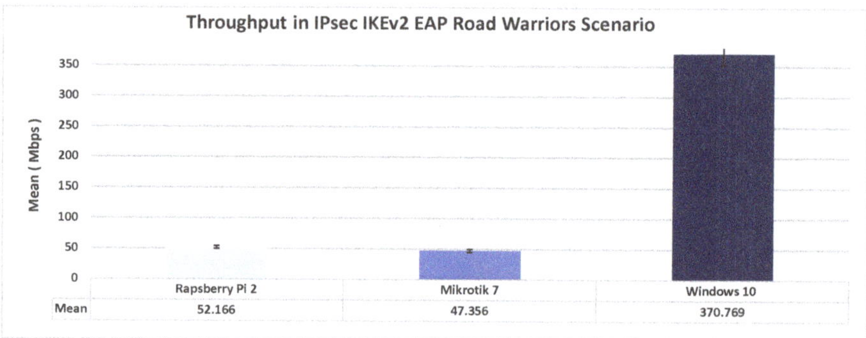

Figure 8. Comparison of the average Throughput in IKEv2-EAP Road Warriors scenario.

5.6. IPsec IKEv2 X509 Road Warriors Scenario

This method uses IKEv2 without EAP also called "Machine Certificate"-based authentication. Supported clients are:

- libreswan
- Windows 7 and up
- Windows Phone (requires the latest firmware)
- OS X and iOS
- Android with strongswan client.

Figure 9 shows the comparison of the average throughput and standard deviation in IKEv2 Cert Road Warriors scenario, wired (on the left) and wireless (on the right). As seen from the figure, Windows 10 as an Ikev2-EAP client turns out to be the best performing, while Mikrotik and Raspberry Pi are almost equivalent. The virtual machine running the demo version of Mikrotik is locked to Ethernet 100 Mbps). Using IKEv2 with certificates, the performance of OpenWRT (client) and the Raspberry Pi 2 are noticeably close. As for the wireless clients, the Android, this time, turns out to be slightly better performing than iPhone 11 and iPhone 8.

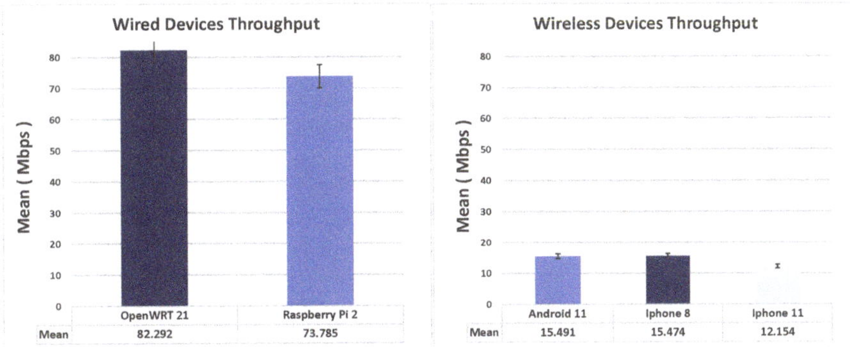

Figure 9. Comparison of the average Throughput in IPsec IKEv2 X509 Road Warriors scenario.

5.7. OpenConnect (Aka Cisco Anyconnect) Road Warriors Scenario

The OpenConnect protocol is designed to create a channel that makes use of UDP packets and secondarily over TCP if UDP traffic is not allowed. The protocols used by Openconnect are: TLS [RFC8446], Datagram TLS [RFC6347] and HTTP [RFC2616]. The session begins with an HTTP connection over TLS on a known port, after which the client authentication begins. Once this is done, the client initiates an HTTP CONNECT command to establish a VPN channel over TCP and, if possible, a secondary VPN channel over UDP, depending on the response from the server. Once these negotiations are complete, IP packets can travel over the VPN. Figure 10 shows the comparison of the average throughput and standard deviation in OpenConnect Road Warriors scenario, wired (on the left) and wireless (on the right). As seen from the figure, OpenWRT as an OpenConnect client turns out to be the best performing, followed by Windows 10, which uses a third-party application to allow the connection but with a ratio of almost 5:1 and finally from the Raspberry Pi 2, with a nearly 10:1 ratio. As for the wireless clients, the iPhones show inferior performance to Android.

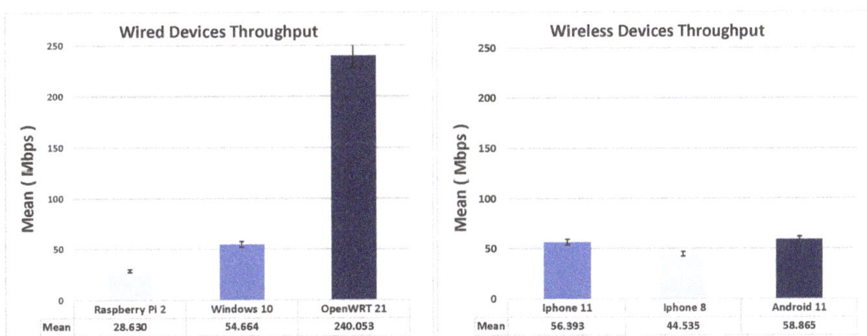

Figure 10. Comparison of the average Throughput in OpenConnect Road Warriors scenario.

5.8. SSTP Road Warriors Scenario

SSTP, the successor to PPTP-based VPNs, was introduced by Microsoft with the Windows Vista system. This VPN service is still trusted in Windows 7, 8, 10, and 11 versions. SSTP uses 256-bit AES encryption, thus providing reasonable security in data transit while also providing a reasonable speed for encrypted tunnel communications. It is mainly used in the Microsoft environment, but implementations are available on Linux (client/server mode), on OpenWRT (client mode) and on the Mikrotik platform (client/server mode). Figure 11 shows the comparison of the average throughput and standard deviation in Wireguard Road Warriors scenario, wired (on the left) and wireless (on the right). As seen from the figure, OpenWRT as an SSTP client, performs less than

Windows 10, but it still has almost double performance compared to Mikrotik in a virtual machine; both use a dedicated kernel module to establish SSTP connections. Finally, the Raspberry Pi 2 also has noteworthy performance. Windows 10, of course, has the best performance with native support for this technology. As for the wireless clients, the iPhones are incompatible with this technology unless one purchases specific apps.

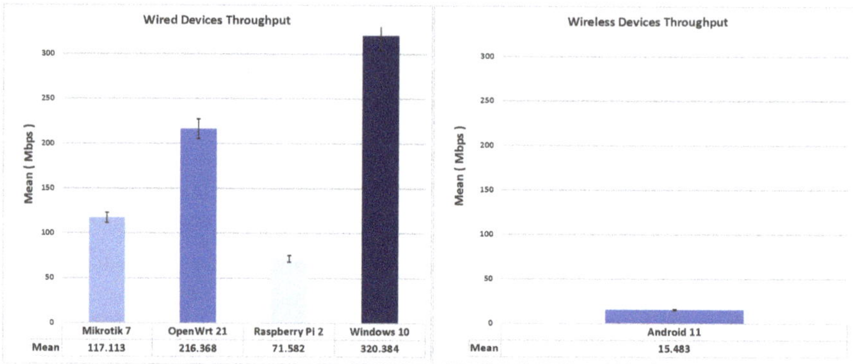

Figure 11. Comparison of the average Throughput in SSTP Road Warriors scenario.

5.9. *OpenVPN Road Warriors Scenario*

OpenVPN, a project managed by the "OpenVPN community Team" allows the creation of IP network tunnels using virtual ethernet adapters on TCP/UDP ports. The encryption, authentication, and certification functions are provided by default by the OpenSSL library or by the Gnu-TLS library, and this guarantees a high standard of protection of private network traffic on the public network. OpenVPN allows two authentication modes, one with a static key (PSK) and the other through SSL / TLS certificates that regulate authentication and key exchange using a customized security protocol. Figure 12 shows a comparison of the average throughput and standard deviation in the OpenVPN Road Warriors scenario, wired (on the left) and wireless (on the right). As can be seen from the figure, OpenWRT as an OpenVPN client turns out to be the best performing, followed by Windows 10, which uses a third-party application to allow the connection. Mikrotik only supports TCP-based connections, but guarantees decent throughput. When it comes to wireless clients, Android falls between the two iPhone generations.

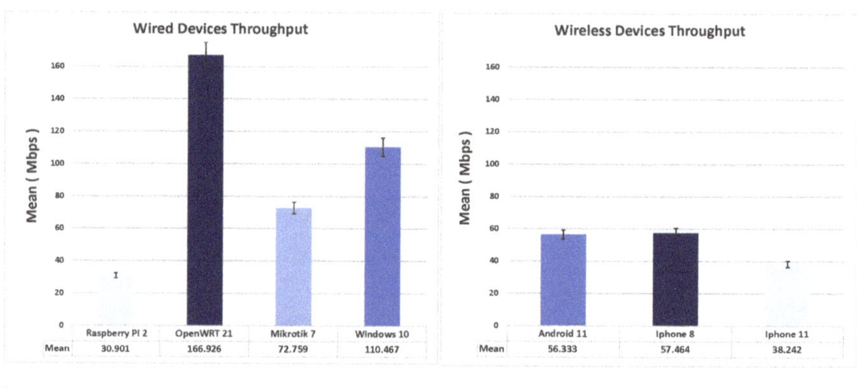

Figure 12. Comparison of the average Throughput in OpenVPN Road Warriors scenario.

5.10. *Wireguard Road Warriors Scenario*

WireGuard [39] is a VPN protocol designed to be at the forefront of speed, management, and scalability, with the goal of being leaner than the main contenders, IPSEC and

OpenVPN. It is designed as a general-purpose VPN protocol that can be deployed on both extremely powerful and extremely limited hardware with excellent performance. It is multiplatform and runs both as a client, and as a server on Windows, macOS, BSD, iOS, Android, and Linux (on which it was developed) and is easily distributable. Despite being a young VPN protocol, its qualities, as described in the specific RFC (RFC7539) [40], have already led it to be a reference for the IT communications of the near future. Figure 13 shows a comparison of the average throughput and standard deviation in the Wireguard Road Warriors scenario, wired (on the left) and wireless (on the right). As seen from the figure, OpenWRT as a Wireguard client turns out to be the best performing, followed by Windows 10, which uses a third-party application to allow the virtualized connection. Mikrotik guarantees decent throughput, even if it is worse than the Raspberry Pi 2. Regarding wireless clients, Android falls between the two iPhone generations.

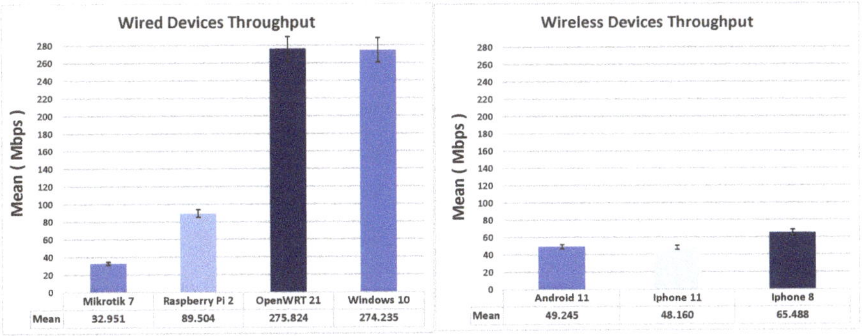

Figure 13. Comparison of the average Throughput in Wireguard Road Warriors scenario.

5.11. Per Device Analysis

The data analysis performed on each VPN as the devices vary, described in the previous paragraphs, provided helpful information to identify the devices with better throughput, given a pre-existing infrastructure. A complementary analysis of what was described in the preceding paragraphs of this section is addressed in this paragraph. The study, in this case, allows people to highlight, for each device examined, the VPN technology if supported, which guarantees better performance. From Figures 14–20, the following graphs group the average throughput and throughput of the STD data by collecting them by device rather than by VPN protocol type. From a first analysis, it is possible to verify that only a few devices allow the use of all the VPN technologies analyzed, in some cases because the software that implements the technology is missing, in others for reasons of architectural choices of the suppliers. This analysis also gives people a clear view of which VPNs are the best performing and helps the network designer make targeted choices, even in architectures already provided by third parties.

Figures 14 and 15 show the data collected on the average throughput for two different generations of iPhone devices as the VPN established for the iPhone 8 and the iPhone 11 varies. From the graphs, we can see how there are variations in performance between the different VPNs supported. One can observe that for the iPhone 8, the Wireguard and OpenVPN VPNs are the best-performing VPNs, while for the iPhone 11, this is no longer true. Our analysis shows that the IKEv1-L2TP VPN maintains the same performance on average for both iPhone devices tested. Furthermore, from Figure 16, which shows the average VPN throughput per Android 11 smartphone, we can conclude that IKEv1-L2TP VPN is the technology that maintains the highest throughput levels on average across all tested wireless devices. For wired devices, whose average throughput is shown in Figures 17–20. The data collected show a much more heterogeneous situation. For example, from the comparison between Figures 17 and 18, where the average throughput for the Mikrotik 7 device and the Raspberry Pi 2 is shown respectively as the VPN varies, one can conclude that the choice to connect these two endpoints via a VPN would fall on SSTP technology. While the others perform optimally on one of the two endpoints. Similarly, if we were to connect two endpoints, such as OpenWRT 21, whose average throughput is represented in Figure 19, and the Windows 10 endpoint, whose average throughput is depicted in Figure 20, the choice would have to fall between a VPN between the first three VPN, that have highest throughputs. In our case, the choice would fall on the Wireguard VPN.

Figure 14 shows the comparison of the average throughput and standard deviation in a Road Warriors scenario, using an iPhone 8 client. The best results in terms of throughput are obtained here using Wireguard.

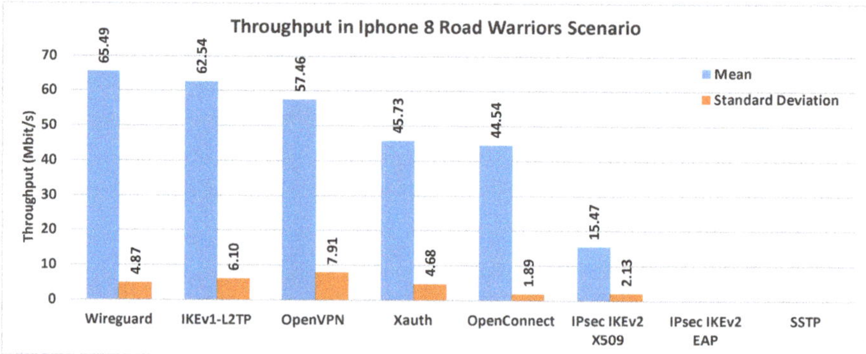

Figure 14. Comparison of Throughputs on iPhone 8 in all scenarios.

Figure 15 shows the comparison of the average throughput and standard deviation in a Road Warriors scenario, using an iPhone 11 client. The best results in terms of throughput are obtained here using IKEv1-L2TP.

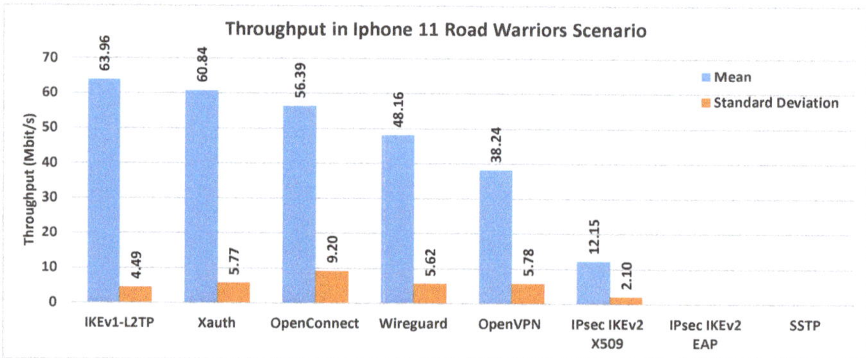

Figure 15. Comparison of Throughputs on iPhone 11 in all scenarios.

Figure 16 shows the comparison of the average throughput and standard deviation in a Road Warriors scenario, using an Android 11 client. The best results in terms of throughput are obtained here using IKEv1-L2TP.

Figure 17 shows a comparison of the average throughput and standard deviation in a Road Warriors scenario, using a Mikrotik 7. x endpoint. The best results in terms of throughput are obtained here using SSTP.

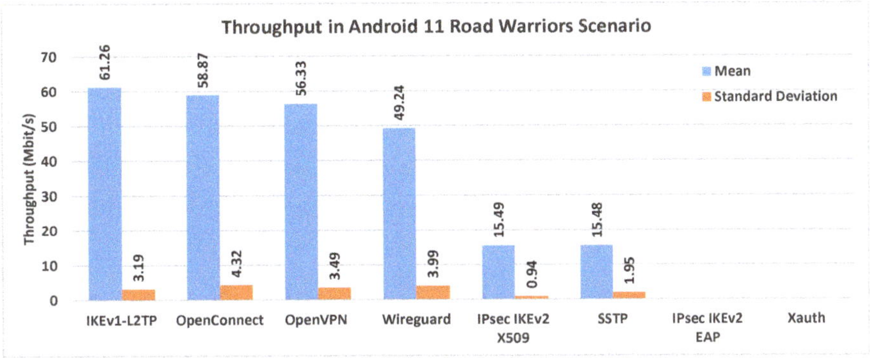

Figure 16. Comparison of Throughputs on Android 11 in all scenarios.

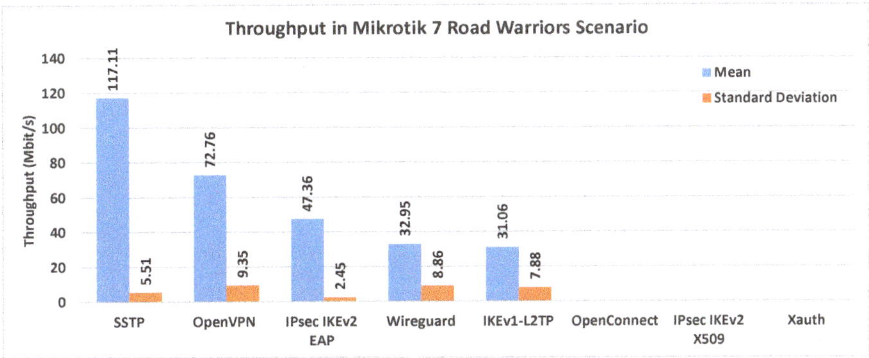

Figure 17. Comparison of Throughputs on Mikrotik 7 in all scenarios.

Figure 18 shows the comparison of the average throughput and standard deviation in a Road Warriors scenario, using a Raspberry Pi 2 endpoint. The best results in terms of throughput are obtained here using Wireguard.

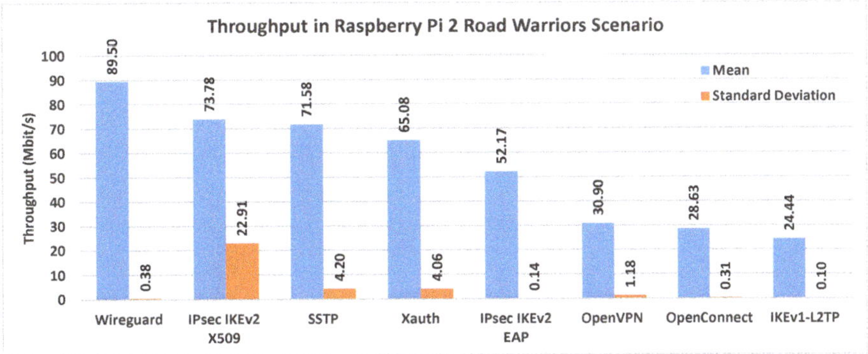

Figure 18. Comparison of Throughputs on Raspberry Pi 2 in all scenarios.

Figure 19 shows the comparison of the average throughput and standard deviation in a Road Warriors scenario, using an OpenWRT 21 endpoint. The best results in terms of throughput are obtained here using IKEv1-L2TP.

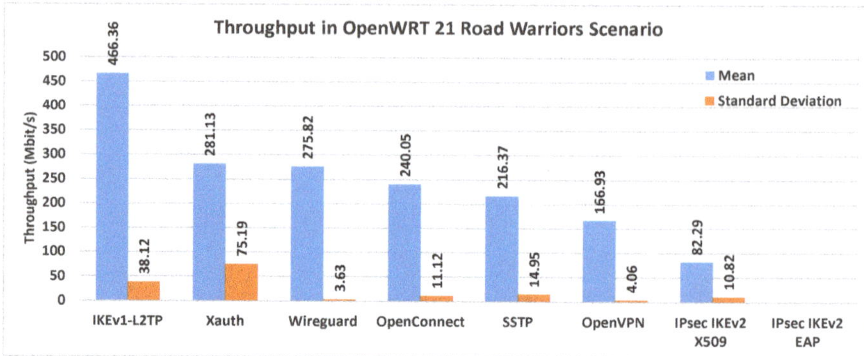

Figure 19. Comparison of Throughputs on OpenWRT 21 in all scenarios.

Figure 20 shows the comparison of the average throughput and standard deviation in a Road Warriors scenario, using a Windows 10 endpoint. The best results in terms of throughput are obtained here using IKEv2-EAP.

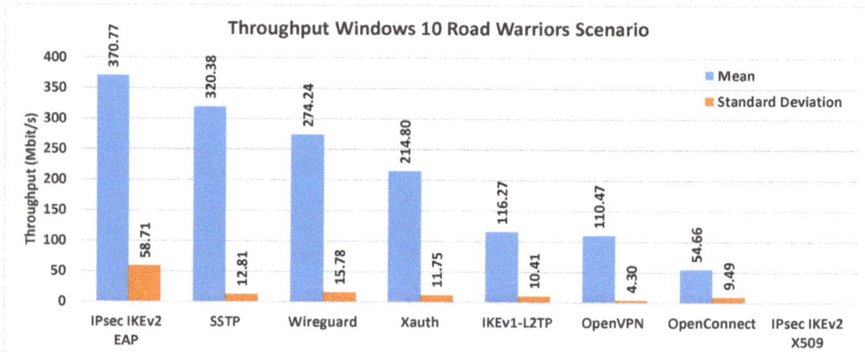

Figure 20. Comparison of Throughputs on Windows 10 in all scenarios.

5.12. Experimental Evaluation of Different VPN Protocols on the Same Hardware

Concerning the experimental evaluation of VPN protocols with different security levels on the same hardware, the choice was OpenVPN on the OpenWRT Router. OpenVPN in its most recent versions sets up two connections: a low-bandwidth "control channel" that negotiates network parameters and keys for the "data channel" and uses TLS to protect packets passing through it, and a "data channel", on which the actual VPN traffic travels, is encrypted with keys traded on the control channel. The version installed on OpenWRT uses TLS 1.3, the most recent, and the related data encryption algorithms were evaluated. In particular, the following testbeds were carried out, the results of which are highlighted in the following two graphs and summarized in Table 5.

Table 5. OpenVPN using different encryption algorithms configuring TLS-cipher suites and data-ciphers.

TLS-CIPHER	DATA-CIPHER 1	DATA-CIPHER 2	DATA-CIPHER 3
TLS-AES-128-GCM-SHA256	AES-128-GCM	AES-256-GCM	CHACHA20-POLY1305
TLS-AES-256-GCM-SHA384	AES-128-GCM	AES-256-GCM	CHACHA20-POLY1305
TLS-CHACHA20-POLY1305-SHA256	AES-128-GCM	AES-256-GCM	CHACHA20-POLY1305

Figure 21 shows the throughput of the OpenVPN using different encryption algorithms. Figure 22 shows the RTT of the OpenWRT VPN using different encryption algorithms.

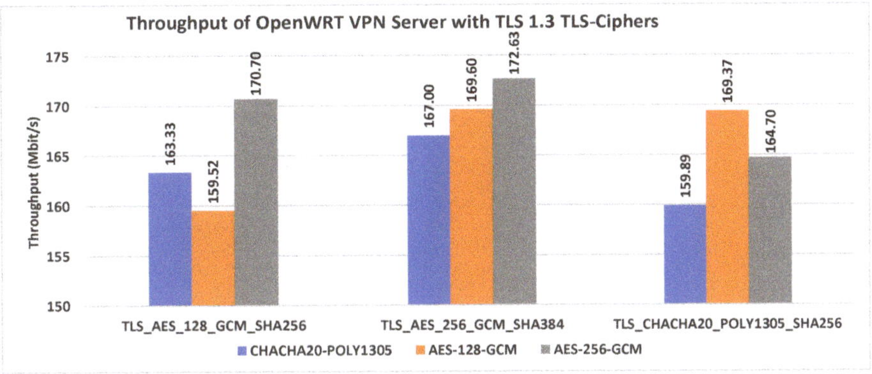

Figure 21. Throughput of OpenVPN using different encryption algorithms.

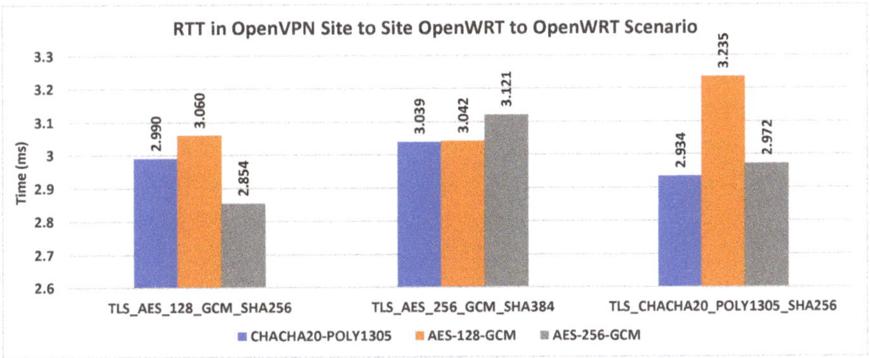

Figure 22. RTT of the OpenWRT VPN using different encryption algorithms.

In Figure 23 the PHY RATE data collected during the VPN connection tests with wireless devices are shown.

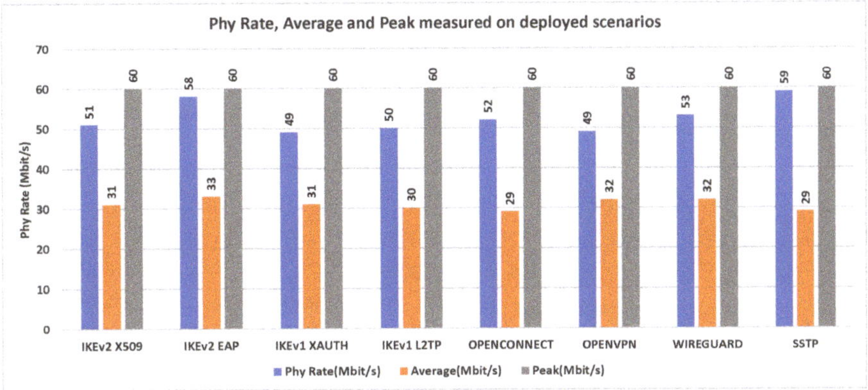

Figure 23. Phy Rate, Average and Peak measured on deployed scenarios.

In Figure 24 the RTT data collected during the VPN connection tests with wired devices are shown.

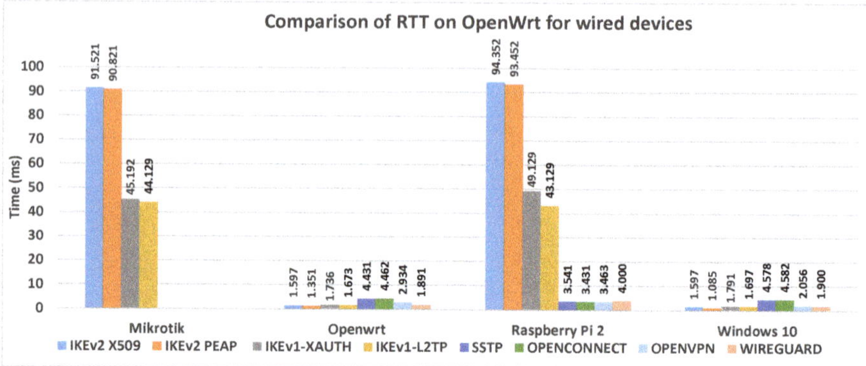

Figure 24. RTT data collected during the VPN connection tests with wired devices.

In Figure 25 the RTT data collected during the VPN connection tests with wireless devices are shown. Both test batteries expected 200 packets sent from one endpoint to another.

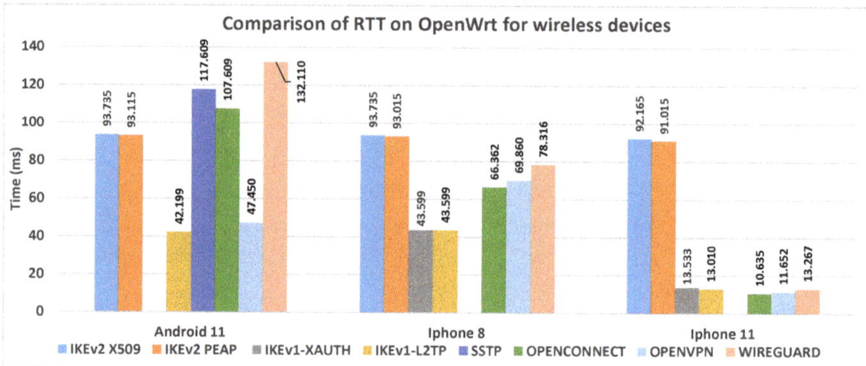

Figure 25. RTT data collected during the VPN connection tests with wireless devices.

From the comparative analysis of Figures 21 and 22, AES-256-GCM, using TLS-ciphers 1,3, turns out to be the best data cipher in terms of throughput and RTT, followed by AES-128-GCM and CHACHA20-POLY1035. From the analysis of Figure 23, OpenWRT results have an average Phy Rate ranging between 29 and 33 Mbps on the TP-LINK TL-WR841N device. From the comparative analysis of Figures 24–27, OpenWRT is the best device/OS for wired connections, while iPhone 11 is the best RoadWarrior wireless client.

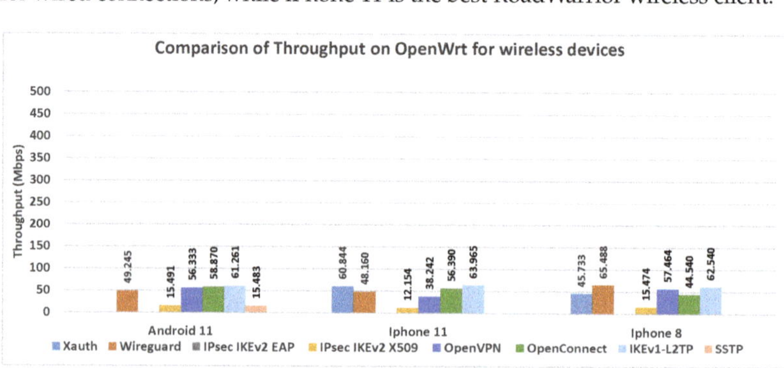

Figure 26. Throughputs data collected during the VPN connection tests with wireless devices.

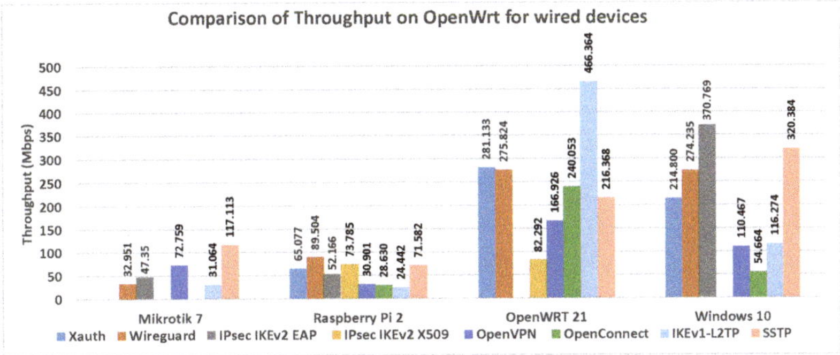

Figure 27. Throughputs data collected during the VPN connection tests with wired devices.

5.13. *Summary of Experimental Data Obtained for Deployed Scenarios*

In the deployed scenarios the practical goodness of choice based on OpenWRT was evaluated.

Making a grouping of the considered VPN deployed, we can make the following considerations: IKEv2-X509, IKEv2-PEAP and IKEv1-L2TP, having the best throughput performance, can represent the first implementation option. Then, the second group is represented by Wirecard, SSTP, and OpenConnect, and the third group is composed of OpenVPN and IKEv1-XAUTH. These considerations are made on the basis of the HW used for our experiments and shown in Table 3. Figure 23 illustrates Phy Rate, Average Phy Rate, and Phy Rate Peak measured on scenarios implemented for all wireless VPN clients, taking into account what is the reference Router (TP-LINK TL-WR841N), which uses 802.11n. Figure 24 illustrates RTT data collected during the VPN connection tests with wired and Figure 25 wireless devices. Tables 6 and 7 summarizes the result obtained for best data transmission efficiency algorithms of deployed scenarios and VPN algorithms which provide the best compatibility with the currently deployed infrastructures.

Table 6. Best data transmission efficiency algorithms of deployed scenarios.

VPN TYPE	PHASE 1 CIPHERS	PHASE 2 CIPHERS
IKEv1-L2TP	aes256-sha1-sha256-sha384-modp2048	aes128-aes256-sha1-sha256-modp2048-modp4096
IKEv2-X509	sha512-modp3072-aes256-modp2048s256	sha512-modp3072-aes256-modp2048s256
IKEv2-PEAP	sha512-modp3072-aes256-modp2048s256	sha512-modp3072-aes256-modp2048s256

Table 7. VPN algorithms which provide the best compatibility with the current infrastructures.

VPN TYPE	CLIENT NUMBERS	CIPHERS
IKEv1-L2TP	7 DIFFERENT VENDORS	aes256-sha1-sha256-sha384-modp2048 aes128-aes256-sha1-sha256-modp2048-modp4096
WIREGUARD	8 DIFFERENT VENDORS	ChaCha20 / Curve25519
OPENVPN	8 DIFFERENT VENDORS	CHACHA20-POLY1305 / AES-128-GCM

In Figure 28 one shows the comparison of throughput on OpenWrt in all scenarios. From Figure 28 it is possible to view the IKE VPN software has the best throughput values. Moreover, that IKEv1 which has lighter cryptography guarantees the best throughput. Also, the ranking shown in Figure 28 confirms that this firmware is a valid option for creating secure, robust networks, even with low-cost devices.

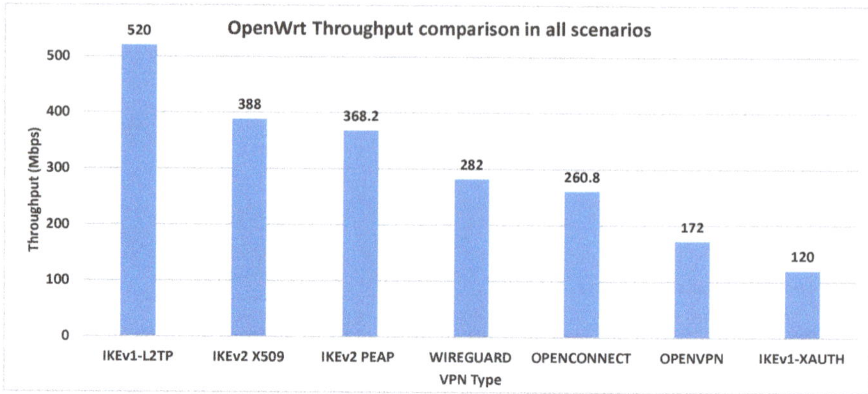

Figure 28. Comparison of Throughputs on OpenWrt in all scenarios.

6. Conclusions

To achieve the objectives of this work, OpenWRT open firmware has been chosen This firmware has shown great potential, used in "enterprise" contests with custom HW such as card APU ("APU BOARDS ", https://openwrt.org/toh/pcengines/apu2 (accessed on 10 June 2022)), and in "home" environments, with standard routers that support open firmware. These potentials are especially useful in outdoor environments, with limited connectivity and in areas difficult to wire (rural/mountain areas), and with the need to manage encrypted connections (often VPN) that cover more or less extensive sensor networks, even of a different nature. Using specific HW and installing dedicated SW, it is possible to directly interconnect these routers as active members of the network using different protocols such as local Zigbee networks interfaced through a Raspberry Pi gateway, like Lora/Lorawan networks, also linked to LTE/4G/5G networks. This firmware, loading a light version of the Linux system, allows having both a flexible operating system (the same router can be both multi VPN concentrator and MQTT broker, for example) and costs reduction by allowing the maximum exploitation of constrained HW. Various implemented scenarios demonstrate the quality of the throughput of these devices as well as their wide flexibility in the choice of the VPN protocols to be implemented. They can be both IPsec initiators/responders and members of TLS MESH over OpenVPN, OpenConnect, and Wireguard networks, achieving good performance in data exchange and resource use. In the scenarios created, by using Iperf, it was evaluated the practical goodness of choice based on OpenWRT, and from Figures 6–13 it turns out that both in client/initiator mode and server/responder mode, it continually ranks in the first three places for Transfer rate. This ranking confirms that this firmware is a valid option for creating secure networks, even with low-cost devices. Future implementations could foresee using more advanced platforms than APUs to increase this system's functionality. It may become a hypervisor, also, to create failover/load balancing policies in addition to that already present infrastructure to make the services perform better without upsetting the basic architecture. OpenWRT is the tool we use to answer all three goals of this paper as it supports all the VPN protocols and implementation deploys examined and it is itself the answer to reaching our third goal. As for the first goal of our work one aims to find the best algorithms that guarantee the best data transmission efficiency on constrained devices for each type of VPN deployed. The constrained devices considered are the TP-LINK OpenWRT router, the Mikrotik router, and the Raspberry Pi 2. The best data transmission efficiency algorithms, from the experiments carried out, appear to be those related to: IKEv1-L2TP, IKEv2-X509, and IKEv2-PEAP. As for the second goal of our work, VPN algorithms that provide best compatibility with the current infrastructures, from the experiments carried out, appear to be those related to: OpenVPN, Wireguard, and IKEV1-L2TP. For that concern VPN, deployed with different security levels on the same hardware, the experimentation proves that OpenVPN, using

TLS 1.3 with different data encryption algorithms, on OpenWRT router results in the best choice, as shown in the discussion of experimentation results. Further analyses with all shown VPNs could be the subject of future works.

Author Contributions: Conceptualization, A.F.G., F.D.R. and D.M.; methodology, M.T.; software, A.F.G.; validation, A.F.G., D.M. and E.G.; data curation, A.F.G. and D.M.; writing—original draft preparation, A.F.G., M.T. and E.G.; writing—review and editing, A.F.G. and M.T.; supervision, F.D.R. All authors have read and agreed to the published version of the manuscript.

Funding: This research received no external funding.

Data Availability Statement: Not applicable.

Conflicts of Interest: The authors declare no conflict of interest.

Abbreviations

The following abbreviations are used in this manuscript:

AES	Advanced Encryption Standard
AH	Authentication Header
EAP	Extensible Authentication Protocol
ESP	Encapsulating Security Payload
GDPR	General data protection regulation
HW	Hardware
IETF	Internet Engineering Task Force
IKE	Internet Key Exchange
IoT	Internet of Things
IPsec	IP Security
ISAKMP	Internet Security Association and Key Management Protocol
MQTT	Message Queue Telemetry Transport
NAT	Network Address Translation
OTP	On-Time Password
PSK	Pre Shared Key
PSTN	Public Switched Telephone Network
RTT	Round Trip Time
RW	Road Warriors
SAD	Security Association Database
SPD	Security Policy Database
SSL	Secure Sockets Layer
SW	Software
TLS	Transport Layer Security
UTP	Unshielded Twisted Pair
VPN	Virtual Private Network

References

1. Khanvilkar, S.; Khokhar, A. Virtual private networks: An overview with performance evaluation. *IEEE Commun. Mag.* **2004**, *42*, 146–154. [CrossRef]
2. Alshalan, A.; Pisharody, S.; Huang, D. A survey of mobile VPN technologies. *IEEE Commun. Surv. Tutor.* **2015**, *18*, 1177–1196. [CrossRef]
3. Gentile, A.F.; Fazio, P.; Miceli, G. A Survey on the Implementation and Management of Secure Virtual Private Networks (VPNs) and Virtual LANs (VLANs) in Static and Mobile Scenarios. *Telecom* **2021**, *2*, 430–445. [CrossRef]
4. Pudelko, M.; Emmerich, P.; Gallenmüller, S.; Carle, G. Performance analysis of VPN gateways. In Proceedings of the 2020 IFIP Networking Conference (Networking), Paris, France, 22–26 June 2020; pp. 325–333.
5. Elezi, M.; Raufi, B. Conception of Virtual Private Networks using IPsec suite of protocols, comparative analysis of distributed database queries using different IPsec modes of encryption. *Procedia-Soc. Behav. Sci.* **2015**, *195*, 1938–1948. [CrossRef]
6. Ullah, S.; Choi, J.; Oh, H. IPsec for high speed network links: Performance analysis and enhancements. *Future Gener. Comput. Syst.* **2020**, *107*, 112–125. [CrossRef]

7. Mao, H.; Zhu, L.; Qin, H. A Comparative Research on SSL VPN and IPSec VPN. In Proceedings of the 2012 8th International Conference on Wireless Communications, Networking and Mobile Computing, Shanghai, China, 21–23 September 2012; pp. 1–4. [CrossRef]
8. Sun, S.H. The advantages and the implementation of SSL VPN. In Proceedings of the 2011 IEEE 2nd International Conference on Software Engineering and Service Science, Beijing, China, 15–17 July 2011; pp. 548–551. [CrossRef]
9. Chen, F.; Wu, K.; Chen, W.; Zhang, Q. The Research and Implementation of the VPN Gateway Based on SSL. In Proceedings of the 2013 International Conference on Computational and Information Sciences, Ho Chi Minh City, Vietnam, 24–27 July 2013; pp. 1376–1379. [CrossRef]
10. Tropea, M.; Spina, M.G.; De Rango, F.; Gentile, A.F. Security in Wireless Sensor Networks: A Cryptography Performance Analysis at MAC Layer. *Future Internet* **2022**, *14*, 145. [CrossRef]
11. De Rango, F.; Potrino, G.; Tropea, M.; Fazio, P. Energy-aware dynamic Internet of Things security system based on Elliptic Curve Cryptography and Message Queue Telemetry Transport protocol for mitigating Replay attacks. *Pervasive Mob. Comput.* **2020**, *61*, 101105. [CrossRef]
12. De Rango, F.; Lentini, D.C.; Marano, S. Static and dynamic 4-way handshake solutions to avoid denial of service attack in Wi-Fi protected access and IEEE 802.11 i. *EURASIP J. Wirel. Commun. Netw.* **2006**, *2006*, 1–19. [CrossRef]
13. Fazio, P.; Tropea, M.; Voznak, M.; De Rango, F. On packet marking and Markov modeling for IP Traceback: A deep probabilistic and stochastic analysis. *Comput. Netw.* **2020** *182*, 107464. [CrossRef]
14. Lammle, T. *Virtual Private Networks (VPNs)*; Sybes: Dublin, CA, USA, 2020; pp. 433–450. [CrossRef]
15. Mohsin Abdulazeez, A.; Salim, B.; Zeebaree, D.; Doghramachi, D. Comparison of VPN Protocols at Network Layer Focusing on Wire Guard Protocol . *Int. J. Interact. Mob. Technol.* **2020**, *14*, 157–177. [CrossRef]
16. Thomson, M.; Turner, S. Using TLS to Secure QUIC. Internet-Draft draft-ietf-quic-tls-31, Internet Engineering Task Force. Work in Progress. Available online: https://datatracker.ietf.org/doc/rfc9001/ (accessed on 10 June 2022).
17. Ezra, P.; Misra, S.; Agrawal, A.; Jonathan, O.; Maskeliunas, R.; Damaševičius, R. Secured Communication Using Virtual Private Network (VPN). *Cyber Secur. Digit. Forensics* **2022**, *73*, 309–319. [CrossRef]
18. Wood, C.A.; Enghardt, R.; Pauly, T.; Perkins, C.; Rose, K. A Survey of Transport Security Protocols. Internet-Draft draft-ietf-taps-transport-security-05, Internet Engineering Task Force, Work in Progress. Available online: https://datatracker.ietf.org/doc/draft-ietf-taps-transport-security/02/ (accessed on 10 June 2022).
19. Pereira, R.; Beaulieu, S. Extended Authentication within ISAKMP/Oakley (XAUTH). Internet-Draft draft-ietf-ipsec-isakmp-xauth-06, Internet Engineering Task Force. Work in Progress. Available online: https://datatracker.ietf.org/doc/draft-ietf-ipsec-isakmp-xauth/ (accessed on 10 June 2022).
20. Smyslov, V.; Weis, B. Group Key Management Using IKEv2. Internet-Draft draft-ietf-ipsecme-g-ikev2-06, Internet Engineering Task Force. Work in Progress. Available online: https://datatracker.ietf.org/meeting/105/materials/slides-105-ipsecme-group-key-management-using-ikev2-00 (accessed on 10 June 2022).
21. Cicirelli, F.; Gentile, A.F.; Greco, E.; Guerrieri, A.; Spezzano, G.; Vinci, A. An Energy Management System at the Edge based on Reinforcement Learning. In Proceedings of the 2020 IEEE/ACM 24th International Symposium on Distributed Simulation and Real Time Applications (DS-RT), Prague, Czech Republic, 14–16 September 2020; pp. 1–8. [CrossRef]
22. Ajiya, A.A.; Idriss, U.S.; Jerome, M.G. Performance Evaluation of IPSEC-VPN on Debian Linux Environment. *Int. J. Comput. Appl.* **2019**, *975*, 8887.
23. Mahmmod, K.F.; Azeez, M.M.; Ahmed, M.A. IPsec Cryptography for Data Packets Security within VPN Tunneling Networks Communications. In Proceedings of the 2020 International Conference on Electrical Engineering and Informatics (ICELTICs), Aceh, Indonesia, 27–28 October 2020; pp. 1–8. [CrossRef]
24. Wouters, P. Deprecation of IKEv1 and Obsoleted Algorithms. Internet-Draft draft-ietf-ipsecme-ikev1-algo-to-historic-06, Internet Engineering Task Force. Work in Progress. Available online: https://www.ietf.org/id/draft-ietf-ipsecme-ikev1-algo-to-historic-06.html (accessed on 10 June 2022).
25. Patel, D.B.V.; Aboba, D.B.D.; Dixon, W.; Zorn, G. Securing L2TP Using IPSEC. Internet-Draft draft-ietf-pppext-l2tp-security-05, Internet Engineering Task Force. Work in Progress. Available online: https://www.rfc-editor.org/rfc/rfc3193 (accessed on 10 June 2022).
26. Mavrogiannopoulos, N. The OpenConnect VPN Protocol Version 1.1. Internet-Draft draft-mavrogiannopoulos-openconnect-01, Internet Engineering Task Force. Work in Progress. Available online: https://datatracker.ietf.org/doc/draft-mavrogiannopoulos-openconnect/02/ (accessed on 10 June 2022).
27. Gont, F. Layer 3 Virtual Private Network (VPN) Tunnel Traffic Leakages in Dual-Stack Hosts/Networks. RFC 7359. 2014. Available online: https://www.rfc-editor.org/info/rfc7359 (accessed on 15 June 2022).
28. Sanchez, D.; García, M.A. A Simple SCCP Tunneling Protocol (SSTP). Internet-Draft draft-sanchez-garcia-SSTP-v1r0-00, Internet Engineering Task Force. Work in Progress. Available online: https://datatracker.ietf.org/doc/draft-sanchez-garcia-SSTP-v0r2/ (accessed on 10 June 2022).
29. Friel, O.; Barnes, R.; Pritikin, M.; Tschofenig, H.; Baugher, M. Application-Layer TLS. Internet-Draft draft-friel-tls-atls-02, Internet Engineering Task Force. Work in Progress. Available online: https://datatracker.ietf.org/doc/draft-friel-tls-atls/02/ (accessed on 10 June 2022).

30. Haga, S.; Esmaeily, A.; Kralevska, K.; Gligoroski, D. 5G Network Slice Isolation with WireGuard and Open Source MANO: A VPNaaS Proof-of-Concept. In Proceedings of the 2020 IEEE Conference on Network Function Virtualization and Software Defined Networks (NFV-SDN), Leganes, Spain, 10–12 November 2020. [CrossRef]
31. Aung, S.T.; Thein, T. Comparative Analysis of Site-to-Site Layer 2 Virtual Private Networks. In Proceedings of the 2020 IEEE Conference on Computer Applications(ICCA), Yangon, Myanmar, 27–28 February 2020; pp. 1–5. [CrossRef]
32. Libreswan. Available online: https://libreswan.org/ (accessed on 20 June 2022).
33. Strongswan. Available online: https://www.strongswan.org/ (accessed on 20 June 2022).
34. Accel-PPP. Available online: https://accel-ppp.org/ (accessed on 20 June 2022).
35. Openwrt. Available online: https://openwrt.org/ (accessed on 20 June 2022).
36. Mikrotik. Available online: https://mikrotik.com/ (accessed on 20 June 2022).
37. Mazon-Olivo, B.; Pan, A. Internet of Things: State-of-the-art, Computing Paradigms and Reference Architectures. *IEEE Lat. Am. Trans.* **2022**, *20*, 49–63. [CrossRef]
38. Kubernetes. Available online: https://kubernetes.io/it/docs/concepts/overview/what-is-kubernetes/ (accessed on 20 June 2022).
39. Wireguard. Available online: https://www.wireguard.com/ (accessed on 20 June 2022).
40. Nir, Y.; Langley, A. ChaCha20 and Poly1305 for IETF Protocols. RFC 7539. 2015. Available online: https://www.rfc-editor.org/info/rfc7539 (accessed on 10 June 2022).

future internet

Article

Attacks on IoT: Side-Channel Power Acquisition Framework for Intrusion Detection

Dominic Lightbody *, Duc-Minh Ngo, Andriy Temko, Colin C. Murphy * and Emanuel Popovici

Electrical and Electronic Engineering, University College Cork, T12 K8AF Cork, Ireland;
120220051@umail.ucc.ie (D.-M.N.); e.popovici@ucc.ie (E.P.)
* Correspondence: 121100122@umail.ucc.ie (D.L.); colinmurphy@ucc.ie (C.C.M.)

Abstract: This study proposes the wider use of non-intrusive side-channel power data in cybersecurity for intrusion detection. An in-depth analysis of side-channel IoT power behaviour is performed on two well-known IoT devices—a Raspberry Pi 3 model B and a DragonBoard 410c—operating under normal conditions and under attack. Attacks from the categories of reconnaissance, brute force and denial of service are applied, and the side-channel power data of the IoT testbeds are then studied in detail. These attacks are used together to further compromise the IoT testbeds in a "capture-the-flag scenario", where the attacker aims to infiltrate the device and retrieve a secret file. Some clear similarities in the side-channel power signatures of these attacks can be seen across the two devices. Furthermore, using the knowledge gained from studying the features of these attacks individually and the signatures witnessed in the "capture the flag scenario", we show that security teams can reverse engineer attacks applied to their system to achieve a much greater understanding of the events that occurred during a breach. While this study presents behaviour signatures analysed visually, the acquired power series datasets will be instrumental for future human-centred AI-assisted intrusion detection.

Keywords: IoT hardware security; IoT side-channel power analysis; attack signatures; simple power analysis; cyber attacks on IoT; IoT time-series power data

Citation: Lightbody, D.; Ngo, D.-M.; Temko, A.; Murphy, C.C.; Popovici, E. Attacks on IoT: Side-Channel Power Acquisition Framework for Intrusion Detection. *Future Internet* **2023**, *15*, 187. https://doi.org/10.3390/fi15050187

Academic Editors: Christoph Stach and Clémentine Gritti

Received: 24 April 2023
Revised: 16 May 2023
Accepted: 19 May 2023
Published: 21 May 2023

1. Introduction

With the ever-evolving global attack surface, recent years have seen an explosion in the volume of cyberattacks. Reports show that 2020 saw a 358% increase in malware attacks compared to the previous year [1]. From 2020, cyberattacks continued to increase globally well into 2021 and, in the first half of 2022 alone, approximately 236.1 million ransomware attacks occurred across the globe.

The World Economic Forum (WEF) estimates that if all cybercrime was amalgamated under the same flag, this country would rank as the world's third-largest economy. Cybercrime caused damages totalling USD eight trillion in 2022 alone [2]. There is no evidence that this acceleration in the growth of cyber-criminality will slow any time soon; in fact, the opposite is proving true.

As mentioned, approximately 236 million ransomware attacks occurred in the first half of 2022 alone. Ransomware [3], a type of malware which encrypts a victim's hard-drive and holds it for 'ransom' unless demands are met, is a seemingly new menace plaguing the headlines. However, ransomware has been a threat for quite a long time. In May 2017, the WannaCry [4] ransomware first made international headlines. WannaCry was unique for its colossal impact. The WannaCry attack infected over 300,000 computers spanning 150 countries, with total damages estimated to be billions of USD [5]. The most destructive threat that WannaCry ushered in was not the prospect of significant ransom demands but rather a new cyber-terrorism trend: to hold hospitals, schools and universities to ransom, with little care if the victims pay the ransom or not. The primary goal of this attack is to cause massive disruption to critical-infrastructure.

In fact, for the year 2023, the Cybersecurity and Infrastructure Security Agency (CISA) revealed that their priority sectors are "water, hospitals and K-12" (K-12 being kindergarten to 12th grade in the United States). These sectors are resource-poor with massive attack surfaces and are heavily targeted by ransomware [6]. In 2022 Emsisoft, a cybersecurity vendor, recorded at least 25 ransomware attacks on "hospitals and multi-hospital health systems", affecting approximately 290 hospitals across the US [7]. In 2021, almost 1 million students countrywide were negatively affected by 67 ransomware attacks against K-12 schools. The estimated cost due to the downtime was USD 3.5 billion [8].

However, this epidemic affecting the health and education sectors is not confined to the United States. Ireland has become a victim of large-scale attacks on these sectors recently. In May 2021, the Irish Health Service Executive (HSE) fell victim to a massive ransomware attack [9]. This attack was the most significant cyber attack on an Irish state agency in history and caused mass disruption to the health service.

The education sector in Ireland has, like the US, seen many attacks in recent years. Third-level institutions, such as the National University of Ireland (NUI) Galway [10], National College of Ireland (NCI) Dublin [11], and Technological University Dublin (TU Dublin) [12], have fallen victim to ransomware attacks which have greatly affected the availability of systems, leading to the temporary closure of education facilities. The most recent of these attacks was a ransomware attack, which led to the closure of Munster Technological University (MTU) for approximately one week [13].

Ransomware is not the only concern in the security field currently. With the massive explosion in the installation of Internet-of-Things (IoT) devices worldwide, the global attack surface continues to grow significantly. This growth in popularity is thanks to the innate ability of this technology to enable communication between (smart) edge devices and the Internet, thus improving the quality of human life [14] or optimising industrial processes. A forecast made by the International Data Corporation (IDC) projects that there will be 55.7 billion IoT devices by 2025 [15]. With such a large deployment, attacks on the IoT have the potential to cause mass disruption, exposing every sector equally to ransomware. Growing concerns regarding IoT security may stop many from adopting this technology. These concerns mainly affect financial technology, healthcare, industry, transportation and education, which have already begun IoT adoption [16].

In September 2017, one of the the largest ever recorded Distributed Denial of Service (DDOS) attacks was performed using the Mirai botnet [17]. This malware is estimated to have infected over 380,000 IoT devices, such as home routers, to create a massive botnet. This botnet was used to target victims with unprecedented levels of traffic. Brian Krebs' website, krebsonsecurity.com, was hit with traffic of 620 gigabits per second (Gbps), one of the largest on record. Later that month, the botnet was used to target the French web host OVH in an attack which shattered all previous records with an estimated 1.1–1.5 terabits per second (Tbps) of traffic [17]. If only 380,000 devices can cause such disruption, with a max pool of possibly 55.7 billion by 2025, attacks like this will most likely become more prevalent in the future.

Another notable attack involving IoT devices was the attack on the Ukrainian power grid in 2015. While much of this attack targeted traditional computing, one central aspect of this coordinated attack was targeting breakers, Serial-to-Ethernet devices, and critical servers' Uninterruptible Power Supplies (UPSs). These can all be considered attacks on IoT. This cyber attack on Ukraine was considered one of the first in history with a quantifiable loss of human life. Over 225,000 customers were left without power, including hospitals providing critical care [18].

Aside from these massive national-infrastructure-level attacks, IoT devices have been the victim of other impactful attacks. In 2017, the US Food and Drug Administration (FDA) announced they had uncovered a massive vulnerability in pacemakers manufactured by St. Jude Medical. These pacemakers could communicate with external services after installation in the patient. Once the attackers could access this communications channel, they would have the ability to deplete the battery, change the functionality and reportedly have the potential to subject the patient to fatal shocks [19].

In 2015, a team from IBM demonstrated a crucial vulnerability in the onboard software of the Jeep SUV. Once exploited, this vulnerability would allow attackers to gain control of the vehicle remotely, affecting the speed of the vehicle and even the steering, which could lead to the vehicle veering off the road [19].

1.1. Contribution

While many parts make up the attack surface in the world today, the remainder of this study will focus on IoT devices. It is important to remember that while this study focuses on IoT devices, the attacks performed and the knowledge gained apply to all cybersecurity fields, from low-power IoT devices to High-powered Computing (HPC).

This work seeks to promote using non-intrusive side-channel power data in cybersecurity. Power data, a key fundamental piece of primary data which every device has, regardless of power level, can reflect all operations running on a device. We believe that by looking at cyberattacks from as many angles as possible, we can defend against them more effectively. We hope the implication of this study will be the broad adoption of side-channel data in security. By studying this type of power-data series, a much greater understanding of cyberattacks and how to defend against them will be gained. One fundamental drawback of many works is that they must analyse the side-channel power data they use. As such, the reader cannot learn anything about the attacks used in these works. To promote the further use of power data in security, in this study we:

- Create two realistic IoT testbeds, each with a different core IoT device.
- Study the normal side-channel behaviour of these testbeds before they are attacked, then subject these testbeds to common attacks in cybersecurity from the fields of reconnaissance, denial of service and brute force, analysing the side-channel signatures obtained.
- Subject the testbeds to capture-the-flag scenarios combining most of the attacks studied into one continuous scenario and discuss automated detection methodologies based on the findings of the analysis of the side-channel power data.

1.2. Organisation

The remainder of this study is organised as follows. Section 2 reviews current security methods for IoT devices and the use of power data in security. Section 3 discusses the testbed setup and data-collection methodology for the replicability of this work. Section 4 studies common attacks, applies them to the testbed and studies the resulting power behaviour. Section 5 combines the previous section's attacks into one attack scenario where the attacker attempts to obtain a secret file. In Section 6, we discuss the intuitions gained from studying this data and what this means for security in the future. Finally, future work and conclusions of this study can be found in Section 7.

2. Related Works

With the massive explosion in the deployment of IoT devices globally, the security of these devices has become a critical concern. IoT devices are resource-constrained by nature. This means conventional security practices tend to be impossible, or impractical, to implement on these devices.

The majority of the work focusing on IoT devices in security only considers the network traffic to and from the device. By inspecting this data, it is possible to infer if a device is under attack or has been attacked. Much research has been conducted in the field of Intrusion Detection Systems (IDS) for IoT devices. These IDS can either be network-based (NIDS), which are usually located on the IoT gateway, or host-based (HIDS), which are implemented on the device itself.

To combat the limited security most IoT devices offer, Passban was created [20]. Passban is an intelligent IDS which can protect IoT devices to which it is directly connected.

Passban is a lightweight solution which can be deployed on cheap, resource-constrained IoT gateways as a HIDS. Trained on the normal behaviour of a device, Passban can detect a wide range of malicious traffic such as brute-force, SYN flood and port-scanning attacks. Passban was designed with scalability in mind, meaning this framework can dynamically scale to new threat definitions without requiring hardware upgrades.

In [21], a deep recurrent-neural-network-based IDS for fog security is developed. Fog computing extends cloud services nearer to IoT devices. It acts as a medium between traditional cloud computing and edge devices, such as the IoT. The proposed framework implemented in the fog-computations layer, comprises a traffic processing engine and a classification engine consisting of a recurrent artificial neural network (ANN). The proposed framework is trained and evaluated using a balanced version of the NSL-KDD [22] dataset and shows high accuracies of 98.27% against denial-of-service attacks, one of the more pervasive attacks against IoT devices.

In [23], a convolutional-neural-network (CNN)-based anomaly-detection IDS framework for IoT is proposed. This framework takes advantage of the strengths of IoT devices and can examine traffic across the broad scope of the IoT. The proposed model can detect a wide range of intrusions and anomalous traffic behaviour and was trained on the Bot-IoT [24] and NID [25] datasets, achieving high accuracies of 92.85% and 99.51%, respectively. This work also presents a framework to incorporate IDS as a program within IoT networks and a strategy to preserve the integrity of IoT networks while seamlessly maintaining availability for legitimate users.

In our previous work, we introduced HH-NIDS [26]—a Heterogeneous Hardware-Based Network IDS framework for IoT security. Using hardware accelerators, HH-NIDS implements anomaly-based IDS approaches for IoT devices. Supervised-learning methodologies on the IoT-23 [27] and UNSW-NB15 [28] datasets were trained to generate lightweight ANN models for anomaly detection, achieving high accuracies of 99.66% and 98.57% for these datasets, respectively. These models were evaluated from a performance and resource-usage perspective on the CPU, GPU and FPGA and implemented on the MAXIM 78000 microcontroller.

A common theme amongst these works is that they focus on the network traffic of the IoT device to detect malicious activity. One key security area that has been largely overlooked is the use of power data. The side-channel power data of a device is a fundamental primary source of data which every device, regardless of computing resources, has. The following works focus more on using power data in security. The volume of works such as these is insufficient, and we argue that much more research must be performed to integrate power data into security.

Earlier works proposed the use of side-channel power data as an attack on the device itself. Work by Kocher et al. [29], presented in 1999, examines specific methods for analysing power consumption measurements to uncover secret keys from tamper-resistant devices. These methods, dubbed "Simple Power Analysis" and "Differential Power Analysis", broke DES encryption, thereby allowing attackers to discern private keys. They also discuss approaches for building cryptosystems which securely operate in insecure hardware that leaks information.

Instead of using the side-channel power data as an attack, other works have been completed which monitor the device power data to detect intrusions. These works are very scarce, however. Some notable examples are listed below.

In our other previous work [30], a lightweight convolutional neural network (CNN) Host-Based IDS (HIDS) for anomaly detection in the power data of an IoT device was created. This CNN framework was implemented on the MAX78000 microcontroller on the edge. The proposed framework is scalable and generic, making it relevant for any device. However, a fundamental limitation of that work was the absence of actual attack examples in the power data. As such, synthetic anomalies were relied on.

WattsUpDoc [31] utilises the side-channel power consumption of medical devices to allow for run-time malware detection. During experimentation, WattsUpDoc performed with an accuracy of 94%, when presented with previously known malware examples, and 85% accuracy regarding unseen malware examples on multiple embedded devices. This framework's non-intrusive methodology, which monitors the side-channel power data from the device, allows for the detection of malware with no software, hardware or network modification requirements of the existing system in place.

DeepPower [32] is another approach which detects malware on IoT devices by analysing their non-intrusive side-channel power signals. This framework utilises deep learning to detect anomalies in the power data. DeepPower initially filters the raw side-channel power data to find suspect power traces. A fine-grained analysis is then performed on these traces to determine which activities they correspond to on the device. The DeepPower framework can detect malicious activity with high accuracy while maintaining a non-intrusive nature, meaning no modifications need to be made to the monitored devices.

In a departure from the theme of IoT devices, the work presented in "Catch Me if You Can" [33] demonstrates how the side-channel power data obtained from High-Powered Computing Platforms (HPCs) can be used to determine what programs are running on a machine and, thus, if any un-authorised programs are running. Using a variety of scientific benchmarks, the proposed framework was tested on an HPC rack at Lawrence Berkeley National Laboratory. This framework can detect if specific programs are running with a recall of up to 95% and a precision of 97%. This work is essential, as it illustrates that using side-channel power data is not simply confined to the IoT field but applies to the entire security sector.

This study proposes a data acquisition framework for non-intrusive side-channel power data for IDS to detect attacks on edge devices. To this end, we create a testbed and subject it to common cyber attacks. The data is then collected and analysed. A unique capture-the-flag scenario is subsequently enacted, which combines the individually studied cyber attacks into a single scenario in which the testbeds are compromised, and a secret file is stolen. One fundamental area for improvement with all the side-channel power-related works presented here is that there needs to be an in-depth explanation of the power signatures they are studying. We seek to demonstrate that a broader understanding of common attack signatures in the power spectrum is crucial. The insight and intuitions gained from analysing the side-channel power data are valuable and should provide insight into how the IoT and other fields can implement more effective IDS frameworks. To the best of the authors' knowledge, for most of the security sector, this study will be the first time they see these common cyber attacks represented in any form other than a packet capture.

3. Materials and Methods

Two separate IoT devices were chosen for this testbed: a Raspberry Pi 3 Model B [34] and a DragonBoard 410 [35]. Based on the Qualcomm Snapdragon 410 processor [36] and manufactured by Arrow Electronics, the DragonBoard 410c comes equipped with advanced processing power, Bluetooth and Wi-Fi. The Snapdragon 410 was an extremely popular processor of which Qualcomm shipped over 200 million units, with a large amount of integration with the mobile-phone market in the mid-2010s. With over 200 million units shipped [37], any signatures obtained for the side-channel power data of this chip can be considered highly valuable. The Raspberry Pi 3 Model B was chosen for a similar reason to the Snapdragon-based system, its extreme popularity. The Raspberry Pi is one of the most popular single-board computers on the market today. Indeed, since the launch of its first version over ten years ago, the Raspberry Pi has sold over 40 million units [38].

The DragonBoard and the Raspberry Pi feature Quad Core 1.2GHz 64-bit CPUs and 1.2GB of RAM. In the case of the DragonBoard, this quad-core CPU is integrated within the Snapdragon 410 along with other components, such as a Qualcomm APQ8016E application processor and a Qualcomm Adreno 306 graphics chip. Regardless of their different manufacturers, manufacturing periods and usual use cases, both systems have startlingly similar specifications. These choices in the device will give a broader understanding of what common attack signatures look like, when viewed in the device current across different devices. Part of this study investigates whether common attacks' signatures are general. While an exact match in the signature cannot be expected due to the different hardware specifications, any generalisation in signatures could greatly assist in understanding and detecting these attacks against IoT devices and networks comprised of various devices. With approximately 245 million devices (combined, between the DragonBoard 410c and the Raspberry Pi), extensive coverage of devices can be assumed during this study. The testbed was kept as simple as possible to maximise explainability while maintaining realism.

Very little functionality was enabled on each board. The Raspberry Pi, running PI OS [39], came with crucial features, such as Bluetooth, installed. These features needed to be installed on the DragonBoard, on which Linaro Debian [40], a specifically designed Debian distribution for the Qualcomm Snapdragon 410, was installed.

Both boards were connected to the internet over Wi-Fi, and the SSH protocol was enabled to broaden the attack surface. SSH [41] or "Secure Shell" is a network protocol allowing users to communicate between computers, share data and remotely log into other devices. SSH would regularly be enabled on IoT devices for maintenance purposes, and, in some cases, normal operating behaviour would require users to log in via SSH to view sensor data or interact with components. Recent research has revealed that 15% of IoT device owners do not change the default passwords. As a result, five passwords can give an attacker access to 10% of all IoT devices worldwide [42]. With an estimated 55 billion IoT devices installed worldwide by 2025, this number of devices with default credentials will be over 5 billion [15].

Bluetooth was enabled on each board to communicate with a TI CC2650STK Sensortag [43]. These Sensortags come equipped with ten low-power MEMS sensors. A simple Python script [44] running on the testbed requested the sensor data over Bluetooth and interpreted the packets received once every 5 s. The Dragonboard and the Raspberry Pi were connected to an Agilent technologies N6705A DC Power Analyser [45]. This device supplied the power to the testbed while also collecting side-channel device current data. This method allows for the non-intrusive monitoring of the side-channel IoT power behaviour. Figure 1 describes a proposal for a generic data acquisition framework for IoT.

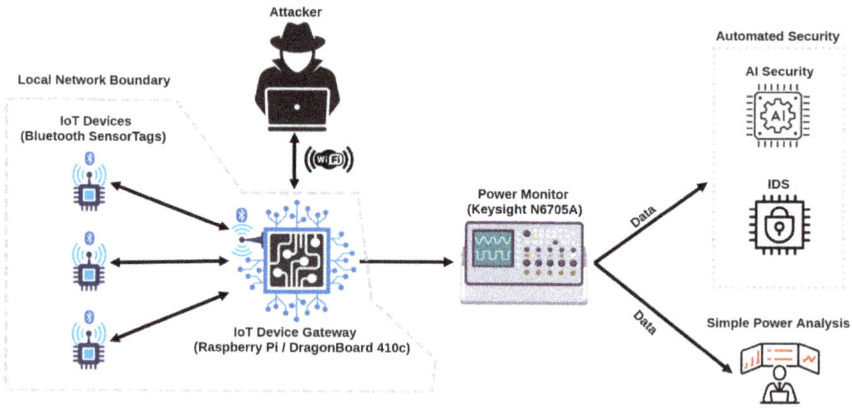

Figure 1. IoT-power data acquisition framework.

3.1. Data Collection

Using the Agilent N6705A, the testbed power was supplied and monitored. Using the data-logging function of the DC power analyser, the current of the testbed was logged at a rate of approximately 50 ksps (one sample every 0.02048 ms). While the Agilent N6705A can sample at much higher rates, 50 ksps is the maximum value for the data-logging function, allowing for much longer data captures.

The data featured in this paper undergoes no pre-processing. While many different pre-processing methodologies were investigated as this work progressed, it was decided to present the raw data, as seen by the power analyser.

Many other power and current sampling options exist, and using an expensive piece of equipment such as that featured in this work is not a strict requirement. For example, the work in [33] shows great success monitoring the power of their High-Powered-Computing (HPC) behaviour with an inexpensive and non-intrusive device which is inserted between the plug and the appliance.

3.2. Normal Power Behaviour

The components of the testbed have already been briefly described, but not how they are used in the regular operation of the device. IoT devices tend to have elementary operations: an automated task they perform many times, looping repeatedly. In the case of this testbed, the devices were each connected to a single Bluetooth SensorTag. Using a Python script, the sensor data was requested every five seconds over Bluetooth. The script then interpreted the incoming Bluetooth packets to display the sensor values. This behaviour is continuous and realistic, as simple functionality is prevalent for simple IoT sensor systems used commercially and in people's homes.

Figure 2 demonstrates three five-second segments of normal side-channel device current from each device. The time periods 0–5 s, 35–40 s and 70–75 s were chosen to represent part of the overall dataset. This normal behaviour is expected throughout the entirety of this work. Noticeable similarities can be seen in the normal-behaviour signature between both boards, even though the Raspberry Pi was drawing approximately double the current of the DragonBoard (0.4 A versus 0.2 A). A remarkably similar signature shape for each spike of Bluetooth communication behaviour can also be observed. Time-wise, both boards complete the Bluetooth communications within the same time frame.

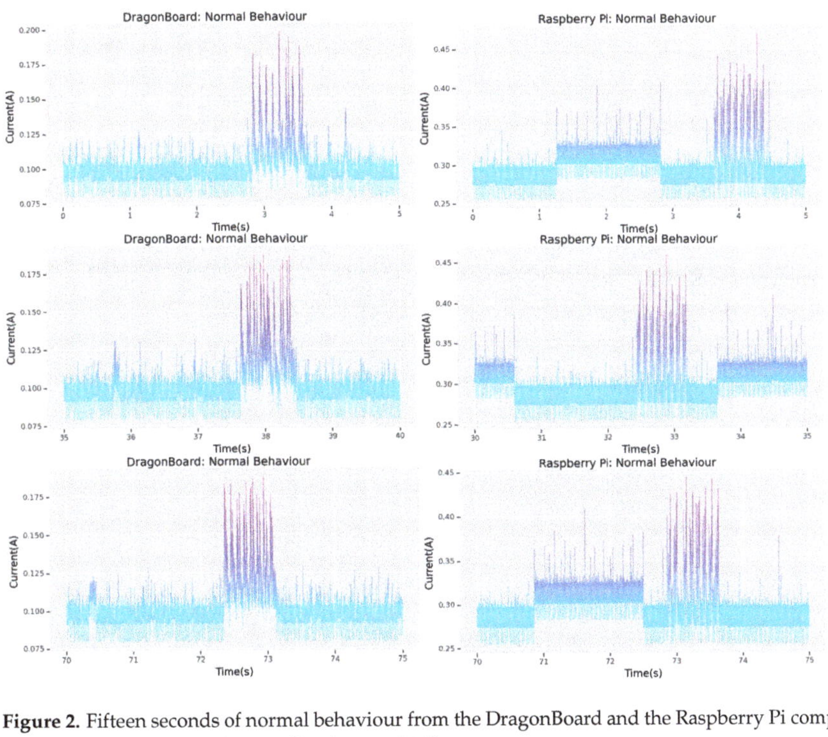

Figure 2. Fifteen seconds of normal behaviour from the DragonBoard and the Raspberry Pi compared (data represented using the "cool" colormap [46]).

The normal behaviour of the PI is not as smooth as the DragonBoard, with evident deviations in the background device current. These deviations will make some minor attack signatures harder to detect, as they can be masked by these slight device current fluctuations. This will be studied in more detail later in this work. In Figure 2 as well as the remaining figures in this study, the side-channel power data is represented using colormaps of increasing intensity [46]. This scale is based on the CISA National Cyber Incident Scoring System (NCISS) [47] with blue colours representing baseline conditions and black colours representing an emergency.

4. Attacks

In this study, the testbed was subjected to three separate fields of attack, each covering a key security concept. These attack fields cover the three core areas of the CIA triad [48] confidentiality, integrity and availability. Furthermore, the attacks are broken into three sub-categories from these core areas: reconnaissance, brute force and denial of service.

4.1. *Reconnaissance*

Reconnaissance attacks focus on the confidentiality of the target. The fundamental goal of this attack type is to gather as much information as possible about a target. This information can be in the form of revealing services operating on a target, the version of the software running on a target, users with accounts on a target or possible peer connections of that target. This is only a small fraction of the information attackers seek, as any information on a target can be used as an attack vector. The "port-scanning attack" was applied to the testbed from this attack category.

4.1.1. Port Scanning

A "port scan" is technique hackers and security teams use to identify weak points in network security. These weak points may be open ports which hackers can exploit to gain entry to a network or a system. A port scan identifies these open ports and gives the attacker valuable information on the system they are attacking, such as if the system or organisation is using a firewall or any IDS or, for example, if ports are allowed to send and receive packets. Any outdated network-facing software running on the target can also be exposed. Outdated softwares regularly have well-documented vulnerabilities and are easily exploited.

The tool Nmap [49] was used to perform this attack on the testbed. Nmap is one of the most popular port-scanning tools and comes preinstalled on many security-focused Linux distributions, for example, Kali [50]. Nmap sends packets and analyses the responses to discover hosts and services on a network. Nmap features many options for probing networks for information; however, only the timing and performance settings were focused on during this work. The settings used are as follows:

- Insane (T5) speed scan, for use on an extremely fast network.
- Aggressive (T4) speed scans, for use on a reliable and relatively fast network.
- Normal (T3), the default speed.
- Polite (T2), slows down the scan to use less bandwidth and use less target machine resources.
- Sneaky (T1), Intrusion-Detection-System evasion with greatly increased delays between actions.
- Paranoid (T0), Intrusion-Detection-System evasion with even further increased delays between actions.

Figure 3 compares the side-channel current signature across the timing settings of the Nmap tool. Similarities in the side-channel current signature between the upper settings of "T" can be seen on the two boards. More specifically, the upper settings of "T3", "T4", and "T5" share a common signature across the two boards. The current level is the only key difference in this signature between the two boards. This behaviour has a much higher amplitude for the Raspberry Pi, at around 0.6 A, versus 0.2 A for the Dragonboard.

As "T5" and "T4" are listed as "Aggressive" and "Insane" in the Nmap documentation, one would expect these settings to exhibit a much larger current signature than their less aggressive counterparts. However, this is not the case, as viewed in Figure 3. These more aggressive settings have a remarkably similar signature to that of "T3"—the default setting. This is because both "T5" and "T4" assume the attacker is operating on a reliable and reasonably fast network. The Raspberry Pi and the DragonBoard could not handle such a bombardment due to their resource-constrained nature and, thus, the signature is closer to that of "T3", a less resource-hungry setting. If these devices were more powerful and the network more reliable, one could reasonably assume much more variation in the side-channel current signatures between "T3", "T4", and "T5".

The latter half of Figure 3 investigates the "Polite" and IDS evasion settings, "T2", "T1", and "T0". One of the core ways the Nmap tool attempts to be "Polite" or avoid IDS detection is by changing delays within the tool's operation. These delays in operation spread the larger signatures witnessed in "T5"–"T3" over much more extended periods. Instead of attempting to scan all 1000 TCP ports on a system in approximately two seconds (the average time "T3"–"T5" take to complete), "T2" will take approximately ten minutes, "T1" over an hour and "T0" tens of hours. These values will increase or decrease based on the target system's resources and the complexity of the network being mapped.

As seen in Figure 3, there is no clear, recognisable signature for settings "T2"–"T0", apart from a slight spike in current seen in setting "T2" for the DragonBoard. While these settings lack the Nmap spike signature witnessed for "T5"–"T3", a long-duration perturbation of the normal behaviour is visible due to the spreading out of the attack over a more extended period. These disturbances, like small bumps in the background current, are less frequent with a decrease in the "T" setting. As seen in Figure 3, this disturbance is quite noticeable for "T2" on the DragonBoard. This disturbance continues for the length of the attack. The "T2" setting on the Raspberry Pi seems less noticeable, and no clear individual signatures can be distinguished when studying Figure 3. However, looking over a more extended period, it is clear that the entire current signature of the board shifts upwards from 0.25 to approximately 0.30 A. Like the DragonBoard, this disturbance persists for the duration of the attack.

Settings "T1" and "T0" can be used for IDS evasion and are very difficult to spot in the side-channel power data. These settings spread out the operations of the tool Nmap to minimise the signature. These delays make the attack last much longer, however. Setting "T1" causes minimal disturbance to normal behaviour. Setting "T0" spreads the operations to a somewhat extreme extent, so much so that there is little disturbance in normal behaviour. Using SPA and DPA methodologies, it should be possible to detect the change in side-channel power due to these attacks, as any behaviour in addition to what is expected will be reflected in the power data, regardless of how slight the change is.

4.2. Brute Force

Brute force attacks, like reconnaissance attacks, affect the confidentiality of the target. Instead of attempting to gain information on a target, however, brute-force attacks attempt to exploit some of the information obtained during the reconnaissance stage, usually to gain unauthorised entry to the target. In essence, brute-force attacks seek to overwhelm the specific security mechanisms they target. An excellent example of a brute-force attack is password cracking. The "SSH Brute-Force" attack was deployed against the testbed from this category.

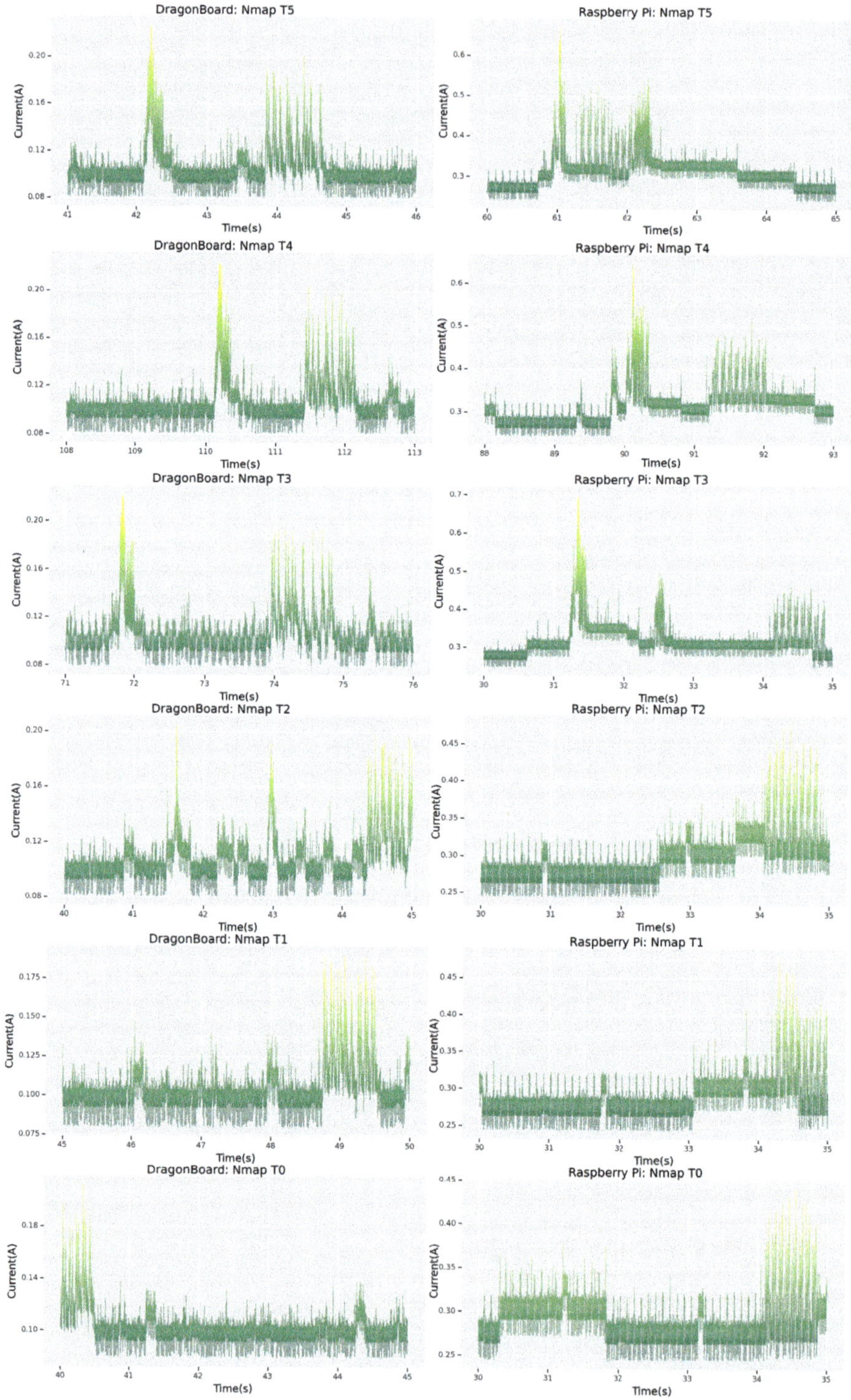

Figure 3. Each Nmap T-Switch behaviour for DragonBoard versus Raspberry Pi (data represented using the "summer" colormap [46]).

4.2.1. SSH Brute-Force

As briefly described, brute-force attacks seek to overwhelm the security mechanisms in place on a target. In the case of the testbed featured in this work, the weak security mechanism is a default username–password combination for the target device. Recent research has revealed that 15% of IoT device owners do not change default passwords. As a result, five passwords can give an attacker access to 10% of all IoT devices worldwide [42].

The port-scan reconnaissance attack revealed that the SSH service [41] was running on the target. SSH is a service which allows users to access remote computer systems securely. Equipped with this knowledge, the attacker will move to crack the SSH password.

This form of brute force is akin to trial and error, where the attacker attempts many passwords, from a password list, against a known username. If the username is unknown, the attack can use two lists simultaneously, one with many possible usernames and the other with many possible passwords. Lists such as these can be readily found on the internet and come pre-installed on some security-focused Linux distributions such as, for instance, Kali.

Once the password-cracking tool guesses the correct password, the attacker will have remote access to the device over SSH. If the user account compromised has admin or "sudo" permissions, this can have disastrous consequences. With such permissions, the attacker will have unhindered access to all files, access other users' files, change system settings and possibly gain entry to other devices on the same network that use the same username–password combination. Even on a standard user account, an attacker can exploit outdated and vulnerable programs on the device in a "privilege-escalation" attack to gain admin permissions.

The parallelised network login cracker Hydra [51] was used to perform the SSH brute-force attack.

The settings of the tool Hydra were varied to investigate the effect on the SSH brute-force side-channel power signature. Figure 4 shows the variation in the side-channel signature over various settings of "t" for both boards. In this case, the "t" setting controls the number of connections in parallel per target. A higher value of "t" and, thus, more connections in parallel, should result in a more significant signature for the attack. Other settings for the tool remained at their default. For these attacks, a chunk from the "wordlist" [52] "Rockyou" [53] was used. In cases where it was intended for Hydra not to find the correct password, the password to the testbed was not included in the wordlist. The correct password was included in the wordlist for experiments where the valid password was to be found.

From studying Figure 4, it can be seen that the current signature, as expected, increases in amplitude and duration with higher settings of "t". Across "t" settings, the amplitude of the attack is greater or equal to that of the Bluetooth signature. With more than one connect in parallel, the brute-force signature completely dwarfs the normal behaviour for both devices, making the attack very obvious in the side-channel power data. It would be unlikely for an attacker to use one "T1", however, as this would take a considerable amount of time to complete. Notably, the signatures share many key features between the two boards, such as the shape of the attack signature and the duration of the attack. This makes an SSH brute-force attack easily recognisable, regardless of against which device it is being performed.

The more powerful Raspberry Pi can sustain higher settings of "t" for longer than the DragonBoard (See "t16" and "t32" in Figure 4). This is likely because the Dragonboard is more optimised for much lower power applications, whereas the Raspberry Pi has fewer limitations on its performance. It should be noted that Hydra warns users that settings above "t4" will likely not work as expected, as many systems will be configured to disallow more than four parallel connections or might not be able to handle them due to resource constraints.

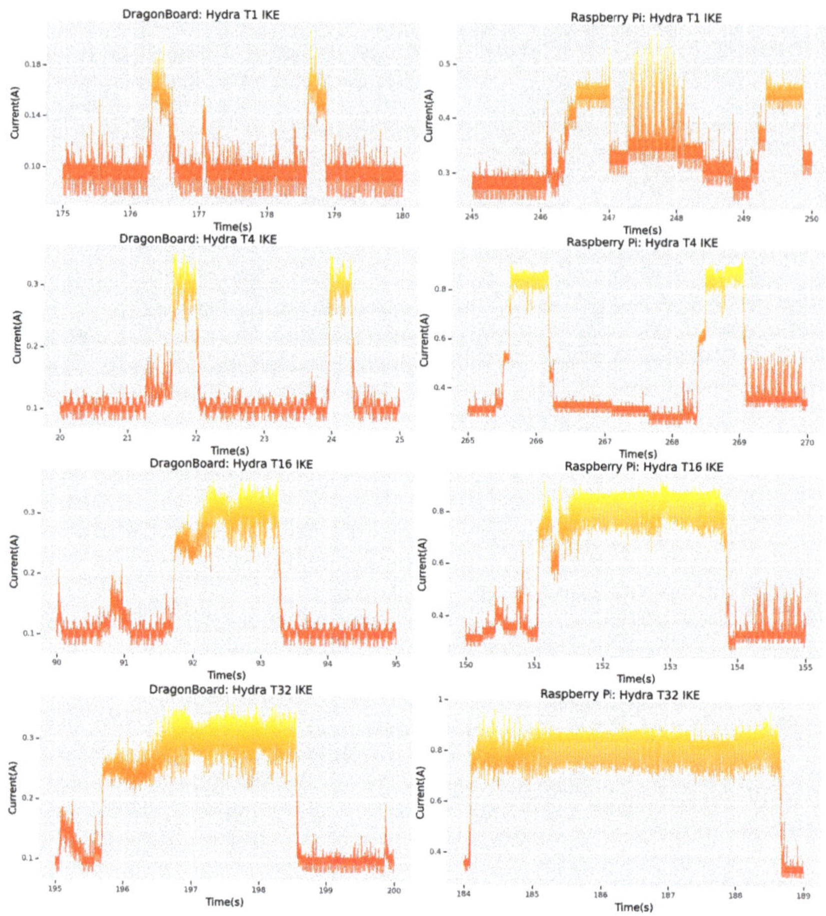

Figure 4. Incorrect Key Events (IKE) across all Hydra t switches compared for DragonBoard and Raspberry Pi (data represented using the "autumn" colormap [46]).

4.2.2. Correct-Key Event

After studying the signature of this attack, it was discovered that there was a unique signature in the event of the correct password being attempted by the tool Hydra. This will be referred to as the Correct-Key Event (CKE) for the remainder of the paper. The opposite case is the Incorrect-Key Event (IKE), where no correct password is found. In fact, in some cases, while watching the data-analyser scope in real-time, it was clear the password had been successfully cracked moments before the tool Hydra informed the attacker of a successful password cracking. This intuition is very similar to how the concepts of simple power analysis (SPA) and differential power analysis (DPA) work [29]. These concepts are both attacks utilising side-channel power analysis approaches to discern if an encryption key has been cracked successfully. In this case, rather than encryption keys being attempted, it is simply a list of passwords being attempted.

The signature of the CKEs varied negligibly regarding the setting "t". In Figure 5, the final attempt for setting "t1"—the correct attempt, can be seen milliseconds before the CKE on each board. As the signature occurs far too quickly after the attempt, this suggests that the CKE is not a network-related response but, rather, a cryptographic response from the target system itself, due to the correct password being attempted. Interestingly, regardless of the Hydra settings used, each board's CKE signature remains the same. To our knowledge, this is the first time this specific event has been documented.

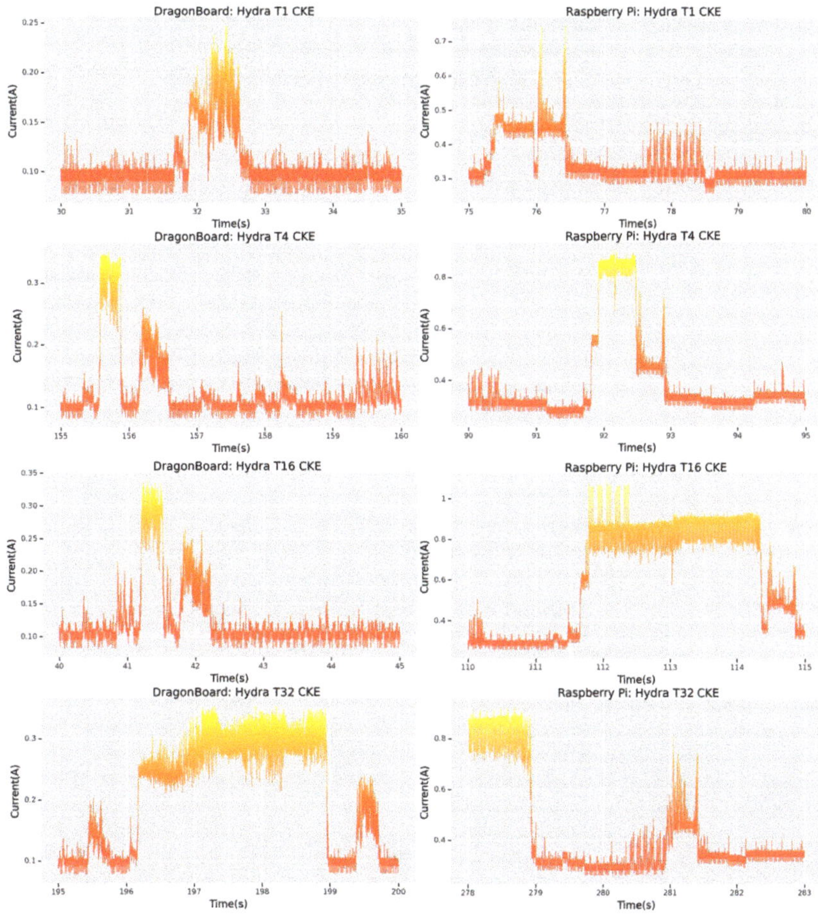

Figure 5. Correct Key Events (CKE) across all Hydra t switches compared for DragonBoard and Raspberry Pi (data represented using the "autumn" colormap [46]).

The impact of the CKE is enormous. For a forensics team, being able to tell precisely when the password was cracked, when responding to a breach, is very valuable. This can allow a forensics team to create a more detailed and accurate report on what happened during the attack, leaving much less time unaccounted for and assisting in countermeasures against future breaches. For instance, if a team sees that the attacker cracked the password very quickly, it would suggest that a weak password was used. This could direct teams to change policies in the organisation to improve security.

An attacker who wishes to hide their attacks can tell Hydra to continue to the end of the password file even after the correct key is found. They may do this to mislead the forensics team, giving the attacker hours, while the forensic team thinks the attack ended when the large IKE signatures stopped. However, if a forensics team attempted to crack their password and look at the CKE signature, they would be able to match the signature with their logged data to ascertain exactly, within milliseconds, when the password was cracked. This is one way in which security and forensics teams can reverse engineer the events that occurred to the victim to better understand what happened during a breach.

4.3. Denial of Service

Denial-of-service (DOS) attacks attempt to deny the availability of a service. These attacks tend to overwhelm the target service with massive traffic levels or exploit vulnerabilities in the target service to deny its availability, while requiring as few resources as possible on the attacker's side. The "SYN flood" denial-of-service attack was carried out from this attack category against the testbed.

SYN Flood

Also known as a half-open attack, an SYN flood attack is a network attack which bombards a server with as many connection requests as possible but does not respond to the acknowledgement of each connection request. Through this methodology, a server is inundated with many half-open TCP requests, which consume the target's resources, while minimal resources are required on the attackers' side. This affects the functionality of the target in serving new requests from legitimate connections or serving existing connections correctly. The SYN flood attack was chosen for this study as it is a remarkably lightweight DOS attack with few resources required on the attacker's side. This is the logical DOS attack against the testbed, a resource-constrained IoT device. To carry out this attack against the testbed, the tool "Hping3" was used [54]. Hping3 is a versatile network tool used by attackers and network auditors alike.

Figure 6 shows the first 10 s of the SYN Flood attack on both devices. Evident destruction of normal behaviour is visible in both cases. The Raspberry Pi can sustain higher numbers of SYN requests than the DragonBoard, resulting in a large signature for the SYN flood attack. Interestingly, the maximum current draw of 0.7 A during the SYN flood attack does not reach the maximum values witnessed during the SSH brute force. This could likely be attributed to the attack surpassing the resources of the Pi, meaning more SYN requests could not be made.

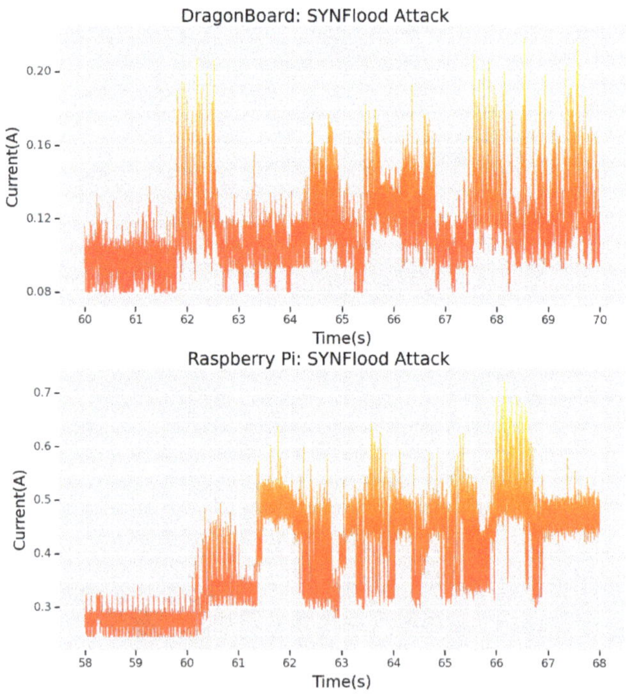

Figure 6. Comparison of SYNFlood attacks on DragonBoard versus Raspberry Pi (data represented using the "autumn" colormap [46]).

In the case of the DragonBoard, a much lower current draw overall can be seen, likely due to the low-power optimised device only being able to handle a few half-open requests. The complete destruction of normal behaviour is evident, nonetheless, with sustained high current draw overall. With these sustained current draws, an attacker may employ such an attack on an IoT device to drain its battery much more swiftly, leading to the device being forced offline.

5. Capture-the-Flag Scenario

To tie together all of the work on this testbed, a "capture-the-flag" scenario (CTF) was carried out against both devices. CTFs in cybersecurity are scenarios where a secret file or "flag" is hidden on a target system. The goal of this scenario is for attackers to hack into the system to obtain this hidden flag. This scenario is often the theme of large competitions where many teams of hackers compete against one another to obtain these hidden flags first. In the case of this work, a file was hidden on the testbed, and the testbed was compromised by combining most of the attacks featured in Section 4, with the hidden flag subsequently being stolen by the attacker.

The entire process of the hacker's first contact with the target to the exfiltration of the hidden flag is all carried out without interruption, in a realistic scenario. This scenario mimics how an attacker would hack a device with minimal security in the field. This scenario is a notable contribution to the field of cybersecurity, as the entire process of the compromised device can be easily represented visually using the side-channel power of the testbed. The CTF scenario is broken up into the key areas of reconnaissance, brute force, infiltration, exploration, exfiltration and withdrawal.

During the CTF scenario study, the data presented will be cross-referenced with the findings from Section 4. Using this methodology, it will be possible to ascertain what occurred during the CTF without prior knowledge of the actual sequence of attacks used in the CTF. A forensics team could employ a methodology such as this after they have had a breach: with limited knowledge of the hackers' actions (for example, they know the attacker infiltrated the system over SSH), apply some well-known attacks to their own devices, obtain the side-channel power data and, finally, compare the signatures obtained with those captured during the breach. Cross-referencing these signatures allows the team to establish a much clearer picture of what happened during the breach.

5.1. Reconnaissance

The CTF scenario begins with attacks on the confidentiality of the testbed. As mentioned in the reconnaissance-attack section, before an attacker begins any attack, they must gather information on their target. Information such as open ports, if the device accepts pings, or even the operating system running on the target can be key.

5.1.1. Ping

Before the attacker commences the port scan, it is prudent to "ping" the target to confirm if it is "live" and allowing ping requests. Ping (Packet Internet or Inter-Network Groper) [55] is an extremely well-known program in computing and used for many purposes. These purposes include verifying if a destination IP exists and is accepting requests. The program works similarly to the concept of a radar ping, where a pulse is sent out, or packet in this case, and returns once it bounces off its destination or, in this case, is returned by the destination.

A "ping" is not actually an attack, but the presence of pings can indicate the initial contact between the attacker and the victim. In cases where a device should see no external communications at all but only a highly constrained range of normal behaviour, pinging could represent a much higher degree of danger. At the discretion of the security teams, specific security mechanisms could be put in place for their particular scenario.

Figure 7 shows the first five seconds of the targets being pinged in the CTF scenario. On the Dragonboard, small signatures belonging to the pings can be seen soon after the normal Bluetooth behaviour. These small pings will continue as long as the attacker pings the target. While the pings themselves may have a minimal signature on this testbed, their detectability only increases with an increase in the duration of the attack. These pings create quite the perturbation in normal behaviour, making this attack obvious even over this short period. For the Raspberry Pi, detecting the pings is slightly more complex. Figure 7 shows no clear individual ping signatures can be distinguished. However, the entire current signature of the board increases from 0.25 A to 0.30 A. This increase continues for the duration of the ping attack. Using traditional power analysis methods, such as simply obtaining the average of the power in a segment, the pings should be detectable on the Raspberry Pi.

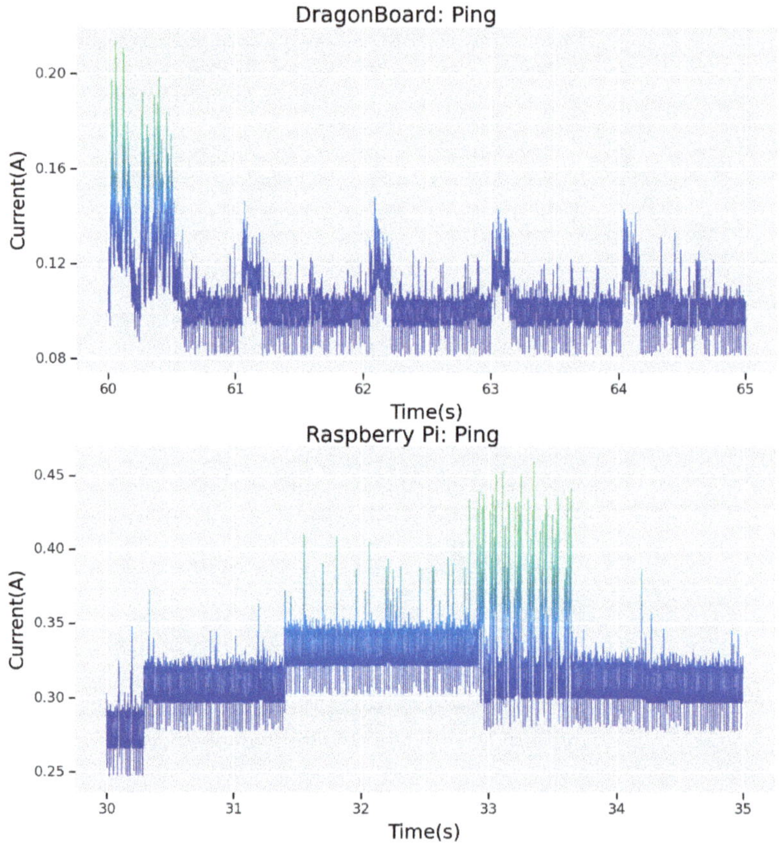

Figure 7. Ping behaviour compared for Dragonboard versus Raspberry Pi (data represented using the "winter" colormap [46]).

5.1.2. Port Scanning

The Nmap tool was used to ascertain what processes were running on the target device; cross-referencing the signatures from Section 4.1.1, one can study Figure 8 and clearly see that this is indeed the signature for a port-scan attack. In fact, for both the Raspberry Pi and the Dragonboard, a clear similarity between Figure 8 and settings "T3"–"T5" shown in Figure 3 can be seen. In both cases, the signatures bear more similarities to the "T3" setting which was, in fact, the setting the devices were subjected to during the CTF.

This methodology of matching signatures can be very powerful for forensics teams after a breach. By subjecting their device to a range of attacks similar to what they think the attacker may have done, they can obtain a signature match which tells them precisely the operations the attacker went through when attacking their device.

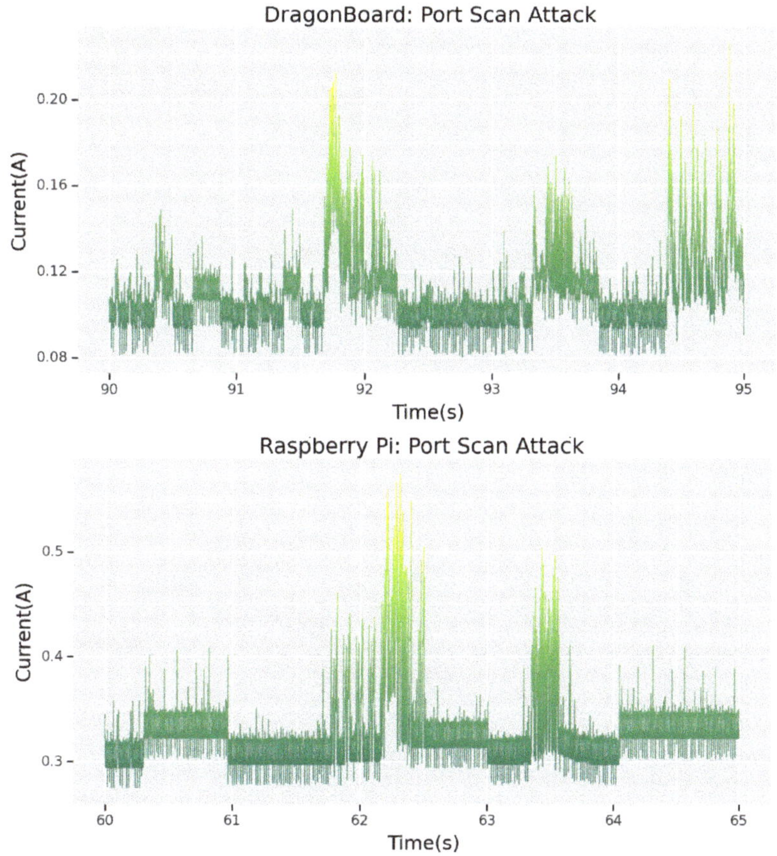

Figure 8. Nmap behaviour during CTF compared for DragonBoard versus Raspberry Pi (data represented using the "summer" colormap [46]).

5.2. Brute Force

The port-scan attack revealed that the SSH service was running on the target; thus, port 22 was open for connection. This directs the attacker's subsequent actions. As they wish to gain entry to the device to steal critical information, this attack vector will be exploited as an SSH brute force. The forensics team can confirm that an SSH brute-force attack has indeed occurred by cross-referencing the data obtained in the CTF with Section 4.2, the matching power signatures.

However, very different signatures are visible between the Dragonboard and the Raspberry Pi. Studying the start of each SSH attack (Figure 9) gives a clearer picture of what occurs in each case. By cross-referencing Figure 9 with Figure 4, it is clear that the Dragonboard has been subjected to setting "t16" on Hydra, while the Raspberry Pi was attacked with only "t4", hence the more significant initial signature for the DragonBoard.

By working backwards, in a method which is similar to reverse engineering, a forensics team can perform many iterations of the same attack type with different settings and, by matching the signatures they obtained with those from the attack, infer with a high degree of certainty not only the attack type but possibly even the settings of the tool the attacker used.

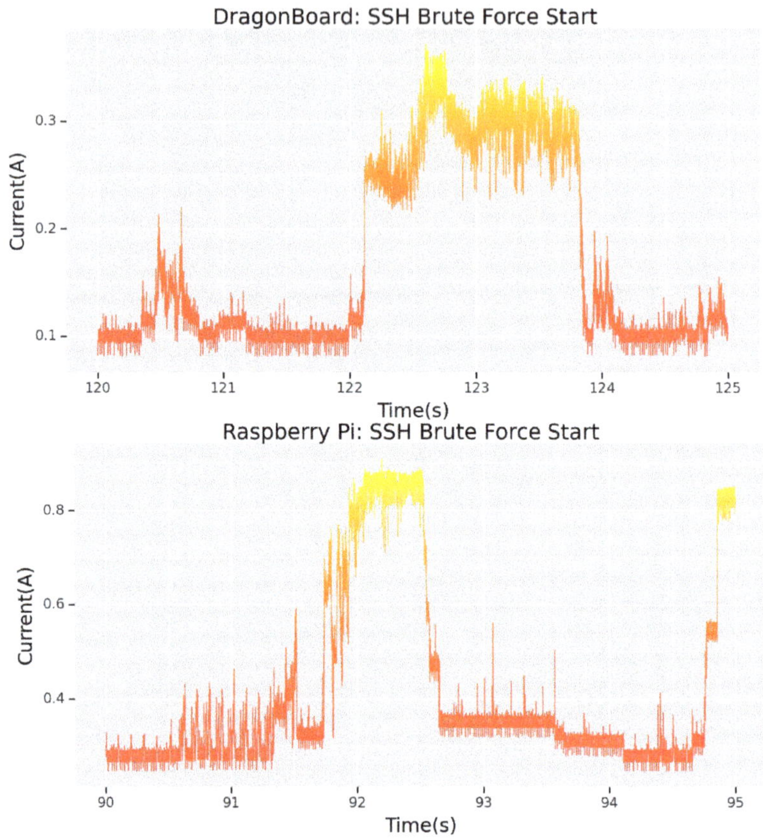

Figure 9. Start of the SSH brute force during CTF, DragonBoard versus Raspberry Pi (data represented using the "autumn" colormap [46]).

In both cases, the CKE is visible in Figure 10 and matches all the expected signatures found in Section 4.2. This is precious information, as now any forensics team dealing with a scenario such as this one can tell, with considerable accuracy, to the nearest millisecond, when the password for their device was cracked.

5.3. Infiltration

After the SSH password is cracked, the target is infiltrated over SSH. This is a notable escalation in danger within this scenario. Now that the attacker has gained entry, they can explore the device with whatever permissions the hacked account has. If the compromised account has admin or "sudo" permissions, this can lead to catastrophic consequences as the attacker will have unhindered access and ability. Even with a regular user account, the attacker can perform a "privilege-escalation attack" to temporarily, or permanently, gain admin permissions. This can lead to the total destruction of the confidentiality, integrity and availability of this device and, possibly, many or all devices on this network which share the same admin password. As the attacker has compromised the admin account in the case of this testbed, they will not need to perform a privilege-escalation attack.

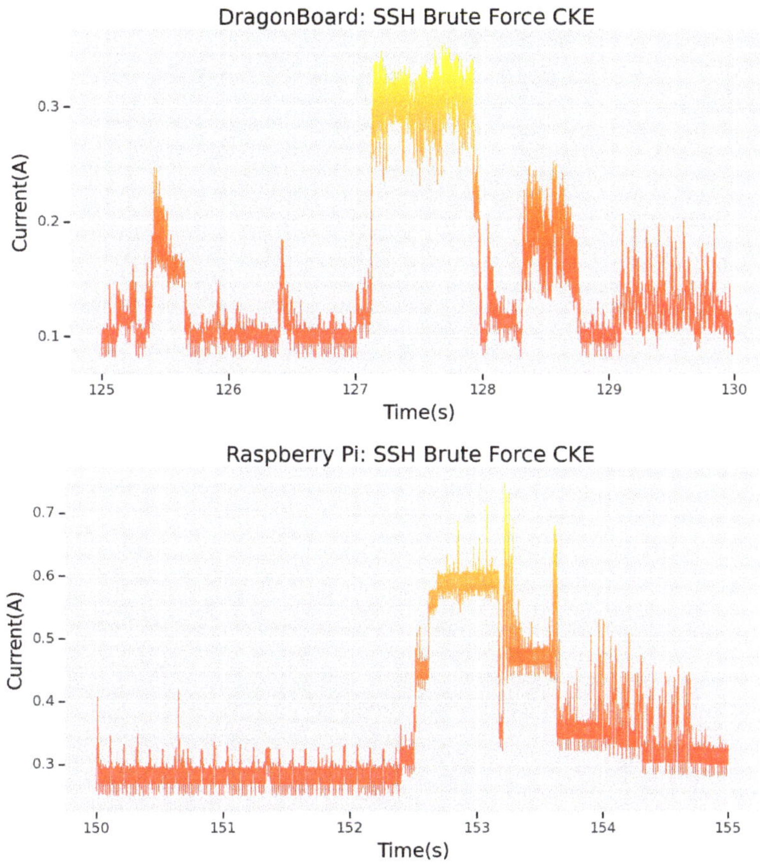

Figure 10. Comparison of the Correct-Key Events during the CTF, DragonBoard versus Raspberry Pi (data represented using the "autumn" colormap [46]).

Interestingly, the signature in Figure 11 is an exact match for that of the SSH Brute force "t1" CKE of each board. The signature for "t1" in Figure 5 shows one final password attempt and the response to it being the correct password, the CKE. This gives a much clearer insight into the operation of the password-cracking tool used against the device.

Of course, looking at the code for Hydra, the tool's operation will be transparent; however, understanding this functionality without looking at the code but, rather, studying its side-channel power data is very valuable. By making such observations, forensics teams can ascertain the inner workings of the tools used to attack their device.

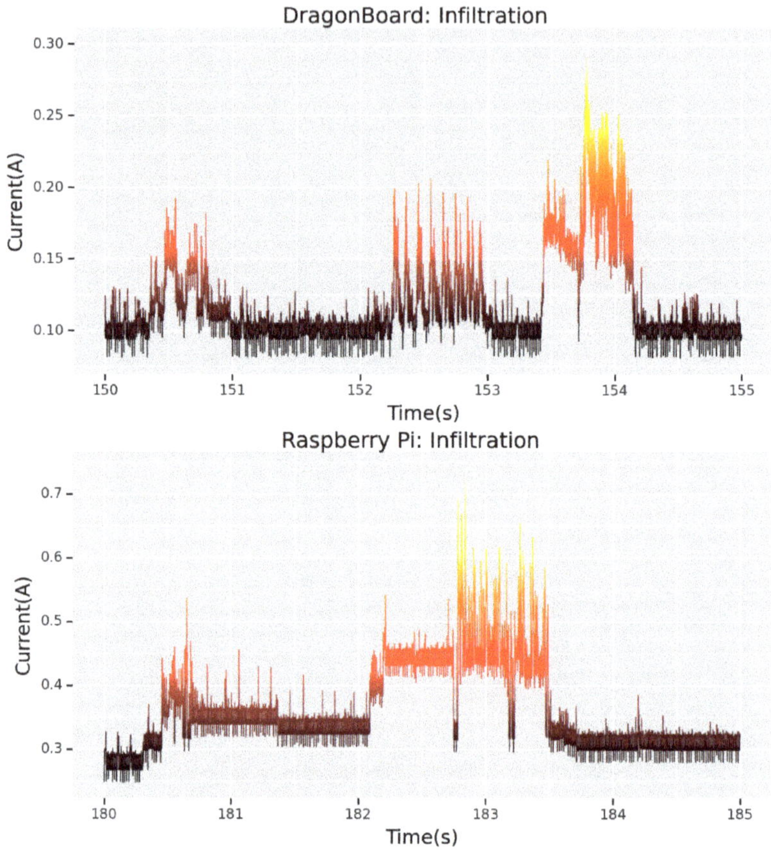

Figure 11. Infiltration during CTF, DragonBoard versus Raspberry Pi (data represented using the "hot" colormap [46]).

5.4. Exploration

After the target is infiltrated, it is explored so that the attacker can find directories and files of interest. An attacker will also check what permissions they and other users have on this device. If they do not have valid admin permissions, they will check for any out-of-date or unsafe software and services on the device. If found, they can then exploit these vulnerabilities in what is known as a "privilege-escalation" attack. This attack can take a standard user account with no permissions and temporarily or permanently give that account admin control. To this end, it is essential that only essential software is installed on IoT devices. In the case of this CTF scenario, the attacker has already obtained an admin account; thus, there is no need for a privilege-escalation attempt.

The exploration of the device causes a significant disruption to normal behaviour. Figure 12 shows a massive perturbation of the normal behaviour for both boards. In the case of the Raspberry Pi, the attacker can be seen in Figure 12 to use the "CAT" tool to open and read a file on the target system. For the Dragonboard, the attacker is simply exploring the directories of the target device in an attempt to find the secret file for which they are searching.

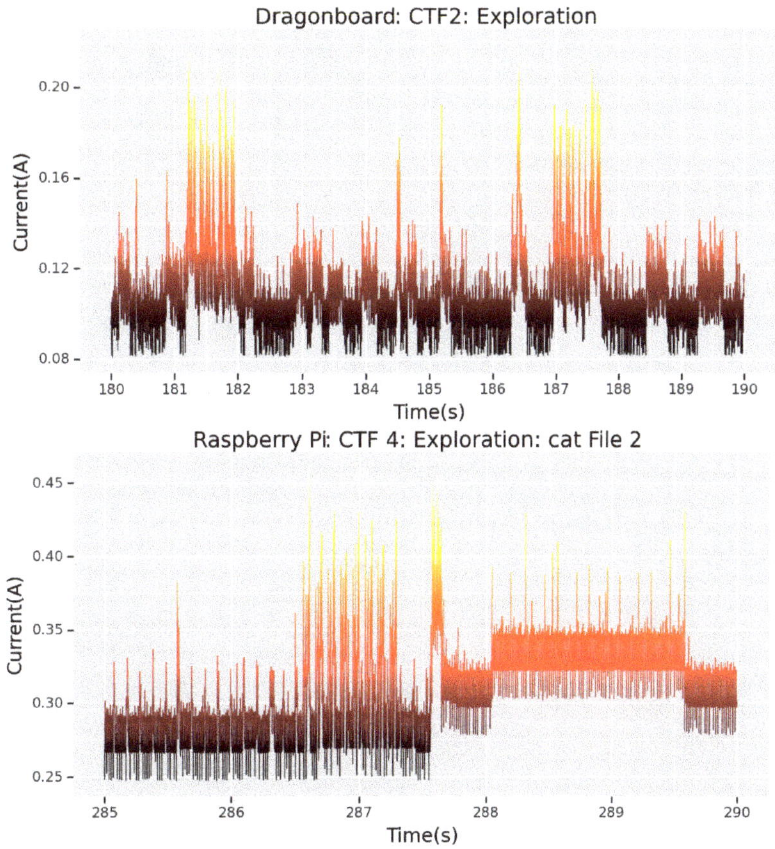

Figure 12. Comparison of attacker exploring target device, DragonBoard versus Raspberry Pi (data represented using the "hot" colormap [46]).

5.5. Exfiltration

The most critical part of this CTF scenario is the exfiltration of confidential data from the device. Exfiltration is the transfer of information from a system without authorisation. It is the complete destruction of the confidentially of the target device. This is the worst-case scenario in this testbed and was the original goal of the attacker—to steal that specific hidden file, the "flag". This flag, in a real-world scenario, could be personally identifying information (PII) of the device owner, a username and password file—which the attacker can try to crack later, or information on the peers to which the device is connected. It should be recognised that regardless of the data exfiltrated, this event should be regarded as being of the highest degree of severity.

Studying Figure 13, it is clear that the tool used to copy the file over SSH is comprised partially of the SSH login event and the SSH logout event (this will be studied in the next section). Knowing the signatures of both the login and logout events, the forensics team can ascertain that a file has been taken and its size. This is possible by studying the extra power signatures between the login and logout events, which correspond to the file itself being copied remotely. As well as this, by studying these signatures, the forensics team would be able to infer that the tool "SCP" [56] or "secure copy", which is used to copy files over SSH, was used.

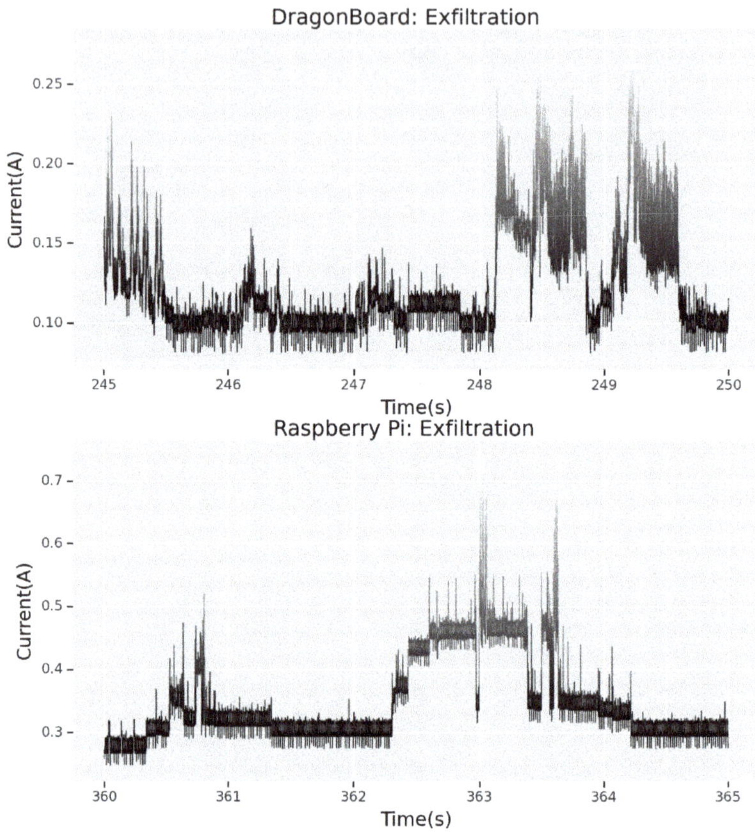

Figure 13. Exfiltration during CTF DragonBoard versus Raspberry Pi (data represented using the "gist_gray" colormap [46]).

5.6. Withdrawal

After the attacker has completed their tasks, they close up communications over SSH. If they remained connected, this would expose them during any investigations, or their continued presence might alert users to their attack. Other than simply logging out, while it does not occur in this CTF scenario, the attacker can try to erase their presence by deleting or amending log files. This would be an attack on the integrity of the device. Studying Figure 14, and refering back to the exfiltration signature, it is evident that the exfiltration signature comprises a login event, followed by a logout event. This means that the exfiltrated file's signature will be the extra power signature in between.

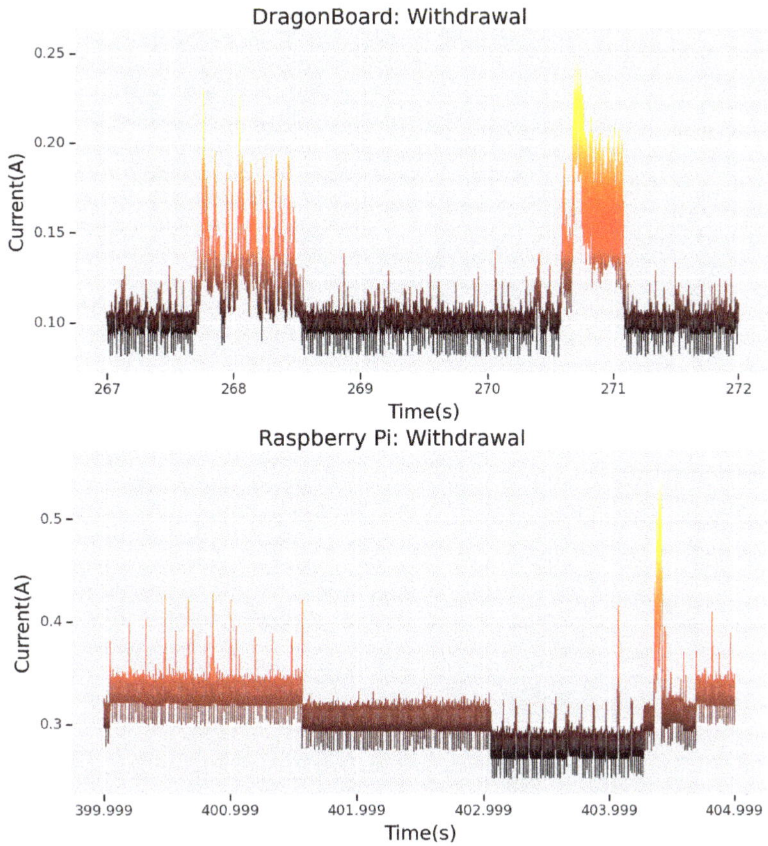

Figure 14. Withdrawal of attacker after CTF completed DragonBoard versus Raspberry Pi (data represented using the "hot" colormap [46]).

6. Discussions

It is essential to consider the detection of the attacks featured in this study from the perspective of signature-based and anomaly-based machine-learning approaches. This study mainly discourages using signature-based methodologies using the side-channel power data for a core reason: the availability of the attacks' signatures and the required quantity. To train a signature-based machine-learning model to classify if a segment is normal or anomalous, sufficient examples of the signatures of each attack would be necessary. If there is one key takeaway from this study, it is that each attack type has a vastly different signature from any other, and each attack itself has a signature which mutates greatly depending on the settings of the tool used to perform the attack. This means that a team hoping to develop a signature-based classifier for their device must obtain many signature examples for each attack and the many settings possible on the tool. This does not even consider that many different tools can perform the same attacks, all of which may have slightly different signatures. These factors make developing a fit-for-purpose signature-based model far too challenging to achieve using side-channel power data.

This is different for anomaly-detection-based machine-learning approaches, however. While clear signatures for both normal and attack classes are required for signature-based methodologies, only a clear signature for normal behaviour is necessary for anomaly-based approaches. These approaches tend to be either semi-supervised, with one labelled normal class, or completely unsupervised, with no labelled classes, and use their logic to cluster the data into the classes they perceive. This study demonstrates that a semi-supervised approach, which leverages the natural abundance of normal data, would be the most effective way to build an automated detector for these attacks. Recent studies such as [57] demonstrate the effectiveness of semi-supervised methodologies such as Bagging-Random Miners (BRM), One-Class K-means with Randomly projected features Algorithms (OCKRA), Isolation Forests (ISOF), and One-Class Support Vector Machines (ocSVM) on the task of anomaly detection. Another approach, [58], utilises unsupervised autoencoders and Gaussian mixture models (GMM) for anomaly detection for network security. As such, an IDS framework which uses a semi-supervised approach such as one of these would be most appropriate for IoT intrusion detection. Semi-supervised methods rely not on the anomaly class for training but on the normal class. Normal data is easily obtained; the more normal examples obtained, the better the model. In this way, a model trained only on normal data can determine if a segment is anomalous simply by the deviations it perceives from the normal behaviour.

While this study discourages using the raw side-channel power signatures of the attacks for signature-based approaches to intrusion detection, it encourages the study of these signatures. Understanding what these attacks look like in specific devices' power data is immensely valuable. This information can be used to educate those in the security sector and influence the design and manufacture of IoT devices. Without a complete understanding of an attack and what it looks like from "all angles", it is impossible to protect against it. A broader understanding of these signatures is also essential from the perspective of reverse engineering attacks. If forensics teams are able to more accurately ascertain what happened during an event, they can better approach protecting against further attacks.

It is crucial to consider far simpler approaches than machine learning to detect these attacks. Deploying a machine-learning approach is not appropriate in every scenario. The resources needed to train and deploy such a model are absent in many settings, especially for IoT devices. While some devices can host a machine-learning-based intrusion-detection system, such as the Raspberry Pi, for instance, excess power usage cannot be allowed for most IoT devices. This means these models must be implemented on IoT gateways instead, with the devices communicating their monitored side-channel data to the gateway for inference.

At a minimum, simple approaches to detecting these attacks must be deployed. With a clear normal class, simple techniques such as thresholds on the side-channel current draw can effectively detect an intrusion from the excess current draw over normal behaviour. If the normal class tends to infrequently surpass this threshold, simply having another threshold on the number of times the current draw can exceed a certain amount within a period would effectively reduce false positives. Excessive false positives can pose a significant problem. Similar to how an Intrusion Prevention System (IPS) works, a fully automated IDS will halt regular operation whenever an anomaly is detected, leading to too much interruption to normal behaviour. For a semi-automated IDS, with a human in the loop, while the human can discern if it is a false positive, too many false positives will lead to an excessive workload for the human. This will likely lead them to disable the IDS entirely. When intrusions do not exceed the normal behaviour's current draw, comparing a data segment's average current draw with a pre-defined baseline can effectively detect anomalies. While these approaches may seem simple, even this bare minimum to detect attacks is not currently being deployed to protect IoT devices in the field. This approach could save companies USD millions in the case of an attack. Where traditional packet-based detection approaches may fail, detecting a simple spike in the power draw of a critical piece of equipment could prevent catastrophic damage.

An essential aspect of this work is its non-intrusive nature. This means that systems can be retrofitted with side-channel monitoring systems without affecting the system's original operation.

Another key takeaway of this study is that, for most of those working in the security sector, this will be the first time they see attacks they are familiar with represented in anything other than packer captures. This study investigates a far-too-overlooked key piece of basic information which every device has, its power data. It demonstrates how security teams in the field can use side-channel power data.

7. Conclusions

This study investigated a far-too-overlooked key piece of fundamental information which every device has, the side-channel power data, and demonstrates how it could be used by security teams in the field. This work highlights the existence of common signatures for attacks across different devices, as well as demonstrating how these signatures mutate with simple changes in attack settings. Some key aspects to consider when developing machine-learning-based methodologies to detect these attacks are also discussed. This work uncovers the "Correct Key Event", when a clear power signature for the password of the device being cracked is obtained. This work also presents a simple reverse-engineering framework which could be employed to ascertain what occurred during a breach, by applying similar attacks and matching the signatures obtained to those captured during the breach. This work highlights the effectiveness of power data in security and promotes its further usage in a supplementary manner to existing methodologies. The work featured in this study is only the initial investigation of this dataset. For future work on this topic, we will apply the machine-learning practices mentioned to develop an effective anomaly-detection-based IDS for IoT devices using non-intrusive side-channel power data and incorporate this into a human-in-the-loop framework. We will release this dataset along with future publications.

Author Contributions: Methodology, D.L., D.-M.N. and E.P.; Software, D.L.; Formal analysis, D.L., D.-M.N., C.C.M. and E.P.; Writing—original draft, D.L. and E.P.; Writing—review and editing, A.T., C.C.M. and E.P.; Supervision, C.C.M., E.P.; Project administration, E.P. All authors have read and agreed to the published version of the manuscript.

Funding: This research is supported in part by a grant from Science Foundation Ireland INSIGHT Centre for Data Analytics (Grant number 12/RC/2289-P2) which is co-funded under the European Regional Development Fund.

Data Availability Statement: The data presented in this study are available on request from the corresponding author.

Acknowledgments: The authors acknowledge the Insight SFI Research Centre for Data Analytics at University College Cork, Ireland. We would also like to acknowledge support from Qualcomm and Dell.

Conflicts of Interest: The authors declare no conflict of interest.

References

1. Griffiths, C. The Latest 2023 Cyber Crime Statistics (Updated March 2023). Available online: https://aag-it.com/the-latest-cyber-crime-statistics/ (accessed on 6 April 2023).
2. Forum, W.E. Partnership against Cybercrime, Insight Report 2020. Available online: https://www.weforum.org/reports/partnership-against-cybercrime/ (accessed on 6 April 2023).
3. Cybersecurity Infrastructure Security Agency. Stop Ransomware | CISA. Available online: https://www.cisa.gov/stopransomware/ (accessed on 6 April 2023).
4. National Cybersecurity and Communications Integration Center. What Is Wannacry/Wanacrypt0r? Available online: https://www.cisa.gov/sites/default/files/FactSheets/NCCICICS_FactSheet_WannaCry_Ransomware_S508C.pdf (accessed on 6 April 2023).
5. Chappell, B.; Neuman, S. U.S. Says North Korea 'Directly Responsible' For WannaCry Ransomware Attack. Available online: https://www.npr.org/sections/thetwo-way/2017/12/19/571854614/u-s-says-north-korea-directly-responsible-for-wannacry-ransomware-attack (accessed on 6 April 2023).
6. Kapko, M. CISA's Priority Sectors for 2023: Water, Hospitals, K-12. Available online: https://www.cybersecuritydive.com/news/CISA-water-schools-healthcare/634657/ (accessed on 6 April 2023).

7. Zacharakos, A. No Relief in Sight for Ransomware Attacks on Hospitals. Available online: https://www.techtarget.com/searchsecurity/feature/No-relief-in-sight-for-ransomware-attacks-on-hospitals (accessed on 6 April 2023).
8. Fowler, B. Ransomware Cost US Schools 3.56 Billion in 2021, Study Says. Available online: https://www.cnet.com/tech/services-and-software/ransomware-cost-us-schools-3-56-billion-in-2021-study-says/ (accessed on 6 April 2023).
9. National Cyber Security Centre. Ransomware Attack on Health Sector—UPDATE 2021-05-16. Available online: https://www.ncsc.gov.ie/pdfs/HSE_Conti_140521_UPDATE.pdf (accessed on 6 April 2023).
10. McGrath, P. NUIG IT Systems Remain Offline after Attempted Cyber Attack. Available online: https://www.rte.ie/news/2021/0930/1249912-nuig-cyber-attack/ (accessed on 6 April 2023).
11. Dwyer, O. IT Services Remain Disrupted at Two Colleges after Ransomware Attacks. Available online: https://www.thejournal.ie/tu-dublin-ransomware-attack-ongoing-5403034-Apr2021/ (accessed on 6 April 2023).
12. Daly, A. TU Dublin's Tallaght Campus Investigating 'Significant' Ransomware Attack. Available online: https://www.thejournal.ie/tu-dublin-ransomware-attack-5401763-Apr2021/ (accessed on 6 April 2023).
13. Munster Technological University. MTU Cyber Attack Update. Available online: https://www.mtu.ie/cyber-attack/ (accessed on 6 April 2023).
14. Kumar, S.; Tiwari, P.; Zymbler, M. Internet of Things is a revolutionary approach for future technology enhancement: A review. *J. Big Data* **2019**, *6*. [CrossRef]
15. International Data Corporation. Future of Industry Ecosystems: Shared Data and Insights. Available online: https://blogs.idc.com/2021/01/06/future-of-industry-ecosystems-shared-data-and-insights/ (accessed on 6 April 2023).
16. Sagu, A.; Gill, N.S.; Gulia, P.; Singh, P.K.; Hong, W.C. Design of Metaheuristic Optimization Algorithms for Deep Learning Model for Secure IoT Environment. *Sustainability* **2023**, *15*, 2204. [CrossRef]
17. Cybersecurity Infrastructure Security Agency. Heightened DDoS Threat Posed by Mirai and Other Botnets. Available online: https://www.cisa.gov/news-events/alerts/2016/10/14/heightened-ddos-threat-posed-mirai-and-other-botnets (accessed on 6 April 2023).
18. Cybersecurity Infrastructure Security Agency. Cyber-Attack Against Ukrainian Critical Infrastructure. Available online: https://www.cisa.gov/news-events/ics-alerts/ir-alert-h-16-056-01 (accessed on 6 April 2023).
19. Kilpatrick, H. 5 Infamous Iot Hacks and Vulnerabilities. Available online: https://www.iotsworldcongress.com/5-infamous-iot-hacks-and-vulnerabilities/ (accessed on 6 April 2023).
20. Eskandari, M.; Janjua, Z.H.; Vecchio, M.; Antonelli, F. Passban IDS: An Intelligent Anomaly-Based Intrusion Detection System for IoT Edge Devices. *IEEE Internet Things J.* **2020**, *7*, 6882–6897. [CrossRef]
21. Almiani, M.; AbuGhazleh, A.; Al-Rahayfeh, A.; Atiewi, S.; Razaque, A. Deep recurrent neural network for IoT intrusion detection system. *Simul. Model. Pract. Theory* **2020**, *101*, 102031. Modeling and Simulation of Fog Computing. [CrossRef]
22. Tavallaee, M.; Bagheri, E.; Lu, W.; Ghorbani, A.A. A detailed analysis of the KDD CUP 99 data set. In Proceedings of the 2009 IEEE Symposium on Computational Intelligence for Security and Defense Applications, Ottawa, ON, Canada, 8–10 July 2009; pp. 1–6. [CrossRef]
23. Saba, T.; Rehman, A.; Sadad, T.; Kolivand, H.; Bahaj, S.A. Anomaly-based intrusion detection system for IoT networks through deep learning model. *Comput. Electr. Eng.* **2022**, *99*, 107810. [CrossRef]
24. Koroniotis, N.; Moustafa, N.; Sitnikova, E.; Turnbull, B. Towards the development of realistic botnet dataset in the Internet of Things for network forensic analytics: Bot-IoT dataset. *Future Gener. Comput. Syst.* **2019**, *100*, 779–796. [CrossRef]
25. Bhosale, S. Network Intrusion Detection. Available online: https://www.kaggle.com/datasets/sampadab17/network-intrusion-detection (accessed on 9 May 2023).
26. Ngo, D.M.; Lightbody, D.; Temko, A.; Pham-Quoc, C.; Tran, N.T.; Murphy, C.C.; Popovici, E. HH-NIDS: Heterogeneous Hardware-Based Network Intrusion Detection Framework for IoT Security. *Future Internet* **2023**, *15*, 9. [CrossRef]
27. Parmisano, A.; Garcia, S.; Erquiaga, M.J. *A Labeled Dataset with Malicious and Benign Iot Network Traffic*; Stratosphere Laboratory: Praha, Czech Republic, 2020.
28. Moustafa, N.; Slay, J. UNSW-NB15: A comprehensive data set for network intrusion detection systems (UNSW-NB15 network data set). In Proceedings of the 2015 Military Communications and Information Systems Conference (MilCIS), Canberra, Australia, 10–12 November 2015; pp. 1–6.
29. Kocher, P.; Jaffe, J.; Jun, B. Differential power analysis. In Proceedings of the Advances in Cryptology—CRYPTO'99: 19th Annual International Cryptology Conference, Santa Barbara, CA, USA, 15–19 August 1999; Springer: Berlin/Heidelberg, Germany, 1999; pp. 388–397.
30. Lightbody, D.; Ngo, D.M.; Temko, A.; Murphy, C.; Popovici, E. Host-Based Intrusion Detection System for IoT using Convolutional Neural Networks. In Proceedings of the 2022 33rd Irish Signals and Systems Conference (ISSC), Cork, Ireland, 9–10 June 2022; pp. 1–7. [CrossRef]
31. Clark, S.S.; Ransford, B.; Rahmati, A.; Guineau, S.; Sorber, J.; Xu, W.; Fu, K. WattsUpDoc: Power Side Channels to Nonintrusively Discover Untargeted Malware on Embedded Medical Devices. In Proceedings of the 2013 USENIX Workshop on Health Information Technologies (HealthTech 13), Washington, DC, USA, 12 August 2013; USENIX Association: Washington, DC, USA, 2013.
32. Ding, F.; Li, H.; Luo, F.; Hu, H.; Cheng, L.; Xiao, H.; Ge, R. DeepPower: Non-Intrusive and Deep Learning-Based Detection of IoT Malware Using Power Side Channels. In Proceedings of the Proceedings of the 15th ACM Asia Conference on Computer and

Communications Security, ASIA CCS '20, Taipei, Taiwan, 5–9 October 2020; Association for Computing Machinery: New York, NY, USA, 2020; pp. 33–46. [CrossRef]
33. Copos, B.; Peisert, S. Catch Me If You Can: Using Power Analysis to Identify HPC Activity. *arXiv* **2020**, arXiv:2005.03135.
34. Raspberry Pi. Raspberry Pi 3 Model B. Available online: https://www.raspberrypi.com/products/raspberry-pi-3-model-b/ (accessed on 12 April 2023).
35. Qualcomm. DragonBoard 410c Development Board. Available online: https://developer.qualcomm.com/hardware/dragonboard-410c (accessed on 12 April 2023).
36. Qualcomm. Snapdragon 410 Processor. Available online: https://www.qualcomm.com/products/mobile/snapdragon/smartphones/snapdragon-4-series-mobile-platforms/snapdragon-processors-410 (accessed on 12 April 2023).
37. Hildenbrand, J. Qualcomm's Snapdragon 410 Used in over 550 Designs and They've Shipped More Than 200 Million Units. Available online: https://www.androidcentral.com/qualcomms-snapdragon-410-used-over-550-designs-and-theyve-shipped-more-200-million-units (accessed on 12 April 2023).
38. Collins, S. The Life of Pi: Ten Years of Raspberry Pi. Available online: https://www.cam.ac.uk/stories/raspberrypi (accessed on 12 April 2023).
39. Pi, R. Raspberry Pi OS. Available online: https://www.raspberrypi.com/software/ (accessed on 12 April 2023).
40. Team, Q.L. Linaro Releases. Available online: https://releases.linaro.org/96boards/dragonboard410c/linaro/debian/21.12/ (accessed on 12 April 2023).
41. die.net. ssh(1)—Linux Man Page. Available online: https://linux.die.net/man/1/ssh (accessed on 12 April 2023).
42. Cimpanu, C. 15 Percent of All IoT Device Owners Don't Change Default Passwords. Available online: https://www.bleepingcomputer.com/news/security/15-percent-of-all-iot-device-owners-dont-change-default-passwords/ (accessed on 12 April 2023).
43. Texas Instruments. TIDC-CC2650STK-SENSORTAG. Available online: https://www.ti.com/tool/TIDC-CC2650STK-SENSORTAG#overview (accessed on 12 April 2023).
44. Harvey, I. sensortag.py-repository. Available online: https://github.com/IanHarvey/bluepy/blob/master/bluepy/sensortag.py (accessed on 12 April 2023).
45. Keysight. N6705A DC Power Analyzer. Available online: https://www.keysight.com/us/en/product/N6705A/dc-power-analyzer-modular-600-w-4-slots.html (accessed on 12 April 2023).
46. Matplotlib Development Team. Choosing Colormaps in Matplotlib. Available online: https://matplotlib.org/stable/tutorials/colors/colormaps.html (accessed on 10 May 2023).
47. Cybersecurity Infrastructure Security Agency. National Cyber Incident Scoring System. Available online: https://www.cisa.gov/sites/default/files/2023-01/cisa_national_cyber_incident_scoring_system_s508c.pdf (accessed on 10 May 2023).
48. Center for Internet Security. Election Security Spotlight—CIA Triad. Available online: https://www.cisecurity.org/insights/spotlight/ei-isac-cybersecurity-spotlight-cia-triad (accessed on 12 April 2023).
49. Nmap.org. Nmap: The Network Mapper. Available online: https://nmap.org/ (accessed on 12 April 2023).
50. OffSec. Kali Linux. Available online: https://www.kali.org/ (accessed on 12 April 2023).
51. Marc van Hauser Heuse. *hydra*. Version 9.2. March 2021. https://github.com/vanhauser-thc/thc-hydra (accessed on 1 March 2023).
52. Kali. wordlists | Kali Linux Tools. Available online: https://www.kali.org/tools/wordlists/ (accessed on 12 April 2023).
53. Burns, W.J. Common Password List (rockyou.txt). Available online: https://www.kaggle.com/datasets/wjburns/common-password-list-rockyoutxt (accessed on 12 April 2023).
54. die.net. hping3(8)—Linux Man Page. Available online: https://linux.die.net/man/8/hping3 (accessed on 12 April 2023).
55. die.net. ping(8)—Linux Man Page. Available online: https://linux.die.net/man/8/ping (accessed on 12 April 2023).
56. die.net. scp(1)—Linux Man Page. Available online: https://linux.die.net/man/1/scp (accessed on 12 April 2023).
57. Villa-Pérez, M.E.; Álvarez Carmona, M.A.; Loyola-González, O.; Medina-Pérez, M.A.; Velazco-Rossell, J.C.; Choo, K.K.R. Semi-supervised anomaly detection algorithms: A comparative summary and future research directions. *Knowl.-Based Syst.* **2021**, *218*, 106878. [CrossRef]
58. An, P.; Wang, Z.; Zhang, C. Ensemble unsupervised autoencoders and Gaussian mixture model for cyberattack detection. *Inf. Process. Manag.* **2022**, *59*, 102844. [CrossRef]

future internet

MDPI

Article

Protecting Sensitive Data in the Information Age: State of the Art and Future Prospects

Christoph Stach [1,*], Clémentine Gritti [2], Julia Bräcker [3], Michael Behringer [1] and Bernhard Mitschang [1]

1 Institute for Parallel and Distributed Systems, University of Stuttgart, Universitätsstraße 38, 70569 Stuttgart, Germany; michael.behringer@ipvs.uni-stuttgart.de (M.B.); bernhard.mitschang@ipvs.uni-stuttgart.de (B.M.)
2 Department of Computer Science and Software Engineering, University of Canterbury, Christchurch 8041, New Zealand; clementine.gritti@canterbury.ac.nz
3 Institute of Biochemistry and Technical Biochemistry, University of Stuttgart, Allmandring 5B, 70569 Stuttgart, Germany; julia.braecker@lc.uni-stuttgart.de
* Correspondence: christoph.stach@ipvs.uni-stuttgart.de; Tel.: +49-711-68588-433

Abstract: The present information age is characterized by an ever-increasing digitalization. Smart devices quantify our entire lives. These collected data provide the foundation for data-driven services called smart services. They are able to adapt to a given context and thus tailor their functionalities to the user's needs. It is therefore not surprising that their main resource, namely data, is nowadays a valuable commodity that can also be traded. However, this trend does not only have positive sides, as the gathered data reveal a lot of information about various data subjects. To prevent uncontrolled insights into private or confidential matters, data protection laws restrict the processing of sensitive data. One key factor in this regard is user-friendly privacy mechanisms. In this paper, we therefore assess current state-of-the-art privacy mechanisms. To this end, we initially identify forms of data processing applied by smart services. We then discuss privacy mechanisms suited for these use cases. Our findings reveal that current state-of-the-art privacy mechanisms provide good protection in principle, but there is no compelling one-size-fits-all privacy approach. This leads to further questions regarding the practicality of these mechanisms, which we present in the form of seven thought-provoking propositions.

Keywords: smart service; privacy techniques; location-based services; health services; voice-controlled digital assistants; image analysis; food analysis; recommender systems; DNA sequence classification

Citation: Stach, C.; Gritti, C.; Bräcker, J.; Behringer, M.; Mitschang, B. Protecting Sensitive Data in the Information Age: State of the Art and Future Prospects. *Future Internet* **2022**, *14*, 302. https://doi.org/10.3390/fi14110302

Academic Editor: Wei Yu

Received: 30 August 2022
Accepted: 18 October 2022
Published: 22 October 2022

1. Introduction

In 1991, Mark Weiser envisioned the computer for the 21st century [1] as a pervasive system that ubiquitously surrounds us, constantly adapting to its context and our current needs. Although this vision did not fully materialize, the *Internet of Things* (*IoT*) is a major step in this direction. Here, various sensors are integrated into everyday objects, enabling them to monitor their surroundings and react to it. Furthermore, all of these IoT-enabled devices, often referred to as *smart devices*, are interconnected and can communicate with each other. Thus, smart devices have a virtually unlimited data stock at their disposal [2].

The full potential of the gathered data can only be exploited if they are interlinked and comprehensively analyzed [3]. Yet, these data, which are labeled *big data*, are generated at high *velocity* and in high *volume*. Profound data processing is therefore not possible on the mostly lightweight smart devices, as they do not have the necessary resources and computing power to do this at an adequate speed and scale. Moreover, there is a high *variety* in the data in terms of schemata and data formats. Thus, extensive data preparation is required to merge these data, which also exceeds the capacities of smart devices [4].

The processing of the captured raw data therefore usually takes place in a powerful backend system. There, the raw data are initially refined. That is, they are cleansed—i.e., missing or erroneous data are treated—and transformed—i.e., different schemata and formats are harmonized [5]. Subsequently, the refined data can be interlinked, further preprocessed, and analyzed [6]. In analogy to the concepts of the knowledge management, three stages can be differentiated. *Data* refers to the unprocessed and unfiltered raw data collected by the smart devices. The refined and interlinked data are termed *information*. *Knowledge* is generated only when the information is analyzed, and the findings are interpreted [7].

Such knowledge provides the foundation for so-called *smart services*. A smart service is any kind of data-driven digital service that is able to adapt to the data and thus offering users the greatest possible utility in their current situation. Smart services can be very small scale, e.g., an adaptive application on a smartphone, as well as highly complex, e.g., when several smart devices interact with each other via actuators [8]. Examples of smart services can be found in the public, industrial, and private sectors, e.g., in the context of *smart cities* [9], *Industry 4.0* [10], and *eHealth* [11].

While such smart services are very appealing since they significantly facilitate the lives of their users in a wide variety of situations, they also pose a serious threat. As we are constantly surrounded by smart devices, they are able to quantify all aspects of their users' lives as well as the lives of innocent bystanders. The thereby gained knowledge provides deep insights into the privacy of these individuals [12]. Data protection laws, such as the *European General Data Protection Regulation* (*GDPR*) [13], therefore entail principles, such as *data minimization* (Article 5(1)(c)), and mandate *data protection by design* (Article 25).

However, this leads to what is known as the *privacy paradox*—although users crave the best possible protection of their privacy, they still do not want to refrain from using smart services, which in turn requires them to share their private data with these services [14]. Therefore, this paradox must be resolved by concealing sensitive knowledge patterns contained in the data without significantly impairing the general data quality. Privacy measures that are too restrictive would render the data unusable, while measures that are too shallow would jeopardize privacy [15]. In order for this balancing act to succeed, however, privacy measures must be tailored to the data and how they are processed.

Therefore, we investigate whether state-of-the-art privacy mechanisms for smart services are up to this task. Our research goal is to systematically analyze the state of the art in this domain. To this end, we identify the privacy threats posed by today's smart services as well as the strengths and weaknesses of the available privacy mechanisms. By comparatively reviewing these two dimensions, we elaborate which issues are already effectively covered by today's privacy mechanisms and which open research questions still need to be addressed as part of future work. With this in mind, we make the following three contributions in our paper:

1. We present modern-day smart services from seven application domains. For each of them, we analyze which data they capture, which types of processing are used to extract information from them, and which knowledge can be derived. The selected application scenarios serve to illustrate the general data processing requirements and the privacy concerns inherently associated with such smart services.
2. We discuss state-of-the-art privacy measures for the identified data types and forms of processing. Hereby, we provide an overview of the current state in terms of the protection of sensitive data when dealing with smart services.
3. We identify, based on our findings, open privacy issues in the context of smart services that need to be overcome in order to comply with the data protection by design principle.

The remainder of this paper is structured as follows: In Section 2, we analyze seven real-world application scenarios of smart services with respect to the involved data processing and highlight the opportunities offered by these smart services as well as the inherent privacy threats. Based on our findings concerning data quality requirements and potential privacy risks, we discuss appropriate state-of-the-art privacy measures and work out their strengths and weaknesses in Section 3. In Section 4, we map the opportunities and threats against the strengths and weaknesses to reflect the current state of privacy mechanisms for smart services. Based on this contrasting juxtaposition, we identify future prospects with regard to privacy issues in smart services and how to overcome them in Section 5 before concluding the paper in Section 6.

2. Analysis of Modern-Day Smart Services

In a first step, we look at seven real-world application scenarios, which require the comprehensive processing of highly sensitive data. This provides insights into the types of data that are involved and the requirements regarding data refinement. In addition, we discuss for each application scenario which opportunities are offered by it and which privacy threats it poses. The selection of the application scenarios is based on the IoT and smart service topics which are currently predominant in the literature. This includes *location-based services* (Section 2.1), which originally had a primarily military background but have long since arrived in the private sector, *health services* (Section 2.2), in which the IoT enables remote health monitoring, *voice-controlled digital assistants* (Section 2.3), which are an important pillar of any smart home, and surveillance services driven by *image analysis* (Section 2.4) [16]. Moreover, end-to-end food processing monitoring to ensure food safety is becoming increasingly common, which is enabled by IoT-supported *food analysis* (Section 2.5) [17]. However, even long-established services, such as *recommender systems* (Section 2.6) gain new capabilities due to the IoT [18]. As the available computing power is also steadily increasing, extensive analysis such as *DNA sequence classification* (Section 2.7) can also be realized with the help of the IoT [19]. In those application domains in general, there are particularly severe privacy threats, such as data leakage or data tampering [20]. We summarize the findings from the study of these seven application scenarios in Section 2.8 and identify knowledge needs to be derived in each scenario as a result of the analysis and what adverse insights can be gained from the data in the process.

2.1. Location-Based Services

Location-based services (*LBS*) are an application domain in which large amounts of data have to be processed. An LBS is a service in which the geographic location of entities is significantly tied in. In addition to tracking stationary entities, e.g., certain locations that are relevant for the service, mobile entities can also be tracked, e.g., shipments that are in delivery [21]. In addition to such non-human entities, the location data of human users are also being tracked by LBS. This is driven in particular by the increased use of smartphones (and affiliated technologies, such as smartwatches) [22]. This kind of device is not only permanently close to its user but also *always on* and *always online* [23]. In addition to a GPS receiver, which enables very accurate positioning, mobile phone tracking also facilitates sufficiently good positioning of users via their GSM cellular location [24]. While satellite-based tracking is limited to outdoor areas, indoor tracking is also possible with standard smart devices. For instance, inertial approaches use built-in accelerometers and gyroscopes to determine the location relative to a given starting point by means of the direction of movement and movement speed. Other approaches are based on the Earth's magnetic field. A built-in magnetic sensor detects the radiated fields, which can be used to determine the current position by means of triangulation [25].

Basically, three different types of location capturing can be found in LBS. The most basic method for static entities is to hardcode their position. For mobile entities, the current location can be captured as singular location information with the help of the aforementioned tracking technologies. If several such singular locations are captured in a temporal sequence, they can be used to form trajectories that describe the movement path of an entity [26]. The latter in particular requires thorough data refinement, e.g., to eliminate outliers [27] or to compensate for inaccuracies in the location information [28].

Figure 1 illustrates three data refinement steps. A problem regarding the capturing of trajectories is that they are only recorded pointwise. Single points can be distorted due to poor positioning signals, e.g., when an entity is passing urban canyons. If the recorded location deviates strongly from the actual location, noise filtering can be used to rectify the trajectories. An outlier can be detected if a point of the trajectory deviates too much from its predecessor and successor. In that case, the corrupted point can be detected, deleted and, e.g., replaced by means of interpolation. If the sampling rate is higher, i.e., if more data are available for data refinement, the results of the cleansing become more accurate [29].

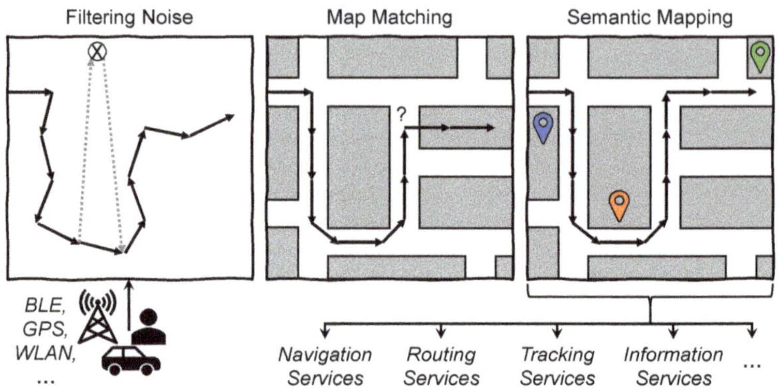

Figure 1. Examples of Data Refinement Steps for Trajectory Data.

While noise filtering can be used to correct significant deviations from single points, smaller deviations can also be corrected by adding complementary knowledge. For instance, if an entity can only move on predefined paths, e.g., a parcel on a parcel conveyor or a car on a road, the trajectory can be projected onto a model of these possible paths. In the depicted map matching, the white paths are roads, and the entity is a car. The denoised trajectory can then be smoothed, since it can be assumed that the car follows the road [30].

As can be seen, however, such a smoothing has its limitations. For instance, it is not possible to determine where the trajectory ends. It is obvious that the car cannot drive off the road. However, the distance to the road above and the one below is identical. If further semantic knowledge is added to the model, even such errors can be corrected. For instance, if some locations are known that are relevant to the entity, then it can be inferred whether it is more likely to use the upper or lower road. In the shown semantic mapping, the green marker is the entity's place of residence. Thus, it can be assumed that the captured trajectory ends there, and the refinement of the trajectory can be made accordingly [31].

Such refined data enable a variety of LBS. For instance, they enable the search for stationary objects—e.g., POI in a guide service [32]—or mobile objects—such as drivers or passengers in a ride-sharing service [33]. Smart navigation systems are also made possible by such data. These navigation systems not only take the planned route into consideration but also contextual knowledge such as congestion information provided by other entities [34]. Social networks in particular are increasingly relying on IoT data from other smart services. Here, these data are merged with social media, whereby new insights are gained. Location data play a significant role in this context, as they are not only a key factor in determining the current context of a person but also because they reveal further connections between users [35]. For instance, in location-based social networks, social groups are identified not only based on mutual friends and common interests but also on matching movement patterns [36]. Another cutting-edge use case is the development of autonomous vehicles that do not require any human intervention to find and follow their route [37]. While in the aforementioned use cases, it is only bothersome if the data refinement is inadequate and thus the tracking of entities is not accurate, in autonomous vehicles, this might endanger the physical integrity of humans. Therefore, it is mandatory to delve into the data. Yet, data processed by LBS provide insights into behavioral patterns. Whenever humans can be linked to the data (e.g., the owner of a smart device), this poses a severe privacy threat [38].

The inherent opportunities and privacy threats of LBS can be broken down as follows:

Opportunities:

- The technical ability to locate smart devices (and thus their owners) with high accuracy enables many services, such as navigation services or location-based information services.
- The location can also be used to derive a lot of additional information about the data subject, e.g., which places the data subject visits frequently.
- If this information is enriched with additional data, such as temporal aspects (How long and when does a data subject stay at a certain place?) or supplementary geographic data (What can be found at that place?), a very precise profile of the data subject can be created. This makes LBS the foundation of many other context-based services since location is a key parameter in context recognition.

Privacy Threats:

- The current location of a data subject is continuously disclosed by an LBS. This enables long-term surveillance of data subjects.
- The data refinement methods described above make it easy to correct even hardware- or software-related inaccuracies, enabling very precise location determination.
- Furthermore, LBS can be used to find out much more about a data subject than might initially appear. For instance, they can be used to determine activities and, in the case of long-term use, to draw conclusions about hobbies and social contacts.

2.2. Health Services

The healthcare sector can also benefit from comprehensive data analyses. The term *eHealth* covers all kinds of services that facilitate the treatment and long-term care of patients and involve the use of modern information technologies [39]. This brings many benefits to the table, as treatment costs can be reduced, the patients' quality of life can be improved, and the workload of physicians can be reduced [40]. To achieve this, however, it is crucial to include the patients fully in the process and overcome technical hurdles [41]. This can be achieved in particular by means of so-called *mHealth*. Here, everyday mobile devices such as smartphones are used for medical treatment [42]. Due to mHealth, there is a mobile application for virtually any health-related use case nowadays [43].

An example of such an mHealth application for people suffering from a chronic disease is an interactive questionnaire with which they can assess their condition on a daily basis. Depending on the answers, follow-up questions are asked. In this way, only essential questions can be asked in a systematic manner without overwhelming patients [44]. In addition to a local evaluation to determine the appropriate question catalog, the results are also forwarded to a backend for further analysis. If a critical situation is detected based on the answers, the patient is advised to see a physician [45]. The physician receives a summary of the analyzed questionnaires, which facilitates his or her daily work and allows him or her to focus on emergency cases [46].

Such applications are especially useful in more rural areas or developing countries where people have no immediate access to a physician. Thus, because of the low-cost hardware required, mHealth can ensure that people in such areas still receive healthcare [47]. Today's smart devices, however, enable far more powerful heath services due to their built-in sensors and connectivity [48]. The so-called *quantified self movement* motivates people to use self-tracking to capture vast amounts of health-related data about themselves, such as blood pressure information, on a permanent basis [49]. While ordinary smart devices come with many sensors that support this kind of self-tracking, the IoT opens up further capabilities [50]. For instance, a smartphone can be turned into a health data hub to which all sorts of IoT-enabled specialized heath meters send their measurement data for storage and processing [51]. As a result, mHealth approaches provide not only a sufficient amount of data but also useful data for more comprehensive medical analyses [52].

In addition to simple questionnaires, the self-assessment of patients can be supported by such health-related measurements. In the case of diabetics, a continuous glucose monitoring device can be coupled with a smartphone to track the blood glucose level in a mobile application. Moreover, external circumstances that might influence the measurements (e.g., stress factors such as noise) can also be captured via the smartphone's built-in sensors. In this way, analyses enable a 360-degree health view on the patients [53]. As these data are tracked continuously, more complex analyses of long-term trends are also feasible [54].

Figure 2 shows the potential of such analyses. First, *descriptive analytics* can be applied to the large amounts of historical data collected about a patient. By looking at a health value over time, it is possible to understand what exactly has happened, e.g., how a patient responded to taking a particular pharmaceutical, and use this knowledge to adjust the medication. Yet, as the data are captured in real time, the current situation can also be monitored by means of *real-time analytics*. For instance, this knowledge can be used to operate a smart insulin pump that supplies a patient with the appropriate amount of insulin based on the current situation. In addition, a model can be trained using historical data, which enables *predictive analytics*. Predicting how a health value will change in the future enables counteracting an adverse trend at an early stage [55]. The potential of such services is therefore virtually unlimited, including education, diagnostics, and treatment [56].

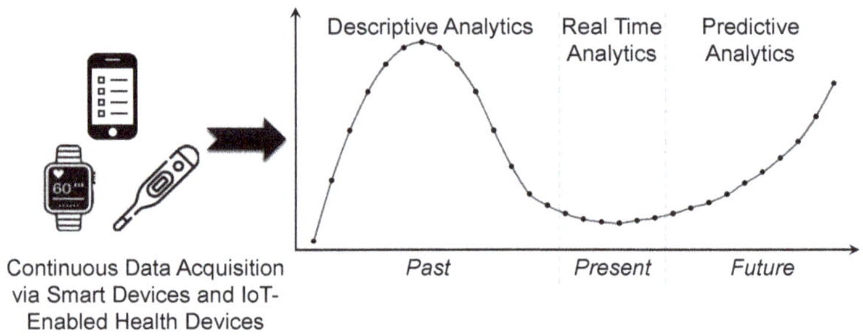

Figure 2. Three Types of Time-Series Analyses Applied in Health Services.

In such an application scenario, the accuracy of the data is key, as it affects the health of patients. Thus, the data must be refined assiduously in order to achieve the highest possible accuracy and eliminate measurement errors [57]. However, such services also raise privacy concerns, since in addition to the health data, which intrinsically contain a vast amount of sensitive information, a vast amount of knowledge about, e.g., lifestyle, environment, and social life is disclosed [58]. As smart devices have become an integral part of our lives, users no longer even notice their presence. Sensors are embedded in everyday objects and enable a comprehensive and almost invisible permanent monitoring of everyone—both the users of the smart devices or accidental bystanders [59]. For this reason, particularly high privacy standards must be applied in such an application scenario [60].

The inherent opportunities and privacy threats of smart health services can be broken down as follows:

Opportunities:

- IoT technologies enable the self-quantification of health-related values, which means that frequently recurring health checks in particular can be performed at home by the patients themselves. This relieves the burden on both patients and physicians.
- The non-intrusive nature of the smart devices allows the permanent monitoring of patients without disturbing their daily routine. This enhances safety, as no health measurements can be forgotten, and health problems can be detected at an early stage.

- Since the smart devices that feed smart health services with data are ubiquitous and capable of capturing a variety of health values, they can be used to provide 360-degree health views on the patient.

Privacy Threats:

- Health data are among the most sensitive data, so the large-scale collection and processing in itself is a privacy threat.
- As smart devices are ubiquitous, data subjects are no longer aware that health data are collected permanently, which makes them unaware of the privacy threat.
- In addition to inferences about diseases, the collected health data also allow insights into other aspects, such as unhealthy behaviors, e.g., whether the data subject is a smoker or carries out little physical activities.

2.3. Voice-Controlled Digital Assistants

In addition, the increasingly popular *Voice-Controlled Digital Assistants* (*VDA*) depend on comprehensive data refinement and subsequent data processing. VDA refers to any hardware and software infrastructure that enables users to request information or give instructions by means of human speech. In addition to IoT devices that serve as microphones and speakers, a software agent is required that permanently listens for a predefined keyword. As soon as this keyword is received, the agent wakes up and records everything that is subsequently spoken. A speech analysis is then performed on the recordings, to interpret the spoken commands semantically and process them logically. Depending on the command, either speech synthesis is used to formulate a response or a machine-processable command is sent to a corresponding IoT-capable device [61]. The most popular VDA are *Alexa* (see https://alexa.amazon.com/, accessed on 30 August 2022), *Siri* (see https://www.apple.com/siri/, accessed on 30 August 2022), *Cortana* (see https://www.microsoft.com/en-us/cortana/, accessed on 30 August 2022), and *Google Assistant* (see https://assistant.google.com/, accessed on 30 August 2022) [62].

Such VDA offer users a variety of benefits: First, the easy accessibility via the voice interface provides a convenient way to give tasks to the VDA. Second, this type of interaction also satisfies a hedonistic desire, as the user has command over a (virtual) assistant that can perform any assigned task. This is further amplified by the fact that the possession of such "future technology" also has a symbolic value, since its usage has a favorable impression on others. Third, VDA also have a positive social aspect, as natural language communication with them overcomes many impediments regarding the perception of the technology—the VDA is rather perceived as a person than a technical object [63]. Due to this variety of benefits, it is therefore not surprising that the popularity and demand for such VDA is growing constantly. This is further enhanced by the fact that more and more IoT devices are compatible with VDA, i.e., everyday objects can be used as input device and can be operated by a VDA. So, the VDA is virtually available at any time and any place [64].

The schematic architecture of the VDA ecosystem is outlined in Figure 3. A compatible input client with a microphone is required on the user side. The software agent of the respective VDA service runs on this client. This can be a conventional personal computer with Internet access, a smart device such as a smartphone, or a dedicated VDA-enabled device such as *Amazon Echo* (see https://www.amazon.com/smart-home-devices/b?node=9818047011, accessed on 30 August 2022). On this client itself, however, neither the voice processing nor the processing of the actual request is carried out. The client only recognizes the specified keyword and records the subsequent instructions. It sends the recording to its processing backend. There, voice processing first retrieves and interprets the request [65]. The request is then processed by means of machine learning approaches. Feedback loops are included so that if the user labels a response incorrect or unsatisfactory, the respecting request is used to train the models further and thereby refine them continuously. To put it simply, with every mistake the VDA makes, the system becomes smarter. IoT-enabled output devices are needed to render the results, e.g., smart speakers, a smartphone application, or compatible third-party devices such as smart light bulbs. To trigger these devices, a VDA backend has various adapters, e.g., an adapter to generate natural language to answer questions verbally or adapters by third-party vendors to control their IoT devices [66].

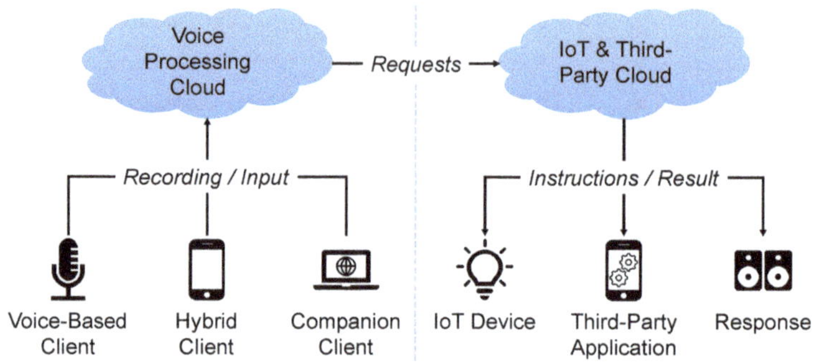

Figure 3. Ecosystem and Architecture of a Voice-Controlled Digital Assistant.

No matter which service is requested by a user, however, an important aspect of VDA is that the misinterpretation of his or her commands is minimal. If this is not achieved, a breach of immersion occurs and user acceptance decreases [67]. To this end, it is imperative that in the backend, models are trained based on many inputs from different users. Furthermore, the incoming requests must be thoroughly refined in order to be able to successfully interpret what is being said. In addition to monitoring the quality of the VDA, a feedback loop is required that involves human operators listening to selected recordings [68].

It is obvious that VDA also raise privacy concerns. The VDA providers are able to identify all of their users. Thus, they are able to link the recordings to the users. Taking this into account, the fact that VDA are able to wiretap users permanently is a particularly disturbing concern. Users do not know what data are actually captured and forwarded for processing. Even third parties can gain access to the data, e.g., for quality control or to provide an affiliated service. So, the privacy concerns regarding VDA are reasonable [69].

The inherent opportunities and privacy threats of voice-controlled digital assistants can be broken down as follows:

Opportunities:

- VDA allow voice-based control of smart devices, making them particularly helpful, e.g., for people with motor disorders.
- The natural language approach of VDA reduces the technical hurdle for people who are less tech-savvy.
- The knowledge that a VDA (theoretically) has at its disposal is almost unlimited. That is, VDA can be used to access required information quickly and easily in almost any situation in life.

Privacy Threats:

- Since a VDA waits for its specific keyword, it is never completely off. That is, all conversations are permanently recorded.
- If a VDA is activated using the keyword and the voice recording is forwarded for processing, it is not possible for data subjects to trace who has access to it.
- For third parties, a VDA is indistinguishable from conventional loudspeakers. Therefore, they are completely unaware that their conversations are also being recorded.

2.4. Image Analysis

There are also many benefits offered by computer-aided image analysis. While the human eye has physical limitations in terms of detail perceptibility, the capabilities of computational image processing are virtually limitless. Therefore, such data processing techniques are applied in a variety of application domains today. For instance, in the food industry, computer vision can be used to reliably detect foreign objects in food products, medical diagnoses can be supported by the automatic interpretation of computed tomography images, and in the context of defense and homeland security, suspects can be recognized on video recordings [70].

In particular in the field of face recognition, remarkable progress has been made in recent years. While in the early days of this research area in 1964, only a few characteristics of the face could be recognized on images, today, the techniques are so reliable and accurate that they are used in a variety of commercial, industrial, legal, and governmental applications [71]. End-users also come into contact with face recognition services, e.g., in *online social networks* (*OSNs*) such as *Facebook* (see https://www.facebook.com/, accessed on 30 August 2022) [72]. Here, the so-called *DeepFace* algorithm is used to identify users on images shared in the network. This algorithm is almost as good as human identification, except that, unlike humans, DeepFace can process large amounts of data in next to no time [73].

Figure 4 illustrates the workflow of face recognition. Initially, all faces on an image must be detected. For instance, characteristic facial features can be identified such as skin color which contrasts with the background, or a model can be trained to detect human faces [74]. Having detected the face, the image needs to be preprocessed, e.g., it has to be cropped to the relevant part of the face. Furthermore, facial landmarks such as the eyes, mouth, and nose have to be marked [75]. For the actual recognition, a comparison with a face database is conducted. To this end, a multinomial classifier is trained using the face database. This classifier can determine with which person from the face database the person on the image has the best match [76]. Various machine learning techniques can be used for this purpose. While in the early days primarily *linear discriminant analysis* was used, today, *support vector machines* and *convolutional neural networks* are used [77].

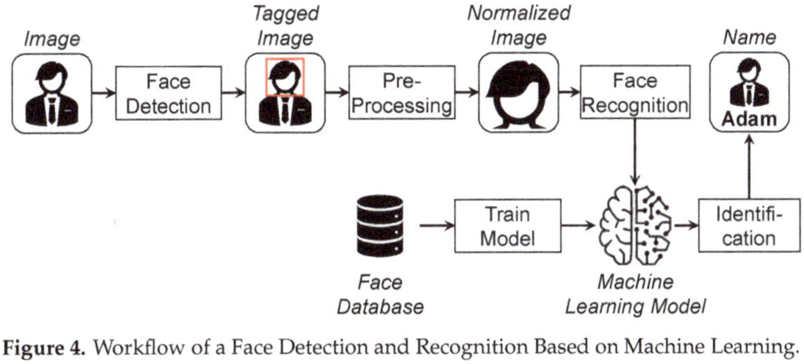

Figure 4. Workflow of a Face Detection and Recognition Based on Machine Learning.

The application of such image analysis techniques naturally also raises privacy concerns. With today's smartphones, IoT-enabled cameras have become ubiquitous in our everyday lives. Moreover, considering that smartphones can tag captured photos with a time code as well as location information, image analysis can be used to determine not only which people are in a picture together but also when and where that meeting has happened. While this may be intended by the main data subject of the photo, it also affects any bystanders. Due to this potential exposure, measures are needed to restrict such privacy-intrusive scenarios [78]. An additional problem associated with image analysis is the bias in data selection. Such a bias leads to inadequate models, which cause errors in face recognition, resulting in users being incorrectly linked to a photo. Thus, it is important to take measures to prevent bias when preparing the data corpus for training the models [79].

The inherent opportunities and privacy threats of image analysis can be broken down as follows:

Opportunities:

- As social media becomes more and more prevalent in people's lives, image analysis is becoming increasingly relevant for them as well. This allows people to be identified and tagged in images, enabling the automatic linking of people with their social contacts as well as with places and activities.
- Comprehensive image analysis enables novel search functionalities, e.g., if users want to find all images of themselves (or other users) that are available in a social network.
- Image analysis is also a key factor in law enforcement and security today, as it can be used to identify suspects rapidly in video recordings.

Privacy Threats:

- When an image is analyzed, locations or activities can be identified in addition to the people depicted in it. By linking this information, a lot of knowledge about the data subject can be derived. Furthermore, by combining all available images, a comprehensive insight into the lives of the depicted persons can be gained.
- The algorithms are subject to certain probabilities of error. If people are incorrectly identified, they may be assigned to the wrong circles of acquaintances or interests, unnoticed by the data subject, which in the worst case can have damaging consequences for their reputation.
- Third parties can also be tagged on images without their knowledge, allowing the algorithm to learn their faces. As a result, they can also be identified in pictures, which means that parts of their private lives are revealed completely without their knowledge.

2.5. Food Analysis

The aim of food chemistry is to investigate and research all biological and non-biological components of foods. This research discipline is fully committed to consumer protection and is intended to ensure the safety of all food products. To this end, chemical analyses are used to determine, e.g., whether sufficient quantities of a valuable ingredient are included in a food product or whether food producers comply with the maximum permitted quantities of unhealthy substances [80]. Here, the use of biosensors can contribute significantly to food safety, as contaminations or toxins in food can be reliably detected [81]. For this purpose, the vast number of proteins in a food sample must be split into peptides in order to detect even the smallest traces of substances that can be unsafe to consume. However, this results in a significantly larger number of analytes. Special tools and techniques are therefore needed to handle this huge amount of data [82].

In simple terms, it is necessary to search for specific patterns in the source material [83]. To cope with the vast amount of analysis data, machine learning is becoming increasingly popular. This enables a much more thorough analysis than a manual analysis. Models are trained from examples. These models generalize the data from the given samples. For new samples, it must be determined to which of the training samples the new sample has the best match. This allows one to analyze even very large datasets, such as complete DNA sequences, for contained patterns [84].

An analysis process in the food chemistry domain is shown in Figure 5. A food sample is analyzed to determine whether it contains certain allergens. Such allergens must be explicitly declared by food producers to protect consumers who have a certain food allergy. In this example, a food chemist determines whether the sample contains glutamate. Yet, the same procedure applies to other allergens as well, such as gluten, crustaceans, or nut seeds. For this purpose, the sample is analyzed with a mass spectrometer to determine the mass-to-charge ratio of all peptides present in a sample. For the separation of contained peptides, liquid chromatography is used. From a proteomics and genomics database for foods, e.g., *UniProt* (see https://www.uniprot.org/, accessed on 30 August 2022), comparison samples of the allergens of interest can be obtained. Then, it can be cross-checked whether sufficiently similar patterns can be recognized in the peptides obtained from the food sample [85]. For this comparison and all subsequent analyses, however, a thorough refinement of the sample data is necessary, e.g., to filter out inaccurate or irrelevant data to be able to focus on the relevant aspects in the data, only [86]. This way, it is possible to detect whether allergens are present in a food sample in any decomposing stage [87].

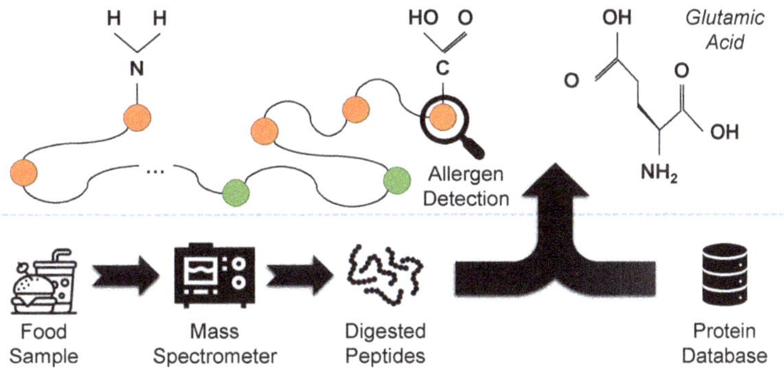

Figure 5. Schematic Representation of Food Analysis to Detect Allergens in a Food Sample.

Such a pattern recognition in a data stream, as it is generated by the measurement devices during the analysis, is also known from the domain of *complex event processing* (*CEP*). A pattern consists of individual facts—in the context of food analysis for instance peptides—which are in a given sequential context [88]. The patterns can be defined by means of pattern templates. These templates provide a level of abstraction to the actual data, as they specify the target sequences based on pairs of measurement values and constraints on those values. A CEP engine then searches for these patterns in the data stream [89]. In the context of food analysis, however, it is not sufficient to find only exact matches but also reasonably good matches. For instance, post-translational protein changes can cause deviations from the entries in the protein sequence database. To this end, data mining techniques such as *k*-nearest neighbor analysis can be used to find similarity matches [90].

It is obvious that privacy does not have to be observed in this application scenario, as the food samples do not have any privacy rights. However, with the help of these analytical methods, it is possible not only to detect allergens in food products but also to determine which ingredients are present in the product. That is, deep insights into the product and manufacturing process can be gained with today's sensors and data processing technology [91]. As a result, it is possible to identify a manufacturer's secret ingredients, which represents a competitive advantage in the food market. Therefore, maintaining confidential business information is a necessary tool in the food chemical sector [92].

The inherent opportunities and privacy threats of IoT-supported food analysis can be broken down as follows:

Opportunities:

- With IoT-supported food analysis, food samples can be analyzed much more efficiently and effectively.
- Due to the increasing number of people suffering from a food allergy, it is important that food products are correctly labeled and that this labeling is also thoroughly verifiable.

- Due to a predominantly automated processing of food products, a thorough inspection of these products is required in order to detect any foreign substances or contaminants at an early stage.

Privacy Threats:

- In this application scenario, there are no privacy threats, but there are confidentiality threats, as food analysis can provide deep insights into the food product, revealing specific ingredients or preparation methods, possibly leading to a loss of competitive advantage.

2.6. Recommender Systems

With the rise of services such as *YouTube* (see https://www.youtube.com/, accessed on 30 August 2022) or large online shopping platforms such as *Amazon* (see https://www.amazon.com/, accessed on 30 August 2022), we have entered an age of oversupply. Users cannot obtain a complete overview of all available items (e.g., video clips or products). Therefore, in all these services and shopping platforms, a mechanism is needed to provide users with information about the items that are particularly relevant to them. Recommender systems represent such a mechanism. In simple terms, a recommender system is a tool that predicts a user's interests based on his or her previous interactions with the service or shopping platform. Such interactions can be, e.g., watching a video clip or buying a product. In this way, information overload can be minimized by presenting a user with only the most relevant information tailored to his or her needs [93]. Although e-resource services such as YouTube or e-commerce platforms such as Amazon are the most prominent uses of recommender systems, such systems have become indispensable in almost all modern e-services [94].

However, if recommender systems would base their suggestions only on the previous interaction of the user in question, no new items could be recommended to him or her. The advantage of these systems is that they not only have knowledge about a single user but can also draw on the interactions of a large number of users. By means of so-called *market basket analysis*, they can determine which combinations of items users frequently interact with, e.g., which video clips a user has on his or her watchlist or which products a user buys at the same time. From such item lists, *association rules* can be derived that describe which item combinations are encountered regularly. If a user is interested in a subset of the items of an association rule, the remaining items can be recommended to him or her [95].

In Figure 6, it is shown how a recommender system can proceed to be able to make tailored recommendations. Basically, there are two different approaches to this end: *content-based filtering* and *collaborative filtering*. In content-based filtering, only the items are considered. For this purpose, similar items are clustered. The clusters are homogeneous within themselves—i.e., the items of a cluster are mutually highly similar with regard to the relevant properties—and heterogeneous to each other—i.e., the items of different clusters are as different as possible with regard to the relevant properties. If a user interacts with an item (or has interacted with it in the past), the remaining items of the same cluster can be recommended to the user, since it is likely that they are also of interest to him or her [96].

Yet, this approach completely neglects social aspects. By including knowledge about the users themselves as well as social relationships between them, considerably better recommendations can be made [97]. In collaborative filtering, therefore, users are also clustered, for instance, based on common interests. This has two effects: On the one hand, the recommendation set can be reduced significantly. It is no longer necessary to recommend all similar items but only those that were also popular among other users from the same cluster. On the other hand, new recommendations can also include distinct items if other users from the cluster have interacted with them [98]. In addition to these two main types of recommender systems, however, there is a variety of hybrid approaches that mix aspects of content-based filtering and collaborative filtering [99].

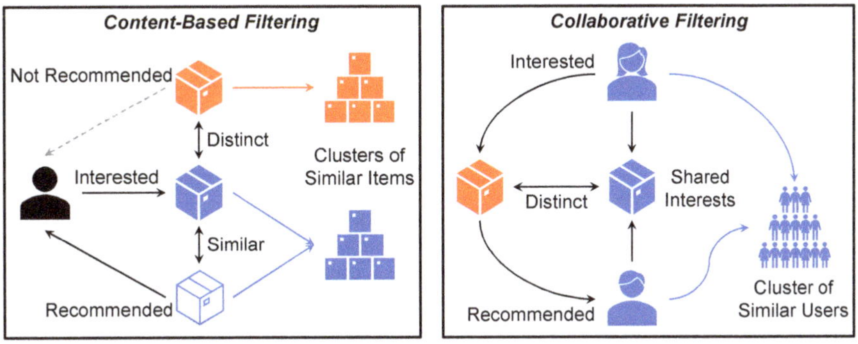

Figure 6. Two Fundamentally Different Modes of Operation of Recommender Systems.

There are two issues in this regard: First, third parties, e.g., content creators or manufacturers of a product, might try to increase the relevance of their items by submitting fake data to be prioritized in the recommendations. So, raw data always have to be refined to purge such misrepresentations. Second, recommender systems rely on knowledge about the community. Only if large amounts of information about users' preferences and usage behavior are available can meaningful recommendations be made. Even if an individual's opinion cannot be inferred directly from the recommendations, such information is present in the base data. Moreover, statistical inferences can also be derived from the processed and accumulated data, which reveal information about the preferences of individuals. Therefore, privacy issues have to be addressed in recommender systems as well [100].

The inherent opportunities and privacy threats of IoT-supported recommender systems can be broken down as follows:

Opportunities:

- IoT-supported recommender systems are able to provide search results that are tailored to the user (e.g., product recommendations) based on contextual information.
- With the help of collaborative filtering, users can also be presented with completely new recommendations, which can expand their horizons, as they were previously unaware that they might be interested in the suggested items.
- Searches become much more efficient, as irrelevant items can be excluded early on, and more effective, as relevant items can be suggested even if they were not directly included in the search query.

Privacy Threats:

- The recommender systems have to collect and analyze a lot of data about a user's interests in order to make suitable suggestions. As a result, they also gain privacy-relevant insights into the life of the data subject.
- In collaborative filtering, the data of several users are combined, and profiles are created, which can be used to derive additional information about a data subject. For instance, knowledge about a data subject can be transferred to the other data subjects in the same cluster with a certain probability.
- A recommender system can also deliberately influence users by making one-sided recommendations.

2.7. DNA Sequence Classification

The analysis of human DNA samples has many useful applications. While in the early days, DNA analysis was mainly used in forensics to uniquely identify individuals—via the so-called *genetic fingerprint*—other areas of application have emerged over time as analytical methods have become more advanced. For instance, it can be used to establish paternity in custody and child support litigation or to trace ancestry over hundreds of thousands of years to study population genetics. Furthermore, DNA analysis can also be used in the medical domain to diagnose inherited disorders and human diseases [101].

Epigenetic modifications, such as DNA methylation, can be used to detect common diseases. For instance, aberrant DNA methylation of imprinted loci can often be observed in connection with certain types of cancer. In addition, autoimmune diseases, metabolic disorders, and neurological disorders are closely related to DNA methylation. Medicine could therefore make significant progress if a deeper understanding of epigenetic mechanisms leading to these diseases is available [102]. Although insights into the mechanisms of DNA demethylation have already been gained, there are still many correlations that have been observed but not yet systematically studied. To this end, however, extensive analyses are required to comprehend all relevant factors [103].

Instead of striving to explain the complex correlations, a reliable classification is already sufficient for a medical diagnosis. For instance, normal and malignant tissue samples from human liver and lung can be studied in order to be able to classify future tissue samples into "normal" and "malignant" based on these findings [104]. Instead of an analysis by a medical expert, artificial intelligence techniques can be used for DNA sequence classification. For this purpose, a large number of labeled samples are collected and used to train a model. This model recognizes the relevant genome sequences that are indicative of a particular disease [105].

Figure 7 is a simplified illustration of how this can be completed. First, suitable target data are selected. Preprocessing steps are applied to these data to structure them and prune errors or missing values. Then, the features relevant for the analysis are extracted. In the given example, sections of the DNA helix of the samples, i.e., sequences of base pairs. With these data, a *convolutional neural network* (*CNN*) is trained. In a CNN, there are multiple hidden layers in addition to an input and output layer. During the training phase, weights are allotted to describe how much certain nodes from one layer contribute to the inputs of the subsequent layer (and thus, eventually, to the final outcome). This allows one to classify new unlabeled data. The advantage of a CNN is that pooling allows one to discard redundant information early on, whereby even large amounts of data can be processed efficiently. This makes it also more robust against overfitting, i.e., it is possible to detect variants of a disease. Furthermore, a CNN can subsequently adjust itself to respond to shifts in circumstances [106].

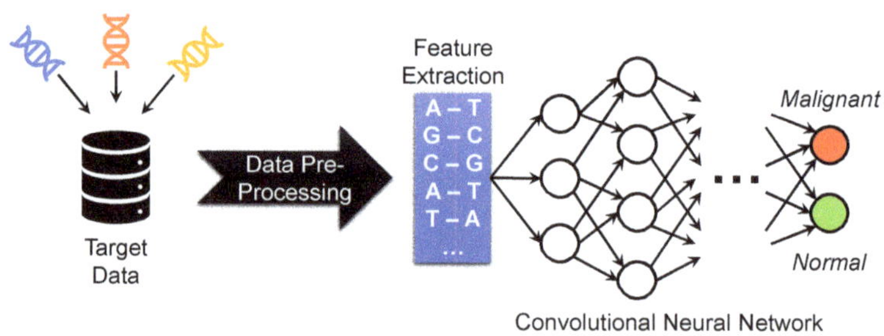

Figure 7. Application of a Convolutional Neural Network for the Classification of DNA Samples.

However, the training of such a CNN requires the establishment of large repositories of annotated genomic data. Although this may represent a key element for future discoveries in human disease, it inevitably also raises a variety of privacy concerns. Even if the metadata of the samples are fully anonymized, the DNA itself represents a unique fingerprint. All knowledge gained from a sample—e.g., ancestry, parenthood, pre-existing conditions, and genetic disorders—can therefore be uniquely traced back to a data subject [107]. Another inherent issue with CNN is its black-box approach. The reason for a certain decision cannot be explained even by domain experts [108]. The layers of a CNN consist of hundreds of millions of parameters that are totally detached from the real-world problem that is modeled by the CNN. Thus, it is not possible to trace which data have influenced the model in what way and therefore were decisive for a certain classification [109].

The inherent opportunities and privacy threats of IoT-supported DNA sequence classification can be broken down as follows:

Opportunities:

- IoT-supported DNA sequence classification enables comprehensive automatic detection of diseases, for instance.
- By using CNN, automatic adaptations to data shifts are facilitated.
- A CNN can learn novel correlations in the DNA structures.

Privacy Threats:

- Training a CNN requires a very large DNA pool (i.e., highly sensitive data). DNA is a unique fingerprint, which means that the collected samples can always be linked to a person.
- Through the DNA analysis as well as the comparison with other samples, additional correlations can be identified (e.g., relatives or hereditary diseases), which reveal a lot of private information.
- The CNN itself or the decisions made by it cannot be explained. Decision making is therefore entirely based on full and blind trust in the CNN.

2.8. Synopsis

The seven application scenarios discussed above serve only to illustrate the general data processing requirements and the privacy or confidentiality concerns involved in big data analytics. The scenarios are representative examples of the most relevant kinds of data and processing types required in today's smart services. In Table 1, the main insights in this regard are summarized. The following section discusses in more detail how to address each of the identified privacy and confidentiality concerns by technical means.

Table 1. Synopsis of the Main Findings Regarding the Processing of Data in Smart Services.

Application Scenario	Required Data Processing	Privacy or Confidentiality Concerns
Location-Based Services	In addition to discrete location information, movement trajectories must be analyzed.	A lot of knowledge can be derived from frequent whereabouts, e.g., place of residence, workplace, interests, and even social contacts.
Health Services	In addition to individual measured values, in particular, temporal progressions of health data must be analyzed.	Health data are particularly sensitive as they reveal not only information about the health condition but also about the lifestyle.
Voice-Controlled Digital Assistants	The recordings must be analyzed to interpret the verbal commands.	The continuous recording enables exhaustive spying on users.
Image Analysis	The contents of the images must be analyzed in order to identify the shown objects.	By identifying the portrayed individuals, it is possible to reconstruct who was where and when with whom; even bystanders can be exposed.

Table 1. *Cont.*

Application Scenario	Required Data Processing	Privacy or Confidentiality Concerns
Food Analysis	Patterns indicating, e.g., allergens must be detected in food samples.	Other patterns reveal secret ingredients, thereby disclosing trade secrets.
Recommender Systems	Large amounts of data from many individuals must be analyzed to make appropriate recommendations.	Although the trained models do not disclose information about individuals, the underlying data do.
DNA Sequence Classification	Neural networks have to be trained based on a comprehensive DNA database to detect new correlations.	DNA data contain sensitive information; hence, third parties must not have full access to the complete dataset.

3. State-of-the-Art Privacy Measures

In this section, we study which technical measures can be used to conceal information in order to comply with privacy requirements. In this context, it is important to reduce the information content in line with processing requirements, i.e., the knowledge required by a data consumer must still be derivable from the data in an adequate quality. If this is not feasible, the technical measures are not applicable in practice—this would be equivalent to withholding the data.

In this context, it is first of all necessary to understand which data must be protected at all in order to be able to apply the privacy measures in a target-oriented manner and to minimize the impact on data quality. The GDPR stipulates that only *personal data* have to be protected. These are data that can be unambiguously traced to a date subject (Article 4(1)). Without such a linkage, no special privacy measures are required. However, most smart services across all domains require an authentication of their users. For instance, it must be possible to link measured values to the correct user, or it must be ensured that a user is authorized to execute certain commands. As part of this authentication, users are identified. This can be avoided by outsourcing the authentication to a trusted third party. This results in a segregation of the identification data and the payload data. The trusted third party only forwards pseudonymized subsets of the identifying attributes to the smart service provider [110]. As a result, the smart service provider cannot directly link the data to a natural person but only to a pseudonymous entity. Yet, if one has access to both the identification data and the payload data, these data can be joined again via the pseudonymized references. From the perspective of the GDPR, these are therefore identifiable entities, i.e., the personal data can be indirectly linked to a natural person. Such data must be protected in the same way as data which can be directly linked to a natural person. Thus, privacy measures are required by law for almost all smart services. However, even with (apparently) fully anonymized data, there is a risk to the privacy of users, which is why additional privacy measures should be taken in any case [111].

Generally speaking, there are three approaches to share data in a privacy-friendly way. First, the base data can be minimized by removing entire data items based on certain properties prior to processing [112]. For instance, in the health care sector, a dataset of patient records can be reduced by removing all medical records of a certain patient in order to preserve the privacy of that patient. In formalized terms, this is the application of a *selection operator* σ known from relational algebra. It is defined as:

$$\sigma_{\varphi}(R) = \{t : t \in R, \varphi(t)\} \tag{1}$$

Let R be a relation and φ be a propositional formula describing which tuples are to be included in the result set, i.e., the tuples t_i for which $\varphi(t_i)$ evaluates to TRUE.

Second, certain attributes of the data items can be excluded from processing [113]. In our example, e.g., the age can be concealed in each patient record. In formalized terms, this is the application of a *projection operator* Π known from relational algebra. It is defined as:

$$\Pi_{a_1,\ldots,a_n}(R) = \{t[a_1,\ldots,a_n] : t \in R\} \tag{2}$$

Let R again be a relation and $a_1,\ldots,a_x$ be the attributes in it. Then, let $t[a_1,\ldots,a_n]$ be the restriction of the tuple t to the first n attributes.

Third, data can be condensed prior to processing [114]. In our example, e.g., not every medical record is shared, but only aggregations of selected values. For instance, the records can first be grouped by diseases, and then, the average age of the patients for each disease can be shared. In formalized terms, this is the application of an *aggregate operator* $\mathcal{G}$ known from relational algebra. It is written as follows:

$$_{G_1,G_2,\ldots,G_m}\mathcal{G}_{f_1(A_1),f_2(A_2),\ldots,f_n(A_n)}(R) \tag{3}$$

Let R again be a relation. Then, let $G_1, G_2, \ldots, G_m$ be the attributes in R to group by, while each f_i is an aggregate function applied to the attribute A_i of the relation schema, e.g., SUM, COUNT, AVERAGE, MAXIMUM, or MINIMUM.

Figure 8 graphically illustrates the functional principle of these three operators.

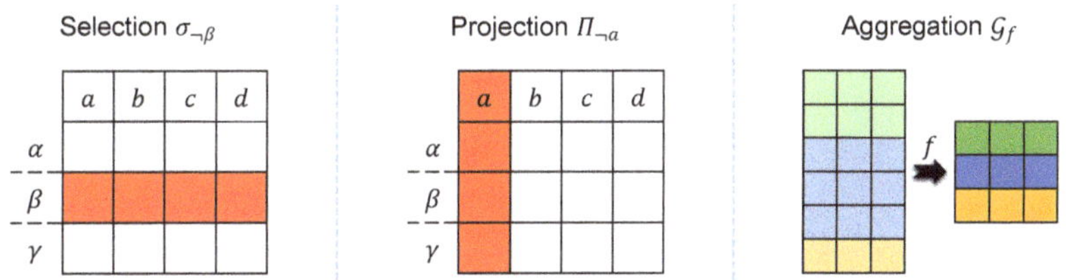

Figure 8. Graphical Illustration of the Effect of a Selection, Projection, and Aggregation on a Dataset.

The undeniable advantage of these three generic filtering techniques is that they can be applied to any type of data to preserve privacy. However, they are highly restrictive. Since entire data items or attributes are excluded from the processing, the amount of data—and thus the effective data quality—is greatly impaired. Yet, today's data refinement methods for deriving knowledge require large amounts of data to be effective. Therefore, perturbing the data typically leads to better results without compromising privacy. The goal is to tamper with specific aspects of the data only while ensuring that the data as a whole remain useful for data consumers [115]. However, there is no one-size-fits-all approach in this regard. Rather, specialized methods are needed that are tailored to the respective type of processing as well as the type of data involved. Only then can sensitive aspects of the data be concealed in a fine-grained manner without limiting the overall data quality [116].

In the following, we therefore discuss four filtering techniques in Sections 3.1–3.4 that are tailored to specific types of data: namely, location data, time series data, audio data, and image data. Section 3.5 then deals with a privacy approach that reorders the data in order to conceal certain patterns. In Section 3.6, statistical methods for privacy protection are addressed. Finally, Section 3.7 outlines how distributed processing can contribute to privacy. Due to the large number of research activities in the field of privacy, only a few representatives can be covered in each category. These seven privacy strategies are mapped to the application scenarios introduced in the corresponding subsections of Section 2. For each privacy strategies, we give a summary of its key strengths and weaknesses at the end of each subsection. To recap our review of the related work, Section 3.8 highlights the key findings.

3.1. Location Privacy

When it comes to protecting location data, two types of use cases must be distinguished: the protection of isolated snapshots and the protection of continuously captured trajectories [117]. The most straightforward approach to conceal the location is to mask it with a random location or to add an arbitrary offset to the actual location [118]. In this approach, however, one of these fake locations might be implausible, e.g., a location in the middle of the ocean. To avoid this and still keep the genuine location private, fixed predefined locations can be used instead of random generated ones, for which it has been ensured that they appear plausible in the context of the data subject [119]. Other approaches combine these two strategies by generating dummy locations following specific rules. In this way, realistic dummy locations can be obtained [120]. It is also possible to hide the real location among many fake ones. For this, a large number of generated dummy locations are shared in addition to the real one. This way, third parties cannot tell which of the location data actually belong to the respective data subject. When using an LBS, it has only to be ensured that all responses for the dummy locations are filtered out [121].

As shown in Section 2.1, such single fake locations would be cleansed during the data refinement process if contiguous trajectories are analyzed. Therefore, other measures must be taken when dealing with trajectories. *Spatial cloaking* takes advantage of the fact that continuous trajectories are composed of discrete locations. A circular cloaking area is set at each of these vertices. Instead of sharing the actual locations, the corresponding cloaking areas are shared. This way, third parties only obtain a rough idea of the path a data subject has traversed. The radius of the cloaking areas determines how accurate a localization can be [122]. The *path confusion* pursues a similar strategy as the dummy locations. Here, however, entire dummy trajectories are generated. In simple terms, phantom paths of non-existent data subjects are made available for data processing among which the actual user trajectories are hidden [123]. Sometimes, the trajectories do not contain any confidential knowledge but rather their temporal correlations. If a data subject travels from A to C via B, this might not be meaningful initially. However, if the temporal relationship is considered, it can be seen whether B was merely passed or whether the data subject stayed there for a longer period of time. Likewise, the temporal component of the trajectories can be used to determine whether two data subjects frequently travel together. To conceal such private information, *temporal cloaking* can be used. Here, the time stamps of the vertices are altered. Thereby, the traveled path of a data subject can still be traced but not when the data subject was there or for how long [124].

These three approaches to conceal trajectories are shown in Figure 9. The examples used to illustrate the privacy techniques in the figure are aligned with Section 2.1. Furthermore, combinations of these approaches are also feasible, e.g., *spatio-temporal cloaking*, in which both the spatial and temporal dimensions of the vertices of a trajectory are perturbed [125]. In addition, there are many other approaches dedicated to a specific task in the context of location-based services. For instance, there are approaches that address the privacy proximity test problem, i.e., how to determine whether an instance is in spatial proximity to another instance (e.g., two users or a user and a POI) without revealing the exact location of either instance [126].

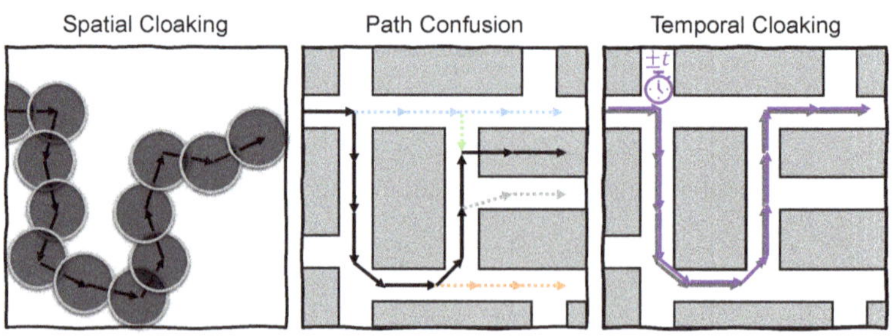

Figure 9. Visualization of Spatial Cloaking, Path Confusion, and Temporal Cloaking.

The key strengths and weaknesses of location privacy approaches are the following:

Strengths:

- There are special techniques that allow to conceal single locations as well as whole trajectories or temporal sequences of trajectories.
- The data quality of the other aspects can be largely maintained.

Weaknesses:

- The techniques are subject to many restrictions regarding the credibility of certain location information or trajectories (e.g., a person will most likely not be in the middle of the ocean), which limits their scope of action.
- Due to the sensor technology available in smart devices, more data sources are available to draw conclusions about location. This makes it easy to debunk a dummy location or a dummy trajectory.

3.2. Privacy-Preserving Time-Series Data

As outlined in Section 2.2, the processing of time-series data is always about identifying temporal patterns and forecasting trends. Privacy-preserving measures for these data must therefore not prevent such processing techniques. Rather, the aim is to conceal particular values or short sections of the time series that contain a particularly high level of sensitive information. This can be either rule-based (e.g., all values above or below a certain threshold) or time-based (e.g., all data recorded within a certain time window). There are basically two strategies to achieve this. On the one hand, the amount of data can be reduced in order to reveal less knowledge. On the other hand, the amount of data can be amplified by additional fake data in order to hide the actual values [127].

Figure 10 shows these two opposing strategies. To achieve data reduction, single values can be deleted from the series if only a few and widely distributed data points are concerned. The resulting gaps can be filled by *interpolation* so that the progression is still coherent [128]. To close gaps in time series where interpolation reaches its limits, e.g., due to the complexity of the progression or as the gaps are too large, machine learning techniques can be used to logically fill the missing parts [129].

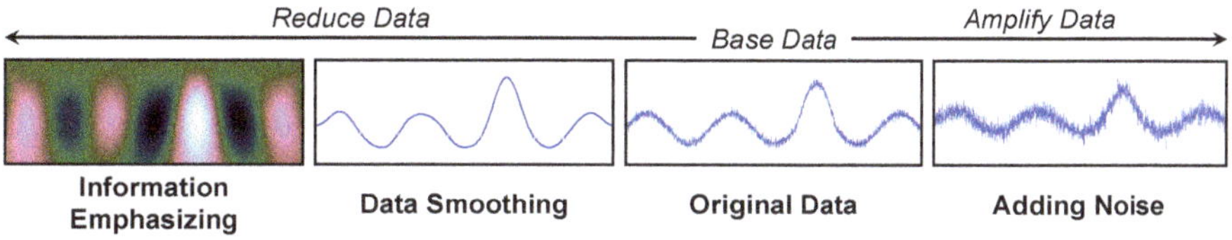

Figure 10. Strategies for Protecting Private Information in Time-Series Data.

If details in general shall be removed from a time series, *data-smoothing* approaches can be used. Originally, these approaches were primarily developed to compress data by removing less relevant parts. This removes noise at the same time. In the context of time-series data, the *discrete cosine transform* is used for this purpose in particular. An input signal is transformed into a finite sum of weighted trigonometric functions with different frequencies, representing a close approximation of the original data. As a result, the time series is smoothened [130]. With regard to privacy, however, this has another advantage, as all details of the single measuring points are wiped, and only the temporal progression remains. By using a *continuous wavelet transform*, the time-series data can be compressed even further [131]. *Information emphasizing* is thereby achieved, i.e., only a chronological sequence of high points and low points is recognizable.

A contrary strategy to protect privacy is *adding noise* to the time series. For instance, an artificial *Gaussian noise* can be generated, which is used to blanket the time-series data.

The parameters of the Gaussian random variable can be used to control how much impact the noise has on the time-series data. In this way, not only each discrete data point can be distorted but also, in the case of particularly strong noise, the progression of the time series itself [132]. While this very simple approach sounds promising at first, it has a significant drawback, as noise filtering is a de facto standard in the data refinement of time-series data. With the help of artificial neural networks, noise can be reliably eliminated on the fly [133]. One reason why such noise can be easily detected and even removed as an anomaly is that its frequency is completely different from that of the time series. Using a sequence of high-pass filters and low-pass filters, a time series can be decomposed into individual frequency bands. The bands affected by noise can be removed, and the remaining bands can be composed to the denoised time series [134]. However, there are also algorithms such as *SNIL* (*spread noise to intermediate wavelet levels*), which distribute the noise over all frequency bands of a time series and are therefore robust against such denoising [135].

The key strengths and weaknesses of privacy approaches for time-series data are the following:

Strengths:

- Privacy techniques can be used to conceal both individual data points as well as data histories in time-series data.
- The data quality of certain aspects (e.g., temporal trends or relevant data points) can be maintained.

Weaknesses:

- When applying the privacy techniques, there must be knowledge about the intended use of the data. An incorrect privacy filter would completely destroy the utility of the data.
- For some data protection techniques, the applications that process the data must be adapted accordingly. Information emphasizing, for instance, provides only maximums and minimums instead of a continuous data stream.

3.3. Voice Privacy

In the context of VDA, as presented in Section 2.3, a variety of privacy concerns arise. First, an imposter could operate the VDA. Since the VDA client is linked to a specific user, all actions of the imposter would be associated with that user. Second, the VDA is able to spy on the user continuously without the user's knowledge. The VDA client can secretly capture any spoken word and sound even if the specified keyword has not been retrieved and forward all recordings to its backend. Third, even if the user has said the keyword, from a privacy point of view the recordings should not always be forwarded to the VDA backend, since unaware bystanders could be heard in them or sensitive knowledge about the user could be derived from his or her voice. Due to the large number of threats, there is not a single privacy approach but rather a framework of protective measures [136]. A pipeline of privacy filters to protect against these threats is shown in Figure 11.

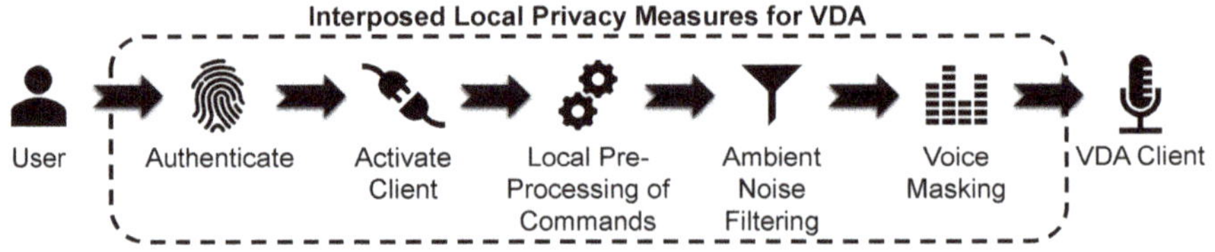

Figure 11. A Pipeline to Ensure Voice Privacy when Dealing with VDA.

A first protective measure consists of an authentication mechanism that uniquely identifies the proper user. In order not to disrupt the immersion when operating the VDA, such authentication can also be accomplished in a voice-based manner. Via characteristic features in their voice, users can be identified quite effectively using a smart speaker [137]. In order to ensure that the voice is not a recording, the environment can be scanned for suspicious magnetic fields, which would be generated by a loudspeaker. This can be achieved by a common smartphone via its built-in magnetometer [138].

Even if the speaker is an authorized user, it must be ensured that the VDA client only becomes active when it has received its keyword. To this end, an obfuscation signal can be generated in the frequency range of human speech. This ensures that the VDA client is permanently busy analyzing this signal whether it contains the keyword. This is the equivalent to a *denial-of-service attack*. Only when the user has said the keyword, the obfuscation signal is deactivated and the VDA client can accept new inputs [139].

Instead of forwarding all recordings to the VDA backend by default, it can first be attempted to handle the commands locally. This can be completed in the edge (e.g., a computer at the user's site, which serves as a hub for all of his or her VDA clients) [140] as well as on the VDA clients themselves [141]. When doing so, ambient noise can also be filtered out of the recording, which may reveal privacy-sensitive information [142]. For instance, if pets or technical devices can be heard in the background, then it can be assumed that they belong to the user. Additionally, conversations of bystanders in the background can also be obfuscated or be filtered out altogether [143]. To determine who is a bystander and who is an active user in a recording, the authentication mechanisms mentioned above can be used.

If the recording has to be sent to the backend for processing (e.g., due to insufficient on-site processing power or if third-party cloud services are required), the privacy of the active user can still be protected. From the voice, a lot of sensitive information can be inferred, e.g., gender, age, intoxication, mood, physical or mental disorders, just to name a few [144]. By voice masking, the user's voice can be distorted in such a way that these aspects are concealed, yet the speech can still be processed by a machine efficiently [145].

The key strengths and weaknesses of voice privacy approaches are the following:

Strengths:

- ⊞ There is a large range of voice privacy approaches, which can also be combined according to privacy requirements.
- ⊞ The voice privacy approaches take different privacy aspects into account, e.g., the protection of unknowing bystanders.

Weaknesses:

- ⊟ The techniques partly require additional hardware or adaptations to the installed hardware.
- ⊟ In some cases, the techniques only relocate the analysis of the data. That is, the sensitive knowledge is merely transferred to another—possibly more trustworthy—provider.

3.4. *Image Privacy*

As described in Section 2.4, image analysis can be used to identify people in photos. This raises a lot of privacy concerns. In order to make privacy-sensitive parts of an image obliterated, it seems to be a straightforward approach to blur these parts extensively. However, this has a negative effect on the quality of the image (and thus on its processability), especially if the blurring affects large parts of the image. Yet, if only small areas are blurred, this obfuscation can be undone in the data refinement process [146]. Even a full redaction of individual objects in the image does not reliably protect privacy. For instance, faces (i.e., exactly the kind of objects that are particularly privacy-sensitive) can still be recognized by means of artificial intelligence [147]. Special obfuscation techniques are therefore needed to conceal the privacy-sensitive areas of an image in a manner that is robust against reconstitution. To this end, *Singular Value Decomposition* is used to create a mask for the areas in question. This mask not only blurs the area but also adds features of other pictures to that area. This way, the result has similarities to several other images. Therefore, the blurred object can no longer be uniquely identified [148].

However, there are two problems from a data subject point of view. First, the manual selection of objects on an image that need to be protected is very cumbersome and time consuming. Second, guidance is needed in selecting the appropriate obfuscation method. For instance, if images shall be shared on social media, the obfuscation techniques must have as little impact on the visual quality as possible. Figure 12 shows what a computer-aided method for solving these two problems looks like.

Figure 12. A Process for Making Privacy-Friendly Versions of Portraits with Sensitive Content.

Initially, the original image is analyzed using deep learning. In this process, the deep features are extracted, i.e., the components that are significant for the recognition of the depicted objects in the underlying model. Based on these deep features, the image is then partitioned into a set of semantic object regions. These regions describe connected areas on the image, e.g., a person, an animal, or an object, while background objects are ignored. Using a classification, all of these regions are then assigned to a semantic meaning, e.g., bystander or child. For these classes, a privacy sensitivity is determined, which indicates how much the respective object needs to be protected [149]. In our example, bystanders receive the protection level orange, while children receive the protection level red. For persons, the semantic region can also be restricted to the face since this usually has the highest privacy sensitivity. To this end, face detection can be used to determine the region to be masked. Depending on the level of protection required, different techniques can be used that affect the visual quality of the image to a greater or lesser extent, e.g., blanking, scrambling, or blurring [150]. This way, a privacy-friendly version of the image suitable for social media can be generated.

The key strengths and weaknesses of image privacy approaches are the following:

Strengths:

- ⊞ Sensitive content can be concealed specifically and according to individual privacy requirements.
- ⊞ The data quality of the main components of an image is fully preserved by the image privacy approaches.

Weaknesses:

- ⊟ Privacy is a highly personal experience. In image privacy approaches, however, the owner or provider of the image decides which privacy requirements apply to the persons visible in an image.
- ⊟ Deep learning is used for the initial image analysis. This means, however, that the original, unaltered image is thoroughly analyzed, and knowledge is generated. This means that much more sensitive knowledge is generated than would otherwise be the case.

3.5. Pattern-Based Privacy

In Section 2.5, we presented that large amounts of data can be processed effectively using CEP. Here, a sequence of isolated data items is interpreted as a stream of events. A CEP engine scans the stream for predefined patterns. As the occurrence of such patterns can be confidential information, there are also several approaches for CEP to protect privacy. For instance, the data stream can be preprocessed locally at the user's site, and the user decides for each event whether to answer truthfully or incorrectly, i.e., whether to feed an event untampered into the stream or to add noise to it. Furthermore, the fact that in the data refinement process, the data are aggregated anyway when converted to knowledge can be exploited. By means of dedicated aggregation techniques applied directly at the user's site, a *zero-knowledge privacy guarantee* (see the work by Gehrke et al. [151]) can be achieved [152].

In doing so, it is disregarded, however, that an isolated event typically entails only little privacy-sensitive information. The information patterns targeted by CEP engines are only revealed by the occurrence of several events in a certain sequence. An arbitrary manipulation of single events is therefore not efficient and deteriorates the quality of the base data unnecessarily. Therefore, other approaches focus in particular on sequences of events. For this purpose, patterns are defined that have to be kept secret. These patterns are then purged from the data stream [153].

Although this approach is considerably less restrictive, it still overly compromises data quality [154]. The problem with this approach is that it only takes into account what has to be concealed but not what knowledge is needed by data consumers. Therefore, two types of patterns are needed to preserve data quality: *private patterns*, which have to be concealed, and *public patterns*, which are required by data consumers. Data producers and data consumers specify which knowledge must not be disclosed or respectively which knowledge is required. This is then mapped to patterns at the data level [155]. When concealing the private patterns, it has to be ensured that no public patterns are blurred in the process or additional erroneous public patterns are created by manipulating the data stream. To this end, a quality metric is used that weights the privacy (i.e., the concealment of private patterns) against the *false negatives* (i.e., public patterns that have been concealed) and *false positives* (i.e., public patterns that have been artificially added). The obfuscation of the private pattern must be conducted in a manner that optimizes this quality metric [156].

To maximize the quality metric, a privacy mechanism has to make use of different obfuscating techniques. Three such techniques are shown in Figure 13. In this example, the private pattern $B \rightarrow C$ has to be concealed, while the public pattern $A \rightarrow B$ should be recognizable. Let $A \rightarrow B \rightarrow C$ be the data stream. A simple technique is to drop an event from the stream or to inject a new one. In the example, the event C is removed, and the event D is artificially created. This conceals the private pattern, while the public pattern remains recognizable. Removing event B would conceal the private pattern as well, but the public pattern would also not be recognizable anymore. Another technique is to tamper with the events. For instance, the attribute values of data item C can be manipulated so that it is recognized as a different event C'. This also conceals the private pattern without affecting the public pattern. Lastly, the events in the stream can be reordered. By swapping B and C, the private pattern is also concealed—albeit, in this case, the public pattern is concealed as well. Using a combination of these techniques, even more complex patterns that contain conjunctions and negations in addition to sequences can be obfuscated [157].

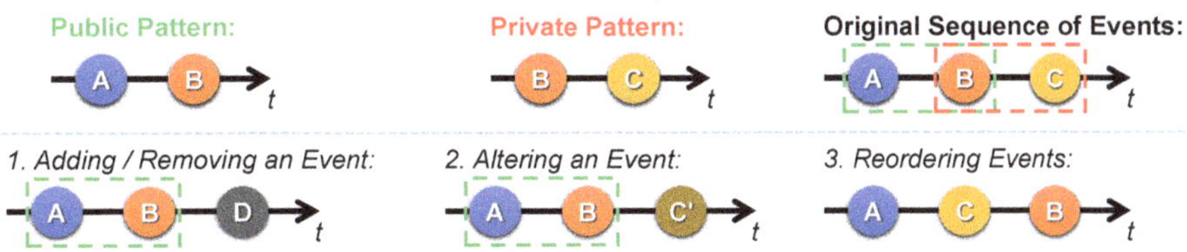

Figure 13. Application of Obfuscation Techniques to Conceal Private Patterns in a Data Stream.

The key strengths and weaknesses of pattern-based privacy approaches are the following:

Strengths:

- Pattern-based privacy does not degrade the data quality of the measurement data.
- Due to the public and private patterns, sensitive information can be filtered out in a target-oriented manner.

Weaknesses:

- The computation of an optimal configuration, i.e., the maximization of the quality metric, is very complex.
- A pattern-based privacy approach requires full control over incoming and outgoing data streams of a data processing system in order to effectively apply the required obfuscation techniques.

3.6. Differential Privacy

The previously discussed privacy techniques focused on data from individual users, which are distorted before processing. When transforming the data to knowledge, the base data of multiple users are often merged. This can be seen, e.g., in the recommender systems presented in Section 2.6. Here, the models that constitute the basis for the recommendations are trained on the preferences of all users. To this end, Dwork [158] introduces *differential privacy* as a measure for assessing the privacy threat to an individual when he or she shares his or her data. The goal is to maximize the accuracy of statistical information about a data collection (e.g., the models of the recommender systems) while minimizing the potential privacy risks for individuals. This can be illustrated by a simple example: In collaborative filtering, let there be a cluster consisting of two persons. Even if the interests of the individual users cannot be derived directly from the trained models, person A can be sure that if a new item is recommended to him or her, person B is interested in this very item. Thus, the privacy of person B has still been violated.

The basic idea of differential privacy is straightforward, as shown in Figure 14. Database 1 contains data about the four users. If the red user wants to know whether her privacy is disclosed when a statistic is computed on these bases data using the function f (respectively, a model is trained on these data), a *neighboring database* Database 2 can be created, which contains all datasets of Database 1 except the data about the user in question. If the function f is applied to Database 2 and the results are similar, the privacy of the red user has not been exposed. So, formally speaking, the following must apply:

$$\mathbb{P}[f(\mathcal{D}_1) \in \mathcal{R}] \leq exp(\epsilon)\mathbb{P}[f(\mathcal{D}_2) \in \mathcal{R}] \tag{4}$$

A function f is *ϵ-differentially private* if its results $\mathcal{R}$ for all of database $\mathcal{D}_1$'s neighboring databases $\mathcal{D}_2$ only differ by an insignificant ϵ. Since there are different privacy requirements depending on the data and processing context, there are approaches that adjust the degree of applied obfuscation according to the level of exposure. That is, there is not a single ϵ that describes the privacy requirements but rather many such measures that are applied depending on the current situation [159].

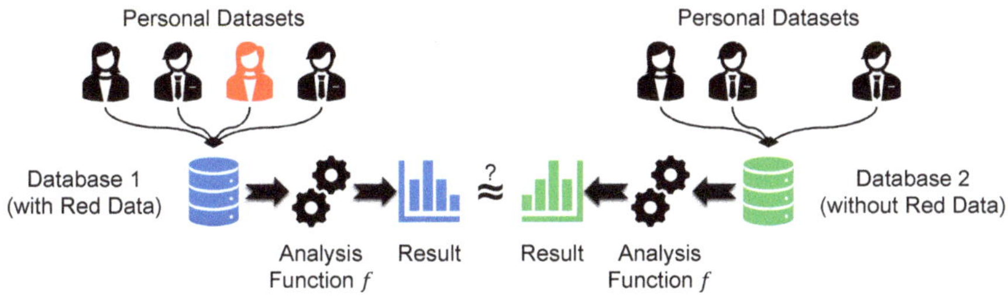

Figure 14. Simplified Representation of the Central Idea of Differential Privacy.

While this sounds promising in theory, it turns out that differential privacy is rarely applied in practice. One reason for this is that there is a lack of awareness of differential privacy principles in the development of algorithms, and in particular, the handling of complex types of data proves to be difficult [160]. In research, however, there is a variety of noise algorithms that establish the ϵ-differentially private properties without impairing the quality of the outcomes excessively [161]. Especially for application scenarios such as recommender systems, differential privacy is therefore suitable [162].

The key strengths and weaknesses of differential privacy approaches are the following:

Strengths:

- Differential privacy approaches allow statistical analysis while preserving the privacy of each individual involved.
- In order to guarantee the differential privacy property, the method is not restricted to any particular technique, which means that an appropriate obfuscation technique can be chosen depending on the base data.

Weaknesses:

- Differential privacy approaches can only be applied when large amounts of data from many different individuals are analyzed.
- Ensuring the differential privacy property is difficult depending on the base data requires the use of destructive noise algorithms. As a result, potentially relevant aspects in the data are lost.

3.7. Federated Learning

Finally, we look at *federated learning*, which is an approach that was not originally intended to provide privacy. Along with the increased application of machine learning in all kinds of domains, the need for data to train the corresponding models has also increased. While the IoT provides an almost unlimited number of smart devices as data sources, organizational problems arise in terms of the communication overhead required to continuously receive the captured data as well as the central computing resources required to process the data. As IoT-enabled devices become increasingly powerful, the federated learning approach shifts many of the data refinement tasks from the central data processor to the distributed data producers. This includes the entire data preparation phase and large parts of the data processing phase. That is, data producers generate information patterns which are gathered by the data processor. The data processor uses these patterns to create or update global knowledge models that contain the information patterns of all data producers. In other words, federated learning takes aggregation to the next level as data producers exchange only knowledge instead of data with the data processor. From the global models of the data processor, knowledge can also be fed back to the local workers in order to improve their local models [163]. This procedural concept is illustrated in Figure 15.

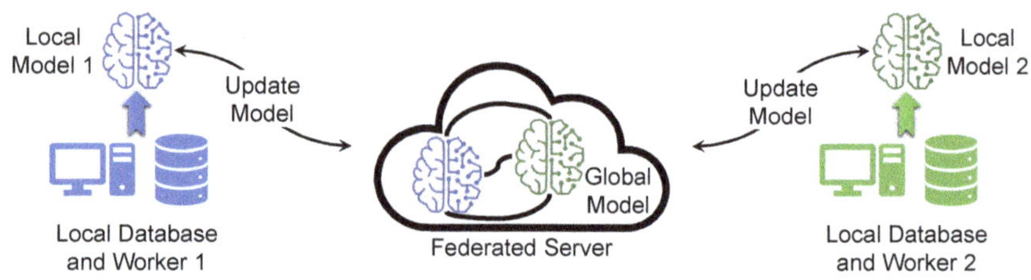

Figure 15. A Distributed Training of a Global Knowledge Model via Federated Learning.

However, federated learning can also contribute to privacy-preserving machine learning. Data privacy is generally regarded as the right of an individual to freely decide how much knowledge he or she wants to disclose about him or herself. This is exactly what federated learning does, as data producers are free to decide which of their data they want to use in the data refinement process and which information patterns they want to include in the knowledge model [164]. By applying privacy filters to the data, differential privacy can be preserved not only for the isolated local models but also for the shared global model. Thereby, a stronger quantifiable privacy protection can be achieved [165]. This protection combined with its distributed computing renders federated learning well-suited for large-scale analyses, such as the DNA sequence classifications presented in Section 2.7 [166].

The key strengths and weaknesses of federated learning approaches from a privacy perspective are the following:

Strengths:

- Federated learning is primarily used to efficiently run complex machine learning processes. The preservation of privacy is a beneficial side effect that comes at no additional cost.
- Federated learning enables data subjects to incorporate their data into global machine learning model but to carry out the necessary processing of their private data locally, i.e., under their full control.

Weaknesses:

- Due to the complex and non-explanatory nature of the trained models, it is not possible for data subjects to understand what knowledge about them is incorporated into the global model by means of their locally computed models.
- The use of federated learning is limited to certain algorithms and algorithm classes.

3.8. Key Findings

Our key findings regarding the means of privacy protection as well as the effects of the aforementioned privacy techniques are summarized in Table 2.

It is evident that there are dedicated privacy filters for different types of data, namely for location data, time-series data, voice data, and image data. Furthermore, there are techniques to conceal private patterns in a stream of data items. Thereby, it is possible to filter sensitive information in a fine-grained manner. Using these and similar obfuscation techniques, the ϵ-differentially private properties can also be established, which ensures that statistical computations are performed on a plethora of personal datasets—including the extraction of information patterns that are the foundation of knowledge—without exposing the privacy of any individual.

Table 2. Summary of the Key Lessons Learned from the Review of the Privacy Techniques.

Privacy Approach	Means of Privacy Protection	Effects of the Measures
General Privacy Measures	The three relational algebra operators—selection, projection, and aggregation—can be applied to base data.	Entire data items or certain attributes can be concealed, and the base data can be condensed.
Location Privacy	Fake locations can be used, and spatial cloaking, path confusion, and temporal cloaking can be applied.	Individual locations or entire trajectories as well as their temporal correlations can be concealed.
Privacy-Preserving Time-Series Data	The data can either be compressed to reduce details or they can be amplified by fake data.	Only temporal progressions can be observed but no details on single data points.
Voice Privacy	The VDA can be jammed, data are preprocessed locally, and the recordings are filtered.	A VDA cannot spy on its users, and the information shared with the VDA backend is minimized.
Image Privacy	Blanking, scrambling, or blurring can be used to mask certain areas of an image.	Objects on an image can be obfuscated in a fine-grained manner based on their privacy sensitivity.
Pattern-Based Privacy	Data items can be added, removed, altered, or reordered.	Private patterns in terms of data sequences can be concealed.
Differential Privacy	In statistical calculations, noise ensures ϵ-differential privacy.	No knowledge about single individuals is disclosed to third parties.
Federated Learning	Data processing is primarily performed locally by data producers.	Data processors only gain insight into highly aggregated knowledge.

However, these dedicated privacy techniques can only be applied systematically if a certain underlying structure is present in the data. For generic raw data, only general privacy measures can be applied. These include, e.g., the three operators of relational algebra mentioned initially: namely, selection, projection, and aggregation. The dedicated privacy filters are therefore significantly better suited for backend systems after comprehensive data refinement measures have been carried out, and processable information has already been retrieved. Meanwhile, on lightweight smart devices, which are more likely to handle raw data, the general privacy measures should be preferred as they are easier to apply. Yet, these general measures have to be used purposefully, as they are not capable of filtering private aspects specifically and thus tend to have a more significant impact on data quality.

Federated learning represents a completely different approach. Here, data preprocessing is carried out in a distributed manner under the supervision of the data subjects. Only highly aggregated information in the form of machine learning models is forwarded to the backend system. In this way, each data subject can decide independently which privacy measures are applied locally to the data. However, data subjects often lack the necessary knowledge to do this in a systematic manner that is tailored to the intended purpose of the models. Moreover, federated learning is only suited for very certain machine learning algorithms and not for general purpose data refinement.

In this context, it is arguable whether it is even necessary to apply privacy measures locally on the smart devices of the users. A local application of privacy filters is primarily reasonable if the data processor to which these smart devices forward the collected data is not trustworthy. If this is not the case, it is more advisable to apply privacy measures on the side of the data processor. Such a global approach can be more targeted and results in stronger privacy protection for the same scope of data filtering—i.e., the trade-off between privacy and utility is significantly better [167]. Data processors in the service domain are commonly assumed to be *semi-honest-but-curious*. That is, they largely carry out their tasks in a trustworthy manner, and they will not deliberately expose the privacy of a user [168].

It is therefore sufficient to provide data processors with means to express their privacy requirements and to verify whether they are respected. So, in this context, federated learning is rather an exception that can be used if highly sensitive data are involved.

Now, the question arises as to what extent these state-of-the-art privacy measures are suitable for the deployment in the application scenarios discussed in Section 2. In the following section, we therefore map the strengths and weaknesses of these approaches against the opportunities offered by and privacy threats posed by the application scenarios in order to answer this question.

4. Assessment of the State of Privacy Mechanisms for Smart Services

After discussing the opportunities and privacy threats of smart services in the most relevant application domains in Section 2 and identifying the strengths and weaknesses of state-of-the-art privacy technologies for these types of smart services in Section 3, we now assess whether the available privacy measures are adequate. To this end, we apply a systematic analysis technique adapted from strategic planning. The so-called *SWOT analysis* is originally used to determine the market position and strategy development of companies. SWOT stands for *strengths*, *weaknesses*, *opportunities*, and *threats*. In this process, internal and external factors are first identified and then juxtaposed with helpful and harmful aspects of a product or a company strategy. From the four resulting intersections (internal factors vs. helpful aspects, internal factors vs. harmful aspects, external factors vs. helpful aspects, and external factors vs. harmful aspects), strengths, weaknesses, opportunities, and threats can be derived. Therefore, SWOT analysis is an important foundation for strategic audits, as it enables a systematic market and environmental assessment [169]. In addition to the original focus on companies, the SWOT analysis is meanwhile used in adapted form in many other domains, for instance, in the education sector when new teaching methods are to be introduced or in the health sector to determine risk factors for patients [170].

For our assessment of the state of privacy mechanisms for smart services, we also use an approach based on the SWOT analysis. We develop an individual SWOT matrix for each of the seven application scenarios. Unlike in the classic SWOT analysis, however, we do not compare internal and external factors with helpful and harmful aspects but rather the strengths and weaknesses of the privacy mechanisms with the opportunities and privacy threats present in the respective application scenarios. In this way, we obtain a good understanding of the extent to which state-of-the-art privacy measures are suitable for the deployment in the application scenarios and which open questions still need to be resolved in this context. The result of our analysis is presented in Figure 16.

	Opportunities of Smart Services	Privacy Threats of Smart Services
Strengths of Privacy Measures	Since either individual locations or trajectories can be concealed, LBS can be provided with the required data.	Excessive spying can be prevented by particularly restrictive privacy techniques that, e.g., blur both locations and trajectories.
Weaknesses of Privacy Measures	As LBS are deceived by fake data, the user experience suffers due to incorrect information.	As LBS permanently track the location, the scope for privacy techniques is limited and private information can still be inferred.

(**a**) Location-Based Services

	Opportunities of Smart Services	Privacy Threats of Smart Services
Strengths of Privacy Measures	By using proper privacy techniques, certain aspects can be concealed without impairing the relevant health data.	Data subjects determine which aspects should be hidden, so that no compromising knowledge is revealed.
Weaknesses of Privacy Measures	Privacy techniques falsify certain aspects of the data, which can have grave consequences in the health context.	As data subjects are not aware of what can be inferred from the collected data, they cannot choose the appropriate privacy technique.

(**b**) Health Services

	Opportunities of Smart Services	Privacy Threats of Smart Services
Strengths of Privacy Measures	Voice-based control of smart devices is not restricted by the privacy measures.	By using additional hardware, a VDA can be fully controlled.
Weaknesses of Privacy Measures	The low entry hurdle of VDA is undermined by the need for additional hardware.	Compliance with the privacy requirements still has to be outsourced to third parties.

(**c**) Voice-Controlled Digital Assistants

	Opportunities of Smart Services	Privacy Threats of Smart Services
Strengths of Privacy Measures	OSNs reveal many private photos—using image privacy techniques, this can be done privacy-aware.	The privacy techniques can be used to blur specific objects in the images.
Weaknesses of Privacy Measures	The overall quality of the image suffers as a result of excessive blurring.	Third parties instead of the affected data subjects decide which objects should be blurred.

(**d**) Image Analysis

	Opportunities of Smart Services	Privacy Threats of Smart Services
Strengths of Privacy Measures	By defining public pattern, pattern-based privacy has negligible impact on the actual analysis.	By defining private pattern, no confidential information is disclosed.
Weaknesses of Privacy Measures	Since the configuration causes a lot of overhead, this affects the cost of an analysis.	Food producers must not have control over the data analysis, which is mandatory for pattern-based privacy.

(**e**) Food Analysis

	Opportunities of Smart Services	Privacy Threats of Smart Services
Strengths of Privacy Measures	Differential privacy does not limit the statistical analysis of recommender systems.	Since individuals are hidden in the mass of all users, the data analysis does not pose a privacy threat.
Weaknesses of Privacy Measures	Properties of extraordinary individuals have to be blurred, i.e., this knowledge is lost.	Through the use of destructive noise, it is possible to deliberately influence users by not recommending certain items.

(**f**) Recommender Systems

	Opportunities of Smart Services	Privacy Threats of Smart Services
Strengths of Privacy Measures	The quality of the global model (and thus of the analysis) is not impaired by federated learning.	Sensitive data is processed locally (and thus under the control of data subjects) and is only forwarded in aggregated form.
Weaknesses of Privacy Measures	Federated learning cannot be applied to any kind of analysis.	The locally computed models cannot be explained, i.e., data subjects do not know what knowledge about them is propagated.

(**g**) DNA Sequence Classification

Figure 16. Results of our SWOT Analysis Broken Down by the Respective Application Scenario.

Across all application scenarios, it can be observed that there is an appropriate privacy technique for each privacy threat. It can also be noted that the privacy techniques are capable of operating in a target-oriented manner. That is, when protecting sensitive data, it is also ensured that the general utility of the base data is not unnecessarily impaired. This is a fundamental requirement, as otherwise, the use of smart services would be severely disrupted or outright prevented, rendering the privacy techniques worthless.

So, in principle, this is a highly promising result, as there are tailored techniques for all types of data and forms of processing. However, our analysis also reveals inherent problems with state-of-the-art privacy technologies for smart services that need to be addressed in order to enable effective data protection by design. Problems arise in particular regarding the selection and configuration of the privacy techniques, since this requires comprehensive technical and domain knowledge, which a normal user does not possess. As a result, the privacy measures are often too restrictive in terms of impairing the quality of the smart service and at the same time not effective enough in terms of protecting all sensitive data. Moreover, each privacy technology is always designed for a specific use case. Comprehensive privacy protection therefore requires the integration of many heterogeneous approaches, which leads to further adjustment problems. Since many privacy approaches rely on a trusted third party to protect sensitive data, such an agglomeration of many different approaches leads to the situation that many third parties gain access to the private data (one party per approach). Instead of disclosing less private information, the data are therefore disclosed to even more parties. Finally, uncontrolled tampering with the data also leads to further security vulnerabilities, which can result in financial or material damage, which is why the choice of the appropriate privacy method must also not be left solely to the inexperienced user.

Based on these insights, we derive seven such open research questions and briefly outline how we believe these questions can be addressed in the following section.

5. Future Prospects

As our assessment of state-of-the-art privacy measures reveals, there is no one-size-fits-all solution when dealing with smart devices. Rather, a privacy approach has to be found which is fully geared toward the intended purpose. Only then is it possible to ensure an adequate data protection. While efficient island solutions for concealing specific sensitive information exist, there are open questions regarding the applicability of such privacy filters. In the following, we therefore discuss seven open research questions that emerge in the context of data protection in the information age.

(a) Privacy Requirements Elicitation. First, it is crucial to identify the privacy risks posed by a smart service, i.e., what sensitive knowledge is exposed. Only when these risks are identified explicitly is it possible for data subjects to give informed consent with regard to the processing of their data. Depending on the threat potential of a service, data access has to be adjusted accordingly.

The *System–Theoretic Process Analysis for Security* (*STPA-Sec*) is a top–down approach that can be used to systematically identify security problems in complex and dynamic systems. For this purpose, a holistic view of the system is obtained. For each component and the communication flow between components, vulnerabilities are annotated in order to identify all potentially insecure control actions. Based on the results of this analysis, the system can be redesigned by means of a security-driven design process [171]. This process can also be adapted to privacy aspects (*STPA-Priv*) [172].

In practice, however, a conceptual problem in STPA-Priv is encountered. Whereas both data producers and data consumers are equally interested in the security of a system, privacy is usually a one-sided interest on the part of data producers—the data consumers (e.g., smart services), meanwhile, are primarily interested in obtaining as much data as possible. The results of such a privacy analysis are therefore less useful for the redesign of a smart service. However, they can be used to determine privacy requirements in the context of a smart service [173].

In this regard, it is important to study how a holistic view of a smart service can be compiled. In the context of the IoT, there are many heterogeneous data sources that can be dynamically added or removed at any time. Furthermore, smart services often use a data backend as a data source. For a privacy analysis of a smart service, it is also necessary to assess its data backends and all their data sources. However, such a comprehensive view of all data sources of a smart service typically cannot be provided.

(b) Holistic Privacy Platform. Even if a user is fully aware of all privacy risks associated with a smart service, s/he also needs to be enabled to take effective countermeasures. The privacy measures discussed in Section 3 certainly contribute to protecting privacy, but they are no out-of-the-box solutions. Rather, they have to be applied at the right place in the data flow. To become a usable privacy measure, the filters must be integrated into a privacy platform. Via such a platform, users can apply the necessary privacy measures appropriately. The advantage of such a central privacy platform is that it can be constantly extended with new privacy filters. Meanwhile, there is already a wide range of privacy filters for some application areas, e.g., the area of behavioral data recorded by smart devices is still largely unexplored [174]. A privacy platform could be easily upgraded once efficient privacy filters for behavioral data have been developed. In this way, the protection of private data is constantly in line with the latest technological advances. There are research approaches for such platforms that, e.g., provide a selection of privacy filters that can subsequently be applied to a data stock [175]. Such a solution primarily addresses backend systems. However, there are also approaches that address the source systems that feed such a data backend, e.g., smart devices [176].

When dealing with smart services, it is important to determine how such a privacy platform can reliably monitor them and control their data access. While a privacy platform for a backend system can restrict access to that data backend and, e.g., apply privacy filters to certain data before they are shared, it cannot be ensured that a smart service does not use multiple data backends. The data obtained from each backend individually might not be very exposing. However, by combining all the gathered data, it might still be possible to derive sensitive knowledge patterns. Reliable privacy protection requires a holistic end-to-end approach from the data source to the data sink, i.e., the smart service. In particular, this means that a smart service must be completely isolated from any data source and can only obtain all data via the privacy platform.

(c) Configuration of Privacy Filters. A user is generally able to specify certain knowledge patterns that are particularly private or confidential at a high abstraction level. However, users are not familiar with the data sources from which these patterns can be derived let alone privacy filters via which they can be concealed efficiently. The configuration and parameterization of such a holistic privacy platform is therefore far too complex for users. Concepts are therefore required to enable them to configure the platform using rather high-level descriptions of their privacy requirements. This requires models that accurately represent which information can be obtained from which data sources and what knowledge can be derived from it. There are metamodels for this purpose to describe such correlation [177]. Furthermore, data subjects are often not at all aware of their privacy requirements. They intuitively have a rough idea but are often unable to fully express it. Yet, using approaches based on collaborative filtering, sensitive knowledge patterns can be recommended to them that may also be relevant to them [178].

Such an approach can be applied successfully for isolated applications, but for complex smart services and their data infrastructure, two major problems arise: On the one hand, in addition to a holistic view on the smart service, domain experts are needed who have the necessary experience to determine which knowledge can be derived from which data sources. Such experts are also needed, e.g., to analyze the smart services using STPA-Priv. Therefore, it is advisable to study whether the model of the data–information–knowledge relationships can be derived from the STPA-Priv results or to what extent this process analysis approach needs to be adapted for this purpose. On the other hand, a recommender system for privacy requirements depends on the assumption that other users have previously been able to formulate their requirements. This results in a chicken-and-egg problem. Since the GDPR mandates a *data protection by default* (Article 25), it should also be studied whether the privacy requirements identified by STPA-Priv can be translated into some sort of basic configuration for the privacy filters representing an advisable baseline for any user.

(d) Deployment of Privacy Filters. Once a configuration has been found, the privacy filters must be deployed appropriately. There are basically two options: the privacy filters are applied either directly to the data sources [179] or to the data backend [180]. However, this decision has a significant impact on privacy and data quality. If a privacy filter is applied very close to the source, i.e., on the user's side, the unfiltered raw data never leave the user's control. That is, the data are impurified before they are forwarded to the data backend. Yet, at this point, not all information about the further usage of the data is available; e.g., a data backend can also feed several smart services, for which different privacy requirements may apply. Therefore, privacy measures cannot be applied in a target-oriented manner, which means that they turn out to be either too restrictive or insufficient. Whereas, when applying privacy filters in the data backend, the data are no longer within the user's sphere of influence and the user must completely trust that his or her privacy requirements are respected by the backend, which represents a major psychological hurdle [181].

In simple terms, the privacy platform must therefore find a deployment plan for which a utility metric is maximized. At an abstract level, this metric looks like this:

$$Utility = Data\ Quality + Privacy \tag{5}$$

The utility of a deployment plan is defined by how well it preserves privacy and how little it impairs data quality. A high-level definition, analogous to the configuration based on knowledge patterns, could look like this:

$$Utility = \sum_i Public_i * w_{Public_i} - \sum_j False_j * w_{False_j} - \sum_k Private_k * w_{Private_k} \tag{6}$$

Here, the data quality is described by how many public knowledge patterns—i.e., non-confidential knowledge patterns—can be detected despite the use of privacy filters ($\sum_i Public_i$), minus all *false positives*, i.e., public knowledge patterns that were falsely recognized due to the use of privacy filters ($\sum_j False_j$). The privacy of a deployment plan is determined by how few private knowledge patterns it exposes ($\sum_k Private_k$). Additionally, a penalty weight w can be assigned to each of these components to prioritize them differently depending on the specific use case. The research question here is how to efficiently determine the one with the best utility from all possible deployment plans.

(e) Secure Data Management. Since the application of privacy filters can be time consuming, they should not be applied as data flow operators over and over for every data access. Instead, frequently used data should be stored in the data backend at different privacy levels, i.e., after different privacy filters have been applied. Zone-based data lakes are suitable for managing big data in different processing stages. In addition to filtered raw data, higher-value information, for instance in the form of machine learning models trained by means of federated learning, can also be stored in such data lakes [182]. For this, the data backend acts as a *data marketplace*. Similar to a marketplace for material goods, customers, e.g., smart services, can pick the data they want [183]. This requires extensive metadata that characterizes the available data so that smart services can also find relevant data [184]. In contrast to conventional marketplaces, however, privacy constraints must be observed in data marketplaces; i.e., not every customer may access every data item [185]. Moreover, since the existence of a data item exposes certain information, even the visibility of the items has to be regulated in this case. In this regard, it must be ensured that the operator of the data marketplace cannot abuse this. Since s/he has sovereignty over the data, s/he could maliciously withhold certain data items to the disadvantage of a data subject or a smart service. By using blockchain technologies this can be prevented, a trustworthy data sharing is enabled [186]. Yet, the use of a blockchain raises further privacy issues, e.g., due to its immutability or the fact that in a data marketplace, there are legitimate reasons why a certain party is not allowed to see a specific data item [187].

Future research therefore has to address how existing metadata models need to be extended to include aspects such as applied privacy measures in addition to data quality or data origin. Furthermore, access policies must be developed to ensure that certain data items are not visible to selected smart services. However, it is also important to ensure that these mechanisms cannot be misused to the disadvantage of data subjects or smart services. Unlike blockchain technologies, all concepts developed in this context must comply with current data protection regulations by design.

(f) Proof of Data Possession. Such a data marketplace must inevitably be able to provide proofs of data possession. It must be verifiable for data subjects whether such a data provider stores the data of the data subject in the agreed form, e.g., at different privacy levels. Although such a verification must be publicly available, it must not jeopardize data protection. That is, a third party must not be able to derive any information by verifying whether certain data about a data subject are retained [188]. To this end, *Proof of Retrievability and Reliability* (*PoRR*) approaches are applied in cloud-based data stores. These approaches verify that a cloud provider faithfully manages the entrusted data in the agreed number of replicas. For this purpose, a so-called *Verifiable Delay Function* (*VDF*) is applied to the data and all replicas. A VFD is slow to compute but easy to verify [189]. The cloud provider is regularly challenged with respect to this function. If the response to this challenge takes too long, it is confirmed that the provider does not have the data at rest but needs to compute the VFD on the fly. In cloud-based data stores, such a procedure can also be applied efficiently [190].

While at first sight there seem to be many similarities between data marketplaces and such cloud-based data stores, there are also decisive differences. In particular, the data marketplace does not manage identical replicas of the source data but rather several variants of them, which are available in different processing stages and to which different privacy filters have been applied. Furthermore, it is also possible that some raw data reside on the smart devices, and the data marketplace only acts as an intermediary between them and the smart services. Therefore, the data infrastructure here is distributed and heterogeneous. Furthermore, some components are computationally weak, namely the smart devices, which is why a mechanism to provide proof of data possession must be lightweight. It is therefore important to investigate to what extent a PoRR approach can be applied to such a data marketplace and what adaptations are necessary to this end.

(g) Prevention of Misinformation. Finally, the spread of misinformation is a major problem in the information age. In the context of a data hub such as a data marketplace, two types of misinformation must be distinguished: On the one hand, data subjects can deliberately manipulate their data using privacy filters in order to gain an advantage [191]. For instance, if they provide health data to a smart service of their health insurance company to obtain a better rate, it must be ensured that unhealthy habits, such as being a smoker, cannot be specifically filtered out. This can be restricted by means of attribute-based authentication of the data sources. Thereby, certain conditions can be specified—e.g., a certain privacy filter is not applied—which the sources must comply with in order to authenticate successfully. However, such information can also reveal a lot about the data. For instance, when certain privacy filters are applied, it can be inferred what kind of knowledge the data subject wants to conceal. Hence, the authentication process also needs to be privacy-friendly. For this purpose, a trusted intermediary can be used, which pre-validates the complete attributes of a source and forwards only those attributes that do not reveal sensitive information to the data recipient for authentication [192].

However, this approach is based on digital signatures. Due to the advent of quantum computers, asymmetric cryptography, which provides the foundation for digital signatures, can no longer be considered secure. Future research therefore needs to explore which post-quantum cryptography approaches can be used instead and to what extent they are suitable for the usage on smart devices, which have to generate the signatures. Furthermore, the question arises how an efficient and trustworthy key management can be implemented in such a scenario. This concerns both the key generation and the key provisioning.

On the other hand, misinformation about a data subject can also be disseminated by third parties. This is well known in the context of OSNs such as Facebook and commonly referred to as *fake news*. Here, as well, information is disseminated by means of word-of-mouth and not directly from a data subject to the intended recipient. For OSNs, there are approaches that can be used to restrain the dissemination of misinformation. Contradictory information, i.e., misinformation and credible information, is identified, and its propagation in the network is monitored [193]. Using a greedy approach, users in the network can be identified who are considered trustworthy and can therefore be assumed to contribute to minimize the spread of misinformation by sharing only credible information in the network. In the case of contradictory information, this allows one to determine which version represents the misinformation—namely, the one which is not shared by a trustworthy user—and then to remove it from circulation [194].

However, this procedure was developed specifically for social networks. It must therefore be studied whether it can also be applied to the data infrastructure of smart services. In particular, it must be assessed to what extent this approach has to be adapted and extended in order to be able to deal with heterogeneous raw data from a wide variety of domains instead of purely textual information. Furthermore, the IoT is a much more dynamic structure than a social network. Whereas in the latter case, users usually remain part of the network for a long time, in the IoT, new data sources are constantly being added or removed. Therefore, it is also important to research how the approach can be adapted to such an ever-changing environment.

We consider these seven challenges to be the most critical ones related to the protection of sensitive data in the information age that need to be addressed in future work.

6. Conclusions

In the modern information age, we are accustomed to smart services facilitating our everyday lives. We use these digital assistants in public, industrial, and private domains. Such data-driven services are so handy, as they adapt their behavior based on the current context and thus always provide an optimized user experience. However, this convenience does not come without a price. Smart services rely on permanent access to a vast amount of data. They analyze these data comprehensively to derive knowledge about the current situation of their users. Yet, this often involves highly personal or confidential data, allowing data processors to obtain sensitive information. For this reason, there are a variety of privacy measures that conceal certain sensitive knowledge patterns in the data without impairing the quality of the data. That is, they try to protect the privacy of data subjects and at the same time maintain the utility of their data for the respective smart services.

For this reason, this paper addresses the question of whether state-of-the-art privacy mechanisms are prepared to meet this challenge. To this end, we carry out a SWOT-like assessment, in which we initially analyze the seven most relevant application scenarios for smart services. It is evident that users can benefit from these smart services in all situations of life. Yet, smart services also pose a high privacy threat due to the personal data they process. Although this statement generally applies to all smart services, it becomes apparent that the different application scenarios involve heterogeneous types of data. There is therefore no universal privacy threat but rather smart services-specific threats. Therefore, we also study the strengths and weaknesses of privacy approaches tailored to smart services. By comparing these two dimensions (opportunities and privacy threats of smart services on the one side and strengths and weaknesses of privacy approaches on the other side), we discover that there is an effective answer to every relevant privacy threat. However, there are fundamental problems with modern privacy technologies for smart services. Users are overwhelmed by the selection and configuration of privacy techniques. These techniques are always tailored to a specific use case, which is why a great deal of domain knowledge is required to apply them effectively. Otherwise, their protective effect is insufficient, and the negative impact on data quality is too high. The latter can lead to further security vulnerabilities if the data tampering is not target-oriented. Therefore, a trusted third party is often required for the application of the privacy techniques. From these findings, we derive pertinent open research questions and give our opinion on how they can be overcome. These research questions deal in particular with concepts for privacy requirements elicitation, a holistic privacy platform, the deployment of privacy filters, the configuration of privacy filters, a secure data management, proofs of data possession, and prevention of misinformation. Only when these issues are fully addressed is a privacy-by-design approach for smart devices feasible.

Author Contributions: Conceptualization, C.S.; methodology, C.S.; software, C.S.; validation, C.S.; formal analysis, C.S.; investigation, C.S.; resources, J.B., C.G. and B.M.; data curation, M.B., J.B., C.G. and C.S.; writing—original draft preparation, C.S.; writing—review and editing, M.B., J.B., C.G., B.M. and C.S.; visualization, C.S.; supervision, C.S.; project administration, C.S.; funding acquisition, B.M. All authors have read and agreed to the published version of the manuscript.

Funding: This research received no external funding.

Institutional Review Board Statement: Not applicable.

Informed Consent Statement: Not applicable.

Data Availability Statement: Not applicable.

Conflicts of Interest: The authors declare no conflict of interest.

Abbreviations

The following abbreviations are used in this paper:

CEP	complex event processing
CNN	convolutional neural network
DNA	deoxyribonucleic acid
e-commerce	electronic commerce
e-resource	electronic resource
e-service	electronic service
eHealth	electronic health
GDPR	general data protection regulation
GPS	global positioning system
GSM	global system for mobile communications
IoT	internet of things
LBS	location-based service
mHealth	mobile health
OSN	online social network
POI	point of interest
PoRR	proof of retrievability and reliability
SNIL	spread noise to intermediate wavelet levels
STPA-Sec	system–theoretic process analysis for security
STPA-Priv	system–theoretic process analysis for privacy
SWOT	strengths, weaknesses, opportunities, and threats
UniProt	universal protein resource
VDA	voice-controlled digital assistant
VDF	verifiable delay function

References

1. Weiser, M. The computer for the 21st century. *Sci. Am.* **1991**, *265*, 94–104. [CrossRef]
2. Presser, M. The Rise of IoT–why today? *IEEE Internet Things Newsl.* **2016**, *12*, 2016
3. Jesse, N. Internet of Things and Big Data: The disruption of the value chain and the rise of new software ecosystems. *AI Soc.* **2018**, *33*, 229–239. [CrossRef]

4. Hariri, R.H.; Fredericks, E.M.; Bowers, K.M. Uncertainty in big data analytics: Survey, opportunities, and challenges. *J. Big Data* **2019**, *6*, 44. [CrossRef]
5. Stach, C.; Bräcker, J.; Eichler, R.; Giebler, C.; Mitschang, B. Demand-Driven Data Provisioning in Data Lakes: BARENTS — A Tailorable Data Preparation Zone. In Proceedings of the 23rd International Conference on Information Integration and Web Intelligence (iiWAS), Linz, Austria, 29 November–1 December 2021; ACM: New York, NY, USA, 2021; pp. 187–198.
6. Stach, C.; Behringer, M.; Bräcker, J.; Gritti, C.; Mitschang, B. SMARTEN — A Sample-Based Approach towards Privacy-Friendly Data Refinement. *J. Cybersecur. Priv.* **2022**, *2*, 606–628. [CrossRef]
7. Liew, A. Understanding Data, Information, Knowledge And Their Inter-Relationships. *J. Knowl. Manag. Pract.* **2007**, *8*, 134.
8. Stöhr, C.; Janssen, M.; Niemann, J.; Reich, B. Smart Services. *Procedia Soc. Behav Sci.* **2018**, *238*, 192–198.
9. Kashef, M.; Visvizi, A.; Troisi, O. Smart city as a smart service system: Human-computer interaction and smart city surveillance systems. *Comput. Hum. Behav.* **2021**, *124*, 106923. [CrossRef]
10. Lee, J.; Kao, H.A.; Yang, S. Service Innovation and Smart Analytics for Industry 4.0 and Big Data Environment. *Procedia CIRP* **2014**, *16*, 3–8. [CrossRef]
11. Pramanik, M.I.; Lau, R.Y.; Demirkan, H.; Azad, M.A.K. Smart health: Big data enabled health paradigm within smart cities. *Expert Syst. Appl.* **2017**, *87*, 370–383. [CrossRef]
12. Nissenbaum, H. Protecting Privacy in an Information Age: The Problem of Privacy in Public. *Law Philos* **1998**, *17*, 559–596. [CrossRef]
13. European Parliament and Council of the European Union. Regulation on the protection of natural persons with regard to the processing of personal data and on the free movement of such data, and repealing Directive 95/46/EC (Data Protection Directive). Legislative Acts L119. Off. J. Eur. Union 2016. Available online: https://eur-lex.europa.eu/eli/reg/2016/679/oj (accessed on 17 October 2022).
14. Gerber, N.; Gerber, P.; Volkamer, M. Explaining the privacy paradox: A systematic review of literature investigating privacy attitude and behavior. *Comput. Secur.* **2018**, *77*, 226–261. [CrossRef]
15. Dewri, R.; Ray, I.; Ray, I.; Whitley, D. Exploring privacy versus data quality trade-offs in anonymization techniques using multi-objective optimization. *J. Comput. Secur.* **2011**, *19*, 935–974. [CrossRef]
16. Ramson, S.J.; Vishnu, S.; Shanmugam, M. Applications of Internet of Things (IoT) – An Overview. In Proceedings of the 2020 5th International Conference on Devices, Circuits and Systems (ICDCS), Coimbatore, India, 5–6 March 2020; IEEE: Manhattan, NY, USA, 2020; pp. 92–95.
17. Dias, R.M.; Marques, G.; Bhoi, A.K. Internet of Things for Enhanced Food Safety and Quality Assurance: A Literature Review. In Proceedings of the International Conference on Emerging Trends and Advances in Electrical Engineering and Renewable Energy (ETAEERE), Bhubaneswar, India, 5–6 March 2020; Springer: Singapore, 2021; pp. 653–663.
18. Nawara, D.; Kashef, R. IoT-based Recommendation Systems – An Overview. In Proceedings of the 2020 IEEE International IOT, Electronics and Mechatronics Conference (IEMTRONICS), Vancouver, BC, Canada, 9–12 September 2020; IEEE: Manhattan, NY, USA, 2020; pp. 1–7.
19. Huffine, E.; Kumar, A.; Kashyap, A. Attaining State of the Art in DNA Tests. In *Handbook of DNA Forensic Applications and Interpretation*; Kumar, A., Goswami, G.K., Huffine, E., Eds.; Springer: Singapore, 2022; pp. 11–23.
20. Zainuddin, N.; Daud, M.; Ahmad, S.; Maslizan, M.; Abdullah, S.A.L. A Study on Privacy Issues in Internet of Things (IoT). In Proceedings of the 2021 IEEE 5th International Conference on Cryptography, Security and Privacy (CSP), Zhuhai, Chinal, 8–10 January 2021; IEEE: Manhattan, NY, USA, 2021; pp. 96–100.
21. Junglas, I.A.; Watson, R.T. Location-Based Services. *Commun. ACM* **2008**, *51*, 65–69. [CrossRef]
22. Raper, J.; Gartner, G.; Karimi, H.; Rizos, C. Applications of location–based services: A selected review. *J. Locat. Based Serv.* **2007**, *1*, 89–111. [CrossRef]
23. Agre, P.E. Welcome to the always-on world. *IEEE Spectr* **2001**, *38*, 10–13. [CrossRef]
24. D'Roza, T.; Bilchev, G. An Overview of Location-Based Services. *BT Technol. J.* **2003**, *21*, 20–27. [CrossRef]
25. Obeidat, H.; Shuaieb, W.; Obeidat, O.; Abd-Alhameed, R. A Review of Indoor Localization Techniques and Wireless Technologies. *Kluw. Commun.* **2021**, *119*, 289–327. [CrossRef]
26. Dey, A.; Hightower, J.; de Lara, E.; Davies, N. Location-Based Services. *IEEE Pervasive Comput.* **2010**, *9*, 11–12. [CrossRef]
27. Bhatti, M.A.; Riaz, R.; Rizvi, S.S.; Shokat, S.; Riaz, F.; Kwon, S.J. Outlier detection in indoor localization and Internet of Things (IoT) using machine learning. *J. Commun. Netw.* **2020**, *22*, 236–243. [CrossRef]
28. Ezzat, M.; Sakr, M.; Elgohary, R.; Khalifa, M.E. Building road segments and detecting turns from GPS tracks. *J. Comput. Sci.* **2018**, *29*, 81–93. [CrossRef]
29. Zheng, Y. Trajectory Data Mining: An Overview. *ACM Trans. Intell Syst. Technol.* **2015**, *6*, 1–41. [CrossRef]
30. Krumm, J. Trajectory Analysis for Driving. In *Computing with Spatial Trajectories*; Zheng, Y., Zhou, X., Eds.; Springer: New York, NY, USA, 2011; pp. 213–241.
31. Chen, C.C.; Chiang, M.F. Trajectory pattern mining: Exploring semantic and time information. In Proceedings of the 2016 Conference on Technologies and Applications of Artificial Intelligence (TAAI), Hsinchu, Taiwan, 25–27 November 2016; IEEE: Manhattan, NY, USA, 2016; pp. 130–137.

32. Teng, X.; Trajcevski, G.; Kim, J.S.; Züfle, A. Semantically Diverse Path Search. In Proceedings of the 2020 21st IEEE International Conference on Mobile Data Management (MDM), Versailles, France, 30 June–3 July 2020; IEEE: Manhattan, NY, USA, 2020; pp. 69–78.
33. Stach, C.; Brodt, A. vHike — A Dynamic Ride-Sharing Service for Smartphones. In Proceedings of the 2011 IEEE 12th International Conference on Mobile Data Management (MDM), Luleå, Sweden, 6–9 June 2011; IEEE: Manhattan, NY, USA, 2011; pp. 333–336.
34. Ceikute, V.; Jensen, C.S. Vehicle Routing with User-Generated Trajectory Data. In Proceedings of the 2015 16th IEEE International Conference on Mobile Data Management (MDM), Pittsburgh, PA, USA, 15–18 June 2015; IEEE: Manhattan, NY, USA, 2015; pp. 14–23.
35. Salim, S.; Turnbull, B.; Moustafa, N. Data analytics of social media 3.0: Privacy protection perspectives for integrating social media and Internet of Things (SM-IoT) systems. *Ad Hoc Netw.* **2022**, *128*, 102786. [CrossRef]
36. Li, N.; Chen, G. Analysis of a Location-Based Social Network. In Proceedings of the 2009 International Conference on Computational Science and Engineering (CSE), Vancouver, BC, Canada, 29–31 August 2009; IEEE: Manhattan, NY, USA, 2009; pp. 263–270.
37. Liu, S.; Li, L.; Tang, J.; Wu, S.; Gaudiot, J.L. *Creating Autonomous Vehicle Systems*, 2nd ed.; Morgan & Claypool: San Rafael, CA, USA, 2020.
38. Primault, V.; Boutet, A.; Mokhtar, S.B.; Brunie, L. The Long Road to Computational Location Privacy: A Survey. *Commun. Surveys Tuts.* **2019**, *21*, 2772–2793. [CrossRef]
39. van Gemert-Pijnen, L.; Kelders, S.M.; Kip, H.; Sanderman, R. (Eds.) *eHealth Research, Theory and Development*; Routledge: London, UK, 2018.
40. Grady, A.; Yoong, S.; Sutherland, R.; Lee, H.; Nathan, N.; Wolfenden, L. Improving the public health impact of eHealth and mHealth interventions. *Aust. N. Z. J. Public Health* **2018**, *42*, 118–119. [CrossRef] [PubMed]
41. Kreps, G.L.; Neuhauser, L. New directions in eHealth communication: Opportunities and challenges. *Patient Educ. Couns.* **2010**, *78*, 329–336. [CrossRef]
42. Marcolino, M.S.; Oliveira, J.a.A.Q.; D'Agostino, M.; Ribeiro, A.L.; Alkmim, M.B.M.; Novillo-Ortiz, D. The Impact of mHealth Interventions: Systematic Review of Systematic Reviews. *JMIR Mhealth Uhealth* **2018**, *6*, e23. [CrossRef]
43. Siewiorek, D. Generation smartphone. *IEEE Spectr.* **2012**, *49*, 54–58. [CrossRef]
44. Bitsaki, M.; Koutras, C.; Koutras, G.; Leymann, F.; Steimle, F.; Wagner, S.; Wieland, M. ChronicOnline: Implementing a mHealth solution for monitoring and early alerting in chronic obstructive pulmonary disease. *Health Inform. J.* **2017**, *23*, 179–207. [CrossRef]
45. Guo, S.; Guo, X.; Zhang, X.; Vogel, D. Doctor–patient relationship strength's impact in an online healthcare community. *Inf. Technol. Dev.* **2018**, *24*, 279–300. [CrossRef]
46. Ball, M.J.; Lillis, J. E-health: Transforming the physician/patient relationship. *Int. J. Med. Inform.* **2001**, *61*, 1–10. [CrossRef]
47. Iyengar, S. Mobile health (mHealth). In *Fundamentals of Telemedicine and Telehealth*; Gogia, S., Ed.; Academic Press: London, UK; San Diego, CA, USA; Cambridge, MA, USA; Oxford, UK, 2020; Chapter 12; pp. 277–294.
48. Rocha, T.A.H.; da Silva, N.C.; Barbosa, A.C.Q.; Elahi, C.; Vissoci, J.a.R.N. mHealth: Smart Wearable Devices and the Challenges of a Refractory Context. In *The Internet and Health in Brazil*; Pereira Neto, A., Flynn, M.B., Eds.; Springer: Cham, Switzerland, 2019; pp. 347–367.
49. Lupton, D. *The Quantified Self*; Polity: Cambridge, UK; Malden, MA, USA, 2016.
50. Swan, M. Sensor Mania! The Internet of Things, Wearable Computing, Objective Metrics, and the Quantified Self 2.0. *J. Sens. Actuator Netw.* **2012**, *1*, 217–253. [CrossRef]
51. Stach, C.; Steimle, F.; Franco da Silva, A.C. TIROL: The Extensible Interconnectivity Layer for mHealth Applications. In Proceedings of the 23rd International Conference on Information and Software Technologies (ICIST), Druskininkai, Lithuania, 12–14 October 2017; Springer: Cham, Switzerland, 2017; pp. 190–202.
52. Swan, M. The Quantified Self: Fundamental Disruption in Big Data Science and Biological Discovery. *Big Data* **2013**, *1*, 85–99. [CrossRef]
53. Chao, D.Y.; Lin, T.M.; Ma, W.Y. Enhanced Self-Efficacy and Behavioral Changes Among Patients With Diabetes: Cloud-Based Mobile Health Platform and Mobile App Service. *JMIR Diabetes* **2019**, *4*, e11017. [CrossRef]
54. Piccialli, F.; Giampaolo, F.; Prezioso, E.; Camacho, D.; Acampora, G. Artificial intelligence and healthcare: Forecasting of medical bookings through multi-source time-series fusion. *Inform. Fusion* **2021**, *74*, 1–16. [CrossRef]
55. Deshpande, P.S.; Sharma, S.C.; Peddoju, S.K. Predictive and Prescriptive Analytics in Big-data Era. In *Security and Data Storage Aspect in Cloud Computing*; Springer: Singapore, 2019; pp. 71–81.
56. Noar, S.M.; Harrington, N.G. *eHealth Applications: Promising Strategies for Behavior Change*; Routledge: New York, NY, USA, 2012.
57. Ben Amor, L.; Lahyani, I.; Jmaiel, M. Data accuracy aware mobile healthcare applications. *Comput. Ind.* **2018**, *97*, 54–66. [CrossRef]
58. Thapa, C.; Camtepe, S. Precision health data: Requirements, challenges and existing techniques for data security and privacy. *Comput. Biol. Med.* **2021**, *129*, 104130. [CrossRef]
59. Kumar, T.; Liyanage, M.; Braeken, A.; Ahmad, I.; Ylianttila, M. From gadget to gadget-free hyperconnected world: Conceptual analysis of user privacy challenges. In Proceedings of the 2017 European Conference on Networks and Communications (EuCNC), Oulu, Finland, 12–15 June 2017; IEEE: Manhattan, NY, USA, 2017; pp. 1–6.

60. Braghin, C.; Cimato, S.; Della Libera, A. Are mHealth Apps Secure? A Case Study. In Proceedings of the 2018 IEEE 42nd Annual Computer Software and Applications Conference (COMPSAC), Tokyo, Japan, 23–27 July 2018; IEEE: Manhattan, NY, USA, 2018; pp. 335–340.
61. Hoy, M.B. Alexa, Siri, Cortana, and More: An Introduction to Voice Assistants. *Med. Ref. Serv. Q.* **2018**, *37*, 81–88. [CrossRef]
62. López, G.; Quesada, L.; Guerrero, L.A. Alexa vs. Siri vs. Cortana vs. Google Assistant: A Comparison of Speech-Based Natural User Interfaces. In Proceedings of the AHFE 2017 International Conference on Human Factors and Systems Interaction (HFSI), Los Angeles, CA, USA, 17–21 July 2017; Springer: Cham, Switzerland, 2018; pp. 241–250.
63. McLean, G.; Osei-Frimpong, K. Hey Alexa … examine the variables influencing the use of artificial intelligent in-home voice assistants. *Comput. Hum. Behav.* **2019**, *99*, 28–37. [CrossRef]
64. Porcheron, M.; Fischer, J.E.; Reeves, S.; Sharples, S. Voice Interfaces in Everyday Life. In Proceedings of the 2018 CHI Conference on Human Factors in Computing Systems (CHI), Montreal, QC, Canada, 21–26 April 2018; ACM: New York, NY, USA, 2018; pp. 1–12.
65. Lei, X.; Tu, G.H.; Liu, A.X.; Li, C.Y.; Xie, T. The Insecurity of Home Digital Voice Assistants – Vulnerabilities, Attacks and Countermeasures. In Proceedings of the 2018 IEEE Conference on Communications and Network Security (CNS), Beijing, China, 30 May–1 June 2018; IEEE: Manhattan, NY, USA, 2018; pp. 1–9.
66. Chung, H.; Park, J.; Lee, S. Digital forensic approaches for Amazon Alexa ecosystem. *Digit. Investig.* **2017**, *22*, S15–S25. [CrossRef]
67. Lopatovska, I.; Rink, K.; Knight, I.; Raines, K.; Cosenza, K.; Williams, H.; Sorsche, P.; Hirsch, D.; Li, Q.; Martinez, A. Talk to me: Exploring user interactions with the Amazon Alexa. *J. Libr. Inf. Sci.* **2019**, *51*, 984–997. [CrossRef]
68. Han, S.; Yang, H. Understanding adoption of intelligent personal assistants: A parasocial relationship perspective. *Ind. Manag. Data Syst.* **2018**, *118*, 618–636. [CrossRef]
69. Bolton, T.; Dargahi, T.; Belguith, S.; Al-Rakhami, M.S.; Sodhro, A.H. On the Security and Privacy Challenges of Virtual Assistants. *Sensors* **2021**, *21*, 2312. [CrossRef] [PubMed]
70. Khan, M.J.; Khan, H.S.; Yousaf, A.; Khurshid, K.; Abbas, A. Modern Trends in Hyperspectral Image Analysis: A Review. *IEEE Access* **2018**, *6*, 14118–14129. [CrossRef]
71. Adjabi, I.; Ouahabi, A.; Benzaoui, A.; Taleb-Ahmed, A. Past, Present, and Future of Face Recognition: A Review. *Electronics* **2020**, *9*, 1188. [CrossRef]
72. Hazelwood, K.; Bird, S.; Brooks, D.; Chintala, S.; Diril, U.; Dzhulgakov, D.; Fawzy, M.; Jia, B.; Jia, Y.; Kalro, A.; et al. Applied Machine Learning at Facebook: A Datacenter Infrastructure Perspective. In Proceedings of the 2018 IEEE International Symposium on High Performance Computer Architecture (HPCA), Vienna, Austria, 24–28 February 2018; IEEE: Manhattan, NY, USA, 2018; pp. 620–629.
73. Taigman, Y.; Yang, M.; Ranzato, M.; Wolf, L. DeepFace: Closing the Gap to Human-Level Performance in Face Verification. In Proceedings of the 2014 IEEE Conference on Computer Vision and Pattern Recognition (CVPR), Columbus, OH, USA, 23–28 June 2014; IEEE: Manhattan, NY, USA, 2014; pp. 1701–1708.
74. Kumar, A.; Kaur, A.; Kumar, M. Face detection techniques: A review. *Artif. Intell. Rev.* **2019**, *52*, 927–948. [CrossRef]
75. Taskiran, M.; Kahraman, N.; Erdem, C.E. Face recognition: Past, present and future (a review). *Digit Signal Process* **2020**, *106*, 102809. [CrossRef]
76. Kortli, Y.; Jridi, M.; Al Falou, A.; Atri, M. Face Recognition Systems: A Survey. *Sensors* **2020**, *20*, 342. [CrossRef]
77. Li, L.; Mu, X.; Li, S.; Peng, H. A Review of Face Recognition Technology. *IEEE Access* **2020**, *8*, 139110–139120. [CrossRef]
78. Senior, A.W.; Pankanti, S. Privacy Protection and Face Recognition. In *Handbook of Face Recognition*; Li, S.Z., Jain, A.K., Eds.; Springer: London, UK, 2021; pp. 671–691.
79. Wang, M.; Deng, W. Deep face recognition: A survey. *Neurocomputing* **2021**, *429*, 215–244. [CrossRef]
80. Nielsen, S.S. (Ed.) *Food Analysis*, 5th ed.; Springer: Cham, Switzerland, 2017.
81. Mishra, G.K.; Barfidokht, A.; Tehrani, F.; Mishra, R.K. Food Safety Analysis Using Electrochemical Biosensors. *Foods* **2018**, *7*, 141. [CrossRef]
82. Korte, R.; Bräcker, J.; Brockmeyer, J. Gastrointestinal digestion of hazelnut allergens on molecular level: Elucidation of degradation kinetics and resistant immunoactive peptides using mass spectrometry. *Mol. Nutr. Food Res.* **2017**, *61*, 1700130. [CrossRef]
83. Berrueta, L.A.; Alonso-Salces, R.M.; Héberger, K. Supervised pattern recognition in food analysis. *J. Chromatogr. A* **2007**, *1158*, 196–214. [CrossRef]
84. Deng, X.; Cao, S.; Horn, A.L. Emerging Applications of Machine Learning in Food Safety. *Annu. Rev. Food Sci. Technol.* **2021**, *12*, 513–538. [CrossRef]
85. Bräcker, J.; Brockmeyer, J. Characterization and Detection of Food Allergens Using High-Resolution Mass Spectrometry: Current Status and Future Perspective. *J. Agric. Food Chem.* **2018**, *66*, 8935–8940. [CrossRef]
86. Mafata, M.; Brand, J.; Medvedovici, A.; Buica, A. Chemometric and sensometric techniques in enological data analysis. *Crit. Rev. Food Sci.* **2022**, 1–15. [CrossRef]
87. Bianco, M.; Ventura, G.; Calvano, C.D.; Losito, I.; Cataldi, T.R. A new paradigm to search for allergenic proteins in novel foods by integrating proteomics analysis and in silico sequence homology prediction: Focus on spirulina and chlorella microalgae. *Talanta* **2022**, *240*, 123188. [CrossRef]
88. Giatrakos, N.; Alevizos, E.; Artikis, A.; Deligiannakis, A.; Garofalakis, M. Complex event recognition in the Big Data era: A survey. *VLDB J.* **2020**, *29*, 313–352. [CrossRef]

89. Alakari, A.; Li, K.F.; Gebali, F. A situation refinement model for complex event processing. *Knowl.-Based Syst.* **2020**, *198*, 105881. [CrossRef]
90. Cardoso, D.R.; Andrade-Sobrinho, L.G.; Leite-Neto, A.F.; Reche, R.V.; Isique, W.D.; Ferreira, M.M.C.; Lima-Neto, B.S.; Franco, D.W. Comparison between Cachaça and Rum Using Pattern Recognition Methods. *J. Agric. Food Chem.* **2004**, *52*, 3429–3433. [CrossRef]
91. Şen, G.; Medeni, İ.T.; Şen, K.Ö.; Durakbasa, N.M.; Medeni, T.D. Sensor Based Intelligent Measurement and Blockchain in Food Quality Management. In *Digitizing Production Systems: Selected Papers from ISPR2021, 7–9 October 2021, Online, Turkey*; Durakbasa, N.M., Gençyılmaz, M.G., Eds.; Springer: Cham, Switzerland, 2022; pp. 323–334.
92. Nielsen, K.M. Biosafety Data as Confidential Business Information. *PLOS Biol.* **2013**, *11*, e1001499. [CrossRef] [PubMed]
93. Bobadilla, J.; Ortega, F.; Hernando, A.; Gutiérrez, A. Recommender systems survey. *Knowl.-Based Syst.* **2013**, *46*, 109–132. [CrossRef]
94. Lu, J.; Wu, D.; Mao, M.; Wang, W.; Zhang, G. Recommender system application developments: A survey. *Decis. Support Syst.* **2015**, *74*, 12–32. [CrossRef]
95. Maske, A.R.; Joglekar, B. An Algorithmic Approach for Mining Customer Behavior Prediction in Market Basket Analysis. In Proceedings of the Sixth International Conference on Innovations in Computer Science and Engineering (ICICSE), Hyderabad, India, 17–18 August 2018; Springer: Singapore, 2019; pp. 31–38.
96. Lops, P.; de Gemmis, M.; Semeraro, G. Content-based Recommender Systems: State of the Art and Trends. In *Recommender Systems Handbook*; Ricci, F., Rokach, L., Shapira, B., Kantor, P.B., Eds.; Springer: Boston, MA, USA, 2011; pp. 73–105.
97. Carrer-Neto, W.; Hernández-Alcaraz, M.L.; Valencia-García, R.; García-Sánchez, F. Social knowledge-based recommender system. Application to the movies domain. *Expert Syst. Appl.* **2012**, *39*, 10990–11000. [CrossRef]
98. Afoudi, Y.; Lazaar, M.; Al Achhab, M. Collaborative Filtering Recommender System. In Proceedings of the International Conference on Advanced Intelligent Systems for Sustainable Development (AI2SD), Tangier, Morocco, 12–14 July 2018; Springer: Cham, Switzerland, 2019; pp. 332–345.
99. Thorat, P.B.; Goudar, R.M.; Barve, S.S. Survey on Collaborative Filtering, Content-based Filtering and Hybrid Recommendation System. *Int. J. Comput. Appl.* **2015**, *110*, 31–36.
100. Resnick, P.; Varian, H.R. Recommender Systems. *Commun. ACM* **1997**, *40*, 56–58. [CrossRef]
101. Saad, R. Discovery, development, and current applications of DNA identity testing. In *Baylor University Medical Center Proceedings*; Taylor & Francis: New York, NY, USA, 2005; Volume 18, pp. 130–133.
102. Jin, Z.; Liu, Y. DNA methylation in human diseases. *Genes Dis.* **2018**, *5*, 1–8. [CrossRef]
103. Onabote, O.; Hassan, H.M.; Isovic, M.; Torchia, J. The Role of Thymine DNA Glycosylase in Transcription, Active DNA Demethylation, and Cancer. *Cancers* **2022**, *14*, 765. [CrossRef]
104. Li, X.; Liu, Y.; Salz, T.; Hansen, K.D.; Feinberg, A. Whole-genome analysis of the methylome and hydroxymethylome in normal and malignant lung and liver. *Genome Res.* **2016**, *26*, 1730–1741. [CrossRef]
105. Ahmed, I.; Jeon, G. Enabling Artificial Intelligence for Genome Sequence Analysis of COVID-19 and Alike Viruses. *Interdiscip Sci.* **2021**, 1–16. *Online ahead of print.* [CrossRef]
106. Wang, G.; Pu, P.; Shen, T. An efficient gene bigdata analysis using machine learning algorithms. *Multimed. Tools Appl.* **2020**, *97*, 9847–9870. [CrossRef]
107. Schwab, A.P.; Luu, H.S.; Wang, J.; Park, J.Y. Genomic Privacy. *Clin. Chem.* **2018**, *64*, 1696–1703. [CrossRef]
108. Rudin, C. Stop explaining black box machine learning models for high stakes decisions and use interpretable models instead. *Nat. Mach. Intell.* **2019**, *1*, 206–215. [CrossRef] [PubMed]
109. Angelov, P.; Soares, E. Towards explainable deep neural networks (xDNN). *Neural Netw.* **2020**, *130*, 185–194. [CrossRef]
110. Almusaylim, Z.A.; Jhanjhi, N. Comprehensive Review: Privacy Protection of User in Location-Aware Services of Mobile Cloud Computing. *Wireless Pers. Commun.* **2020**, *111*, 541–564. [CrossRef]
111. Finck, M.; Pallas, F. They who must not be identified—Distinguishing personal from non-personal data under the GDPR. *Int. Data Priv. Law* **2020**, *10*, 11–36. [CrossRef]
112. Rassouli, B.; Rosas, F.E.; Gündüz, D. Data Disclosure Under Perfect Sample Privacy. *IEEE Trans. Inf. Forensics Secur.* **2020**, *15*, 2012–2025. [CrossRef]
113. Al-Rubaie, M.; Chang, J.M. Privacy-Preserving Machine Learning: Threats and Solutions. *IEEE Secur. Priv.* **2019**, *17*, 49–58. [CrossRef]
114. Dou, H.; Chen, Y.; Yang, Y.; Long, Y. A secure and efficient privacy-preserving data aggregation algorithm. *J. Ambient Intell. Humaniz. Comput.* **2022**, *13*, 1495–1503. [CrossRef]
115. Liu, B.; Ding, M.; Shaham, S.; Rahayu, W.; Farokhi, F.; Lin, Z. When Machine Learning Meets Privacy: A Survey and Outlook. *ACM Comput. Surv.* **2021**, *54*, 31:1–31:36. [CrossRef]
116. Alpers, S.; Oberweis, A.; Pieper, M.; Betz, S.; Fritsch, A.; Schiefer, G.; Wagner, M. PRIVACY-AVARE: An approach to manage and distribute privacy settings. In Proceedings of the 2017 3rd IEEE International Conference on Computer and Communications (ICCC), Chengdu, China, 13–16 December 2017; IEEE: Manhattan, NY, USA, 2017; pp. 1460–1468.
117. Jiang, H.; Li, J.; Zhao, P.; Zeng, F.; Xiao, Z.; Iyengar, A. Location Privacy-Preserving Mechanisms in Location-Based Services: A Comprehensive Survey. *ACM Comput. Surv.* **2021**, *54*, 4:1–4:36. [CrossRef]

118. Ardagna, C.A.; Cremonini, M.; Damiani, E.; De Capitani di Vimercati, S.; Samarati, P. Location Privacy Protection Through Obfuscation-Based Techniques. In Proceedings of the 21st Annual IFIP WG 11.3 Working Conference on Data and Applications Security (DBSec), Redondo Beach, CA, USA, 8–11 July 2007; Springer: Berlin/Heidelberg, Germany, 2007; pp. 47–60.
119. Alpers, S.; Betz, S.; Fritsch, A.; Oberweis, A.; Schiefer, G.; Wagner, M. Citizen Empowerment by a Technical Approach for Privacy Enforcement. In Proceedings of the 8th International Conference on Cloud Computing and Services Science (CLOSER), Funchal, Madeira, Portugal, 19–21 March 2018; SciTePress: Setúbal, Portugal, 2018; pp. 589–595.
120. Kido, H.; Yanagisawa, Y.; Satoh, T. An anonymous communication technique using dummies for location-based services. In Proceedings of the 2005 International Conference on Pervasive Services (ICPS), Santorini, Greece, 11–14 July 2005; IEEE: Manhattan, NY, USA, 2005; pp. 88–97.
121. Hara, T.; Suzuki, A.; Iwata, M.; Arase, Y.; Xie, X. Dummy-Based User Location Anonymization Under Real-World Constraints. *IEEE Access* **2016**, *4*, 673–687. [CrossRef]
122. Siddiqie, S.; Mondal, A.; Reddy, P.K. An Improved Dummy Generation Approach for Enhancing User Location Privacy. In Proceedings of the 26th International Conference on Database Systems for Advanced Applications (DASFAA), Taipei, Taiwan, 11–14 April 2021; Springer: Cham, Switzerland, 2021; pp. 487–495.
123. Ma, Y.; Bai, X.; Wang, Z. Trajectory Privacy Protection Method based on Shadow vehicles. In Proceedings of the 2021 IEEE International Conference on Parallel & Distributed Processing with Applications, Big Data & Cloud Computing, Sustainable Computing & Communications, Social Computing & Networking (ISPA/BDCloud/SocialCom/SustainCom), New York, NY, USA, 30 September–3 October 2021; IEEE: Manhattan, NY, USA, 2021; pp. 668–673.
124. Khazbak, Y.; Fan, J.; Zhu, S.; Cao, G. Preserving personalized location privacy in ride-hailing service. *Tsinghua Sci. Technol.* **2020**, *25*, 743–757. [CrossRef]
125. Li, C.; Palanisamy, B. Reversible spatio-temporal perturbation for protecting location privacy. *Comput. Commun.* **2019**, *135*, 16–27. [CrossRef]
126. He, Y.; Chen, J. User location privacy protection mechanism for location-based services. *Digit. Commun. Netw.* **2021**, *7*, 264–276. [CrossRef]
127. Stach, C.; Bräcker, J.; Eichler, R.; Giebler, C.; Gritti, C. How to Provide High-Utility Time Series Data in a Privacy-Aware Manner: A VAULT to Manage Time Series Data. *Int. J. Adv. Secur.* **2020**, *13*, 88–108.
128. Pourahmadi, M. Estimation and Interpolation of Missing Values of a Stationary Time Series. *J. Time Ser. Anal.* **1989**, *10*, 149–169. [CrossRef]
129. Ramosaj, B.; Pauly, M. Predicting missing values: A comparative study on non-parametric approaches for imputation. *Computation Stat.* **2019**, *34*, 1741–1764. [CrossRef]
130. Thomakos, D. Smoothing Non-Stationary Time Series Using the Discrete Cosine Transform. *J. Syst. Sci. Complex* **2016**, *29*, 382–404. [CrossRef]
131. Rhif, M.; Ben Abbes, A.; Farah, I.R.; Martínez, B.; Sang, Y. Wavelet Transform Application for/in Non-Stationary Time-Series Analysis: A Review. *Appl. Sci.* **2019**, *9*, 1345. [CrossRef]
132. Dwork, C.; Kenthapadi, K.; McSherry, F.; Mironov, I.; Naor, M. Our Data, Ourselves: Privacy Via Distributed Noise Generation. In Proceedings of the 24th Annual International Conference on the Theory and Applications of Cryptographic Techniques (EUROCRYPT), St. Petersburg, Russia, 28 May–1 June 2006; Springer: Berlin/Heidelberg, Germany, 2006; pp. 486–503.
133. Gao, Q.; Zhu, L.; Lin, Y.; Chen, X. Anomaly Noise Filtering with Logistic Regression and a New Method for Time Series Trend Computation for Monitoring Systems. In Proceedings of the 2019 IEEE 27th International Conference on Network Protocols (ICNP), Chicago, IL, USA, 8–10 October 2019; IEEE: Manhattan, NY, USA, 2019; pp. 1–6.
134. Moon, Y.S.; Kim, H.S.; Kim, S.P.; Bertino, E. Publishing Time-Series Data under Preservation of Privacy and Distance Orders. In Proceedings of the 21th International Conference on Database and Expert Systems Applications (DEXA), Bilbao, Spain, 30 August–3 September 2010; Springer: Berlin/Heidelberg, Germany, 2010; pp. 17–31.
135. Choi, M.J.; Kim, H.S.; Moon, Y.S. Publishing Sensitive Time-Series Data under Preservation of Privacy and Distance Orders. *Int. J. Innov. Comput. Inf. Control* **2012**, *8*, 3619–3638.
136. Cheng, P.; Roedig, U. Personal Voice Assistant Security and Privacy–A Survey. *Proc IEEE (Early Access)* **2022**, 1–32. [CrossRef]
137. Mhaidli, A.; Venkatesh, M.K.; Zou, Y.; Schaub, F. Listen Only When Spoken To: Interpersonal Communication Cues as Smart Speaker Privacy Controls. *Proc. Priv. Enhanc. Technol.* **2020**, *2020*, 251–270. [CrossRef]
138. Chen, S.; Ren, K.; Piao, S.; Wang, C.; Wang, Q.; Weng, J.; Su, L.; Mohaisen, A. You Can Hear But You Cannot Steal: Defending Against Voice Impersonation Attacks on Smartphones. In Proceedings of the 2017 IEEE 37th International Conference on Distributed Computing Systems (ICDCS), Atlanta, GA, USA, 5–8 June 2017; IEEE: Manhattan, NY, USA, 2017; pp. 183–195.
139. Gao, C.; Chandrasekaran, V.; Fawaz, K.; Banerjee, S. Traversing the Quagmire That is Privacy in Your Smart Home. In Proceedings of the 2018 Workshop on IoT Security and Privacy (IoT S&P), Budapest, Hungary, 20 August 2018; ACM: New York, NY, USA, 2018; pp. 22–28.
140. Saade, A.; Dureau, J.; Leroy, D.; Caltagirone, F.; Coucke, A.; Ball, A.; Doumouro, C.; Lavril, T.; Caulier, A.; Bluche, T.; et al. Spoken Language Understanding on the Edge. In Proceedings of the 2019 Fifth Workshop on Energy Efficient Machine Learning and Cognitive Computing—NeurIPS Edition (EMC2-NIPS), Vancouver, BC, Canada, 13 December 2019; IEEE: Manhattan, NY, USA, 2019; pp. 57–61.

141. He, Y.; Sainath, T.N.; Prabhavalkar, R.; McGraw, I.; Alvarez, R.; Zhao, D.; Rybach, D.; Kannan, A.; Wu, Y.; Pang, R.; et al. Streaming End-to-end Speech Recognition for Mobile Devices. In Proceedings of the 2019 IEEE International Conference on Acoustics, Speech and Signal Processing (ICASSP), Brighton, UK, 12–17 May 2019; IEEE: Manhattan, NY, USA, 2019; pp. 6381–6385.
142. Tiwari, V.; Hashmi, M.F.; Keskar, A.; Shivaprakash, N.C. Virtual home assistant for voice based controlling and scheduling with short speech speaker identification. *Multimed. Tools Appl.* **2020**, *79*, 5243–5268. [CrossRef]
143. Perez, A.J.; Zeadally, S.; Griffith, S. Bystanders' Privacy. *IT Prof* **2017**, *19*, 61–65. [CrossRef]
144. Hernández Acosta, L.; Reinhardt, D. A survey on privacy issues and solutions for Voice-controlled Digital Assistants. *Pervasive Mob. Comput.* **2022**, *80*, 101523. [CrossRef]
145. Qian, J.; Du, H.; Hou, J.; Chen, L.; Jung, T.; Li, X.Y. Hidebehind: Enjoy Voice Input with Voiceprint Unclonability and Anonymity. In Proceedings of the 16th ACM Conference on Embedded Networked Sensor Systems (SenSys), Shenzhen, China, 4–7 November 2018; ACM: New York, NY, USA, 2018; pp. 82–94.
146. Tian, C.; Fei, L.; Zheng, W.; Xu, Y.; Zuo, W.; Lin, C.W. Deep learning on image denoising: An overview. *Neural Netw.* **2020**, *131*, 251–275. [CrossRef]
147. Oh, S.J.; Benenson, R.; Fritz, M.; Schiele, B. Faceless Person Recognition: Privacy Implications in Social Media. In Proceedings of the 14th European Conference on Computer Vision (ECCV), Amsterdam, The Netherlands, 11–14 October 2016; Springer: Cham, Switzerland, 2016; pp. 19–35.
148. Fan, L. Practical Image Obfuscation with Provable Privacy. In Proceedings of the 2019 IEEE International Conference on Multimedia and Expo (ICME), Shanghai, China, 8–12 July 2019; IEEE: Manhattan, NY, USA, 2019; pp. 784–789.
149. Yu, J.; Zhang, B.; Kuang, Z.; Lin, D.; Fan, J. iPrivacy: Image Privacy Protection by Identifying Sensitive Objects via Deep Multi-Task Learning. *IEEE Trans. Inf. Forensics Secur.* **2017**, *12*, 1005–1016. [CrossRef]
150. Sarwar, O.; Rinner, B.; Cavallaro, A. A Privacy-Preserving Filter for Oblique Face Images Based on Adaptive Hopping Gaussian Mixtures. *IEEE Access* **2019**, *7*, 142623–142639. [CrossRef]
151. Gehrke, J.; Lui, E.; Pass, R. Towards Privacy for Social Networks: A Zero-Knowledge Based Definition of Privacy. In Proceedings of the 8th Conference on Theory of Cryptography (TCC), Providence, RI, USA, 28–30 March 2011; Springer: Berlin/Heidelberg, Germany, 2011; pp. 432–449.
152. Quoc, D.L.; Beck, M.; Bhatotia, P.; Chen, R.; Fetzer, C.; Strufe, T. PrivApprox: Privacy-Preserving Stream Analytics. In Proceedings of the 2017 USENIX Annual Technical Conference (USENIX ATC), Santa Clara, CA, USA, 12–14 July 2017; USENIX Association: Berkeley, CA, USA, 2017; pp. 659–672.
153. Li, F.; Wang, N.; Gu, Y.; Chen, Z. Effective Privacy Preservation over Composite Events with Markov Correlations. In Proceedings of the 2016 13th Web Information Systems and Applications Conference (WISA), Wuhan, China, 23–25 September 2016; IEEE: Manhattan, NY, USA, 2016; pp. 215–220.
154. Churi, P.P.; Pawar, A.V. A Systematic Review on Privacy Preserving Data Publishing Techniques. *J. Eng. Sci. Technol. Rev.* **2019**, *12*, 17–25. [CrossRef]
155. Stach, C.; Mitschang, B. ACCESSORS: A Data-Centric Permission Model for the Internet of Things. In Proceedings of the 4th International Conference on Information Systems Security and Privacy (ICISSP), Funchal, Madeira, Portugal, 22–24 January 2018; SciTePress: Setúbal, Portugal, 2018; pp. 30–40.
156. Palanisamy, S.M.; Dürr, F.; Tariq, M.A.; Rothermel, K. Preserving Privacy and Quality of Service in Complex Event Processing through Event Reordering. In Proceedings of the 12th ACM International Conference on Distributed and Event-Based Systems (DEBS), Hamilton, New Zealand, 25–29 June 2018; ACM: New York, NY, USA, 2018; pp. 40–51.
157. Palanisamy, S.M. Towards Multiple Pattern Type Privacy Protection in Complex Event Processing Through Event Obfuscation Strategies. In *Data Privacy Management, Cryptocurrencies and Blockchain Technology: ESORICS 2020 International Workshops, DPM 2020 and CBT 2020, Guildford, UK, 17–18 September 2020, Revised Selected Papers*; Garcia-Alfaro, J., Navarro-Arribas, G., Herrera-Joancomarti, J., Eds.; Springer: Cham, Switzerland, 2020; pp. 178–194.
158. Dwork, C. Differential Privacy. In Proceedings of the 33rd International Colloquium on Automata, Languages, and Programming (ICALP), Venice, Italy, 10–14 July 2006; Springer: Berlin/Heidelberg, Germany, 2006; pp. 1–12.
159. Psychoula, I.; Chen, L.; Amft, O. Privacy Risk Awareness in Wearables and the Internet of Things. *IEEE Pervasive Comput* **2020**, *19*, 60–66. [CrossRef]
160. Machanavajjhala, A.; He, X.; Hay, M. Differential Privacy in the Wild: A Tutorial on Current Practices & Open Challenges. In Proceedings of the 2017 ACM International Conference on Management of Data (SIGMOD), Chicago, IL, USA, 14–19 May 2017; ACM: New York, NY, USA, 2017; pp. 1727–1730.
161. Jain, P.; Gyanchandani, M.; Khare, N. Differential privacy: Its technological prescriptive using big data. *J. Big. Data* **2018**, *5*, 15. [CrossRef]
162. Zhu, T.; Li, G.; Zhou, W.; Yu, P.S. Differentially Private Recommender System. In *Differential Privacy and Applications*; Springer: Cham, Switzerland, 2017; pp. 107–129.
163. Li, T.; Sahu, A.K.; Talwalkar, A.; Smith, V. Federated Learning: Challenges, Methods, and Future Directions. *IEEE Signal Process. Mag.* **2020**, *37*, 50–60. [CrossRef]
164. Yang, Q.; Liu, Y.; Cheng, Y.; Kang, Y.; Chen, T.; Yu, H. *Federated Learning*; Morgan & Claypool: San Rafael, CA, USA, 2019.
165. Wu, X.; Zhang, Y.; Shi, M.; Li, P.; Li, R.; Xiong, N.N. An adaptive federated learning scheme with differential privacy preserving. *Future Gener. Comput. Syst.* **2022**, *127*, 362–372. [CrossRef]

166. Rieke, N.; Hancox, J.; Li, W.; Milletarì, F.; Roth, H.R.; Albarqouni, S.; Bakas, S.; Galtier, M.N.; Landman, B.A.; Maier-Hein, K.; et al. The future of digital health with federated learning. *NPJ Digit. Med.* **2020**, *3*, 119. [CrossRef] [PubMed]
167. Wang, H.; Zhao, Q.; Wu, Q.; Chopra, S.; Khaitan, A.; Wang, H. Global and Local Differential Privacy for Collaborative Bandits. In Proceedings of the Fourteenth ACM Conference on Recommender Systems (RecSys), Rio de Janeiro, Brazil, 22–26 September 2020; ACM: New York, NY, USA, 2020; pp. 150–159.
168. Chai, Q.; Gong, G. Verifiable symmetric searchable encryption for semi-honest-but-curious cloud servers. In Proceedings of the 2012 IEEE International Conference on Communications (ICC), Ottawa, ON, Canada, 10–15 June 2012; IEEE: Manhattan, NY, USA, 2012; pp. 917–922.
169. Piercy, N.; Giles, W. Making SWOT Analysis Work. *Mark. Intell. Plan.* **1989**, *7*, 5–7. [CrossRef]
170. Benzaghta, M.A.; Elwalda, A.; Mousa, M.M.; Erkan, I.; Rahman, M. SWOT Analysis Applications: An Integrative Literature Review. *J. Glob. Bus. Insights* **2021**, *6*, 55–73. [CrossRef]
171. Young, W.; Leveson, N.G. An Integrated Approach to Safety and Security Based on Systems Theory. *Commun. ACM* **2014**, *127*, 31–35. [CrossRef]
172. Shapiro, S.S. Privacy Risk Analysis Based on System Control Structures: Adapting System-Theoretic Process Analysis for Privacy Engineering. In Proceedings of the 2016 IEEE Security and Privacy Workshops (SPW), San Jose, CA, USA, 22–26 May 2016; IEEE: Manhattan, NY, USA, 2016; pp. 17–24.
173. Mindermann, K.; Riedel, F.; Abdulkhaleq, A.; Stach, C.; Wagner, S. Exploratory Study of the Privacy Extension for System Theoretic Process Analysis (STPA-Priv) to elicit Privacy Risks in eHealth. In Proceedings of the 2017 IEEE 25th International Requirements Engineering Conference Workshops, 4th International Workshop on Evolving Security & Privacy Requirements Engineering (REW/ESPRE), Lisbon, Portugal, 4–8 September 2017; IEEE: Manhattan, NY, USA, 2017; pp. 90–96.
174. Hanisch, S.; Cabarcos, P.A.; Parra-Arnau, J.; Strufe, T. Privacy-Protecting Techniques for Behavioral Data: A Survey. *CoRR* **2021**, *abs/2109.04120*, 1–43.
175. Wu, X.; Zhang, Y.; Wang, A.; Shi, M.; Wang, H.; Liu, L. MNSSp3: Medical big data privacy protection platform based on Internet of things. *Neural Comput. Applic* **2022**, *34*, 11491–11505. [CrossRef]
176. Stach, C.; Gritti, C.; Mitschang, B. Bringing Privacy Control Back to Citizens: DISPEL — A Distributed Privacy Management Platform for the Internet of Things. In Proceedings of the 35th ACM/SIGAPP Symposium on Applied Computing (SAC), Brno, Czech Republic, 30 March–3 April 2020; ACM: New York, NY, USA, 2020; pp. 1272–1279.
177. Shapiro, S.S. Time to Modernize Privacy Risk Assessment. *Issues Sci. Technol.* **2021**, *38*, 20–22.
178. Stach, C.; Steimle, F. Recommender-based Privacy Requirements Elicitation—EPICUREAN: An Approach to Simplify Privacy Settings in IoT Applications with Respect to the GDPR. In Proceedings of the 34th ACM/SIGAPP Symposium On Applied Computing (SAC), Limassol, Cyprus, 8–12 April 2019; ACM: New York, NY, USA, 2019; pp. 1500–1507.
179. Stach, C. How to Deal with Third Party Apps in a Privacy System — The PMP Gatekeeper. In Proceedings of the 2015 IEEE 16th International Conference on Mobile Data Management (MDM), Pittsburgh, PA, USA, 15–18 June 2015; IEEE: Manhattan, NY, USA, 2015; pp. 167–172.
180. Beierle, F.; Tran, V.T.; Allemand, M.; Neff, P.; Schlee, W.; Probst, T.; Pryss, R.; Zimmermann, J. Context Data Categories and Privacy Model for Mobile Data Collection Apps. *Procedia Comput. Sci.* **2018**, *134*, 18–25. [CrossRef]
181. Stach, C.; Alpers, S.; Betz, S.; Dürr, F.; Fritsch, A.; Mindermann, K.; Palanisamy, S.M.; Schiefer, G.; Wagner, M.; Mitschang, B.; et al. The AVARE PATRON - A Holistic Privacy Approach for the Internet of Things. In Proceedings of the 15th International Joint Conference on e-Business and Telecommunications (SECRYPT), Porto, Portugal, 26–28 July 2018; SciTePress: Setúbal, Portugal, 2018; pp. 372–379.
182. Stach, C.; Giebler, C.; Wagner, M.; Weber, C.; Mitschang, B. AMNESIA: A Technical Solution towards GDPR-compliant Machine Learning. In Proceedings of the 6th International Conference on Information Systems Security and Privacy (ICISSP), Valletta, Malta, 25–27 February 2020; SciTePress: Setúbal, Portugal, 2020; pp. 21–32.
183. Busch-Casler, J.; Radic, M. Personal Data Markets: A Narrative Review on Influence Factors of the Price of Personal Data. In Proceedings of the 16th International Conference on Research Challenges in Information Science (RCIS), Barcelona, Spain, 17–20 May 2022; Springer: Cham, Switzerland, 2022; pp. 3–19.
184. Driessen, S.W.; Monsieur, G.; Van Den Heuvel, W.J. Data Market Design: A Systematic Literature Review. *IEEE Access* **2022**, *10*, 33123–33153. [CrossRef]
185. Spiekermann, S.; Acquisti, A.; Böhme, R.; Hui, K.L. The challenges of personal data markets and privacy. *Electron Mark* **2015**, *25*, 161–167. [CrossRef]
186. Stach, C.; Gritti, C.; Przytarski, D.; Mitschang, B. Trustworthy, Secure, and Privacy-aware Food Monitoring Enabled by Blockchains and the IoT. In Proceedings of the 2020 IEEE International Conference on Pervasive Computing and Communications Workshops (PerCom Workshops), Austin, TX, USA, 23–27 March 2020; IEEE: Manhattan, NY, USA, 2020; pp. 50:1–50:4.
187. Bernal Bernabe, J.; Canovas, J.L.; Hernandez-Ramos, J.L.; Torres Moreno, R.; Skarmeta, A. Privacy-Preserving Solutions for Blockchain: Review and Challenges. *IEEE Access* **2019**, *7*, 164908–164940. [CrossRef]
188. Gritti, C.; Chen, R.; Susilo, W.; Plantard, T. Dynamic Provable Data Possession Protocols with Public Verifiability and Data Privacy. In Proceedings of the 13th International Conference on Information Security Practice and Experience (ISPEC), Melbourne, VIC, Australia, 13–15 December 2017; Springer: Cham, Switzerland, 2017; pp. 485–505.

189. Boneh, D.; Bonneau, J.; Bünz, B.; Fisch, B. Verifiable Delay Functions. In Proceedings of the 38th International Cryptology Conference (Crypto), Santa Barbara, CA, USA, 17–19 August 2018; Springer: Cham, Switzerland, 2018; pp. 757–788.
190. Gritti, C.; Li, H. Efficient Publicly Verifiable Proofs of Data Replication and Retrievability Applicable for Cloud Storage. *Adv. Sci. Technol. Eng. Syst. J.* **2022**, *7*, 107–124. [CrossRef]
191. Chow, R.; Golle, P. Faking Contextual Data for Fun, Profit, and Privacy. In Proceedings of the 8th ACM Workshop on Privacy in the Electronic Society (WPES), Chicago, IL, USA, 9 November 2009; ACM: New York, NY, USA, 2009; pp. 105–108.
192. Gritti, C.; Önen, M.; Molva, R. Privacy-Preserving Delegable Authentication in the Internet of Things. In Proceedings of the 34th ACM/SIGAPP Symposium on Applied Computing (SAC), Limassol, Cyprus, 8–12 April 2019; ACM: New York, NY, USA, 2019; pp. 861–869.
193. Litou, I.; Kalogeraki, V.; Katakis, I.; Gunopulos, D. Real-Time and Cost-Effective Limitation of Misinformation Propagation. In Proceedings of the 2016 17th IEEE International Conference on Mobile Data Management (MDM), Porto, Portugal, 13–16 June 2016; IEEE: Manhattan, NY, USA, 2016; pp. 158–163.
194. Litou, I.; Kalogeraki, V.; Katakis, I.; Gunopulos, D. Efficient and timely misinformation blocking under varying cost constraints. *Online Soc. Netw. Media* **2017**, *2*, 19–31. [CrossRef]

future internet

MDPI

Article

A Blockchain Self-Sovereign Identity for Open Banking Secured by the Customer's Banking Cards

Khaled A. M. Ahmed *, Sabry F. Saraya, John F. Wanis and Amr M. T. Ali-Eldin

Computer Engineering and Control Systems Department, Faculty of Engineering, Mansoura University, Mansoura 35516, Egypt; drsfsaraya@mans.edu.eg (S.F.S.); jfzaki@mans.edu.eg (J.F.W.); amr.ali-eldin@mans.edu.eg (A.M.T.A.-E.)
* Correspondence: khalida.whab@gmail.com

Abstract: Open finance is evolving and extending open banking. This creates a large context that implies a financial and identity data exchange paradigm, which faces challenges to balance customer experience, security, and the self-control over personal identity information. We propose Self-Sovereign Banking Identity (SSBI), a Blockchain-based self-sovereign identity (SSI) to secure private data sharing by utilizing trusted customer's banking cards as a key storage and identity transaction-signing enclave. The design and implementation of the SSI framework is based on the Veramo SDK and Ethereum to overcome the limitation of signing curve availability on the current banking Java Cards needed for Hyperledger Indy. SSBI uses the elliptic curve SECP256K1 for transaction signing, which exists for several payment cards in the market. SSBI enables automated financial services and trust in the service provider communication. This work analyzes the flow and framework components, and evaluates the usability, integration, and performance in terms of throughput, latency, security, and complexity. Furthermore, the proposed approach is compared with related solutions. The presented prototype implementation is based on a test Ethereum network and signing transactions on the banking card. The preliminary results show that SSBI provides an effective solution for integrating the customer's banking cards to secure open banking identity exchange. Furthermore, it allows the integration of several scenarios to support trusted open banking. The Blockchain layer settings need to be scaled and improved before real-world implementation.

Keywords: digital identity; Blockchain; self-sovereign identity; banking card; open banking

Citation: Ahmed, K.A.M.; Saraya, S.F.; Wanis, J.F.; Ali-Eldin, A.M.T. A Blockchain Self-Sovereign Identity for Open Banking Secured by the Customer's Banking Cards. *Future Internet* **2023**, *15*, 208. https://doi.org/10.3390/fi15060208

Academic Editors: Christoph Stach, Clémentine Gritti and Claude Chaudet

Received: 8 April 2023
Revised: 16 May 2023
Accepted: 5 June 2023
Published: 8 June 2023

1. Introduction

In the modern transformation of digital banking, the customer does not have control of the sharing of their personal and financial data with the increasing number of authorized financial service providers. Open finance [1] is an emerging field within open banking, in which financial institutes, regulated websites, and financial applications obtain access to transactional details, for example, customer financial data such as savings, insurance, and customer credit (or part of it) from the user's bank accounts and payment services, usually via the Application Programming Interface (API) and based on customer consent. The purpose is to not only recommend cheaper services, but also to provide advice and customized product recommendations, similar to open banking but with a kind of 'read' data permission [2]. Open finance comes with 'write' permissions to execute cost savings on behalf of the customer, for example, transferring an additional USD 150 to the user's saving account with the aim of tailoring the best financial decision for the customer.

The banking sector has a significant stake in the digital trust era, including via customers' personal data, Know Your Customer (KYC) procedures, policy regulations, and secure authentication. There is a strong potential for banks to invest in identity management and expand payment card services and digital payment credentials with digital identity capabilities based on the strong position of trust and security measures. Banks, as identity actors, can act as an identity issuer, deliver digital identity services, and provide credentials and authentication services.

In such a sensitive data-sharing context, many challenges appear, such as banks trying to adapt to a higher volume of transactions, data polices, and governmental regulations to ensure proper security and privacy controls for users to manage their own data between financial pillars. To avoid privacy leakage in such a paradigm, recalling for instance the Facebook–Cambridge Analytical data scandal [3], customers want to control and grant permission to each sharing transaction. In our proposed prototype, SSBI provides self-sovereign identity (SSI) features based on Veramo version 3 APIs [4] and strengthens the security of managing customer private keys using industry-trusted banking cards, and related form factors, assuming at least one banking account is available before the registration phase. Banks and other parties cannot share the data without permission from the customer in a decentralized framework. These data can be limited to credentials and proofs instead of disclosing the detailed personal identity of financial information. This attempts to address the issues of centralized or cloud-based identity systems, including leakage of immutability, traceability, and third-party control over individual data.

In this paper, we propose a framework to enable users to control their identities and privacy in the context of open finance and open banking, using a Blockchain-based SSI platform to improve the trust between customers and involved financial parties. SSBI integrates the banking card to improve the security of signing identity creation and sharing transactions, and considers a seamless user experience. In addition, this work is validated against a typical use case scenario and possible results are discussed to evaluate the proposal in contrast to related solutions.

During the demonstrated implementation, we designed and implemented an SSI framework based on the Veramo SDK with an underlying Ethereum Blockchain to extend the model, as suggested in [5]. Our contributions include the integration of the customer's banking cards to hold the identity secret keys, the signing of pairwise DIDs (Decentralized Identifiers), and verifiable proofs to balance the security and customer experience. This improves the user's trust by controlling the identity enrolment and claims of several financial services provided by open banking institutions. Our work overcomes the limitation of signing curve availability on banking Java Cards that are required to sign Hyperledger Indy [6] transactions, as stated by [7]. The Ethereum signing curve exists for several market banking cards, which we have utilized for this work, to issue ethr DID and VCs (Verifiable Credentials). Furthermore, by storing the private key and performing the transaction signing on the trusted smart card, the level of trust is increased compared to manipulating these keys on a mobile device's memory.

This work applies the proposed framework to a typical open banking case, and describes the integration and the method used to implement the card personalization flow to ensure feasible deployment of the identity applet for the banking Java Card. SSBI implements the SDK calls to integrate with a mobile app, which in turn can be part of a mobile banking app and the Internet banking portals. The proposed work allows identity key recovery by splitting the secret key seeds and passphrase between the trusted bank (card issuer) and the identity owner (the customer).

This paper is organized as follows: Section 2 presents the underlying background. In Section 3 we present related work in contrast to SSBI. Section 4 describes the proposed solution, design objectives, platform layers, and architecture, and breaks down the components. Section 5 provides the proposed flow and related phases. Section 6 describes the implementation and the tools used, in addition to discussing key recovery and integration of customer banking cards for SSI operations. Section 7 presents the discussion of the security and performance evaluation, in addition to comparative analysis. Section 8 provides the conclusion and discussion of future work.

2. Background

The user-centric model, and now the SSI approach, are consequences of the revolution of identity management from centralized silos to a federated model. These imply the principals of returning the control of identity data to their owners, as stated by Christopher Allen and Kim Cameron [8]. The self-sovereign identity principals [8] include:

- Users must have an independent existence;
- Users must control their identities;
- Users must have access to their own data;
- Systems and algorithms must be transparent;
- Identities must be long-lived;
- Information and services about identity must be transportable;
- Identities should be as widely used as possible;
- Users must agree to the use of their identity;
- Disclosure of claims must be minimized.

SSI systems based on Blockchain have the means to deploy the concept of identity and benefit from the rich characteristics provided by Blockchain, such as immunity, cryptographic data storage, distributed key management, and distributed storage [9]. SSI returns the control over identity information storage and selective disclosure rights to the identity owner. It enables the decentralized architecture to minimize the probability of data leakage that may be caused by compromising centralized identity repositories. In addition, combining SSI with Blockchain provides data consistency, availability, and privacy, and reduces correlation by creating a unique identifier for each service.

The aim of Blockchain is to store cryptographic transactions in a public ledger based on a decentralized consensus mechanism, which is hard to break or modify, according to the investigation results from [10]. Our proposed platform utilizes the Ethereum public ledger to record the identity transaction hashes, to prevent hacking, denial of service attacks, and identity theft or loss. In addition, it enables smart contracts, which allow multiple benefits such as automation of approved financial services and integration with decentralized applications.

uPort [11,12] is an identity system implementation based on Decentralized Identifier (DID) specifications and W3C [13], and relies on Ethereum Blockchain to deploy several smart contracts to manage identities, attributes, and attestations of attributes. It allows selective disclosure and sharing of identity attributes between the user and related parties. Public data are stored in the Interplanetary File System (IPFS) [14]; the public key is bound to a hash link that points to public profile data in the IPFS of Ethereum Blockchain. Only users who own the private key and the data hash on the IPFS, which ensures the integrity, can modify this link. This work does not use the uPort app since it was decommissioned in June 2021 and replaced by the Veramo framework [4].

Veramo [4,15,16] is a new and evolving iteration of uPort. It provides open source libraries for a modular API for SSI and verifiable data, creating and managing interoperable DID and Verifiable Credentials (VCs) [17] without relying on third-party vendors or centralized systems. It supports many platforms, such as Node, browsers, and React Native. Veramo is fitted to our framework and banking use case as a middle layer to control DID and VCs.

3. Related Work

KYC2 [18] is a framework for client onboarding based on the SSI model and Hyperledger Indy [6]. The model was discussed in terms of the SSI concept and General Data Protection Regulation (GDPR) [19] rules. It advised the need for key management to secure the secret agent keys and recovery conditions. This work proposes using banking smart cards for cryptographic operation handling and deploys the SSI concept based on Ethereum Blockchain; specifically, the curve exists for several market banking Java Cards and was validated in our proof of concept.

The authors in [20] provide a survey and overview of the SSI concept and available solutions in terms of the architecture and actors in the approach to identifier-registry and claim-registry models. Blockchain enables these decentralized registries, although it is not a necessity, and the storage of both identity and claim registries can also be decentralized. The paper presents on-chain and off-chain storage with the advantages and disadvantages for each option. In our work, we store private information off-chain while the integrity and authenticity proofs are stored on-chain.

SCARAB [21] is a model used for decentralized and secure access control, in which the accountability issue of SSI Blockchain-based systems is resolved by registering every data access request in public mode. It introduced on-chain secret verifiable sharing to collective management using the Byzantine protocol, and also introduced identity skipchains to control access policies to manage the identity and enable user's self-sovereign identity management [21]. SCARAB stores everything on-chain, which affects the scaling and performance, making sensitive identity data available for the public, and meaning that it is not possible to erase it.

BBM [5] is a Blockchain-based SSI model for open banking and enables secure communication between involved parties. In contrast to BBM, it is proposed here to use the banking card, which is trusted and conveniently used by customers for daily purchases, to act as tamper-resistant storage for the identity private key, and to perform the cryptography required for verifiable credential issuance and sharing with banks and third-party agencies. Instead of relying on additional devices, such as hardware tokens, we utilize the banking card to balance security and usability. This work proposes the implementation of a prototype to assess the feasibility and possibility of integration with the existing payment card issuance flow.

The SSI framework has been used to prevent banking scam calls [7]. The authors proposed an SSI model based on Hyperledger Indy [6] and utilization of the user's banking card with the aim of preventing banking scam calls. In contrast, in our framework, we leverage the use case and overcome the limitation of the signing curve on Java Card ED25519, which is required for Indy transactions. The proposed framework is based on Veramo, which relies on Ethereum Blockchain, where the required SECP256K1 signing curve exists for several banking cards used for this work, to issue ethr DIDs and VCs. This fits within the open banking use case; the proposed framework allows smart contracts that automate several online financial services.

In [22], the authors proposed a protocol to reach a high level of assurance (LoA) for a mobile-phone-based SSI identity wallet with the main focus on achieving a high LoA. It considered key tamper-resistant storage requirements to achieve the high LoA via combining a software wallet key part and a YubiKey 5Ci token [23] for the hardware key part. This protocol is applicable for our implementation, since we rely on the banking card for the tamper-resistant storage of the key materials to secure the SSI wallet instead of the mobile secure element, which does not exist in some market mobile devices. In contrast, in our work we leverage the banking use case and balance the usability and strong security without relying on an additional device for banking customers. Furthermore, our implementation supports encrypted strong authentication and VC manipulation for the open banking services, and not only the identity wallet.

The authors in [24] proposed "Casper", a mobile identity wallet that enables inter-banking Know Your Customer (KYC) based on SSI and Rahasak Blockchain to manage smart contracts. In contrast, in our work, we attempt to combine high LoA as a market transition step to build on the customer's trust in the ecosystems of banks and financial institutes. This increases the trust and identity control via the sharing of a similar payment experience using banking cards and modern wearables. Our work can be extended to several business values, such as smooth KYC, password-less login to the Internet and mobile banking, and loyalty programs. We rely on Ethereum Blockchain, and benefit from the Veramo SDK to handle DID and VC-related smart contracts, in addition to the key recovery mechanism, by combining a secret part at the customer's bank and the passphrase that is held by the customer.

PriFob [25] is a privacy-aware fog-enhanced Blockchain online global credential management solution. It uses a reliable encryption scheme with public Blockchain, allowing digital signatures and zero-knowledge proofs. Furthermore, it deploys two consensus algorithms (Proof of Authority (PoA) and Signature of Work (SoW)) for efficient verifiable credential handling. PriFob has been evaluated in terms of security and performance metrics, such as throughput and latency, based on the simulation and emulation of the framework. It showed better performance compared to other Blockchain-based solutions. In our work, we introduce the usage of the banking card to improve the security of key management and to achieve a high level of assurance, and aim to present the SSI benefits to customers, financial agencies, and banks for integration in the specific field of open banking. Similar to PriFob, our work is based on Ethereum Blockchain, and our framework can be improved to enhance the performance by applying the adopted reliable encryption and efficient consensus algorithms.

4. Proposed Solution

This section provides an overview of the components and actors used to build a demonstrative implementation, Figure 1 shows the SSI-based Blockchain identity systems components. The aim of utilizing banking cards is to balance the trusted industry security behind the day-to-day usage of personal payment cards and the user's experience, and to provide ease of use without relying on additional hardware tokens. As an extension, the proposed architecture supports smart cards with biometric fingerprints and other form factors such as wearable tags and banking bracelets. This section describes the designed flow, framework components, and interaction between the parties. This implementation shows the feasibility and attractive values for adoption by banks and other financial institutes to secure and improve open banking integrations, and to provide convenience to customers, who can control their identity and sensitive information sharing.

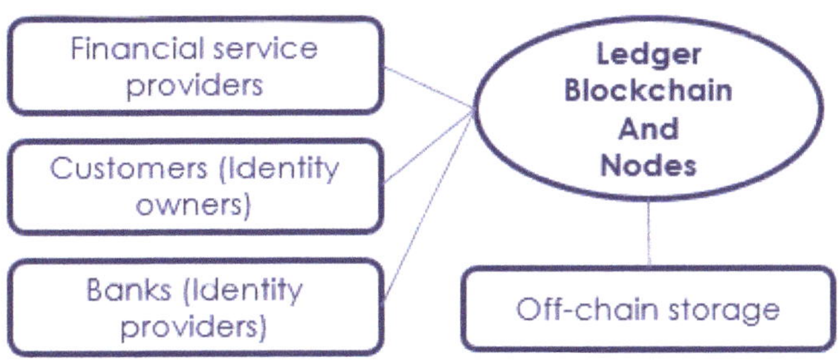

Figure 1. Blockchain SSI-based identity system components.

This work proposes an architecture based on the SSI and Blockchain implementations, Ethereum with Veramo and smart contracts. Specifically, a decentralized framework is used that stores private information off-chain while the integrity and authenticity proofs are stored on-chain. Our proposal provides a friendly interface as a mobile application or via web access integrated with a payment card for a seamless user experience, which allows third parties to verify the validity of identity data. The Veramo SDK layer provides public profile data to be stored on the IPFS; this means the user has to interactively manage the sharing of their private identity with other parties using web links or QR codes. Mnemonics and QR code technology allow presenting the secret keys in mnemonic terms to help with memory retention. Seed phrases are analogies for mnemonic words to ease portability and retrieve the key parts.

Furthermore, new banking cards in the market are equipped with biometric fingerprints [26], as shown in Figure 2, which improve the user experience by eliminating PIN entry for each identity transaction.

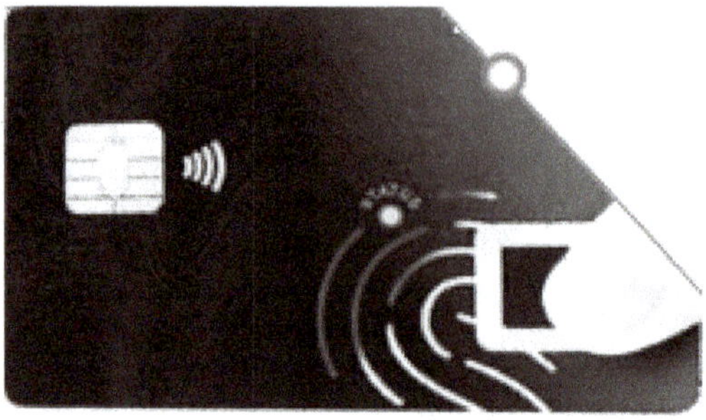

Figure 2. NFC banking card sample equipped with a fingerprint.

4.1. *Requirements and Design Goals*

The SSI concept emphasizes that personal data are subject to governance and are not dependent on any external factor, as listed in the below characteristics [27]:

- Complete control over private data;
- Ensuring the privacy and security over users' data;
- Trust in any central institution is not required;
- Data portability;
- Data integration;
- Personal data transparency.

This concept is important as it reduces the chance of data leakage incidents since information is not stored in central locations [28]. A high level of assurance is assessed based on specifications [22] compiled from two standards that control digital identity: (1) The ISO 29115 [29] standard; this standard covers the four main phases of (I) enrolment, (II) electronic identification mean management, (III) authentication, and (IV) management and organization; and (2) NIST (National Institute of Standards and Technology (NIST), 2020) [30]. We include these standards' requirements in our work to ensure a high level of assurance by extending the banking card usage as a hardware-based second factor authenticator.

4.2. *Design Objectives*

The design of SSBI aims to address and satisfy the following list of requirements within our proposed implementation:

- Prevent financial data leakage when third-party agencies promote financial products or KYC services, and eliminate user's fraud calls and phishing messages by trusting DID and VCs.
- Trusted identity owner since we consider the active account bank user, which means the KYC process is to completed in advance with a unique DID.
- Achieve a high level of assurance using Veramo key management combined with tamper-resistant secret key storage, by relying on a payment card to prevent theft and unauthorized usage, with potential enhancement to use a smart card with embedded biometric fingerprint authentication.
- Ensure validity of identity and proofs by asserting the revocation status on-chain using Veramo APIs.
- Strong and multifactor authentication against attacks, such as replay, Man-in-The-Middle, and theft attacks, through the usage of a banking card, PIN, and biometric card.
- User-controlled identity and data sharing. The user is able to manage one or more identities, including their attributes, using an open system that prohibits third parties

to restrict its usage. The user is able to select and authorize related parties to access these identities by enabling access to subsets of identity attributes. User consent can be given, and the related party is able to request access to the identity attribute of an end user. The attributes must be retrievable without a direct communication channel between the user and the party.
- The identity data owner is able to revoke access at any time.
- A secure and seamless user experience is achieved by combining the trusted issued banking cards for cryptographical identity operations. Furthermore, we propose new form factors such as stickers, tags, and biometric payment cards.
- An identity and secret key recovery mechanism is proposed where the key part is saved by the issuing bank while the passphrase is memorized by the user.

4.3. SSBI Layers

The framework consists of four layers, as presented in Figure 3: a banking applications layer, a Veramo identity layer, a Blockchain layer, and a customer data layer, as shown in Figure 3. Veramo APIs provide identity-related functionality, and we store data hashes on Blockchain (smart contracts) while customer-sensitive data are kept off-chain to maintain users' privacy. Ethereum Blockchain is used for strong cryptography DIDs. Public keys and VC transactions are stored on the Ethereum Blockchain; however, the VC itself is stored off-chain, such as in a database on a PC or in the mobile device's memory.

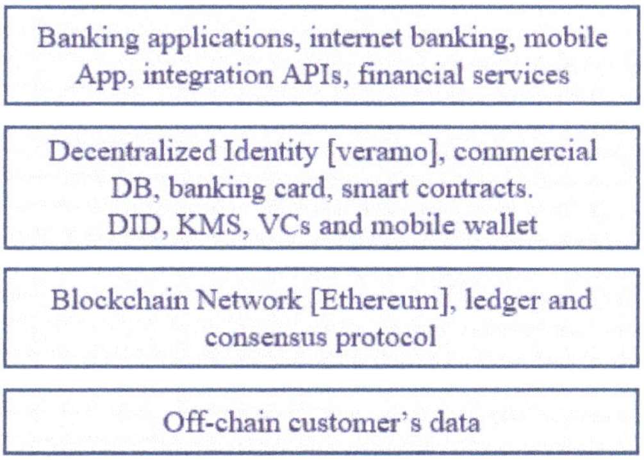

Figure 3. Proposed SSBI layers.

The banking application layer provides a means for third-party agencies and financial service providers to integrate and communicate through a list of APIs to obtain access to customers' account data and business functionalities.

The Veramo layer consists of backend software to manage communication, DIDs, DKMS, and VC-related functionality, in addition to the mobile application interface and database storage. Non-sensitive data, such as the public key, are stored on Blockchain. An Android mobile application is used to connect with agencies or other financial service providers through a QR scan.

The Blockchain layer and the SSBI framework using Veramo relies on Ethereum Blockchain and four smart contracts: the controller contract, the proxy contract, the registry contract, and the application contract, which provide consistency, security, and smart contract features.

The proposed architecture for SSBI, main actors and system functionalities are presented in Figure 4.

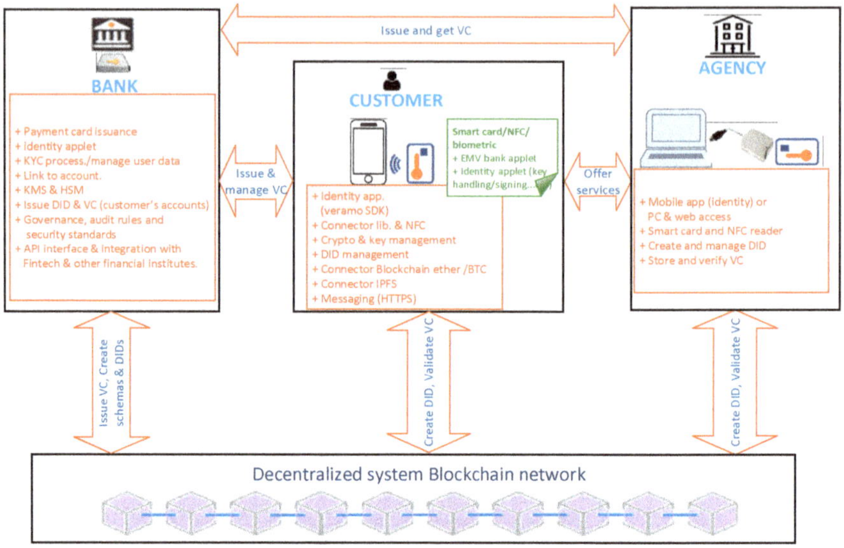

Figure 4. Proposed architecture customized for SSBI, extended from the model in [7].

4.4. *Sub-Systems (Actors and Features)*

The objective is to provide an easy-to-use open banking service, powered by SSI to reduce the risk of fraud, provide transparency, and provide verifiable and publicly resolvable identity information for an efficient KYC and open banking paradigm, as follows:

1. The contributing parties are based on Blockchain, which underlies Ethereum, to provide a decentralized infrastructure that enables smart contracts to manage identity information and provide immune storage for VCs and proofs to allow cryptographical verifications.
2. The banking card is used by the customer as an extended feature to enable secure signing operations and maintain secure private key storage.
3. An information management platform with web and mobile app interfaces allows identity data management flow, creation, revoking, verification, and selective data sharing.
4. DIDs [13] (Decentralized Identifiers) are linked to their owner, and represent the user, customer, financial institutes, third-party providers, and verifier parties. In our case, the Ethereum address is able to register, issue, query, and verify proofs and VCs.
5. Smart contracts [31–34] interact with Blockchain directly and execute the on-chain application logics among the system including the Ethereum *DIDRegistry.sol* to manipulate the DIDs, the Ethereum *EthereumClaimsRegistry.sol* to track claims, VCs, and proofs, and *Verifier.sol* to implement verification and check validity of VCs and proofs.
6. System admin and auditor roles.

4.5. *SSBI Components*

The main platform components are:

- Mobile app and agent operations;
- Server-side application to handle communication between parties, access to Blockchain, and smart card connections;
- Applet to install on the banking card;
- Veramo layer for DID and VC management.

Figure 5 shows the UML diagram presenting the designed classes to implement the SSBI platform. The off-chain class handles the storage of verifiable credentials and identity attributes, in addition to allowing the interface with agent application calls and providing one instance per entity. For the customer, this application is deployed as a smartphone app, although it can be implemented as a PC application. The identity operations include DID operations and connectivity with Blockchain during registration, modification, and DID revocation. The entity class has an interface to the pairwise DID class. The JavaCardApplet class provides several functions to perform cryptographic operations, such as signing DIDs and VCs, and holding and generating keys, in addition to state and PIN management. The connector provides handling and exchange of the APDU command (main application protocol data unit) between the app and the card through NFC connections or a USB smart card reader.

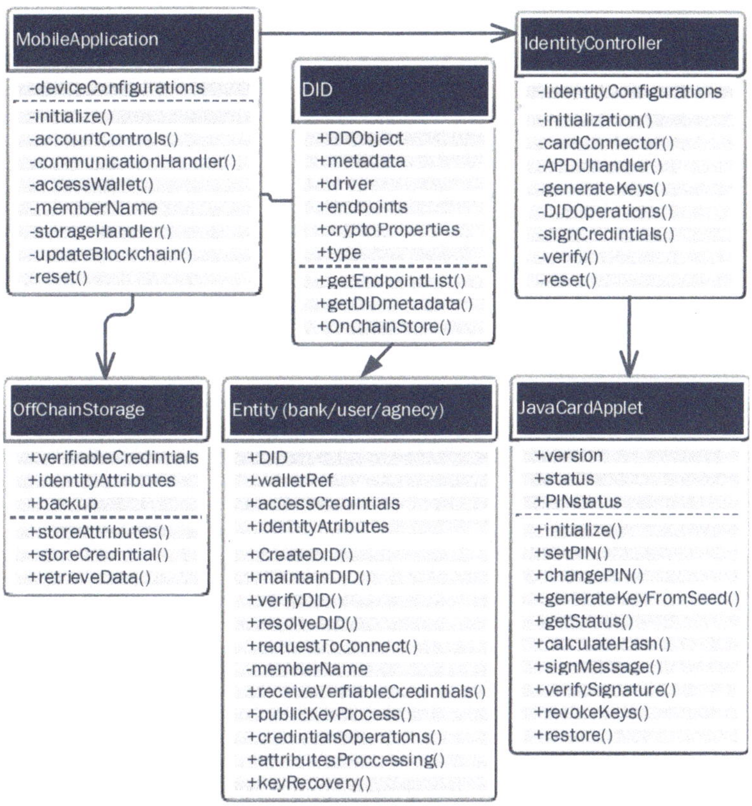

Figure 5. UML class diagram adapted from [7].

According to Lopez [35], the content of a certified document must include:

- URI for a unique identity credential (a set of claims by the issuer about a subject);
- such a credential is associated with a DID.
- URI to identify the issuer (issuer DID).
- URI for credential type.
- URI for protocols and terminology to help parties read the credentials.
- Cryptographic proof of the issuer.
- Claims data, and the statement or characteristics about a subject (metadata).
- Issuance date.
- Expiration date.
- Location of the credential status (such as a smart contract reference).

The format of VC and certificate exchange has to be in a protected and secure digital channel such as JSON-LD, as proposed by European Blockchain technical specifications [36] and according to the DID standards of W3C [17], as follows:

- The context of data description inside JSON-LD to ensure common understanding between parties.
- Identifiers—according to W3C, a DID should be used to identify issuers, credential subjects, persons, things, and verifiable IDs or attestations.
- Types, such as W3C (Verifiable Credential), in addition to other types.
- Issuer—the public institution, government entity, or an organization that is trusted to issue a specific type of VC, typically a bank or financial institute for our case.
- Credential subject, which is a DID to identify a person, thing, or organization.
- Claims, which constitute a verifiable ID or attestation for a semantic statement to a certain value about an attribute of a DID subject such as (.dateOfBirth).
- Issuance date is the date the claim becomes valid, and it is not mandatory that it be the same as the issuance date of the VC. Expiration date is the date when the claim should become false or expire. Status is the current claim status, such as active, suspended, or revoked.
- Proofs—at least one proof mechanism should exist to allow the verify operations.
- Optional extensibility—as per W3C, extensions of the data model can be used to add more attributes.

Moreover, Lopez [35] considers it a mandatory feature to allow public querying and resolving in the Blockchain of all involved parties' DIDs.

4.6. DID and Registry Using Veramo SDK

We propose managing DID and related identity operations using Veramo, which is one of the popular DID platforms implemented for Ethereum, and is based on smart contracts. Veramo is the next generation of uPort [11,12], which uses a more powerful and modular architecture and can run on backend servers, and web or mobile devices.

The Veramo platform consists of identity management and messaging protocols that are interoperable with any distributed ledgers for decentralized web and SSI models [4,16]. The platform uses the DID standard ERC1056 [34], which is an Ethereum pattern designed to create and update identities based on a lightweight identifier that utilizes smart contracts applicable for off-chain usage. The underling DID method allows any key pair account, for instance, a SECP256K1 public key, or any Ethereum smart contract to be considered as a valid identifier without additional registration. Attributes such as 'service end point' are retrieved by resolving an ERC1056 smart contract deployed in the network and listed in the registry repository. Any identity is mapped to a single Ethereum address, which is controlled by its owner. However, the advanced ownership model is managed by multi-signature contracts. EIP-1812 [37] is used by Veramo as a standard for claiming registry. It allows off-chain management of VCs by storing claims off the chain, and verifies them using on-chain smart contracts that implement either a stat channel or off-chain libraries. This is in contrast to other standards such as ERC-735 and ERC-780 [32], which rely on storing on-chain claims, and, as such, do not comply with GDPR rules regarding the storage of personal information on public databases. ERC1056 is compliant with W3C DID recommendations and allows identities to have multiple associated attributes and delegations for on-behalf identity operation.

5. Proposed Flow

5.1. Typical Initialization on the Mobile App

As presented in Figure 6:

1. The bank instructs the customer to download the mobile app from an authorized store.
2. The customer creates a DID and inserts his basic information to generate a secret key on the banking card and chooses the recovery mechanism.
3. The customer performs the first KYC for enrolment and creates a bank account.
4. The bank issues a new payment card and links it to the customer account. In addition, the bank installs an identity applet to serve the identity and open banking data sharing.

5. The bank confirms second factor authentication, e.g., to verify a one-time password (OTP) on the registered mobile number.
6. The customer performs initial configuration, sets a PIN, and inserts seeds for the master key.
7. The bank deploys an identity applet for an agency smart card and links it to the agency ID.
8. The agency performs identity card applet initialization, sets the PIN, and inserts master key seeds following the bank's instructions, and installs the corresponding application.

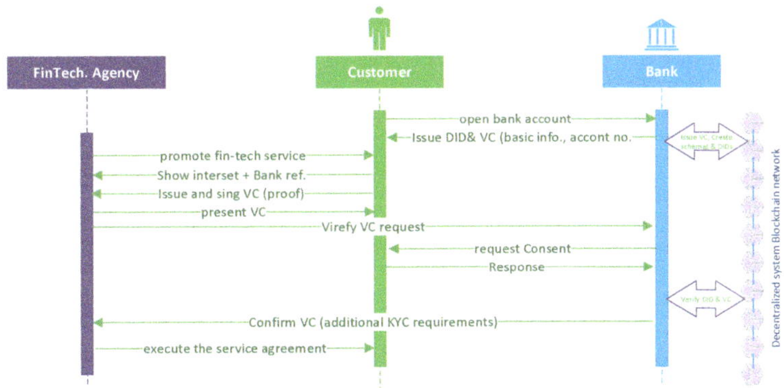

Figure 6. UML representation of the initialization phase.

5.2. Authorization Phase

As presented in Figure 7:

1. The user uses the mobile application to scan the QR code for the Fintech agency DID.
2. The user triggers verification of the DID (ether DID).
3. The user triggers connection with the fintech agency and creates a new pairwise DID.
4. The fintech agency requests a VC.
5. The customer selects which information to share.
6. The bank authenticates the user and picks the corresponding schema.
7. The bank signs the user claim using a Hardware Security Module (HSM) private key.
8. The bank sends the VC to the customer.
9. The user shares the VC, and it is signed/confirmed by requesting that the user presents an NFC banking card (from Bank A) and PIN or the biometrics for the card authentication.
10. The fintech agency validates the submitted VC by querying the Ethereum distributed ledger (DL), validates the digital signature to verify the user credentials, and validates the bank signature.
11. [Optionally request additional VC/proofs.]
12. The fintech agency signs a proof with the proposed financial service details (KMS/HSM key).
13. The user confirms and generates a signed proof by signing the transaction using his/her banking card.
14. The fintech agency executes a manual digital contract or, in automated mode, triggers a smart contract to the user.

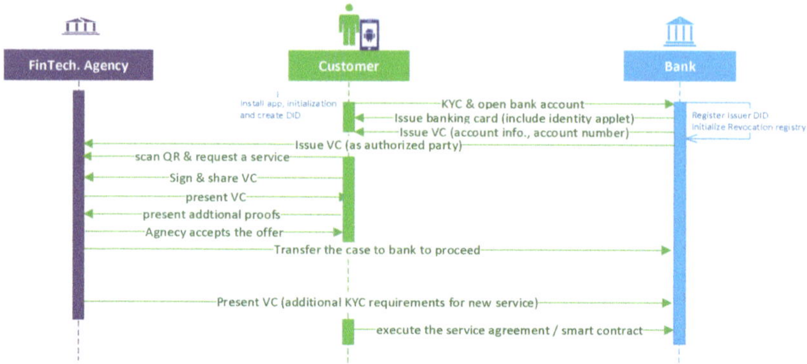

Figure 7. UML representation if the authorization phase.

6. Demonstrative Implementation

We implemented a mobile phone application to issue and manage the user's DID, proofs, and VCs based on the Veramo SDK and connector to utilize the banking smart card for secret key management through NFC communication. This is instead of saving the key encrypted on mobile memory. In addition, we use a webserver to act as a registry to handle the storage of DIDs and VCs, and to allow communication with other framework parties such as banks and financial agencies. The customer's bank issues the VC as proof of the account and initializes the identity applet with a secret key as part of normal banking card issuance. We simulate the interaction and basic flow functionality using a web mock service for indirect actors.

The mobile application for Android and IOS is based on React Native communication with an NFC banking card to set the PIN, generate secret keys, and sign the transaction to create pairwise DIDs and sign VCs using the connector library and mobile NFC reader.

6.1. Set of Tools Used in Our Demonstrated Implementation

- **Veramo** [4]—SDK to handle verifiable data and related SSI functionalities;
- **React Native** [38]—a framework for creating the mobile and web applications;
- **Node.js** [39]—the runtime environment;
- **Express.js** [40]—a web framework used to create the API;
- **Infura** [41]—used as access point for Ethereum networks;
- **Firestore** [42]—used to save data in a cloud database;
- **TypeScript** [43]—programming language;
- **Expo** [44]—used for building and debugging the application;
- **Mocha** [45]—used for unit testing.

6.2. DID and DID Document

SSBI uses the DID method *did:ethr*, which is a representation of the Ethereum address linked to the ERC1056 smart contract, and its DID document [13] stored off the chain, such as in the IPFS.

Sample DID Document

```
{
    "@context": [http://www.w3.org/2019/did/v1],
    "id": "did:ethr:0xdf0d ... de1aebd143e",
    "publicKey": [
    {
      "did:ethr:0xdf0d ... de1aebd143e#key-1",
      "type": "EcdsaSecp256k1VerficationKey2019",
```

"publicKeyHex": "035bc987531e4823 ... 46821d763ea36",
"controller": " did:ethr:0xdf0d ... de1aebd143e"
}
],
"authentication": [
"did:ether:ysktz-uyzt-82re-d#key-1"
]
}

6.3. *Claims and Verifiable Credential*

The VC is formatted as JSON-LD, which contains the following information in the form of a Uniform Resource Identifier (URI), to define the means and protocol for parties to communicate: credential type, issuer, date, the credential and its subject, cryptographic proof of the issuer, optional metadata, and revocation condition.

Sample of Input Data for a VC

[
{
"@context": "http://schema.org",
"@type": "bankingAccount",
"agent": { "did": "did:ethr:0xdf0d ... de1aebd143e" },
"bankingCredential": {
"issuer": { "name": "Bank A" },
"description": "customer valid account",
"accountName": "User A",
"accountNo": "012345678",
"status": "Active", //or "Revoked"
"time": "2021-11-29T08:00:00.000-07:00"
}
}
]

The application allows authorities, such as authorized banks in our case, to issue claims and VCs and store them on the IPFS. The authorized entities, such as central banks, regularly monitor banks and apply rules policies. Based on this, we consider the bank as a trusted authority to issue VCs. Off-chain there are two GUI interfaces: the first allows the bank to create pairwise DIDs and issue VCs; the second enables verifiers and other open banking agencies to validate the customer's VCs.

The following is a list of SSBI subset functionalities to handle DIDs and VCs:

- Create_DID;
- Create_VC;
- Resolve_DID;
- Verify_VC;
- List_existing_DIDs;
- List_existing_VCs;
- Retrieve_VC;
- Send_signed_VC.

The on-chain component is part of a front-end user interface (Android or iOS app hosting identity operations) and React Native application (customer and open banking agents). The functionality includes creation of DIDs, removal of DIDs, and VC manipulation for issuance and verification, in addition to underlying connectivity with the banking card and wearables to perform key management operations and Ethereum transaction signing. Sample snapshots to show SSBI sub-functionalities presented in Figure 8. The flow was validated and tested using Ethereum Blockchain. The off-chain part is based on Java script and a Node.js application that communicates with the client through web3.js and API implementation. The encrypted data are stored in an SQLite3 database [46] and are accessible via a JDBC driver. The open banking API layer includes financial operations such as Account API, Payment, Money transfer, and Card life cycle.

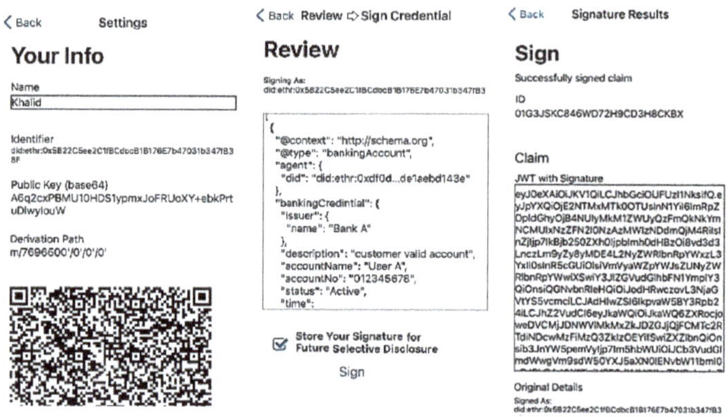

Figure 8. Sample snapshots present SSBI sub-functionalities to register and *create_DID* and fill in banking VC details, followed by signing and storing the VC by calling the *Create_VC* API with the underlying presence of the customer card to perform Ethereum transaction signing.

6.4. Key Recovery

The bank issues the payment card including the applet and initial key with seeds, and personalizes it for the user based on the Primacy Account Number (PAN) and the bank (issuer) private key to derive the seed and corresponding key. The operation is managed through the bank's HSM and strong key management system (KMS). The trusted bank complies with, and is audited for, security standards such as the Payment Card Industry Data Security standards PCI-DSS [47], policies, and monitored operations. During the initialization phase, the customer injects his passphrase. The pairwise DID key is generated based on both the key and the phrase. To recover the key in the case that it or the payment card is lost, the bank issues a new card with the same derived key and the customer inserts his passphrase to recover the key. The DID is then restored and all corresponding VCs and proofs can be requested again or restored from the encrypted local database.

An alternative option is to apply the mechanism suggested in the BBM model [5], by storing a recovery network in a smart contract; then, during the recovery on the new device, the bank verifies the issuance of a new card based on triggering the recovery smart contract.

6.5. Using Customer's Banking Smart Card to Secure Identity Private Keys and Sign Blockchain Transactions

6.5.1. Java Card for Banking and Identity Operations

SSBI introduces the utilization of the customer banking card to store identity secret keys and perform transaction signing to balance security and usability. The majority of banking payment cards support signature implementation for the SECP256K1 curve-based ECDSA, which is compatible with the Veramo platform and Ethereum Blockchain.

Banking customers commonly use payment smart cards for day-to-day purchases and money withdrawals. This relies on strong industry standards and security rules, such as EMV [48] and PCI-DSS [47], with advanced deployment of strong cryptographic keys and certificates. In addition to usually maintaining a secure PIN code or biometric verification to perform payment transactions, customers trust the issuer (banks) and rely on security rules that control this industry. We extend this level of trust and user experience to apply the financial identity sharing paradigm to open banking by using the same card for private key storage and signing the underlying Blockchain transaction for DID and VC operations, as suggested by [7].

Java smart cards are designed to be tamper-resistant and follow Java Card Platform specifications [49], where the developed applet can be installed once and run on several card's virtual machines; this applet is deployed on the card's operating system layer. In contrast to many studies that recommend a hardware wallet [50,51], we utilize the banking card to manage identity secret keys. This was also proposed by [7], where the applet is installed separately from the banking applet, with programmed functions to initialize master keys, generate sub-keys, and sign transactions securely on the card for DID and VC operations.

6.5.2. Java Card and Identity Libraries

The libraries required to interact with the banking Java Card for identity cryptographic operations include:

- APDU handling to construct and exchange the APDU commands according to context calls such as verify_PIN and get_status;
- Applet client for the interaction with the card applet methods, building the logic command, and the input data, in addition to handling the response from the applet, e.g., the signed hash;
- Card connector to establish the hardware handler with the smart card and manage the secure channel between the reader and the card;
- Java Card identity applet, which is the installed and personalized applet (cap) file on the card;
- Hashing and key operations, where the corresponding key methods include generating the keys and setup recovery methods.

6.5.3. Generating Identity Applet Master Key

During the initialization phase, the master key for the identity applet on the banking card is generated with the following steps:

- The applet performs pseudo-random code generation;
- The user inserts mnemonic words;
- The root seed (bank key part) + phrase (customer part) are generated;
- The applet performs the HMAC-SHA512 function;
- The resulting hash is used to generate the master private key and master chain code for the HD key root [52].

6.5.4. Initialization

This section describes the logical flow to initialize and issue the banking card for the identity applet next to the normal payment applet.

On the card issuer (bank) side:

1. Include applet installation (*.cap file) during the pre-personalization step;
2. Personalize the payment applet and link the card to the customer account;
3. The bank performs chip key rotation to secure the card and prevent unauthorized changes to installed applets

On the customer side:

1. The customer receives the payment card (instantly or by mail);
2. During the card activation process, the customer is then instructed to install the identity software application from a safe repository using a web link provided by the bank;
3. As part of initializing the identity application, the customer connects his card, usually via an NFC interface, to set the owner PIN, generate seed words, back it up for recovery in the future, and generate the master private key.

6.5.5. DID and VC Signing

In operation mode, the logical flow to sign DID and VC cryptographic operations using the Java Card is as follows:

On the customer side:

1. Identity application—build transaction message;
2. Request user to present banking card to NFC mobile device;
3. PIN authentication or biometric verification;
4. Transaction data signed on the card and the APDU responds with hash;
5. The application handles the Blockchain transaction broadcasting and waits for mining to complete;
6. Using the Veramo SDK, generate the DID or sign the VC according to the requested identity operation;
7. Store the created DID or VC in the case of proofs;
8. Present the VC to the user and ask for consent to share with a financial third party.

The applet supports three different categories of APDU commands that follow the ISO 7816 [53] smart card standard: initially, external authentication techniques such as PIN setting and verifying instructions; secondly, status and key generation; finally, the identity and Blockchain commands, such as signing and verifying transaction hashes.

7. Discussion and Evaluation

The implementation in this work is based on leveraging SSI, Blockchain, and tamper-resistant key management, with the aim of simplifying and securing the banking identity in decentralized mode. In the design, we considered security and easy integration with existing banking platforms, with benefits arising from security standards such as PCI-DSS and from the use of HSM. Each customer will be able to control and gain from sharing his identity and financial data without relying on a centralized system or third parties. The Veramo SDK and Ethereum Blockchain allow the management of DIDs, which are sets of characteristics and claims that uniquely identify persons, entities, or simply things. In addition, users can share claims and attested proofs with others with minimal attributes required for each case. This ensures high privacy requirements are met with the concept of zero-knowledge proof and selective disclosure, which exists in our work by relying on the Veramo framework. Our work aims to control and maintain the trust between participants in the open banking context based on SSI implementation, and achieves privacy, security, and a high level of trust, while balancing the user experience, by utilizing the customer banking cards to securely manage the private keys and on-chain identity operations. In addition, it is able to manipulate the required characteristics, such as key recovery and portability, via the use of a client mobile app, enables interoperability by relying on DID and VC specifications, and addresses the ability to be integrated with the banking ecosystem.

In contrast to using hardware tokens, such as in the work presented in [22], to achieve a high level of assurance we rely on existing banking cards, which are usually part of every banking customer's life, and provide the required level of trust and security (controlled by PCI-DSS and related standards), while remaining portable and easy to use. This can be extended for many other valued services, such as fast KYC and password-less access to banking mobile applications and online banking, on top of secure interaction with open banking and financial services.

7.1. SSBI Evaluation

In this section, we evaluate SSBI according to the stated design objectives, and in terms of the fulfillment of the requirements and features of the framework.

Prevent financial data leakage in open banking: SSBI helps to prevent identity and financial data leakage during open banking operations. Third-party agencies can promote financial products and perform KYC services on behalf of banks via the trusted SSI platform where customers and agencies can verify each other. By providing trust in DIDs and VCs, this eliminates fraudulent calls and phishing.

Trusted identity issuer: The proposed solution capitalizes on having banks as a trusted party. SSBI considers the active bank account customer, where the KYC process is performed in advance with a unique DID.

High level of assurance: SSBI combines Veramo key management with a tamper-resistant secret key storage, which is a payment card, to prevent unauthorized access to an identity.

Digital identity verification: The approach relies on Veramo SSI APIs to ensure identity proof, VC validity, and revocation status through the access to Ethereum Blockchain.

Strong authentication and secure key handling: Multifactor authentication is enabled through the usage of a banking card for sensitive key operations and transaction signing. Moreover, banking cards use PIN verification and have recently begun to use embedded biometric fingerprint authentication.

Identity ownership and SSI: The user is able to control several identity proofs and related attributes, request and store the VC and proofs from the identity issuer (banks in our case), and control the sharing of the VC or selective attributes with other parties.

Ability to revoke the proofs: The identity owner is able to revoke the proofs and VC at any time, by calling the corresponding Veramo API to revoke the Ethereum smart contract.

Seamless user experience: SSBI combines the trusted issued banking cards, which are used in day-to-day payments and money withdrawals, for identity cryptographic operations. It is more convenient for the customer to use modern form factors such as payment stickers and miniTags. The mobile app communicates through NFC to simplify the process.

Key recovery: Identity and private key recovery are achieved by combining two key parts and performing the generation action, where the key part is saved by the issuing bank while the passphrase is memorized by the user.

7.2. Security Analysis

The proposed SSI framework uses the Veramo (uPort) SDK and integration with a banking card. To evaluate the security of Veramo, we applied the threat model from [54]. SSBI does not rely on the official uPort mobile app; however, we utilize the same SDK functions:

1. The private key in uPort is stored on the mobile secure enclave, where the SSBI stores the key on the banking card, which in turn cannot be extracted because the threat score is reduced significantly.
2. To counter the potential threat of stealing or gaining unauthorized access to the smartphone, SSBI isolates the private key store on external devices (the banking card), which prevents attacker identity operations.
3. uPort does not include an additional layer of protection such as a PIN or biometrics. We implement access control to identity the app using a PIN or by presenting the smart card.
4. Recovery mode in uPort is available with a social mechanism; however, we applied a different mechanism by involving the trusted bank to store the keys (HSM), while the user is responsible for the passphrase. The key cannot be regenerated without having the two components.
5. The action time stamp exists and can cause false revocation. This is possibly mitigated by forcing the SSBI app to fetch a governance timeserver.
6. Denial identity actions due to stealing or losing the mobile device can be mitigated via the recovery mechanism.
7. uPort app deletion means erasing the private keys. In contrast to the SSBI app, the deleted key is still safe inside the user's banking card.

7.3. Security Challenges of Using the Banking Card for Identity Transactions

SSBI introduces the banking card to sign identity transactions. This raises additional security concerns related to the applet and APDU data exchange. Here, we present a set of these concerns alongside their corresponding mitigations:

1. Malware attack by capturing the PIN or passphrase during customer entry—this is possibly mitigated using a secure keypad instead of a standard device keypad.
2. Presenting the banking card to sign the identity transaction is vulnerable to a change in the data package sent to the card by an attacker—this can be mitigated by establishing a secure messaging layer [55] based on a session key and securing all APDU data exchange.
3. The identity applet installed on the banking card is vulnerable to unauthorized updates or re-installation—this is mitigated by the default use of access to a chip key, which is a 3DES key securely managed by the chip manufacturer and shared only with the issuer bank. Additionally, the issuing script performs key rotation to lock the Java Card state to avoid further installations.
4. Another layer of authentication could be a smart card with an embedded biometric fingerprint sensor to achieve a better user experience and security.

7.4. Performance Analysis

To analyze the general performance of the SSBI prototype implementation, we implemented different test scenarios and stress testing with recorded customer groups. The complete flow of the payment card issuance and applet installation was performed using the desktop personalization tool at an early step, followed by card initialization, setting the customer PIN, and key generation. During the Create_DID process, the card is presented to perform the signing operation, with the average time to respond being 3.2 s, assuming that the customer presents the physical card to the mobile app to eliminate the human action time of fetching the card. Moreover, to be able to perform real-world stress testing on the system throughput and latency, we isolated the card signing time during the stress testing and prepared a set of pre-signed transactions (10, 25, 50, and up to 300) to push the network.

The testing environment was as follows:

- Number of organizations: two banks and one agency;
- Access to Ethereum network: Infura;
- Consensus mechanism: Proof of Authority (PoA);
- Nodes: four nodes (one validator and three peer nodes);
- Customer device: Samsung Galaxy x4 and iPhone 6;
- The system ran on: Intel(R) Core(TM) i7-10850H CPU @ 2.70 GHz;
- OS: Windows 10;
- State database: Firestore;
- Mocha tool for unit testing;
- Caliper benchmarking tool;
- Solidity to write smart contracts.

To perform the transaction on the Ethereum network, the computation cost is expressed in the term of gas expenses. The gas amount is the parameter of computation complexity. There is a significant cost of the operations that deploy smart contracts such as VC and DID creation. The function of SSBI was validated for general comparison and the evaluation results were gathered for the system throughput and latency, followed by assessment of the complexity and consistency of the Blockchain interaction operations. The test was conducted based on a peer network of four nodes: one node was configured as a validity node to perform PoA consensus and sign the transaction; the remaining controller nodes sent the transactions to the validation node.

7.4.1. Throughput and Latency

The throughput and time per transaction achieved for different sets of recorded customers, from 100 to 500, are shown in Figure 9. In addition, the performance when sending a concurrent set of calls, of 10, 25, 50, and up to 300 transactions per second, was assessed, as presented in Figures 10 and 11.

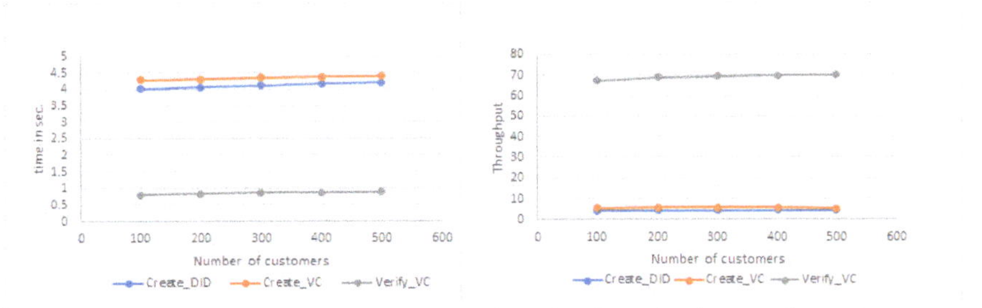

Figure 9. Transaction processing time (**left** side) and throughput (**right** side) for create and verify operations for different recorded sets of customers.

To test the throughput performance, we used sets of customers of 100, 200, and up to 500 as a small-scale implementation. The send rate varied from 10 to 300 transactions per second. The signing operation by the Java Card depends on the speed of APDU command exchange and the chip characteristics to process the SECP256K1 signing function. For the sample card used for our testing, we observed an average time for the card crypto-operation of ~3.2 s. The broadcasting, mining, and publishing to the Blockchain are all included in the transaction time duration. We were able to measure the time including the signing by the card for the case of one transaction at a time, while for stress testing, we pre-signed a set of 10, 25, 50, and up to 300 transactions to assess the performance. It was observed that key generation on the card takes a relatively long duration, of about 10 s; however, it is done only once during the registration phase for each customer, and hence it is still acceptable behavior.

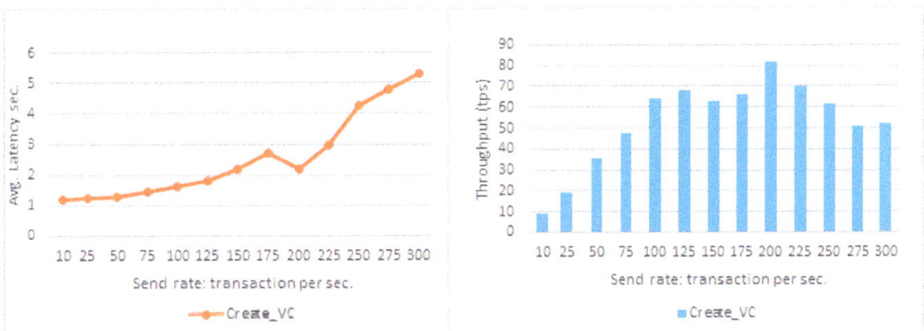

Figure 10. Latency (**left** side) and throughput (**right** side) for Create_VC (write operation) under different sending rates of 10, 25, 50, and up to 300 tps.

The observed behavior for sending rates of 10, 25, 50, and up to 300 transactions per second (tps) for Create_VC calls, which in terms of Blockchain access is a write operation, is shown in Figure 10, which indicates the throughput is relatively increasing. However, after 125 tps the throughput starts to slightly decrease, then significantly increases at 200 tps, before gradually decreasing again. The average throughput of write transactions is 53.07 tps.

The latency increases up to the sending rate of 175 tps, then drops at 200 tps, before the trend gradually increases again. The observed average latency for the write operation was 2.55 s.

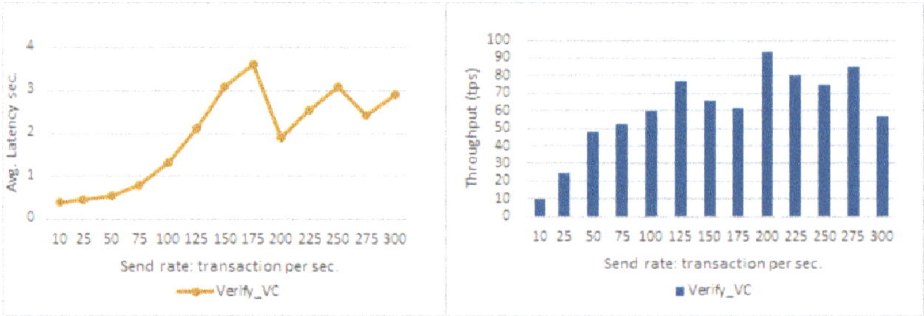

Figure 11. Latency **left** side) and throughput (**right** side) for Verify_VC (read operation) under different sending rates of 10, 25, 50, and up to 300 tps.

For Verify_VC calls, which are read operations, as shown in Figure 11, the transaction throughput increases until starting to decrease after 125 tps; furthermore, it oscillates and relatively increases at 200 tps and 275 tps, and then the trend goes down. The average throughput of read transactions was 60.69 tps.

The transaction latency increased with higher sending rates up to 175 tps, then dropped at 200 tps, before the trend gradually increased again, although it started to decrease after 250 tps. The observed average latency for the read operation was 1.93 s.

Throughput significantly varies depending on the function performed, such as Create_VC, which implies smart contracts; the throughput is much lower than that of read operations such as Verify_VC. We tested the performance with a set of customers ranging from 100 to 500 and sending rates of 10, 25, 50, and up to 300 tps. The latency and throughput for this testing show that optimum performance happened at 200 tps, with a throughput of 82 tps in writing operations and 93 tps in reading operations, while the latency observed was 2.2 s for writing and 1.9 s for reading. However, these preliminary evaluation metrics are highly impacted by the capacity and Blockchain setup configuration, such as the size of the network and the number of nodes, in addition to the expected load (transaction sending rate). Although the system showed a stable throughput, which complies with our preliminary proof-of-concept implementation, it does not satisfy the real-world banking use case.

The customer expects a much faster transaction. Hence, we will validate other options in future work to improve the throughput and speed, such as the proposed solution in [25] and other private Blockchain deployments, with the aim of achieving a maximum of 15 s per transaction. As per our discussions with SMEs from different banks, the expected time for customer convenience should not exceed 15 s per transaction at peak load, and the system should be capable of concurrently handling 1000 transactions in the pilot phase, with the ability to scale based on pilot performance analysis.

7.4.2. Proof of Authority PoA Consensus Mechanism

Proof of Authority (PoA) consensus is considered an efficient deviation from Proof of Stake (PoS), where the assigned validators sign and confirm the transactions. PoA includes a governance penalty system to eliminate malicious behavior and can provide faster transaction rates compared to PoW, excluding the mining time. The PoA mechanism reduces the utilization of computational resources, which is preferred in our case as it leads to a relatively stable transaction time

In our work, we use Proof of Authority (PoA), which is efficient and reasonable for verifiable credential handling, since this kind of platform requires only a small number of sealers. To ensure audit trails of access to records in this financial service ecosystem, it is necessary to have the participation of banks and regulated authorities.

7.4.3. Consistency

Distributed ledger consistency implies that no two nodes are in agreement, to ensure that they decide differently [56]. This is hard to achieve in Blockchain-based applications for both agreement on the valid blocks and the order of those blocks. Our proposed work relies on uPort and Ethereum, which has a default solution by hardening the puzzle difficulty. This leads to the generation of one block every several minutes, which is enough to propagate the new block through the Blockchain network.

7.4.4. Complexity

To assess the complexity of the SSBI system, we considered the work of [57], in which the authors suggested using the push–pull mooring effect to extend the complex theory, explaining this to the field experts, and proposing that they use the solution. We attempted to adapt the mechanism for our use case and conducted an interview with senior banking business managers to propose the idea and explain the features proposed to extend the usage of payment cards for identity operations. The interview included a three-slide presentation, a guide on how to use the demo, a presentation of the demo, and a request for feedback regarding risks and their intention to use.

The collected feedback from four representatives in four different banks, in addition to three expert engineers who worked in banking card issuance consultants, is summarized as the following:

- The idea seemed to be promising and extends the trust in banks for identity operations.
- The integration with bank systems was theoretically possible but still needed to be validated by a bank architect.
- A question was asked about the permission of the payment scheme to add the identity applet to the payment app, which should be accepted. However, it needs consultation regarding the specific payment scheme.
- After trying a demo using the card signing to share the identity with another bank, five participants out of seven, saw it as being easy to use but difficult to register.
- Bank representatives saw it as an opportunity for KYC enrolment and updates.
- Card issuance representatives thought that it has good potential and value for the customers.
- A general comment was made about linking Blockchain to crypto-currency and an explanation was provided that the scope is different.
- The average complexity level as collected from participants was 2 out of 10.

7.5. Analysis and Comparison

KYC2 [18] advised the need for key management, secret handling, and recovery options. This work proposes using banking smart cards for identity cryptographic operations and deploying the SSI concept based on Ethereum Blockchain; in particular, the curve exists for multiple-market banking Java Cards, which we validated in this proof of concept.

SCARAB [21] stores everything on-chain. This affects the scaling and system performance; furthermore, it stores identity-sensitive data on public servers, with no possibility to erase them.

BBM [5] is a Blockchain SSI model for open banking; however, SSBI uses the banking card, which is trusted and available to use by the customer for daily purchases. This work extends the use of the card to store and manipulate the identity transaction signing and allows the sharing of VCs with third parties through the Veramo SSI platform. This achieves the balance between usability and assurance without additional hardware tokens. This work attempted to implement a prototype, and to assess its feasibility and possibility of being integrated with the existing banking card issuance process.

The SSI framework prevents banking scam calls [7]. We leverage its use case and overcome the limitation of the signing curve on Java Card ED25519, which is required for Indy transactions. Instead, the framework is based on Veramo and Ethereum Blockchain, for which the required SECP256K1 signing curve exists in several banking cards, to issue ethr DIDs and VCs. This fits within open banking use cases, and the proposed framework supports smart contracts to enable automated online financial services.

In [22], the authors proposed a protocol to achieve a high level of assurance (LoA) by combining a software wallet key part and a YubiKey 5Ci token [23] hardware key part. In contrast, in SSBI we leverage the banking use case and balance the usability and strong security without additional devices. Moreover, the framework supports strong authentication and VC exchange for open banking services, and not only the identity operations.

The authors in [24] proposed "Casper", a mobile identity wallet that enables inter-banking Know Your Customer (KYC) based on SSI. In this work, we combine high LoA using banking cards to improve the identity control and share a payment experience that is similar to that of using banking cards to support extension to several business values, such as smooth KYC, password-less login to mobile banking, and much more. We rely on Ethereum Blockchain, and the Veramo SDK to handle DIDs and VCs. In addition, we use a key recovery mechanism that combines a secret part from the bank with a known passphrase held by the customer.

PriFob [25] is a global online credential management solution that provides privacy, which is fog-enhanced and based on a publicly permissioned Blockchain. The efficiency and performance improved due to the adaptation of PoA and SoW consensus protocols. Although it is not applied directly for the banking use case, we focus on improving security and the customer experience by involving banking cards in the flow of identity sharing. However, the reliable encryption scheme and consensus algorithm in PriFob can be utilized in our future improvements to achieve better scalability.

A comparative analysis between SSBI and the previous work is summarized in Table 1.

Table 1. SSBI comparative analysis.

	KYC2 [18]	**SCARAB** [21]	**Onename.io** [58]	**PriFob** [25]	**Casper** [24]	**SSI Bank Scam Calls** [7]	**BBM** [5]	**SSBI**
Banking KYC exchange	Yes	No	No	No	Yes	No	Yes	Yes
Utilize banking card for identity operations	No	No	No	No	No	Yes	No	Yes
Recoverability	No	No	No	?	No	No	Yes	Yes
Flexibility	No	Yes	No	Yes	Yes	Yes	Yes	Yes
Portability	Yes	Yes	No	Yes	Yes	No	Yes	Yes
Smart contract	Yes	No	?	Yes	No	No	Yes	Yes
Interoperability	Yes	Yes	No	Yes	No	Yes	Yes	Yes
Privacy protection	Yes	Yes	Yes	Yes	Yes	Yes	Yes	Yes
Integrity and confidentiality	Yes	No	Yes	Yes	Yes	Yes	Yes	Yes
Accountability	Yes	Yes	No	Yes	Yes	Yes	Yes	Yes
Selective disclosure	Yes	No	No	Yes	Yes	Yes	Yes	Yes
Achieve high LoA	No	No	No	No	No	Yes	No	Yes
Blockchain	Hyperledger Indy	Public ledger	Blockstack	Publicly permissioned Blockchain	Rahasak	Hyperledger Indy	Ethereum	Ethereum
Consensus	Plenum	Byzantine protocol	Stacks	PoA/SoW	PoW	Plenum	PoW	PoA

8. Conclusions and Future Work

This work is a step toward adopting and emphasizing the SSI identity-sharing concept for the open banking paradigm. The proposed platform relies on usability and customer trust in using a payment card as tamper-resistant secure key storage for identity cryptographic operations and transactions performed on Ethereum Blockchain. SSBI provides a mobile app and a web-based trace component to manage communication and reporting to entities and controllers including third-party agencies, banks, and financial institutes. Personal data are stored in the customer's mobile device while proofs are stored on Blockchain. Verifiable credentials are shared between the customers and the prover entities.

SSBI benefits from the trusted banking card industry and security measures such as PCI-DSS standards. The payment card secures the issuance process and the audited life cycle ensures attestation, genuine smart chip keys, and certified installed applets. We assume that the customer holds at least one active bank account in the initial phase. The platform eliminates fraudulent banking calls and financial identity theft due to the strongly encrypted credentials and proofs.

We reviewed the related solutions, proposed a solution, and described the framework architecture, layers, and components. In addition to the design objectives, we also describes the flow, including the initialization and authorization phases. This work provides a demonstrative proof-of-concept implementation based on banking Java Cards to sign the identity transactions on Ethereum Blockchain. We presented the integration of the banking card, personalization, applet development, and the flow to integrate with the banking issuance process. We presented a performance evaluation and security analysis, in addition to comparisons to related solutions.

The preliminary evaluation of our implemented prototype, which exploits the use of a banking card with a Blockchain-based SSI solution, is highly impacted by the capacity and Blockchain configuration, such as size of the network, the number of nodes, and the expected load. The system showed a stable throughput, which complies with the requirements of our preliminary proof-of-concept implementation, but does not satisfy the real-world banking use case. The customer expects faster transaction execution.

Hence, in future work, we will validate other options to improve the performance in terms of throughput and speed, with the aim of achieving a maximum of 15 s per transaction and a 1000 tps scale, which can fit with pilot real-world implementations. Furthermore, we will integrate the proposed approach with an existing bank system to implement practical settings and real test cases, in addition to validating end-to-end encrypted data sharing and cost analysis for identity operations.

Author Contributions: Conceptualization, K.A.M.A. and A.M.T.A.-E.; methodology, K.A.M.A.; software, K.A.M.A.; validation, J.F.W. and S.F.S.; formal analysis, K.A.M.A., A.M.T.A.-E. and J.F.W.; writing—original draft preparation, K.A.M.A.; writing—review and editing, K.A.M.A., J.F.W. and A.M.T.A.-E.; supervision, S.F.S. and A.M.T.A.-E. All authors have read and agreed to the published version of the manuscript.

Funding: This research received no external funding.

Data Availability Statement: Not applicable.

Conflicts of Interest: The authors declare no conflict of interest.

References

1. Open Finance. Available online: https://fastpayltd.co.uk/blog/what-is-open-finance (accessed on 25 March 2021).
2. Remolina, N. *Open Banking: Regulatory Challenges for a New Form of Financial Intermediation in a Data-Driven World*; SMU Centre for AI & Data Governance Research Paper No. 2019/05; SSRN: Rochester, NY, USA, 2019. [CrossRef]
3. Confessore, N. Cambridge Analytica and Facebook: The Scandal and the Fallout So Far. *The New York Times*. Available online: https://www.nytimes.com/2018/04/04/us/politics/cambridge-analytica-scandal-fallout.html (accessed on 17 May 2022).
4. Veramo. Available online: https://veramo.io/ (accessed on 22 December 2022).
5. Dong, C.; Wang, Z.; Chen, S.; Xiang, Y. BBM: A Blockchain-Based Model for Open Banking via Self-sovereign Identity. In Proceedings of the International Conference on Blockchain, Third International Conference, Held as Part of the Services Conference Federation, SCF 2020, Honolulu, HI, USA, 18–20 September 2020; Lecture Notes in Computer Science. Springer: Berlin/Heidelberg, Germany, 2020; Volume 12404, pp. 61–75. [CrossRef]
6. Hyperledger Indy. Available online: https://www.hyperledger.org/projects/hyperledger-Indy (accessed on 10 April 2021).
7. Ahmed, K.A.M.; Saraya, S.F.; Wanis, J.F.; Ali-Eldin, A.M.T. A Self-Sovereign Identity Architecture Based on Blockchain and the Utilization of Customer's Banking Cards: The Case of Bank Scam Calls Prevention. In Proceedings of the 2020 15th International Conference on Computer Engineering and Systems (ICCES), Cairo, Egypt, 15–16 December 2020. [CrossRef]
8. Allen, C. The Path to Self-Sovereign Identity. 2016. Available online: http://www.lifewithalacrity.com/2016/04/the-path-to-self-soverereign-identity.html (accessed on 13 February 2022).

9. Tobin, D.R.A. The Inevitable Rise of Self-Sovereign Identity. Sovrin Foundation Technical Report. 2018. Available online: https://sovrin.org/wp-content/uploads/2018/03/The-Inevitable-Rise-of-Self-Sovereign-Identity.pdf (accessed on 15 March 2022).
10. Xu, J.J. Are blockchains immune to all malicious attacks? *Financ. Innov.* **2016**, *2*, 25. [CrossRef]
11. uPort.me. Available online: https://www.uport.me (accessed on 10 April 2021).
12. uPort Whitepaper. Available online: https://whitepaper.uport.me/uPort_whitepaper_DRAFT20170221.pdf (accessed on 10 April 2021).
13. Reed, M.S.D. Decentralized identifiers (dids) v0.12. In *Community Group Report*; W3C: Cambridge, MA, USA, 2019.
14. IPFS. Available online: https://ipfs.io (accessed on 10 April 2021).
15. Veramo Agents. Available online: https://veramo.io/docs/veramo_agent/introduction/ (accessed on 10 March 2022).
16. Veramo Specifications. Available online: https://identity.foundation/didcomm-messaging/spec (accessed on 22 December 2022).
17. W3C. Available online: https://www.w3.org/TR/vc-data-model/ (accessed on 15 April 2021).
18. Soltani, R.; Nguyen, U.T.; An, A. A New Approach to Client Onboarding Using Self-Sovereign Identity and Distributed Ledger. In Proceedings of the IEEE International Conference on Internet of Things (iThings) and IEEE Green Computing and Communications (GreenCom) and IEEE Cyber, Physical and Social Computing (CPSCom) and IEEE Smart Data (SmartData), Halifax, NS, Canada, 30 July–3 August 2018; pp. 1129–1136. [CrossRef]
19. GDPR, E.U. GDPR. Available online: http://eugdpr.org/eugdpr.org.html (accessed on 12 June 2021).
20. Mühle, A.; Grüner, A.; Gayvoronskaya, T.; Meinel, C. A survey on essential components of a self-sovereign identity. *Comput. Sci. Rev.* **2018**, *30*, 80–86. [CrossRef]
21. Kokoris-Kogias, E.; Alp, E.C.; Siby, S.D.; Gailly, N.; Jovanovic, P.; Gasser, L.; Ford, B. Hidden in Plain Sight: Storing and Managing Secrets on a Public Ledger. *IACR Cryptol. ePrint Arch.* **2018**, *2018*, 209.
22. Abraham, A.; Schinnerl, C.; More, S. SSI strong authentication using a mobile-phone based identity wallet reaching a high level of assurance. In Proceedings of the 18th International Conference Security Cryptography, SECRYPT 2021, No. Secrypt, Online, 6–8 July 2021; Volume 1, pp. 137–148. [CrossRef]
23. Yubikey. Available online: https://www.yubico.com/at/product/yubikey-5ci (accessed on 15 March 2021).
24. Bandara, E.; Liang, X.; Foytik, P.; Shetty, S.; De Zoysa, K. A Blockchain and Self-Sovereign Identity Empowered Digital Identity Platform. In Proceedings of the 2021 International Conference on Computer Communications and Networks (ICCCN), Athens, Greece, 19–22 July 2021; Volume 2021. [CrossRef]
25. Baniata, H.; Kertesz, A. PriFoB: A Privacy-aware Fog-enhanced Blockchain-based system for Global Accreditation and Credential Verification. *J. Netw. Comput. Appl.* **2022**, *205*, 103440. [CrossRef]
26. Biometric Card. Available online: https://www.thalesgroup.com/en/markets/digital-identity-and-security/banking-payment/cards/emv-biometric-card (accessed on 12 April 2022).
27. Abraham, A. Self-Sovereign Identity—Whitepaper about the Concept of Self-Sovereign Identity including Its Potential. 2017. Available online: https://technology.a-sit.at/en/whitepaper-self-sovereign-identity (accessed on 10 March 2021).
28. Ogawa, A. What Is the Self-Sovereign Identity? The New Potential of Blockchain. Info-Com T & S World Trend Report, No. 346. 2018.
29. *ISO/IEC 29115:2013*; Information Technology—Security Techniques—Entity Authentication Assurance Framework. International Organization for Standardization: Geneva, Switzerland, 2013.
30. *NIST SP 800-63*; Digital Identity Guidelines. National Institute of Standards and Technology: Gaithersburg, MD, USA, 2020.
31. Solidity. Available online: https://solidity-by-example.org/app/ (accessed on 15 May 2021).
32. EIP-780. Available online: https://github.com/ethereum/EIPs/issues/780 (accessed on 15 March 2021).
33. Solidity Language. Available online: https://docs.soliditylang.org/en/v0.8.4/solidity-by-example.html (accessed on 15 March 2021).
34. eip-1056. Available online: https://eips.ethereum.org/EIPS/eip-1056 (accessed on 14 January 2021).
35. López, M.A. *Self Sovereign Identity: The Future of Identity: Self-Sovereignity, Digital Wallets, and Blockchain*; Technical Report; Inter-American Development Bank: Washington, DC, USA, 2020.
36. Infrastructure, E.B.S. EBSI's Technical Specification. Available online: https://ecas.ec.europa.eu/ (accessed on 10 July 2022).
37. EIP-1812. Available online: https://eips.ethereum.org/EIPS/eip-1812 (accessed on 1 March 2022).
38. Reactnative. Available online: https://reactnative.dev/ (accessed on 18 March 2021).
39. Nodejs. Available online: https://nodejs.org/en/ (accessed on 15 March 2021).
40. Expressjs. Available online: https://expressjs.com/ (accessed on 15 March 2021).
41. infura.io. Available online: https://infura.io/ (accessed on 15 March 2021).
42. Firestore. Available online: https://firebase.google.com/docs/firestore (accessed on 10 April 2022).
43. Typescript. Available online: https://www.typescriptlang.org/ (accessed on 10 May 2021).
44. Expo. Available online: https://docs.expo.dev/ (accessed on 13 November 2021).
45. mocha.js. Available online: https://mochajs.org/ (accessed on 12 August 2021).
46. Sqlite. Available online: https://sqlite.org/index.html (accessed on 12 April 2021).
47. PCI. Available online: https://www.pcisecuritystandards.org/ (accessed on 10 May 2022).
48. EMV. Available online: https://www.emvco.com/ (accessed on 15 April 2021).

49. Java Card Plaftorm. Available online: https://download.oracle.com/otndocs/jcp/java_card_kit-2.2.2-fr-oth-JSpec/ (accessed on 19 March 2022).
50. Fritsche, J.E.M.R.V.; Palma, L.M. Recommendations for implementing a Bitcoin Wallet Using Smart Card. Dep. Informática e Estatística—Univ. Fed. St. Catarina (UFSC), Campus Univ. Trindade Cx.P. 476/CEP 88040—Florianóp.—SC—Brazil 2018. Available online: https://repositorio.ufsc.br/bitstream/handle/123456789/192174/TCC%20Ricardo%20Fritsche%20Final.pdf?sequence=1 (accessed on 5 March 2021).
51. Bamert, T.; Decker, C.; Wattenhofer, R.; Welten, S. BlueWallet: The Secure Bitcoin Wallet. In *Lecture Notes in Computer Science*; Springer International Publishing: Berlin/Heidelberg, Germany, 2014; pp. 65–80.
52. BIP-32. Available online: https://github.com/bitcoin/bips/blob/master/bip-0032.mediawiki (accessed on 18 March 2021).
53. *ISO/IEC 7816-8:2021*; Identification Cards—Integrated Circuit Cards—Part 8: Commands and Mechanisms for Security Operations. International Organization for Standardization: Geneva, Switzerland, 2021. Available online: https://www.iso.org/obp/ui/#iso:std:iso-iec:7816:-8:en (accessed on 1 March 2021).
54. Grüner, A.; Mühle, A.; Lockenvitz, N.; Meinel, C. Analyzing and comparing the security of self-sovereign identity management systems through threat modeling. *Int. J. Inf. Secur.* **2023**, 3. [CrossRef]
55. Hölzl, M.; Asnake, E.; Mayrhofer, R.; Roland, M. A password-authenticated secure channel for App to Java Card applet communication. *Int. J. Pervasive Comput. Commun.* **2015**, *11*, 374–397. [CrossRef]
56. Kertesz, H.B.A. Consistency analysis of distributed ledgersin fogenhanced blockchains. In Proceedings of the European Conference on Parallel Processing, Lisbon, Portugal, 1–3 September 2021.
57. Sun, W.; Dedahanov, A.T.; Shin, H.Y.; Li, W.P. Using extended complexity theory to test SMEs' adoption of Blockchain-based loan system. *PLoS ONE* **2021**, *16*, e0245964. [CrossRef]
58. OneName.io: The Bridge Between Physical & Digital Identity & Blockchain for the Billions. WordPress.com 2015. Available online: https://rywalk.wordpress.com/2015/02/13/onename-the-bridge-between-physical-digital-identity (accessed on 12 August 2021).

future internet

Article

Blockchain-Based Loyalty Management System

André F. Santos, José Marinho and Jorge Bernardino *

Polytechnic of Coimbra, Coimbra Institute of Engineering (ISEC), Rua Pedro Nunes, 3030-199 Coimbra, Portugal; a21280412@isec.pt (A.F.S.); fafe@isec.pt (J.M.)
* Correspondence: jorge@isec.pt

Abstract: Loyalty platforms are designed to increase customer loyalty and thus increase consumers' attraction to purchase. Although successful in increasing brand reach and sales, these platforms fail to meet their primary objective due to a lack of incentives and encouragement for customers to return. Along with the problem in originating sales, they bring excessive costs to brands due to the maintenance and infrastructure required to make the systems feasible. In that sense, recent blockchain technology can help to overcome some of these problems, providing capabilities such as smart contracts, which have the potential to reinvent the way loyalty systems work and solve current problems. Although blockchain is a relatively new technology, some brands are already investigating its usefulness and rebuilding their loyalty systems. However, these platforms are independent and linked directly to a brand. Thus, there is a need for a generic platform capable of creating and managing different loyalty programs, regardless of the size of the business. This paper explores the shortcomings of current loyalty programs identified through the literature review, and proposes a loyalty management system with blockchain integration that allows any retailer to create and manage their loyalty programs and have customers interact directly with multiple programs in a single application.

Keywords: loyalty programs; blockchain; smart contracts; loyalty management

Citation: Santos, A.F.; Marinho, J.; Bernardino, J. Blockchain-Based Loyalty Management System. *Future Internet* **2023**, *15*, 161. https://doi.org/10.3390/fi15050161

Academic Editors: Christoph Stach and Clémentine Gritti

Received: 18 March 2023
Revised: 21 April 2023
Accepted: 25 April 2023
Published: 27 April 2023

1. Introduction

A loyalty program is a marketing strategy used by companies to reward their most loyal customers. Its other goals include attracting new customers, encouraging repeat purchases, and collecting information about customers. Currently, there are a wide range of loyalty programs; this is an issue because different applications are needed and the rewards do not suit all customers. Moreover, there is no interaction between the different programs, which makes it almost impossible to keep track of the status of all of them, and means users have to provide their information multiple times.

Some loyalty programs have found a way to operate with different businesses, but there are many difficulties in tracking information outside of their scope, for example, when a partner provides information for an external service.

A number of loyalty management systems have emerged as a way of overcoming these problems. These loyalty management systems provide new possibilities to build loyalty programs, and the tools to monitor them, faster. At the same time, they offer a way to create an ecosystem of loyalty programs that supports the user experience. While they solve drawbacks such as user experience, they still have problems related to third-party tools and incomplete connection between different systems.

Therefore, blockchain may be a viable option to address these challenges. The concept of blockchain consists of a chain of blocks that allows the storage of information. Each block stores a set of transactions, and these blocks are part of a network that any user can access. This technology introduces a new paradigm based on three core ideas: security, decentralization, and disintermediation [1].

For maximum security, this technology uses distributed consensus algorithms that ensure that every transaction is trustworthy. Additionally, it utilizes mechanisms of cryptography and digital signatures to ensure only the owner can access its data [1]. A final security-related concept is persistence, which means the data cannot be changed. Concerning decentralization and disintermediation, these two properties play a crucial role. Decentralization allows the user to control and to be the owner of the information it produces, as opposed to the current reality of large companies such as Meta and Google. Another property of a decentralized network is the trustless environment, which means no one needs to know or trust another entity, and each member has a copy of the data. This environment prevents anyone from corrupting the data and optimizes the distribution of resources [2].

Since blockchain does not use third-party tools to analyze and verify transactions, it uses the concept of smart contracts. Smart contracts are self-executing programs stored on the blockchain. These contracts are triggered when the established conditions are met. Usually, they are used to automate the execution of an agreement between entities so that all of them can be aware of the expected result. This mechanism brings benefits such as performance and efficiency, trust and transparency, and security, as it only works as long as the conditions are met and stored later in the immutable blockchain. Additionally, blockchain may allow people to save money as it avoids delays, fraud, and intermediaries [3].

This paper explores the shortcomings of current loyalty programs, identified through a literature review, and how the blockchain technologies might overcome them. Therefore, an innovative blockchain-based loyalty system is proposed, with the following contributions:

- A fully decentralized platform;
- Management of multiple loyalty programs of different types on a single platform;
- Integration of meta transactions without including personal information, providing a seamless experience for any level of customer expertise;
- The ability to handle multiple tokens within a single contract through the use of an ERC-1155 interface;
- Reduction of financial and maintenance costs for retailers by facilitating near real-time transactions and eliminating the need for intermediaries and proprietary infrastructures;
- Provision of better exposure to retailers and a better user experience for customers through a single marketplace in which all loyalty programs coexist.

The rest of this paper is structured as follows. Section 2 presents a literature review, which includes a background on loyalty programs, blockchain, smart contracts, and meta transactions; it also discusses related work. Section 3 describes and explains the proposed loyalty management system. Section 4 discusses the proposed system and compares it with other systems. Finally, Section 5 presents our conclusions and future work.

2. Literature Review

This section first provides an overview of the current state of loyalty programs and the challenges they face. In addition, it clarifies how blockchain technology has the potential to revolutionize these programs by introducing technical concepts that can facilitate a more effective approach. Second, this section presents a comprehensive analysis of the current state of the art regarding the use of blockchain in various domains and contexts associated with loyalty programs.

2.1. Background

Loyalty programs are one of the main options for brands to increase their accessibility and attractiveness. However, these systems have shortcomings that can influence interaction with users. Blockchain is an innovative technology capable of transforming the behavior of these systems and solving current problems. For better understanding and context, this subsection describes the issues of current loyalty programs and presents how blockchain technology complements these programs. The concepts of smart contracts and meta transactions are also explained.

2.1.1. Traditional Loyalty Programs

Currently, traditional loyalty programs are based on a variety of mechanisms, and it is possible to classify them based on several categories: stamp cards, points, cashback, tiers, and coalition [4]. However, despite being a dominant choice and offering several tools to attract new customers and retain them, these programs have problems related to customer and company perspectives [5].

Starting with the customer perspective, the main current difficulties are as follows:

- The first step in joining a new loyalty program is usually to provide the system with personal information, which nowadays, with unquestionable security concerns, may imply a constraint for some users. Additionally, customers feel that they belong to too many loyalty programs, since each brand has its own program [6];
- With traditional approaches, the customer does not own assets such as points, tiers, and others. Instead, the program controls and stores the assets inside its database. Additionally, the value of the assets is changeable due to the company having the power to decide when to inflate or deflate it. These programs naturally have limits due to the way they are implemented. An example is that rewards only have value within the scope of the application and cannot be transferred to another system [5];
- The customer experience of traditional loyalty programs is not satisfactory. Despite their popularity, customers are less motivated to continue interacting with the system due to the effort–reward relationship. Most programs set an expiration date for rewards, which indirectly requires the customer to spend them before the deadline, even if he/she does not intend to. Another factor that reduces the customer experience is the lack of customization in the rewards for customers, with their needs and desires not usually considered (each customer is different, but compensation is the same for all) [7].

From the perspective of companies, the main difficulties consist of the following:

- Companies are responsible for protecting all the information of customers. Moreover, to process the transactions, they rely on third party entities, which makes them very costly. Apart from the need for intermediaries, a company is also responsible for infrastructure management, which adds overwhelming complexity and increases costs;
- Building partnerships with the traditional approach is complex due to the nature of the programs. Usually, brands conceive of loyalty programs with the intention of expanding their own business, and therefore, said program becomes restricted to its original scope. Whenever a company decides to build a partnership with another company, the typical solution consists of creating a new loyalty program. Consequently, another issue arises: that of who owns the assets (since with the traditional approach, the customers do not own them).

2.1.2. Blockchain

Blockchain is a recent technology that has introduced a new paradigm for the secure storage and sharing of information. This technology essentially consists of a distributed database of a chain of blocks, where each block stores a transaction. To guarantee security, blockchain utilizes several mechanisms, such as cryptography, consensus algorithms, and digital signatures [1]. All transactions are performed in a decentralized way, removing the need for intermediaries, and once inside a block, the transaction cannot be deleted. This modern technology is based on the following core principles [8]:

- Decentralization—one of the main goals of blockchain is to remove an entity's ability to have absolute control of all data. Hence, this technology is based on a network in which every user has equal power and the users themselves validate the transactions [3];
- Immutability—every node in the network holds a copy of the digital ledger. To add a new transaction, all nodes must check the transaction, and most of them must validate it. Validated records become irreversible and cannot be changed. In a scenario of a sudden change in one node, all nodes check the validity of the new state. When all

nodes finish the verification, a voting process is started, and if the majority rejects the change, the new state is rejected. Then, a copy of a trusted digital ledger is sent to that node [8];
- Transparency—any user can access any record by accessing the node and consulting the timestamp [1];
- Security—when interacting with the blockchain, it is mandatory for a user to enter the key every time s/he wants to create a transaction. Each record is individually encrypted. The information inside the blockchain is hashed cryptographically, which means that each piece of data has its own unique identifier [8];
- Disintermediation—in order to avoid the use of third-party entities, blockchain takes advantage of smart contracts. These contracts are used to facilitate the negotiation process but still provide security and reliability to the process.

Although blockchain is a relatively new technology, there is growing interest in and research into its applicability across a range of sectors. In [1], the authors describe what distinguishes blockchain from currently known paradigms, and characterize their specifications from fundamentals to architecture. After specifying the capabilities of the technology, the authors present recent studies on the incorporation of blockchain in certain areas. The domains involved in these studies are financial activities, health, information systems, wireless networks, IoT (Internet of Things), smart grids, government and military services, and defense. For each specified domain, this survey presents some examples of applicability and their respective benefits, limits, and challenges.

2.1.3. Loyalty Programs Complemented with Blockchain

Blockchain integration will reshape the way current loyalty programs work and add some benefits. First, companies will not need to spend more money on infrastructures to manage the assets and the system. Furthermore, with blockchain integration, third parties are removed due to the characteristics of decentralization, disintermediation, and consensus, leading to financial savings. Data security is now guaranteed by blockchain mechanisms, leaving the company less burdened.

It will become simpler to create partnerships with other companies because they are in a shared ecosystem (blockchain) in which there are no restrictions, as any asset is valid.

For end users/customers, there are also some benefits. The first of these is the elimination of any physical support for managing rewards, such as stamp cards. Secondly, users do not need to provide any personal information to interact with the blockchain. The user just needs a virtual wallet. In addition, with the possibility for companies to partner with each other more easily, the number of loyalty programs the user belongs to will decrease. The user will also be able to trade assets between loyalty programs or even with other users. Finally, the user will be the owner of her/his assets in these loyalty programs, which is the opposite of the traditional approach.

2.1.4. Smart Contracts

Smart contracts are self-executing computer programs stored on the blockchain. The primary purpose of these contracts is to provide a more secure way to produce transactions without having to rely on third parties and question whether the information has been altered for personal benefit. These contracts contain the logic to proceed with transactions, with this logic being defined by programmers. Each contract, when stored in the blockchain, receives a unique address, which is the entry point for the interaction.

To interact with smart contracts, every user needs a virtual wallet, as every interaction with the blockchain requires a fee payment due to processing effort. The hosted blockchain's native cryptocurrency is used to pay this fee. During the interaction with a smart contract, the user sends data to the contract that will be processed based on its logic.

The benefits of smart contracts are speed and efficiency due to automatization processes. Once a given condition is met, the execution of the associated contract is triggered, and the transaction is performed without errors or any kind of bureaucracy. Associated with automatization, these contracts offer trust and transparency, since there are no third-party entities, and the blockchain encrypts every transaction and shares it across the nodes. Additionally, with a consensus mechanism it is nearly impossible to modify transactions for personal benefit. As a result of their nature, smart contracts help entities save money and time by eliminating intermediaries and the associated delays and expenses.

Currently, the adaptability and reusability of smart contracts is an open research issue. In [9], the author proposed a design pattern to implement the reusability of verification rules across multiple contracts, which allows reconfigurability at runtime. The study was conducted using permissioned blockchains and Java as the programming language. The proposed design pattern exploits polymorphism and manages inheritance through sealed classes, and includes two layers: abstract and concrete. The abstract layer includes an abstract contract class with a single abstract method ("checkSC"), as well as an interface for rule implementation. The concrete layer contains all the contracts, which are classes that extend the abstract contract class and therefore implement its single abstract method (i.e., the contract behaviour). Furthermore, the concrete layer includes the definition of rules, where each rule is encapsulated by a class that implements the rule interface. Although a simple pattern, the proposal in [9] reduces code redundancy and allows configuration at runtime.

In the same context, the paper [10] details a design pattern, called the proxy pattern, that allows smart contracts to be updated without losing data. OpenZeppelin provides an extensible plugin based on this pattern. It comes with extensive documentation explaining its functionality [11]. Unlike a single contract that combines both data and business logic, the proxy pattern involves splitting the contracts into a proxy contract and a business logic contract. The proxy contract acts as a dedicated storage and delegates calls to the business logic contract, while the business logic contract handles all the business logic operations. In the proposed design pattern, the read and write operations performed by the delegated contract have an impact on the storage of the proxy contract.

2.1.5. Meta Transactions

Every day, thousands of smart contracts are deployed on existing blockchains. Smart contracts can offer significant advantages, but they can also present challenges for new users in the blockchain space. The primary issue is related to user onboarding, which requires the use of native cryptocurrencies to interact with the contracts and can have an impact on the overall user experience. Users who are not familiar with the blockchain environment will need to pre-purchase cryptocurrency to use the loyalty application.

Meta transactions are a mechanism that allows anyone to interact with a blockchain regardless of their level of knowledge on this topic, and to make transactions without fee payments. This is achieved by decoupling the owner of the data to be sent and the entity that will pay the fee [12].

The meta transactions' flow consists of the following three steps [12]:

1. The user who wants to interact with a smart contract creates a transaction request. This transaction is equivalent to a usual smart contract transaction, except that it will be signed with the user's private key and additional parameters to "craft" the transaction;
2. Following the transaction request creation and signature phase, the request is sent to an entity that acts as a relayer and validates the transactions according to a specification (whitelist for example). Once validated, the relayer sends the transaction to the blockchain network in which the targeted smart contract is hosted;
3. Finally, the smart contract receives the transaction and performs the intended action.

Meta transactions are a design pattern that offers a seamless experience to users, in which they do not have to spend money to engage with the blockchain.

2.2. Related Work

This section presents an analysis of the current state of the art regarding the use of blockchain in various domains and contexts associated with loyalty programs.

In [13], the authors propose and describe an implemented loyalty program that is based on blockchain. The authors identified issues related to the mechanisms of customer dissatisfaction and loyalty. Afterward, they explain the solution they propose, which consists of a smart contract created in the Ethereum network [14] and capable of producing tokens. These tokens are called TECH tokens (the name given by the authors) and are similar to cryptocurrencies. TECH tokens are stored in Ethereum wallets and can be transferred between wallets to perform payments or exchange goods in the system. Since the Ethereum network requires a fee for each transaction and the cost is expressed in the ether cryptocurrency, the authors implemented and tested the system on Rinkeby, an Ethereum-compatible test network. The implementation consisted of three distinct phases. First, the smart contract is written and deployed on the blockchain. Then, the token is created according to ERC20 standards [15], and finally, the frontend is developed in React.js for user interaction. The system has integrated Web3.js as a way to allow the frontend to communicate with the blockchain.

The authors in [5] propose a universal loyalty platform. Before presenting the solution, they detail how blockchain can improve a loyalty program from both the point of view of customers and companies. The proposed solution consists of an application in which customers can use points to redeem rewards, and a website for companies to interact with and manage their business. The system uses React Native and React for mobile and website frontend, respectively. The system also uses Stellar and Hyperledger Fabric Blockchains to manage users' information and assets. At the customer level, upon registration, a Stellar account is created. This account allows the user to trade any asset type, being able to exchange points between loyalty programs. At the company level, upon registration, a Stellar account is created. During the registration process, the company can choose other companies to collaborate with. Once this entire process is over, the smart contract allocated in the Hyperledger blockchain is invoked to create and connect the inherent token to that company and link it to the Stellar account.

Two studies from 2019 [16,17] aimed to explore how blockchain can impact loyalty programs and understand their relation. Both studies presented four theoretical foundations: (1) self-determination theory (SDT) [18]; (2) customer perceived value; (3) loyalty programs' design in loyalty program efficiency; and (4) blockchain application. The authors use SDT to define four dimensions of customer motivation: economy, autonomy, competence, and relatedness. To better understand the link between SDT and customer perceived value, the authors assess customers' perceived value in three domains that they proposed: economic utility, psychological self-fulfillment, and social interaction. For the last two theoretical foundations, the authors focus more on loyalty programs, specifically on design effectiveness and blockchain infrastructure. Both papers chose BubiChain as the target blockchain infrastructure, due to its success in the loyalty program market, to undertake the evaluation. They also chose an exploratory case study on a loyalty points-based system. Although the works are analogous, the depth of the study differs between them.

The authors of the first paper [16] only assess the main characteristics of blockchain integration and its impact on each SDT domain. The second paper [17] evaluates the BubiChain infrastructure itself and how it works, the main elements of the loyalty points-based system, how the code is structured, and, finally, the impact of BubiChain on participation in loyalty programs. While the exploratory case with blockchain has proven to be more effective than existing traditional schemes to satisfy customers, both papers state that more future research is needed to explore the relationship between blockchain design for loyalty programs and customer behaviors.

The work described in [7] aims to review the potential and challenges of blockchain in marketing areas for building brand loyalty. The authors first describe blockchain technology and its unique features. Then, an overview of the characteristics of brand loyalty and current failures is presented. The authors of the paper also explain how integration with blockchain infrastructure can eventually solve and improve the quality of loyalty programs, making them more effective and affordable.

The authors in [19] propose an implementation of a loyalty program for a mobile platform based on the Waves blockchain. The choice for this blockchain platform is due to its lightweight and fast features when compared to Ethereum, and to the premise that the cell phone is an essential commodity today. The proposed solution consists of three key components: a token creator in Waves, a web API, and a database to store accounts' information.

In [6], the authors give a simple introduction to blockchain and aim to analyze its impact on the retail industry and customer loyalty. Although the authors focus both on the supply chain and loyalty areas, their emphasis is on the need for a loyalty program as the best solution to retain customers. According to them, customers remain dissatisfied even though numerous companies are now investing in loyalty programs. To support this claim, the authors provide statistics from a survey in which they list several problems pointed out by the respondents. They also suggest that blockchain integration could potentially address these issues and provide a lower cost system with better security and transparency.

In [20], the author studies the importance of blockchain technology in enhancing loyalty programs. The scope of this paper is restricted to the airline business. Regardless of the study domain, the author explains how blockchain can impact and revolutionize a company's experience. Integrating blockchain technology from a business perspective could reduce maintenance costs and enhance customer service. Blockchain can foster better customer relationships and captivate customers by providing the benefits of traditional loyalty programs and better customer service. From the point of view of the transaction process, blockchain also makes all the validation, storage and sharing of transactions easier, since intermediaries would no longer be needed. For better context, the author explains the five basic principles of blockchain. This study consists of a survey that was carried out using two sources of information: the primary source, using 450 questionnaires, and the secondary source, the literature. The 450 questionnaires were distributed among customers and airline managers, and a high-efficiency response rate was achieved. The use of blockchain technology was emphasized by the surveyed airlines. They concluded that the degree of effective use of customer loyalty programs is related to blockchain technology, and the results showed that loyalty systems were effectively improved by blockchain integration.

In [21], the authors propose a blockchain-based framework to enhance loyalty programs in the fast-moving consumer goods (FMCG) industry. This framework aims to revolutionize the conventional process of exchanging coupons by introducing a token called promotion asset exchange (PAX). Instead of customers purchasing products and retaining coupons for later trading with merchants, followed by the cumbersome process of merchants trading numerous coupons with the manufacturing company, which can lead to counting errors and result in confusion, customers will scan a QR code on the product packaging, and the promotion (PAX token) will be stored in their wallet. Once the customer has accumulated enough PAX tokens, they can be traded with the merchant, which will trigger a transfer to the merchant's wallet. To receive payment, the merchant can return the PAX token to the manufacturing company, which will refund it to the customer based on the number of tokens. The authors emphasize that blockchain technology enables the secure tracking of all transactions, enabling manufacturing companies to better understand customer needs while ensuring no coupons or payments are lost due to human error.

The authors of [22] propose a blockchain-based platform for coalition loyalty program management. The primary objective of the platform is to provide customers with the capability to exchange their points across distinct existing blockchain loyalty programs. The system consists of different companies that operate their loyalty programs, customers who accumulate points through these programs, and a platform that facilitates token exchange. The token exchange platform incorporates smart contracts that define the exchange rates between the tokens of each company. To ease this process, the companies are required to negotiate the rates and establish smart contracts with the rates. To enable interoperability, the authors utilize sidechain technology for conducting transactions across different blockchains. In order to encourage customer engagement, the authors allow customers to participate in the blockchain consensus protocol and receive rewards in return. The authors have selected the proof of stake (PoS) protocol due to its low computational requirements, minimal delay, and potential to attract more customers. The proposed system ensures that customers can effortlessly exchange their points with other companies, and companies can easily join the coalition/partnership without the need for a system overhaul.

The authors of [23] propose a blockchain-based platform that facilitates the exchange of loyalty points among customers that is similar to the work described in [22]. The proposed system leverages a many-to-many matching method based on the call auction approach used in the stock exchange market, thereby eliminating the need for exchange rate negotiations. The call auction method increases the likelihood of orders being executed, prevents points from going unused, and incentivizes customers to accumulate more points. In contrast to the system proposed in [22], in which companies are responsible for handling rate negotiations and defining the corresponding smart contracts, this system [23] uses the call auction method to simplify the points exchange process, making it more efficient and eliminating the need for partnerships and negotiations. Both systems aim to enhance the current blockchain-based loyalty programs and improve customer experience.

A study on the impact of a blockchain-based loyalty program on customer loyalty in the renewable energy sector was conducted by the authors in [24]. More specifically, this paper addresses the problem of green electricity tariffs (GETs), where companies often fail to generate the promised amount of renewable energy, leading to dissatisfaction, mistrust, and customers considering switching to competitors. First, the authors present a theoretical study assessing how blockchain can increase institutional trust and decrease institutional distrust. They then propose a loyalty program that allows users to track electricity generation data, monitor their electricity consumption, and optimize their consumption by setting rules for their smart devices. The proposed loyalty program will provide customers with tokens that can be redeemed to reduce the price of GETs, donate to charity, use other utility services such as electric scooters or car sharing, or reinvest in renewable energy shares. While affordability is a system requirement in [24], it is not clear whether customers are responsible for paying blockchain fees, which may be impractical due to high transaction prices on the Ethereum network. This problem could be solved by introducing meta transactions. In addition, the smart contract architecture of the proposed system is not demonstrated, and it is unclear whether the system utilizes multiple contracts or any interface for token creation. Although the authors have achieved the desired outcome, the results are specific to the renewable energy sector. Furthermore, the proposal in [24] is supported by a qualitative analysis that includes interviews and a literature review, but lacks quantitative analysis.

Although there are already some proposals for blockchain integration, such as [5,13,19,21], they present a poor user experience with respect to blockchain onboarding, mainly due to the need for user knowledge, the authentication process, the input of personal information, and the lack of meta transactions. Moreover, they contradict the premise of blockchain decentralization using centralized database systems. Systems such as [22,23] propose innovative concepts and solutions to address the challenges faced by existing loyalty programs that rely on blockchain technology. All these systems use blockchain to manage the loyalty programs' infrastructure.

3. Proposed Loyalty Management System

In order to properly design a loyalty management system, first of all, we have to understand the problem and consider some factors. These factors include understanding blockchain capabilities and how they fit into the proposed system, addressing current customer and retailer concerns, and providing a better overall user experience.

The proposed system is a mobile application capable of managing several loyalty programs through interaction with a blockchain. Regardless of the knowledge and role of the user (i.e., retailer or customer), the application is designed to completely abstract users from the underlying blockchain.

3.1. System Description and Features

The system is a self-managed loyalty platform with which retailers of all sizes can create loyalty programs in which their customers can participate. The platform has two distinct users (customers and retailers), and the loyalty programs can include a variety of mechanisms. Despite the system requiring the use of blockchain technology, it abstracts and helps users in onboarding by providing each one with a wallet. These wallets are connected to the users themselves and are private. Furthermore, customers are not required to pay any transaction fee generated by the blockchain interaction, since the resulting costs are intended to be supported by the retailers.

This system includes two main components: a mobile application and smart contracts. Most blockchain applications are web applications, which limits interaction and provides an entry barrier for some less tech-savvy users. On the contrary, a mobile solution improves accessibility, taking advantage of the ubiquity of smartphones in everyday life. Despite having two types of users, the mobile application is the same for both of these types. The objective of smart contracts is to manage the loyalties' mechanisms and to ensure that the assets of the users are secure.

One of the most critical aspects of loyalty programs is the need for security measures, which is especially relevant today, since security is a major global concern. With the objective of ensuring customer privacy, the proposed system does not store any personal information. On the other hand, it uses Web 2.0 authentication methods, which allow users to use their previously created accounts, thus preventing them from providing any personal information whenever they access the loyalty management system. Although the data stored on the used blockchain are publicly accessible and can be viewed by anyone, the underlying consensus protocols make it possible to validate transactions and identify any potential risk, which makes the blockchain a useful technology for ensuring data security. Furthermore, by implementing appropriate smart contracts, addresses not whitelisted can be excluded, and any malicious changes detected during transactions can be immediately reversed, thus ensuring data integrity.

Despite the mobile application being unique, both types of users have different use cases. Customers have access to a marketplace in which all the loyalty programs are within the scope of the application. For each loyalty program, a customer can view the information related to the business and a description of the loyalty program itself. Additionally, the customer can enroll in loyalty programs to start participating and earning rewards. Once the customer subscribes to a program, he/she can begin to spend and earn rewards. The system helps customers to track all the assets related to the various loyalty programs so that they can check and be aware of their progress.

Retailers have different use cases, as their goal is to attract more customers, thus increasing their business. Retailers can create a loyalty program by providing their business information and by choosing a loyalty mechanism along with a brief description. Additionally, during this process, the retailer can upload images to give a better perception of the loyalty program or business. Since the main goal is to increase brand reach, the retailer can access the analysis and statistics data of the loyalty program. As the retailer must pay transaction fees, he/she can also buy the required cryptocurrency in the app.

The mobile application provides a transaction system based on a QR code approach to trading assets with retailers/customers, thereby enhancing the overall user experience. All transactions require the existence of two individuals, and the rewards are defined by the retailers during the process. The two individuals must be one retailer and one customer, since customer-to-customer and retailer-to-retailer transactions are not possible.

Figure 1 gives an overview of the proposed loyalty management system, illustrating the main components and interactions. Customers and retailers, which are the only actors in the system, use their personal smartphones running the mobile application to access the system.

A customer accesses the system to interact and discover new retailers, while a retailer must create its loyalty program and exchange rewards with customers. The mobile application employs Web 2.0 credentials for user login. To enable customer interaction with the underlying blockchain network at no financial cost, each transaction originated by a customer first passes through a relayer, which then redirects it to the respective smart contract. On the contrary, a retailer interacts directly with the smart contracts. For the execution of asset transfers or rewards, the receiving entity displays the QR code while the other agent reads it. Upon completion of the QR code-reading process, the transfer of assets from the QR code reader to the receiver is triggered through a transaction sent to the smart contract. Depending on whether the transfer is being performed by the retailer or the customer, it will go directly to the smart contract or through the relayer, respectively.

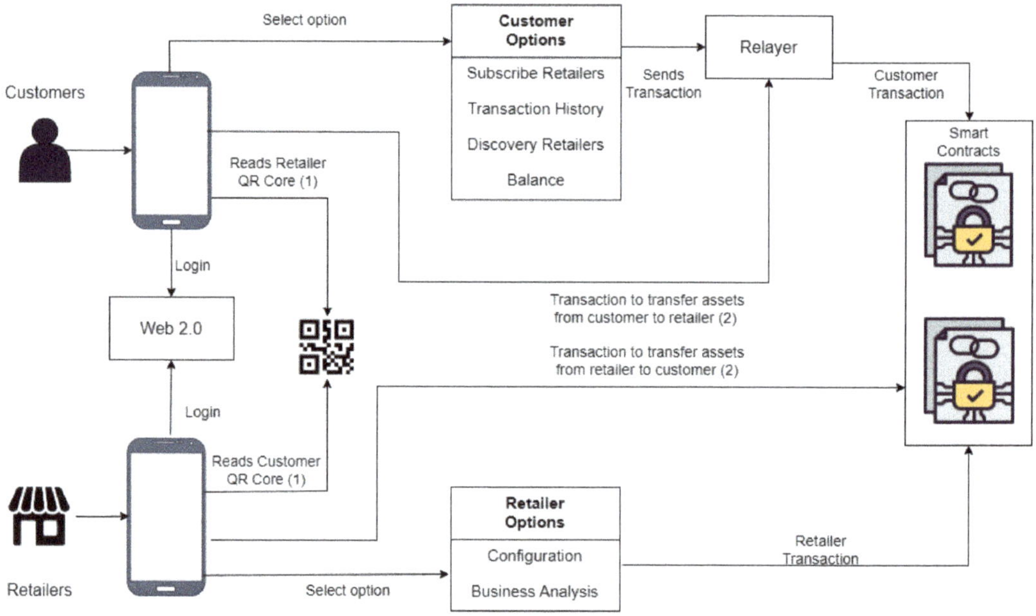

Figure 1. Proposed loyalty management system overview.

Figure 2 presents three screenshots of the mobile user interface. The first screenshot shows the customer's home page, which provides a concise summary of the retailers that the customer is subscribed to, along with their five most recent transactions. The second screenshot shows the retailer's page, which provides comprehensive information about the retailer and its loyalty program, including the rewards available. Finally, the third screenshot shows the QR code that the user will need to scan in order to redeem their rewards.

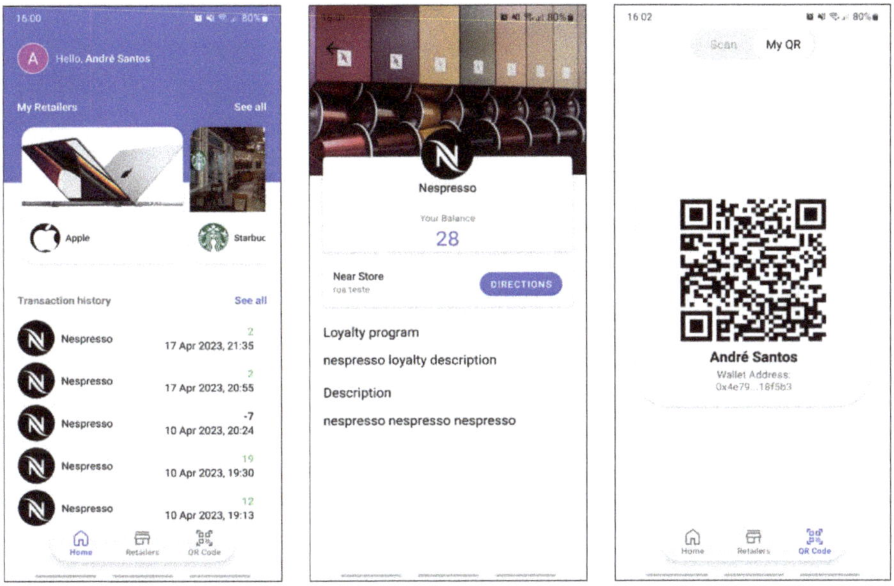

(**a**) (**b**) (**c**)

Figure 2. Mobile app screenshots. (**a**) customer profile; (**b**) retailer page; (**c**) QR code.

Following an agile methodology, Tables 1–3 summarize the system user stories mentioned previously.

Table 1. Common user stories.

User Story	Description	Name
US-1	As a user, I need to login so that I can use the app	Login
US-2	As a user, I need to have a QR code identifier so that I can receive transactions	Display QR code
US-3	As a user, I need to scan QR codes so that I can send transactions	Read QR code

Table 2. Customer user stories.

User Story	Description	Name
US-4	As a customer, I need to check my loyalty balance so that I can be aware of my progress	Check loyalty balance
US-5	As a customer, I need to access my transaction history so that I can track all my transactions	Transaction history
US-6	As a customer, I need a profile page so that I can see my system information	Profile page
US-7	As a customer, I need a discovery page so that I find new businesses to subscribe to	Discovery page
US-8	As a customer, I need to subscribe to a retailer so that I can start earning rewards	Subscribe retailers

Table 3. Retailer user stories.

User Story	Description	Name
US-9	As a retailer, I need to configure my profile so that I can create my loyalty program	Retailer configuration
US-10	As a retailer, I need to check my business status so that I can gain insight into my success	Analysis of the business
US-11	As a retailer, I need to exchange money into cryptocurrency so that I can pay for blockchain interactions	Currency exchange

3.2. Blockchain Selection

One of the most critical components within the proposed system is the choice of the underlying blockchain platform. Currently, there are numerous blockchain networks with distinct characteristics and purposes. For this specific type of loyalty management system, some key factors are essential, such as the number of transactions per second, cost per transaction, ability to support smart contracts, scalability, and a good source of information and documentation. Since this system will support several loyalty programs and each customer can subscribe to several programs, high transaction volumes per second are essential to ensure good performance. Considering the expected number of transactions, the cost of each transaction must be minimal so that retailers are not penalized and the platform becomes more affordable.

Since the purpose of the system is to support and manage different types of loyalty schemas and loyalty programs, smart contracts are mandatory. Therefore, the chosen blockchain must support them. Due to the need for the system to handle platform growth, scalability is required. Therefore, the blockchain network must permit scaling. The blockchain system must also have good documentation to ensure the best use of blockchain resources and capabilities.

The choice of a set of blockchain network candidates took all the mentioned factors into account. The chosen networks were Avalanche, Ethereum, Hyperledger Fabric, Polygon, and Solana. These networks are currently the most popular and are distinct from each other. Table 4 summarizes their main characteristics for comparison. The information presented was obtained from the available documentation, except for the transaction price. Transaction prices for Polygon, Ethereum, and Avalanche have been determined utilizing the following specialized websites, which are designed for scanning transactions provided by the blockchains: PolygonScan [25], EthereumIo [26], and Avascan [27] for Avalanche. Solana's price is available on their official website [28]. It is important to note that these prices are subject to change, as they depend on several factors, such as the value of the cryptocurrency at any given time and the volume of transactions processed over the course of a day. Concerning Hyperledger, the transaction price is zero, as it is a private network.

Table 4. Blockchain network comparison.

	Avalanche	Ethereum	Hyperledger	Polygon	Solana
Environment execution	EVM [1]-AVM [2]	EVM [1]	Docker	EVM [1]	LLVM
Smart contract language	Solidity	Solidity	Java/GO	Solidity	C/Rust
Transaction per second	4500	15–27	3500	10,000	65,000
Permission type	Public/Private	Public	Private	Public	Public
Transaction price	<$0.01	≈$4	$0	<$0.01	$0.00025
Layer	Layer 1	Layer 1	Layer 1	Layer 2	Layer 1

[1] EVM—Ethereum virtual machine; [2] AVM—Avalanche virtual machine.

Despite Ethereum being the most popular among the five solutions, it presents a weakness regarding the number of supported transactions per second. Additionally, the cost for each transaction is too high, which can lead to platform rejection. Hyperledger Fabric (designated Hyperledger in Table 4) is a private network with a smart contract environment and language distinct from the usual Ethereum Virtual Machine (EVM) and Solidity language, which can cause implementation problems. Even though Solana supports a higher number of transactions per second and results in a lower cost, it suffers from the same problems as Hyperledger.

The Avalanche and Polygon networks are both compatible with smart contracts developed for the Ethereum environment, since they use EVM. The main differences between these two solutions are the level of popularity, the supported volume of transactions per second, and the type of layer. Polygon is more popular and allows more transactions per second. It is also a layer-two network, while Avalanche is a layer-one. This means Avalanche is a network that works independently, while Polygon works in parallel with a main network (i.e., a layer one network such as Ethereum and Avalanche). In the Polygon case, the main network is Ethereum, but in the future, it will also be compatible with other main networks.

In conclusion, the preferred blockchain network for the proposed loyalty system is Polygon, due to its increasing popularity and capability of providing a fair price and a good volume of transactions per second. In addition, Polygon is compatible with Ethereum, which allows the transfer of smart contracts to the Ethereum network at any time.

3.3. Polygon Blockchain

Polygon is a layer-two scaling solution designed to improve existing blockchains. Polygon's objective is to solve scalability issues while not compromising on decentralization. This is accomplished by using sidechains and leveraging the developer community and the existing ecosystem. Each blockchain network is made up of multiple networks. These networks consist of two distinct environments, one for testing and another for production. The production network is entitled the mainnet. Sidechains are "copies" of the mainnet that support asset transfers between them and the mainnet. They are an alternative to a mainnet network that create a new blockchain with its own mechanisms. Currently, Polygon is only compatible with Ethereum. Its key features are the following [25]:

- Scalability—fast, low-cost, and secure transactions;
- High throughput—it offers up to 10,000 transactions per second on a single sidechain. It is possible to add more sidechains to increase the horizontal scaling;
- Security—Polygon validators are also stakeholders in the proof-of-stake system, required to hold and lock a certain amount of cryptocurrency as a security deposit. This ensures that validators have a vested interest in maintaining network security and integrity. Validators who act maliciously or make errors that result in a loss of funds will be penalized by losing their security deposit. As a result, validators have a strong incentive to follow the rules and act in the best interests of the network;
- Public sidechains—sidechains are naturally public, permissionless, and capable of supporting multiple protocols.

3.4. System Architecture

After taking into consideration all factors mentioned previously, it is essential to define a detailed architecture with all the different software components, the interaction with external services, and the communication protocols. The architecture is composed of three major modules: (i) the mobile application the users will interact with; (ii) external services for facilitating user integration with blockchain and providing a set of required features, namely InterPlanetary File System (IPFS), Torus Web3Auth, Ramp Exchange, and Openzeppelin; and (iii) smart contracts that contain the system's business logic. Figure 3 represents the system architecture.

As the system allows for different loyalty mechanisms, to facilitate the entire loyalty program management process, each smart contract represents a specific loyalty mechanism. Since numerous retailers with a specific loyalty mechanism can exist, each smart contract will inherit the ERC-1155 interface [29] through the contract available through Openzeppelin. ERC-1155 is a token standard interface that facilitates the smart contract representing multiple tokens simultaneously, and includes management functions that efficiently help in reducing the fees. This interface, along with the already pre-built ERC-1155 contract, simplifies the whole management process and diminishes the effort needed to create smart contracts from scratch. Furthermore, since ERC-1155 allows multiple tokens to be managed within a single contract, the proposed loyalty management system can significantly reduce the number of smart contracts required, unlike current systems that require separate ERC-20 contracts for each token. Given the large number of tokens involved in our system, the use of ERC-1155 thus results in a more efficient and scalable solution that requires a minimum number of contracts: a single smart contract is required for delegating calls and storing data unrelated to loyalty programs, and a separate smart contract inheriting the ERC-1155 contract is required for each loyalty mechanism.

The proposed loyalty system includes an intelligent contract manager to simplify blockchain calls, which is responsible for receiving transaction requests and delegating their execution to a contract running a specific loyalty mechanic. The manager is also responsible for storing all generic data, such as which retailers are defined in the system, customer subscriptions, and smart contract addresses for loyalty mechanics. Through this approach, calls to interact and obtain information from blockchain are simplified. Furthermore, in order to prevent any unauthorized data manipulation, all data are directly associated with the user's unique wallet address. This approach guarantees that only the users themselves can modify their data, thus preventing any malicious user from tampering with other users' information.

In order to streamline the development of smart contracts for each mechanism and store them in the loyalty manager component of the proposed system, a loyalty mechanism interface was created (see Figure 3). This interface defines four methods that each mechanism contract must implement to properly define its behavior. The loyalty manager stores all the mechanism contracts with a map, upon which the keys are an "enum" and the values are references to the loyalty mechanism interface type that corresponds to contracts responsible for delegating the received calls to the corresponding contracts.

Torus Web3Auth is a pluggable authentication infrastructure for Web3 wallets and applications. It provides seamless onboarding by providing web 2.0 login flows, such as Facebook, Gmail, and Twitter. Despite the fact that the user is abstracted from wallet creation, he/she is always in control of ownership. Torus Web3Auth also manages the key infrastructure of the wallets (thus taking the responsibility away from the system) and can perform direct integration through a software development kit (SDK) for native Android, which provides a group of tools to enable the implementation of authentication. This authentication infrastructure allows for better user integration with blockchain, with all abstraction layers provided.

Despite the proposed loyalty management system not being intended to store personal information, retailers can publish images about their business and other business information. All these data are stored in the IPFS module in a JSON-like file, in which the file is also encrypted and assigned a hash in order to access it. The hash is a set of randomly generated characters that is created based on the content of the file. IPFS is a distributed system for storing and accessing files and a peer-to-peer hypermedia protocol. Since inserting information into the blockchain has an economic cost, non-relevant and sensitive information related to users and the system can be left off the chain. IPFS is appropriate for this scenario. The only way of accessing the data is through the unique hash code, which will be stored on the blockchain and linked to the retailer's information.

Ramp is a service that allows exchange between cryptocurrency and physical money. The need for this service comes from the retailer use case, as retailers need to pay transaction fees to interact with the blockchain. So, to simplify this activity, the platform directly supports the required money exchange option.

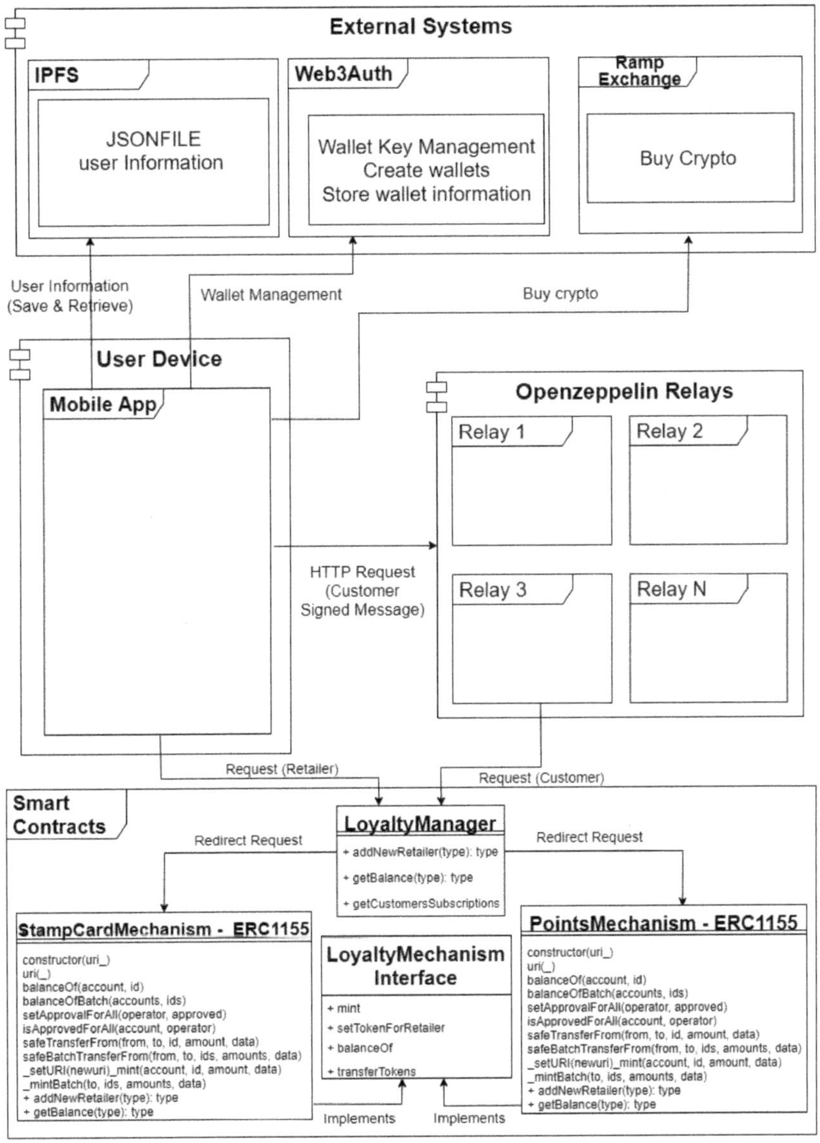

Figure 3. Loyalty system architecture.

Meta transactions are essential to provide a more satisfactory onboarding process for any user, regardless of their degree of expertise. Openzeppelin relays will be employed to implement meta transactions, acting as intermediaries that will receive transactions from customers, thus preventing them from paying fees. These transactions will be sent via HTTPS and signed by the client that created them. In the current system architecture, all requests are routed to the loyalty manager or, in the case of customer requests, to the relay, which are then routed to the loyalty manager contract. However, as the number of customer requests may increase in the future, it may be necessary to assign multiple relays to the loyalty manager to prevent any relayer from being overwhelmed with requests and causing system blockages, and to ensure a smoother flow of operations.

Summarizing the aforementioned set of external services, Web3Auth is used for wallet/key infrastructure management, IPFS is used to store any data in off-chain mode, Ramp is used by retailers to purchase cryptocurrency to pay fees, and Openzeppelin relays are for meta transactions.

The mobile application is the interaction center in which users will execute their requests. In the case of retailers, the application communicates directly with the blockchain and external services. In the case of the customers, an HTTPS request is redirected to a smart contract, and is sent to a relayer whenever it is necessary to interact with the blockchain.

Figure 4 illustrates a temporal sequence diagram of a blockchain transaction in the proposed loyalty management system. Both types of users, i.e., customer and retailer, must first log in to interact with the system, as shown in the diagram. Then, if the user is a retailer, invoked transactions are directly sent to the blockchain, whereas if it is a customer, transactions go through a relay to avoid fees. After completing the login process, the retailer queries the blockchain to retrieve their data, such as the IPFS hash. Once the data are returned, a second request is sent to IPFS to retrieve the remaining information associated with the hash. When the transaction reaches the smart contracts, it firstly contacts the loyalty manager contract that is responsible for delegating the transaction to the right contract loyalty type.

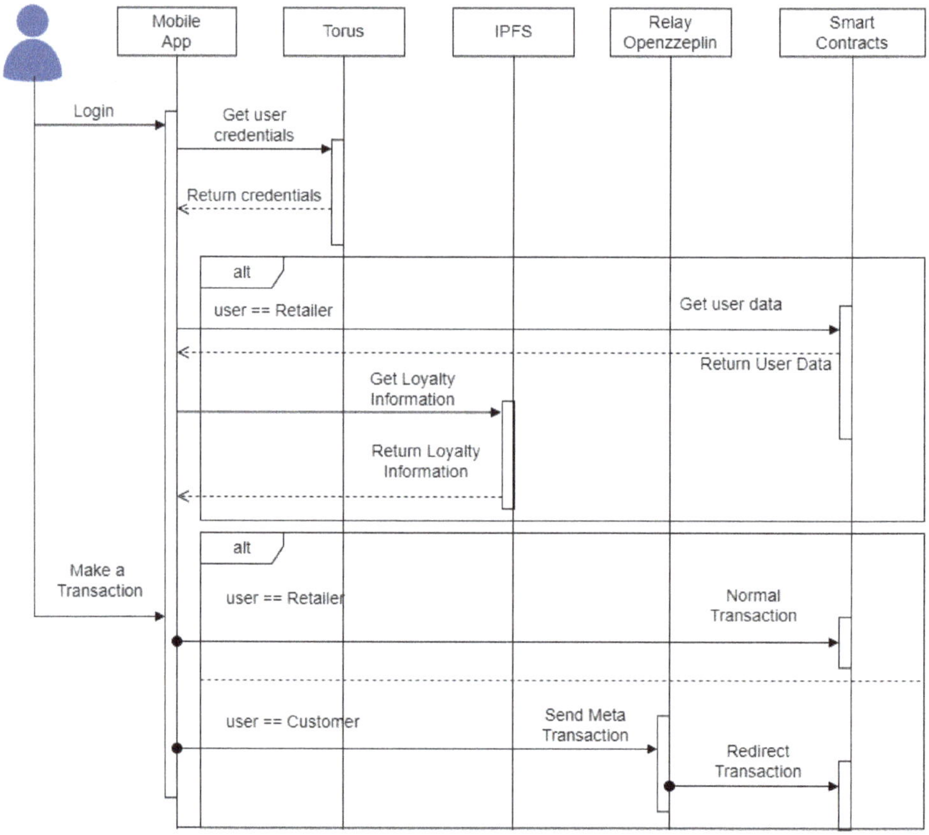

Figure 4. Sequence diagram.

3.5. Smart Contract Design

Smart contract design is a critical aspect of a fully decentralized system. As the sole source of project execution for data-related operations, smart contract design plays a pivotal role in determining the efficiency and cost-effectiveness of the system.

In order to streamline the implementation of multiple loyalty mechanisms, an interface has been created that must be implemented by each contract representing such mechanisms. Figure 5 shows the loyalty mechanism interface, which defines four core methods that form the basis of each contract, as they inherit the capabilities of the ERC-1155 contract. In addition, the interface defines two standard events that must be included in each mechanism contract to ensure consistency and standardization across the system.

```
interface LoyaltyMechanism {
    function mint(address to, uint amount) external;
    function setTokenForRetailer(address retailer) external;
    function balanceOf(address account, address retailerAddress) external view returns(uint);
    function transferTokens(address from, address to, uint amount, bool isSenderRetailer) external;


    event Mint(address indexed from, uint tokenId, uint amount);
    event Transfer(address indexed from, address indexed to, uint tokenId, uint amount, uint256 timestamp);
}
```

Figure 5. Loyalty mechanism interface.

Figure 6 highlights the points mechanism contract, which is responsible for managing the entire loyalty programs for this specific mechanism. The main functions of this contract include assigning a unique token ID to each merchant, allowing only the owner to mint tokens for their respective merchants, and performing transfers. Figure 7 shows the mint function, which is responsible for creating tokens for the retailer. To prevent unauthorized minting by any entity, only the token owner has the ability to generate tokens for their designated loyalty program. In addition, when a new retailer registers its loyalty points program, the smart contract assigns a token ID and stores it within a hash table (see Figure 8). This combined mechanism of token ID assignment and owner-restricted minting capability ensures secure and controlled token generation.

```
contract LoyaltyProgramPoints is ERC1155, LoyaltyMechanism {

    uint tokenIDIncrementor;
    mapping (address => uint) tokenOwners;

    constructor() ERC1155("") {
        tokenIDIncrementor = 0;
    }
```

Figure 6. Loyalty program points contract.

```
function mint(address to, uint amount) external {
    uint tokenId = getRetailerTokenId(to);
    require(ownerOf(tokenId, to), string.concat("Not the owner of the Token", Strings.toString(tokenId)));
    _mint(to, tokenId, amount, "");

    emit Mint(to, tokenId, amount);
}
```

Figure 7. Loyalty program points—function mint.

```
function setTokenForRetailer(address retailer) external{
    uint tokenId = getRetailerTokenId(retailer);
    if (tokenId == 0) {
        tokenIDIncrementor = tokenIDIncrementor + 1;
        tokenOwners[retailer] = tokenIDIncrementor;
    }
}
```

Figure 8. Loyalty program points—function setTokenForRetailer.

Finally, the loyalty manager contract, shown in Figure 9, acts as a central hub for all incoming requests in the loyalty system. It is also designed to support meta transactions, allowing customers to transact without incurring transaction fees. To fulfil this requirement, the contract inherits from the ERC2771Context contract and requires a forward contract as a parameter to validate whether a transaction should be executed or ignored. This forward contract and the ERC2771Context contract are standard contracts provided by OpenZeppelin and have not been implemented by us.

Similar to the approach presented in [9], where contracts store rules with references to interfaces, our Loyalty Manager contract follows a similar pattern. This allows for a more dynamic code implementation and the ability to add new contracts during runtime. To achieve this, the contract uses mapping, as shown in Figure 9, with an *enum* as the key, which is converted to integers at compilation time, and uses the contract references as the values. In the future, the *enum* can be removed, and contracts can be defined directly with dynamic incrementing integers.

```
contract LoyaltyManager is ERC2771Context {
    uint retailersRegisterdCounter;

    //adress = retailer address
    mapping (address => uint) retailersKey;
    //uint reatailer Key
    mapping (uint => RetailerInfo) registeredRetailers;

    //adress = customer address
    mapping (address => RetailerInfo[]) customersSubscriptions;

    mapping (LoyaltyMechanismType => LoyaltyMechanism) loyaltyMechanismContracts;

    enum LoyaltyMechanismType {POINTS, STAMP_CARDS}

    struct RetailerInfo {
        address retailerAddress;
        string ipfsHash;
        LoyaltyMechanismType loyaltyMechanismType;
    }

  constructor(MinimalForwarder forwarder, LoyaltyProgramPoints loyaltyProgramPoints)
     ERC2771Context(address(forwarder))
  {
     retailersRegisterdCounter = 0;
     loyaltyMechanismContracts[LoyaltyMechanismType.POINTS] = loyaltyProgramPoints;
  }
```

Figure 9. Loyalty manager contract.

By using references to the loyalty mechanism interface, the loyalty manager contract only recognizes the four defined methods, providing enhanced security and preventing access to other functions that may be defined by the contracts themselves. This approach also makes the code more dynamic, eliminating the need for nested 'if' statements to determine the correct contract to invoke. Figures 10 and 11 show how requests can be redirected to the appropriate contract in a straightforward manner.

```
/**
* Mint Retilaers token
*
* - 'amount' amount of tokens to create
*
* This funtion only works if the caller is a retailer
*
* First obtains the retailer loyalty mechanism to call the correct contract
* At same time verifies if the caller is a retailer; If not the transaction is reverted
* Makes a call to the contract mint function and passes the sender and amount as arguments
* The sender must be a Retailer
*/
function mintRetailersToken(uint amount) external {
    LoyaltyMechanismType loyaltyMechanismType = getLoyaltyMechanismType(_msgSender());
    loyaltyMechanismContracts[loyaltyMechanismType].mint(_msgSender(), amount);
}
```

Figure 10. Loyalty manager—function mint.

```
* Transfer tokens
*
*  - 'to' address of the receiver of the tokens
*  - 'amount' amount of tokens to transfer
*
* Checks if at least one retailer is present
* Makes a call to the contract transferTokens function and passes the sender of the order as an argument
* The sender can be a Retailer or Customer
* The 'to' can be a Retailer or Customer
*/
function transferTokens(address to, uint amount) external {

    bool isReceiverRetailer = checkIfAddressIsRetailer(to);
    bool isSenderRetailer = checkIfAddressIsRetailer(_msgSender());

    require(isReceiverRetailer || isSenderRetailer , "Both users are Customers. One Retailer has to be present");

    LoyaltyMechanismType loyaltyMechanismType;

    if (isReceiverRetailer) {
        require(checkIfCustomerIsSubscribedToRetailer(to, _msgSender()), "The customer is not subscribed to retailer");
        loyaltyMechanismType = getLoyaltyMechanismType(to);
    }
    else {
        require(checkIfCustomerIsSubscribedToRetailer(_msgSender(), to), "The customer is not subscribed to retailer");
        loyaltyMechanismType = getLoyaltyMechanismType(_msgSender());
    }

    loyaltyMechanismContracts[loyaltyMechanismType].transferTokens(_msgSender(), to, amount, isSenderRetailer);
}
```

Figure 11. Loyalty manager- function transferTokens.

As discussed in Section 2.1.4, Openzeppelin provides a plugin that allows the proxy pattern to be used without having to implement it manually. This plugin allows the use of the proxy pattern and the modification of the business logic of the contracts while preserving the data. All of the above contracts are publicly available in the GitHub repository [30].

4. Discussion

The proposed system aims to manage various loyalty programs with different mechanisms. Even though the blockchain solves significant problems, it is impossible to address concerns related to rewards and personal preferences due to inherent human characteristics.

The proposed loyalty management system offers limitless opportunities and can reshape how traditional loyalty systems are implemented. It provides new paradigms and possibilities to users with any level of knowledge about technology to create and interact with the blockchain without having the inconvenience of learning it. This way, users can easily participate and enjoy the loyalty programs of their favorite brands.

The system works on the Polygon chain, which implies the use of Solidity as the smart contract programming language. Polygon also includes a network for testing purposes (which is useful while developing the system) and offers ways to track transactions.

A mobile application is available to interact with the blockchain. This application abstracts its users by setting up all the requirements for them. Additionally, the way smart contracts are designed allows for a more manageable infrastructure of loyalty programs.

With the ERC-1155 token interface standard, each retailer can mint its own tokens and be responsible for their generation, thus gaining better control and ownership. Whenever a customer purchases something, the retailer mints the desired number of tokens and gives them to the client. In order for the customer to be able to purchase goods or services from a retailer, he/she needs to have a certain number of tokens from that retailer.

The designated external services are intended to produce a better user experience. Web3Auth is used to grant the user the ability to join the system with any credentials they have from Web 2.0, while managing the wallet and keys for user interaction. Ramp service is used to accommodate users when they need to buy cryptocurrencies. It offers an easy access point for currency exchange without leaving the application. Relays are used to diminish entry barriers for customers, as blockchain requires mastery, experience, and cryptocurrency to interact with. Since customers do not adhere to applications for which constant payment is required, the system acceptance would fail without the use of relays.

With the proposed loyalty management system, the users apply their previous Web 2.0 credentials and do not have to enter new information. Furthermore, the system does not include any database, which means that personal data will not be stored.

4.1. Comparison with Other Systems

There are several advantages that can be highlighted when comparing the proposed loyalty system with the related works discussed in Section 2. The universal blockchain system [5] supports multiple loyalty programs and partnerships, which our system can incorporate. However, this system limits the loyalty mechanism to points and uses a backend server for the users' application to communicate. Additionally, it stores user data in a database. In contrast, the proposed system eliminates the need to store personal information. Moreover, it does not rely on any backend system and achieves a more decentralized approach.

The system proposed in paper [13] has similarities with the universal blockchain system, since both employ the points mechanism using tokens. However, the proposed system incorporates additional types of loyalty mechanics and abstracts the users from the concept of tokens, enabling them to interact with points in a manner similar to non-blockchain-based platforms. Another distinguishing factor is that the author in [13] employs the ERC 20 interface for token creation, which requires that each token exists as a smart contract, whereas the proposed system uses ERC-1155 for managing multiple tokens. Additionally, the expectation that users possess knowledge of Ethereum wallets in [13] may present challenges for novice users.

The loyalty program that utilizes the Waves blockchain and a mobile application [19] employs a token creator that generates a unique token for each retailer. However, this system also stores personal information and potentially creates a token contract for each retailer, similar to the aforementioned system [13].

The reward system in [21] uses tokens that can be converted into points, and the platform is specifically designed for the FMCG industry, limiting its scope to this domain. However, this system resembles our transaction process, in that it employs QR codes to initiate the transfer of assets.

Although the systems proposed in [22,23] do not directly aim to manage loyalty programs, they improve these programs by facilitating the exchange of assets between different loyalty programs. Currently, our loyalty management system lacks features that allow for partnerships between retailers; we plan to implement these features in the future.

Since the existing loyalty systems discussed in Section 2 do not specify their authentication process, it is assumed that they have created one and currently manage it. This results in an additional step for user onboarding, as they have to enter new information to create a new account. In contrast, the proposed system employs Web3Auth, which uses Web 2.0 credentials to streamline the authentication process. Additionally, a significant advantage of the proposed system is the use of meta transactions, which provide a superior user experience by eliminating the need for users to pay blockchain fees.

4.2. System Benefits

The designed system architecture offers countless benefits for loyalty program ecosystems; it helps to reshape the traditional approach for a more modern, intuitive, and secure model.

The main benefits of the system are the following:

- The platform fully embraces decentralization and leverages the power of the ERC-1155 token to efficiently manage tokens and reduce the number of contracts required.
- The system provides near real-time transaction speed, allowing users to redeem their rewards faster without having to wait too long;
- Due to the behavior of smart contracts, third-party tools and intermediaries are no longer necessary;
- The system helps to reduce costs. As intermediaries and third-parties are abolished, retailers can save money. Furthermore, the infrastructure needed to manage the loyalty program is transferred to the blockchain;
- Blockchain offers a secure infrastructure;
- A single digital wallet can be used to store all retailers' tokens, eliminating the problem of managing rewards from multiple brands. It also gives complete control to the user;
- The platform accommodates all distinct levels of user knowledge, allowing more experienced users to use their wallet providers, and inexperienced people to have a smoother first contact with it;
- The platform is capable of managing several types of loyalty mechanisms, increasing the diversity of loyalty programs. In addition, the platform welcomes any retailer of any business size. This makes it easier for new retailers to join and customers with more options to choose from;
- It provides a marketplace with all available loyalty programs. On the one hand, the marketplace helps retailers gain greater exposure to attract more customers. On the other hand, customers have a better access point to all available loyalty programs, meaning they are able to discover new businesses;
- The system takes advantage of the fact that smartphones are increasingly present in people's lives, thus enhancing the overall experience.

4.3. System Limitations

The proposed system has great advantages. However, it also has some limitations:

- Due to human nature, it is almost impossible to solve concerns about rewards and personal preferences;
- Even though the system does not require any personal information, some information is required at the retailer's business level;
- The blockchain is a new subject, and this may affect outcomes. Because it is so new, it may be subject to impactful changes at any time, which can affect the system as it is currently designed, making it unsuitable for the current purpose;
- Exchanging of tokens between two customers is not allowed;
- Retailers cannot form any partnerships, which limits the possibility of trading retailers' assets;
- The transaction process requires the physical presence of the retailer and the customer. The application does not have an item or service available for purchase;

- Transactions can only be executed in person between a retailer and a customer;
- Remote transactions are not currently possible within this system.

5. Conclusions and Future Work

In this paper, a proposed loyalty management system that aims to manage different types of loyalty programs and mechanisms is described. This proposal includes several challenges in the technical domain, since the blockchain introduces new paradigms.

The system aims to overhaul the way in which traditional loyalty programs operate, and attempts to eliminate several entry barriers and infrastructure issues. It aims to replace conventional physical cards, instead adopting the strategy of using the smartphone, an increasingly ubiquitous object in everyday life. The platform also includes a universal system that aggregates loyalty programs, thereby simplifying the user experience.

The usability of this system allows users to have better interaction with the brands that they appreciate the most, alongside a better idea of and ownership over their rewards. Additionally, the system provides a marketplace with all available loyalty programs in a single space, removing the need to commit to having multiple applications and allowing management of all loyalty schemes at once. The system allows retailers to customize their own loyalty program without restrictions on the type of mechanism implemented. Furthermore, it exposes brands to new audiences, increasing the accessibility and reachability of the business.

The proposed loyalty management system reduces the problem of managing the loyalty program. It removes the responsibilities from the retailers' side, giving them more freedom to focus on satisfying customers and improving their loyalty program and business. It also removes concerns about the privacy of personal information, since the system does not require it. From the customers' perspective, it offers a new way to explore and know new businesses without the burden of having to download new apps and repeatedly introduce their data. It offers an opportunity to have a first encounter with the blockchain without worrying about learning and understanding the process. Overall, the platform prioritizes the user experience more.

Although the system solves significant problems, it has also some limitations. It cannot control human aspects, such as preferences for rewards. The blockchain is still extremely new and is constantly changing, which can impact the platform. The platform presents limitations about partnerships among retailers and the transaction process. The transaction process only considers trades between a retailer and a customer, and needs both to be present in the same room.

In general, the use of the proposed system is intuitive and simple. Its primary objective is to design a platform capable of managing different loyalty programs with distinct mechanisms. It presents a scalable solution, such that it is possible to add and expand its functionalities.

In terms of future work, we will be continuing with the implementation of the proposed system, particularly with regard to the remainder of the loyalty mechanism and improvements to the mobile app. It is also relevant to continue studying new blockchain networks and technologies, as they could potentially be better options. Furthermore, it is necessary to expand transaction possibilities in order to give users better control of assets. Regarding partnerships, it is crucial to review the architecture of smart contracts. It might be beneficial to make the application more interactive through the addition of a functionality that allows the purchase of products and services without the necessity of physically going to the store.

Author Contributions: Conceptualization, J.B.; Methodology, A.F.S. and J.B.; Software, A.F.S.; Validation, A.F.S., J.M. and J.B.; Formal analysis, A.F.S., J.M. and J.B.; Investigation, A.F.S.; Resources, A.F.S.; Data curation, A.F.S.; Writing—original draft preparation, A.F.S.; Writing—review and editing, A.F.S., J.M. and J.B.; Supervision, J.B. and J.M.; Project administration, J.B. and J.M.; Funding acquisition, J.B. All authors have read and agreed to the published version of the manuscript.

Funding: This research received no external funding.

Data Availability Statement: The data presented in this study are openly available in GitHub, reference number [30].

Conflicts of Interest: The authors declare no conflict of interest.

References

1. Krichen, M.; Ammi, M.; Mihoub, A.; Almutiq, M. Blockchain for Modern Applications: A Survey. *Sensors* **2022**, *22*, 5274. [CrossRef] [PubMed]
2. AWS—Amazon. What Is Decentralization in Blockchain? Available online: https://aws.amazon.com/blockchain/decentralization-in-blockchain/ (accessed on 30 October 2022).
3. IBM. What Are Smart Contracts on Blockchain? Available online: https://www.ibm.com/topics/smart-contracts (accessed on 30 October 2022).
4. Agrawal, D.; Jureczek, N.; Gopalakrishnan, G.; Guzman, M.N.; McDonald, M.; Kim, H. Loyalty Points on the Blockchain. *Bus. Manag. Stud.* **2018**, *4*, 80–92. [CrossRef]
5. Agrawal, M.; Amin, D.; Dalvi, H.; Gala, R. Blockchain-based Universal Loyalty Platform. In Proceedings of the 2019 International Conference on Advances in Computing, Communication and Control (ICAC3), Mumbai, India, 20–21 December 2019; pp. 1–6. [CrossRef]
6. Hader, M.; Elmhamedi, A.; Abouabdellah, A. Blockchain technology in supply chain management and loyalty programs: Toward blockchain implementation in retail market. In Proceedings of the 2020 IEEE 13th International Colloquium of Logistics and Supply Chain Management (LOGISTIQUA), Fez, Morocco, 2–4 December 2020; pp. 1–6. [CrossRef]
7. Antoniadis, I.E.; Kontsas, S.; Spinthiropoulos, K. Blockchain and Brand Loyalty Programs: A Short Review of Applications and Challenges. *Procedia Econ. Bus. Adm.* **2019**, *5*, 8–16. [CrossRef]
8. GeeksforGeeks. Features of Blockchain. Available online: https://www.geeksforgeeks.org/features-of-blockchain/ (accessed on 7 December 2022).
9. Górski, T. Reconfigurable Smart Contracts for Renewable Energy Exchange with Re-Use of Verification Rules. *Appl. Sci.* **2022**, *12*, 5339. [CrossRef]
10. Kim, K.; Ryu, J.; Lee, H.; Lee, Y.; Won, D. Distributed and Federated Authentication Schemes Based on Updatable Smart Contracts. *Electronics* **2023**, *12*, 1217. [CrossRef]
11. OpenZeppelin. Proxy Upgrade Pattern. OpenZeppelin Docs. 2023. Available online: https://docs.openzeppelin.com/upgrades-plugins/1.x/proxies#upgrading-via-the-proxy-pattern (accessed on 18 April 2023).
12. Polygon. Meta Transactions. 2022. Available online: https://wiki.polygon.technology/docs/develop/meta-transactions/meta-transactions/ (accessed on 6 January 2023).
13. Sönmeztürk, O.; Ayav, T.; Erten, Y.M. Loyalty Program using Blockchain. In Proceedings of the 2020 IEEE International Conference on Blockchain (Blockchain), Rhodes, Greece, 2–6 November 2020; pp. 509–516. [CrossRef]
14. Ethereum Network. 2015. Available online: https://ethereum.org/ (accessed on 15 January 2023).
15. Vogelsteller, F.; Buterin, V. ERC-20 Token Standard. 19 November 2015. Available online: https://eips.ethereum.org/EIPS/eip-20 (accessed on 15 January 2023).
16. Wang, L.; Luo, X.; Hua, Y.; Wang, J. Exploring how blockchain impacts loyalty program participation behaviors: An exploratory case study. In Proceedings of the 52nd Hawaii International Conference on System Sciences, Maui, HI, USA, 8–9 January 2019; pp. 1–10. [CrossRef]
17. Wang, L.; Luo, X.; Lee, F. Unveiling the interplay between blockchain and loyalty program participation: A qualitative approach based on Bubichain. *Int. J. Inf. Manag.* **2019**, *49*, 397–410. [CrossRef]
18. Ryan, R.M.; Deci, E.L. Overview of self-determination theory: An organismic dialectical perspective. In *Handbook of Self-Determination Research*; University of Rochester Press: Rochester, NY, USA, 2002; Volume 2, pp. 3–33.
19. Perez, L.J.D.; Ibarra, L.; Alejandro, G.-F.; Rumayor, A.; Lara-Alvarez, C. A loyalty program based on Waves blockchain and mobile phone interactions. *Knowl. Eng. Rev.* **2020**, *35*, e12. [CrossRef]
20. Udegbe, S. Impact of blockchain technology in enhancing customer loyalty programs in airline business. *Int. J. Innov. Res. Adv. Stud.* **2017**, *4*, 257–263.
21. Bülbül, Ş.; İnce, G. Blockchain-based Framework for Customer Loyalty Program. In Proceedings of the 2018 3rd International Conference on Computer Science and Engineering (UBMK), Sarajevo, Bosnia and Herzegovina, 20–23 September 2018; pp. 342–346. [CrossRef]
22. Nguyen, C.T.; Hoang, D.T.; Nguyen, D.N.; Pham, H.-A.; Tuong, N.H.; Dutkiewicz, E. Blockchain-based Secure Platform for Coalition Loyalty Program Management. In Proceedings of the 2021 IEEE Wireless Communications and Networking Conference (WCNC), Nanjing, China, 29 March–1 April 2021; pp. 1–6. [CrossRef]
23. Tu, S.-F.; Hsu, C.-S.; Wu, Y.-T. A Loyalty System Incorporated with Blockchain and Call Auction. *J. Theor. Appl. Electron. Commer. Res.* **2022**, *17*, 56. [CrossRef]
24. Utz, M.; Johanning, S.; Roth, T.; Bruckner, T.; Strüker, J. From ambivalence to trust: Using blockchain in customer loyalty programs. *Int. J. Inf. Manag.* **2023**, *68*, 102496. [CrossRef]

25. PolygonScan. PolygonScan. 2023. Available online: https://polygonscan.com/ (accessed on 8 April 2023).
26. EthereumIo. EthereumIo. 2023. Available online: https://etherscan.io/ (accessed on 8 April 2023).
27. Avascan. AvalancheScan. 2023. Available online: https://avascan.info/ (accessed on 8 April 2023).
28. Solana. Solana Website. 2023. Available online: https://solana.com/ (accessed on 8 April 2023).
29. Radomski, W.; Cooke, A.; Castonguay, P.; Therien, J.; Binet, E.; Sandford, R. ERC-1155: Multi Token Standard. Ethereum Improvement Proposals, 17 June 2018. Available online: https://eips.ethereum.org/EIPS/eip-1155 (accessed on 18 April 2023).
30. Smart Contracts Repository. Available online: https://github.com/AndreFSantos00/loyalty-smart-contracts (accessed on 21 April 2023).

future internet

Article

Blockchain, Quo Vadis? Recent Changes in Perspectives on the Application of Technology in Agribusiness

Geneci da Silva Ribeiro Rocha [1,*], Diego Durante Mühl [1], Hermenegildo Almeida Chingamba [2], Letícia de Oliveira [1,3,*] and Edson Talamini [1,3]

1 Interdisciplinary Center for Studies and Research in Agribusiness—CEPAN, Universidade Feral do Rio Grande do Sul—UFRGS, Porto Alegre 90040-060, Brazil
2 Graduate Program in Production Engineering—PPGEP, Universidade Federal Rio Grande do Sul—UFRGS, Porto Alegre 90035-190, Brazil
3 Department of Economics and International Relations—DERI, Faculty of Economics—FCE, Universidade Federal do Rio Grande do Sul—UFRGS, Porto Alegre 90040-000, Brazil
* Correspondence: geneci.6813.srr@gmail.com (G.d.S.R.R.); leticiaoliveira@ufrgs.br (L.d.O.)

Abstract: Information technologies such as blockchain are developing fast, overcoming bottlenecks, and quickly taking advantage of their application. The present study analyzes recent changes concerning the benefits, disadvantages, challenges, and opportunities of blockchain applications in agribusiness. Interviews were conducted with and a questionnaire was applied to professionals working in the development and application of blockchain technology in agribusiness, to compare their perception of the recent advances. The results showed that the importance of blockchain technology to improve governance and information flow along supply chains has increased, and this is the main perceived benefit. The main disadvantages were removing intermediaries and the high cost of implementing the technology. The absence of a widely accepted platform in blockchain operations is the leading and growing challenge, while patterns for blockchain technology seem to be being overcome. The integration of blockchain with new technologies, and the competitiveness provided by the technology, are seen as the main and growing opportunities. Despite the study limitations, we conclude that the benefits and opportunities associated with blockchain application in agribusiness outweigh the challenges and disadvantages in number and importance, and are becoming more relevant.

Keywords: trust; transparency; immutability; profitability; sharing data; business models; symmetry of information; traceability; transaction costs; smart contracts

Citation: Rocha, G.d.S.R.; Mühl, D.D.; Chingamba, H.A.; de Oliveira, L.; Talamini, E. Blockchain, Quo Vadis? Recent Changes in Perspectives on the Application of Technology in Agribusiness. *Future Internet* **2023**, *15*, 38. https://doi.org/10.3390/fi15010038

Academic Editors: Christoph Stach and Clémentine Gritti

Received: 6 December 2022
Revised: 10 January 2023
Accepted: 11 January 2023
Published: 16 January 2023

1. Introduction

Blockchain is a relatively recent technology that is rapidly developing. In its trajectory, several obstacles have been overcome to enable its application in different sectors of society. Initially, this section recalls relevant aspects of blockchain technology's background and presents the research question that guided the present study.

1.1. Background

In 1991, Haber and Stornetta [1] proposed a method for date and time stamping digital documents. However, digital documents cannot be time-stamped or authenticated in the same way as paper documents. At the time, the authors proposed a basic structure consisting of hash cryptography (algorithm) that organizes immutable information into linked blocks. An authentication service organizes the documents or blocks of information sequentially, according to date and time.

The spreading of encrypted information among different clients (companies or people) guarantees that recorded data cannot be altered without leaving traces [2]. In this flow of information, first, if a client wants to falsify previously registered information, the authentication service can check the previous record and identify the fraud attempt. Second, suppose the authentication service itself conspires with a forger to try to falsify previously recorded information, in that case, he/she cannot do it without leaving traces, since the information is distributed among the other clients that will be able to audit the forgery [1]. In other words, even if a client and the authentication service are dishonest and act together to create fraud, the other clients can identify it. So, the technology works like a logbook that can keep permanent data records about decisions and transactions [3]. Since it is composed of blocks of information in a chain, the technology started to be called the 'blockchain'.

Blockchain technology is becoming efficient, robust, and cost-effective for managing interactions between multiple participants in a network, in a very reliable and decentralized manner [4,5]. Likewise, it facilitates the use of information by subsidizing demands and providing the real-time updating of records. It also increases supply chain transparency and product or service traceability by allowing the exchange of transactional data between two or more actors in the chain [6]. This is because data stored in the blockchain is reliable and cannot be deleted, altered, or corrupted, without most parties involved in the transaction being aware of the changes.

Blockchain is considered a significant technological trend that could influence business in general, as it is a disruptive solution for several business segments capable of increasing the trust of bilateral or multilateral relationships [6]. Thus, blockchain could create competitive advantages in many business sectors, especially agribusiness. Rocha et al. [7] point out several blockchain applications in agribusiness activities, such as finance, energy, logistics, environmental management, agriculture, livestock, and agro-industry. According to the authors, blockchain can be successfully implemented, and applications for logistics can result in energy, financial and environmental advantages.

Agribusiness comprises long supply chains involving various economic sectors, and fundamental activities such as providing food and energy for the human and animal populations. As end-consumers seek more information about product quality, origin, and mode of production, company managers, supply chains, and governments, are pressured to adopt solutions that ensure product transparency and traceability [8,9].

In agribusiness, blockchain technology can be applied in solutions to ensure the security and traceability of products in a system in which information is transmitted between all links and actors along the value chain. Information sharing allows for building trust and transparency between actors. Consumers can trace products to check their origin and production mode, hindering the possibilities of product counterfeiting and adulteration [7,10]. Thus, a traceability system based on blockchain brings benefits and information to all those involved in the supply chain, including the final consumers [11–13], since it enables the use and exchange of data in a secure, transparent, and effective manner [14].

By integrating blockchain, IoT, and wireless sensor networks, the end consumer can track the product, verifying its origin, harvest date, mode of transport, and certification, increasing consumer confidence in the product they are purchasing. Consequently, technology can increase the demand for these products among those consumers who value such information [15,16].

The information recorded by the blockchain is secure and can be publicly or privately accessible, depending on the preferred strategy, and most importantly, it is immutable. This allows the information to be accessible to retailers, auditors, governments, consumers, producers, and other stakeholders along the supply chain. Moreover, it provides for monitoring and auditing the data, assessing economic and environmental performance, or other indicators of interest [17]. Additionally, blockchain has the advantage of eliminating dependence on a central authority. Its decentralized nature elevates the importance of the network effect. The technology's potential benefits increase as the network's size expands, making it more robust against external attacks [18].

Finally, the interoperability of blockchain and the use of mobile devices, from computers to smartphones, put the power of decision-making in the hand of consumers [19].

However, the benefits and opportunities associated with using this technology and the advantages and disadvantages it provides to stakeholders influence managers' decisions to adopt blockchain [20]. Furthermore, business models need to rebuild existing systems, popularize the technology, and train staff to adapt to the new management process [21,22].

1.2. Research Question

Due to the numerous advantages of blockchain technology, its development and applications have advanced rapidly. For example, Arooj et al. [23] identified that in 2017 blockchain technology was at the crossing chasm stage when the industry was adopting proofs of concept, funding was being provided by venture capital, and issues related to scalability, sustainability, and throughput, were resolved. Three years later, in 2020, blockchain reached the adaption movement stage, characterized by the definition of standards and protocols, an explosion in the use of blockchain, funding by specialized IT companies, the use associated with IoT for the development of smart contracts, and use in e-government solutions. At the current stage, blockchain is in use for major banking application shifts, forcing changes in banking infrastructure, IoT applications, and AI-based smart contracts.

Therefore, in a short time, blockchain technology has evolved from proofs of concept into practical applications in several sectors. As a result, several authors address the adoption of blockchain in agribusiness [24,25]. Still, more studies analyzing the perception of actors involved in the application of blockchain in agribusiness over time are needed.

For this reason, the present study analyzes this recent evolution from the following research question: What changed in the perception of professionals involved in the application of blockchain in agribusiness regarding the benefits, disadvantages, challenges, and opportunities of the technology in recent times? Thus, the objective is to analyze the recent changes in the benefits, disadvantages, challenges, and opportunities of blockchain applications in agribusiness from the perspectives of experts.

The adoption and impacts of blockchain technology on agribusiness supply chains follow the same pace as other sectors and can be affected by successive events in a short time. Thus, the perception of benefits, disadvantages, challenges, and opportunities can change relatively briefly. Identifying the benefits and opportunities that have increased their relative importance may indicate remarkable stimuli to promote technology development and adoption along supply chains.

On the other hand, disadvantages whose relative importance has increased signal weaknesses of the technology that need to be strengthened to make blockchain affordable and competitive. Identifying opportunities whose relative importance has increased reinforces aspects that can be prioritized in the search for gains related to the development and use of blockchain technology. Therefore, by achieving the objective and answering the central question of this research, we hope to bring contributions to the various actors interested in the development and adoption of blockchain technology along supply chains, such as developers, farmers, industrial managers, logistics operators, financial system, regulators, and consumers.

2. Materials and Methods

In answering the research question proposed in this study, the methodology was divided into two phases and classified into two approaches. The first phase considers exploratory and qualitative research (Figure 1). In this stage, interviews were conducted with Brazilian experts who are somehow involved with the development and application of blockchain technology in agribusiness. The interviews sought to understand the perception of experts about the benefits, disadvantages, challenges, and opportunities of blockchain technology in order to qualify the objective answers.

A semi-structured interview script was used to collect information from experts and researchers on blockchain technology (See Supplementary Material S1). The script was designed based on the concepts found in the literature reviewed. Interviewees needed to understand blockchain technology, its structure, and the agribusiness sector. The semi-structured interviews allowed the interviewer to steer the questions in the direction best suited to the agribusiness context.

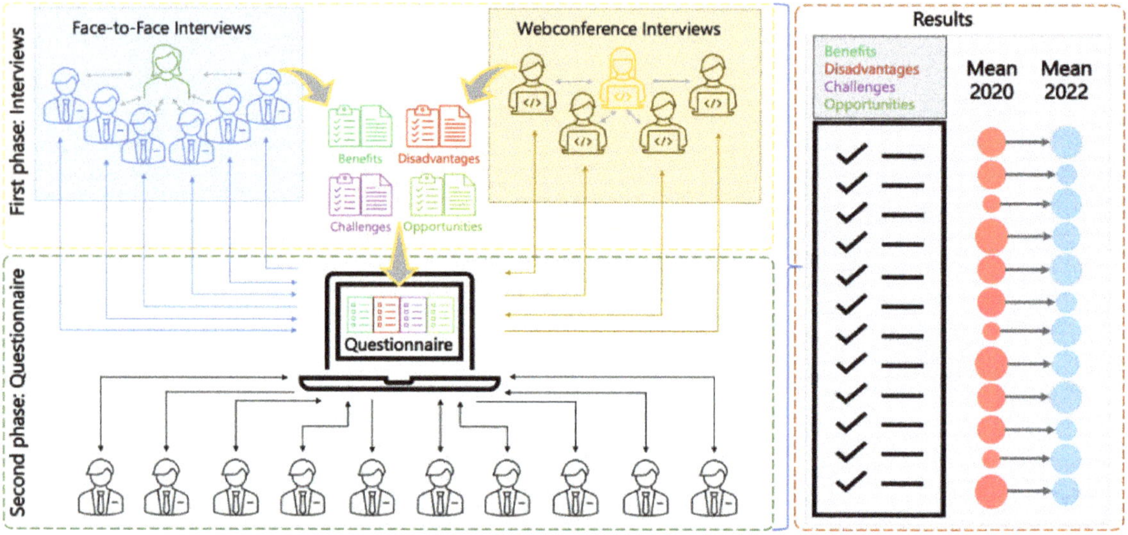

Figure 1. The framework of methodological phases and procedures carried out in the research. Source: elaborated by the authors.

The interviews were conducted with 10 specialists, six in person and four remotely via web conference (Skype), in January 2020, with an average duration of 50 min. After the interviewees' permission, the interviews were recorded and later transcribed in a text editor to be analyzed. The respondents are managers, coordinators, developers, or researchers with knowledge of programming, technology, and business models. They were collaborators from many companies and universities. All interviewees were involved in blockchain technology development or implementation activities or were university researchers with research projects on the topic (see Table 1).

Table 1. Profile of the professionals interviewed in the first phase of the research.

Expert	Education and Background	Working on ...	... at the Organization
1	Electrical Engineer, MSc and PhD in Computer Science	Computer networking and security	Universidade Federal do Rio Grande do Sul—UFRGS, Institute of Informatics and Graduate Program in Computing
2	Economist and MSc in Business Management. Background in Risk Management, Foreign Trade and International Business, Alternative Energy	Support to a multidisciplinary group of specialists	Blockchain Collaborative Institute—ICoLab
3	Accountant, MSc and PhD in Administration—Systems Management and Information Technology	Accounting and information system	Universidade Federal do Rio Grande do Sul—UFRGS, Accounting Department
4	Economist, PhD in Economic Development, Space and Environment	Applied economics.	Universidade Federal do Paraná—UFPR, Departament of Economics

Table 1. *Cont.*

Expert	Education and Background	Working on ...	... at the Organization
5	Lawyer, MSc and PhD in Law Science	Public International Law, International Economic Law, Economic Criminal Law, and International Criminal Law	Ulbra University and Graduate Program in Law at UniRitter University
6	B.A. and MSc in Computational Mathematics. PhD in Computation Science	Blockchain researcher	IBM (São Paulo, SP)
7	B.A. in Data Processing, MSc in Applied Informatics	Innovation and technology entrepreneur	Park Hub and ONE Percent Software Innovation Studio
8	Accountant. MSc and PhD in Administration—Information Systems and Decision Support	Costs and management information systems research	Universidade Federal do Rio Grande do Sul—UFRGS, Faculty of Economics
9	B.A., MSc, and PhD in Administration	Digital transformation, enterprise mobility, cryptocurrencies, and Blockchain	UNISINOS, UniRitter, UFRGS, Feevale, ICoLab e ONE Percent Software Innovation Studio
10	Ecologist, MSc in Civil Engineering (Water Resources), PhD in Economics, and Post-doc in Cryptocurrency Design and Alternative Economic Systems for Planet Regeneration	Collaborating professor at the Institute of Economics at Unicamp University and entrepreneur	Unicamp University and Satisfied Vagabonds (Costa Rica)

Source: elaborated by the authors based on research data.

The interviews were analyzed using content analysis techniques. First, the content of the interviews was grouped into categories related to the benefits, disadvantages, challenges, and opportunities of applying blockchain technology in agribusiness. Subsequently, this information was used in preparing a questionnaire with a Likert scale to be applied in the second methodological stage of the study.

The second phase of the research was quantitative, and the information collected via interviews was organized into a questionnaire with a Likert scale. The questionnaire was developed in a digital platform and sent to the personal e-mail of the 10 experts who had participated in the interviews in the first stage (January 2020). The response rate was 60%, considering six experts answered the questionnaire in October 2022. In addition to the experts who participated in the interviews in the first phase of research (interviews), the questionnaire was also forwarded to other experts who were somehow involved with developing and applying the technology in agribusiness, and seven responses were obtained.

All the participants (first and second phases) were invited to express their perceptions of the benefits, disadvantages, challenges, and opportunities related to blockchain applications in agribusiness. For each benefit, disadvantage, challenge, and opportunity proposed in the questionnaire, all the participants should assign a Likert value considering the importance of the respective item at the beginning of 2020 and another value (1 = lower, 3 = equal, or 5 = higher) considering the importance of the same item at the time of completing the questionnaire (October 2022).

The purpose of the questionnaire was to quantify the experts' responses by capturing the change in their perception regarding the benefits, disadvantages, challenges, and opportunities of applying blockchain technology in agribusiness. Thus, we calculated the average of each item of the four axes under analysis: benefits, disadvantages, challenges, and opportunities, in order to compare the averages between the years 2020 and 2022. Based on the mean values, figures with the main results were prepared, presented, and analyzed in the next section.

It is emphasized that, in Brazil in general, and in agribusiness in particular, the existence of experts and companies working in the development and applications of blockchain technology is still restricted. Nevertheless, the experts were identified and invited to participate in the research on their evidence in blockchain research and development, availability of contacts, and willingness to contribute, justifying the low number of participants.

3. Results and Discussion

3.1. The Realms of the Benefits, Disadvantages, Challenges, and Opportunities of Blockchain

Experts were asked about blockchain benefits, disadvantages, challenges, and opportunities for agribusiness. The interviewees presented different points of view according to their areas of expertise. Thus, the interviews were analyzed, and the essential items were highlighted in Table 2.

Table 2. Experts' opinions about the benefits, disadvantages, challenges, and opportunities of blockchain technology.

	Item	Why? Experts' Opinions about the Items
Benefits	Reduction of transactions costs	The authenticity and trust provided by the blockchain reduce the agents' insecurity and consequently minimize transaction costs.
	Better governance and information flow	Blockchain allows governance by convenience. In the first step, each party involved does not need to share all their information or data, but only what matters and with each need.
	Smart digital contracts	Smart contracts can perform operations, such as executing financial transactions or authenticating documents in a common legal agreement. This will make everyday business less bureaucratic, as it will decentralize operations.
	Trust, transparency, and immutability	Blockchain technology makes production chains more transparent, as information is recorded and auditable in a decentralized manner.
	Optimization of resources and processes	The adoption of the blockchain platform will speed up bureaucratic processes, which is reflected in the optimization of resources overall.
	Greater profitability	Adopting the blockchain platform will give speed to processes, higher quality in records, reduced transaction costs, resource optimization, and increased profitability.
	Information on the network visible to everyone	Blockchain improves trust in contractual relationships as long as the information on the network is verified by all participants, eliminating the need for a trusted third party.
	More security throughout the production chain	Supply chains need to be more integrated. So, blockchain will help the whole process, integrating all actors, from the producer to the consumer.
Disadvantages	High cost of technology implementation	Currently, there is a high cost to develop and implement blockchain technology.
	High maintenance cost	There is a high cost of maintaining and storing information for operationalizing blockchain.
	Elimination of intermediaries	The intermediaries' elimination could break the links in the supply chain by monopolizing production and excluding many people from the production process.
	Absence of a single blockchain technology standard for different business models	No robust, global platform can be found that stands out. As a result, different solutions are being developed to address specific problems.
	Lack of a winning platform	The solutions developed for specific problems have not allowed for establishing a widely accepted platform, code, or universal solution.
Challenges	Collect, process, and store data	As blockchain technology becomes widespread, vast amounts of data will be generated. Moreover, many copies of the information are distributed among users, requiring big storage and processing capacities.
	New business model generation	Creating decentralized businesses is still counterintuitive. Nevertheless, blockchain enables the creation of decentralized and autonomous networks, directly integrating the end consumer and the farmer.

Table 2. *Cont.*

	Item	Why? Experts' Opinions about the Items
Challenges	Break with old paradigms	Breaking with old paradigms will be a challenge because it means changing society and the way people are used to doing things.
	Skilled labor	Specific software engineers are vital players in the advancement and spread of blockchain.
	Lack of a winning platform	Many pilot projects are in the early stages of application, such as IoT Blockchain, Bitcoins Blockchain, Rapper Ledger Blockchain, and Corda Blockchain. An integrative and widely accepted platform is not available.
	Intermediate elimination	The farmer can transfer the product directly to the final consumer so that the technology will exclude intermediary actors from the supply chains, which could create serious social problems.
	Quality of information entered on the network	The information inserted incorrectly in the network, either by some fault or by bad faith, may be out of touch with reality.
	Absence of blockchain technology standards	There are various attempts and initiatives, each with positive features and challenges.
Opportunities	Integration with new technologies	Blockchain has the integrated potential to store contact information, financial information, and logistics information securely and audibly.
	Product traceability and certification	Transparency about the products' origin, processes, and inputs can be updated in real-time, allowing the consumer access to product information.
	Digitization of the production chain	The possibility of improving the efficiency of overall processes can leverage the digitization of end-to-end supply chains.
	Information symmetry	All actors involved may have access to important information. However, all info may be available according to a strategy agreed upon by traders.
	New domestic and foreign markets	Blockchain guarantees the final product in terms of transparency and certification of its origin. In addition, higher levels of confidence allow new buyers to access markets.
	Competitive edge	Innovation in agribusiness has already proven to bring benefits. The blockchain platform will be able to automate many processes, increasing standardization and quality and leading to competitive advantages.
	Product differentiation	At first, safety and traceability will be a differentiator for consumers.
	Product standardization	The exchange of information and the automation of processes will lead to greater standardization of the production process.

Source: elaborated by the authors based on research data.

The benefits relate to the advantages or good things that blockchain can do for agribusiness and society. Similarly, the opportunities are more related to favorable circumstances for developing and implementing blockchain in agribusiness. On the other hand, the challenges represent the impediments to blockchain development. Finally, the disadvantages are problems that blockchain might generate.

Many of the explanations made by the experts are in line with data found in scientific literature. For example, the possibility of reducing transaction costs occurs because blockchain may solve many supply and demand matching issues. This will be made possible by the symmetry of information between consumers and suppliers [26].

In this same sense, smart contracts can significantly contribute to the governance of organizations. With blockchain, deals can become decentralized, and transactions can be executed with smart contracts and virtual currencies [27]. In other words, contracts and payments can be executed and settled automatically according to parameters agreed upon in advance by the parties. Human action will be required only at the time of negotiations.

Following the same perspective, trust and transparency among negotiating actors can lead to more efficient use of resources. Blockchain may secure information sharing while maintaining the privacy of those involved [28]. This integration improves efficiency in resources such as time, people, and others. Thus, even if the efficiency gains are short-lived, the new technology allows rethinking processes associated [29].

However, sometimes the experts were at variance with the literature. According to one expert, adopting technology will have a chance of reducing gains from some actors or eliminating them from the supply chain. The intermediaries' elimination could break the links in the supply chain by monopolizing production and excluding many people from the production process. On the other hand, since blockchain simplifies processes by dispensing with intermediaries involved in contracts for goods and services, parties can control tangible or intangible material damage by sharing access data [30]. So, depending on the point of view, eliminating intermediaries can be an advantage or a problem. A technology that is good for business may not necessarily be good in social terms.

Regarding the disadvantages and challenges, using blockchain may require sharing data by partners in a supply chain. In this sense, some partners may feel insecure when sharing information, and data loss and hacking by those involved may also occur. Furthermore, it may imply that anyone can have access to private information, leading to the non-participation of some actors in a blockchain network [31].

These results are in line with the findings in the study by [32], who highlighted important points that may affect the adoption of blockchain in the agricultural sector, such as lack of regulatory guidelines adopted by the government, security issues of technology, lack of awareness among actors, blockchain complexity, resistance to change by collaborators, trust between parties involved in the network, and high investments in technology development.

As blockchain technology is recent, it still requires large investments in infrastructure and maintenance, given the lack of specialized labor for its development. Given this, there is a need for employee training for blockchain adoption.

Blockchain has also faced several security challenges, mainly in digital assets blockchain systems, which the experts did not point out. However, according to the literature, Sybil attacks are becoming frequent and challenging to prevent, with several fake nodes being possible on a blockchain network [33]. Such a situation can lead to a non-acceptance of technology on a large scale. Iqbal and Matulevičius [33] explored this subject in more detail and proposed a framework based on the domain model for security risk management to explore Sybil and double-spending risks in blockchain systems. The model illustrates the assets protected or to be protected and the classification of threats that the attacker can unleash using the Sybil attack, which causes double spending on blockchains.

Gopalakrishnan et al. [34] argue that implementing a blockchain platform can involve high development costs, depending on the strategy and the intended solution. However, the opportunity to reduce transactional costs can make the technology viable. For example, Niforos [35] estimates that the operational costs of supply chains account for two-thirds of the final cost of goods. Therefore, processors, distributors, and consumers can benefit from reducing these costs via blockchain implementation.

On the other hand, some agribusiness services and activities may cease to exist since automation may remove intermediaries. Consequently, the blockchain follows principles like those applied in implementing Bitcoins, where open access and removing intermediaries are critical features of the technology [6].

3.2. Changes in Experts' Perceptions about Blockchain from 2020 to 2022

In general, the experts believe that most benefits have become more important over almost three years (Figure 2A). Only the benefit of more security throughout the production chain has maintained its importance between 2020 and 2022, and this is the benefit with the lowest importance score. The benefits with the highest importance scores in 2020 were reduced transaction costs and improved trust, transparency, and data immutability. Reduced transaction costs remained the benefit with the highest importance score in 2022, accompanied by enhanced governance and information flow. However, the main change in perspective concerns the perception that blockchain technology enables better governance and information flow along supply chains, whose perceived importance score increased by +15.4% (3.9 to 4.5). The second and third largest relative changes in perceived importance relate to achieving greater profitability in activities (+11.1%) and the possibility of drawing up smart digital contracts between actors in supply chains (+10.2%).

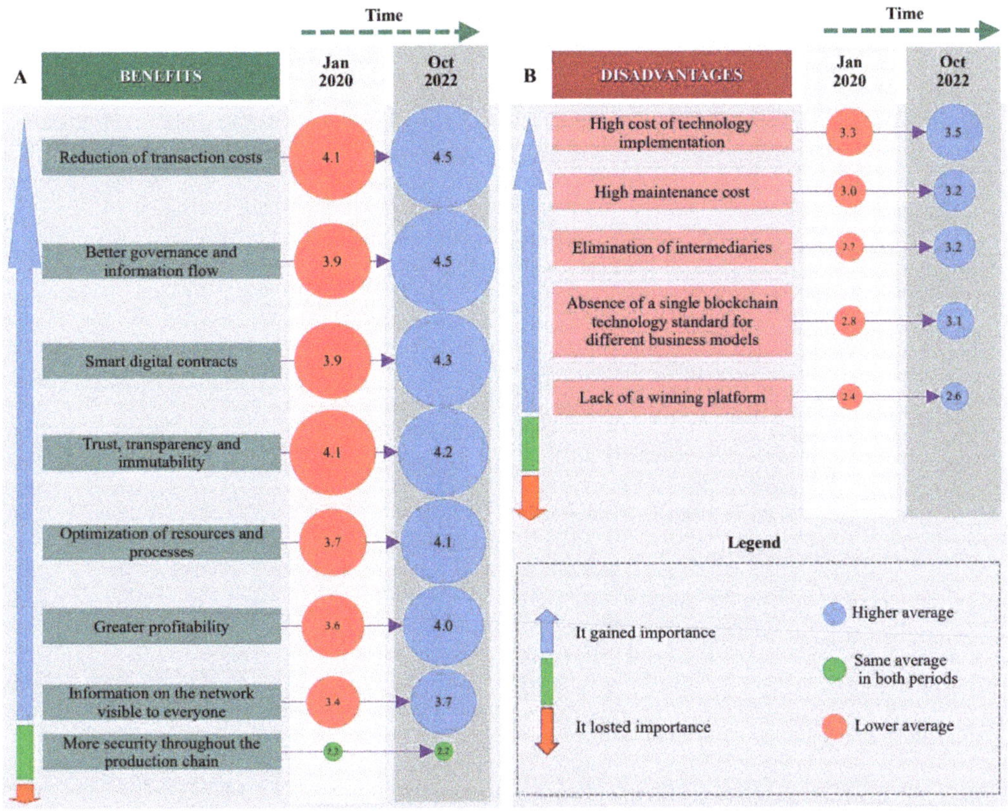

Figure 2. Changes in perceptions of the benefits (**A**) and disadvantages (**B**) of applying blockchain technology in agribusiness. Source: prepared by the authors based on survey data.

All disadvantages gained importance in the period. Blockchain implementation and maintenance costs were pointed out as the main disadvantages of the technology. The elimination of intermediaries as a negative result of adopting the technology had the highest percentage growth, 18.5 percent (score from 2.7 to 3.2).

The experts listed more and attributed more importance to the benefits than the disadvantages in both periods. During the interview, the experts also pointed out that blockchain technology will reach farmers through public, private, or cooperative initiatives. The interviewees also pointed out that farmers may seek to collaborate to drive the modernization of activities, showing optimism about the adoption of blockchain in agribusiness.

The benefits (Figure 2A) suggest more secure and transparent relationships among supply chain members, including final consumers. Experts believe that direct connections between producers and consumers can be established, reversing the current commercial logic in which intermediary actors have great power of governance in supply chains. In other words, aspects of supply and demand will become more transparent, and market speculation movements may be accessible to all actors in the supply chain, making markets fairer and less speculative.

Besides benefits and disadvantages, there are challenges and opportunities for adopting blockchain in agribusiness. Therefore, there are bottlenecks to be overcome and incentives to be explored in the technology development and adoption process. According to the experts' perception, all opportunities have become more critical in the three years (Figure 3B). The possibility of integrating blockchain with new emerging technologies was highlighted as the foremost opportunity. Traceability, certification, and supply chain digitization are among the top opportunities for technology adoption. On the other hand, the opportunity with the most significant variation in perception was a competitive advantage, with an increase of +17% (score from 3.0 to 3.5). In the same direction, information symmetry and product differentiation were the second and third opportunities with the most significant variation in perceived importance.

As the transactions carried out in a supply chain were placed on a blockchain, the information would be symmetric among actors since they could access it in real time. Therefore, a blockchain can be seen as an opportunity to record data with better reliability. In this sense, the competitive advantage is provided by technology's possibility of total and rigorous control of all chain stages. Furthermore, it promises to establish the automation of production processes, configure greater transparency, providing agribusiness sectors with greater business opportunities with new international markets. In the same direction, information symmetry and product differentiation were the second and third opportunities with the most significant variation in perceived importance.

Data collection, processing and storage, and the creation of new blockchain-based business models ranked as the top challenges for the technology (Figure 3A). They were the second and third items with the most significant change in perspective (11%, from 3.7 to 4.1). Meanwhile, the leading absolute change in perspective was related to the lack of a dominant or winning platform, with a 17% increase (2.3 to 2.7). Additionally, breaking old paradigms is also considered important for technology development and use. Finally, the possibility of eliminating intermediaries in business and the quality of the information entered into the systems are challenges that have remained stable during the period.

The experts pointed out that supply chain actors must restructure how they record information. At a later stage, current business models may be replaced, and information recording may become a major challenge. Sharing sensitive information requires the collaboration of many players, and some intermediaries may be hampered by technology. Influential intermediaries may hinder technology adoption or try to monopolize information to preserve their position, maintain profitability, and sustain their competitive advantage. In the future, the way of doing business will change, and some players may no longer be needed, as some interviewees reported.

On the other hand, the current supply chain cycle is considered long and inefficient, leading to the continuous deterioration of product quality and food safety [36,37]. From addressing this weakness arise many opportunities for digitizing information with reduced costs and significant improvement in input and food flows. Among several opportunities, the potential application of blockchain is associated with the ability to keep reliable records available and auditable to all stakeholders, including end consumers. Furthermore, proper blockchain design makes it impossible to alter recorded data without the consent of all involved. Therefore, blockchain can dramatically reduce transaction costs, as insecurity is almost wholly removed. Blockchain technology also makes supply chains more transparent and more efficient information flow between links. Future scenarios indicate that blockchain solutions are expected to be widely adopted and supply chains computerized to meet the aspirations of Industry 4.0.

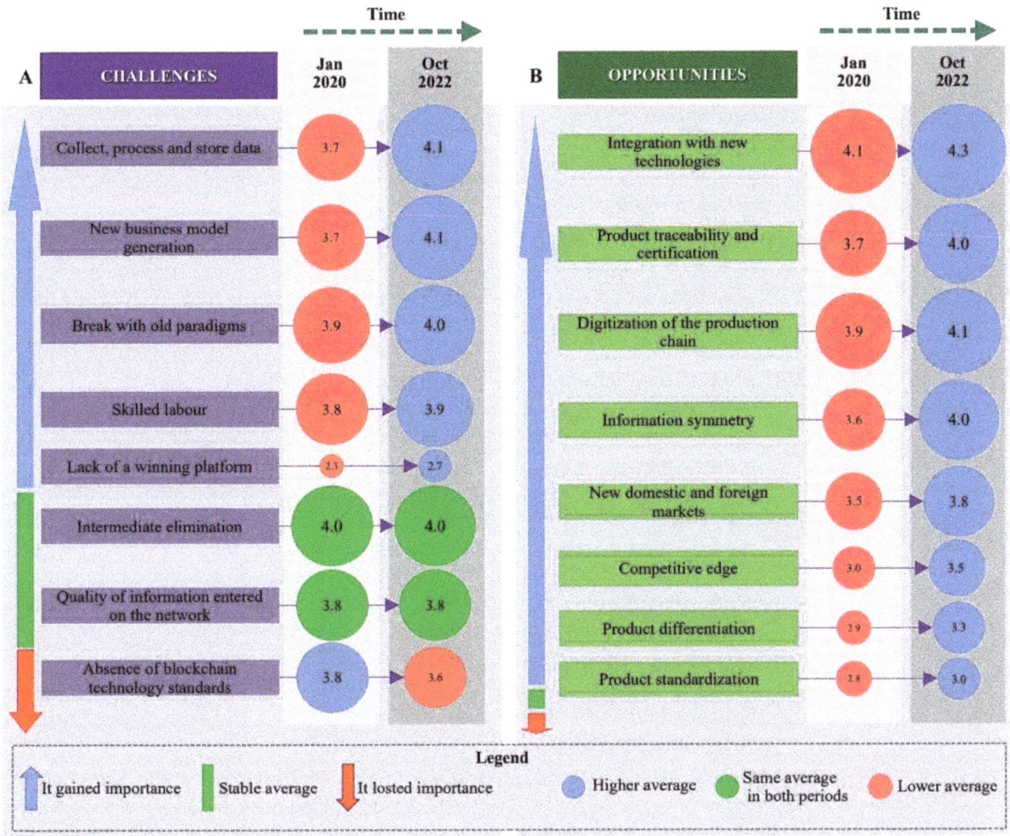

Figure 3. Changes in perceptions of the challenges (**A**) and opportunities (**B**) of applying blockchain technology in agribusiness. Source: prepared by the authors based on survey data.

Responses from the interviewees indicate that in 2020 blockchain technology was still in its early stages of development and its impact on the economy is not yet noticeable. However, it is believed that advances in the development of technology will change the way economic agents relate to each other, transforming the flow and exchange of information and contracts both domestically and globally and transforming the economy as a whole [38]. Our findings in the present study already signal that advances have occurred in this brief, almost three-year period.

Blockchain will be like a decentralized database. Everyone involved in the supply chain will be responsible for feeding, storing, and maintaining real-time information to allow transactions to be made and completed simultaneously [39]. The decentralized nature of the technology elevates the importance of the network concept, and the potential benefits of the technology increase as the size of the network expands [26]. In this context, business models need to rebuild existing systems, popularize the technology, and train personnel to adapt to the new management process [22,40].

Furthermore, interviewees perceive that the novelty of blockchain technology may impact the trust of the parties involved in a smart contract. However, human-computer interaction techniques can help build this trust [34]. The globalization of trade has forced supply chains to expand their nodes and relationships, involving diverse entities and complex transactions that cover a wide range of geographic places. As a result, farmers, entrepreneurs, consumers, logistics companies, media, financial institutions, industry associations, and governmental regulators, could be connected by blockchain through smart contracts, automating records, reducing transaction costs, and reducing risks of contract fraud [41].

4. Conclusions

At the end of the analysis, the following were the main changes observed in the perception of professionals regarding the development and application of blockchain technology in agribusiness:

(a) Experts believe that blockchain technology has been consolidating the promise of delivering benefits for supply chain management since it improves governance and information flow, facilitates the creation of smart contracts, reduces transaction costs between actors, and increases trust, transparency, and immutability in the sharing of information.

(b) Eliminating intermediary agents in transactions and high costs for the implementation and maintenance of blockchain have been the main disadvantages, signaling that the level of competitiveness in the blockchain market still offers opportunities for new entrants.

(c) Generating new business models and collecting, processing, and storing data are challenges to be met by overcoming old paradigms, especially concerning information sharing. In addition, other challenges are being overcome, such as establishing standards for technology while eliminating intermediaries, qualifying the information added to the network and the workers' skills, and developing a platform capable of integrating the other blockchain initiatives that seem to be constantly advancing.

(d) Finally, integrating blockchain technology with other emerging technologies, especially the IoT, digitalizing supply chains' information, guaranteeing symmetrical access to information, and enabling to trace and certify products are seen as increasing opportunities.

Therefore, we conclude that blockchain and its application to agribusiness present the fast development characteristic of any emerging information technology, with several potential uses. In almost three years, the perceptions of professionals working in this field are that most of the benefits of blockchain have grown in importance. In opposition, fewer disadvantages have been identified, with a slight increase in perceived importance. Thus, the benefits of blockchain outweigh the disadvantages in number and importance and are perceived as more relevant.

Similar perceptions were observed regarding opportunities and challenges. At least three challenges were perceived as less important in 2022 than in 2020, suggesting the perception that the development of blockchain technology has been able to overcome some bottlenecks perceived as critical not so long ago. On the other hand, all the opportunities considered in the study have increased their perceived importance. We can conclude, therefore, that the perception is that: first, the development of the technology has been overcoming the challenges; and second, the possibilities of blockchain applications in agribusiness have strengthened the belief that they can generate many opportunities.

Our conclusions are limited by the number of professionals participating in the study and cannot be generalized. Blockchain is still an emerging technology, and its development and implementation depend on several players and variables. At the current stage of blockchain development and adoption, it is impossible to predict the real impacts that technology may have on agribusiness. Future studies can focus on mapping the fundamental variables for technology development and their effects on supply chains, business, and the possibility of removing actors from supply chains. Finally, major changes can occur in agribusiness due to technologies such as blockchain, and researchers can explore these transformations by proposing appropriate solutions and avoiding adverse developments. The threat of cyber-attacks, for example, was not addressed in this study but deserve attention in future studies.

Supplementary Materials: The following supporting information can be downloaded at: https://www.mdpi.com/article/10.3390/fi15010038/s1, Supplementary Material S1: Script for the interviews with Blockchain experts.

Author Contributions: Conceptualization, G.d.S.R.R., D.D.M., H.A.C., L.d.O. and E.T.; methodology, G.d.S.R.R., H.A.C., L.d.O. and E.T.; software, D.D.M.; validation, G.d.S.R.R. and L.d.O.; formal analysis, G.d.S.R.R. and L.d.O.; investigation, G.d.S.R.R., D.D.M. and L.d.O.; resources, G.d.S.R.R., D.D.M. and L.d.O.; data curation, G.d.S.R.R. and L.d.O.; writing—original draft preparation, G.d.S.R.R., H.A.C., D.D.M. and L.d.O.; writing—review and editing, G.d.S.R.R., L.d.O. and E.T.; visualization, G.d.S.R.R. and E.T.; supervision, G.d.S.R.R. and L.d.O.; project administration G.d.S.R.R. and L.d.O.; funding acquisition, G.d.S.R.R. and E.T. All authors have read and agreed to the published version of the manuscript.

Funding: This research was funded by the National Council for Scientific and Technological Development under Grant Number 303956/2019-4 and Grant Number 140931/2022-8. Coordination for the Improvement of Higher Education Personnel-CAPES-Scholarship and Aid Control System-SCBA Process Number: 88887.614568/2021-00 and Process Number: 88887.642788/2021-00.

Data Availability Statement: Not applicable.

Acknowledgments: The authors would like to thank all the professionals who dedicated their time and knowledge to participate in this research, helping us to build knowledge about blockchain technology.

Conflicts of Interest: The authors declare no conflict of interest.

References

1. Haber, S.; Scott Stornetta, W. How to time-stamp a digital document. In *Advances in Cryptology-CRYPTO' 90. CRYPTO 1990. Lecture Notes in Computer Science*; Springer: Berlin/Heidelberg, Germany, 1991; Volume 537, pp. 437–455. [CrossRef]
2. Nasu, H.; Kodera, Y.; Nogami, Y. A Business-to-Business Collaboration System That Promotes Data Utilization While Encrypting Information on the Blockchain. *Sensors* **2022**, *22*, 4909. [CrossRef] [PubMed]
3. Chod, J.; Trichakis, N.; Tsoukalas, G.; Aspegren, H.; Weber, M. On the financing benefits of supply chain transparency and blockchain adoption. *Manag. Sci.* **2020**, *66*, 4378–4396. [CrossRef]
4. Nesarani, A.; Ramar, R.; Pandian, S. An efficient approach for rice prediction from authenticated Blockchain node using machine learning technique. *Environ. Technol. Innov.* **2020**, *20*, 101064. [CrossRef]
5. Chaer, A.; Salah, K.; Lima, C.; Ray, P.P.; Sheltami, T. Blockchain for 5G: Opportunities and challenges. In Proceedings of the 2019 IEEE Globecom Work GC Wkshps, Waikoloa, HI, USA, 9–13 December 2019. [CrossRef]
6. Tapscott, D.; Tapscott, A. La revolución blockchain Descubre descubre cómo esta nueva tecnología transformará la economía global. *Ed. Deusco.* **2017**, *1*, 439.
7. Delgado-von-Eitzen, C.; Anido-Rifón, L.; Fernández-Iglesias, M.J. Blockchain applications in education: A systematic literature review. *Appl. Sci.* **2021**, *11*, 11811. [CrossRef]
8. Leng, K.; Bi, Y.; Jing, L.; Fu, H.C.; Van Nieuwenhuyse, I. Research on agricultural supply chain system with double chain architecture based on blockchain technology. *Future Gener. Comput. Syst.* **2018**, *86*, 641–649. [CrossRef]
9. Lin, Y.P.; Petway, J.R.; Anthony, J.; Mukhtar, H.; Liao, S.W.; Chou, C.F.; Ho, Y.F. Blockchain: The evolutionary next step for ICT e-agriculture. *Environments* **2017**, *4*, 50. [CrossRef]
10. Aung, M.M.; Chang, Y.S. Traceability in a food supply chain: Safety and quality perspectives. *Food Control* **2014**, *39*, 172–184. [CrossRef]
11. Galvez, J.F.; Mejuto, J.C.; Simal-Gandara, J. Future challenges on the use of blockchain for food traceability analysis. *TrAC Trends Anal. Chem.* **2018**, *107*, 222–232. [CrossRef]
12. Islam, S.; Manning, L.; Cullen, J.M. Systematic assessment of food traceability information loss: A case study of the Bangladesh export shrimp supply chain. *Food Control* **2022**, *142*, 109257. [CrossRef]
13. Tian, F. A supply chain traceability system for food safety based on HACCP, blockchain & Internet of things. In Proceedings of the 14th International Conference on Service Systems and Service Management ICSSSM 2017, Dalian, China, 16–18 June 2017. [CrossRef]
14. Vilas-Boas, J.L.; Rodrigues, J.J.P.C.; Alberti, A.M. Convergence of Distributed Ledger Technologies with Digital Twins, IoT, and AI for fresh food logistics: Challenges and opportunities. *J. Ind. Inf. Integr.* **2022**, *31*, 100393. [CrossRef]
15. Baralla, G.; Pinna, A.; Corrias, G. Ensure traceability in european food supply chain by using a blockchain system. In Proceedings of the 2019 IEEE/ACM 2nd International Workshop on Emerging Trends in Software Engineering for Blockchain, WETSEB 2019, Montreal, QC, Canada, 27 May 2019; pp. 40–47. [CrossRef]
16. Huang, H.; Zhou, X.; Liu, J. Food Supply Chain Traceability Scheme Based on Blockchain and EPC Technology. In Proceedings of the Smart Blockchain, Second International Conference, SmartBlock 2019, Birmingham, UK, 11–13 October 2019; Volume 11911, pp. 32–42. [CrossRef]

17. Janssen, M.; Weerakkody, V.; Ismagilova, E.; Sivarajah, U.; Irani, Z. A framework for analysing blockchain technology adoption: Integrating institutional, market and technical factors. *Int. J. Inf. Manag.* **2020**, *50*, 302–309. [CrossRef]
18. Wang, W.; Lyu, G. Sequential product positioning on a platform in the presence of network effects. *Int. J. Prod. Econ.* **2020**, *229*, 107779. [CrossRef]
19. Hardjono, T.; Lipton, A.; Pentland, A. Toward an Interoperability Architecture for Blockchain Autonomous Systems. *IEEE Trans. Eng. Manag.* **2020**, *67*, 1298–1309. [CrossRef]
20. Oguntegbe, K.F.; Di Paola, N.; Vona, R. Behavioural antecedents to blockchain implementation in agrifood supply chain management: A thematic analysis. *Technol. Soc.* **2022**, *68*, 101927. [CrossRef]
21. Zhao, S.; Li, S.; Member, S.; Yao, Y. Blockchain Enabled Industrial Internet of Things Technology. *IEEE Trans. Comput. Soc. Syst.* **2019**, *6*, 1442–1453. [CrossRef]
22. Xu, J.; Guo, S.; Xie, D.; Yan, Y. Blockchain: A new safeguard for agri-foods. *Artif. Intell. Agric.* **2020**, *4*, 153–161. [CrossRef]
23. Arooj, A.; Farooq, M.S.; Umer, T. Unfolding the blockchain era: Timeline, evolution, types and real-world applications. *J. Netw. Comput. Appl.* **2022**, *207*, 103511. [CrossRef]
24. Krithika, L.B. Survey on the Applications of Blockchain in Agriculture. *Agriculture* **2022**, *12*, 1333.
25. Antonucci, F.; Figorilli, S.; Costa, C.; Pallottino, F.; Raso, L.; Menesatti, P. A Review on blockchain applications in the agri-food sector. *J. Sci. Food Agric.* **2019**, *99*, 6129–6138. [CrossRef]
26. Liu, L.; Li, F.; Qi, E. Research on Risk Avoidance and Coordination of Supply Chain Subject Based on Blockchain Technology. *Sustainability* **2019**, *11*, 2182. [CrossRef]
27. Wright, A.; De Filippi, P. *Decentralized Blockchain Technology and the Rise of Lex Cryptographia*; SSRN: Rochester, NY, USA, 2015. [CrossRef]
28. Yermack, D. Corporate governance and blockchains. *Rev. Financ.* **2017**, *21*, 7–31. [CrossRef]
29. Ali, R.; Barrdear, J.; Clews, R.; Southgate, J. The Economics of Digital Currencies. Rochester, NY, USA. 2014. Available online: https://papers.ssrn.com/abstract=2499418 (accessed on 10 January 2023).
30. Nofer, M.; Gomber, P.; Hinz, O.; Schiereck, D. Blockchain. *Bus. Inf. Syst. Eng.* **2017**, *59*, 183–187. [CrossRef]
31. Bumblauskas, D.; Mann, A.; Dugan, B.; Rittmer, J. A blockchain use case in food distribution: Do you know where your food has been? *Int. J. Inf. Manag.* **2020**, *52*, 102008. [CrossRef]
32. Nagariya, R.; Mukherjee, S.; Baral, M.M.; Patel, B.B.; Venkataiah, C. The Challenges of Blockchain Technology Adoption in the Agro-Based Industries. Available online: https://ijmems.in/cms/storage/app/public/uploads/volumes/59-IJMEMS-22-0063-7-6-949-963-2022.pdf (accessed on 10 January 2023).
33. Iqbal, M.; Matulevicius, R. Exploring Sybil and Double-Spending Risks in Blockchain Systems. *IEEE Access* **2021**, *9*, 76153–76177. [CrossRef]
34. Gopalakrishnan, P.K.; Hall, J.; Behdad, S. Cost analysis and optimization of Blockchain-based solid waste management traceability system. *Waste Manag.* **2021**, *120*, 594–607. [CrossRef]
35. States, U.; Alliance, G.; Facilitation, T. *Beyond Fintech: Leveraging Blockchain for More Sustainable and Inclusive Supply Chains*; International Finance Corporation: Washington, DC, USA, 2017.
36. Yao, Q.; Zhang, H. Improving Agricultural Product Traceability Using Blockchain. *Sensors* **2022**, *22*, 3388. [CrossRef]
37. Tsoukas, V.; Gkogkidis, A.; Kampa, A.; Spathoulas, G. Enhancing Food Supply Chain Security through the Use of Blockchain and TinyML. *Information* **2022**, *13*, 213. [CrossRef]
38. Waldo, J. A Hitchhiker' s Guide to the Blockchain Universe. *Commun. ACM* **2019**, *62*, 38–42. [CrossRef]
39. Staples, M.; Chen, S.; Falamaki, S.; Ponomarev, A.; Rimba, P.; Tran, A.B.; Weber, I.; Xu, S.; Zhu, J. Risks and Opportunities for Systems Using Blockchain and Smart Contracts. Available online: https://publications.csiro.au/rpr/download?pid=csiro:EP175103&dsid=DS2 (accessed on 10 January 2023).
40. Zhao, G.; Liu, S.; Lopez, C.; Lu, H.; Elgueta, S.; Chen, H.; Boshkoska, B.M. Blockchain technology in agri-food value chain management: A synthesis of applications, challenges and future research directions. *Comput. Ind.* **2019**, *109*, 83–99. [CrossRef]
41. Perboli, G.; Musso, S.; Rosano, M. Blockchain in Logistics and Supply Chain: A Lean Approach for Designing Real-World Use Cases. *IEEE Access* **2018**, *6*, 62018–62028. [CrossRef]

future internet

Article

Comparison of Distributed Tamper-Proof Storage Methods for Public Key Infrastructures

Fabian Honecker, Julian Dreyer and Ralf Tönjes *

Faculty for Engineering and Computer Sciences, University of Applied Sciences, 49076 Osnabrück, Germany
* Correspondence: r.toenjes@hs-osnabrueck.de

Abstract: Modern Public Key Infrastructures (PKIs) allow users to create and maintain centrally stored cryptographic certificates. These infrastructures use a so-called certificate chain. At the root of the chain, a root Certification Authority (CA) is responsible for issuing the base certificate. Every verification and certification step within the chain is based upon the security of said root CA. Thus, its operation security is of great concern. Since the root certificates are stored locally on the root CA, any Denial of Service (DoS) attack may render the whole certificate chain, which is based on of the attacked root CA, inoperable. Therefore, this article evaluates different approaches to a decentralized data storage system that is based on the Distributed Ledger Technology (DLT). To show the real-world potential of the proposed approaches, we also evaluate the different technologies using a novel PKI mechanism called Near Field Communication Key Exchange (NFC-KE). The results indicate that modern distributed data storage solutions such as Interplanetary Filesystem (IPFS) and SIA can have significant performance and decentralization benefits in comparison to purely Blockchain-based technologies like Hyperledger Fabric. However, they lack any Smart Contract functionality, which requires a software developer to implement verification mechanisms in centralized software solutions.

Keywords: Public Key Infrastructures; decentralized data storage; blockchain data storage; Certification Authority; decentralized trust; public key cryptography

Citation: Honecker, F.; Dreyer, J.; Tönjes, R. Comparison of Distributed Tamper-Proof Storage Methods for Public Key Infrastructures. *Future Internet* **2022**, *14*, 336. https://doi.org/10.3390/fi14110336

Academic Editors: Christoph Stach and Clémentine Gritti

Received: 28 October 2022
Accepted: 16 November 2022
Published: 18 November 2022

1. Introduction

Establishing a trust relationship between two parties that never met before is an inherently hard task. Nevertheless, this task is executed each time a user visits, e.g., a website that is secured using the Hypertext Transfer Protocol Secure (HTTPS). Here, the user needs to be confident that the response of the server is authentic and has not been tampered with. Otherwise, a malicious server impersonator could send unauthentic messages and thereby compromise the communication between the user and the server. To remedy this fundamental problem, modern cryptographical digital signature schemes are used in conjunction with certificate chains.

These signature schemes mostly rely on asymmetric public key algorithms to sign data packets using a private/secret key (SK_{serv}), e.g., of the server, and an accompanying public key (PK_{serv}) that is stored within a cryptographic certificate. Now, when using this mechanism on its own, again, an attacker could craft a malicious certificate and thereby impersonate the real server. Therefore, multiple intermediary parties are introduced within the system. These are called *Certification Authority* (*CAs*). Their purpose is to issue and maintain user certificates and thereby act as a trusted intermediary. To enhance the cryptographic trust between the Certification Authorities (CAs) and the certificate owners, multiple CAs are chained together, forming a certificate chain. This chain is based on a root CA that issues a root Certificate and stores its corresponding private key securely on the client. Additional CAs are added by issuing new certificates that are signed by the root CA and are only capable of issuing and verifying new certificates. Most commonly, these certification chains have a length of two to four CAs and thus allow for a robust and secure digital trust relationship that forms the basis of modern web traffic.

This system has three significant problems that are under current research [1–3]: First, the chain of trust is based on the trust for the root CA. This trust is artificially generated, e.g., by the developers of modern Web browsers and operating systems that store the root certificates in the default keychain of the system. These certificates need to be trusted implicitly. Second, the maintenance effort for establishing and running a custom certification chain is beyond the scope of competencies of most Small and Medium Enterprises (SMEs). These enterprises often do not own a proper IT department to establish such a certification chain and are therefore depending on a proper solution provided by external contractors. The third problem is an inherent flaw in the design of the root CA and its centralized, on-device keystore. If a root CA has been compromised, the attacker has potential access to the file system and all of its files and certificates. This allows, in the worst case, for tampering of the certificate contents or maliciously uploaded certificates that are not properly issued. In cases when a formerly well-intended CA turns out to be fraudulent, the same above problems arise.

The Distributed Ledger Technology (DLT) offers a solution for problems one and three by using its decentralized execution and data storage concepts and, when combined with a lightweight Public Key Infrastructure (PKI) approach, also a potential real-world applicable solution for problem two. Therefore, we chose to adapt the Near Field Communication Key Exchange (NFC-KE) PKI scheme, which offers a lightweight public key exchange scheme. This scheme is particularly suited for SMEs since it does not require a certification chain for its operation.

Most research on the general topic of DLT-based CAs is focusing on novel protocol approaches for diversifying the realm of new signature verification techniques, aided by, e.g., Blockchain platforms [4–6]. A general pattern evolved among these works, showing that popular DLT frameworks like Hyperledger Fabric (HLF) [5] or Ethereum [6] are valid candidates for successfully implementing those novel PKI approaches. These approaches rely on the Smart Contract capabilities of said platforms, thereby neglecting distributed data storage systems such as the Interplanetary Filesystem (IPFS), which also use the DLT at its core.

Therefore, our works aims to reason the feasibility of implementing novel DLT-based PKI schemes using distributed file storage technologies. To show the advantages of a tamper-proof storage system that is based on the DLT for PKIs, this article evaluates a broad spectrum of modern DLT solutions in regards to their specific traits and benefits, as well as their performances under the given application scenario. The main contributions can be summarized as follows:

- Concept proposals for decentralized tamper-proof storage algorithms for PKIs, based on *private IPFS*, *public IPFS* and *Sia*.
- Real-world application and adaptation to the lightweight NFC-KE PKI scheme [7]
- A performance evaluation of the system using standardized Key Performance Indicators (KPIs), e.g., *execution times*

The remainder of this article is structured as follows: First, in Section 2, we provide a brief overview of other approaches to decentralize modern PKI schemes and use case scenarios for the decentralized IPFS. In the following Section, Section 3, we explain the new concepts for the decentralized and tamper-proof public key stores for each DLT framework by first introducing the selected DLTs and introducing the NFC-KE and its intricacies. After this, the concepts are implemented in Section 4 and evaluated for their performances afterward in Section 5. The performance evaluation uses the same KPIs as previously used in [7] to establish a common basis for comparison. The results of the performance analysis and its implications are further discussed in Section 6. Finally, Section 7 concludes this article. All abbreviations used in this article are listed and explained in the abbreviations table at the end of this article.

2. Related Work

The previously mentioned drawbacks of modern PKIs have been addressed previously. In their research paper [1], Kfoury et al. proposed a new methodology of creating a fully functional PKI that leverages the Ethereum Blockchain platform and its Smart Contract capabilities. Among other contributions, their proposed system offers a proper Preshared-Key (PSK) exchange and keystore concept that is decentralized and thus offers enhanced tamper-resistance. By using an Application Management Platform in conjunction with the public Ethereum Blockchain, the authors achieved an implementation that showed the real-world application potential of the proposed system. Their approach is, therefore, capable of replacing common CA implementations for their system.

Trust and usability issues of common PKIs have also been identified by Hepp et al. [2] in their extensive exploration of potentials and challenges of Blockchain-based PKIs. The authors research four distinct usage scenarios that may benefit from the introduction of the DLT or the Blockchain in particular. Concepts such as the PKI, Pretty Good Privacy (PGP), and X.509 are shown to be capable of benefiting from the introduction of a Blockchain-backed approach for fixing common security problems and trust issues. The authors conclude that, at the state of writing, the Blockchain-based PKIs offers a large number of potentials, although there are major challenges to solve, especially concerning speed and scalability.

Additional Blockchain-based concepts for a decentralized PKI have been proposed by Papageorgiou et al. [3]. Their approach is focussing on being particularly applicable for the Internet of Things (IoT) and its special demands. The IoT is characterized, among other facets, by the dynamic joining and leaving of new or known network participants. Common PKI solutions, as the authors argue, are not always capable of coping with the corresponding scalability demands. Furthermore, the common monolithic PKI systems establish a single source of failure. In case of a PKI system outage, no trusted communication is possible anymore. To remedy these issues, the authors analyzed the abstract inner routines of common CA architectures and identified singular building blocks that can be decentralized. Using this method, they identified the verification component and swapped it out for a decentralized authorization component that offers a distributed information basis. To show the real-world applicability of the concept, an implementation using the HLF Blockchain Platform is provided. The results indicate that this solution is indeed applicable for IoT networks and is thus capable of replacing common monolithic PKI solutions.

In practice, the IPFS is often used in combination with a Blockchain-based Smart Contract platform for handling logic. Ramesh et al. [8] use the IPFS together with a private Ethereum-based application for secure data management that is especially suited for the IoT. Their approach is based upon a common IoT network structure, comprised of data consumers and producers. The exchanged data, which themselves shall be encrypted using specialized security measures, is stored decentralized on the IPFS. To allow for encryption and data verification, the Ethereum Smart Contracts allow for a custom programmability for the desired use case. In their specific setup, the authors used a Trusted Platform Module (TPM) for local key storage and verification operations. They were able to show that their system offers a throughput performance of eight transactions per second using a private Ethereum setup, thereby proving the real-world applicability. Though, the performance itself is mainly influenced by the data verification on behalf of the data consumers and producers.

Another application example of the IPFS being employed in combination with the Ethereum Blockchain is provided by Uddin et al. [9]. In their research paper, the authors propose a decentralized and secure data sharing system that is based upon the IPFS for handling the decentralized data storage and an Ethereum-based encryption and decryption mechanism. Furthermore, the Ethereum Smart Contracts are used for data identification and client authentication. Since their setup is not fundamentally new, the authors compare their proposed solution to previous works by Aslam et al. [10], Hasan et al. [11], Nizamuddin et al. [12], and Park et al. [13], which all have unique drawbacks in terms of security and performance. Since their system is intended to be used on the public Ethereum blockchain, the authors also provide a cost estimation based on Ethereum transaction/gas prices. Their approach has shown to be real-world applicable and offers new security perspectives for similar use cases.

3. Concepts and Evaluation Methods

The following subsections will give a brief overview of the several technologies that are used for the concept of a tamper-proof data storage for PKIs. First, the DLTs used within this article are described. Furthermore, this section also includes a statement on how these concepts are evaluated later. For that, the main lightweight PKI scheme NFC-KE will also be explained and discussed.

IPFS Private

The IPFS [14] is a distributed Peer-To-Peer (P2P) network intended for storing and allowing access to data, websites, and applications. In contrast to centralized systems that manage the data access via location-based addressing, IPFS implements a content addressing scheme [15] for accessing data stored within it. Using this scheme, a distributed download of one single file via multiple peers in different locations is possible. Said peers are called *IPFS Peers* in the following.

The decentralized storage of files offers, beyond others, some noteworthy advantages:

1. **Redundancy:** Centrally stored files are lost, if the central file server is out of operation, e.g., after a fire hazard. Using IPFS, data redundancy can be achieved since data stored on one given peer is also replicated to multiple peers within the network. Therefore, the data availability is enhanced.
2. **Performance:** The data query can be accelerated because, using a custom routing scheme within the IPFS network, files can be queried from geographically close IPFS peers. This routing scheme is facilitated by distributed hash tables (DHTs) and is based on a P2P basis [16]. Thus, a file that has been uploaded in Brazil can be queried within a few milliseconds by a client located in Germany (Note: After complete file replication within the network and if there are IPFS peers closely located).
3. **Tamper-proof design:** For third parties, manipulation or deletion of a file stored within the IPFS is made almost infeasible since the data redundancy and content addressing schemes cryptographically and physically deny any attempt to alter data.

The content addressing works by cryptographically hashing the given files using the *SHA-256* hashing algorithm. Thus, two identical files will always produce the same data hash (called *Content identifier (CID)*). Furthermore, each file is split up into chunks of the same size to allow for an even more extended spreading of data. This is made possible by using Directed Acyclic Graphs (DAGs) and Merkle-DAGs in particular [17].

Fundamentally, the IPFS is a homogeneous network comprised of peers that are storing the desired files on their local hardware. Each peer is therefore responsible for storing and managing the data, as well as communicating with other network participants. In general, IPFS is a publicly available implementation that can be used either in a private deployment or using public Application Programming Interfaces (APIs) and frameworks, although the inner workings of the technology are common for every setup.

When using a public IPFS setup, each peer needs a financial incentive to store the data on its local hard drive. Within a private setup, this requirement is not given since every peer is set up on-demand and is well-intended by default.

Public IPFS-Filecoin

IPFS can be implemented in both a private and a public setup. Each paradigm involves different intricacies that need to be considered. On the one hand, public IPFS solutions offer a broad decentralization and may spread all around the globe while, on the other hand, private setups do not require a financial incentive to be run and also allow for generally better performance.

Filecoin [18] is a Blockchain-based P2P network that uses such a financial incentive to guarantee a persistent and reliable data storage. By using Filecoin, each client will pay a small fee for the data storage to the peers hosting the storage space. Those peers are essentially storage servers connected to the underlying IPFS network. After registering to the Filecoin network, they can offer their free available storage space to any client willing to pay them. Thus, a digital marketplace for decentralized data storage, which is backed by a Blockchain keeping track of the Filecoin transactions, is established.

Additionally, storage providers can promote special benefits like lower cost, higher redundancy, or enhanced access speed. After a client chose to store his data on a given provider's hardware, the provider needs to execute a *Proof-of-Replication*. This scheme, being beyond the scope of this article, is responsible for eliminating the *double-spending* problem and essentially ensures a proper storage operation that is provably correct.

Sia

Sia is a decentralized cloud storage platform that implements a different storage scheme than Filecoin or IPFS in general. Instead of lending storage space from a central provider, Sia allows clients to lend storage space to each other. The underlying Sia Blockchain only stores the storage contracts the clients agreed on. Each contract has a storage interval in which the data storage provider needs to guarantee the storage of the data. To prove this, the provider executes a *Proof-of-Storage* algorithm, thereby validating the contract.

Additionally, the storage providers get compensated for each successful proof using the native *SIA Token*. In return, the data client will have to spend some of his SIA Tokens for the data to be stored on the network. After the process is complete, chunks of the data will be replicated to 30 different peers within the Sia network and, using error correcting code (ECC), can be reconstructed using only ten unique peer responses [19].

Hyperledger Fabric

Hyperledger Fabric is a private/permissioned Blockchain platform that offers custom programmability using Smart Contracts and a fully custom network setup. In contrast to other, mostly public Blockchain platforms, the network structure of Fabric is heterogeneous. It is comprised of different nodes each with a different role, alhough each peer role may occur multiple times within the network.

1. Channels

On the top level, each custom Fabric network is based around a *Channel* [20]. It hosts a custom Blockchain together with a so-called *World State* database. The Blockchain is responsible for keeping the transactions logged and persisted within the network while the World State contains all the software-defined objects and their connected states. Fundamentally, it acts as a distributed database that is kept in sync using an ordering mechanism.

2. Organizations

Each Channel can be organized within several *Organizations* [21]. They allow for a more business-oriented network structure and encapsulate the clients within it from the rest of the network. Each Organization can host its own CA, which is responsible for issuing new client certificates. Additionally, Smart Contracts can be deployed on an organizational basis.

3. Peers

The Peers are physical entities within the network [22]. The Channel and Organization concepts in themselves are purely logical while the Peers are separate clients within the network. Each Peer hosts a full copy of the shared ledger, in this case, the Fabric Blockchain and the World State. To keep these in sync with other Peers and thus reach a consensual basis, each Peer will communicate with other Peers within the same Organization. Furthermore, the peers are responsible for executing and validating Smart Contracts being installed onto them. During network setup, an administrator may choose to configure some Peers as *Endorsement Peers*. Those Peers are on duty of executing, checking, and endorsing Smart Contracts while non-Endorsement Peers only hold and maintain a copy of the Blockchain and the World State.

4. Smart Contracts

In the Fabric domain, Smart Contracts are called *Chaincodes* since they are code fragments, written in high-level languages like Java or GoLang [23] and run on-chain using the underlying Fabric Blockchain. They are the fundamental way to alter the World State by adding, deleting, updating, and reading data from it. Thus, the Smart Contracts allow for decision-based data storage for data that can be stored within the World State database. Methods like cryptographical signature verification, decryption, and permission handling are all possible using the Smart Contract functionality of Fabric.

5. Ordering Service

After a Peer issues a new transaction, it needs to be stored on the Blockchain. Since there might be multiple transactions being issued at the same time within the network, a protocol for reaching a consensus must be employed. For this, the role of an *Orderer* has been created [24]. Using a Byzantine-Fault-Tolerant algorithm, the (multiple) Ordering nodes, which themselves are distinct clients within the network, are responsible for receiving the new transactions from the peers, sorting them by their timestamps and other metrics, and creating a new block for the Blockchain. The new block will hold the newly collected transactions and is sent to every peer within the network. Eventually, every peer will receive the new Block and append it to its local copy of the Blockchain and update the World State accordingly.

NFC-Key Exchange and Evaluation Methods

Modern PKI hierarchies and certification chains are often too cumbersome to set up and maintain, especially for SMEs. Some scenarios within the Industry 4.0 domain require the use of trusted communication between the involved parties. Mostly, these scenarios are based on sensor networks and low-cost devices. Thus, large server environments and multiple CA instances are too expensive in terms of resources and maintenance effort for these specific scenarios.

Nevertheless, modern cryptographic signature schemes are more often used in these communication scenarios. This process is also fostered by more modern and potent hardware on the sensor side that allows for faster signing of data packets. The authenticity of data is crucial in most industrial scenarios since no malicious sensor nodes shall be able to intervene within the network, e.g., by using unauthenticated data requests or by acting as a Man-in-the-Middle (MitM) and manipulating well-intended data packets.

Therefore, Dreyer et al. proposed a novel public key exchange scheme based on the Near Field Communication (NFC) as a transport medium [7]. The main idea of the concept assumes that each sensor is capable of signing a given data packet using its own secret key SK_S. The resulting signature can be verified using the corresponding public key PK_S. In practice, this basic concept works until a malicious actor joins the network and begins signing his own data packets. Therefore, the proposed NFC-KE concept involves a central PKI-like entity, called *Signing Hub*, that is responsible for verifying access to the network and thus allowing for communication. Since only previously authenticated publickeys will be added to the system, this process is called an *authentic public key exchange*.

The Signing Hub is comprised of a TPM connected, in this case, to a Raspberry Pi microcontroller. Additionally, it has an NFC-Reader attached to its serial port allowing for NFC tags to be read and written. The main challenge–response protocol of the NFC-KE is depicted in Figure 1.

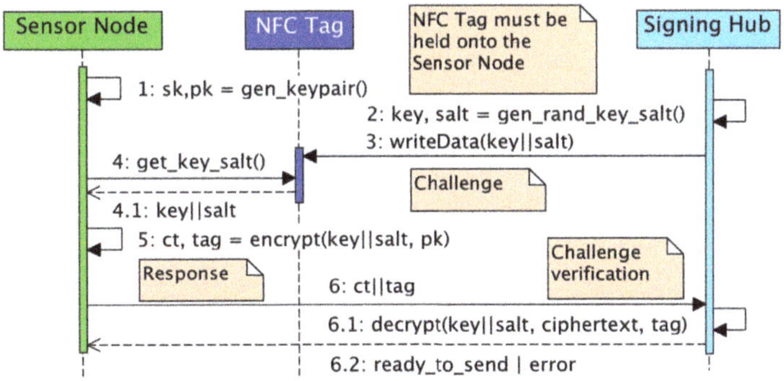

Figure 1. NFC-KE Phase 1: Fundamental public key exchange scheme [7].

The fundamental working principle of the NFC-KE requires a physical Proof-of-Possession scheme to be performed between the Sensor Node and the Signing Hub. Using the NFC, which requires a close physical distance for communication, a cryptographic token can be exchanged between the two parties, thereby implicitly proving authorization due to the low likelihood of an unauthorized NFC-communication. For further details on the NFC-KE, the reader is referred to the original paper [7].

Due to its lightweight setup and its ability to be implemented using low-cost hardware, the NFC-KE scheme is a valid candidate for low-resource PKI replacement in Industry 4.0 and Industrial Internet of Things (IIoT) environments in general.

The NFC-KE in its basic variant is susceptible to arbitrary file-tampering if an attacker is able to exploit a potentially unknown software vulnerability, e.g., by infiltrating the Raspberry Pi microcontroller. The properly exchanged authentic public keys of the Sensor Nodes are stored as certificate files on the Signing Hub. If an attacker had access to the filesystem, arbitrary upload and manipulation of said files could be possible. In the worst case, an attacker could be able to inject a nonauthorized public key and thus circumvent the whole authorization mechanism. This, therefore, creates a general single point of failure, as well as a setup vulnerable to public key tampering.

To remedy this problem, Dreyer et al. also proposed an extension to the former NFC-KE protocol that is using the HLF Blockchain for a more decentralized data storage of the public keys [25,26]. The new approach essentially replaces the standard filesystem-based storage approach of the NFC-KE scheme in favor of a Blockchain-based approach. Instead of storing and implicitly validating the newly exchanged Sensor Node's public key on the Signing Hub device itself, this functionality is outsourced to an external Blockchain Platform, in this case, HLF. The authors showed that using their approach, the NFC-KE experiences, on the one hand, a decrease in performance while, on the other hand being more tolerant against system outages, tampering attempts and unauthorized access due to the higher redundancy and tamper resistance of the HLF platform.

For their new approach, the authors introduced two new network participants, one being the Application Server and the other being the HLF instance (s. Figures 2 and 3). Generally, the Application Server is responsible for validating the requests originating from the Signing Hub on a network level and passing them on to the HLF instance. The latter is responsible for performing upper-level validations such as permission control and signature verification. Since HLF offers a fully programmable Smart Contract runtime, every signature verification can be performed on-chain, thus enabling an enhanced distributed trust. Additionally, each Sensor Node's public key, after being successfully exchanged, is stored within the World State of the HLF platform. This eliminates the need for a distinct file for each new key and thereby reduces the chances of arbitrary file tampering. For further details on the inner workings of the general extension scheme, the reader is referred to [25].

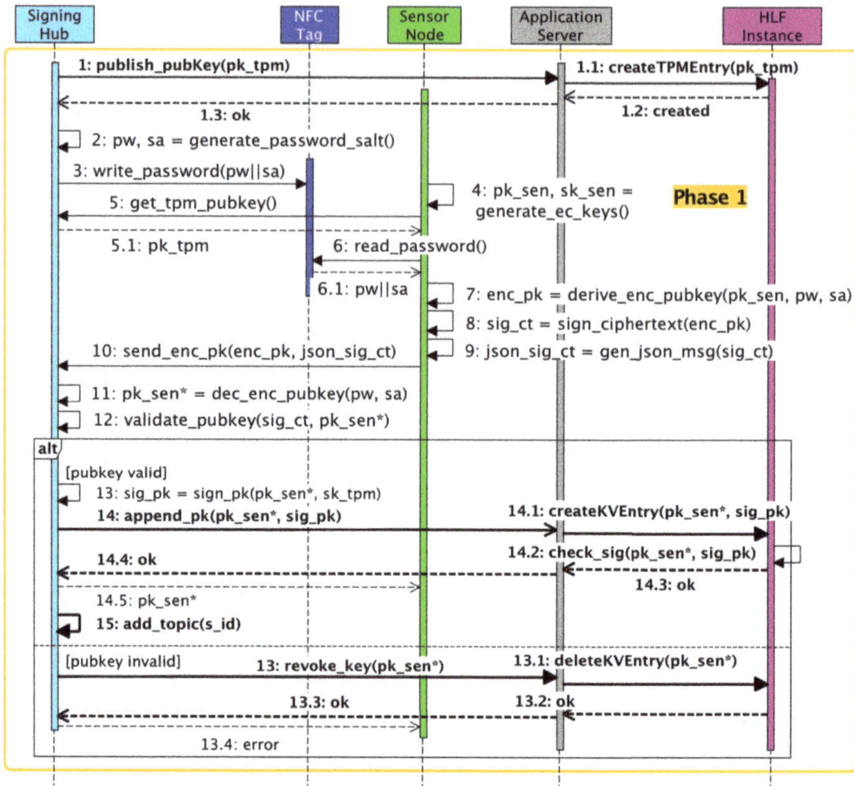

Figure 2. NFC-KE Phase 1 HLF extension [25].

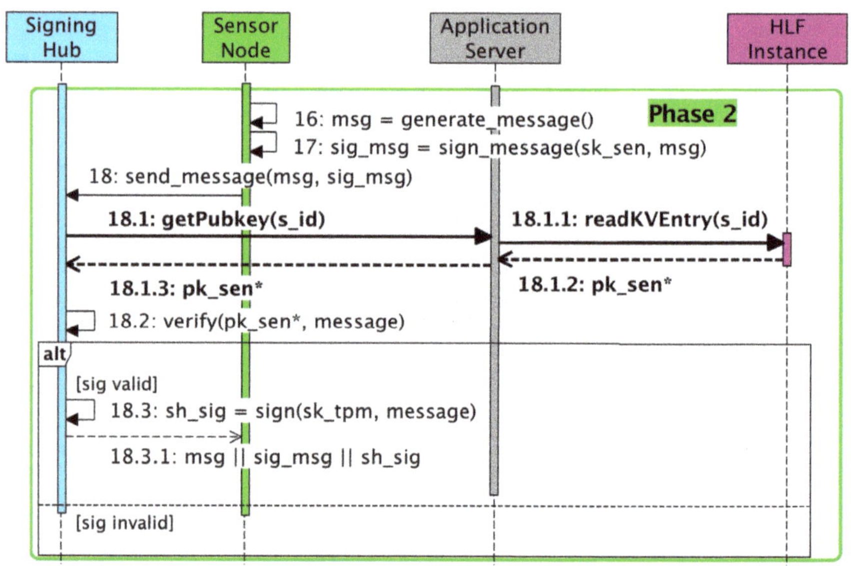

Figure 3. NFC-KE Phase 2 HLF extension [25].

The basic idea of outsourcing the filesystem keystore to an external DLT-based solution is also pursued by Papageorgiou et al. [3]. Their approach also includes a preceding analysis phase of the system in regards to potential parts that may benefit from being outsourced to a DLT. Thus, for this article, we identified the filesystem in particular and analyzed how different DLT solutions can handle the same functionality as the HLF-based approach of [25].

4. Alternative Implementations and Concept Analysis

One goal of this article is to present an alternative approach for a DLT-based data/file storage that is suited for the given use case as a replacement for the filesystem keystore in the NFC-KE scheme. Apart from the file storage per se, the implemented NFC-KE extension using HLF is also capable of executing Smart Contracts, e.g., for verifying cryptographical signatures. The other DLTs mentioned in Section 3 lack such a feature entirely and focus mainly on the plain storage aspect of the data.

Therefore, a custom verification mechanism is required to map the Smart Contract functionality to a centralized entity. For all concepts explained in the following sections, the following assumptions are set:

- Each Sensor Node has a unique identificator ID_S
- All Sensor Node public keys will be called PK_S
- The corresponding file for a given PK_S will be named $< ID_S >$ *.json*

4.1. Application Server

Since a stateless Application Server component is already present in the NFC-KE extension (s. Figure 2), it can be extended to support the verification of incoming signatures for new public keys. We chose to implement the Application Server using Node.js and JavaScript to achieve a universal basis for each DLT evaluation. Each of the proposed DLTs provide a custom NodeJS library and/or API endpoints for interacting with them [27–29]. Additionally, to reach enhanced modularity and exchangeability of the different technologies, object-oriented programming paradigms were leveraged, enabling a loosely coupled system.

When a new public key needs to be added to the system, the Application Server will perform a validation of the corresponding signature, following the NFC-KE scheme. For this, it has to query the publicly available public key of the TPM PK_{TPM}, which has to be stored immutably in the given DLT. The following subsections will describe this particular process in more detail. After querying PK_{TPM}, the Application Server can use the key for signature verification of the newly sent Sensor Node's public key PK_S. If the verification is successful, the key can be passed onto the given DLT and stored permanently.

Another problem arises when querying the newly added Sensor Node's public key since the presented DLT platforms do not offer a *Key-Value* mapping like HLF does. In the presented NFC-KE extension, the unique Sensor Node identifier ID_S is used as a key in the World State database and PK_S as the corresponding value. To map this scheme to the other DLT platforms, new concepts are needed and further described in the following subsections. Naive implementations could make use of the filename, e.g., Sensor_X.pem, and use it as an identifier. Since the presented DLTs are using content-based addressing, the unique content hash (called CID) of the file is used for addressing. Therefore, a mapping between the CID and the sensor id is required, resulting in the same fundamental problem.

Private IPFS

The fundamental IPFS system allows for file deletion using a garbage collection mechanism. Each file has a property called *Pinned*. If a file is pinned, the internal IPFS garbage collector will not delete the file. Else, if the file is unpinned, eventually the garbage collection will delete and remove the file from the network. Depending on the network size and speed, this process can take some time.

The pinning request has some intricacies that allow a direct *SensorID-Publickey* mapping, just like HLF. Each *Pin* has some metadata attached to it that can be arbitrarily altered by the user. Therefore, when uploading a file to the IPFS the Application Server will pin the file accordingly and add the given Sensor ID ID_S to its metadata. Using integrated network functionalities, the IPFS can be queried for all (or specific) pins. The resulting list can then act as a direct ID_S-PK_S mapping.

The functionality of adding the TPM key to the IPFS instance also works similarly. The required pin for the TPM public key PK_{TPM} has a special identifier, e.g., TPM_{PK} and is programmed into the application server. This pin cannot be altered after being inserted.

Public IPFS

The previous concepts apply specifically for the private IPFS setup and the provided IPFS Software Development Kit (SDK) [27]. For the public IPFS variant, *Filebase* [30] offers a free to use API that exposes all IPFS functionalities. Though, the ID_S-PK_S mapping can be handled easier by leveraging the Filebase API. To query a given file, a custom GET request needs to be crafted that contains the following file name as a parameter. On request, the API will return the complete object stored under that file name. Alternatively, if the CID of the file is known, the file can also be queried using public IPFS gateways [31].

An alternative way of accessing the public IPFS is the *Web3.Storage* API. In the background, Web3.Storage uses the IPFS with a financial incentive concept called *Filecoin*. Using this incentive system, data can be guaranteed to be persisted for the long term, since each data hoster will be rewarded with Filecoin tokens. Whenever a user uploads a new file to the network, he will receive a unique CID for said file. This can then be used to query the data from the network. For users of the Web3.Storage API, this feature is free to use [32].

Adding a new Sensor's public-key to the IPFS using the Web3.Storage API differs, again, in its way of mapping the ID_S to its corresponding PK_S. Since the user receives a unique CID of the file and can use this CID to query it, a direct search for the key is not trivially possible. For this purpose, the API offers a method to query all files uploaded via a given Web3.Storage account. This feature can be exploited to receive a full list of all files. Since by definition the uploaded files will follow the naming scheme $< ID_S > .json$, the returned list can be trivially searched for a given ID_S. The list, therefore, also contains a direct ID_S-CID mapping, allowing the public key storing concept to work.

To add the TPM's public key to the IPFS using the Web3.Storage API, the same methodology is used as in the previous approach using the private IPFS implementation. The TPM's public key is uploaded under a specific identifier which is programmed into the application server's logic.

Sia Skynet

In its abstract form, Sia offers the same functionalities as plain IPFS storage systems when using its *Skynet* API. For public key storage, Skynet offers a method to upload arbitrary data using a file and a corresponding filename. Again, following set conventions, the filename will be using the naming scheme $< ID_S.json >$. After uploading a file to the Skynet network, the user receives a so-called *Skylink*, which can be compared to the concept of a CID that is used in IPFS. Essentially, a Skylink allows direct access to the given file.

To support the query of an uploaded public key, an alternative approach needs to be proposed, since the previously described concepts that apply for private and public IPFS solutions do not apply to Sia Skynet. Other than Web3.Storage, for example, Skynet does not offer a direct API endpoint to query all uploaded files. Therefore, we propose an alternative approach using the available *SkyDB* [33] that is offered by the Sia Blockchain. Other than the Skynet system, the SkyDB is directly stored on the Sia Blockchain and thus only allows for small datasets to be stored. By adding a mapping file, called *devices.json* for this case, a mapping can be created for each new ID_S-PK_S pair using the Filename and the returned Skylink ID. On creation, the *devices.json* file will be stored on-chain and can be queried, edited, and updated at any time. After a successful query of the mapping file, the desired ID_S-PK_S mapping for the given Sensor Node can be extracted.

To support the special storing of the TPM's public key, this approach also follows the same principle as described in the previous section for IPFS.

4.2. Further Considerations

Every presented approach has some particularly minor accompanying problems, that require consideration.

Since the public APIs all rely on a properly working internet connection, increased network latency can be expected in comparison to private deployments. Public DLT setups offer enhanced global decentralization and therefore a more resilient data availability. As stated previously, data can be queried on a P2P basis and is thus more directly available. On the one hand, using the financial incentive system, e.g., Filecoin, data is guaranteed to be pinned/stored persistently on the public IPFS. There is still a remaining yet small chance of the uploading peer to disconnect from the network during initial upload. Then, the given file will not be pinned or uploaded. This financial incentive, on the other hand, renders the corresponding solutions more expensive to use since the public APIs all offer free plans that allow for a limited amount of data to be stored, e.g., 5GiB per day, or a decreased performance (e.g., latency, bandwidth, etc.) in comparison to a paid plan.

All presented alternative approaches to replace the central NFC-KE on-device public key storage lack the Smart-Contract-based verification functionality that enables a decentralized, fully autonomous signature verification. Since this feature is missing from all of the presented technologies, it is not part of the following performance comparison. It shall be noted that this is a major drawback in comparison to the previously proposed NFC-KE extension.

Nevertheless, current PKI setups also do not rely on a decentralized verification system, therefore legitimating this comparison of storage-only technologies.

5. Performance Comparison

To show the real-world benefits and problems of the previously proposed concepts, this section will contribute the following points:

- Uniform comparison basis for decentralized storage systems
- Real-world performance analyses of *IPFS <Private | Public>*, *Sia Skynet* and *HLF*.
- General and technology-specific implementation advice, based on performance results

When testing the private deployments, in this case, IPFS private and HLF, the setup can be individually created based on requirements or limitations, e.g., by the physical hardware. For this evaluation, the IPFS private setup is tested using two, four, and eight peers, respectively, and HLF in its setup with four orderers, two logical organizations, and two peers per organization, thus four peers in total. The latter configuration is chosen to establish a basis for comparison between the results of this analysis and the performance evaluation in [25].

To establish an experimental baseline, the same Signing Hub setup was chosen with the exact configuration as previously setup in [7,25], comprised of a Raspberry Pi 3B+ with an Infineon SLB9670 TPM 2.0 attached to it. Since the network in which the different Sensor Nodes and the Signing Hub operate is also a point of concern, this performance evaluation will consider a basic WiFi-based network using 802.11ac and an Commercial off-the-shelf (COTS) industrial access point. No special Quality of Service (QoS) presets or other performance-enhancing preferences were set or changed. All created peer instances were set up in a virtualized container environment using Docker. Finally, the Application Server was built and run using the following hardware specifications:

- **OS:** Arch Linux 64-Bit, Kernel Version 5.19.3-arch1-1
- **CPU:** AMD Ryzen 9 3900X, 12 Cores, 24 Threads, 3.80–4.60 GHz
- **RAM:** 31.3 GiB DDR4-3200, CL14-14-14-34
- **Network**: 1 GiB Ethernet Link

Since the NFC-KE can be separated into two distinct phases that each have different requirements in terms of filesystem access and availability, the performance analyses will also differentiate between *Phase 1* and *Phase 2*, where Phase 1 mainly consists of the authentic public key exchange (cf. Figure 2) and Phase 2 covers the following authentic message exchange between the Signing Hub and the Sensor Node (cf. Figure 3).

5.1. Authentic Key Exchange-Phase 1

First, every previously described DLT and, if applicable, their different setups are compared in terms of execution times. Using the Boxplot representation shown in Figure 4 of the collected data, the mean values and standard deviations of the datasets can be extracted directly.

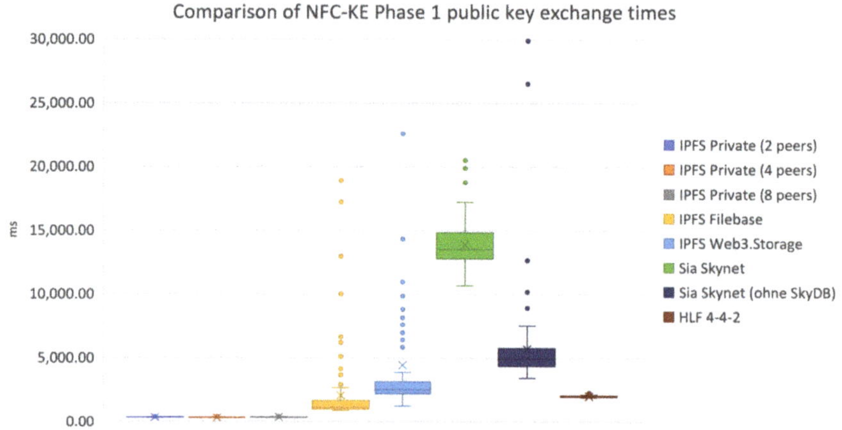

Figure 4. NFC-KE Phase 1 execution time comparison.

When compared with the other technologies, the *IPFS private* setup is executing the NFC-KE process the fastest. On average, the public key exchange happens within 286 to 334 ms, thereby not being influenced by the number of peers within the network. Additionally, the distribution of data points and the number of significant outliers is also lower when compared to the other DLT setups. IPFS Filebase is also executing the NFC-KE Phase 1 faster than the previously proposed HLF extension, although this setup experiences more outliers due to increased network jitter and API restrictions. On average, the authentic public key exchange has been completed within 992 and 1654 ms. The public IPFS setup based on Web3.Storage is significantly slower in comparison to the Filebase setup. With a mean completion time within 1921 and 2353 ms, the Web3.Storage setup is the slowest IPFS-based storage solution and also experiences the most outliers in direct comparison to the Filebase and private IPFS setups. The Sia Skynet setups share the last places among all tested setups in terms of execution times. Since the proposed setup uses the integrated SkyDB for public key query, two different API endpoints need to be contacted. To get a deeper understanding of the different execution times, we also included a setup without the SkyDB and thus without the public key query functionality. Using this setup, only one public key can be stored and queried which is only useful for testing. The results show that even without the SkyDB functionality, the Skynet-based approach is significantly more unstable and slower in comparison to all other DLT setups. With a mean execution interval of 4377 to 5746 ms, the Skynet solution does not offer any in-time storage solution. By adding the SkyDB backup functionality, and thus enabling the main NFC-KE Phase 1, the execution times range between 12.8 to 14.8 s. When compared to other setups, the outliers are less scattered in the latter setup case.

The stark performance degradation between the public and private setups is to be expected since the external APIs are highly reliant on the network performance. Furthermore, another main result of this particular comparison is the comparably fast execution times of the private IPFS solution in comparison to the HLF execution times. Even the public Filebase approach is, on average, faster than the proposed HLF NFC-KE extension, although being more prone to outliers in the execution times and thus less reliable.

5.2. Authentic Message Sending-Phase 2

The second phase of the NFC-KE is, in terms of real-world impact, the most important since the corresponding execution times will directly influence the responsiveness of the system. Generally, Phase 2 is querying the file storage system for the given Sensor Node's public key to verify the message signature, originating from said Sensor Node. Therefore, lower execution times will allow for faster message processing. A significant differentiation needs to be made between the two phases because Phase 1 must be executed only once for each Sensor Node, whereas Phase 2 is executed repeatedly afterward. Thus, the impact of particularly bad Phase 1 execution times is less significant than comparatively high execution times in Phase 2.

For the performance comparison in Phase 2, the number of simultaneously sending Sensor Nodes influences the general performance of the Signing Hub. Therefore, the following evaluation will, additionally, include performance numbers for one to six simultaneously sending Sensor Nodes that are virtualized and sending data within an interval of one to two seconds. The following Figure 5 shows the mean execution times of each DLT scenario and also differentiates between each Sensor Node configuration.

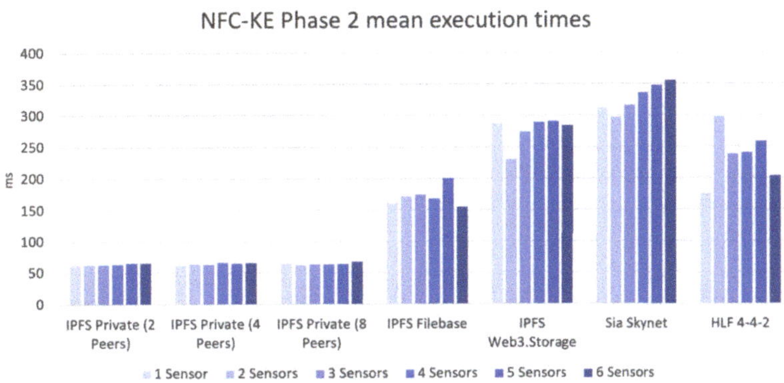

Figure 5. NFC-KE Phase 2 mean execution time comparison.

As can be seen, by the resulting mean execution times, the number of Sensor Nodes does, in this case, not influence the performance numbers in a significant way. In the case of the IPFS-based solutions, these are almost neglectable. Again, the private IPFS deployments offer the lowest execution times at almost 60 ms constantly. The Filebase setup is considerably faster than the HLF approach with a 150 ms mean execution time. The general pattern that already set itself apart in Section 5.1 does repeat itself in this case, since the Sia Skynet solutions also take the most time for Phase 2 execution with over 310 ms mean execution time. Note that for this comparison leaving out the SkyDB functionality like previously described in Phase 1 (cf. Section 5.1) does not apply here, since the message-sending part is inherently reliant on the public key query functionality. When comparing the IPFS-based solutions, only the Web3.Storage approach falls behind the Phase 2 execution times of the HLF-based approach.

To get a deeper understanding of the Sensor Node number's influences, Figure 6 shows the particular execution times of the IPFS private scenario using eight peers within the network.

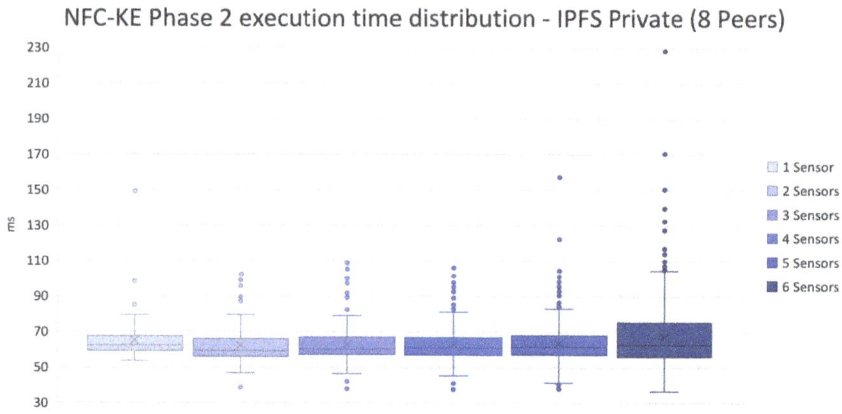

Figure 6. NFC-KE Phase 2 execution time distribution-IPFS private-8 Peers.

Here, the mean and median execution times remain almost constant among all tested Sensor Node numbers. The standard deviation and the number of outliers increase proportionally with increasing numbers of Sensor Nodes. Thus, the reliability of the system is slightly decreasing. This effect is only noticeable when visualizing the performance numbers for the private IPFS setup. Other DLT setups do not show the given behavior, due to other effects being more influential to the generally longer execution times such as latency, network bandwidth, and congestion.

6. Discussion

The results presented in Section 5 indicate that modern distributed file-storage systems, based on the DLT, are viable candidates for replacing centralized filesystem public key stores. Some technologies, especially those which can be set up in a private configuration, are capable of executing transactions significantly faster than, e.g., their public counterparts. Therefore, some specific technologies may not be suitable for every desired use case due to a lack of required performance.

In general, the performance numbers lead to the conclusion that the IPFS private setup is the most reliable and, in terms of performance, the most responsive solution for a high-performance system. Other implementation candidates on the IPFS basis have shown to be capable of being a proper alternative to the existing HLF NFC-KE extension because their performance numbers are either lower or equal to each other. Only the Sia Skynet solutions did not prove to be a valid replacement alternative for the HLF approach due to significantly higher execution times among all test scenarios. There are some additional points to the HLF performance results that need acknowledgement. First, the HLF implementaion requires a consensus among all participants to synchronize the underlying Blockchain state. Only after this process is complete, the stored data can be queried by external parties. Furthermore, HLF internally performs multiple cryptographical operations such as signing, verifying, and hashing transaction data, thereby increasing performance numbers. These points all add up to form the resulting performance of HLF.

When discussing the real-world application potential of the presented DLTs for the use within PKIs, the level of decentralization needs to be considered as well. The private variants of the presented DLTs are generally less decentralized and autonomous than the public variants. This is due to the restricted access to the given network since a private DLT instance is mostly created by one single administrator and then spread to known clients within the network, e.g., a local data center or multiple office computers. Therefore, decentralization is a point to consider since these network clients are, again, connected within the same network and may be located geographically close to each other. Nevertheless, a network administrator may be able to geographically decentralize private

IPFS peers, e.g., by using different data centers or renting a virtual server. This would, in return, require a higher maintenance effort and would not scale dynamically because of a single authority needing to manage all peers simultaneously. Due to their public access nature, the public DLTs offer global access to the underlying network and, thus, allow for increased redundancy and outage resilience.

There are also some points, that require attention when it comes to choosing a DLT for the PKI use case. DLT Platforms such as HLF or Ethereum [34] offer custom programmability and can be used as a decentralized, trusted execution environment that can enhance the trustworthiness of the given PKI. Each operation can be cryptographically signed and later verified by the DLT network. Enabling this distributed consensus renders the whole network more tamper-resistant and less prone to external intervention. In this particular case, the previously proposed NFC-KE extension using HLF used the integrated Smart Contract functionality to verify and store the exchanged public keys. This functionality had to be moved away from the DLT because the evaluated distributed file storages do not support a custom Smart Contract functionality, leaving the centralized Application Server the duty of verifying the received public keys. Therefore, the HLF approach offers increased security while not offering a true global decentralization or, in some cases, acceptable performance numbers.

Other security considerations arise when focusing on the internal structure of the private IPFS network in particular. Since the network peers are configured on a P2P basis and inform each other about changes in the internal network structure, any attacker who is able to infiltrate one IPFS peer is also capable of reconfiguring the entire private IPFS network. By using advanced concepts that are under current development, these issues can be remedied by employing different roles and responsibilities among the different IPFS peers. Therefore, further network and Operations Security (OpSec) measures need to be employed when setting up a private IPFS deployment.

In this article, only public APIs were considered for interacting with the IPFS or Sia network respectively. These technologies also offer the option to set up a custom peer instance which can connect to the given public network. For this, a distinct wallet containing cryptographic material needs to be stored and maintained, thereby adding new maintenance effort to the overall architecture. Though, by using such a configuration, latency and execution times are presumably lower since the connection to the custom peer, and thus to the whole P2P network, can be established without further routing. Additionally, by setting up one or more custom public IPFS peers, a pinning service would be required to take advantage of a public IPFS network. If such a pinning service is not used, no third-party IPFS peer will replicate the data due to lack of incentive.

Furthermore, the evaluation approach chosen in this article has been carried out in a controlled environment. In particular, the same hardware and physical network configurations were used for testing all aforementioned approaches. Additionally, the evaluation only focuses on examining the most significant KPIs that are relevant for a real-world application, in particular the execution times. It can be argued, that other indicators such as scalability, ease of use, availability or read/write latency times of the DLT are also of concern. Evaluating these KPIs was beyond the primary scope of this article.

7. Conclusions

This article compares four different distributed file storage technologies in different network configurations with each other and evaluates them in terms of performance and applicability for use within PKIs. The comparison is intended to assist developers and researchers alike to better decide on which technology is best suited for the intended use case. The most notable points to consider, when choosing a distributed file storage system are the content addressing paradigm as well as the lack of custom programmability.

The technical comparison of the storage system was conducted by comparing integrity, availability, persistency, and performance limitations of the several DLTs. The integrity of the data is warranted by either the CIDs or Skylinks. In this case, the private IPFS setup has shown to be the most promising candidate for the use, e.g., within modern Industry 4.0 applications because of its custom setup, its significantly better performance in comparison to the public setup or other technology and, generally, its improved resilience. Further means to increase the overall network security must be employed by the network administrators.

The availability of the public DLT variants has shown to be considerably higher than the private counterparts, given the assumption that a private DLT instance is hosted in a geographically small space, e.g., a data center. Though, in terms of persistency, public DLT solutions require an incentive for each peer within the given network to persistently store the desired files. Since data storage is a paid resource on the networks, deletion of files is possible. The private DLT counterparts do not require such an incentive and, therefore, attain a higher level of data persistency.

The performance analyses of the different DLTs also indicate that the private IPFS setup is, at least for the intended use within PKIs, the most suited candidate if high performance is mandatory. Since it lacks any custom programmability, its feature set is limited to only the decentralized data store which may reduce the trustworthiness of the overall system. This feature is offered by the already existing NFC-KE HLF extension, although the performance of this approach is significantly lower than the private IPFS approach.

Future research may focus on combining the decentralized Smart Contract functionality of, e.g., HLF and decentralized file storage such as IPFS and evaluating the resulting performance. Following this approach, a PKI would benefit from the distributed consensus and the decentralized, redundant file storage. Additionally, focusing on purely DAG-based DLT technologies such as *IOTA*[35] can also benefit the overall scope of evaluating new DLT-based PKIs by changing the fundamental data structure of the filesystem. Since this article focused on the lightweight operation and low maintenance requirements of the overall system, future research may also evaluate the performance of our setup using the aforementioned custom public IPFS and Sia peers respectively.

Author Contributions: Conceptualization, F.H. and J.D.; data curation, F.H.; formal analysis, J.D.; funding acquisition, R.T.; investigation, F.H.; methodology, F.H.; project administration, R.T. and J.D.; resources, F.H. and J.D.; software, F.H. and J.D.; validation, R.T. and J.D.; visualization, F.H. and J.D.; writing—original draft, F.H.; writing—review and editing, J.D. All authors have read and agreed to the published version of the manuscript.

Funding: This research received no external funding.

Data Availability Statement: All datasets can be found on Github via https://github.com/JulianD267/Tamper-Proof-Storage-Methods accessed on 27 October 2022.

Conflicts of Interest: The authors declare no conflict of interest.

Abbreviations

The following abbreviations are used in this manuscript:

API	Application Programming Interface
CA	Certification Authority
CID	Content identifier
COTS	Commercial off-the-shelf
DAG	Directed Acyclic Graph
DHT	distributed hash table
DLT	Distributed Ledger Technology
ECC	error correcting code
HLF	Hyperledger Fabric
HTTPS	Hypertext Transfer Protocol Secure
IIoT	Industrial Internet of Things

IoT	Internet of Things
IPFS	Interplanetary Filesystem
KPI	Key Performance Indicator
MitM	Man-in-the-Middle
NFC	Near Field Communication
NFC-KE	Near Field Communication Key Exchange
OpSec	Operations Security
P2P	Peer-To-Peer
PGP	Pretty Good Privacy
PKI	Public Key Infrastructure
PSK	Preshared Key
QoS	Quality of Service
RTT	Round Trip Time
SDK	Software Development Kit
SME	Small and Medium Enterprise
TPM	Trusted Platform Module

References

1. Kfoury, E.; Khoury, D. Distributed Public Key Infrastructure and PSK Exchange Based on Blockchain Technology. In Proceedings of the 2018 IEEE International Conference on Internet of Things (iThings) and IEEE Green Computing and Communications (GreenCom) and IEEE Cyber, Physical and Social Computing (CPSCom) and IEEE Smart Data (SmartData), Halifax, NS, Canada, 30 July–3 August 2018; pp. 1116–1120. [CrossRef]
2. Hepp, T.; Spaeh, F.; Schoenhals, A.; Ehret, P.; Gipp, B. Exploring Potentials and Challenges of Blockchain-based Public Key Infrastructures. In Proceedings of the IEEE INFOCOM 2019-IEEE Conference on Computer Communications Workshops (INFOCOM WKSHPS), Paris, France, 29 April–2 May 2019; pp. 847–852. [CrossRef]
3. Papageorgiou, A.; Mygiakis, A.; Loupos, K.; Krousarlis, T. DPKI: A Blockchain-Based Decentralized Public Key Infrastructure System. In Proceedings of the 2020 Global Internet of Things Summit (GIoTS), Dublin, Ireland, 3 June 2020; pp. 1–5. [CrossRef]
4. Schaerer, J.; Zumbrunn, S.; Braun, T. Veritaa-IoT: A Distributed Public Key Infrastructure for the Internet of Things. In Proceedings of the 2022 IFIP Networking Conference (IFIP Networking), Catania, Italy, 13–16 June 2022; pp. 1–9. [CrossRef]
5. Melo, W.; Machado, R.C.S.; Peters, D.; Moni, M. Public-Key Infrastructure for Smart Meters using Blockchains. In Proceedings of the 2020 IEEE International Workshop on Metrology for Industry 4.0 & IoT, Roma, Italy, 3–5 June 2020; pp. 429–434. [CrossRef]
6. Singla, A.; Bertino, E. Blockchain-Based PKI Solutions for IoT. In Proceedings of the 2018 IEEE 4th International Conference on Collaboration and Internet Computing (CIC), Philadelphia, PA, USA, 18–20 October 2018; pp. 9–15. [CrossRef]
7. Dreyer, J.; Fischer, M.; Tönjes, R. NFC Key Exchange—A light-weight approach to authentic Public Key Exchange for IoT devices. In Proceedings of the 2021 IEEE 7th World Forum on Internet of Things (WF-IoT), New Orleans, LA, USA, 14 June–31 July 2021; pp. 374–379. [CrossRef]
8. Ramesh, V.K.C.; Kim, Y.; Jo, J.Y. Secure IoT Data Management in a Private Ethereum Blockchain. In Proceedings of the 2020 IEEE 44th Annual Computers, Software, and Applications Conference (COMPSAC), Madrid, Spain, 13–17 July 2020; pp. 369–375. [CrossRef]
9. Uddin, M.N.; Hasnat, A.H.M.A.; Nasrin, S.; Alam, M.S.; Yousuf, M.A. Secure File Sharing System Using Blockchain, IPFS and PKI Technologies. In Proceedings of the 2021 5th International Conference on Electrical Information and Communication Technology (EICT), Khulna, Bangladesh, 17–19 December 2021; pp. 1–5. [CrossRef]
10. Aslam, F.; Javaid, N. Blockchain-based secure data sharing platform for research data rights management over the Ethereum network. In *This Work Is Submitted as Major Assignment for the Fulfilment of the Graduate Course, Research Methodology in Information Technology (RMIT)*; 2019. Available online: https://www.researchgate.net/publication/339989757_Blockchain-Based_Secure_Data_Storage_for_Distributed_Vehicular_Networks (accessed on 27 October 2022).
11. Hasan, H.R.; Salah, K. Proof of Delivery of Digital Assets Using Blockchain and Smart Contracts. *IEEE Access* **2018**, *6*, 65439–65448. [CrossRef]
12. Nizamuddin, N.; Hasan, H.; Salah, K.; Iqbal, R. Blockchain-based framework for protecting author royalty of digital assets. *Arab. J. Sci. Eng.* **2019**, *44*, 3849–3866. [CrossRef]
13. Park, J.S.; Youn, T.Y.; Kim, H.B.; Rhee, K.H.; Shin, S.U. Smart contract-based review system for an IoT data marketplace. *Sensors* **2018**, *18*, 3577. [CrossRef] [PubMed]
14. Protocol Labs, Inc. Interplanetary Filesystem. Technical Report. 2022. Available online: https://docs.ipfs.tech/ (accessed on 11 October 2022).
15. Protocol Labs, Inc. IOTA Wiki. Technical Report. 2022. Available online: https://wiki.iota.org/learn/about-iota/an-introduction-to-iota (accessed on 11 October 2022).
16. Protocol Labs, Inc. IPFS: Distributed Hash Tables (DHTs). Technical Report. 2022. Available online: https://docs.ipfs.tech/concepts/how-ipfs-works/#distributed-hash-tables-dhts (accessed on 11 October 2022).

17. Protocol Labs, Inc. IPFS: Directed Acyclic Graphs (DAGs). Technical Report. 2022. Available online: https://docs.ipfs.tech/concepts/how-ipfs-works/#directed-acyclic-graphs-dags (accessed on 11 October 2022).
18. Protocol Labs, Inc. Filecoin Docs. Technical Report. 2022. Available online: https://docs.filecoin.io/get-started/overview/ (accessed on 12 October 2022).
19. Filebase, Inc. What Is the Difference Between IPFS and Sia? Technical Report. 2022. Available online: https://docs.filebase.com/storage-networks/what-is-the-difference-between-ipfs-and-sia (accessed on 12 October 2022).
20. Linux Foundation. Channels. Technical Report. 2018. Available online: https://hyperledger-fabric.readthedocs.io/en/latest/channels.html (accessed on 12 October 2022).
21. Linux Foundation. Organization. Technical Report. 2021. Available online: https://hyperledger-fabric.readthedocs.io/en/latest/glossary.html#organization (accessed on 12 October 2022).
22. Linux Foundation. Peers. Technical Report. 2021. Available online: https://hyperledger-fabric.readthedocs.io/en/latest/peers/peers.html (accessed on 12 October 2022).
23. Linux Foundation. Smart Contracts and Chaincode. Technical Report. 2020. Available online: https://hyperledger-fabric.readthedocs.io/en/latest/smartcontract/smartcontract.html (accessed on 12 October 2022).
24. Linux Foundation. The Ordering Service. Technical Report. 2022. Available online: https://hyperledger-fabric.readthedocs.io/en/latest/orderer/ordering_service.html (accessed on 12 October 2022).
25. Dreyer, J.; Tonjes, R.; Aschenbruck, N. Decentralizing IoT Public- Key Storage using Distributed Ledger Technology. In Proceedings of the 2022 International Wireless Communications and Mobile Computing (IWCMC), Dubrovnik, Croatia, 30 May–3 June 2022; pp. 172–177. [CrossRef]
26. Dreyer, J.; Tönjes, R.; Aschenbruck, N. Towards securing Public-Key Storage using Hyperledger Fabric. In Proceedings of the 2022 IEEE International Conference on Blockchain and Cryptocurrency (ICBC), Shanghai, China, 2–5 May 2022; pp. 1–3. [CrossRef]
27. Filebase Team. API Documentation. Technical Report. 2022. Available online: https://docs.filebase.com/api-documentation (accessed on 13 October 2022).
28. Protocol Labs, Inc. API Documentation. Technical Report. 2022. Available online: https://github.com/ipfs/js-ipfs (accessed on 13 October 2022).
29. Skynet Devs. Skynet Developer Guide. Technical Report. 2021. Available online: https://docs.skynetlabs.com/ (accessed on 13 October 2022).
30. Filebase Team. Introducing Support for IPFS, Backed by Decentralized Storage. Technical Report. 2022. Available online: https://filebase.com/blog/introducing-support-for-ipfs-backed-by-decentralized-storage/ (accessed on 13 October 2022).
31. Protocol Labs, Inc. IPFS Public Gateway Checker. Technical Report. 2022. Available online: https://ipfs.github.io/public-gateway-checker/ (accessed on 13 October 2022).
32. Protocol Labs, Inc. Web3.Storage Docs. Technical Report. 2022. Available online: https://web3.storage/docs/ (accessed on 18 October 2022).
33. Vorick, D. SkyDB: A Mutable Database for the Decentralized Web. Technical Report. 2020. Available online: https://blog.sia.tech/skydb-a-mutable-database-for-the-decentralized-web-7170beeaa985 (accessed on 18 October 2022).
34. Wackerow, P. Ethereum Development Documentation. Technical Report. 2022. Available online: https://ethereum.org/en/developers/docs/ (accessed on 18 October 2022).
35. IOTA Foundation. The Complete Reference for IOTA. Technical Report. 2022. Available online: https://wiki.iota.org/#core-libraries (accessed on 18 October 2022).

future internet

Article

Users' Perceptions of Key Blockchain Features in Games

Iikka Paajala [1,*], Jesse Nyyssölä [1], Juho Mattila [1,2] and Pasi Karppinen [1]

[1] Faculty of Information Technology and Electrical Engineering, University of Oulu, Pentti Kaiteran katu 1, 90570 Oulu, Finland
[2] Ikune Labs, Loukkutie 11 A1, 90540 Oulu, Finland
* Correspondence: iikka.paajala@oulu.fi; Tel.: +35-84-4010-1802

Abstract: The blockchain is an emerging technology that has the potential to revolutionize the gaming industry among a wide range of different business fields. So far, only a few studies have been conducted about blockchain gaming. This study introduces a mobile game utilizing blockchain asset tokens and smart contracts. It was developed for research purposes and used to demonstrate blockchain-based games using semi-structured interviews. This study follows the exploratory research paradigm, which aims to map research of little-known areas. This study focuses on how participants perceived blockchain attributes such as trust, transparency, and user-generated content and how this affected engagement and their willingness to play the game again. Based on our evaluation, generating blockchain assets positively impacted player retention. According to the results, providing genuine asset ownership through the blockchain contributes to environmental engagement and self-engagement, as well as player retention. Another positive blockchain feature discovered from the interview data is user-generated content implementation into games.

Keywords: blockchain; games; non-fungible token; qualitative study

Citation: Paajala, I.; Nyyssölä, J.; Mattila, J.; Karppinen, P. Users' Perceptions of Key Blockchain Features in Games. *Future Internet* **2022**, *14*, 321. https://doi.org/10.3390/fi14110321

Academic Editors: Christoph Stach and Clémentine Gritti

Received: 10 October 2022
Accepted: 31 October 2022
Published: 4 November 2022

1. Introduction

Blockchain is an emerging technology that has the potential to revolutionize the gaming industry among a wide range of different business fields [1–3]. Information systems research about the blockchain has remained scarce, and the need for more academic studies has been notable in recent years [3,4].

Blockchain games have not yet been able to leave a mark in the gaming industry; nevertheless, the fusion of games and blockchain shows immense potential [5,6]. So far, only a very few studies have been conducted about the usefulness of the blockchain in gaming [3]. The most well-known blockchain game Cryptokitties has been studied at least by Min et al. [3] and Jiang & Liu [7]. A systematic literature review on blockchain-based applications by Casino et al. [8] and Ali et al. [1] did not address blockchain games at all. Only a few years ago, the majority of blockchain studies were about Bitcoin [4]. Nowadays, research has also been done in sectors such as finance, government, manufacturing, and health [1,8]. Blockchain is still a relatively novel technology, and research on other topics will gradually emerge [2], and the expectations of blockchain will increase the demand for blockchain applications [9].

The use of the blockchain has several benefits [10]. For example, trust and transparency are argued to be inherent blockchain features [11]. In the gaming context, transparent blockchain data enables the players or third-party stakeholders to audit the smart contract and related game rules, enhancing the game's trustworthiness [3]. With the use of blockchain token technology, users generating their own content for games can be granted ownership and compensation via smart contracts [10]. This can encourage players to participate in content creation in several ways [3,10].

Player retention is crucial to the success of a mobile game [12] since not all mobile gamers return to the game after trying it shortly after its installation [13]. Player engagement in game analytics aims to measure a user's behavior related to a particular game.

Bouvier et al. [14] described engagement in terms of attention, immersion, involvement, presence, and flow.

This study introduces a mobile game, IkuneRacers, utilizing blockchain asset tokens and smart contracts. The game was developed for research purposes at the University of Oulu. This IT artifact was used as demonstration material when conducting semi-structured interviews. Based on the data analysis, key themes and categories were recognized. The aim of this study is to evaluate trust, transparency, and user-generated content through artifact demonstration and interviews. Furthermore, we examine whether blockchain implementation can influence player retention and engagement.

The research questions are:

RQ1: How do players perceive transparency, trust, and user-generated content in blockchain games?

RQ2: Does blockchain game implementation influence users' perception of their engagement and player retention?

The structure of the paper is as follows. First, we introduce blockchain research, how blockchain assets are related to games, and studies on player retention and engagement. After that, we explain how our blockchain game (demonstration material) was designed, the research methods, and how the data was collected. The results are presented in Chapter 4. Chapter 5 includes a discussion and draws conclusions, and presents future research opportunities and limitations of the study.

The presented data were collected for a master's thesis by one of the authors.

2. Background

Thus far, the blockchain has been primarily studied in practitioner literature in computer science and cryptography studies, but theory and empirically driven IS research on the blockchain is gradually emerging [15]. Rossi et al. [15] assess existing research on blockchains and provide a framework for blockchain research in information systems (IS). In their framework, the authors [15] provide research agendas for several fields in IS, from which 'Agenda for behavioral IS research on the blockchain' is the most relevant for this study. Rossi et al. [15] call for more studies on the blockchain to be done either on the protocol level, application level, or in between. On an application level, one crucial issue is blockchain adoption [15]. Although several promising blockchain applications have been proposed, widespread adoption is rare [2]. Rossi et al. [15] believe privacy, security, and scalability are key factors, and research on these topics could make both academic and practical contributions [15].

2.1. Blockchain Assets and Games

A blockchain game can offer the potential of true ownership of crypto assets, as the asset ownership can be stored with a smart contract to the blockchain. Also, smart contracts enable the transfer of assets without third-party involvement [16]. Pillai, Biswas, and Muthukkumarasamy [17] define crypto assets as digital assets that are stored on the blockchain and utilize various techniques (e.g., cryptography, distributed consensus, peer-to-peer network, smart contract) to create, transact, and verify in a decentralized manner. They presented three categories of crypto assets: asset-tokens, crypto-coins, and utility-tokens. In their framework, the assets can be classified based on the fungibility and tangibility of the asset [17]. Table 1 illustrates the characteristics of different digital tokens, and Table 2 further explains them. In the context of the blockchain, a token is tangible if it represents something with a tangible existence, such as company shares or digital art. Tangible tokens can also represent ownership, e.g., in a physical world like property or a car [17]. If the asset is unique, it cannot be traded directly into similar tokens (aka a non-fungible token, NFT). An asset is fungible if it is exchangeable using a common standard, value, and characteristics like crypto coins (i.e., cryptocurrencies such as Bitcoin and Ether), which are fungible and intangible assets used as a means of exchange. Lastly, utility tokens are intangible and non-fungible and represent access to a product or service and also serve paying functions [17,18]. One example of utility tokens is the ERC-20 token standard used in the Ethereum network [18]. Asset-tokens can be fungible or non-fungible. Pillai, Biswas, and Muthukkumarasamy [17] use the CryptoKitties game as an example of tangible and non-fungible asset-tokens. It has similar principles to what was used when designing this game. In Cryptokitties, players "breed" digital cats 'CryptoKitties', which are non-fungible tokens (ERC-721 token standard), as each is unique [19].

Table 1. Types of digital assets [17].

	Tangible	**Intangible**
Non-Fungible	Asset-token	Utility-token
	ERC-21 token passport	Service
Fungible	Asset-token	Cryptocoins
	ERC-20-token Car	Bitcoin Ether

Table 2. Further explanation of crypto assets. Table modified from Pillai, Biswas and Muthukkumarasamy [17].

		Crypto Assets	
	Crypto Coins	**Asset Tokens**	**Utility Tokens**
Fungibility	Fungible	Fungible or non-fungible	Non-fungible
Tangibility	Intangible	Tangible	Intangible
Represents	Digital object, medium of exchange	Object with tangible characteristic	Digital object providing access rights
Example	Bitcoin, ETH	Car, property, NFT	Service, subscription

Min et al. [3] categorized blockchain-based games and divided them into four groups based on their utilized blockchain characteristics: rule transparency, asset ownership, (cross-platform) asset reusability, and user-created content. Curran [5] added to that list decentralized asset ownership and exchange, fast and secure payment networks, and the ability for developers to monetize their creations. The most popular blockchain-based benefits for games are rule transparency (i.e., gambling) and asset ownership (i.e., collecting in-game items) [5]. Standardized cross-platform asset reusability could benefit games utilizing asset ownership [5]. The ownership of an individual crypto asset can be validated using smart contract standards, which adds trust to the system and therefore gives the users confidence to invest in it [5].

2.2. Blockchain Trust, Transparency, and User-Generated Content

According to Queiroz and Wamba [11], both trust and transparency are built-in blockchain features. In fact, blockchain systems tend to be safer and more transparent than regular databases [8]. In the blockchain environment, transparency also covers asset trading [16]. As blockchain-based public ledgers are transparent, trust in the system increases [10,11]. In the gaming environment, this allows the players or third-party stakeholders to audit the smart contract and related game rules [3]. A further aspect of the blockchain, the reduced need for intermediaries, can also benefit the developers as they can build and distribute their products directly to the customer via smart contracts [5].

User-generated content can be used to create and update game elements, which can be crucial in retaining veteran players [3]. However, this topic can include built-in development opportunities for users, especially in blockchain game environments, and the created material can even be preserved and shared among multiple games. Furthermore, blockchain can store the ownership information of user-generated material. Blockchain and smart contracts have also been proposed as a method for paying for contributions to intellectual property [10]. These characteristics suggest that blockchain implementation can encourage players to participate in content creation in several ways [3,10]. As an example, in the game Last Trip, the developers provide story frames whereby the players contribute content to the game [3].

Although blockchain technology enables co-content creation by developers and non-developers alike, this study concentrates on asset ownership within the possibilities of blockchain technology. Asset ownership was implemented on the application on the first (and current) iteration.

2.3. Retention and Engagement

Predicting player retention is crucial to the success of a mobile game [12]. The reason for this is that a considerable number of mobile gamers do not return to the game after trying it shortly after its installation [13]. Player retention can be measured in numerous ways, one being the 'Retention profile,' which provides valuable information for the developers on how different user segments and cohorts play the game and how long they will keep playing [20]. Drachen et al. [12] analyzed variables that affect player retention in a free-to-play mobile game with a dataset of over 130,000 players. The key findings of the study were that the most positive predictors of retention included variables that tracked long and consistent participation. An interesting finding by Drachen et al. [12] was that good results in the early stages resulted in lower retention, which suggests that the players found the games too easy and lost interest. However, according to Weber et al. [21], games should not be too hard to play either. The blockchain-related issue of the gas fee has been proven to affect player retention [22], but this paper does not attempt to study that phenomenon further.

Bouvier et al. [14] described engagement with the terms: attention, immersion, involvement, presence, and flow. When a player's attention is towards game content, but the consciousness is in the real world, the state is engagement. When the player's consciousness shifts into the game world, the player feels a presence, and flow is an action-oriented aspect of presence [14]. In addition, gamers eventually abandon the game without proper motivation [23]. Motivation can be divided into intrinsic (playing the game itself) and extrinsic motivations (e.g., rewards and punishments) [24]. Concentrating on intrinsic reasons, motivation can be further divided into human needs, such as competence, autonomy, and relatedness [24]. These needs are positively linked to game enjoyment. The need for competence in games is met through challenges and goals. The need for autonomy is met if the player is given the power to make choices during the gameplay. Relatedness can be satisfied if games offer social or online interactions [24]. In this study, we are not measuring flow, although we do ask for interviewees' opinions on the matter.

3. Research Methods

To find answers to research questions involving such novel technology and topics as blockchain and asset tokens, we build an artifact to demonstrate a blockchain-based game and tokens to support the interviews. The interviewees got familiar with the game and the selected blockchain mechanics prior to the interviews.

This study follows the exploratory research paradigm (e.g., [25], which is described as mapping research of little-known areas of study. It aims to discover the nature of the problem and guide academics to understand it better. Exploratory research is flexible and can use both quantitative and qualitative methods. However, exploratory research does not aim to test a hypothesis based on existing theories and knowledge.

3.1. The Artifact

The artifact developed in this study is a blockchain-based mobile multiplayer game prototype. The game is a turn-based racing game where players tap their cars for them to move forward. The game demonstrates blockchain features, as the cars, including their attributes, such as name, color, and speed, are blockchain items. Additionally, the user accounts are managed via smart contracts. In the current version of the game, players receive a prize in the form of coins for participating in races, which can generate more cars.

Figure 1 illustrates the garage screen view of the designed game. Racer names are generated randomly and work as a seed for the blockchain to calculate attributes for the racer object. The column "Dice" shows the speed value of the car, which consists of three dice-related attributes, which are represented in the format "xDy + z." Here, x refers to the number of dice thrown, whereas D is short for dice, y represents the number of die sides, and z is a constant modifier added to each result. In addition to the dice values, the blockchain stores several other values, like the racer's level, its model and color, the number of wins, and the races it has participated in. The model and color of the racers are values stored in the blockchain, but apart from other attributes, they are represented visually, as shown in a table format on the garage page (Figure 1).

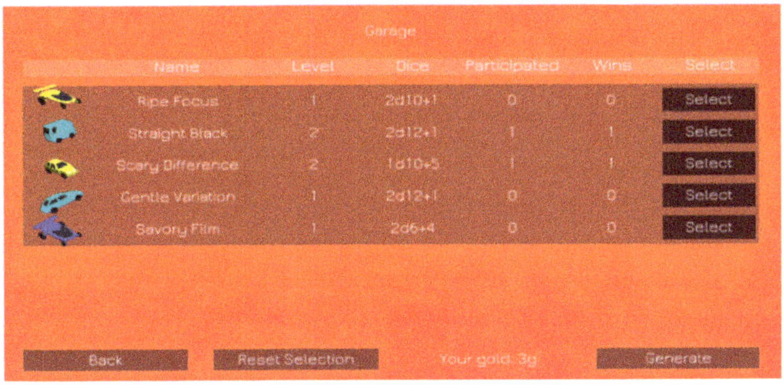

Figure 1. Garage view of the game.

For blockchain implementation, the platform used was Loom, which is meant for decentralized applications and includes a software development kit (SDK). For development, testing, and data gathering purposes, we used the testnet provided by Loom called Extdev instead of the Ethereum or the Loom mainnet. Extdev is free for development purposes and has the same functionality as the mainnet, including a web platform that enables examining the generated blocks. Both the Ethereum and Loom mainnets require tokens that have real-world value.

The smart contracts of the application were written in a Solidity language and built and deployed with Truffle. Smart contracts define the way the data is stored in the blockchain, and the functions used to manage the blockchain and applications built on it. Visual Studio Code (text editor) with a Solidity extension was used for development, while Visual Studio 2019 was chosen for Unity development. Loom's SDK enabled implementing queries to the blockchain as if they were built locally. This made it possible to retrieve relevant data from the blockchain.

As indicated by Rossi et al. [15], designing smart contracts for the artifact plays a key role in blockchain implementation. The smart contracts followed many of the principles by Tonelli et al. [26] to make them modular and service-oriented. The smart contract functionality in the game in this study was based on well-defined autonomous tasks and was, therefore, authority-free, which means that there were no privileges or exclusive rights written in the code of the blockchain interactions. To minimize the risks of smart contract-related logical errors and software bugs, it was decided to have only the racer and account data stored and managed on the blockchain while everything else is handled by the client. Well-defined and autonomous smart contract tasks also help to reduce complexity and therefore reduce the mentioned risks.

The account is identified based on the private key. When creating an account, the "GenerateKeys()" function is called. Based on this key, the client will use a hash function to calculate its address, which the blockchain will use to identify the user. "GetAccountName()" is called without parameters as the contract knows who is calling based on the address.

3.2. Demonstrations and Interviews

Ten game demonstrations, along with semi-structured interview sessions, were conducted either face-to-face or online due to the COVID-19 pandemic in the spring of 2020. During the online meetings, the participants could not try the artifact themselves but had to rely on a thorough gameplay demonstration through screen sharing. Two of the interviews were held in English, and the other eight were in Finnish. Some of the interviewees were recruited based on their participation in a blockchain course at the university, while others were recruited through personal connections.

All the participants were studying or graduated from the University of Oulu or Oulu University of Applied Sciences. Three of the interviewees were studying in or had a degree from a university of applied sciences, and the other seven were studying in or had a degree from a university.

Because the blockchain is a novel technology and could be unfamiliar to the general population, a demonstration of the artifact and explanation relevant to blockchain-related qualities took place prior to the interviews. This is supported by Schultze and Avital [27], as the purpose is not to affect interviewee responses but to hand out more perspectives for interviews. The artifact was demonstrated to 10 participants before being interviewed. The demonstration included a hands-on gameplay session (face-to-face) or a gameplay showcase (online), open discussion on the game and blockchain, and showcasing the generated blocks on the web platform. All the interviews were conducted, recorded, and coded by the second author.

After the game demonstration, the conducted interview consisted of questions focused on the topics presented in the Background section. These include how interview participants regard blockchain-related aspects (such as asset ownership, asset value, immutability, transparency, security, and trust) and how those are linked to video game characteristics like retention and engagement.

The interview question structure contained the following five topics:

1. Background questions
2. What makes assets valuable
3. Asset ownership and the lack of a trusted third party
4. Data on the blockchain is immutable and transparent
5. Security and trust of blockchain systems

The interviews were semi-structured and utilized laddering, and the questions depended on those elements the interviewees found personally meaningful. Interviews followed the method by Schultze and Avital [27] to collect relevant data for this study. For interview perspectives, Schultze and Avital [27] take up a combination of localistic and romantic views. The localistic view allows the interview to be a platform for activities, i.e., political action or impression management, which is affected by different contexts, such as the interviewee's age or gender. In a way, interviews are seen as windows into social reality [27]. In the romantic view, interviews are conversations in which the interviewer can participate and intervene when necessary [27]. Naturally, the interviewer is not fishing for preferred answers but handing out more perspectives [28]. Also, according to Holsten and Gubrium [28], interviews are not only for stating facts but also for constructing meanings.

Schultze and Avital [27] suggest three scientific interviewing methods that can be used for IS research. These are appreciative, laddering, and photo-diary interviewing. For this study, the laddering interviewing method was selected. In practice, questions such as "How do X and Y differ from Z?" are used to make distinctions between elements. On the other hand, questions like "Why is Z important?" aims to create a meaningful connection between elements [27]. Here, the interviews aimed to identify the meaningful connections between video games and blockchain features.

Because blockchain is a novel technology and could be unfamiliar to the general population, a demonstration of artifacts and explanations of relevant blockchain-related qualities took place prior to the interviews. This is supported by Schultze and Avital [27], as the purpose is not to affect interviewee responses but to hand out more perspectives for interviews.

After demonstrating the game, the conducted interview consisted of questions focused on the topics presented in the Background section. These include how interview participants regard blockchain-related aspects (such as asset ownership, asset value, immutability, transparency, security, and trust) and how those are linked to video game characteristics like retention and engagement.

The interview question structure contained the following five topics:

1. Background questions
2. What makes assets valuable
3. Asset ownership and the lack of trusted third party
4. Data on blockchain being immutable and transparent
5. Security and trust of blockchain systems

The interviews were semi-structured and utilized laddering, and the questions depended on those elements the interviewees found personally meaningful. Table 3 presents the interview method and language used in interviews. The interview questions are listed by the topic in the Appendix A.

Table 3. The interview method and language of the interviewee.

Interviewee Number	Interview Method	Interview Language
#1	Face-to-face meeting	English
#2	Face-to-face meeting	English
#3	Online meeting	Finnish
#4	Face-to-face meeting	Finnish
#5	Online meeting	Finnish
#6	Online meeting	Finnish
#7	Online meeting	Finnish
#8	Online meeting	Finnish
#9	Online meeting	Finnish
#10	Online meeting	Finnish

3.3. Data Analysis

The data analysis followed a thematic analysis suggested by Terry et al. [29], and it consisted of six steps. The first step is familiarization, wherein some insights are gained from the data. Here the entire dataset is gone through, notes are taken, and some early analytic ideas are formed. [29] In the second step, the data is reduced and organized into patterns. The data mass becomes an easily organizable set of labels, which are generated based on the relevancy of the research questions. This might indicate that some datasets have several labels and some none [29].

Theme development is a process of pattern identification, and that happens in the third step of the thematic analysis [29]. A central idea or concept supporting a set of codes is developed for a theme. According to Terry et al. [29], in this step, visual aids, such as tables and figures, are crucial to demonstrating different themes and relationships between them, as well as identifying new themes.

In the fourth step, the themes are reviewed so that they work with the labeled data, datasets, and research questions. The fifth step is the phase for defining the theme, whereby the researcher interprets and describes each theme and its representation. Also, "thin" themes are dropped out. The sixth and final step is reporting [29].

4. Results

The research results reported here focus on blockchain issues of transparency, trust, user-generated content, player retention, and engagement. Also, some gameplay (involving the blockchain) improvement suggestions were gathered from the interviews, including a leveling system wherein the players can improve and customize their garages and cars. The interviewees were also interested in generating blockchain objects and were hoping to see more content on that. In the following sections, the participants' (P1–P10) responses are divided into themes in accordance with the research questions. Some of the participants' quotations include comments in parentheses that represent the interviewer's interpretations. Interview topics and identified themes and subthemes are presented in Table 4. The interview data is from the master's thesis of Jesse Nyyssölä [30].

Table 4. Interview topics and corresponding themes with sub-themes.

Interview Topic	Themes	Sub-Themes
General questions	Personal relationship with video games	Personal motivation
		Emotions in video games
	Blockchain-related experience	Cryptocurrencies
		Blockchain games
	Blockchain-related experience	Suggestions for improvement
		Retention on artefact
		Engagement on artefact
What makes assets valuable	What gives value to an asset?	Assets with monetary value
		Assets with non-monetary value
	Giving all assets monetary value (i.e., making them sellable and purchasable)	Financial incentive
		Unfair playing field in video games
Asset ownership and the lack of a trusted third party	Authorities over data in a video game context	Authorities controlling/ exploiting
		High prerequisites for authority-free environment
	Exchange through a central system as opposed to person-to-person	Convenience of centralized system
		Social aspect of person-to-person exchange
	Actual ownership of virtual assets	
Data on the blockchain, immutable and transparent	The effects of transparency in video games and IS	Traceability and accountability
		Privacy concerns
	Third-party involvement in video games	Continued support
		Modding
Security and trust on the blockchain	Trust on the blockchain	

4.1. Retention, Engagement, and Gaming Experience

General questions included basic questions about the artifact and gaming and background information about the participant. For the question, 'Why do participants choose to play video games over other forms of media entertainment,' the major answer included mentions of active participation in games (nearly all participants) or social aspects of games (P3, P7, P10). Answers also included topics such as 'Autonomy over your character' (P4, P6, P8) and the 'Variety of environments and interacting with it,' while one interviewee focused on the role of thinking and decision-making as the following quotation reveals:

> *"Video games are great because you get to use your brain, create things yourself, decide on the course of the events yourself, how it is going, what you are allowed to do, what you want to do. In streaming services, you can choose what to watch."* (P8)

Playing video games can be relaxing and reduce stress by giving something else to think about (P8). Also, the feeling of immersion was mentioned (P3). Most participants noted that playing video games involves a wide range of emotions, including excitement, annoyance, nostalgia, and fearfulness:

> *"If you are looking for alleviating stress, it (the emotion) can be, for example, calmness. For example, in some role-playing games like Witcher, you can see yourself running in the virtual world and hear the wind in your ears, and some music can calm the situation at the same time. It is so versatile in terms of which emotions and senses touch you."* (P6)

The interviewees who stated their interest in video games knew little or nothing about blockchains. Some of them knew about Bitcoin or other cryptocurrencies, but the participants who were interested in the blockchain had more varied background knowledge on the topic. For games in general, they suggested a few blockchain-related characteristics, such as the private key (P5), currency conversion (P6), asset ownership (P9), and asset reusability (P3). Specifically for the artifact, seven participants suggested improvements, including more customization, visual aids, and more refined progression and competition systems.

> *"I like the idea of cross-platform or cross-game related things very much. [. . .] If we imagine that you have a Mario game and they would release four different games where you could get this cross-stuff from each game so that you could shuffle them around with this blockchain-like idea, I would probably buy them all. [29]"* (P3)

One participant (P2) stated that the game was not interesting, but the blockchain implementation seemed interesting from a technical point of view. Another participant (P3) also noted that the genre of the game is not relevant, but the uniqueness of the virtual assets provides an exciting setting. Another participant saw the game just as a technical demonstration but, at the same time, supported the core ideas of the blockchain functionality:

> *"I would play a game like that, but of course not with cars because I think cars are the most boring thing. However, if we were to say that something like Diablo would work that way, that all items are blockchain objects, it would be a really good thing. Having items in a game be reflected in the outer world or other currency, I do not see that it would make any game worse. [29]"* (P8)

Four participants stated that they might want to play again (P2, P4, P6, P10). This was largely attributed to the interest in generating new cars. However, two participants (P2, P10) said that they would only play it a few times because the game lacked meaningful gameplay content.

When asked about the perceived engagement of the artifact, the answers varied. One noted that the current game is not engaging, but the fundamental idea of a car race could be (P9). Two participants noted that the competitive spirit of a car race was an engaging factor (P4, P6). Two participants were keen on generating tokens (P5, P8). For one participant, the engagement was enabled by the possibility of progressing to better vehicles (P7). Re-playability (retention) and engagement by participants are shown in Table 5.

Table 5. Summary of the answers regarding re-playability and engagement.

	Play Again	Engaged
Yes	4	2
No	3	2
No answer	3	1
Partly	-	5

To ensure that the participants could assess the effects of the blockchain regarding assets, it was necessary to establish how the participants perceived the value of their assets in general. All the interviewees gave an example of a valuable item in a video game, and some could distinguish between an item with monetary value and a non-monetary value. One participant gave an example of both types of valuable items:

> *"In Player Unknown's Battlegrounds, I had things that could be converted by selling to Steam currency, and that could be used to buy any games sold on Steam. Additionally, in games like the World of Warcraft, if you have spent a couple of years trying to find a certain item that you have a chance to get only once a week with a drop rate lower than one percent, so even if it is not in any way significant or it does not have monetary value, it still feels good to achieve it."* (P5)

When asked if the instrumental value of the game item could be made exchangeable with the value of money, some respondents feared that (at its worst) one would be able to buy a competitive advantage. The topic was complicated since there are positive ideas of combining real money with games, like being able to sell your items after quitting the game (financial incentive), but also negative aspects like bots and possibly an unfair playing field. The most supportive comment was the following:

> *"It would not bother me even if we could trade genuinely valuable things. That would be fun. I can hear how my money goes down the drain. On the other hand, if everything was for sale, it could mean that using so-called 'honest methods,' it would be really difficult to get those items. That could be one negative drawback affecting people playing the game."*(P3)

One interviewee considered not spending money on video games to be a part of the challenge of the game:

> *"The worst thing you can do is that you buy with money the best item you can get. The game is over at that point. So, you are paying money so that you do not want to play the game anymore."* (P8)

Regarding the preference of trading person-to-person versus through a centralized system, most participants agreed that the option with the central system is better because it is often more convenient. However, there were also respondents (P6, P8) who advocated for person-to-person trading:

> *"Old RuneScape did not have Grand Exchange yet, . . . so people would gather in these trade worlds where thousands of people were in one place shouting what they wanted to sell. Some people wanted money in return, but some traded items for other items. It was a really functional and extremely pleasant social event where people were gathering as if they were at a marketplace."* (P6)

4.2. Transparency, Trust, and User-Generated Content

Transparency of video game data was one of our interview foci. This chapter presents findings regarding blockchain transparency, trust, and user-generated content. Transparency is related to the themes of traceability, accountability, and privacy concerns. One participant (P1) underlined the distinction between transparency and public data:

> *"I think transparency is important for everything. I have been contemplating public and private data for some time. While I believe that even if it is private, it should be transparent, not all private data can be public. Transparent means that you can see a log of the actions, and [so can] the public, [. . .] everybody can also see the data."* (P1)

According to one interviewee (P9), transparency can improve traceability and reduce application abuse. Five participants (P2–5, P6 & P10) pondered privacy issues.

> *"Surely it (transparent data in a system) is not better than private. If I happen to sell something, I do not want everybody to check that 'Oh, they sold that.' [. . .] If it does not enable everyone to see something that a regular guy would like to keep hidden, then it is not necessarily a bad thing."* (P10)

The interviewee's comments on trust in the blockchain were mixed. None of the participants could give a direct "yes" or "no" answer to the question, "From what you have experienced, would you say that you trust the blockchain?" Only three participants indicated at least some trust in the blockchain, but other comments were conditional, for example: "

If it is made in the right hands" or "depends on the implementation." Answers show suspicious attitudes towards novel technology participants are not yet familiar with. One participant clearly states reluctance to adopt blockchain in the first wave:

> *"I would say that with my understanding still, I would not (trust blockchain). [. . .] I have played games that are based on servers, and they have worked for me and everyone else, so I trust those. Maybe for me to trust the blockchain, it would require that it becomes*

a mainstream thing and is proven safe. [...] I am not the first guy who goes to try a new invention." (P10)

Closely related to immutability and transparency, one of the common features in blockchain applications is user-generated content. Since the game designed for this study did not incorporate user-generated content directly into the mechanics of the game, the interviews evaluated more external forms of user-generated content. Specifically, modifications and developing the game further after the original developers quit (i.e., preserving the game). All except one participant (P8) had positive reactions to these ideas.

"The law of supply and demand comes straight away and an old saying that you should not shoot a milk-producing cow. What I mean by that is that when the developers stop developing the game, it is because no one wants to fund it anymore, no one wants to buy it, and it does not have any players." (P8)

However, another interviewee (P6) had quite a different opinion on the matter:

"That (open third-party involvement) would have saved so many good game projects so far. For example, Age of Empires Online, some eight years ago, was a concrete example that there was a game that was playable but ran out of developers. There were players, sure, but no one wanted to update and continue developing the game." (P6)

According to one respondent (P4), the benefit of third-party modding is that mods often fix bugs that the developers have not been bothered to fix. Another participant (P6) states that mods enable playing older games with newer technology. From a business point of view, the relationship between the developer and the community can profit from modding (P5). One participant saw external modification as a failure of the developers:

"I am just a gamer who does not want to use my own thinking time to ponder these modding issues, and I just play what is fun. [...] It is the developers' responsibility to get the people to remain in the game and to direct us players to do what makes us stay with the game." (P8)

Although the questions regarding user-generated content within a game were not in focus, one participant (P3) was aware of its possibilities:

"Another thing that I find intriguing about blockchain is that if you get things through spending a lot of effort, you get rare things, and then you could sell those things for real money. I think that it is a really neat idea. Those guys who built Middle-Earth in Minecraft—they built it for 13 years—I would say that if they could now sell the Middle-Earth as a map, I think that is what they would do." (P3)

5. Discussion

This study examined whether the blockchain adds value to the gaming experience. This was done by creating a blockchain game and demonstrating it to interviewees. This chapter analyzes the results by discussing core findings in the context of relevant literature.

Transparency is one of the core benefits of blockchain-based systems [3,12] and the interviews gave useful feedback on system transparency issues. Many interviewees raised both privacy concerns and possible applications of transparent data. In addition, the traceability of transactions was mentioned by interviewees, which is also emphasized by Casino et al. [8]. We propose researching high-level abstractions, for example, experimental settings where transparency is represented as an independent variable.

Another interview topic was blockchain-related security. The underlying proposal was to determine whether the interviewees trusted blockchain applications. However, not a single participant gave a direct answer to that question. Queiroz and Wamba [11] state that blockchain transparency or trust does not significantly affect blockchain adoption. Based on the findings of this study, blockchain in gaming was not seen as a security threat, although there were some critical replies to transparency and trust issues.

Our findings of player retention are supported partly by previous studies. Drachen et al. [12] argue that games should not be too easy to play to increase retention. According to our findings, some participants stated that the game did not have enough meaningful content to increase their willingness to play it in the future.

Following Drachen et al.'s [12] argument, it would be safe to assume that implementing blockchain into games would not attract individuals to play the game again if the game itself was not perceived as engaging. However, participants who said they would like to try the game again said that the blockchain was the reason, and in more detail, they were interested in creating blockchain assets. Additionally, there were two interviewees who stated that they would want to play the game again, although the game was unengaging for them. This suggests that there is something interesting in the implementation of blockchain rather than in the game itself. The issue of solving player retention can be difficult, as the literature findings show reduced retention for both too-easy and too-difficult games [12,21].

Bouvier et al. [14] presented a model with four types of engagement: environmental engagement, social engagement, self-engagement, and action engagement. The model provided a possibility to analyze the engagement type of those interviewees who identified engaging aspects in the artifact. However, the model by Bouvier et al. [14] defines engagement types on a very conceptual level, while the participants in our interviews identified specific reasons for engagement. That is why we first divided the engagement types into three categories that closely correspond to interview replies: competitive engagement (P4, P6, P9), engaging generation (P5, P8), and engaging progression (P7). From the three categories, engaging generation is the easiest to identify as a blockchain characteristic because the blockchain provides practice for genuine asset ownership. When reflecting on our results of the model by Bouvier et al. [14], the types of engagement would be environmental engagement (curiosity, relaxation, exploration) and self-engagement (customizing and developing a story around the character). Although social engagement did not come up in the answers based on the demonstrated game, the interviews revealed that enabling social engagement (i.e., token trading) would be relevant for a blockchain-based application.

The financial aspects of crypto assets received little attention from the respondents, although blockchains offer room for emerging design solutions for handling monetary issues. Instead, there were varying opinions on what makes assets valuable in the first place (intrinsic versus instrumental values). Regarding exchange, two participants were adamant about the significance of person-to-person trading as opposed to centralized exchange systems. For them, trading is an essential factor in engagement, especially when considering social engagement.

According to our findings, participants were eager to customize assets as soon as they could. It seems that providing genuine asset ownership through blockchain contributes to environmental engagement and self-engagement. The demonstrated game cannot measure how blockchain affects social or action engagement. The social aspect was mentioned several times, which could be one important focus for blockchain game development, for example, enabling token trading.

User-generated content is often mentioned in related literature [3,10]. The interviews included two aspects: modifications and preserving the game. These aspects were primarily seen as a positive possibility. When implementing this feature into games, the developers should consider other related blockchain topics, such as asset valuation and transparency. Adding the possibility for user-generated content directly into the game (e.g., racetracks) seems like a viable choice, according to the interviews and participants' positive views on it.

All the aforementioned key observations and corresponding implications can be seen in Table 6.

Table 6. Key observations from interviews and their implications.

Observations	Implications
Blockchain technology was seen as a novel technology among gamers.	Using blockchain technology in games can provide a huge competitive advantage, since there are not many game companies using it as their games' core technology.
Transparency was perceived as a complex issue.	Even though transparency is one of the fundamental properties of blockchain technology, gamers might not automatically perceive it as a positive element.
Asset ownership was perceived as the most important blockchain feature in games.	Asset ownership can potentially increase player engagement and retention.
User-generated content was seen as a positive possibility.	Adding the possibility for user-generated content directly into the game seems like a viable choice.

From a technical standpoint, implementing token trading is straightforward. Since the tokens are already tied to their respective owners, only the function to transfer the tokens between accounts is needed. One way to achieve this is by implementing a token standard. ERC-721 is a commonly accepted standard for unique tokens or NFTs. The standard would also aid interoperability between systems, for example, in common crypto wallets such as MetaMask. For the user, this would emphasize the point that a crypto token is an asset of value that they have genuine ownership of.

A limitation of this study was that only 3 out of the 10 participants could try out the actual game. This was due to the COVID-19 pandemic and meeting restrictions caused by it. However, the artifact was demonstrated via screen sharing. Naturally, in the cases of actual hands-on experience, the participants seemed to be more engaged with the game. There are only a few examples of blockchain games designed for academic purposes [31–33]. Karapapas et al. created a proof-of-concept implementation of a decentralized and fair baseline system for a trading game [31]. However, they did not involve any end-user participants to evaluate their design [31], and neither did Alefs et al. for their designed blockchain gaming platform [32]. Yilmaz et al. investigated the use of non-fungible tokens as trading mechanisms in Virtual Reality Metaverse settings, with only three participants testing their system [33].

Our experiences encourage using qualitative research methods when the study is explorative and inductive. However, measuring player retention and engagement would benefit from more extended experiments and quantitative data gathering. Additionally, questions related to transparency and trust are somewhat challenging to address when emerging technology like the blockchain is being evaluated. The potential for future research on blockchain-based video games is vast. Some recommended research topics emerged from this study:

Design Science Research study on a system that enables user-created content with the blockchain;

The effects of the blockchain on retention based on quantitative game user data;

The effects of the blockchain on different engagement types based on quantitative data by users.

Also, designing an identical game without any blockchain and testing it with a control group could enhance the results of this study.

6. Conclusions

The objective of this study was to investigate the effects of blockchain implementation (NFTs and user-generated content), especially on player engagement and retention in games, and how players view blockchain features such as trust and transparency. A unique mobile game utilizing blockchain technology was purposefully designed and created as a university project to demonstrate and evaluate the effects of the blockchain in a game. Qualitative data was gathered via semi-structured interviews with either test play or game demonstration sessions. A total of 10 interviews were conducted, either face-to-face or through online meetings. According to the results, participants' interest in generating blockchain assets or interest in the general implementation of the blockchain seems to resonate with a person's willingness to try the game again. Few distinct reasons for engagement were found from the analysis, and from those, environmental engagement and self-engagement were most associated with blockchain asset generation.

In conclusion, this study provides initial suggestions on the effects of the blockchain on player retention and engagement. Some of the identified themes are not exclusive features of blockchains, but one of the goals of this study was to learn to look at the blockchain as a tool that can be applied to specific problems whenever they present themselves.

Author Contributions: Resources, J.M.; Software, J.N.; Supervision, P.K.; Writing—original draft, I.P., J.N., and P.K.; Writing—review & editing, J.M. All authors have read and agreed to the published version of the manuscript.

Funding: This research received no external funding.

Institutional Review Board Statement: Not applicable.

Informed Consent Statement: Informed consent was obtained from all subjects involved in the study.

Data Availability Statement: Not applicable.

Conflicts of Interest: The authors declare no conflict of interest.

Appendix A

Example interview questions by topic:

1. General questions
    - How would you describe your relationship with video games?
    - How would you differentiate video games from other media entertainment?
    - Why are those aspects important?
    - How do you feel when playing video games?
    - How do you feel about the blockchain? Do you feel the pressure to get involved?
    - What do you have to say about the blockchain game prototype?
    - Would you play the game again? Why/why not?
    - If the blockchain functionalities were implemented in an existing game you play, how would you feel about that?
    - Was the game engaging?
    - How did you feel about the responsiveness in the game?
2. What makes assets valuable
    - Have you ever considered any video game asset/item valuable?
    - How would you differentiate the value of money from that item?
    - Why is that important?
    - Do you think these could be unified?
3. Asset ownership and the lack of a trusted third party
    - In the game you played, how does it make you feel that there isn't a higher authority over the assets than yourself? Explain the feelings.
    - How does the lack of a middleman make you feel from a social point of view?
    - Have you ever considered who actually owns your virtual assets/items on traditional games or software platforms?
    - If so, did it concern you? Why/why not?
    - If so, how would you differentiate the concern from concern over losing something you physically own?
4. Data on blockchain, immutable and transparent
    - All of the transactions done in the game are public, and your racers can be identified using your address. How do you feel about that?
    - Do you think there's value in the possibility of multiplayer interactivity after developer support ends? Why?
    - How about 3rd party modding possibilities during development? Why?

5. Security and trust on the blockchain
    - How would you describe your security concerns regarding information systems in general?
    - From what you've experienced, would you say that you trust the blockchain? Why/why not?

References

1. Ali, O.; Jaradat, A.; Kulakli, A.; Abuhalimeh, A. A comparative study: Blockchain technology utilization benefits, challenges, and functionalities. *IEEE Access* **2021**, *9*, 12730–12749. [CrossRef]
2. Iansiti, M.; Lakhani, K.R. The truth about blockchain. *Harv. Bus. Rev.* **2017**, *95*, 119–127.
3. Min, T.; Wang, H.; Guo, Y.; Cai, W. Blockchain games: A survey. In Proceedings of the IEEE Conference on Games (CoG), London, UK, 20–23 August 2019; pp. 1–8. [CrossRef]
4. Yli-Huumo, J.; Ko, D.; Choi, S.; Park, S.; Smolander, K. Where is current research on blockchain technology?—A systematic review. *PLoS ONE* **2016**, *11*, e0163477. [CrossRef] [PubMed]
5. Curran, B. Blockchain Games: The Current State of Blockchain Gaming Technology. 2019. Available online: https://blockonomi.com/blockchain-games/ (accessed on 13 June 2022).
6. Agrawal, G. Examples of Blockchain Games and How They Work. 2019. Available online: https://medium.com/crowdbotics/examples-of-blockchain-games-and-how-they-work-7fb0a1e76e2e (accessed on 17 November 2021).
7. Jiang, X.J.; Liu, X.F. Cryptokitties transaction network analysis: The rise and fall of the first blockchain game mania. *Front. Phys.* **2021**, *9*, 57. [CrossRef]
8. Casino, F.; Dasaklis, T.K.; Patsakis, C. A systematic literature review of blockchain-based applications: Current status, classification, and open issues. *Telemat. Inform.* **2019**, *36*, 55–81. [CrossRef]
9. Hughes, L.; Dwivedi, Y.K.; Misra, S.K.; Rana, N.P.; Raghavan, V.; Akella, V. Blockchain research, practice, and policy: Applications, benefits, limitations, emerging research themes and research agenda. *Int. J. Inf. Manag.* **2019**, *49*, 114–129. [CrossRef]
10. Felin, T.; Lakhani, K. What problems will you solve with blockchain? *MIT Sloan Manag. Rev.* **2018**, *60*, 32–38.
11. Queiroz, M.M.; Wamba, S.F. Blockchain adoption challenges in supply chain: An empirical investigation of the main drivers in India and the USA. *Int. J. Inf. Manag.* **2019**, *46*, 70–82. [CrossRef]
12. Drachen, A.; Lundquist, E.T.; Kung, Y.; Rao, P.; Sifa, R.; Runge, J.; Klabjan, D. Rapid prediction of player retention in free-to-play mobile games. In Proceedings of the Twelfth Artificial Intelligence and Interactive Digital Entertainment Conference, Burlingame, CA, USA, 8–12 October 2016; pp. 23–29.
13. Runge, J.; Gao, P.; Garcin, F. Churn prediction for high-value players in casual social games. In Proceedings of the IEEE Conference on Computational Intelligence and Games, Dortmund, Germany, 26–29 August 2014.
14. Bouvier, P.; Lavoué, E.; Sehaba, K. Defining engagement and characterizing engaged behaviors in digital gaming. *Simul. Gaming* **2014**, *45*, 491–507. [CrossRef]
15. Rossi, M.; Mueller-Bloch, C.; Thatcher, J.B.; Beck, R. Blockchain research in information systems: Current trends and an inclusive future research agenda. *J. Assoc. Inf. Syst.* **2019**, *20*, 1388–1403. [CrossRef]
16. Gao, S.; Li, Y. An empirical study on the adoption of blockchain-based games from users' perspectives. *Electron. Libr.* **2021**, *39*, 596–614. [CrossRef]
17. Pillai, B.; Biswas, K.; Muthukkumarasamy, V. Blockchain interoperable digital objects. In Proceedings of the International Conference on Blockchain, San Diego, CA, USA, 25–30 June 2019; Springer: Cham, Switzerland, 2019; pp. 80–94.
18. Graves, S.; Hussey, M. What Are ERC-20 Tokens, Gas, ETH? *Ethereum's Architecture Explained.* 2022. Available online: https://decrypt.co/resources/what-are-erc-20-gas-ether-ethereum (accessed on 20 September 2022).
19. CryptoKitties White Pa-Purr. Version 2.0. 2017. Available online: https://drive.google.com/file/d/1sooeAaJHzhw_XhFGMJp3VNcQoM43byS/view (accessed on 20 September 2022).
20. Seufert, E.B. *Freemium Economics: Leveraging Analytics and User Segmentation to Drive Revenue*; Elsevier: Amsterdam, The Netherlands, 2014.
21. Weber, B.G.; John, M.; Mateas, M.; Jhala, A. Modeling player retention in madden NFL 11. In Proceedings of the Twenty-Third IAAI Conference, San Francisco, CA, USA, 9–11 August 2011.
22. Jiang, Y.; Fan, S.; Cai, W. Economic analysis of loot box market in blockchain games. In Proceedings of the Fourth ACM International Symposium on Blockchain and Secure Critical Infrastructure, Nagasaki, Japan, 30 May 2022; Association for Computing Machinery: New York, NY, USA; pp. 35–46.
23. Wang, H.; Chen, K.; Xu, D. A maturity model for blockchain adoption. *Financ. Innov.* **2016**, *2*, 12. [CrossRef]
24. Przybylski, A.K.; Rigby, C.S.; Ryan, R.M. A motivational model of video game engagement. *Rev. Gen. Psychol.* **2010**, *14*, 54–166. [CrossRef]
25. Robson, C. *Real World Research: A Resource for Social Scientists and Practitioner-Researchers*, 2nd ed.; Wiley-Blackwell: Hoboken, NJ, USA, 2002.

26. Tonelli, R.; Pinna, A.; Baralla, G.; Ibba, S. Ethereum smart contracts as blockchain-oriented microservices. In Proceedings of the 19th International Conference on Agile Software Development: Companion, Porto, Portugal, 21–25 May 2018; Association for Computing Machinery: New York, NY, USA, 2018.
27. Schultze, U.; Avital, M. Designing interviews to generate rich data for information systems research. *Inf. Organ.* **2011**, *21*, 1–16. [CrossRef]
28. Holstein, J.A.; Gubrium, J.F. *The Active Interview*; Sage: London, UK, 1995; Volume 37.
29. Terry, G.; Hayfield, N.; Clarke, V.; Braun, V. Thematic analysis. In *The SAGE Handbook of Qualitative Research in Psychology*; Willig, C., Rogers, W., Eds.; American Psychological Association: Newbury Park, CA, USA, 2017; pp. 17–36.
30. Nyyssölä, J. Assessing the Effects of Blockchains in Video Games: Case IkuneRacers. Master's Thesis, University of Oulu, Oulu, Finland, 2020.
31. Karapapas, C.; Syros, G.; Pittaras, I.; Polyzos, G.C. Decentralized NFT-based evolvable games. In Proceedings of the 4th Conference on Blockchain Research & Applications for Innovative Networks and Services (BRAINS), Paris, France, 27–30 September 2022; pp. 67–74.
32. Alefs, K.; Hartl, F.; Newman, L.; Ozdeveci, B.; Uriawan, W. Secure Decentralized Online Gaming with Lending Functionalities. *arXiv* **2022**, arXiv:2205.02348.
33. Yilmaz, M.; Hacaloğlu, T.; Clarke, P. Examining the use of non-fungible tokens (NFTs) as a trading mechanism for the metaverse. In *Systems, Software and Services Process Improvement, Proceedings of the 29th European Conference, EuroSPI 2022, Salzburg, Austria, August 31–2 September 2022*; Communications in Computer and Information Science Series; Yilmaz, M., Clarke, P., Messnarz, R., Wöran, B., Eds.; Springer: Cham, Switzerland, 2022; Volume 1646.

future internet

MDPI

Article

Abstracting Data in Distributed Ledger Systems for Higher Level Analytics and Visualizations

Leny Vinceslas [1,2], Safak Dogan [1,*], Srikumar Sundareshwar [3] and Ahmet M. Kondoz [1]

1 Institute for Digital Technologies, Loughborough University London, London E20 3BS, UK
2 Institute of Sound Recording, University of Surrey, Guildford GU2 7XH, UK
3 RegulAItion Ltd., Belmont Business Centre, Lewes BN8 6QL, UK
* Correspondence: s.dogan@lboro.ac.uk

Abstract: By design, distributed ledger technologies persist low-level data, which makes conducting complex business analysis of the recorded operations challenging. Existing blockchain visualization and analytics tools such as block explorers tend to rely on this low-level data and complex interfacing to provide an enriched level of analytics. The ability to derive richer analytics could be improved through the availability of a higher level abstraction of the data. This article proposes an abstraction layer architecture that enables the design of high-level analytics of distributed ledger systems and the decentralized applications that run on top. Based on the analysis of existing initiatives and identification of the relevant user requirements, this work aims to establish key insights and specifications to improve the auditability and intuitiveness of distributed ledger systems by leveraging the development of future user interfaces. To illustrate the benefits offered by the proposed abstraction layer architecture, a regulated sector use case is explored.

Keywords: distributed ledger technology (DLT); blockchain; block explorer; hyperledger fabric; abstraction layer; information visualization; analytics

Citation: Vinceslas, L.; Dogan, S.; Sundareshwar, S.; Kondoz, A.M. Abstracting Data in Distributed Ledger Systems for Higher Level Analytics and Visualizations. *Future Internet* **2023**, *15*, 33. https://doi.org/10.3390/fi15010033

Academic Editors: Christoph Stach, Clémentine Gritti and Paolo Bellavista

Received: 1 November 2022
Revised: 6 January 2023
Accepted: 9 January 2023
Published: 11 January 2023

1. Introduction

Distributed ledger technologies (DLTs) are becoming more widely used. They record operations between multiple parties in an immutable way. They are built on consensus-based decentralized systems, which address the trust issue between the involved parties. Numerous applications of distributed ledgers are currently being developed in various fields of the industry [1], such as agriculture [2], energy [3], finance [4], security [5], intellectual and digital property [6,7] or healthcare [8,9]. Figure 1a describes how a record of operations is maintained in a conventional centralized ledger. For example, a government or a bank may operate as a clearing house with complete control on the ledgers. In comparison, Figure 1b illustrates operation handling within a DLT framework, where each peer is maintaining its own ledger [10].

Motivated by the need to provide secure and decentralized services, DLTs keep track of very large amounts of ever-growing data [11]. By design, ledgers in DLTs persist low-level data that make conducting complex business analysis of the recorded operations challenging. Usually, the record of operations can be accessed by third party applications via querying the ledger [12,13]. This is achieved by employing a native set of application programming interfaces (APIs) where information about transactions, smart contracts or blocks can only be queried by their corresponding cryptographic hash. Although this access method allows for data lookup, such basic APIs are not adequate to devise high-level information from blockchains, often needed for analytics. Consequently, block explorers and similar visualization and analytics tools often only offer limited unintuitive information. These challenges call for an innovative approach for introducing a middleware between the presentation and data query layers that enables more accessible analytics and information visualizations. This can be achieved by employing an abstraction layer, which aggregates data from the ledger, pre-processes it and provides higher level APIs to block explorers and analytics dashboards that can then intuitively present information readily.

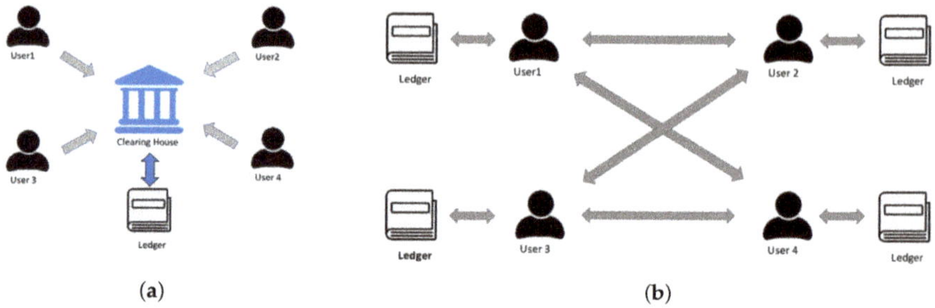

Figure 1. Transaction handling in different ledger architectures [10]. (**a**) Conventional centralized ledger. (**b**) Decentralized ledger.

In data analytics, visual representations are primarily designed to make sense of data and provide insights. They usually model data structures, which help to expand the boundary of individuals' cognitive system [14]. Not only do visual representations support users' reasoning, but also provide a construct to manipulate information. They can be used to both structure information and reduce individuals' cognitive burden by easing external anchoring, information foraging, and cognitive offloading [15].

In DLTs, audits are often conducted through lists and tables of low-level block data that do not easily allow for tracking and tracing digital assets [11]. The introduction of adequate visual representations of the ledgers' data has the capacity to enable a higher level of functionalities, and therefore improve the intuitiveness of the auditing processes.

A 2019 survey showed that time series representation accounted for 53% of all visualization techniques used for DLT representation, followed by basic charts such as bar charts, pie charts and histograms, which accounted for 41%, while tree and graph visualizations represented 38% [16]. The main reason behind this distribution can be explained by the time domain nature of the blockchain components, which contain time-stamped information. However, to represent data within more specific contexts, higher level types of visualizations were not used as frequently. For instance, map-based visualizations and other multi-dimensional representations accounted for only 13% and 7% of all visualization types, respectively.

To build intuitive visualizations or conduct an in-depth analysis of blockchain's recorded operations, data must be readily available in a format that can be consumed by the frontend visual representations. Although most DLTs provide access to the ledger via their software development kits (SDKs), they only offer low-level query interfaces that often lack semantic richness or functionality. For instance, information about transactions, contracts or blocks can usually be queried by their corresponding hashes. Such basic query interfaces are not adequate to map out high-level information or derive visualizations from blockchains' ledgers, and hence make it significantly challenging for users to deduce valuable insights that can serve complex business analysis quickly.

As a result, a few block explorers in the literature and marketplace implement high-level ledger visualizations or analysis functionalities. These limitations drastically reduce the possibility to adapt the abstraction of the displayed information to the targeted audience and application. The level of abstraction needs careful consideration when designing a visualization system to avoid situations where the users are presented with either insufficient or too much low-level data. In visualization and analytics, mismatched levels of granularity can result in both less accurate comprehension of a situation and higher time to complete a given task [17].

Curating information and presenting users with notable insights at the right level of granularity is often the responsibility of data analysts and designers. However, in the relatively new DLTs landscape, there are no such easily accessible solutions to implement the appropriate blockchain visualization tools to address a target audience's needs in a specific scenario [17].

For instance, firms of the regulated sectors willing to address compliance and trust issues might engage in decentralized digital assets trading by means of DLTs [18]. By design, DTLs encourage each participant to operate several peers on the network. For an effective and intuitive auditing process, visualization and analytics tools would need to represent the data at a participant-level, in an abstract and high-level manner. This requires aggregating the data from all the peers operated by a same participant. However, current off-the-shelf solutions typically provide information at a peer-level and are therefore limited to low-level representation of the ledgers' data [19].

The lack of high-level abstractions make the development of visual analytics time consuming for the developers since new functionalities need to be designed and implemented to analyze and aggregate the available low-level data. Additionally, delegating such functionalities to frontend applications can result in high resource-consuming services for the devices rendering such information. Therefore, there is a clear need for a standardized intermediate-level abstraction that provides higher level information from the low-level block data [19].

This article proposes an abstraction layer framework that facilitates better design of business analytics for DLT-based systems and the decentralized applications (DApps) that run on them. The purpose of this work is to establish a higher level of abstraction that improves the auditability and intuitiveness of distributed ledger records and enables further development of future user interfaces including analytics and visualization tools. Based on the analysis of existing initiatives and identification of the relevant user requirements, we infer specifications to improve the auditability and usability of block explorers. An abstraction layer has been designed to bridge the gap between a DLT and a user interface. As a result, the proposed abstraction layer coupled with a new user interface (UI) promotes the ease of analytics such as tracking and tracing of the history of operations, clustering of user addresses, and labeling of entities. Finally, to illustrate the benefits offered by the proposed abstraction layer architecture, we explore an industrial case study based on the RegNet platform [20]. RegNet is an infrastructure for trusted data access that removes the need to explicitly share data through the use of federated learning and tokenization. Its goal is to provide access to analytics in an auditable and privacy preserving manner, while being able to comply with data policies such as General Data Protection Regulation (GDPR) [21].

The remainder of this paper is organized as follows. We firstly review the related work by highlighting notable aspects in existing visualization tools, block explorers and abstraction layer implementations. Secondly, we propose an approach to build account and transaction-oriented abstractions while addressing auditability and intuitiveness issues. Thirdly, we provide an example of application in the RegNet case study. Finally, a discussion of the article is provided with an outlook on the future of the topic.

2. Related Work

In DLTs, the ledger is a global data structure collectively maintained by a set of mutually untrusting participants [22]. Changes to the ledger are organized into transactions which record the identifiers of their creators and beneficiaries. Transactions are hashed and grouped into blocks which are then chained together. Each block is appended via its header pointing to its predecessor. The synchronization of all peers on the state of the blockchain network is achieved using a consensus algorithm. This append-only ledger system provides DLTs with immutable records of transactions and therefore makes the blockchain tamper resistant. In addition to transactions, DLTs can implement smart contracts. Smart contracts are executable scripts that read or write to the ledger and are deployed across peers of the network.

2.1. Visualization Tools

Visualization tools refer to pieces of software developed to represent DLTs' data through infographics. In contrast to block explorer, they do not usually provide extended search capability. These DLT's data representations can be sorted into different task domains [16]. Tools focusing on analyzing patterns of individual blockchain components, i.e., transactions, addresses and blocks can be classified under the transaction detail analysis task domain. For example, Blockchain Explorer proposes to visualize weekly or monthly transaction volumes as a tile map [23]. Ethviewer shows the real-time transaction pool in Ethereum using a node-link diagram to represent blocks and transactions [24]. BitExTract is a collection of visual analytic tools that analyses activities among Bitcoin exchanges, including transactional volume, market share, and connectivity between exchanges [25].

Tools representing information through a network and flows perspective can be classified in the transaction network analysis task domain. This category of representation is usually based on tree or node-link diagrams showing the connectivity among blockchain components. For instance, Daily-Blockchain provides a real-time representation of Bitcoin transactions where the nodes of the network evolve over time [26]. Bitforce5 only shows the most recent transactions [27]. This ensures a constant performance or rendering over time. For more granularity, BlockchainVis can either display the total amount of Bitcoin transactions or a specific address selected by the user [28]. Blockchain.com provides a tree diagram in which users can click through the tree levels to follow the value flow with respect to addresses [29]. Instead of presenting the value flow as a tree structure, BitConeView provides a unique diagram showing the value flow of a seed transaction as it appears in blocks from top to bottom [30]. Unlike the previous static value flow visualizations, BitInfoCharts dynamically represents the flow of transactions over the entire history of a blockchain utilizing a node-link diagram arranged in a linear layout [31].

Tools representing aggregated statistics of the network can be categorized under the network activity analysis task domain. For instance, Blockchain.com provides a long list of time series charts to display a wide range of Bitcoin network statistics, such as the total hash rate, average block size, total transaction fee, and mining difficulty [29]. BitNodes implements map-based visualizations where a node crawler gathers reachable Bitcoin nodes locations to estimate the global distribution [32].

The solutions proposed in the literature as well as by the online tools mainly focus on low-level aspects such as block creation, transactions and simple currency exchange taking place in distributed ledgers. These types of representations provide highly technical insights on network status, which do not facilitate intuitive audits of ledgers and might not be accessible for non-expert users. Moreover, due to their very specialised range of analysis, these tools do not offer comprehensive analytics of the data.

2.2. Block Explorers

Block explorers are tools that allow users to browse through ledgers including blocks, account addresses and transaction data. Block explorers are mostly search tools; however, they recently tend to adopt dashboard oriented layouts and integrate network activity analysis elements. This approach provides them with a more comprehensive design and improves their intuitiveness.

A variety of block explorers have been developed to analyze the transaction details and audit network activity of different distributed ledger platforms [33–37]. These are often based on lists and tables of data which fail to provide DApp-specific information easily. For example, Figure 2a shows the main view of the Hyperledger Explorer (HE) dashboard [33,38]. It presents users with statistical insights regarding the Hyperledger Fabric (HF) network (number of blocks, transactions, nodes and smart contracts) grouped by organizations or averaged over time. It also displays the name of the peers operating on a specific channel and details about the last committed blocks. Figure 2b shows the block explorer view, which allows us to navigate into the network history and retrieve specific blocks. The transaction view displays the transaction history in a similar fashion than for the blocks. Figure 2c shows the transaction detail view which presents users with the details of a specific transaction when its ID number is selected.

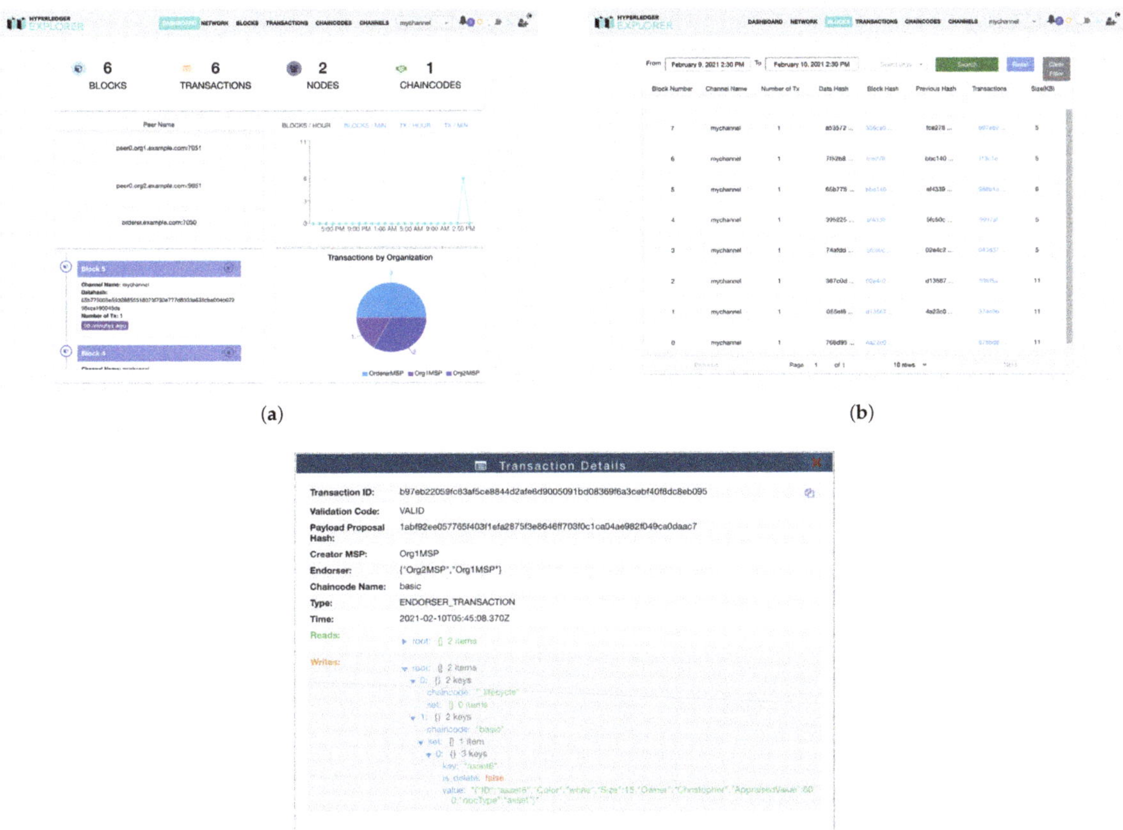

(a) (b) (c)

Figure 2. Screen-shots of Hyperledger Explorer [39]. (**a**) Dashboard main view; (**b**) block explorer view; and (**c**) transaction details pop-up window.

Alethio block explorer provides richer analytics [40]. In addition to the standard browsing history features, it maps interactions between accounts by tracing transactions and evoked smart contracts. These interactions are visually represented using simple node-link diagrams. It also allows us to keep track of account balances, search for information by account alias rather than by block and transaction hashes, and attach diverse social information to account addresses. Its functionalities, such as address tracking, tracing, labeling and data aggregation of DLT data improve auditability.

These block explorers appear to suffer from a lack of granularity in their presentation styles. They present users with very detailed information while failing at providing an overall context or general tendencies.

2.3. Abstraction Services

To overcome these limitations, both above mentioned block explorers implement an additional backend software sitting between the UI and ledger's low-level query interface.

This standardized middleware aims to abstract the complexity of user interactions with blockchains, and is responsible for querying, aggregating, and conditioning the ledger data, so that it can offer higher level analytics more easily.

Several platforms have advocated for the need for abstraction layers. For instance, Ledgerdata Refiner is a ledger data query platform developed for interfacing permissioned DLTs such as HF [19]. It is based on a data analysis middleware, which extracts and synchronizes the ledger data, and then parses the relationship among them. From the queried blocks and transactions, the middleware provides end users with tailored queries to access aggregated ledger information.

Datachain is another example, which is an interoperable framework that eases the extraction of data from different underlying blockchains [41]. It allows users to define specific high-level query abstractions, and perform data requests, extract transactions, manage data assets and derive high-level analytic insights automatically.

More recently, The Graph proposed a decentralized indexing protocol for querying data off Ethereum and InterPlanetary File System (IPFS). In this framework, queries are based on the standardized API language GraphQL [42]. As previously introduced middleware, The Graph is also extracting, synchronizing, parsing and conditioning data from DLTs before returning it in a format that can readily be consumed in applications. The specificity of The Graph is that thanks to its decentralized architecture, it eliminates the trust issues between middleware services and client applications.

The difference between the aforementioned solutions and our requirements is two folds. Firstly, these block explorers and visual tools mostly display raw data from the ledgers with minimal contextualization. They seem to mainly target data analysts. Secondly, it would be counter-productive to employ cloud services for processing data from privacy-preserving DLTs. For these reasons we aim to develop a self-contained middleware that would enable the design of analytics and visualization with higher level of contextualization and accessible to general users.

3. Proposed Approach

We propose two types of visual representations: (i) a transaction-oriented abstraction emphasizing on the time series of the transaction history while allowing tracing and tracking of assets, and (ii) an account-oriented abstraction focusing on interactions between entities of the audited DLTs and providing insights on inter-party behaviors.

The transaction-oriented abstraction uses a directed acyclic graph layout. As shown in Figure 3a, vertices represent transactions while directed edges illustrate the transaction flow between the source outputs and target inputs. This visualization shows the flow of transactions relative to a given transaction. It allows tracking and tracing of assets from their origins to end points across a predefined number of hops. For a low granularity level, only one-hop tracking is displayed with respect to Tx_3, which corresponds to the blue vertices in Figure 3a. For a higher granularity level, a two-hop tracking is represented with blue and grey vertices. Using an adaptive design that displays details on demand, this visualization gives access to a continuum of granularity. In addition to the vertices, the directed edges can be augmented with asset values and transaction timestamps. Informing about the smart contracts that triggered the represented transactions can also provide pertinent insights. The implementation of this transaction-oriented abstraction requires knowledge about the mapping between the transactions of interest. However, this information is not directly available in the ledger and must be obtained through data parsing, aggregation and analysis.

The account-oriented abstraction uses force-directed graphs where different granularity levels are implemented. As shown in Figure 3b, when a macro level is chosen, the square-shaped vertices represent addresses while directed edges are illustrating interactions between accounts. At a lower level, circular vertices denote clusters of accounts, forming entities linked by the directed edges. Entities and directed edges can be of variable sizes, representing the quantity of accounts by cluster and total value or amount of all inter-cluster interactions that occurred during a predefined time period, respectively. By nesting accounts within different cluster sizes, the visualization can efficiently adapt to the required level of detail. The implementation of the account-oriented abstraction requires knowledge on entities and their interactions, which needs clustering and labeling the ledger data.

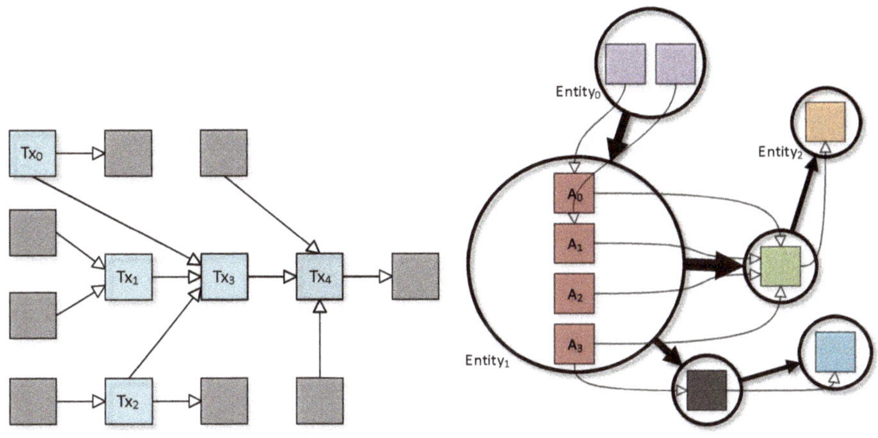

Figure 3. Abstracted visualizations. (**a**) Transaction-oriented abstraction; and (**b**) account-oriented abstraction.

To address the implementation of the two visual representations, we propose a general architecture for building an abstraction layer. The main purpose of this layer is to hide the complex interactions with ledgers. Through a simplified interface, users can connect to underlying ledgers to derive high-level analytic insights or perform high-level requests such as asset tracking and tracing. With this abstraction layer, we establish a framework where different services can be easily integrated to provide a transparent and richer query interface for business analytics. As depicted in Figure 4, it first extracts transaction, block, contract and channel details through blockchain ledger SDKs. At this stage, it is conceivable to query data from different DLTs. The ledger data is then parsed and aggregated to create comprehensive objects with common data structures, easy to manipulate in a given framework. The parsed data is then cached so that both current and historic data can be accessed by the pre-processing services. Pre-processing services aim at deriving high-level information from the low-level cached data. For instance, this is where information is mapped or filtered according to predefined heuristics. The last service of the abstraction layer provides interfaces to query the computed analytics.

- Parsing is the process of converting raw low-level data structures into higher level objects. Blockchain data structures are optimized for transaction validations and data retrieval across a distributed network and thus are not best suited for conducting analysis easily. For instance, to implement the proposed transaction-oriented abstraction, the parsing procedure must first collect the transactions from one or several blocks prior to mapping their inputs to previous transaction outputs. In addition, transactions must be assigned with IDs and timestamps along with the associated addresses of creators and beneficiaries to facilitate retrieval procedures [43] different parsing procedure;
- Aggregation refers to the collection and integration of data from multiple sources into a single storage destination. During this process, the different data sources required to infer higher level information are gathered and stored within a common data structure. For example, the proposed transaction-oriented abstraction needs to establish links between transaction inputs and outputs. To derive this mapping, transaction metadata

of different blocks is aggregated, and corresponding source and destination addresses are matched;

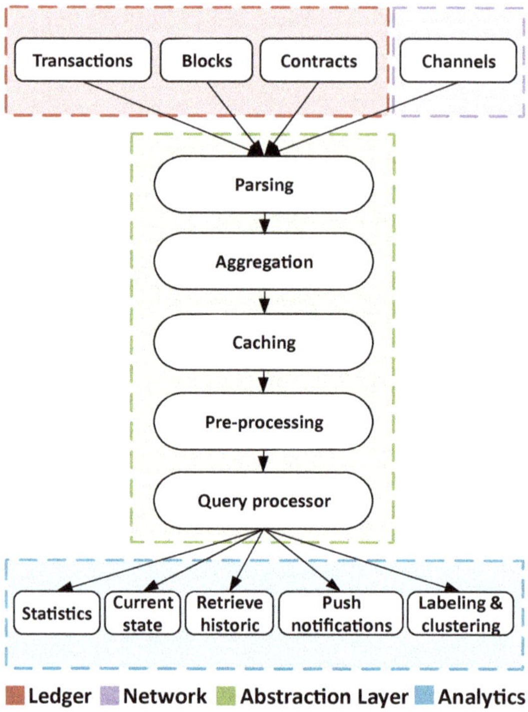

Figure 4. Abstraction layer architecture.

- Caching is the process of storing data resulting from previous computations so that future requests for that data can be executed faster. Both hardware and software used for caching depend on critical requirements such as the data volume, persistence time, access rate, throughput, and format. In the present scenario, the parsed and aggregated data can be cached in a server hard drive and RAM using a regular or graph database. The latter usually provides a better basis for analyzing relationships between entities [44];
- Pre-processing defines the operation of taking the cached data as input to generate the information requested by the query services. For example, this step is necessary to compute statistical insights on the network state, i.e., number of transactions per day. In addition to cached data, the pre-processing operation can also request data from third-party services. In the case of the proposed account-oriented abstraction, a pre-processing service will access the stored aggregated data to cluster addresses based on various possible heuristics [45]. Entities can then be inferred from the clustered accounts. Address clustering is particularly powerful when combined with labeling, i.e., labeling clusters with real-world entity designations [46]. On a small scale, labels can be determined by users through the query services. However, for large-scale labeling, automated scraping of open-source information or access to a third-party service provider is desirable;
- Query processor refers to the interfaces allowing third-party applications or users to query high-level data through a set of predefined instructions. Queries can initiate reading pieces of information collected or generated by the other abstraction layer services. Through a set of rich queries, this service aims to deliver requested data in a readily consumable format. To build the proposed visualizations, the query

services can be implemented using Representational State Transfer (REST) APIs and the JavaScript Object Notation (JSON) file format. The defined set of APIs will allow client applications to remotely execute pre-processing services to submit labels and clustering rules before querying the pre-processed data.

4. Use Case Scenario

4.1. The RegNet Platform

RegNet is a privacy-preserving data-access and data-collaboration platform for the regulated sectors, and addresses data-privacy challenges by combining DLTs, cryptography and machine learning [20]. Participants can request and provide access to each other's data while the resulting sharing agreements are stored in a distributed ledger. RegNet seeks to provide a trusted infrastructure to enable the exchange of data in a way where no sensitive information leaves the data-holders' firewalls. To ensure both security and relevance of exchanged information, RegNet uses privacy enhancing techniques on the data accessed between participants. In addition, RegNet also implements federated learning capabilities [47] that promotes a secure collaborative way to build larger data models together with semi-trusted participants.

RegNet's decentralized capabilities are provided by HF. HF is a permissioned distributed ledger platform targeting enterprise-grade business applications [48–50]. In addition to usual blockchain features, such as a decentralized ledger and tamper-proof data sharing, HF offers a more efficient consensus mechanism with higher throughput [51]. HF also addresses scalability and privacy issues by establishing the concept of channels. Channels allow the chosen data to be shared only among permissioned participants and thus provide a more adaptive data protection structure [52]. Areas of application include a digital vaccine passport [53], anti-counterfeit system [54], privacy-preserving in healthcare [55] and E-Voting System [56].

Since RegNet's decentralized capabilities are based on HF, HE appears as the default endpoint to access the ledger data and provide participants with monitoring and auditing features. However, the API implemented by HE is not entirely complying with the RESTful standard and does not allow the implementation of a suitable auditing layer for the data access applications that run on RegNet. In addition, a clear need for abstraction from HE's data arose. We therefore designed and implemented the discussed abstraction layer to more accessibly provide the interfacing and aggregating functionalities needed for analytics.

4.2. Architecture

The abstraction layer is designed and implemented as a middleware, which sits between the front-end dashboard application and the HE, as shown in Figure 5. The primary role of the abstraction layer is to hide the bad endpoints of the HE API from the dashboard API. The secondary purpose of this layer is to translate requests and responses so that the dashboard API can be compliant with the RESTful standards. This allows the developed applications to gain robustness for deployment while increasing their compatibility for future developments. In addition, this middleware provides the foundations for additional features such as persistence layers, data aggregation and processing, and authentication methods.

The abstraction layer was designed following a microservice architecture [57] using the Micronaut framework [58]. The microservice approach structures an application as a collection of smaller and consistent services. Under this type of architecture, microservices are separated autonomous components of an application, each accountable for a specific functionality and able to communicate with each other to form a coherent entity. The advantages of microservices oriented development are that it provides better maintainability in complex and large systems by enabling the deployment of many independent services, each of which may have a granular and autonomous life-cycle. An additional benefit is that microservices can scale out independently. Instead of having a single monolithic application that must be scaled as a unit, it can alternatively scale specific microservices to the consumer application need and to an extent to the demand on the network. In our network visualization application, each microservice is deployed using Docker container images.

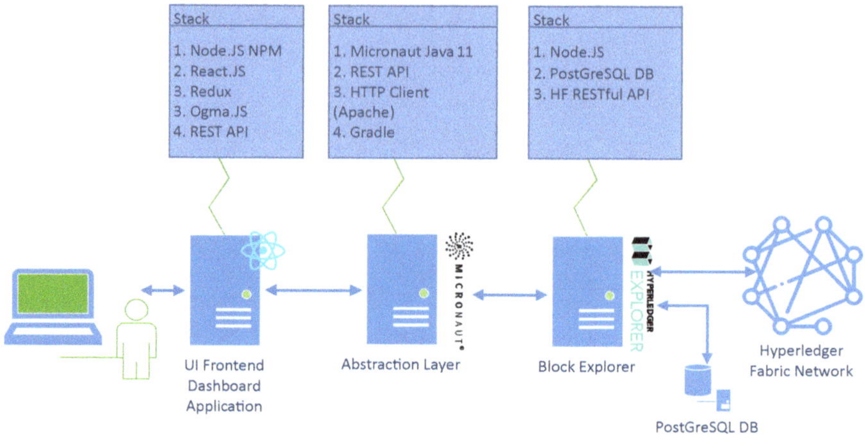

Figure 5. Application architecture and stack structures.

4.3. Consumer-Driven Contract Testing

When implementing microservice architectures, integration points between services can be a source of failure. Consumer-driven contract testing is an approach where the consumer of a service defines a contract and verifications are made against this contract within the provider's test life-cycle [59,60]. Contract testing is also practical to test microservices in isolation before deploying them in a live environment. Depending on the scope and perspective of the testing, there are a number of tools available that can be used to implement contract tests [61,62]. To test integration points between our microservices, we employed the tool Pact [63]. Pact is a set of open-source libraries and frameworks for automating contract-driven testing, and is specifically well-suited for internal provider- and consumer-focused testing [64].

4.4. Queries

In the proposed design, the abstraction layer receives requests from the front-end application. If the requested data are not available in the persistence layer, the middleware queries the HE and external sources for the appropriate data. It is noteworthy that a single query from the front-end often results in a collection of different requests emanating from the middleware. After parsing, aggregating, processing and caching the data, the middleware responds back to the dashboard application providing the requested data in a JSON format.

4.5. Network Visualization

Figure 6 shows the implementation of one such account-oriented dataset access based analytics utility onto the RegNet platform. The visualization is rendered using Ogma, a JavaScript library for interactive graph visualization [65]. The implemented account-oriented graph features nodes that represent organizations and edges that specify the relationships between nodes. Nodes are clustered accounts belonging to the same organization and thus sum up all account activities of a participant. The number of data models made available by each organization is reflected by the node sizes while the node color is indicating the channel on which an organization operates. In HF, channels are separated ledgers that enable the privacy and the scalability of the platform. The width of the directed edges illustrates the quantity of access permission granted between two organizations, which is also numerically displayed. Adaptive granularity is introduced by a tooltip providing a summary and further insights onto an organization when its node is double-clicked.

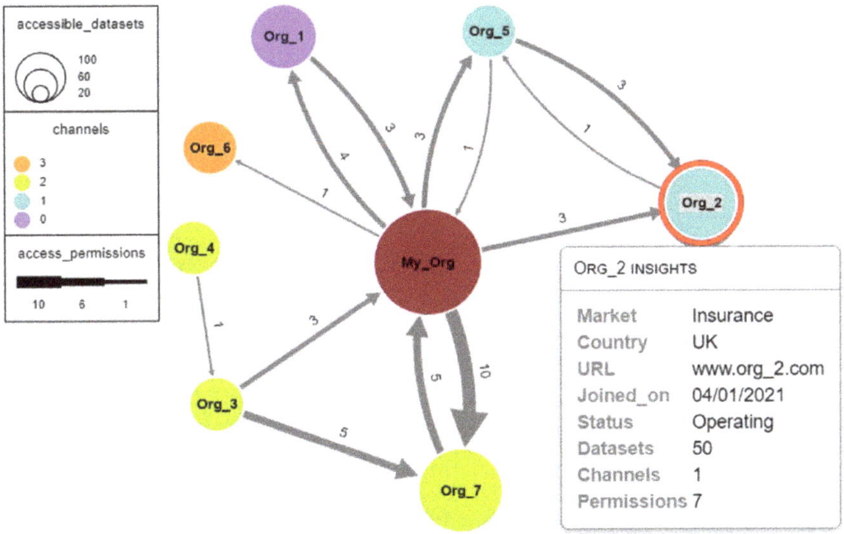

Figure 6. RegNet network visualization. The information used to populate the visualization is based on the data of the HF network deployed as a proof-of-concept in the RegNet case scenario.

4.6. Dashboard Concept

Figure 7 is a mock-up that proposes an integration of the network visualization in a dashboard framework. The design proposition is articulated around two main aspects: a news feed and a node-link diagram. The news feed presents users with the latest events that occurred in the network. New events appear at the top of the column, pushing the oldest ones toward the bottom in real time. Events can be filtered depending on their nature or grouped together for better visibility. The purpose of this presentation design is to highlight the time dimension of the network occurrences and enhance the perception of the dynamic aspect of the RegNet platform.

In this dashboard concept, users are able to interact back and forth with the news feed and the node-link diagram. For instance, when a specific event in the news feed is selected, the corresponding network components are highlighted. Moreover, a dynamic visual transition shows the direction or location of the corresponding action and its relation to other organizations or smart contracts. An additional window on the right side of the dashboard also displays the previous transactions between the highlighted entities and a plot of the number of transactions per day.

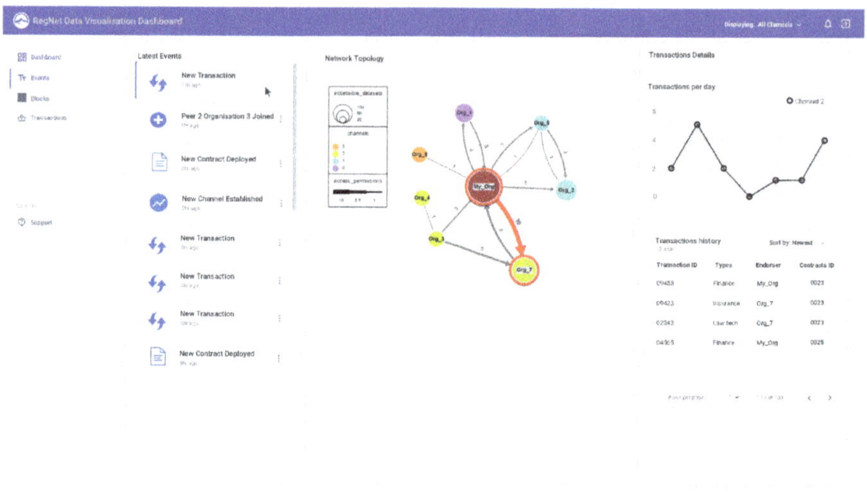

Figure 7. Concept design of the network topology view of the RegNet dashboard. From the left to the right column: page selection, news feed, network topology visualization and transaction insights. The analytics used to populate the dashboard are based on the data of the HF network deployed as a proof-of-concept in the RegNet case scenario.

5. Discussion and Outlook

In data analytics, visual representations are important since they can provide constructs that intuitively assist with inferring new information and can also reduce individuals' cognitive burden. The introduction of adequate visual representations for ledger data enables a higher level of analytics and therefore augments the intuitiveness of auditing processes. Nevertheless, when designing analytics solutions, the level of abstraction needs careful consideration; higher level data can facilitate richer and quicker analytics. To apply these concepts, we designed two distributed ledger visual representations, i.e., transaction-oriented abstraction highlighting transactions history and structure while allowing tracing and tracking of assets, and an account-oriented abstraction focusing on interactions between entities and providing insights on inter-participant behaviors. To enable the implementation of these high-level visual representations, we proposed an abstraction layer architecture. Its main purpose it to provide coherent interfacing and aggregating functionalities allowing the production of readily consumable data. A comparison between the proposed approach and available abstraction services for HF is provided in Table 1. To illustrate the proposed visual concepts and application architecture, a use case based on a dashboard for the regulated sectors has been explored. Employing higher level abstractions to represent HF data enables better comprehension of entities' interactions and improves the auditability of the system in comparison to HE.

As a result of their simplicity, the middleware and microservice architecture enable better maintainability and scalability of the system. The autonomous life-cycles of microservices allow them to be deployed at a relatively fast pace. However, due to their technical heterogeneity, a larger set of skills is required for their development and maintenance. In addition, to comply with the best practices, individual testing of the microservices needs to be performed. Both the necessary skill-set and the additional testing make this type of architecture less cost effective in the short term or at low scale compared to traditional monolithic applications. Yet, as a result of the high scalability and maintainability of the microservice architecture, this extra-cost can be recovered when developing and operating larger systems.

Ultimately, a universal higher level query language could be designed on top of this abstraction layer, which sits on blockchains. In turn, just like what Structured Query Language (SQL) is to Relational Database Management System (RDBMS), such language with its compositional, pragmatic and rich semantics would make business level querying much easier.

Table 1. Comparative table of available abstraction services for Hyperledger Fabric and the proposed approach.

Categories	Features	Proposed Approach	Hyperledger Explorer [38]	Ledgerdata Refiner [19]	Datachain [41]
Architecture	RESTful API	✓			
	Microservice based	✓			
Data management	Processed data persistence	✓		✓	✓
	Ledger parsing & aggregation	✓		✓	✓
	Aggregation of external data	✓			
Low level queries	Block browsing	✓	✓	✓	✓
	Transaction browsing	✓	✓	✓	✓
	Block & transaction search by ID	✓		✓	✓
High level queries	Statistics on transactions	✓	✓	✓	
	Tracking & tracing transactions	✓			
	Ledger operation chronology	✓		✓	
	Network change report	✓			
Enabled Visualizations	Transaction flow & volume	✓			
	Organization & node activity	✓			
	News feed	✓			

Future work will investigate the scalability and interoperability of such system. In addition, a user study could also be conducted to evaluate the usability and effectiveness of the proposed visualizations and dashboard concept.

Author Contributions: Conceptualization, L.V. and S.D.; methodology, L.V. and S.D.; software, L.V.; validation, L.V., S.D. and S.S.; investigation, L.V.; resources, S.D. and A.M.K.; writing, L.V.; review and editing, S.D.; visualization, L.V.; supervision, S.D. and S.S.; project administration, S.D.; funding acquisition, S.D. and A.M.K. All authors have read and agreed to the published version of the manuscript.

Funding: This research was funded by Innovate UK [application number 45079; project number 106159].

Data Availability Statement: Data sharing is not applicable to this article.

Conflicts of Interest: The authors declare no conflict of interest.

References

1. Casino, F.; Dasaklis, T.K.; Patsakis, C. A systematic literature review of blockchain-based applications: Current status, classification and open issues. *Telemat. Inform.* **2019**, *36*, 55–81. [CrossRef]
2. Feng, H.; Wang, X.; Duan, Y.; Zhang, J.; Zhang, X. Applying blockchain technology to improve agri-food traceability: A review of development methods, benefits and challenges. *J. Clean. Prod.* **2020**, *260*, 121031. [CrossRef]
3. Alladi, T.; Chamola, V.; Rodrigues, J.J.; Kozlov, S.A. Blockchain in smart grids: A review on different use cases. *Sensors* **2019**, *19*, 4862. [CrossRef]
4. Ali, O.; Ally, M.; Clutterbuck; Dwivedi, Y. The state of play of blockchain technology in the financial services sector: A systematic literature review. *Int. J. Inf. Manag.* **2020**, *54*, 102199. [CrossRef]
5. Kshetri, N. Blockchain's roles in strengthening cybersecurity and protecting privacy. *Telecommun. Policy* **2017**, *41*, 1027–1038. [CrossRef]
6. Wang, J.; Wang, S.; Guo, J.; Du, Y.; Cheng, S.; Li, X. A summary of research on blockchain in the field of intellectual property. *Procedia Comput. Sci.* **2019**, *147*, 191–197. [CrossRef]

7. Wang, Y.C.; Chen, C.L.; Deng, Y.Y. Authorization mechanism based on blockchain technology for protecting museum-digital property rights. *Appl. Sci.* **2021**, *11*, 1085. [CrossRef]
8. Agbo, C.C.; Mahmoud, Q.H.; Eklund, J.M. Blockchain technology in healthcare: A systematic review. *Healthcare* **2019**, *7*, 56 [CrossRef]
9. Ratta, P.; Kaur, A.; Sharma, S.; Shabaz, M.; Dhiman, G. Application of blockchain and internet of things in healthcare and medical sector: Applications, challenges, and future perspectives. *J. Food Qual.* **2021**, *2021*, 7608296. [CrossRef]
10. Kadam, S. Review of distributed ledgers: The technological advances behind cryptocurrency. In Proceedings of the International Conference Advances in Computer Technology and Management (ICACTM), Pune, India, 23–24 February 2018.
11. Dinh, T.T.A.; Liu, R.; Zhang, M.; Chen, G.; Ooi, B.C.; Wang, J. Untangling blockchain: A data processing view of blockchain systems. *IEEE Trans. Knowl. Data Eng.* **2018**, *30*, 1366–1385. [CrossRef]
12. Zhu, Y.; Zhang, Z.; Jin, C.; Zhou, A.; Qin, G.; Yang, Y. Towards rich Qery blockchain database. In Proceedings of the 29th ACM International Conference on Information & Knowledge Management, Virtual, 19–23 October 2020; pp. 3497–3500.
13. Przytarski, D.; Stach, C.; Gritti, C.; Mitschang, B. Query Processing in Blockchain Systems: Current State and Future Challenges. *Future Internet* **2021**, *14*, 1. [CrossRef]
14. Hegarty, M. Diagrams in the mind and in the world: Relations between internal and external visualizations. In Proceedings of the International Conference on Theory and Application of Diagrams, Cambridge, UK, 22–24 March 2004; pp. 1–13.
15. Liu, Z.; Stasko, J. Mental models, visual reasoning and interaction in information visualization: A top-down perspective. *IEEE Trans. Vis. Comput. Graph.* **2010**, *16*, 999–1008. [PubMed]
16. Tovanich, N.; Heulot, N.; Fekete, J.D.; Isenberg, P. Visualization of blockchain data: A systematic review. *IEEE Trans. Vis. Comput. Graph.* **2019**, *27*, 3135–3152. [CrossRef] [PubMed]
17. Oscar, N.; Mejía, S.; Metoyer, R.; Hooker, K. Towards personalized visualization: Information granularity, situation, and personality. In Proceedings of the 2017 Conference on Designing Interactive Systems, Edinburgh, UK, 10–14 June 2017; pp. 811–819.
18. Polyviou, A.; Velanas, P.; Soldatos, J. Blockchain technology: Financial sector applications beyond cryptocurrencies. *Multidiscip. Digit. Publ. Inst. Proc.* **2019**, *28*, 7.
19. Zhou, E.; Sun, H.; Pi, B.; Sun, J.; Yamashita, K.; Nomura, Y. Ledgerdata Refiner: A Powerful Ledger Data Query Platform for Hyperledger Fabric. In Proceedings of the 2019 Sixth International Conference on Internet of Things: Systems, Management and Security (IOTSMS), Granada, Spain, 22–25 October 2019; pp. 433–440.
20. Treleaven, P.; Sfeir-Tait, S. *Future Data-Driven Regulation*; Technical report; UCL: London, UK, 2020.
21. Pithadia, H.J. Algorithmic Regulation using AI and Blockchain Technology. Ph.D. Thesis, UCL (University College London), London, UK, 2021.
22. Rauchs, M.; Glidden, A.; Gordon, B.; Pieters, G.C.; Recanatini, M.; Rostand, F.; Vagneur, K.; Zhang, B.Z. *Distributed Ledger Technology Systems: A Conceptual Framework*; Cambridge Center for Alternative Finance, Judge Business School: Cambridge, UK, 2018.
23. Kuzuno, H.; Karam, C. Blockchain explorer: An analytical process and investigation environment for bitcoin. In Proceedings of the 2017 APWG Symposium on Electronic Crime Research (eCrime), Phoenix, AZ, USA, 25–27 April 2017; pp. 9–16.
24. Ethviewer. Available online: http://ethviewer.live (accessed on 14 December 2022).
25. Yue, X.; Shu, X.; Zhu, X.; Du, X.; Yu, Z.; Papadopoulos, D.; Liu, S. Bitextract: Interactive visualization for extracting bitcoin exchange intelligence. *IEEE Trans. Vis. Comput. Graph.* **2018**, *25*, 162–171. [CrossRef]
26. Daily blockchain. Available online: https://dailyblockchain.github.io/ (accessed on 14 December 2022).
27. Bitforce5. Available online: https://www.bitforce5.com/ (accessed on 14 December 2022).
28. Bistarelli, S.; Santini, F. Go with the-bitcoin-flow, with visual analytics. In Proceedings of the 12th International Conference on Availability, Reliability and Security, Reggio Calabria, Italy, 29 August–1 September 2017; pp. 1–6.
29. Blockchain.com. Available online: https://www.blockchain.com/explorer/ (accessed on 12 December 2022).
30. Di Battista, G.; Di Donato, V.; Patrignani, M.; Pizzonia, M.; Roselli, V.; Tamassia, R. Bitconeview: Visualization of flows in the bitcoin transaction graph. In Proceedings of the 2015 IEEE Symposium on Visualization for Cyber Security (VizSec), Chicago, IL, USA, 25–26 October 2015; pp. 1–8.
31. BitInfoCharts. Available online: https://bitinfocharts.com/bitcoin/explorer/ (accessed on 14 December 2022).
32. Bitnodes. Available online: https://bitnodes.io/ (accessed on 14 December 2022).
33. Hyperledger Explorer Github. Available online: https://github.com/hyperledger/blockchain-explorer (accessed on 30 November 2022).
34. Etherchain. Available online: https://etherchain.org (accessed on 30 November 2022).
35. Ethplorer. Available online: https://ethplorer.io (accessed on 30 November 2022).
36. Etherscan. Available online: https://etherscan.io/ (accessed on 30 November 2022).
37. Blockscout. Available online: https://blockscout.com (accessed on 30 November 2022).
38. Hyperledger Explorer Documentation. Available online: https://blockchain-explorer.readthedocs.io/en/main/ (accessed on 14 December 2022).
39. Thinkit.co.jp. Available online: https://thinkit.co.jp/article/18190 (accessed on 13 December 2022).
40. Alethio. Available online: https://explorer.aleth.io (accessed on 30 November 2022).

41. Trihinas, D. Interoperable Data Extraction and Analytics Queries over Blockchains. In *Transactions on Large-Scale Data-and Knowledge-Centered Systems XLV*; Springer: Berlin/Heidelberg, Germany, 2020; pp. 1–26.
42. Tal, Y.; Jannis, P.; Brandon, R. The Graph. Available online: https://thegraph.com/ (accessed on 30 November 2022).
43. Kalodner, H.; Möser, M.; Lee, K.; Goldfeder, S.; Plattner, M.; Chator, A.; Narayanan, A. Blocksci: Design and applications of a blockchain analysis platform. In Proceedings of the 29th {USENIX} Security Symposium ({USENIX} Security 20), online, 12–14 August 2020; pp. 2721–2738.
44. Tsoulias, K.; Palaiokrassas, G.; Fragkos, G.; Litke, A.; Varvarigou, T.A. A Graph Model Based Blockchain Implementation for Increasing Performance and Security in Decentralized Ledger Systems. *IEEE Access* **2020**, *8*, 130952–130965. [CrossRef]
45. Fröwis, M.; Gottschalk, T.; Haslhofer, B.; Rückert, C.; Pesch, P. Safeguarding the evidential value of forensic cryptocurrency investigations. *Forensic Sci. Int. Digit. Investig.* **2020**, *33*, 200902. [CrossRef]
46. Harrigan, M.; Fretter, C. The unreasonable effectiveness of address clustering. In Proceedings of the 2016 Intl UIC/ATC/ScalCom/CBDCom/IoP/SmartWorld. IEEE, Toulouse, France, 18–21 July 2016; pp. 368–373.
47. Yang, Q.; Liu, Y.; Chen, T.; Tong, Y. Federated machine learning: Concept and applications. *ACM Trans. Intell. Syst. Technol. (TIST)* **2019**, *10*, 1–19. [CrossRef]
48. Androulaki, E.; Barger, A.; Bortnikov, V.; Cachin, C.; Christidis, K.; De Caro, A.; Enyeart, D.; Ferris, C.; Laventman, G.; Manevich, Y.; et al. Hyperledger fabric: A distributed operating system for permissioned blockchains. In Proceedings of the Thirteenth EuroSys Conference, Porto, Portugal, 23–26 April 2018; pp. 1–15.
49. Hyperledger Fabric Github. Available online: https://github.com/hyperledger/fabric (accessed on 30 November 2022).
50. Hyperledger Fabric Documentation. Available online: https://hyperledger-fabric.readthedocs.io/en/latest/whatis.html (accessed on 14 December 2022).
51. Nasir, Q.; Qasse, I.A.; Abu Talib, M.; Nassif, A.B. Performance analysis of hyperledger fabric platforms. *Secur. Commun. Netw.* **2018**, *2018*. [CrossRef]
52. Valenta, M.; Sandner, P. Comparison of ethereum, hyperledger fabric and corda. *Frankf. Sch. Blockchain Cent.* **2017**, *8*, 1–8.
53. Shih, D.H.; Shih, P.L.; Wu, T.W.; Liang, S.H.; Shih, M.H. An International Federal Hyperledger Fabric Verification Framework for Digital COVID-19 Vaccine Passport. *Healthcare* **2022**, *10*, 1950. [CrossRef] [PubMed]
54. Chen, C.L.; Shang, X.; Tsaur, W.J.; Weng, W.; Deng, Y.Y.; Wu, C.M.; Cui, J. An Anti-Counterfeit and Traceable Management System for Brand Clothing with Hyperledger Fabric Framework. *Symmetry* **2021**, *13*, 2048. [CrossRef]
55. Stamatellis, C.; Papadopoulos, P.; Pitropakis, N.; Katsikas, S.; Buchanan, W.J. A privacy-preserving healthcare framework using hyperledger fabric. *Sensors* **2020**, *20*, 6587. [CrossRef] [PubMed]
56. Díaz-Santiso, J.; Fraga-Lamas, P. E-Voting System Using Hyperledger Fabric Blockchain and Smart Contracts. *Eng. Proc.* **2021**, *7*, 11.
57. Microservices. Available online: https://microservices.io (accessed on 30 November 2022).
58. Micronaut. Available online: https://micronaut.io/ (accessed on 30 November 2022).
59. Meyer, B. Applying'design by contract'. *Computer* **1992**, *25*, 40–51. [CrossRef]
60. Lehvä, J.; Mäkitalo, N.; Mikkonen, T. Consumer-driven contract tests for microservices: A case study. In Proceedings of the International Conference on Product-Focused Software Process Improvement, Barcelona, Spain, 27–29 November 2019; pp. 497–512.
61. Sotomayor, J.P.; Allala, S.C.; Alt, P.; Phillips, J.; King, T.M.; Clarke, P.J. Comparison of runtime testing tools for microservices. In Proceedings of the 2019 IEEE 43rd Annual Computer Software and Applications Conference (COMPSAC), Milwaukee, WI, USA, 15–19 July 2019; Volume 2, pp. 356–361.
62. Sotomayor, J.P.; Allala, S.C.; Santiago, D.; King, T.M.; Clarke, P.J. Comparison of open-source runtime testing tools for microservices. *Softw. Qual. J.* **2022**, 1–33. [CrossRef]
63. Pact. Available online: https://docs.pact.io/ (accessed on 12 December 2022).
64. Ma, S.P.; Fan, C.Y.; Chuang, Y.; Lee, W.T.; Lee, S.J.; Hsueh, N.L. Using service dependency graph to analyze and test microservices. In Proceedings of the 2018 IEEE 42nd Annual Computer Software and Applications Conference (COMPSAC), Tokyo, Japan, 23–27 July 2018; Volume 2, pp. 81–86.
65. Ogma-Linkurious. Available online: https://ogma.linkurious.com/overview (accessed on 30 November 2022).

future internet

Article

SAUSA: Securing Access, Usage, and Storage of 3D Point CloudData by a Blockchain-Based Authentication Network

Ronghua Xu [1], Yu Chen [1,*], Genshe Chen [2] and Erik Blasch [3]

[1] Department of Electrical and Computer Engineering, Binghamton University, Binghamton, NY 13902, USA
[2] Intelligent Fusion Tech, Inc., Germantown, MD 20876, USA
[3] The U.S. Air Force Research Laboratory, Arlington, VA 22203, USA
* Correspondence: ychen@binghamton.edu

Abstract: The rapid development of three-dimensional (3D) acquisition technology based on 3D sensors provides a large volume of data, which are often represented in the form of point clouds. Point cloud representation can preserve the original geometric information along with associated attributes in a 3D space. Therefore, it has been widely adopted in many scene-understanding-related applications such as virtual reality (VR) and autonomous driving. However, the massive amount of point cloud data aggregated from distributed 3D sensors also poses challenges for secure data collection, management, storage, and sharing. Thanks to the characteristics of decentralization and security, Blockchain has great potential to improve point cloud services and enhance security and privacy preservation. Inspired by the rationales behind the software-defined network (SDN) technology, this paper envisions SAUSA, a Blockchain-based authentication network that is capable of recording, tracking, and auditing the access, usage, and storage of 3D point cloud datasets in their life-cycle in a decentralized manner. SAUSA adopts an SDN-inspired point cloud service architecture, which allows for efficient data processing and delivery to satisfy diverse quality-of-service (QoS) requirements. A Blockchain-based authentication framework is proposed to ensure security and privacy preservation in point cloud data acquisition, storage, and analytics. Leveraging smart contracts for digitizing access control policies and point cloud data on the Blockchain, data owners have full control of their 3D sensors and point clouds. In addition, anyone can verify the authenticity and integrity of point clouds in use without relying on a third party. Moreover, SAUSA integrates a decentralized storage platform to store encrypted point clouds while recording references of raw data on the distributed ledger. Such a hybrid on-chain and off-chain storage strategy not only improves robustness and availability, but also ensures privacy preservation for sensitive information in point cloud applications. A proof-of-concept prototype is implemented and tested on a physical network. The experimental evaluation validates the feasibility and effectiveness of the proposed SAUSA solution.

Keywords: Blockchain; smart contract; point cloud; security; privacy preservation; software-defined network (SDN); big data; assurance; resilience

Citation: Xu, R.; Chen, Y.; Chen, G.; Blasch, E. SAUSA: Securing Access, Usage, and Storage of 3D Point CloudData by a Blockchain-Based Authentication Network. *Future Internet* **2022**, *14*, 354. https://doi.org/10.3390/fi14120354

Academic Editors: Christoph Stach and Clémentine Gritti

Received: 1 November 2022
Accepted: 25 November 2022
Published: 28 November 2022

1. Introduction

With the rapid development of three-dimensional (3D) acquisition technologies, 3D sensors are increasingly available and affordable, such as light detection and ranging (LIDAR) sensors, stereo cameras, and 3D scanners. Complemented with two-dimensional (2D) images, 3D data acquired by sensors demonstrate rich geometric, shape, and scale information such that they provide an opportunity for a better understanding of surrounding environments for machines [1]. In general, 3D data can be represented with different formats, such as depth images, point clouds, meshes, and volumetric grids. When compared to other 3D data formats, 3D point cloud representation preserves the original geometric information along with associate attributes in a 3D space without any discretization [1].

Therefore, point clouds have been widely adopted in numerous application fields, including 3D scanning and modeling, environmental monitoring, agricultural and forestry, bio-medical imagery, and so on [2].

Recently, deep learning (DL) on point clouds has been thriving in many scene-understanding-related applications, such as virtual/augmented reality (VR/AR), autonomous driving, and robotics. Nevertheless, the massive amount of point cloud data aggregated from distributed 3D sensors also poses challenges for securing data collection, management, storage, and sharing. By using signal processing or neural network techniques, several efficient point cloud compression (PCC) methods [3] have been proposed to reduce the bandwidth of wireless networks or the storage space of 3D point cloud raw data. However, there are still many efforts to achieve efficient end-to-end data delivery and optimal storage management. From the architecture aspect, conventional point-cloud-based applications rely on centralized cloud servers for data collection and analysis. Such a centralized manner is prone to single-point failures because any successful attacks such as distributed denial-of-service (DDoS) to the control (or data) server may paralyze the entire system. Other than that, a centralized server that manages 3D sensors and stores point clouds under a distributed network environment may lead to performance bottlenecks (PBNs), and it is vulnerable to data breaches caused by curious third parties and security threats in data acquisition, storage, and sharing process.

Because of several key features, such as the separation of the control and data planes, logically centralized control, the global view of the network, and the ability to program the network, software-defined networking (SDN) can greatly facilitate big data acquisition, transmission, storage, and processing [4]. At the same time, Blockchain has been recognized as a promising solution for security and privacy in big data applications [5] with its attractive properties, including decentralization, immutability, transparency, and availability. Therefore, combining SDN and Blockchain demonstrates great potential to revolutionize centralized point cloud systems and address the aforementioned issues.

In this paper, we propose a secure-by-design networking infrastructure called SAUSA, which leverages SDN and Blockchain technologies to secure access, usage, and storage of 3D point clouds datasets in their life-cycle. SAUSA adopts a hierarchical SDN-enabled service network to provide efficient and resilient point cloud applications. Network intelligence based on dynamic resource coordination and SDN controllers ensures optimal resource allocation and network configuration for point cloud applications that demand various QoS requirements. To address security issues in point cloud data collection, storage, and sharing, we design a lightweight and secure data authentication framework based on the decentralized security fabric.

By leveraging a hybrid on-chain and off-chain storage strategy, data owners can store the encrypted meta-data of point clouds into distributed data storage (DDS), which is more reliable than existing solutions [6,7] that use cloud data servers to store audit proofs. In addition, encrypting meta-data on DDS also protects the privacy of data owners. Data owners place the Swarm hash of meta-data and the access control policy on the Blockchain (on-chain storage), while the original point clouds are saved by private storage servers. Thanks to the transparency and auditability properties of Blockchain, data owners have full control over their point cloud data, and authorized users can verify shared data without relying on any trusted third party authority. Hence, the point cloud data integrity verification is more credible in a distributed network environment.

In summary, the key contributions of this paper are highlighted as follows:

(1) The comprehensive architecture of SAUSA is introduced, which consists of a hierarchical SDN-enabled point cloud service network and a decentralized security fabric, and key functionalities for network traffics based on point cloud applications are described;
(2) The core design of the data authentication framework is illustrated in detail, especially for the workflow in data access control, integrity verification, and the structure of hybrid on-chain and off-chain storage;

(3) A proof-of-concept prototype is implemented and tested under a physical network that simulates the case of point cloud data sharing across multiple domains. The experimental results verify the efficiency and effectiveness of our decentralized data access authorization and integrity verification procedures.

The remainder of the paper is organized as follows: Section 2 provides the background knowledge of SDN and Blockchain technologies and reviews the existing state-of-the-art on Blockchain-based solutions to secure big data systems. Section 3 introduces the rationale and system architecture of SAUSA. The details of data the authentication framework are explained in Section 4. Section 5 presents the prototype implementation, experimental setup, performance evaluation, and security analysis. Finally, Section 6 summarizes this paper with a brief discussion on current limitations and future directions.

2. Background and Related Work

This section describes the fundamentals of the point cloud concept and explains key techniques including SDN, Blockchain, and smart contracts. Then we introduce the state-of-the-art on decentralized solutions to secure big data acquisition, storage, and analytic.

2.1. Deep Learning on 3D Point Clouds

By providing a simpler, denser, and more close-to-reality representation, 3D point clouds are prevalent in representing both static and dynamic 3D objects. By definition, a 3D point cloud is a set of points $\{P_i\}_{i=1}^{n}$ embedded in the 3D space and carrying both geometry and attribute information [2]. Given a Cartesian coordination system, the geometry information refers to the point position, which can be expressed as a coordinate tuple $c_i = (x_i, y_i, z_i)$. The attribute information is used to describe the visual appearance of each point, and it may have different formats according to various user cases, such as color value tuple (R, G, B) and normal vectors (n_x, n_y, n_z).

As a dominating technology in artificial intelligence (AI), deep learning on point clouds has been thriving with an increasing number of solutions to 3D point cloud applications, and typical examples are 3D shape classification, 3D object detection and tracking, and 3D point cloud segmentation [1]. Regarding 3D shape classification, the whole point cloud file is used to extract a global shape embedding, which is then input into several fully connected layers of the neural network to achieve the classification task [8]. In 3D object detection scenarios, a 3D object detector firstly processes the point cloud of a frame, and then, it produces a set of detected objects with 3D bounding boxes [9]. As a 3D object detection algorithm can detect the locations of target objects in the first frame, 3D object tracking methods can use the embedded rich information of point clouds to estimate their state in subsequent frames [10]. Given the understanding of the global geometric structure and fine-grained details of each point, 3D point cloud segmentation methods can be classified into three types: semantic segmentation, instance segmentation, and part segmentation [1].

2.2. Overview of SDN

The emergence of the software-defined network (SDN) paradigm has attracted great interest in designing intelligent, flexible. and programmable networks. As defined by the Open Networking Foundation (ONF), SDN refers to an emerging network architecture, where network control policies are decoupled from the forwarding mechanism and are directly programmable [11]. Unlike traditional networks that are vertically integrated, the control and data planes are decoupled in SDN frameworks. As a result, control logic and network intelligence are moved to an external entity called the SDN controller, while network devices simply make forwarding decisions that are flow-based rather than destination-based [12]. The network is programmable through software applications running on top of the SDN controllers that logically control the underlying network infrastructure and interact with the upper-layer management panel.

With its inherent characteristics of decoupling the control and data panels and programmability on the centralized control panel, SDN brings potential benefits in conventional network architecture and operations [11]. SDN can enhance network configuration and management by using the unification of the control panel over heterogeneous network devices; thus, the entire network can be easily configured with programmable controllers and then dynamically optimized according to the global network status. In addition, an SDN controller allows for the centralization of the control logic with global knowledge of the network state, and it is promising to improve network performance with optimal utilization of the underlying infrastructure. Moreover, SDN offers a convenient platform for the validation of techniques and encourages the innovation of next-generation networks.

2.3. Blockchain and Smart Contract

From the system architecture aspect, a typical Blockchain system consists of three essential components: a distributed ledger, a consensus protocol, and smart contracts. Essentially, distributed ledger technology (DLT) is a type of distributed database that is shared, replicated, and maintained by all participants under a P2P networking environment. Each participant maintains a local view of the distributed ledger in the context of a distributed computing environment, and a well-established consensus allows all participants to securely reach an agreement on a global view of the distributed ledger under the consideration of failures (Byzantines or crash faults). Given different consensus algorithms and network models, distributed consensus protocols are categorized into Nakamoto consensus protocols [13] or Byzantine fault-tolerant (BFT) consensus protocols [14]. From the topology aspect, Blockchains can be classified into three types: public (permissionless) Blockchains, private (permissioned) Blockchains, and consortium Blockchains [15].

Thanks to the cryptographic and secure computing schemes, a *smart contract* (SC) brings programmability to the Blockchain by integrating protocols with user interfaces to formalize and secure the relationships of participants over computer networks [16]. Essentially, SCs are programmable applications containing predefined instructions and data, and they are compiled and saved in the addressable storage of the Blockchain. Through exposing a set of public functions or application binary interfaces (ABIs), an SC acts as the autonomous trusted agent between parties to perform predefined business logic functions or contract agreements under specific conditions. Owing to the secure execution of the predefined functional logic, global unique address, and open-access ABIs, the SC is an ideal candidate to implement decentralized applications (Dapps) under dynamic, heterogeneous, and distributed network environments.

2.4. Related Work

By leveraging Blockchain and deep reinforcement learning (DRL), a Blockchain-enabled, efficient data collection and sharing framework is proposed to provide a reliable and safe environment for data collection [17]. A distributed DRL-based scheme aims to achieve the maximum data collection and ratio and geographic fairness in the long term, while the Ethereum Blockchain provides a tamper-proof distributed ledger to ensure the security and reliability of data sharing. The simulation results demonstrated that the proposed scheme can prevent against attacks in data collection and sharing. However, the performance adopting Blockchain has not been evaluated, and the storage overhead by directly storing data on the distributed ledger was not discussed.

To solve the distrust issues of big data sharing on collaborative edges, a Blockchain-based framework was proposed to ensure efficient and reliable data sharing across resource-limited edge nodes [18]. A green consensus mechanism called proof-of-collaboration (PoC) allows edge devices to mine blocks given their collaboration credits, rather than their computational resources. In addition, this work designed a novel futile transaction filer (FTF) algorithm that offloads transactions from the storage to the cache layer to reduce the response time and storage overhead occupied by the Blockchain. Moreover, the smart-contract-based express transaction (E-TX) can support asynchronous validation, and hollow blocks can significantly reduce the redundancy in block propagation. However, transactions encapsulating raw data are still directly stored on the distributed ledger, and this brings privacy concerns.

With the popularity of the edge–fog–cloud computing paradigm, verifying the integrity of data in use has become a challenging problem. Inspired by the smart contract and Blockchain technology, a real-time index authentication for an event-oriented surveillance video query system is proposed to provide a decentralized video stream security mechanism in the distributed network environment [6]. The hash value of video recordings is stored in the Blockchain as immutable evidence, which is used for the authenticity of raw data in the verification process. The experimental results showed that the entire index authentication process incurs marginal computational overhead for service providers.

To solve the issues of the traditional data integrity of cloud servers, a Blockchain-based data integrity verification in P2P cloud storage is proposed, which allows for more open, transparent, and auditable verification of big data [7]. The raw data are divided into several shards, which are stored in private storage, while the digits of the shards construct the hash Merkle trees, which are saved on P2P cloud storage servers for data integrity verification. As the root of a Merkle tree is recorded on the Blockchain before uploading the data, users can verify the integrity of the data without relying on any third party authority.

Combining homomorphic verification tags (HVTs) and the data-auditing Blockchain (DAB), a decentralized big data auditing scheme is proposed for smart city environments [19]. Unlike [6,7], the data owners unload their files and HVTs to cloud service providers (CSPs), while all auditing proofs generated by the CSPs are stored into the blocks of the DAB. As all historical auditing proofs cannot be tampered with, data owners or users can verify the data integrity without relying on third party auditors (TPAs). The comparison shows the lower communication and computational overheads incurred in the auditing process. However, the storage overhead of recording auditing proofs on the DAB was not discussed.

As a decentralized storage platform that aims to address the issue of file redundancy, the Interplanetary File System (IPFS) has been used to solve the problems of centralized big data storage. A Blockchain-based secure storage and access scheme is proposed to provide the security and efficiency of electronic medical record sharing [20]. Attribute-based encryption (ABE) is used to encrypt medical data, and then, the encrypted data are stored in the IPFS. ABE allows only authorized users to decrypt medical data in the IPFS. The hash values (data address of the IPFS) of the medical data are recorded in the Blockchain for data retrieval process and verification. Similar to the scheme in [20], EduRSS [21] combines Blockchain, storage servers, and encryption techniques to manage educational records in a decentralized manner. The encrypted original educational records are saved in distributed off-chain storage servers, while the hash information of the records is stored on the Blockchain. EduRSS utilizes smart contracts to regulate the data storage and sharing process.

To comply with the privacy requirement of the General Data Protection Regulation (GDPR), which allows users to rectify or even erase their own data, several solutions have been proposed to delete and update data on the Blockchain. A redactable Blockchain based on hash function modification is proposed to re-write or compress the on-chain data on the append-only distributed ledger [22]. Due to secret trapdoor information, chameleon hash functions [23] can efficiently find hash collisions, which allow for redactable blocks without breaking the hash chain. To enable redactable off-chain data over the IPFS, a delegated content erasure is proposed to enforce complete content removal across the entire network [24]. The proposed protocol relies on a "proof-of-ownership" to ensure anonymous and censorship-resistant off-chain data storage, such that only a user is allowed to delete its own contents. Unlike the above redactable solutions, a pseudonymization-based approach [25] is proposed to satisfy GDPR as integrating with the Blockchain. The pseudonymization uses cryptographic hash functions for encrypting the date or pseudonymous identities for anonymity. Therefore, only users who have encryption keys of the pseudonymization can decrypt data or even eliminate content.

3. Design Rationale and System Architecture

Aiming at a self-adaptive and secure-by-design service architecture for assurance- and resilience-oriented 3D point cloud applications, SAUSA leverages SDN to achieve efficient resource coordination and network configuration in point cloud data processing and delivery. By combining Blockchain and distributed data storage (DDS) to build a decentralized authentication network, SAUSA is promising to guarantee the security and privacy of data access, usage, and storage in 3D point cloud applications.

Figure 1 demonstrates the SAUSA architecture, which consists of two sub-frameworks: (i) a hierarchical SDN-enabled point cloud service network; (ii) a decentralized security fabric based on Blockchain and DDS.

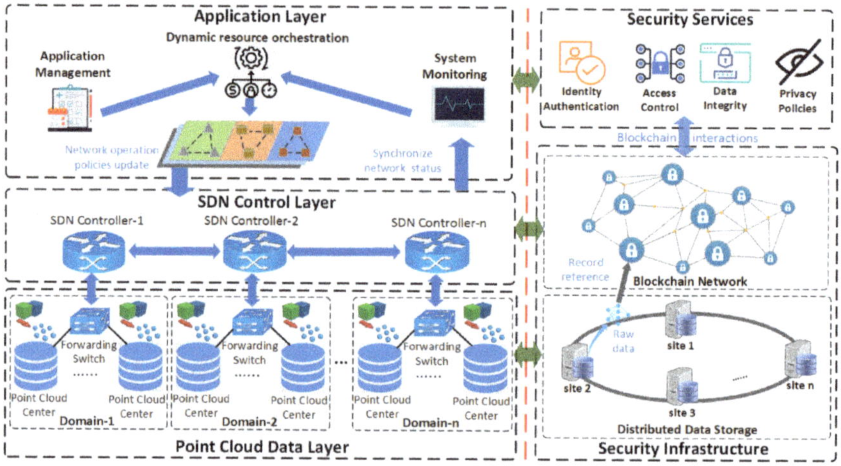

Figure 1. System architecture of SAUSA.

3.1. Hierarchical SDN-Enabled Point Cloud Service Network

As a potential technology to improve network performance and reduce management cost, the rationale of SDN is utilized to design a conceptual network architecture for multi-domain PC applications. Since this paper focuses on the Blockchain-based authentication network architecture, the key components and the workflow of the SDN are briefly described. The detailed SDN designs will be presented in our future work. The left part of Figure 1 shows the hierarchy of a point cloud service network according to point cloud application stage: acquisition, aggregation, and analytic. The point cloud data layer acts as an infrastructure layer including multiple domain networks, which are responsible for raw data collection, processing, and delivery. In each domain, point cloud centers interconnect with each others via forwarding switches. The 3D sensors generate cloud points and send them back to the point cloud centers, which are actually local servers, to process and store the data. Given the decisions made by the SDN controllers, the forwarding switches can forward the data traffic flows efficiently to satisfy the QoS requirements.

The network intelligence and control logic of each domain network are performed by the SDN controller, which can be deployed on fog or cloud computing platforms. By using a pre-defined southbound API, each SDN controller can either update the configuration of forwarding switches to change the network operations or synchronize the status to have the global view of a domain network. Northbound interfaces allow an SDN controller to interact with the upper-level application layer, such as providing domain network status to the system monitoring and accepting the network operation policies' update. Therefore, these SDN controllers construct a control layer, which acts as a broker between point cloud applications and fragmented domain networks, and they can provide network connectivity and data services among heterogeneous domain networks.

The application layer can be seen as a "system brain" to manage the physical resources of the point cloud data layer with the help of SDN controllers. The application management maintains registered users and their service requirements, while the system monitoring can provide the global status of the point cloud ecosystem. Given the inputs from application management and system monitoring, the dynamic resource coordination adopts machine learning (ML) algorithms, which achieve fast resource (e.g., computation, network, and storage) deployment and efficient service re-adjustments with QoS guarantees.

3.2. Decentralized Security Fabric

As the right part of Figure 1 shows, the decentralized security fabric consists of two sub-systems: (i) a security services layer based on the microservice-oriented architecture (MoA); (ii) a fundamental security networking infrastructure atop the Blockchain and DDS. To address heterogeneity and efficiency challenges such as developing and deploying security services in the distributed network environment, our security services layer adopts container technology to implement microservices for PC applications. The key operations and security schemes are decoupled into multiple containerized microservices. As container is loss-coupled from the remaining system with the OS-level isolation, these microservices can be independently updated, executed, and terminated. Each microservice unit (or container) exposes a set of RESTful web service APIs to users of PC applications and utilizes local ABIs to interact with the SCs deployed on the Blockchain.

The Blockchain network acts as a decentralized and trust-free platform for security services, and it uses a scalable PoW consensus protocol to ensure the immutability and integrity of the on-chain data on the distributed ledger if the majority (51%) of the miners are honest. The security mechanisms are implemented by self-executing SCs, which are deployed on the Blockchain by trusted oracles such as system administrators. Thus, the security service layer can provide secure and autonomous microservices in a decentralized manner. To reduce the overheads of directly recording large data on the distributed ledger, we bring DDS into the security infrastructure as off-chain storage, which is built on a Swarm [26] network. Unlike the IPFS, which does not guarantee storage, Swarm maintains content-addressed DHT and relies on data redundancy to offer secure and robust data services. Moreover, the inclusion of incentives makes Swarm more flexible to integrate with the Ethereum Blockchain. The meta-data of point clouds and operation logs that require a heterogeneous format and various sizes are encrypted and then saved into the DDS. Raw data on the DDS can be easily addressed by their references (Swarm hash), which are recorded on the Blockchain for audition and verification. A Swarm hash has a much smaller size (32 or 64 bytes) than its raw data; therefore, it is promising to improve efficiency in transaction propagation and privacy preservation without directly exposing raw data on the transparent Blockchain.

4. Blockchain-Based Lightweight Point Cloud Data Authentication Framework

This section presents the details of the decentralized and lightweight data authentication framework. SAUSA guarantees security and privacy preservation for point clouds collection, storage, and sharing. We firstly introduce the participants and workflow in the framework. Then, we describe the structure of hybrid on-chain and off-chain storage. Finally, we explain the data access authorization and integrity verification procedures.

4.1. Data Access Control and Integrity Verification Framework

Figure 2a shows the framework of secure data access, storage, and usage based on Blockchain and the DDS. In this framework, owners can upload point clouds generated by 3D sensors to their private server, which acts as a service provider for the users of applications. By storing the access control policy and audit proof in the Blockchain, each owner can fully control its data, and the authorized user can verify the data stored on the private server. The overall workflow is divided into three stages according to the 3D point cloud life-cycle.

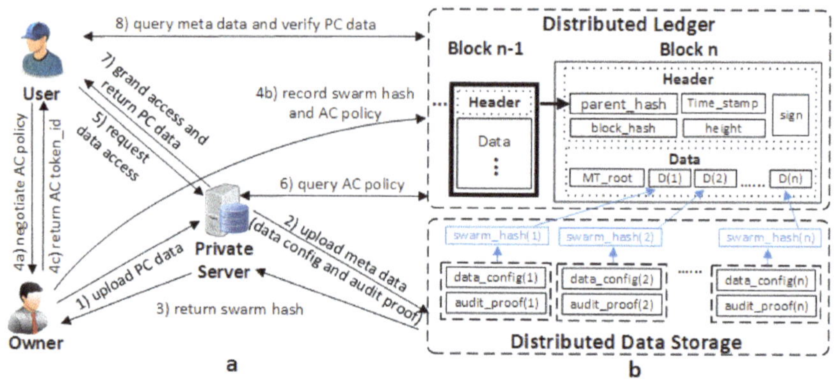

Figure 2. Illustration of Blockchain-based data authentication framework. (**a**) shows the workflow of 3D point cloud data storage, access authorization, and verification. (**b**) shows the structure of the hybrid on-chain and off-chain storage.

- *Data storage*: Owners and their private servers are in the same domain, and they can exchange secret keys via a trustworthy key distribution center (KDC). As a result, an owner and its private server can use shared secret keys to establish a secure communication channel for PC data transmission. In Step 1, the owner uses a shared secret key to encrypt point cloud data PC_i and then sends encrypted data to a private server. After receiving point clouds in Step 2, the private server stores encrypted PC_i into local storage and then records meta-data (e.g., configuration and audit proof) MD_i on the DDS. In a meta-data item, the configuration contains the URL address of a private server and other data properties such as the format and size, and the audit proof consists of an authenticator of raw data and a signature signed by the data owner. In Step 3, a site of the DDS stores the received MD_i and calculates a Swarm hash as a unique reference to address MD_i on the DDS. Finally, the Swarm hash is returned to the private server, and then, the private server transfers the Swarm hash to the data owner.
- *Data access control*: The data access control (AC) process is built on a capability-based access control (CapAC) scheme [27]. In Step 4a, a data user contacts a data owner to negotiate an AC policy for PC data sharing. Then, the data owner verifies the data user's identity and authorizes access rights for the data user given pre-defined AC policies. In Step 4b, the data owner stores the Swarm hash of the meta-data along with the assigned access rights in a distributed ledger (Blockchain). As long as the AC data have been successfully saved in an AC token on the Blockchain, a *token_id* is returned to the data owner. Finally, the data owner sends the *token_id* back to the data user as a notification, as Step 4c shows. In Step 5, a user first sends data access requests to a private server, which stores PC_i. Then, the private server retrieves the AC policy from the Blockchain and checks if the access rights assigned to the user are valid, as Step 6 shows. If the access authentication is successful, the private server uses shared secret keys to decrypt PC_i and return it to the data user, as Step 7 shows. Otherwise, the private server denies the access requests without sharing the data with unauthorized users.
- *Data verification*: To audit the received PC_i from a private server, the user queries the Swarm hash from the Blockchain and then retrieves meta-data MD_i from the DDB accordingly, as Step 8 shows. Because meta-data MD_i contains the audit proof that was submitted by the data owner when it uploaded PC_i, the data user can verify if PC_i satisfies the data properties and consistency of the authenticator and signature. In the data verification process, the user first checks if the properties of PC_i satisfy the configuration in MD_i. Then, it locally calculates the audit proof AP_i' according to

PC_i and compares it with the AP_i recorded in MD_i. If the audit proofs are equal, the data integrity has been guaranteed. Otherwise, the data may be inconsistent with the original version or corrupted during storage or sharing.

4.2. Structure of the Hybrid On-Chain and Off-Chain Storage

In general, a 3D model construction needs multiple segmented point clouds. and each point cloud segment PC_i may have a large data size and demand privacy preservation. Thus, it is impractical to directly store point clouds in a transparent Blockchain for data authentication. To ensure efficient and privacy-preserving data storage and sharing, we adopted a hybrid on-chain and off-chain storage structure in the data authentication framework, as shown in Figure 2b. In the point cloud data collection stage, the meta-data of the point cloud segments are saved in the DDS, while the raw data are managed by private servers. The meta-data MD_i contain the data configuration (e.g., server address and properties), which is relatively small regardless of the size of the original data. In addition, an audit proof consists of the integrity of the authenticator of a point cloud segment and a signature singed by a data owner, which are byte strings with a small length. Therefore, the small size of the meta-data can greatly reduce the communication cost in the verification process. Furthermore, the meta-data are encrypted and then saved on the DDS, and only authorized users are allowed to query and decrypt the meta-data. It is promising to protect the privacy of data owners without exposing sensitive information on the Blockchain and DDS.

In our Swarm-based DDS, each of the stored meta-data has a unique Swarm hash as the addressable reference to the actual data storage, and any change of the stored data will lead to an inconsistent Swarm hash. Therefore, recording the Swarm hash on an immutable distributed ledger provides the non-tamperability property of the meta-data on the DDS. To verify the data integrity of a large point cloud file, the Swarm hash of meta-data MD_i is considered as a digest $D(i)$, which is located on a leaf of the Merkle tree. Then, we use such an ordered list of digests to construct a binary Merkle tree $MT\_root = BMT(D(1), D(2), ...D(N_m))$, where N_m is the number of meta-data. Modifying digests or changing the sequential order will lead to different root hash values MT_root of the Merkle tree. Therefore, MT_root is also stored on the distributed ledger as the data integrity proof of the entire file. In the data verification process, a data user can query digests from the Blockchain and then in parallel validate the integrity of the segment data. Then, it can easily reconstruct the Merkle tree of the digests and obtain MT_root'. Finally, the data integrity of the entire point cloud file can be efficiently verified by comparing MT_root' with MT_root on the distributed ledger.

4.3. Decentralized Data Authentication Procedures

The Blockchain-based data access authorization and integrity verification procedures are presented as pseudo-code in Algorithm 1. Given a list of meta-data M, data owner traverses each meta-data MD_i and uploads them to the DDS, then appends the returned Swarm hash D_i to *ordered_swarm_hash*, as Lines 2–6 show. Following that, the data owner feeds *ordered_swarm_hash* to function $BMT()$, which will construct a binary Merkle tree and output the root hash *mk_root* (Line 7). Finally, the data owner calls the smart contract function $set\_dataAC()$ to record *mk_root* and *ordered_swarm_hash* into the distributed ledger as the public audit proof, which can be uniquely addressed by *token_id* (Line 8).

In the data verification procedure, the data user firstly uses *token_id* as the input to call the smart contract function $query\_dataAC()$, which will return the public audit proof information stored on the Blockchain (Line 10). Regarding token validation, the data user performs function $BMT()$ on the received *ordered_swarm_hash* to recover the root hash mt_root', then checks if mt_root' is consistent with the audit proof *mt_root*. If the validation fails, it directly returns a false result. Otherwise, it goes ahead to the meta-data verification. Given the received *ordered_swarm_hash*, the data user traverses each digest D_i, which is used to download the meta-data MD_i from the DDS. Any wrong digest or corrupted meta-data will lead to a *NULL* result returned by the function *download_data*(). Finally, a valid list of the meta-data is returned only if all meta-data can be successfully retrieved, as Lines 16–23 show.

Algorithm 1 The data access authorization and integrity verification procedures

```
1: procedure: authorize_data(token_id, M)
2:    ordered_swarm_hash = []
3:    for MD_i in M do
4:       D_i ← upload_data(MD_i)
5:       ordered_swarm_hash.append(D_i)
6:    end for
7:    mt_root ← BMT(ordered_swarm_hash)
8:    receipt ← Contract.set_dataAC(token_id, mt_root, ordered_swarm_hash)
9: procedure: verify_data(token_id)
10:   mt_root, ordered_swarm_hash ← Contract.query_dataAC(token_id)
11:   mt_root' ← BMT(ordered_swarm_hash)
12:   if mt_root' ≠ mt_root then
13:      return False
14:   end if
15:   MD = []
16:   for D_i in ordered_swarm_hash do
17:      MD_i ← download_data(D_i)
18:      if MD_i == NULL then
19:         return False, NULL
20:      end if
21:      MD.append(MD_i)
22:   end for
23:   return True, MD
```

5. Experimental Results and Evaluation

In this section, the experimental configuration based on a proof-of-concept prototype implementation is described. Following that, we evaluate the performance of running SAUSA based on the numerical results, which is especially focused on the impact of the Blockchain on the system performance. In addition, a comparative evaluation of the previous works highlights the main contributions of SAUSA in terms of the lightweight Blockchain design, performance improvement, and security and privacy properties. Moreover, we analyze the security properties and discuss potential attacks.

5.1. Prototype Implementation

We used the Python language to implement a proof-of-concept prototype including client and server applications and microservices. A micro-framework called Flask [28] was used to develop RESTful APIs for the applications and microservices. We used standard python library cryptography [29] to develop all security primitives, such as the digital signature, symmetric cryptography (Fernet), and hash function (SHA-256). Solidity [30] was used for smart contracts' implementation and testing, and all SCs were deployed on a private Ethereum test network.

The experimental infrastructure worked under a physical local area network (LAN) environment and included a cloud server and several desktops and Raspberry Pi (Rpi) boards. Figure 3 shows the experimental setup for our prototype's validation. A desktop emulated the private server, which stored the point clouds data managed by the data owner. To evaluate the impact of the hardware platforms on the data user side, both the Rpis and desktops were used to simulate a user client that requests data access. The private Ethereum network consisted of six miners, which are deployed on the cloud server as six containers separately, and each containerized miner was assigned one CPU core, while the other microservice containers that were deployed on the desktops and RPis worked in light-node mode without mining blocks. All participants used Go-Ethereum [31] as the client application to interact with the smart contracts on the private Ethereum network. Regarding the Swarm-based DDS, we built a private Swarm test network consisting of five desktops as the service sites. Table 1 describes the devices that were used to build the experimental testbed.

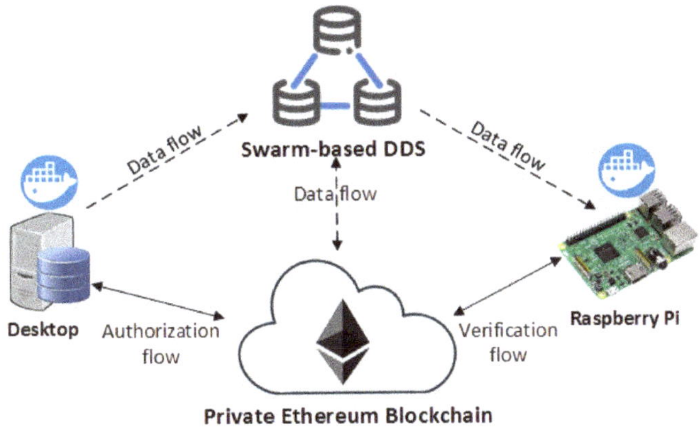

Figure 3. The experimental setup and network configuration.

Table 1. Configuration of experimental nodes.

Device	Cloud Server	Desktop	Raspberry Pi 4 Model B
CPU	Intel(R) Xeon(R) Gold 5220R CPU @ 2.20 GHz (96 cores)	Intel Core TM i5-3470 (4 cores), 3.2 GHz	Broadcom ARM Cortex A72 (ARMv8), 1.5 GHz
Memory	512 GB DDR4	16 GB DDR3	4 GB SDRAM
Storage	4 TB HHD	500 GB HHD	64 GB (microSD)
OS	Ubuntu 20.04	Ubuntu 20.04	Raspbian (Jessie)

5.2. *Performance Evaluation*

This section evaluates the performance of executing the operations in the data authorization and verification. In the data authorization process, the desktop launches a transaction, which encapsulates the Swarm hash of the meta-data in the Blockchain, and then, the states of the SC can be updated until a block containing transactions committedby the miners. Thus, we evaluated the end-to-end latency and gas usage during a successful data authorization operation. According to Algorithm 1, the whole data integrity verification procedure is divided into three steps: (1) the client (Rpi or desktop) queries the data token containing the Swarm hash of the meta-data and the root from the Blockchain; (2) the client validates the Merkle root and Swarm hash in the data token; (3) the client retrieves the meta-data from the DDS and verifies them. Therefore, we evaluated the processing time of the individual steps on different platforms by changing the number of meta-data (N_m). Finally, we analyzed the computational overheads incurred by retrieving the meta-data from the DDS and performing symmetric encryption on the meta-data. We conducted 50 Monte Carlo test runs for each test scenarios and used the averages to measure the results.

5.2.1. End-to-End Latency and Gas Usage by Data Authorization

We scaled up N_m in the data authorization scenarios to evaluate how the size of the ordered list of digests (Swarm hash) impacts the performance. As a transaction's committed time is greatly influenced by the Blockchain confirmation time, we observed that all data authorization operations with different N_m demonstrated almost a similar end-to-end latency (about 4 s) in our private Ethereum network. Regarding the computational complexity and processed data required by the SC, the gas used by the transactions may vary. Figure 4 shows the gas usage by data authorization transactions as N_m increases. The longer the ordered list of digests, the more gas is used per each transaction that stores the data on the Blockchain. Hence, recording the Swarm hash, rather than the meta-data or even the raw data on the distributed ledger, can greatly reduce the gas consumption of the Blockchain transaction.

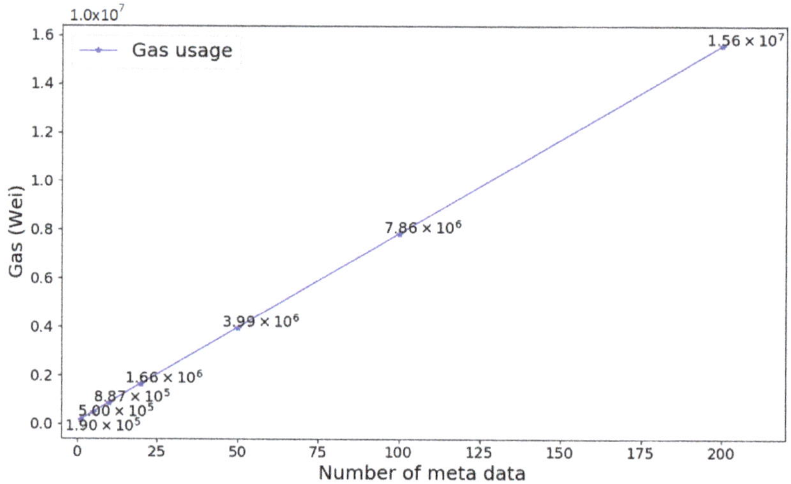

Figure 4. Gas usage in data authorization.

5.2.2. Processing Time by Data Verification

Figure 5 shows the average delays to evaluate how a data token query function of the SC can be successfully handled by the client as N_m increases from 5 to 200. Regarding a larger N_m, the query token procedure of the SC needs more computational resources to process the data on the distributed ledger. Thus, the delays of querying a data token on both platforms scale linearly with N_m with the same gain. Due to different computational resources, the processing time of the data token query on the Rpis is almost double that on the desktops.

Figure 6 shows the computational overheads by validating token the data on the client side as N_m changes. The data token data validation requires reconstructing the binary Merkle tree of the ordered list of Swarm hashes, which results in a traversal complexity of $\mathcal{O}(N_m)$. Then, the root hash can be used as the fingerprint for all the meta-data to check for inconsistencies, which requires a computational complexity of $\mathcal{O}(1)$. Finally, the computational overheads incurred by verifying the token data are scale linearly with N_m. Computing the root hash of the binary Merkle tree demands intensive hash operations such that the computational power of the client machines dominates the performance of the data token validation. Therefore, a larger N_m in the data token validation brings more delays on the Rpis than the desktops. However, the impact was almost marginal in our test scenarios such that $N_m \leq 200$.

Figure 7 shows the processing time of verifying the meta-data on the client side as N_m increases. In the meta-data verification stage, a client uses the Swarm hash list in the data token to sequentially retrieve N_m meta-data from the DDS, which results in a communication complexity of $\mathcal{O}(N_m)$. Regarding the fixed bandwidth of the test network, increasing N_m allows for a larger round-trip time (RTT) and more computational resources in meta-data transmission. As a result, the delays of verifying a batch of meta-data are scale linearly with N_m. Unlike the desktops, the Rpis have limited computational resource to handle each data transmission. Therefore, the Rpis take a longer time to verify the same amount of meta-data than the desktops do.

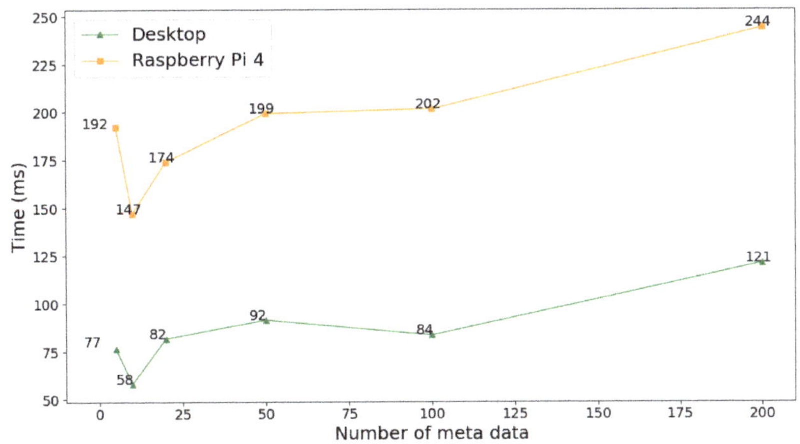

Figure 5. Latency by data token query on different platforms.

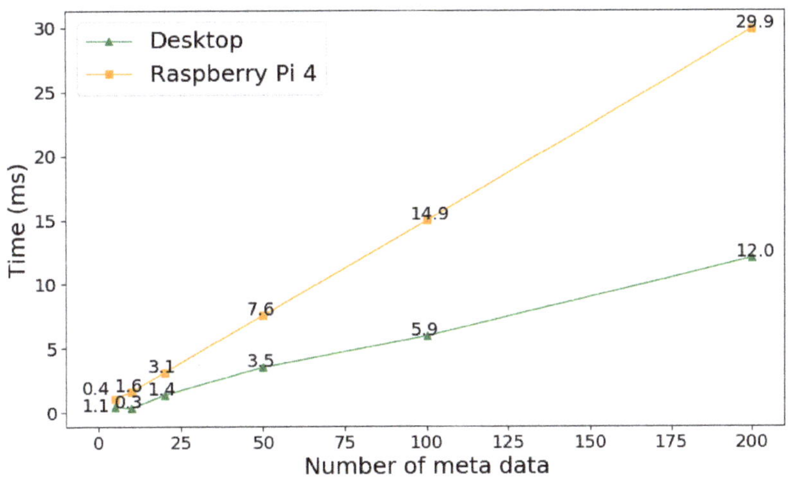

Figure 6. Processing time by data token validation on different platforms.

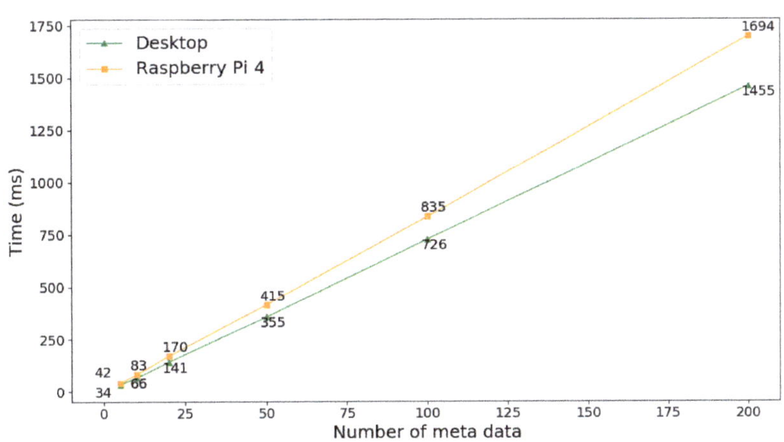

Figure 7. Processing time by meta-data verification on different platforms.

5.2.3. Computational Cost by Preserving Meta-Data Privacy

In our test scenario, the average size of the meta-data file was about 2 KB. Figure 8 shows the processing time of accessing data from (to) the DDS and executing encryption over a meta-data file on the client side. The delays incurred by uploading a meta-data file to the Swarm network and then downloading it from a service site are almost the same on the desktops and Rpis. However, the RPis took longer to encrypt and decrypt the data than desktops did due to the limited computational and memory resources. Compared to the Swarm operations, performing encryption algorithms on meta-data brings extra overheads in the data verification process on both platforms. As a trade-off, using encrypted meta-data to ensure privacy preservation is inevitable at the cost of a longer latency in the service process.

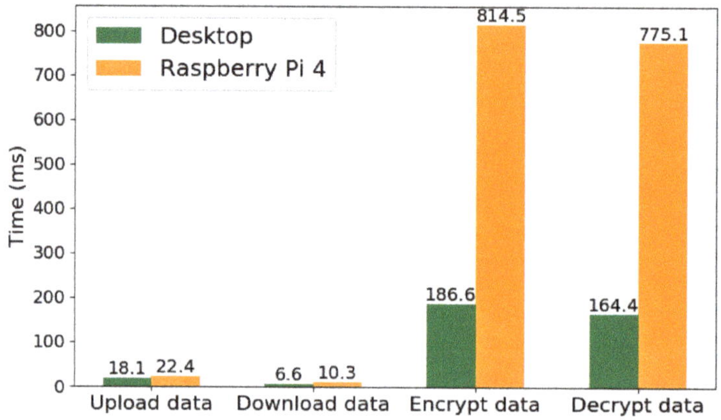

Figure 8. Processing time of meta-data operations: accessing Swarm and symmetric encryption.

5.3. Comparative Evaluation

Table 2 presents the comparison between our SAUSA and previous Blockchain-based solutions to big data applications. The symbol $\surd$ indicates that the scheme guarantees the security properties or implements some prototypes to evaluate system performance or other specifications. The symbol $\times$ indicates the opposite case. Unlike existing solutions, which lack details on the optimal network framework for QoS or evaluations on the impact of applying Blockchain to big data applications, we illustrate a comprehensive system architecture, along with details on the SDN-based service and lightweight data authentication framework. We especially evaluated the performance (e.g., network latency, processing time, and computational overheads) of the Blockchain-enabled security mechanism in the data access authentication and integrity verification process.

Table 2. Comparison among existing Blockchain-based solutions.

Scheme	Blockchain	Storage	Performance	Security	Privacy
[17]	×	DLT	×	√	×
[18]	Green Blockchain	DLT	√	√	×
[6]	Ethereum	Fog Server	√	√	×
[7]	×	DDS	√	√	×
[19]	×	DLT	×	√	×
[20]	×	DDS	√	√	×
[21]	Ethereum	Storage Server	√	√	√
SAUSA	Ethereum	DDS	√	√	√

Regarding storage optimization and privacy preservation for point cloud data sharing, the hybrid on-chain and off-chain data storage structure not only reduces the communication and storage overheads by avoiding directly saving large volumes of raw data or audit proofs in Blockchain transactions, it also protects sensitive information by only exposing the references of encrypted meta-data on the transparent distributed ledger as the fingerprint proof. Unlike existing solutions, which rely on a centralized off-chain storage (e.g., centralized fog server or storage server) to store audit proofs, using a decentralized Swarm network as the off-chain storage is promising to enhance the robustness (availability and recoverability) of point cloud data sharing in multi-domain applications.

5.4. Security and Privacy Analysis

In this section, we first discuss the security and robustness of SAUSA and evaluate the impact of several common attacks on the proposed scheme. Then, we briefly describe the privacy preservation of SAUSA. Regarding the adversary model, we assumed that the capability of attackers is bounded by probabilistic polynomial time (PPT) such that they cannot compromise the basic cryptographic primitives, such as finding hash function collisions or breaking the cipher-text without knowing the secret keys. Moreover, we assumed that an adversary cannot control the majority of miners within the Ethereum network.

5.4.1. Sybil Attack

In a Sybil attack, an adversary can forge multiple fake identities to create malicious nodes. As a result, these malicious nodes can control the DDS network or even the consensus network to some extent. However, in the proposed SAUSA, permissioned network management provides the basic security primitives, such as the PKI and KDC for identity authentication and message encryption. Thus, all nodes with invalid identities are prevented from joining the domain networks. Furthermore, properly defined AC strategies are promising to reduce the impact of Sybil attacks across different application domains.

5.4.2. Collusion Tamper Attack

An adversary can compromise multiple nodes that collude to tamper with the PC data to influence the accuracy of 3D object detection and tracking. The collusion tamper attack could be easily achieved, especially for a small network. Our SAUSA anchors the meta-data of the original PC data to the Ethereum Blockchain. Once transactions encapsulating the meta-data are finalized on the immutable public distributed ledger, it is difficult for an adversary to attempt to revert the transactions or the status of smart contracts by controlling the majority (51%) of the nodes within a public Ethereum network. As the meta-data recorded on the Blockchain can be used as audit proofs for verifying the integrity of data on local private servers, the possibility of collusion tampering is reduced.

5.4.3. DDoS Attack

In conventional cloud-based systems, an adversary can accessmultiple compromised computers, such as using bots, to send huge volumes of TCP request traffic to target cloud servers in a short period of time. As a result, unexpected traffic jams by the DDoS attack overwhelm centralized servers such that service and networking functions become unavailable. Our solution adopts a DDS to achieve efficient and robust meta-data storage and distribution. As the DDS uses a DHT-based protocol to coordinate and maintain meta-data access service sites over a P2P network, it is hard for an adversary to disrupt the meta-data service by launching DDoS attacks to target service sites. Moreover, our data authentication framework relies on SCs deployed on Ethereum to ensure decentralization. Therefore, our approach can mitigate the impact of DDoS attacks better than centralized data auditing methods.

5.4.4. Privacy Preservation of PC Data

In data acquisition, users rely on trusted private servers to protect the raw PC data by AC policies and encryption algorithms. In the data sharing process, only encrypted meta-data along with references are exposed to the public network. The decentralized data authentication framework prevents attackers from violating access privileges or inspecting any sensitive information. However, the prototype of the SAUSA presented in this paper has no integrated privacy protection module to deter data privacy breach by honest or curious users, such as dishonest data users or private servers who attempt to obtain private information from PC data without deviating from pre-defined security protocols. Therefore, a data-privacy-preserving component based on differential privacy or secure multi-party computation is needed to guarantee PC data privacy, and we leave this for our future work.

6. Conclusions and Future Work

This paper presented SAUSA, which combines SDN and Blockchain technology to support efficiency, assurance, and resilience-oriented point cloud applications. The hierarchical SDN-enabled service network can provide efficient resource coordination and network configuration to satisfy the QoS of point cloud applications. A lightweight data authentication framework atop the Blockchain and DDS aims to secure 3D point cloud data access, usage, and storage in a decentralized manner. The experimental results based on a prototype implementation demonstrated the effectiveness and efficiency of our SAUSA. However, there are open questions that need to be addressed before applying SAUSA to real-world 3D point cloud scenarios. We leave these limitations to our future works:

(1) SAUSA uses Ethereum to build a Blockchain network, which ensures security and scalability in open-access networks. However, PoW mining brings unsustainable energy consumption, longer transaction committed latency, and lower throughput. Thus, it is not suitable for time-sensitive applications. Lightweight Blockchain designs, such as Microchain [32], are promising to optimize computational utilization and improve performance in terms of end-to-end latency and transaction throughput. Our on-going efforts include validating SAUSA in a real-world point cloud scenario and the investigation of the integration of Microchain to reduce data authorization latency.

(2) This paper focused on the decentralized security scheme's implementation and validation; however, there are still unanswered questions and challenges about networking service intelligence in point cloud applications. In future work, we will investigate SDN controllers and virtual network functions (VNFs) to efficiently manage network and storage resources within each domain and evaluate the system performance and security properties according to various attack scenarios.

Author Contributions: Conceptualization, R.X., Y.C., G.C. and E.B.; methodology, R.X. and Y.C.; software, R.X.; validation, R.X. and Y.C.; formal analysis, R.X. and Y.C.; funding acquisition, Y.C.; investigation, R.X. and Y.C.; resources, R.X. and G.C.; data curation, R.X.; writing—original draft preparation, R.X. and Y.C.; writing—review and editing, R.X. and Y.C.; visualization, R.X.; supervision, Y.C.; project administration, Y.C. All authors have read and agreed to the published version of the manuscript.

Funding: This research was partially funded by the United State National Science Foundation (NSF) under the grant CNS-2141468.

Institutional Review Board Statement: Not applicable.

Informed Consent Statement: Not applicable.

Data Availability Statement: Not applicable.

Acknowledgments: The views and conclusions contained herein are those of the authors and should not be interpreted as necessarily representing the official policies or endorsements, either expressed or implied, of the Air Force Research Laboratory or the U.S. government.

Conflicts of Interest: The authors declare no conflict of interest.

Abbreviations

The following abbreviations are used in this manuscript:

ABI	Application binary interfaces
AC	Access control
AI	Artificial intelligence
AR	Augmented reality
BFT	Byzantine fault tolerant
DApp	Decentralized app
DDS	Distributed data storage
DDoS	Distributed denial-of-service
DL	Deep learning
DLT	Distributed ledger technology
GDPR	General Data Protection Regulation
IoT	Internet of Things
IPFS	Interplanetary File System
KDC	Key distribution center
LIDAR	Light detection and ranging
ML	Machine learning
MoA	Microservice-oriented architecture
ONF	Open Networking Foundation
P2P	Peer-to-peer
PBN	Performance bottleneck
PC	Point cloud
QoE	Quality-of-experience
QoS	Quality-of-service
RRT	Round-trip time
SC	Smart contract
SDN	Software-defined networking
SPF	Single point of failure
VR	Virtual reality

References

1. Guo, Y.; Wang, H.; Hu, Q.; Liu, H.; Liu, L.; Bennamoun, M. Deep learning for 3d point clouds: A survey. *IEEE Trans. Pattern Anal. Mach. Intell.* **2020**, *43*, 4338–4364. [CrossRef] [PubMed]
2. Cao, C.; Preda, M.; Zaharia, T. 3D point cloud compression: A survey. In Proceedings of the 24th International Conference on 3D Web Technology, Los Angeles, CA, USA, 26–28 July 2019; pp. 1–9.
3. Bui, M.; Chang, L.C.; Liu, H.; Zhao, Q.; Chen, G. Comparative Study of 3D Point Cloud Compression Methods. In Proceedings of the 2021 IEEE International Conference on Big Data (Big Data), Orlando, FL, USA, 15–18 December 2021; pp. 5859–5861.
4. Cui, L.; Yu, F.R.; Yan, Q. When big data meets software-defined networking: SDN for big data and big data for SDN. *IEEE Netw.* **2016**, *30*, 58–65. [CrossRef]
5. Deepa, N.; Pham, Q.V.; Nguyen, D.C.; Bhattacharya, S.; Prabadevi, B.; Gadekallu, T.R.; Maddikunta, P.K.R.; Fang, F.; Pathirana, P.N. A survey on Blockchain for big data: Approaches, opportunities, and future directions. *Future Gener. Comput. Syst.* **2022**, *131*, 209–226. [CrossRef]
6. Nikouei, S.Y.; Xu, R.; Nagothu, D.; Chen, Y.; Aved, A.; Blasch, E. Real-time index authentication for event-oriented surveillance video query using Blockchain. In Proceedings of the 2018 IEEE International Smart Cities Conference (ISC2), Kansas City, MO, USA, 16–19 September 2018; pp. 1–8.
7. Yue, D.; Li, R.; Zhang, Y.; Tian, W.; Peng, C. Blockchain based data integrity verification in P2P cloud storage. In Proceedings of the 2018 IEEE 24th International Conference on Parallel and Distributed Systems (ICPADS), Singapore, 11–13 December 2018; pp. 561–568.
8. De Deuge, M.; Quadros, A.; Hung, C.; Douillard, B. Unsupervised feature learning for classification of outdoor 3d scans. In Proceedings of the Australasian Conference on Robitics and Automation, Sydney, Australia, 2–4 December 2013; University of New South Wales: Kensington, Australia, 2013; Volume 2, p. 1.
9. Caesar, H.; Bankiti, V.; Lang, A.H.; Vora, S.; Liong, V.E.; Xu, Q.; Krishnan, A.; Pan, Y.; Baldan, G.; Beijbom, O. nuscenes: A multimodal dataset for autonomous driving. In Proceedings of the IEEE/CVF Conference on Computer Vision and Pattern Recognition, Seattle, WA, USA, 14–19 June 2020; pp. 11621–11631.

10. Munoz, D.; Bagnell, J.A.; Vandapel, N.; Hebert, M. Contextual classification with functional max-margin markov networks. In Proceedings of the 2009 IEEE Conference on Computer Vision and Pattern Recognition, Miami, FL, USA, 20–25 June 2009; pp. 975–982.
11. Xia, W.; Wen, Y.; Foh, C.H.; Niyato, D.; Xie, H. A survey on software-defined networking. *IEEE Commun. Surv. Tutorials* **2014**, *17*, 27–51. [CrossRef]
12. Kreutz, D.; Ramos, F.M.; Verissimo, P.E.; Rothenberg, C.E.; Azodolmolky, S.; Uhlig, S. Software-defined networking: A comprehensive survey. *Proc. IEEE* **2014**, *103*, 14–76. [CrossRef]
13. Nakamoto, S. Bitcoin: A Peer-to-Peer Electronic Cash System. Available online: file:///C:/Users/MDPI/Downloads/21260-bitcoin-a-peer-to-peer-electronic-cash-system.pdf (accessed on 24 November 2022).
14. Lamport, L.; Shostak, R.; Pease, M. The Byzantine generals problem. *ACM Trans. Program. Lang. Syst. (TOPLAS)* **1982**, *4*, 382–401. [CrossRef]
15. Ferrag, M.A.; Derdour, M.; Mukherjee, M.; Derhab, A.; Maglaras, L.; Janicke, H. Blockchain technologies for the internet of things: Research issues and challenges. *IEEE Internet Things J.* **2018**, *6*, 2188–2204. [CrossRef]
16. Szabo, N. Formalizing and securing relationships on public networks. *First Monday* **1997**, *2*, 9. [CrossRef]
17. Liu, C.H.; Lin, Q.; Wen, S. Blockchain-enabled data collection and sharing for industrial IoT with deep reinforcement learning. *IEEE Trans. Ind. Inform.* **2018**, *15*, 3516–3526. [CrossRef]
18. Xu, C.; Wang, K.; Li, P.; Guo, S.; Luo, J.; Ye, B.; Guo, M. Making big data open in edges: A resource-efficient Blockchain-based approach. *IEEE Trans. Parallel Distrib. Syst.* **2018**, *30*, 870–882. [CrossRef]
19. Yu, H.; Yang, Z.; Sinnott, R.O. Decentralized big data auditing for smart city environments leveraging Blockchain technology. *IEEE Access* **2018**, *7*, 6288–6296. [CrossRef]
20. Sun, J.; Yao, X.; Wang, S.; Wu, Y. Blockchain-based secure storage and access scheme for electronic medical records in IPFS. *IEEE Access* **2020**, *8*, 59389–59401. [CrossRef]
21. Li, H.; Han, D. EduRSS: A Blockchain-based educational records secure storage and sharing scheme. *IEEE Access* **2019**, *7*, 179273–179289. [CrossRef]
22. Ateniese, G.; Magri, B.; Venturi, D.; Andrade, E. Redactable Blockchain–or–rewriting history in bitcoin and friends. In Proceedings of the 2017 IEEE European Symposium on Security and Privacy (EuroS&P), Paris, France, 26–28 April 2017; pp. 111–126.
23. Krawczyk, H.; Rabin, T. Chameleon Hashing and Signatures. 1998. Available online: https://eprint.iacr.org/1998/010 (accessed on 24 November 2022).
24. Politou, E.; Alepis, E.; Patsakis, C.; Casino, F.; Alazab, M. Delegated content erasure in IPFS. *Future Gener. Comput. Syst.* **2020**, *112*, 956–964. [CrossRef]
25. Campanile, L.; Cantiello, P.; Iacono, M.; Marulli, F.; Mastroianni, M. Risk Analysis of a GDPR-Compliant Deletion Technique for Consortium Blockchains Based on Pseudonymization. In Proceedings of the International Conference on Computational Science and Its Applications, Cagliari, Italy, 13–16 September 2021; Springer: Berlin/Heidelberg, Germany, 2021; pp. 3–14.
26. Swarm. Available online: https://ethersphere.github.io/Swarm-home/ (accessed on 30 September 2022).
27. Xu, R.; Chen, Y.; Blasch, E.; Chen, G. Blendcac: A smart contract enabled decentralized capability-based access control mechanism for the iot. *Computers* **2018**, *7*, 39. [CrossRef]
28. Flask: A Pyhon Microframework. Available online: https://flask.palletsprojects.com/ (accessed on 30 September 2022).
29. Pyca/Cryptography Documentation. Available online: https://cryptography.io/ (accessed on 30 September 2022).
30. Solidity. Available online: https://docs.soliditylang.org/en/v0.8.13/ (accessed on 30 September 2022).
31. Go-Ethereum. Available online: https://ethereum.github.io/go-ethereum/ (accessed on 30 September 2022).
32. Xu, R.; Chen, Y.; Blasch, E. Microchain: A Light Hierarchical Consensus Protocol for IoT Systems. In *Blockchain Applications in IoT Ecosystem*; Springer: Berlin/Heidelberg, Germany, 2021; pp. 129–149.

future internet

MDPI

Article

PVPBC: Privacy and Verifiability Preserving E-Voting Based on Permissioned Blockchain

Muntadher Sallal [1], Ruairí de Fréin [2,*] and Ali Malik [2]

[1] Department of Computing and Informatics, Bournemouth University, Dorset BH12 5BB, UK
[2] School of Electrical and Electronic Engineering, Technological University Dublin, D07 EWV4 Dublin, Ireland
* Correspondence: ruairi.defrein@tudublin.ie

Abstract: Privacy and verifiability are crucial security requirements in e-voting systems and combining them is considered to be a challenge given that they seem to be contradictory. On one hand, privacy means that cast votes cannot be traced to the corresponding voters. On the other hand, linkability of voters and their votes is a requirement of verifiability which has the consequence that a voter is able to check their vote in the election result. These two contradictory features can be addressed by adopting privacy-preserving cryptographic primitives, which at the same time as achieving privacy, achieve verifiability. Many end-to-end schemes that support verifiability and privacy have the need for some voter action. This makes ballot casting more complex for voters. We propose the PVPBC voting system, which is an e-voting system that preserves privacy and verifiability without affecting voter usability. The PVPBC voting system uses an effective and distributed method of authorization, which is based on revocable anonymity, by making use of a permissioned distributed ledger and smart contract. In addition, the underlying PVPBC voting system satisfies election verifiability using the Selene voting scheme. The Selene protocol is a verifiable e-voting protocol. It publishes votes in plaintext accompanied by tracking numbers. This enables voters to confirm that their votes have been captured correctly by the system. Numerical experiments support the claim that PVPBC scales well as a function of the number of voters and candidates. In particular, PVPBC's authorization time increases linearly as a function of the population size. The average latency associated with accessing the system also increases linearly with the voter population size. The latency incurred when a valid authentication transaction is created and sent on the DLT network is 6.275 ms. Empirical results suggest that the cost in GBP for casting and storing an encrypted ballot alongside a tracker commitment is a linear function of the number of candidates, which is an attractive aspect of PVPBC.

Keywords: verifiable voting; online voting; Selene; distributed ledger technology

Citation: Sallal, M.; de Fréin, R.; Malik, A. PVPBC: Privacy and Verifiability Preserving E-Voting Based on Permissioned Blockchain. *Future Internet* **2023**, *15*, 121. https://doi.org/10.3390/fi15040121

Academic Editors: Christoph Stach and Clémentine Gritti

Received: 19 January 2023
Revised: 10 March 2023
Accepted: 21 March 2023
Published: 25 March 2023

1. Introduction

Many conventional offline services, such as voting, mail, and payments, are migrating online due to the rapid development of the Internet and information technologies [1]. E-voting is an area of research which is attracting significant attention. The uptake of e-voting has gradually spread throughout European countries –and to non-European countries– with successful outcomes. However, e-voting comes with its own security challenges which need to be overcome to safeguard the pillars of free-and-fair election processes. These challenges include ensuring: (1) strong voter authentication; (2) voter privacy; (3) end-to-end (E2E) verifiability; and, finally, (4) election transparency and integrity [2].

To design and implement a successful e-voting system, several requirements must be met. These requirements are named and described as follows:

1. **Eligibility:** each voter should only be allowed to vote once. Only eligible voters are allowed to vote;
2. **Fairness:** no early results can be declared during the election period which would influence others who have not voted yet;

3. **Individual verifiability:** the ability to check that the vote was correctly counted should be available to individual voters;
4. **Universal verifiability:** every independent person should be able to verify the operations performed at every stage of the election as well as to check that the published results actually sum to the total number of cast votes;
5. **Vote privacy:** how a voter voted should not be disclosed to anyone;
6. **Receipt freeness:** the system should not offer any information which voters can use to prove how they voted. This information is typically expressed in the form of a receipt;
7. **Coercion resistance:** the system should not allow the voter to present evidence to a coercer of how they voted.

We address the following problem. How can blockchain be used in an e-voting framework in a way in which voter usability is preserved, privacy is enforced, and, finally, verifiability is ensured? We hypothesize that a solution addressing this problem statement should consist of (1) a conceptual framework which is based on blockchain; (2) the use of novel cryptographic primitives and mechanisms as part of this framework to enable checks of individual and universal verifiability. Consequently, voters must be able to verify the availability and accuracy of their votes in the final tally and that the published results are correct (sum of all votes); and, finally, (3) safeguarding the voter's privacy during the authentication phase and post-election phase, as well as ballot privacy during the election phase. In our previous work, we put forth the VMV (Verify-My-Vote) voting system [3], which uses the Selene voting scheme as the underlying voting system to preserve verifiability. In this paper, we extend VMV by incorporating privacy and verifiability functionalities into existing e-voting platforms without affecting the voter's usability.

1.1. Motivation

Privacy is a crucial requirement in e-voting systems which ensures that the connection between voters and their vote remains hidden. Many e-voting systems achieve privacy by making use of several cryptographic mechanisms. The aim is to remove the linkability feature between a vote and the caster of the vote during all election phases. Preserving privacy during the voter-identification (ID) phase is a challenging task to achieve. This is because additional cryptographic arrangements must be established to prevent linking a voter's identity to the contents of their vote. Blind signatures and mixed-nets are among the techniques that can be applied to remove the connection between the voter and their ballot [4]. These arrangements may clash with voter-usability requirements, which can negatively affect voter turn-out.

Meeting the verifiable-e-voting requirement is important in reducing the need to trust electronic systems. Systems meeting this requirement afford voters and observers the ability to independently check whether the votes were recorded, counted, and tallied correctly. However, Internet-based voting systems in use today are not sufficiently robust to satisfy the verifiable-election requirement as they do not provide proof, by means of corroborating evidence, that would allow clear, individual and universal determination of the election's verifiability [5]. To ensure verifiability, many E2E methods necessitate voter involvement, such as executing the cut-and-choose action during the vote-casting phase. Cut-and-choose protocols require complex interaction with the voter (e.g., interactive proofs) to reduce the probability of corruption [6]. However, cut-and-choose protocols make the voting process more complicated for voters, negatively affecting their ease of use. Replacing existing systems with a brand new E2E verifiability scheme that supports voters usability presents a risk to businesses.

1.2. Contributions

To conceal the link between the voter and ballot during the authentication phase in such a way that a cryptographic action by voters is not required, an authentication protocol which keeps the voter's identity confidential is introduced in this paper. The protocol uses an access model which is based on the revocable-anonymity concept where incorporating identity revocation in an anonymous communication system is achieved using a permissioned distributed ledger (DL) and smart contract. To ensure resilience, trust, and privacy, e-voting requires a framework for access control which matches the decentralized attribute of e-voting.

To provide E2E verifiability, the Selene e-voting scheme [7] is employed as an e-voting scheme which is designed to support individual and universal verifiability. This means voters and observers can check the validity of votes during the election and post-election phases. Selene simplifies the voter's experience during the vote-casting time by hiding the cryptographic complexity from the voter. The proposed system uses DL technology within Selene to manage cryptographic primitives.

We summarise our contributions as follows:

- We propose an e-voting framework, which is called the PVPBC voting system, which preserves crucial e-voting features: (i) voter privacy and anonymity by making use of a permissioned ledger technology and smart contracts; and (ii) E2E verifiability by making use of the Selene voting scheme and a permissioned DL. Importantly, these crucial features are provided without affecting user experience, as the PVPBC framework does not affect the protocol fulfilled by the voter when they are voting.
- We propose a comprehensive architecture for a capability-based authorization protocol, which can be used to authenticate voters in e-voting platforms. The proposed approach supports dynamic voter authentication in e-voting systems based on an access-control model which supports revocable anonymity, as a result of using permissioned DLs. The protocol includes capability management and access-right validation.

Compared to existing work, the PVPBC voting system we introduce has a number of benefits, which we emphasize here: (i) it supports usability preservation for voters in the sense that voting is the only task required of the voter; (ii) the system integrity and security is supported by a distributed authentication mechanism based on DL technology; (iii) the system is verifiable by all parties; and (iv) the system fulfils vote privacy as well as voter privacy.

1.3. Organisation

The structure of this paper is outlined as follows. We start by reviewing the state-of-the-art in Section 2 and by identifying gaps in the current state-of-technology. In Section 3, we introduce the PVPBC voting system on a system-component level. We outline the assumptions made and discuss how authentication with revocable anonymity is achieved. Section 4 discusses the role of the permissioned distributed-ledger technology (DLT) in the context of the contributed e-voting scheme in terms of the front-end system and authentication. The voting experience/protocol followed by the voter at the pre-election phase, voting phase and also the tracking-number retrieval process is outlined in Section 5. Section 6 describes the technical details involved in the pre-election, election-phase and post-election-phase setup. A thorough analysis, under the headings: eligibility, privacy, integrity, fairness and verifiability, is provided in Section 7. This analysis is supplemented by a performance analysis in Section 8, which evaluates the registration and authentication schemes along with an evaluation of the voting-phase performance. We provide our conclusions and make recommendations for future research in Section 9.

2. Related Work

E-voting is attracting increasing research attention. The aim of this research has been to preserve the key attributes of e-voting which are privacy, integrity, transparency, and E2E verifiability. From the protocol point of view, preserving vote privacy is challenging when e-voting systems are being designed. Several mechanisms have been adopted in the literature to guarantee privacy in e-voting systems. One mechanism is to use blind signature schemes [8]. The distinguishing feature of these schemes is that the voter receives a token which includes a blind signature from the administrator as an indication of the voter's eligibility to vote. In response to this, the voter should send their vote anonymously along with the signed token as a proof of eligibility. An alternative approach is to use homomorphic encryption [9]. In this approach, the voter encrypts their vote. After that, the administrator determines the encrypted tally from the encrypted votes by making use of the encryption algorithm's homomorphic properties. Randomness, which is achieved using mix-nets, is also an exploited mechanism to preserve privacy by mixing up votes so that the connection between voters and their ballot is obscured [10] .

Homomorphic encryption is the foundational principle for a number of e-voting systems. In the system proposed in [11], the voter can encrypt their vote and generate a proof of validity which does not require interaction—this is referred to as a zero-knowledge proof– by making use of an El-Gamal scheme and a set of private and public keys that are shared between parties. A coalition of honest authorities can check the proofs from the voters and then combine all correct encrypted votes in order to decrypt them utilising the proofs. Even though most of the security requirements are fulfilled, the complexity of the proofs as well as the form of the votes affects the scalability of the system.

Several proposals for E2E-verifiable voting systems, which employ common verifiability techniques to support the integrity, and thus the credibility of the election, have been introduced. Selected examples include Prét à Voter [12], Wombat [13], Scantegrity II [14], Helios [11], Belenios [15], Civitas [16], and the Selene protocol [7]. These techniques are suitable for either paper voting or electronic voting. Recent verifiability techniques target remote voting from the voter's device.

The Benaloh Challenge, which is discussed in [17], is a typical mechanism to confirm that a vote is cast according to the voter's intent where a cut-and-choose method is applied to ensure that the vote is accurately constructed. After creating the vote and encrypting it, the voter has two options, either to cast the vote or to audit it. The auditing process involves the disclosure of the vote, without casting it, and offers proof that the vote was accurately constructed. The voter is able to carry out vote auditing several times before submitting an un-audited vote. This action is achieved by using one of the following algorithms, the Helios, Belenios, Civitas or the Wombat algorithm.

Pretty Good Democracy [18] is a verifiable e-voting system which makes use of a Code-Vote approach as a way to obtain individual verifiability. A code sheet is given to the voter in this approach, which contains a voting code for each candidate, as well as a return code. These codes are delivered to the voter using a private channel such as the post. During the voting phase, the voter casts their vote by submitting the code of their candidates. Using the return code, the voter extracts verification that the vote has been correctly received. This is possible because only the election system has knowledge of the voting codes and the return codes.

In recent years, several Internet voting systems have been proposed that either use DL technology or have been launched as DL-based systems [19–22]. Some leading examples of live systems include, Follow My Vote (https://followmyvote.com/, accessed on 18 January 2023) and Democracy.Earth (https://www.democracy.ea, accessed on 18 January 2023). Typically, these proposals view the act of casting a vote as a transaction which should be documented on a blockchain or DL. A systematic review of the literature on blockchain-based solutions for e-voting and an analysis of the techniques reviewed is given in [23]. Blockchain is reviewed from a security perspective in [24]. The authors perform their analysis at three levels: the process, data and infrastructure levels and emphasize the need to consider these levels in light of urgent business and industrial concerns. Occasionally, the act of casting a vote is analogous to transferring a vote "coin" to a specific candidate. Nevertheless, these proposals often fail to tackle concerns related to electronic voting. These concerns include, but are not limited to, ballot secrecy, E2E verifiability, resistance to coercion, confirmation of accurate casting and recording of votes, voter eligibility, and the guarantee that all ballots cast before the end of the election will be counted. In addition, these actions could have unfavorable consequences, such as disclosing real-time running totals or voter turnout during the election, or linking a monetary value to a vote. Limited levels of technical details are available for some of these systems. The technical details of the schemes that are available are difficult to review and to independently assess. On the other hand, the VMV voting system introduced in [3] also uses the Selene voting scheme as the underlying voting system and employs blockchain to preserve verifiability. However, VMV sacrifices the voter-privacy property as the EA has full control of the voter's identity. VMV was added as a verifiability layer on the already-available legacy system.

A solution for the integrity and verifiability problems is commonly sought by applying the DL concept. Adoption of a permissionless DL on public, peer-to-peer networks gives rise to a number of security and performance issues which have not been addressed in these solutions. Firstly, inconsistency in the DL, which might result as the necessary and sufficient conditions for independent consistency verification, is not well-defined or understood. Secondly, due consideration is not given to information propagation delay, which is an on-going concern in public networks that maintain DLs [25,26]. Potential solutions exist in the form of modern machine-learning approaches that aim to minimize variation in propagation times [27,28] and deep-learning-based traffic-classification schemes [29] which seek to utilize network resources better. The net result of these issues is that votes might be absent from the DL at the end of an election. Finally, performance issues, which are related to transaction costs, are high when a public DL is used.

In summary, we compare the shortcomings and strengths of state-of-the-art e-voting systems in Table 1 under the headings Eligibility (El), Privacy (Pr), Verifiability (Vr) and Usability (Us), in order to establish the need for PVPBC.

Table 1. Motivating the need for PVPBC by comparing the strengths and shortcomings of existing e-voting systems under the headings: Eligibility (El), Privacy (Pr), Verifiability (Vr) and Usability (Us).

Voting Scheme	El	Pr	Vr	Us	Limitations	Strengths
Prét à Voter [12]	✓	✓	✓	×	Secrecy of the election is at risk when the election authority is compromised. Uses cut-and-choose protocols which require complex interaction with the voter.	Offers verifiable elections.
Wombat [13]	×	✓	✓	×	Paper-and-cryptographic-based voting system. Usability issues identified during the trials which arise due to the ballot design.	Overcomes the privacy issues associated with on-demand print ballots.
Scantegrity II [14]	×	✓	✓	×	Does not support verifiability in remote voting. Uses cut-and-choose protocols which require complex interaction with the voter.	Individual verifiability is achieved by a cut-and-choose mechanism.
Helios [11]	×	✓	×	×	Does not support verifiability in remote voting. Requires a separated public authentication-mechanism service. It is vulnerable to ballot stuffing.	Offers verifiable online elections.
Pretty Good Democracy [18]	✓	×	✓	×	Privacy is at risk as the election system has knowledge of the voting codes associated with the verifiability process.	Individual verifiability is achieved by making use of a code-vote approach.
Belenios [15]	✓	✓	✓	×	Not suitable for use in high-stake elections as it is not coercion-resistant.	Offers verifiable online elections.

Table 1. *Cont.*

Voting Scheme	El	Pr	Vr	Us	Limitations	Strengths
VMV [3]	✓	×	✓	✓	The voter-privacy property is at risk as the EA has full control of the voter's identity.	Offers verifiable and usable online elections.
DLT systems [19–22]	✓	×	×	×	Fails to tackle concerns related to electronic voting. These concerns include, but are not limited to, ballot secrecy, E2E verifiability, resistance to coercion, confirmation of accurate casting and recording of votes, voter eligibility, and the guarantee that all cast ballots before the end of the election will be counted.	Offers faster Internet voting based on blockchain.

3. Core of the PVPBC Voting System

The key innovation of the proposed Privacy- and Verifiability-Preserving E-voting Based on Permissioned Blockchain (PVPBC) voting system is to introduce a fully verifiable e-voting protocol. The PVPBC voting system should provide the verifiability, voter usability, and integrity properties. The PVPBC voting system is comprised of three components which are: the front-end system; the election authority (EA); and finally, the trusted third party (TTP). It is a requirement that these components are independent of each other. To satisfy the aim of implementing E2E verifiability, the PVPBC voting system uses the Selene e-voting scheme [7]. As a consequence of using the Selene protocol, the front-end system has two distinct, desirable characteristics. Firstly, the voters' cast ballots are gathered and presented on a permissioned distributed ledger in an unencrypted format, and then matched with anonymous tracking numbers. This approach is distinct from many other e-voting systems described in the literature, which only disclose an encrypted or hashed rendition of the votes on the bulletin board (BB), (an area within the e-voting system which includes the final election results). By publishing plaintext votes on the DL, voters can be confident that their vote was accurately recorded and tallied. It follows that voters have no need to trust or possess any specific understanding of the cryptographic methods utilized in the election. Publishing votes in plaintext on the DL guarantees individual verifiability. In effect, any voter can check their intended vote against the plaintext vote recorded in the DL by using their tracking number. Universal verifiability is also guaranteed because the tally can be calculated by anyone using the plaintext votes. The PVPBC voting system also provides mechanisms to safeguard against coercion, as a result of using the Selene protocol. These mechanisms are available to voters if they experience coercion.

To achieve the objective of providing access-control processes which are effective for voting services and information in e-voting systems, the PVPBC voting system offers an anonymous authentication mechanism where the voter's vote can be revoked by the election service provider with the cooperation of a trusted organization. An identity-based capability-token management strategy which is robust is proposed, based on the blockchain network. The use of smart contracts is instrumental in achieving effective registration, propagation and the revocation of access authorization. Consequently, election service providers are able to verify eligible voters without knowing their real identity. Voters only reveal their identity to the trusted organization as a TTP. During the election period, the TTP collaborates with the election service providers to validate voter eligibility without revealing their identity to the election services. This authentication feature is implemented in the PVPBC voting system by making use of a permissioned DL and smart contract. The permission DL is implemented by several trusted authorities within the trust organization, to store and initiate capability access tokens for eligible voters.

3.1. System Component Overview

Figure 1 illustrates the PVPBC voting system's components. Further explanation of how these components work will be provided in Section 4, but a brief overview is provided now in order to provide a working knowledge of the system.

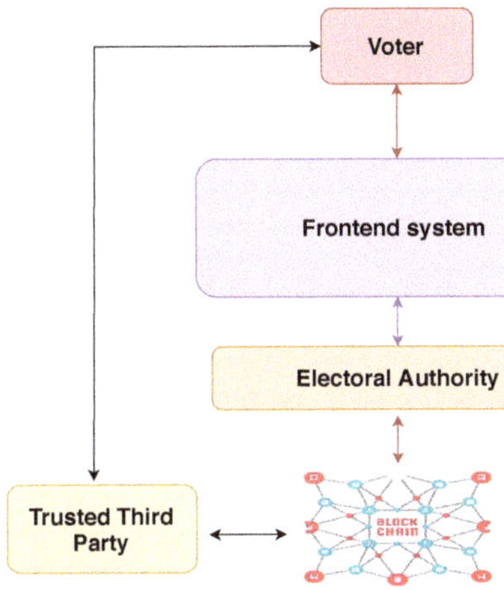

Figure 1. Component overview for the PVPBC voting system and interactions between the election actors: the election authority, the trusted third party, the voter and the front-end system. PVPBC is supported by blockchain.

Trusted third party (TTP): an individual or organisation which ensures that the election is correctly held by making use of DL technologies and smart contract to verify eligible voters.
Front-end system: a self-contained component which provides a verifiability functionality similar to the Selene protocol. A crucial difference from the Selene protocol is that the cryptographic functionality is provided by the front-end system, which includes key generation, signing, and decryption for voters. As a result, these functionalities do not have to be provided by the voters. This is managed independently of the EA. The front-end system also provides the authentication and voting services for voters. Specifically, the front-end system verifies voter-authentication requests and allows eligible voters to access the ballot and to cast their votes.

The front-end system is composed of several components. These components collectively provide secure and verifiable voting.

- **Voter-key management:** This component manages voter keys and their application. This is run, not by the EA, but by an independent trusted party.
- **Tellers:** generation and management of the election threshold key and the cryptographic manipulations required by the Selene protocol is performed by a set of tellers [30].
- **Web bulletin board (WBB):** To distribute trust, the front-end system utilizes a permissioned DL with a group of peers. This process requires a consensus of over two-thirds majority for agreement on the contents of the ledger.

Voters: Voters interact with the registration website offered by the TTP. Other interactions with the voters are typically via email as well as a voting website which is offered by the front-end system. In our proposed system, voters are not required to handle their own keys. Instead, the voter-key management component manages this for the voters, and voters use the credentials provided to access the required cryptographic functions. The voter-key management component is run by trusted parties within the front-end system.
Election authority: the EA is in charge of setting up the election, running it, and verifying voter eligibility in collaboration with the front-end system and the TTP.

Authentication-permissioned DLT: The authentication DLT manages voter authentication by making use of capability tokens, which are embedded in a smart contract. The authentication DLT is maintained by the TTP.

3.2. Assumptions

We make the following assumptions. We assume that:

1. There are authenticated channels from the front-end system to the authentication server of the EA.
2. There is a secure channel from the EA to the TTP.
3. Tellers are trusted parties who collaborate with the EA to run the election.

3.3. Authentication with Revocable Anonymity

The PVPBC voting system is based on a brand-new voter-authentication mechanism which preserves voter privacy and enforces the legal requirements of the election service authorities. To verify voters, the PVPBC voting system uses TTP certification along with DL technology and a smart contract. The TTP issues a capability access token for each eligible voter. The DL technology holds the access tokens securely and releases them when certain requirements have been fulfilled. At least five parties participate in the protocol, they are listed now. The TTP is comprised of two parties, the T_A, which is a trust body within the organization, and the T_m, which is an administrator. The third party is the voter, V. The permissioned DLT is run by two additional parties, an administrator, T_m, and several T_As.
Initialisation phase: Prior to the election, the T_m maintains a list of ID numbers, which belong to eligible voters. It is populated on the DLT by the TTP administrator T_m. Once the ID number related to the voter is verified, a profile for the registered voter is created on the DLT by calling the *Register Smart Contract* (more details are provided in Section 6.1). Each voter profile on the DLT includes the registration information tuples (Voter ID number (VID), Flag—"Not voted"). The VID is a unique virtual ID, which is assigned for each registered voter on the DLT.
Smart-contract deployment: A smart contract, which oversees the capability access token for each eligible voter, must be deployed on the DLT by the TTP. The smart contract can securely manage any algorithmically specified protocol, thanks to cryptographic protocols, when it is deployed over the DLT network. The smart contract supplies a deployAction () ABI (smart-contract interaction interface) for this function. The permission to execute this ABI rests with T_m. When operating this ABI, the T_m must send a transaction which includes the voter information defined in the previous phase, namely, the initialization phase. After synchronizing the permissioned DLT data, any node in the permissioned network, for example, the T_m or the T_A, has the ability to communicate with the smart contract using the contract address provided.
Authorization validation: The authorization-validation process takes place at the election service provider, in response to receiving an authentication request from the voter during the pre-election phase. More precisely, as shown in Figure 2, the election service provider verifies the present condition of the smart contract in the permissioned DLT to obtain the capability token associated with the voter's VID. The smart contract furnishes an accessRequest() ABI for verification of the token. The election service provider can accomplish this ABI by submitting the pertinent information, the voter's VID and the action to be performed using the transaction. The transaction is mined, included into a block and broadcast to the nodes, T_m and T_A, in the TTP DLT network. As part of this process, each node that receives the transaction carries out the ABI to check if the voter possesses the necessary access rights. This assures system users that nodes cannot deceive others with false processing results. The result is a robust and trustworthy access-control system. The election service provider will grant access to the ballot to the voter if the access-right policies and conditional constraints are satisfied and also depending on the

capability-token validation and access-authorization-process result. If these conditions are not satisfied, the voting request is denied.

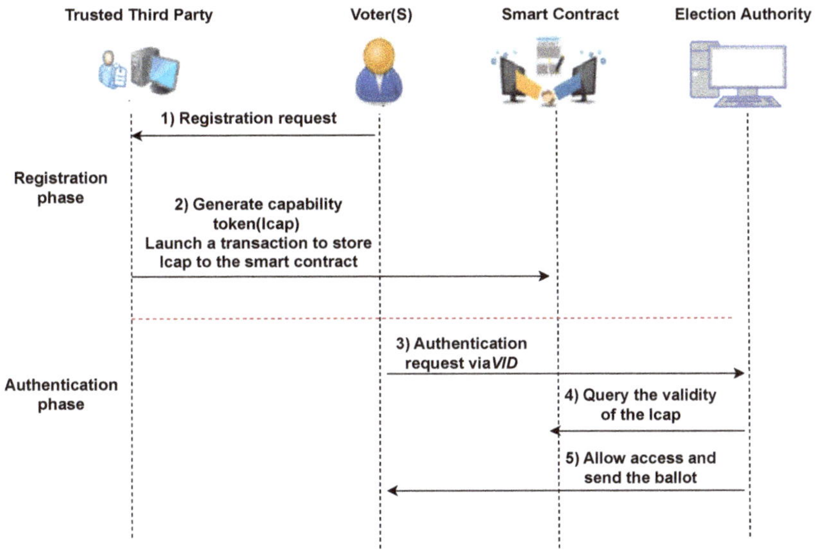

Figure 2. Voter-access control is carried out using two phases: the registration phase and the authentication phase. Effective access-control processes for voting services and information in e-voting systems is achieved by PVPBC by offering an anonymous authentication mechanism where the voter's vote can be revoked by the election service provider with the cooperation of a trusted organization.

Capability Token Structure

The T_m is responsible for maintaining a profile record for eligible voters on the DLT. In the profile record, all registered voters are associated with a virtual identity (VID) which is globally unique (we use two forms of ID. The voter ID is the normal identification required at the registration phase. It is typically a name or an email address. VID is another form of voter ID on the DLT. VID is a key which identifies the eligible voter on the authentication blockchain.), which is used as the prime key for identifying voters'off-chain profile. It is maintained by the TTP. At least one main account belongs to each voter. It is indexed by its VID address in the permissioned DLT. Consequently, the DLT is used to capture the VID for profiling registered voters. Generally, each eligible voter should be assigned a capability access token, which is stored in a smart contract. The capability token specifies the access rights associated with the voting ballot. It also contains awareness information. The capability token uses the following parameters.

- F: a cryptographic hash function which operates in one direction only;
- VID_V: the virtual ID of an eligible voter which solicits access, with the purpose of taking part in the election;
- EXP_R: the token expiration time;
- AR: the set of access rights pertaining to a predetermined set of actions;
- T: a timestamp;
- n_i: the tracking number for registered voters. This tracking number is a sparse selection of integers.

The capability token is then defined as a function of these parameters,

$$Token_R = F(VID_V, EXP_R, AR, T, n_i). \quad (1)$$

In the adopted authentication mechanism, the capability token corresponding to a particular voter is identified by the VID of the voter. Example actions in the set of access rights, AR, are the *vote* and the $NULL$ actions. If the access-rights set is null, $AR = NULL$, the voting operation is not allowed.

4. Role of Permissioned DLTs

The PVPBC voting system establishes two separate permissioned DLTs, each of which is used at a certain stage of the election to fulfill certain features and functions within the proposed system.

4.1. Election Permissioned DLT in the Front-End System

The front-end system makes use of DL technology. This DL is integrated into the Selene protocol as a key element to obtain the necessary assurances ahead of the election, e.g., the first half of commitments to tracking numbers. It is also responsible for storing the election's verifiability evidence as it is generated, and it handles the mixing and decryption of the votes and tracker numbers at the end of the ballot. This is carried out in this manner to ensure that the commitments that were placed in the ledger before the election started do not change after the election. Using the ledger guarantees that the same verifiability information is seen by all observers, and that the verifiability information cannot be changed at a later point in time.

In the front-end system, a permissioned DL is used. It is considered to be the appropriate design choice. This is because the DL is purpose-built for the election. In their current forms, EAs are necessary to manage the election roll. The need for the EA motivates the requirement for having trusted authorities. The EA's role is to start the election and to take responsibility for the election outcome. A benefit of the proposed system is that only the individuals responsible for conducting the election, who are already trusted, need to be granted write access to the ledger. Consequently, malicious nodes do not have the ability to take part in the DL's network. As a result, the security of the proposed system is increased. An additional advantage of permissioned DLs is that they permit faster and lighter consensus than the permissionless DLs. One side effect of the permissionless public-ledgers approach is that they give less control over consistency. This is because they support *eventual consensus*, which means that forks are possible and are eventually resolved. A good example of this is the branch rule of Bitcoin [31,32].

4.2. Authentication Permissioned DLT

The proposed system makes use of a permissioned DLT to manage and to store the authentication information. The authentication DLT is different from the election DLT, which is used to manage the verifiability data within the front-end system. The authentication DLT is maintained by trusted parties within the trusted third party organisation. This means that only those authenticated and trusted parties can join the network and can securely add, delete, and manage all voter information. During the registration phase, a TTP will be responsible for adding the voter information to the DLT as well as for deploying the smart contract related to the capability access token. We use the DLT here to evaluate the access policies, to preserve the access token integrity and to detect token double spending.

5. Voting Experience

The voting experience has three distinct stages: the pre-election phase; the voting phase; and, finally, the tracking-number retrieval phase. Each stage is now described.

5.1. Pre-Election Phase

Our starting point is the assumption that the EA oversees voter registration. It maintains the list of eligible voters. The EA also provides precise instruction, extensive advertisement and distribution, and election assistance. In Figure 3, we illustrate that voters need to remotely register with the TTP before the election. The voter needs to provide an ID key, e.g., national ID, in addition to other information, such as their e-mail address, phone number, and address. In response, the voter is given a unique pseudonyms key (VID) which can be provided during the voting phase for authentication purposes.

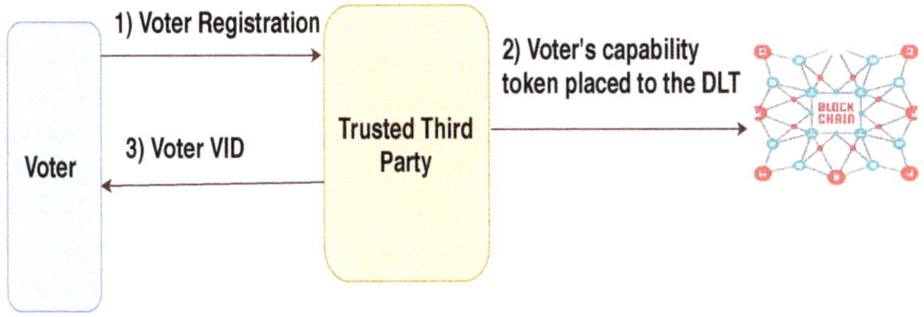

Figure 3. Voter experience during the registration phase: voters need to remotely register with the TTP prior to the election. The voter is given a unique pseudonyms key, which is used to provide authentication during voting.

During the pre-election phase, which is illustrated in Figure 4, voters need to authenticate themselves with the front-end system using their VID, which is generated during the registration step. Specifically, the voter, while interacting with the front-end system, is asked to input their VID, which is linked to the capability token stored in the authentication permissioned DLT. Upon retrieving a valid access token, the voter's cryptographic primitives (voter's signing keys, and voters' trapdoor keys) are created and a tracker commitment is placed on the DLT. We describe the voter's cryptographic primitives and tracker commitment in more detail in Section 6. Once the authentication stage has been successfully navigated, the voter is forwarded to the voting webpage.

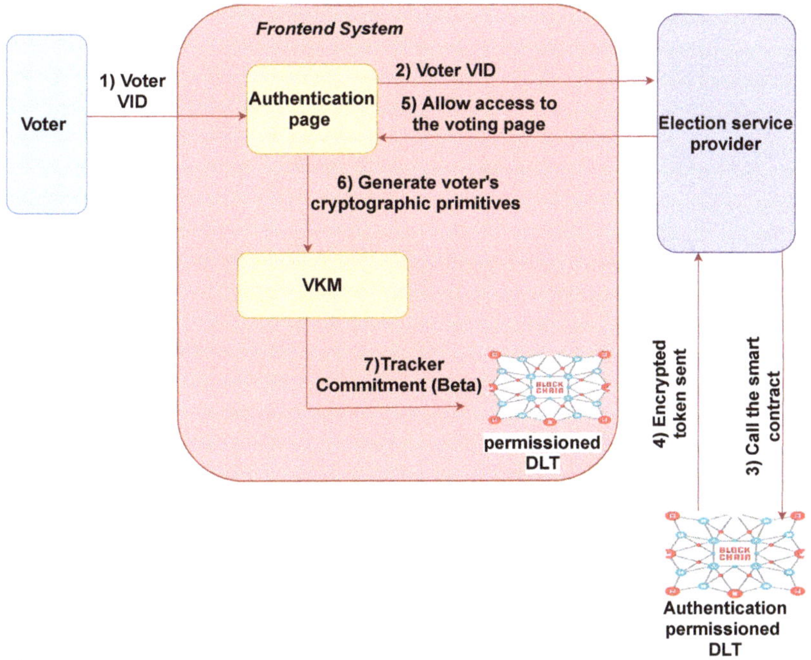

Figure 4. Voter registration is facilitated by the interaction with the front-end system. Subsequent interactions with the election service provider, the authentication permissioned DLT and VKM are instigated by the authentication page in the front-end system.

5.2. Voting Phase

The casting of votes takes place during the voting period. After passing the authentication stage, which is illustrated in Figure 5, the voter is forwarded to the voting webpage. The voter is presented with choices and is prompted to make a selection(s). After the voter has confirmed their selection(s), they may proceed to submit their ballot. A ballot, in encrypted form, which contains the voter's choices, is sent to the authentication server within the EA for checking and signing. The server checks whether the ballot is accurately formatted. An example inspection which is carried out examines if the ballot has the accurate election identifier tag. When the ballot is correctly formed, the authentication server responds to the front-end system with an acknowledgement. This acknowledgement contains EA's signature on the ballot. Ultimately, the signed encrypted ballot is paired with the encrypted tracking number and stored on the DLT within the front-end system.

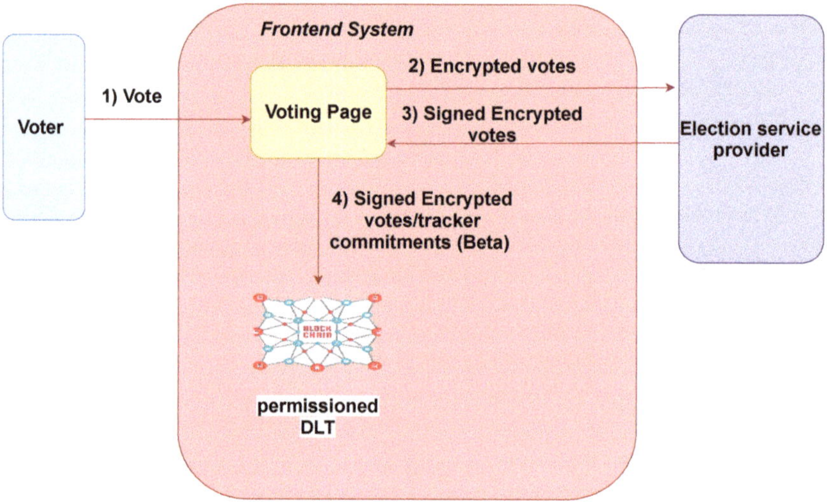

Figure 5. Voter experience during the voting phase: Once authentication is complete, the voter is forwarded to the voting webpage. The voter is shown the choices and is allowed to make their selection(s).

5.3. Post-Election Phase and Tracking-Number Retrieval

Figure 6 illustrates the post-election phase processes. Once the election has ended, the front-end system WBB has the cast votes, in plaintext form, and the corresponding tracking numbers. Email or post is subsequently used by the TTP to send the α (trapdoor key which opens to a tracker β which is already posted on the DL) term to the voter. When the voter receives the α term, if they wish to verify their vote was cast as intended they can open their tracking-number commitment, access the information contained in the WBB and check their own vote in plaintext.

The voter's unique tracker is calculated by a supported web application within the front-end system using the received α term, the public β (a tracker commitment which is publicly assigned to each voter and paired with the encrypted and decrypted votes. α opens to a β, then the voter tracking number can be constructed) term, and the trapdoor key *sk*. The purpose of this action is to support individual verifiability.

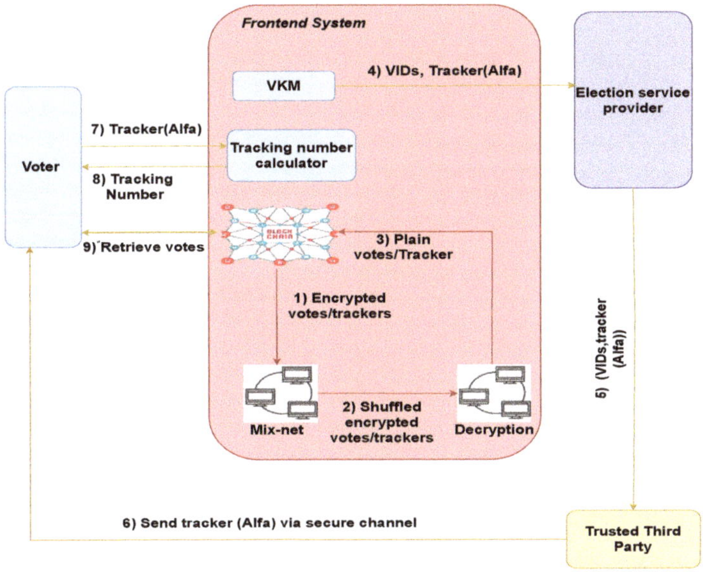

Figure 6. Voter experience during the post-election phase and tracking-number retrieval phase.

6. PVPBC Protocol: Cryptographic Considerations

We will now delve into a technical description of the protocol, paying particular attention to the cryptographic steps. This discussion is arranged into three parts: (1) the pre-election technical setup; (2) the election phase; and, finally, (3) the post-election phase.

6.1. Pre-Election Technical Setup

The pre-election setup phase consists of two main processes, a registration process and an election-key generation process. We describe voter authentication and the voter's cryptographic primitives in this context.

Registration: The voter-registration process is performed by the TTP by making use of a permissoned DLT and smart contract. Anyone with an ID number, for example, a national ID number, which is on the TTP's whitelist of IDs has permission to register for the election. When the ID field on the registration page is completed by the voter, as illustrated in Figure 7, the regApp.js application makes eth.calls to the registrar contract to check that the ID provided is on the whitelist of IDs, and if the associated voter has registered previously. When the outcome of the check is positive, the regApp.js communicates with the registrar contract to store the new voter information. This information consists of the voter's ID, Quorum DL address, and e-mail address. The purpose of this step is to link the user's Quorum DLT address and e-mail address so that they cannot register twice. Once registered, the registrar contract generates an access token for the voter and encrypts it with the private key of the TTP. The voter uses the access token for authentication on the blockchain. The voter is given a VID which is a parameter included in the access token generated by the smart contract (see Section 3.3 to obtain further information on this process).

Voter authentication: The voter interacts with the front-end system, providing their unique virtual identity (VID) to obtain entry to the voting platform, and then casts their ballot. Once voter authentication is complete, the voter should complete the unique virtual-identity (VID) field in the authentication page on the front-end system. After that, the authentication request and the voter's VID is forwarded to an election service provider. This process is illustrated in Figure 4. The election service provider first fetches the capability token from the smart contract by using the voter's VID, and then makes the decision whether or not to grant access to the ballot. The TTP changes the flag field of the voter's profile tuple to "voted" after successful authentication and updates the hashes of the blockchain. This prevents voters from voting more than once. Once voter access has been granted, the voter is directed to the voting page on the front-end system.

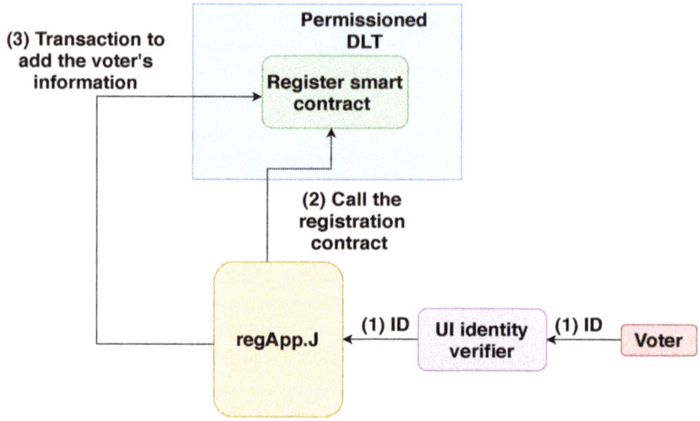

Figure 7. The voter-registration phase is managed by the TTP by making use of a permissoned DLT and smart contract.

Voter's cryptographic primitives: When the voter has been authenticated, their voter-keys management (VKM) component, which is part of the front-end system, generates cryptographic primitives for the corresponding authenticated voter. Cryptographic primitives include the voter's signing keys, and the voters' trapdoor keys. An ElGamal signing key pair, $(sign_i, vk_i)$, where $sign_i$ is the voter's signature key and vk_i is the public verifier key, is generated by the VKM component. The voter, V_i, generates signatures that can be verified via vk_i. It is assumed that a secret key, sk_i, and a public key (voter's trapdoor key), PK_i, can be generated by the VKM componentfor each voter, V_i, where $PK_i = g^{sk_i}$.

We move on to consider the creation of an encrypted tracking number for each authenticated voter. Voters obtain encrypted tracking numbers using the following procedure:

1. The tracking number, n_i, for the registered voter is extracted from the retrieved token.
2. The term, g^{n_i}, is calculated for the imported tracking number, n_i, in order to make sure that the tracking number falls in the appropriate subgroup.
3. Tracking numbers, g^{n_i}, are encrypted using the encryption key, PK_T.
4. The Sako–Kilian protocol [30] is used to re-encrypt and shuffle the set of encrypted tracking numbers, $(g^{n_i})_{pk_T}$. The resulting shuffled list is then assigned to the voters, so that each voter is associated with a unique secret encrypted tracker, $\{g^{n_{\pi(i)}}\}_{PK_T}$, where π is the permutation induced by the shuffle.

At this junction, the calculation of trapdoor commitments that open for a unique tracker occurs. The procedure by which the tracking-number commitment for each voter is generated is outlined.

1. A random number, r_i, is generated for each voter, i, by summing random values, $r_{i,j}$, which were generated by each teller, j, for voter i, where, $r_i = \Sigma_{j=1}^{i} r_{i,j}$.
2. The product, $\{PK_i^{r_i}\}_{PK_T}$, is generated using the component which is generated by each teller, j, $\{PK_i^{r_{i,j}}\}_{PK_T}$, and then by taking the product of these components. Similarly, $\alpha_i = g^{r_i}$ can be computed in a distributed fashion. The value, $\alpha_i = g^{r_i}$, will not be published, but it will be communicated privately to voters at the end of the election. In the PVPBC voting system, it is not computed until the end of the election.
3. The product of $\{PK_i^{r_i}\}_{PK_T}$ and $\{g^{n_{\pi(i)}}\}_{PK_T}$ is computed to obtain $\{PK_i^{r_i}.g^{n_{\pi(i)}}\}_{PK_T}$.
4. Threshold decryption is applied to reveal the commitment $\beta_i = PK_i^{r_i}.g^{n_{\pi(i)}}$. The commitment β_i is published.

The initialization block, which acts as the genesis block for the chain, is used to initialize the DL in the front-end of the PVPBC voting system. This block does not contain votes. All of the election information, for example, the election public key, the set of valid choices the voters can choose, are included. As a result, a DL is linked to a specific election. As the system parameters are integrated into the DL, potential disagreement over the system parameters is prevented. To allow trustees to join the permissioned network and to participate in maintaining the DL, security credentials are generated for them. A tuple of terms is posted to the DL for each voter V_i, before the voting phase starts, in the following manner:

$$(vk_i\ ,\ PK_i\ ,\ \{g^{n_{\pi(i)}}\}_{PK_T}\ ,\ \beta_i). \tag{2}$$

Election-key generation: In the PVPBC front-end system, tellers are responsible for generating the threshold election key, PK_T. This is the same in the Selene protocol. The election public key, PK_T, is created in a distributed way by the election tellers using the key-generation protocol proposed by Pedersen in [33]. In this protocol, every teller T_j has a share $s_j \in Z_q$ of a secret S.

As the values $PK_j = g^{s_j}$ are made public, every teller is committed to these values. The encryption key, PK_T, for the election is computed using

$$PK_T = \Pi_{j=1}^{n} PK_j, \tag{3}$$

where n is the number of tellers. All tellers are sent the calculated encryption public key, PK_T. It is not possible for a single teller to recover the secret key, $S = \sum_{j=1}^{n} s_j$. The authentication server, which is run by the EA, runs the key-generation algorithm of the digital signature scheme to produce the authentication server's public/private (verification/signing) keys. The primary action that takes place during the election phase, e.g., voting, is now described.

6.2. Election Phase

Voting: After gaining access to the voting site, the voter engages with the voting page, which is part of the front-end system, to cast their ballot. The PVPBC front-end system encrypts the cast ballot using the election key, and then signs the encrypted ballot using the voter's signing key. Then, the front-end system submits the ballot, V_i, to the authentication server at the EA using an authenticated communications channel.

If a ballot is received by the authentication server, and it has the correct format –the ballot is tagged with the correct election identifier–, an acknowledgement is sent as a response by the EA, which consists of a signature applied to the ballot V_i. If the ballot does not conform to the correct format, the EA does not output anything. If the voter or front end tries to vote more than once and an acknowledgement has already been issued by the EA, then the EA reissues the initial acknowledgement and does not consider the new vote. After this set of events, the encrypted vote is published to the permissioned DL alongside the encrypted tracking number, voter's public key, and trapdoor commitment. This is the public record of votes received in encrypted form. This means the signed encrypted vote is created at the vote time, and posted on the WBB when it is cast. Carrying out elections using this process results in the provision of public, real-time information regarding the turnout for the election. The voter is not involved with the voter-side cryptographic actions of the Selene protocol in the voting phase. Instead, the voter's task during the voting phase is to create and cast a ballot in the usual way.

Voting-phase technical details: After casting the ballot, the front-end system carries out several steps to produce the final result. For each cast ballot:

1. The ballot is encrypted and signed, obtaining

$$V_i = sign_{v_i}(\{Vote_i\}_{PK_T}). \tag{4}$$

2. A non-interactive zero-knowledge proof of the knowledge of the plaintext is created, Π_i, and it is attached to the encrypted ballot. In addition, this proof also includes the

voter's verification key, aiming to ensure the proof is valid only for this verification key.

$$V_i = sign_{v_i}(\{Vote_i\}_{PK_T}), (\Pi_i). \tag{5}$$

3. The front end, V_i, submits vi's ballot to the authentication server, EA, using an authenticated channel alongside the voter's VID. If the ballot is correctly formed, then the EA signs the ballot and sends the ballot back to the front-end system.
4. The signed encrypted ballot and the NIZKP are posted on the ledger alongside the previously published encrypted tracking number, tracking number commitment, and voter's identity (PK_i). This process is defined mathematically in

$$(vk_i, PK_i, \{g^{n_{\pi(i)}}\}_{PK_T}, \beta_i, sign_{v_i}(\{Vote\}_{PK_T}), \Pi_i). \tag{6}$$

This is the only information published on the ledger. It is important to note that the standard voting information collected by the election system is not published.

6.3. Post-Election Phase

At the end of the voting phase, as in the Selene protocol, the front-end system starts the mix and decryption process. During this process, each encrypted vote and its corresponding encrypted tracking number are mixed, shuffled, and decrypted by the election tellers. At the end of this process, tracking numbers alongside plaintext votes are posted to the permissioned ledger.

After the votes and tracking numbers have been published, the front-end sends α_i and the corresponding VID to the EA, which, in turn, forwards it to the TTP. The TTP uses the VID to locate the identity of registered voters on the authentication blockchain in order to send their α_i term via e-mail. In response to receiving the α term, the voter can access their tracking number via a call to the front-end system to apply their trapdoor key to (α_i, β_i). This enables them to identify their vote on the ledger.

After the election finishes, the front-end system carries out two steps to produce the final result.

Mix and decryption: For each voter, V_i, the encrypted tracking number and encrypted vote are extracted to give a pair of the form:

$$(\{g^{n_{\pi(i)}}\}_{PK_T}, \{vote_i\}_{PK_T}). \tag{7}$$

This pair is put through a verifiable shuffle performed by the mix-nets [34], and then a threshold set of tellers perform a verifiable decryption of these shuffled pairs. Proof of shuffling and correct decryption are uploaded on the WBB. After decryption, the pairs, $(n_{\pi(i)}, vote_i)$, are published on the WBB.

Tracking-number retrieval: Once the grace period has ended, each voter, V_i, is sent the α_i described in Section 6.1. The received α_i can be combined with the β_i term to reconstruct the tracking number encrypted under the voter's public key PK_i.

$$(\alpha_i, \beta_i) = \{g^{n_{\pi(i)}}\}_{PK_i} \tag{8}$$

The voter, V_i, can request that the front-end system use their secret key to decrypt and retrieve the $g^{n_{\pi(i)}}$, and thus obtains the tracker $n_{\pi(i)}$, which they can then use to look up their vote on the ledger.

7. Security Analysis

A security analysis of the system is now provided. It is organized using the headings: eligibility, privacy, ballot privacy, integrity, and, finally, fairness. Finally, we provide some examples of malicious attacks from which e-voting systems suffer, and analyze the resilience of PVPBC with regard to these attack types.

7.1. Security Attributes

We investigate the strengths and weaknesses of PVPBC with respect to key security attributes.

7.1.1. Eligibility

In order for voters to take part in the election, they need to provide a valid VID. Each valid VID needs to reference a valid capability access token in the authentication DLT. The TTP only provides access tokens to authenticated voters that were included in the list of eligible voters, which was compiled during the initialisation phase, and to authenticated voters that have not previously requested to vote. This means only eligible voters can vote. It also means that they can acquire only one eligibility token, and, therefore, they can only cast one valid vote. The eligibility property is "invalidated" when an eligible voter tries to vote more than once. Invalidation is not possible because the TTP changes the flag field of the tuple to "voted" in the DLT after successful authentication, as part of the voter-authentication phase. Therefore, the voter-authentication request will be rejected when the submitted VID references a profile on the authentication DLT which is flagged "voted".

7.1.2. Privacy

We analyze privacy using two subcategories "Voter Privacy" and "Ballot Privacy".

In terms of voter privacy, the front-end and election provider has no access to the voter identities beyond pseudonyms (VID: unique key allocated by TTP). Voters only submit the VID to the front-end as a proof of identity during the authentication phase. They only need to reveal their real identity to the TTP during the registration phase. Apart from the TTP, no party within the proposed system is able to link a VID to its individual voter.

Regarding ballot privacy, the PVPBC voting system guarantees that no party, at any point of the election, is able to reveal how a particular voter voted. While the front-end is collecting votes, the only link between the voter and their vote in plaintext is the VID key which acts as a pseudonymous identity. Mapping a VID key to the corresponding voter can only be carried out by the TTP, which is responsible for registering eligible voters. Therefore, ballot privacy is maintained with respect to the PVPBC voting system in three ways: the front-end has no access to the voters' true identities; election service providers have no access to both voters and votes; and, finally, the TTP has no access to the plaintext votes linked to their VIDs.

Similar to the Selene protocol, the front end provides the assurance that the published verifiability information does not compromise the privacy of the vote, provided that the requirement that each teller node is independent is met. This is maintained by the front end.

7.1.3. Integrity

The PVPBC-voting-system design ensures that even an insider at the EA or an attacker gaining access to the vote cannot invisibly change votes, as any change can be detected by a voter performing a verification. Consequently, the PVPBC voting system no longer requires the election provider to be trusted to ensure election integrity. This is because it can be verified independently. Use of DLT trustees means that the authority alone cannot change the commitments and verifiability parameters, because collusion between a majority threshold of trustees would be required.

7.1.4. Fairness

The PVPBC voting system guarantees that the election result will not be known during the election period. Therefore, voters cannot be influenced. This property is achieved by separating the election phase from the post-election phase. During the election phase, cast votes are published to the DLT in an encrypted form alongside commitments and encrypted tracking numbers. The ballot is only mixed, shuffled, and decrypted during the post-election phase by a set of tellers.

7.1.5. Verifiability

The PVPBC voting system supports individual and universal verifiability features because it incorporates the Selene voting scheme in its design. Specifically, the proposed system enables universal verifiability because all of the required information is published on the ledger. This allows every independent person to verify cryptographic operations performed at every stage of the election. Some examples of the types of verifications performed are listed. They include verifying the requirements that the tracker numbers are unique, and obtaining proof that shuffling and decryption have all been carried out correctly. Individual verifiability is achieved in the proposed system because voters are able to verify their vote on the DLT after receiving their α term and by reconstructing their tracking number. However, this does not offer the same level of individual verifiability provided by the Selene protocol. This is because the front end has control over the voter's private key. This is in contrast to the Selene protocol where voters control their private key. Consequently, they control the α term and are able to generate a fake α. We believe that sacrificing this Selene protocol feature in the PVPBC voting system is a feature worth sacrificing, given that it helps to preserve the voter usability.

7.1.6. Usability

The PVPBC voting system supports voter usability while preserving privacy and verifiability. Specifically, PVPBC preserves voters' privacy and verifiability without requiring any complex actions to be performed by the voter during the pre-election, election, and post election phases. During the pre-election phase, a voter receives a VID from the TTP to be submitted to the PVPBC's frontend. Once the valid VID is submitted to the frontend, the voter is allowed to access the candidates list and cast her vote. All the voter's cryptographic primitives are managed by the front end (e.g. voter-key generation, and vote encryption, etc). During the post-election phase, the voter will receive the α term which helps in reconstructing the tracking number. The tracking-number reconstruction is performed by the front end, meaning the voter does not perform any complex operation to locate the tracking number associated with her vote. To conclude, voter usability is fulfilled across all election phases without affecting the verifiability and privacy features.

7.2. Malicious-Attack Analysis

E-voting systems are subject to several forms of attack which target the availability, integrity, and anonymity of elections. We discuss and provide a qualitative analysis of the susceptibility of PVPBC to these classes of attack.

7.2.1. Malware

Malware, which includes Trojan horses, worms, spyware, and ransomware, typically targets the integrity of an election by preventing the voter's vote from being recorded as intended [35,36]. Distributed denial-of-service (DDOS) attacks can have a significant negative impact on network routing and performance metrics [37]. A DDOS attack aims to disrupt the voting process by slowing down communication in the network, including vote casting, tallying, and auditing. Regarding PVPBC's resilience to these attacks, vote integrity is achieved as the vote is cast through the front-end system which is controlled and run by the election authority as well as the trustees. We assume that there is a secure channel between the voter and the front-end which reduces the likelihood that adversaries can intercept the votes and alter them during the election phase. Regarding intrusions that intensify DDOS attacks, PVPBC offers a high level of resilience as a result of its use of permissioned blockchain. A consequence of this design choice is that participants are pre-defined and authenticated before joining the network. In addition, every DL node maintains a copy of the voting information (e.g., encrypted votes and verification proofs, etc). This helps PVPBC to stay alive; consequently, it is available for voters if a particular node fails due to a DDOS attack. When the node goes live again, it synchronises with other nodes and updates its DL ledger. In summary, PVPBC achieves a high level of resilience due to its use of blockchain technology.

7.2.2. Cross-Site-Scripting (XSS) Attacks

XSS attacks normally target e-voting systems at the application level. The reason for this is that the application has direct interactions with the voter [38]. As a counter-measure for this type of attack, security measures need to be considered at the application level. In PVPBC, insertion of the voter's contributed content is not allowed at the front-end. Consequently, there is no facility for the voters and candidates to perform editing. In addition, the front-end is designed so that voters' entries are checked and validated against the eligible entry format. This ensures non-persistent XSS attacks are avoided.

7.2.3. Collusion between DLT Participants

We adopted a practical Byzantine fault tolerance (PBFT) consensus protocol which supports e-voting resilience and trust [39]. A PBFT consensus ensures that the system stays available and accessible by voters even if a node goes offline due to a malicious attack. This is an advantage in comparison to e-voting systems which adopted a centralised infrastructure, which is vulnerable to a single point of failure. An additional strength of PVPBC is that we designed the PBFT so that the collusion between PVPBC blockchain members is very challenging. In other words, PBFT prevents the system from reaching an agreement if there are faulty or malicious nodes. To reach an agreement in PBFT, nodes vote on the validity of the information (transactions/blocks). This process provides safety because breaching the data integrity would require $(n-1)/3$ faulty nodes out of a total of n, the honest nodes, to collaborate. The PBFT protocol ensures resilience by including the election authority and other trusted parties as known participants in the permissioned group. They cannot compromise the integrity of the ledger unless the number of participating collaborators surpasses a threshold.

7.2.4. Broken Authentication

Attackers try to gain unauthorised access to online applications, in particular e-voting applications [40]. The possibility of successfully obtaining unauthorised access to e-voting systems during the election is a very serious issue which compromises the integrity of the election. In PVPBC, an authentication protocol based on blockchain is implemented to ensure valid and efficient authentication is carried out. The authentication process is maintained by the election authority in collaboration with the TTP by making use of a capability token which is verified and stored on the blockchain. As blockchain offers an immutable history, it is very challenging to forge the capability access token.

8. Performance Analysis

In this section, we investigate the performance of the PVPBC voting system. To carry out this investigation, we implemented a prototype system. Development of this prototype is initially described. We evaluate the gas and time required to deploy and register a smart contract as part of the registerVoter() function. We then evaluate the gas and time associated with the voter-authorisation phase, where the voter submits an access request to the front-end system. We provide a measurement of the authorisation time, which is the time that elapses between when an access request is submitted to the front-end system and when the request is acted upon. We investigate the impact of different voter population sizes on the authorisation time. The access-request transaction latency within the authentication DLT network is also investigated.

8.1. Prototype System: Voter Registration and Authentication Scheme

We evaluated the PVPBC voting system by implementing a proof-of-concept (PoC) prototype using a Quorum blockchain network. The implementation was divided into three phases: Quorum-network configuration, smart-contract development and deployment, and front-end application development using the REST APIs. For the PoC implementation of the PVPBC voting system, we built REST APIs using Node.js and Express. These APIs allowed us to access and to interact with the blockchain and the functions defined in the smart contract from the front-end system. We used Postman, which is an API development tool, which allowed us to send requests to the Rest APIs and to receive responses. Regarding the TTP and the front-end blockchains, a Byzantine fault tolerance consensus was implemented. The block time was maintained at the default value of 50 ms. We developed all smart-contract code in Solidity using the remix IDE, which allowed us to write, to test, and to debug our smart-contract transactions before deploying the contract on the blockchain network. Finally, we used web3.js and Node.js to interact with the Quorum blockchain and smart contract from the client side. We continue by describing the functionalities implemented in more detail. We then describe the performance evaluation carried out for each function. Finally, we evaluate the voting phase. The goal of this final experiment was to verify how the cost of casting a ballot varied as a function of the number of candidates.

Voter registration: The voter-registration process is described in Algorithm 1. The registerVoter() function in the register-smart-contracts component is used to register a voter on the blockchain. During voter registration, the TTP calls the smart-contract registrar to verify that the provided voter's ID ($voter_{id}$) is part of the whitelist. If the check is successfully passed, the voter is allowed to register themselves for the election ($Access_v$). The TTP handles the registration transaction, which adds the voter's information to the authentication blockchain. Subsequently, an access token is placed on the blockchain. The voter's access token ($Token_v$) is entered by the TTP responsible for pushing the contract to the blockchain. It is then stored in the contract state (smartContract).

Algorithm 1: Voter registration.

Input: $voter_{id}$
Output: VID_v,*invalid()*

$Access_v = false;$
$candidates = candidates;$
if $(voter_{id} \in candidates)$ **then**
  $Access_v = true;$
  $Token_v = (timestamp, expDate, Access_v);$
  smartContract= update($Token_v$);
  VID_v= address(smartContract)
  return VID_v;
else
  return $invalid(Access_v)$;
end

Time and gas cost: In order to determine the number of units of gas required by PVPBC to perform voter registration, we measured the gas required to deploy the smart contract as part of voter registration for 10 contracts. We repeated each experiment 30 times for each contract, and report the average time in seconds and cost for each contract. The results are tabulated in Table 2 for each contract. The average, of the average time required for each contract, is 34 s and the average gas cost for all contracts is 181000.

Authorisation: During the voter-authorisation phase, the voter submits an access request to the front-end system. This process is described in Algorithm 2. When an access request is initiated using the voter's VID (VID_v), the EA initially checks whether or not the token data associated with the user's address exists in the local database ($localDatabase$). If the search for token data fails, the EA interacts with the authorisation contract (querySmartContract(VIDv)), which fetches the token data from the TTP DLT network by calling an exposed-contract method and saving the token data to the local database. After retrieving the access token, the accessRequest() checks the current capability status of the token, such as the initialized, isValid, issuedate, and expireddate functions. If the status of any token is not valid, the authorisation process stops and a deny-access request is sent back to the subject. The EA goes through all the access rules specified in the retrieved token to decide whether to grant the access to the ballot, or not (e.g., Flag 'voted'). The process checks whether or not the request made by the voter (REST-ful method) to access the ballot matches the access rules in the retrieved access token.

Table 2. Voter registration: The average contract deployment time for 10 contracts, where the average computed for 30 experiments is given and the cost of gas is reported. The average time for deployment over all trials is 34 s and the associated average cost is 181000.

Contract	Time (s)	Cost (Gas)
1	32	182000
2	35	173000
3	33	185000
4	34	180000
5	35	182000
6	35	179000
7	34	184000
8	33	186000
9	35	175000
10	34	184000
Average	34	181000

Algorithm 2: Authorise Access (Function accessRequest()).

Input: VID_v
Output: $token_{Access}$
Search (VID_v);
if $(VID_v \in localDatabase)$ **then**
    $extractToken = token_v$;
else
    extractToken=querySmartContract(VID_v);
end
verifyTokenStatus(extractToken);
if *isValid()* **then**
    EvaluateAccessRight();
    return $token_{Access}$;
else
    Deny();
end

Time and gas cost: We measured the gas required to deploy the authorisation contract and the associated time. The experiments were repeated 30 times for each contract, and the results obtained are given in Table 3. The average time reported is 33 s, which is similar to the average time report for voter registration; however, the average gas cost reported is significantly larger, e.g., 549719.
Authorisation time: The duration between when an access request is submitted to the front-end system and when the request is granted or acted upon is called the authorisation time. To measure the authorisation time, we used the ConsenSys Quorum blockchain to emulate the authentication DLT. For the blockchain private network, we implemented three nodes, which were distributed over three Linux machines, to mimic the TTP nodes. We developed a PoC front-end system which allows the voter to submit the VID to the EA.

Table 3. Voter authentication: The average deployment time for 10 contracts and the associated costs are given for 30 experiments. The average of these contract deployment times is 33 s and the average cost is 549719.

Contract	Time (s)	Cost (Gas)
1	49	582033
2	25	539234
3	30	524892
4	35	489234
5	28	547834
6	20	542934
7	27	593485
8	42	534584
9	39	558394
10	35	584574
Average	33	549719

To evaluate the authorisation time, we conducted an experiment which measured the time that elapsed between when the voter access request was made and when the access request was granted. Specifically, we registered voters with the TTP using the registerVoter() function. Access tokens and VID's were created and placed on the DLT. The next step started when VIDs were sent at the same time, T_1, via the front-end application. The experiment used the accessRequest() function in the smart-contract authorisation. This function used VID to fetch the associated token data from the TTP DLT, and it checked the current capability status of the token. We measured the time, T_2, which was when the access token was retrieved and access was granted. The authorisation time, T_A, was measured as the time difference between T_2 and T_1,

$$T_A = T_2 - T_1. \tag{9}$$

In order to determine the impact of different voter population sizes, we ran the experiment several times and increased the voter population for each experiment. The voter population sizes used were 100, 150, 200 , 250, 300, 350, 400, 450, 500 voters. Figure 8 illustrates the average authentication time for each voter population size. Figure 8 provides evidence that the authorisation time increases linearly as a function of the voter population size. For example, the average latency associated with 100 voters obtaining access to the election process is on average 20 s. The average latency increases approximately linearly. When the voter population size is 500 voters, the recorded authentication time latency is approximately 80 s. This linear increase suggests that the voter authorisation time scales well over population sizes of 100 to 500 voters.

The access-request-transaction latency within the authentication DLT network is an important performance metric. The transaction latency in the blockchain environment is the time taken to complete information (transaction) verification. When a node receives a transaction, it verifies whether it is valid or not. If the transaction is valid, the node forwards it to its neighbours. Alternatively, invalid transactions are discarded. We conducted the following experiment to characterise this performance metric. The latency incurred when a valid authentication transaction is created and sent in the DLT network is measured. We also recorded the time at which the access token is retrieved. The goal was to understand the time required to send and receive a response to an access request, which was sent by the EA. We repeated the experiment 30 times for each function call using one client system. Figure 9 plots the latency incurred by this process.

The access-request transaction incurs an average latency of 6.275 ms in Figure 9. It is important to note that the block time of Quorum affects the value of the transaction latency in the system. The block time is defined as the time it takes to add a new block to the blockchain. We employed the default block time of 50 ms, which was set for the

RAFT consensus in this experiment. However, the research of [41] has shown that an increase in block time and transaction rate increases the latency of transactions in the system. This change is because transactions will have to wait longer before they are added to a block, thereby increasing the latency. Current blockchain consensus protocols are not well-adapted for e-voting systems, considering the fact that the existing proof- and voting-based consensus mechanisms may affect the performance of the e-voting systems due to latency and transaction throughput challenges. In future work, we will investigate and design a blockchain consensus protocol suitable for e-voting systems, where transactions are validated with lower latency to ensure fair and efficient performance without compromising security. A related line of attack could also involve adapting modern machine-learning approaches that estimate variation in propagation times [27], improving monitoring [42] and deep-learning classification schemes [29] that aim to perform better utilization of the underlying network resources.

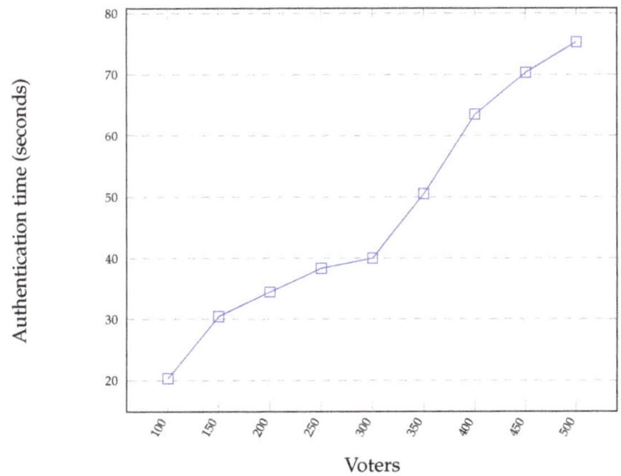

Figure 8. The voter authorisation time is plotted as a function of the number of voters. The authorisation time is an approximately linear function of the voter population size, when it is the range 100 to 500 voters, which underlines the scalability of PVPBC.

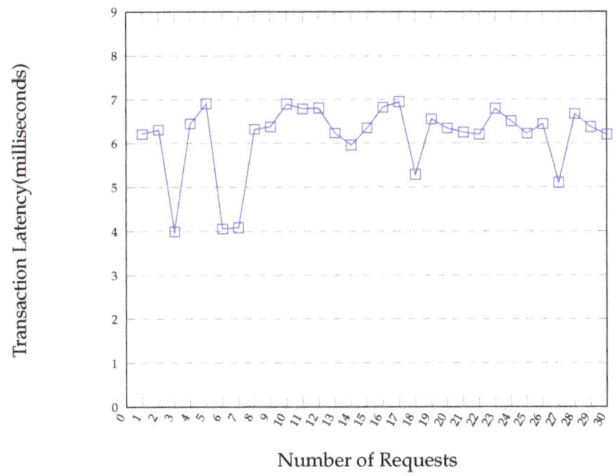

Figure 9. The time required to send and receive a response to an access request is illustrated as a function of the number of requests. The access-request transaction incurs an average latency of 6.275 ms.

8.2. Voting-Phase Performance Evaluation

We deployed our voting-phase implementation, which is described in Section 5.2, on the Quorum private blockchain platform in order to mimic its operation on a real-world, production network. We recruited different numbers of candidates to contest the election, in order to evaluate implementation. The number of candidates was increased from 1 to 7. In this experiment, the blockchain was used as a ballot box during the election phase to store the signed encrypted ballots as well as the tracker commitment (Beta). The goal of these experiments was to verify how the cost for casting a ballot varied with different numbers of candidates. Figures 10 and 11 show the average gas-consumption cost (resp. cost in GBP) for casting an encrypted ballot and tracker commitment based on different numbers of candidates. In this experiment, the cost was calculated in GBP. Based on the data points in Figure 11, it is reasonable to suggest that the costs in GBP increases approximately linearly as the number of candidates increases from 1 to 7. When there are 7 candidates, the cost is GBP 0.6. The cost for 2 candidates is approximately GBP 0.22 . The linear relationship described here to characterise the cost in GBP for casting and storing an encrypted ballot alongside tracker commitment as a function of the number of candidates is an appealing property of PVPBC.

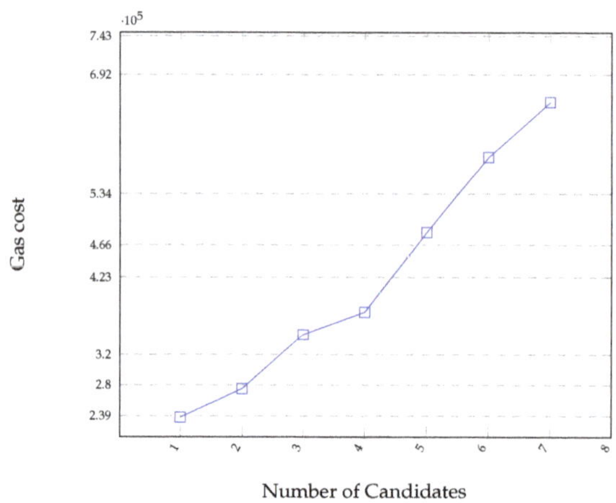

Figure 10. Gas cost for casting and storing an encrypted ballot alongside tracker commitment.

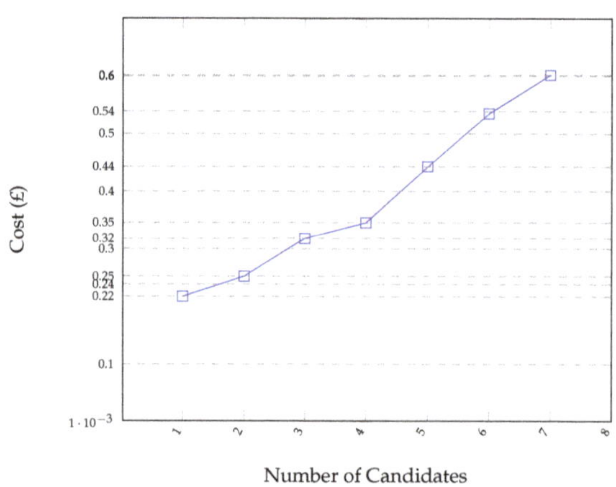

Figure 11. Cost in GBP for casting and storing an encrypted ballot alongside tracker commitment.

9. Conclusions

We proposed an e-voting system based on DLT called PVPBC. The primary advantage of this system is that it preserves voter privacy and verifiability. It achieves this by using DLT to ensure revocable anonymity, and to preserve voter privacy. An additional feature of PVPBC is that it uses advanced cryptographic primitives to support ballot privacy. Voters' verifiability is ensured by using the Selene voting scheme. To demonstrate that the provision of these features is achieved by PVPBC, we provided a detailed security analysis of PVPBC to establish how it preserves privacy, security, verifiability, integrity, fairness, and eligibility. Empirical evidence of the performance characteristics of PVPBC was then provided. We evaluated its performance on a prototype system, by measuring the time and cost required at the registration phase, authentication phase, and voting phase. In summary, the average time required by PVPBC to perform voter registration was 34 s and the associated average cost of gas was 181000 for each contract. During the voter-authorisation phase, the average time to deploy contracts was 33 s and the average cost of gas was 549719. The scalability of the system was then investigated by examining the average authorisation time as the voter population size increased. We provided evidence that the authorisation time was an approximately linear function of the voter population size over the range of population sizes examined. We also reported that access-request transactions incurred an average latency of 6.275 s. The inherent linear relation between various performance metrics and the number of voters and candidates is an appealing property of PVPBC. For example, we suggested that a linear relationship described the cost in GBP for casting and storing an encrypted ballot alongside tracker commitment as a function of the number of candidates. In summation, PVPBC delivers the following key features: privacy, verifiability, integrity, and, finally, eligibility and initial results suggest that the system is scalable.

Author Contributions: M.S.: Conceptualization; Methodology; Software; Validation; Investigation; Resources; Data curation; Writing—original draft preparation; Visualization; Project administration; Writing—review and editing. R.d.F.: Investigation, Conceptualization, Funding acquisition, Project administration Data curation, Writing—original draft preparation, Writing—review & editing. A.M.: Visualization, Data curation, Resources, Writing—original draft preparation, Funding acquisition. All authors have read and agreed to the published version of the manuscript.

Funding: This paper has emanated from research supported in part by a Grant from Science Foundation Ireland under Grant numbers 13/RC/2077_P2 and 15/SIRG/3459.

Data Availability Statement: Not applicable.

Conflicts of Interest: The authors declare no conflict of interest.

References

1. Tarasov, P.; Tewari, H. The future of e-voting. *IADIS Int. J. Comput. Sci. Inf. Syst.* **2017**, *12*, 148–165.
2. Jafar, U.; Aziz, M.J.A.; Shukur, Z. Blockchain for electronic voting system—Review and open research challenges. *Sensors* **2021**, *21*, 5874. [CrossRef] [PubMed]
3. Sallal, M.; Schneider, S.; Casey, M.; Dragan, C.; Dupressoir, F.; Riley, L.; Treharne, H.; Wadsworth, J.; Wright, P. VMV: Augmenting an internet voting system with Selene verifiability. *arXiv* **2019**, arXiv:1912.00288.
4. Kho, Y.X.; Heng, S.H.; Chin, J.J. A Review of Cryptographic Electronic Voting. *Symmetry* **2022**, *14*, 858. [CrossRef]
5. Mursi, M.F.; Assassa, G.M.; Abdelhafez, A.; Samra, K.M.A. On the development of electronic voting: A survey. *Int. J. Comput. Appl.* **2013**, *61*, 1–11.
6. Arapinis, M.; Kashefi, E.; Lamprou, N.; Pappa, A. A comprehensive analysis of quantum e-voting protocols. *arXiv* **2018**. arXiv:1810.05083.
7. Ryan, P.Y.; Rønne, P.B.; Iovino, V. Selene: Voting with transparent verifiability and coercion-mitigation. In *Financial Cryptography and Data Security: FC 2016 International Workshops, BITCOIN, VOTING, and WAHC, Christ Church, Barbados, 26 February 2016, Revised Selected Papers 20;* Springer: Berlin/Heidelberg, Germany, 2016; pp. 176–192.
8. Cruz, J.P.; Kaji, Y. E-voting system based on the bitcoin protocol and blind signatures. *IPSJ Trans. Math. Model. Its Appl.* **2016**, *10*, 14–22.
9. Kardaş, S.; Kiraz, M.S.; Bingöl, M.A.; Birinci, F. Norwegian internet voting protocol revisited: Ballot box and receipt generator are allowed to collude. *Secur. Commun. Netw.* **2016**, *9*, 5051–5063. [CrossRef]

10. Basin, D.; Radomirović, S.; Schmid, L. Dispute resolution in voting. In Proceedings of the 2020 IEEE 33rd Computer Security Foundations Symposium (CSF), IEEE, Boston, MA, USA, 22–26 June 2020.
11. Adida, B. Helios: Web-based open-audit voting. In *Proceedings of the 17th conference on Security symposium (SS'08)*; USENIX Association: Berkeley, CA, USA, 2008; pp. 335–348.
12. Chaum, D.; Ryan, P.Y.A.; Schneider, S. A practical voterverifiable election scheme. In Proceedings of the Computer Security—ESORICS 2005, 10th European Symposium on Research in Computer Security, Milan, Italy, 12–14 September 2005; pp. 118–139.
13. Ben-Nun, J.; Fahri, N.; Llewellyn, M.; Riva, B.; Rosen, A.; Ta-Shma, A.; Wikström, D. A new implementation of a dual (paper and cryptographic) voting system. In Proceedings of the 5th International Conference on Electronic Voting, (EVOTE), Bregenz, Austria, 11–14 July 2012.
14. Carback, R.; Chaum, D.; Clark, J.; Conway, J.; Essex, A.; Herrnson, P.S.; Mayberry, T.; Popoveniuc, S.; Rivest, R.L.; Shen, E.; et al. Scantegrity II Municipal Election at Takoma Park: The First E2E Binding Governmental Election with Ballot Privacy. In Proceedings of the 19th USENIX Security Symposium, Washington, DC, USA, 11–13 August 2010.
15. Cortier, V.; Fuchsbauer, G.; Galindo, D. BeleniosRF: A Strongly Receipt-Free Electronic Voting Scheme. *IACR Cryptol. EPrint Arch.* **2015**, *2015*, 629.
16. Clarkson, M.R.; Chong, S.; Myers, A.C. Civitas: Toward a secure voting system. In Proceedings of the IEEE Symposium on Security and Privacy (S&P 2008), Oakland, CA, USA, 18–21 May 2008; pp. 354–368.
17. Benaloh, J. Simple verifiable elections. In Proceedings of the 2006 USENIX/ACCURATE Electronic Voting Technology Workshop, Vancouver, BC, Canada, 1 August 2006.
18. Ryan, P.Y.A.; Teague, V. Pretty good democracy. In Proceedings of the Security Protocols XVII, 17th International Workshop, Cambridge, UK, 1–3 April 2009; Revised Selected Papers, pp. 111–130.
19. Lee, K.; James, J.I.; Ejeta, T.G.; Kim, H.J. Electronic voting service using block-chain. *J. Digit. Forensics Secur. Law* **2016**, *11*, 8. [CrossRef]
20. Meter, C. Design of Distributed Voting Systems. *arXiv* **2017**, arXiv:1702.02566.
21. Bistarelli, S.; Mantilacci, M.; Santancini, P.; Santini, F. An end-to-end voting-system based on bitcoin. In *Symposium on Applied Computing*; ACM: Rochester, NY, USA, 2017.
22. Faour, N. Transparent Voting Platform Based on Permissioned Blockchain. *arXiv* **2018**, arXiv:1802.10134.
23. Benabdallah, A.; Audras, A.; Coudert, L.; Madhoun, N.E.; Badra, M. Analysis of Blockchain Solutions for E-Voting: A Systematic Literature Review. *IEEE Access* **2022**, *10*, 70746–70759. [CrossRef]
24. Leng, J.; Zhou, M.; Zhao, J.L.; Huang, Y.; Bian, Y. Blockchain Security: A Survey of Techniques and Research Directions. *IEEE Trans. Serv. Comput.* **2022**, *15*, 2490–2510. [CrossRef]
25. Sallal, M.; de Fréin, R.; Malik, A.; Aziz, B. An empirical comparison of the security and performance characteristics of topology formation algorithms for Bitcoin networks. *Array* **2022**, *15*, 100221. [CrossRef]
26. Sallal, M.F. Evaluation of Security and Performance of Clustering in the Bitcoin Network, with the Aim of Improving the Consistency of the Blockchain. Doctoral Dissertation, University of Portsmouth, Portsmouth, UK, 2018.
27. de Fréin, R.; Izima, O.; Malik, A. Detecting Network State in the Presence of Varying Levels of Congestion. In Proceedings of the 2021 IEEE 31st International Workshop on Machine Learning for Signal Processing (MLSP), Gold Coast, Australia, 25–28 October 2021; pp. 1–6. [CrossRef]
28. Izima, O.; de Fréin, R.; Malik, A. A Survey of Machine Learning Techniques for Video Quality Prediction from Quality of Delivery Metrics. *Electronics* **2021**, *10*, 2851. [CrossRef]
29. Malik, A.; de Fréin, R.; Al-Zeyadi, M.; Andreu-Perez, J. Intelligent SDN Traffic Classification Using Deep Learning: Deep-SDN. In Proceedings of the 2020 2nd International Conference on Computer Communication and the Internet (ICCCI), Nagoya, Japan, 26–29 June 2020; pp. 184–189. [CrossRef]
30. Sako, K.; Kilian, J. Receipt-free mix-type voting scheme. In Proceedings of the International Conference on the Theory and Applications of Cryptographic Techniques, Saint-Malo, France, 21–25 May 1995; Springer: Berlin/Heidelberg, Germany, 1995; pp. 393–403.
31. Sallal, M.; Owenson, G.; Salman, D.; Adda, M. Security and performance evaluation of master node protocol based reputation blockchain in the bitcoin network. *Blockchain Res. Appl.* **2022**, *3*, 100048. [CrossRef]
32. Sallal, M.; Owenson, G.; Adda, M. Bitcoin network measurements for simulation validation and parametrisation. In Proceedings of the 11th International Network Conference, Frankfurt, Germany, 19–21 July 2016.
33. Pedersen, T.P. A threshold cryptosystem without a trusted party. In *Workshop on the Theory and Application of of Cryptographic Techniques*; Springer: Berlin/Heidelberg, Germany, 1991; pp. 522–526.
34. Wikström, D. *User Manual for the Verificatum Mix-Net Version 1.4. 0*; Verificatum AB: Stockholm, Sweden, 2013.
35. Estehghari, S.; Desmedt, Y. Exploiting the Client Vulnerabilities in Internet E-voting Systems: Hacking Helios 2.0 as an Example. *EVT/WOTE* **2010**, *10*, 1–9.
36. Iqbal, R.; Butt, T.A.; Afzaal, M.; Salah, K. Trust management in social internet of vehicles: Factors, challenges, blockchain, and fog solutions. *Int. J. Distrib. Sens. Netw.* **2019**, *15*, 1550147719825820. [CrossRef]
37. Abuidris, Y.; Kumar, R.; Yang, T.; Onginjo, J. Secure large-scale E-voting system based on blockchain contract using a hybrid consensus model combined with sharding. *Etri J.* **2021**, *43*, 357–370. [CrossRef]

38. Riadi, I.; Raharja, P.A. Vulnerability analysis of E-voting application using open web application security project (OWASP) framework. *Int. J. Adv. Comput. Sci. Appl.* **2019**, *10*. [CrossRef]
39. Feng, L.; Zhang, H.; Chen, Y.; Lou, L. Scalable dynamic multi-agent practical byzantine fault-tolerant consensus in permissioned blockchain. *Appl. Sci.* **2018**, *8*, 1919. [CrossRef]
40. Sohoel, H.; Jaatun, M.G.; Boyd, C. OWASP Top 10-Do Startups Care? In Proceedings of the 2018 International Conference on Cyber Security and Protection of Digital Services (Cyber Security), IEEE, Scotland, UK, 11–12 June 2018; pp. 1–8.
41. Baliga, A.; Subhod, I.; Kamat, P.; Chatterjee, S. Performance evaluation of the quorum blockchain platform. *arXiv* **2018**, arXiv:1809.03421.
42. de Fréin, R. State Acquisition in Computer Networks. In Proceedings of the 2018 IFIP Networking Conference (IFIP Networking) and Workshops, Zurich, Switzerland, 14–16 May 2018; pp. 1–9. [CrossRef]

future internet

MDPI

Article

Addressing ZSM Security Issues with Blockchain Technology

Michael Xevgenis [1,*], Dimitrios G. Kogias [2], Panagiotis A. Karkazis [3] and Helen C. Leligou [1]

[1] Department of Industrial Design and Production Engineering, University of West Attica, 122 43 Attica, Greece
[2] Department of Electrical and Electronics Engineering, University of West Attica, 122 43 Attica, Greece
[3] Department of Information and Computer Engineering, University of West Attica, 122 43 Attica, Greece
* Correspondence: mxevgenis@uniwa.gr

Abstract: Undoubtedly, we are witnessing a new era of computer networks that aspire to support modern demanding applications by providing the highest Quality of Experience (QoE) to the end user. Next Generations Networks (NGNs) ensure that characteristics such as ultra-low latency, high availability and wide service coverage can be met across the network regardless of the network infrastructure ownership. To accomplish that, beyond the necessary improvements in the radio propagation field, changes have been made in the core network functions which are now characterized as programmable, and software defined. Software Defined Networks (SDNs) and Network Function Virtualization (NFV) are the keystones of the NGNs flexibility. The high expectations of NGNs' performance and the continuous changes in the network conditions lead to the development of new network management frameworks that add elasticity and dynamicity and minimize human intervention. ETSI (the European Standards Organization) presents the Zero-touch Service Management (ZSM) framework that uses hyped technologies such as Artificial Intelligence (AI) and Machine Learning (ML) to achieve full end-to-end automation of the network services' management across one or many different domains. Focusing on multi-domain network service management, there are several security issues identified by the standardization team which mostly derive from the lack of trust among network providers. In the present research, we explore the suitability of blockchain technology adoption for facing these security issues. Blockchain technology inherently addresses security in trustless environments such as the infrastructures defined by the ZSM team. Our contribution is three-fold: (a) we define the architecture of a multi-domain network infrastructure that adopts the ZSM approach and integrates blockchain functionality, (b) we explore the adoption of different blockchain and distributed ledger technologies (DLT) approaches to address ZSM security needs and (c) we provide guidelines to prospective solution designers/implementers on the detailed requirements that this solution has to meet to maximize the offered value.

Keywords: zero touch networks; next generation networks; cross-domain resource management; blockchain/DLT

Citation: Xevgenis, M.; Kogias, D.G.; Karkazis, P.A.; Leligou, H.C. Addressing ZSM Security Issues with Blockchain Technology. *Future Internet* **2023**, *15*, 129. https://doi.org/10.3390/fi15040129

Academic Editors: Christoph Stach and Clémentine Gritti

Received: 17 February 2023
Revised: 24 March 2023
Accepted: 25 March 2023
Published: 28 March 2023

1. Introduction

Next Generation Networks (NGNs) offer network services based on technologies such as Software Defined Networking (SDN) and Network Function Virtualization (NFV). SDN and NFV reshape the nature of modern networks as they support network services via virtualized environments, without the need for a hardware networking device [1,2]. Therefore, the Providers (NPs) can easily trade (virtualized) resources to support modern Network Services (NSs) at different Quality of Service levels without having to resort to optimization of every single algorithm (from routing like [3] to the upper layer). In the new scene, the marketplace of resources grows as new players are entering the market [4–6]. These services are implemented by one or many Virtual Network Functions (VNFs) which are supported by a collection of computational resources in the form of Virtual Machines (VMs) or Containers (i.e., Dockers). The performance of the NSs both in terms of availability and latency affects the Quality of Experience (QoE) of the end-user [7,8]. Therefore, it is crucial to orchestrate the performance of the individual NSs [9,10]. Considering that modern networks must support applications with very different QoS requirements, the ability of flexible and agile provisioning of high availability, ultra-low latency and 100% coverage is of high importance [11,12] and SDN/NFV-enabled networks play a crucial role in this [13,14]. Massive, seemingly infinite capacity, imperceptible latency, ultra-high reliability, personalized services with extreme improvements in customer experience, global web-scale coverage, and support for massive machine-to-machine communication are only a subset of the requirements that these deployments should fulfill. The flexibility offered by the SDN/VNF architecture opens the opportunity for NPs to enhance the utilization of their resources by the dynamic reconfiguration and allocation of workload to the different devices/resources. For given NS demands, NPs can trade resources to support NSs with predefined network characteristics when needed. To maximize the benefits of resource sharing, NPs need a framework for a highly dynamic, self-optimized resource management process. Additionally, to avoid human errors and reduce the response time of the system, the resource management process should require minimum human intervention.

The requirements of NPs have led the research community to the development of the Zero-touch network and Service Management (ZSM) standardization, kicked off by the ETSI (the European Standards Organization) in 2017 [15]. The pivotal deployment of 5/6G and network slicing gave birth to the need for a radical change regarding the management and orchestration of modern networks and services. More specifically, there is a need to handle: (a) the increased overall complexity of networks derived from their transformation into programmable, software-driven, service-based and holistically managed architectures, (b) the unprecedented operational agility (i.e., real-time management of NS) required to support new business opportunities enabled by technology breakthroughs, such as network slicing. The ultimate automation goal is to enable largely autonomous networks which will be driven by high-level policies and rules; these networks will be capable of self-configuration, self-monitoring, self-healing and self-optimization without further human intervention.

Besides the important benefits stemming from ZSM introduction, a set of security challenges are also introduced in this highly dynamic, automated resource management environment, as already pointed out by their proposers [15]. The concerns are related to the untrusted nature of modern networks where large numbers of NPs are involved and the security level of the automated mechanisms (many times powered by Artificial Intelligence (AI) and Machine Learning (ML)). Since these automation mechanisms are responsible to make important decisions regarding the management of the network, the safeguarding of this mechanism is vital for the network's well-being. These mechanisms must not be compromised, and their decisions must not be manipulated or tampered with. Although several solutions have been discussed in the ETSI documents, the complexity of the system increases as multiple different techniques are combined.

On the other hand, Blockchain technology (one of the most hyped technologies in 2022) is adopted in many different use cases. The ability to establish trust in an untrusted environment, the data integrity and the transaction validity ensured in the absence of a trusted third party are the main characteristics that make blockchain attractive. To accomplish that, blockchain solutions run in a decentralized and distributed network of nodes that are characterized as public, private, permissioned and permissionless, offering different degrees of participation control. In the last few years, blockchain has been successfully adopted in several sectors beyond cryptocurrency, such as supply chain management, maritime and gaming with several distributed applications (Dapps) [16–20].

The current work proposes to adopt blockchain technology to address the security concerns mentioned by the ZSM standardization team in [21] and describes how blockchain technology can be used, i.e., presents the architecture of such an implementation and provides the lifecycle of a management service in a ZSM-supported scenario. Next, it qualitatively assesses this proposition to prove that it inherently addresses the security issues identified by the ZSM group. Furthermore, we identify the criteria against which an implemented solution should be evaluated. Considering that any prospective developer will use a basis for the solution of an existing blockchain or DLT platform, we evaluate indicative platforms (adopting Blockchains or Directed Acyclic Graphs (DAGs) structures), against the criteria mentioned earlier. To the best of our knowledge, this investigation and solution proposition has not been yet presented in other related works; they may mention that the use of a blockchain-based approach could be beneficial but the details of its design and implementation and the suitability of currently available blockchain approaches are not discussed.

This paper is organized as follows: Section 2 presents research works related to the use of AI/ML and blockchain in modern networks focusing on the topic of ZSM networks. Section 3 highlights the main requirements and elements of the ZSM framework and examines the case of cross-domain procedures placing emphasis on the security-related open issues involved in this scenario. In Section 4, the architecture of the blockchain-enabled ZSM framework is described in detail followed by an analysis of the main components and entities of this architecture. In Section 5, the criteria against which such a solution should be evaluated in real life are presented so as to guide the prospective developers to select appropriate baseline blockchain/DLT frameworks. Section 6 provides the assessment of the proposed architecture while Section 7 is devoted to the examination of baseline Blockchain/DLT platforms that are candidates for building a ZSM-oriented solution on top of them. Finally, Section 8 concludes the paper and additionally provides the potential and the next steps of this research direction.

2. Related Work

This section of the paper presents the related work with respect to (a) AI/ML techniques in resource management focusing on multi-domain scenarios and (b) blockchain-based solutions in NGNs and ZSM-compliant networks. The introduction of AI/ML for automation purposes succeeds in enhancing the automation level at the expense of introducing security vulnerabilities, which can be rectified by blockchain solutions.

2.1. The Role of AI/ML in the Implementation of the ZSM Concept

Artificial intelligence and Machine Learning techniques have been pursued to support the profiling of a service, the forecasting of the quality a service will experience for a given deployment scenario, and the placement of an NFV among others.

Dalgkitsis et al. [22], examine the use of Reinforcement Learning (RL) and more specifically, they leverage a Deep Deterministic Policy Gradient (DDPG) RL algorithm to solve the NFV placement problem in a scenario that consists of a Data Center (DC) and multiple Mobile Edge Computing (MECs) infrastructures. The goal is to minimize latency for ultra-Reliable Low Latency Communications (uRLLC). Uzunidis et al. [23], focus on the resource management process in NGNs and the proper network service profiling and placement in order to offer high QoE to the end user. In this work, authors present a framework to address the problem of service profiling and to predict the system's "critical points", focusing on complex services running over containers. In [24], the authors focus on the problem of choosing the proper amount of resources to support applications based on VNFs, especially in Mobile Edge Computing (MEC) environments where the computational resources are limited. They argue that AI technology can solve this problem in 5G networks, and they use it to develop a predictive autoscaling mechanism in NFV MANO that could automatically adapt the resources beforehand to the workload used by the application without any human intervention. Authors in this paper leverage Federated Learning (FL) techniques to design deep learning models for predictive Virtual MEC Application Functions (VMAF) autoscaling in a multi-domain setting that can better react to the changing service requirements, optimize the network resource usage, and also comply with data protection policies. Authors in [25] present a framework for Zero Touch Networks that use AI technology and microservices to perform self-orchestration of end-to-end network services. The presented research is part of the European H2020 program called

CHARITY. The goal is to increase the QoE by respecting Key Performance Indicators (KPIs), which are based on NGNs characteristics such as high availability and ultra-low latency. The outcome of this research is an Artificial Intelligence based Resource aware Orchestration (AIRO) framework in Cloud Native Environment that has been tested through simulation. However, the authors do not address the security issues when AI technology is used. This is anticipated to raise new challenges [26] when federated learning approaches enabling different NPs to contribute to the model training will be brought to the scene.

2.2. *Examining Blockchain as a Mean for the Security Enhancement in NGNs*

Authors in [27], present a combination of AI technology and DLTs in order to increase the security and trust in multi-operator mobile/cellular networks. The authors highlight the ability of AI to offer characteristics such as self-adaptation and self-reaction to next-generation networks which are susceptible to changes regarding the network conditions. This research is part of the 5GZORRO project, and its goal is to present a conceptual architecture of a solution that uses AI and DLTs. Another work of the same project [28] proposes the use of Smart Contracts (SCs) coupled with Cloud-Native operational Data Lakes to provide a zero-touch solution for the automated service assurance of multi-domain network slices. The SLAs which define the proper performance of the services are applied in the form of Smart Contracts (SCs) deployed in a blockchain network to increase the transparency of the process and to facilitate the integrity of the agreement. This research presents an architecture for a Smart Contract-based service assurance mechanism for network slices in a multi-domain environment that is SLA-driven. Additionally, this work aims to present a definition of AI-driven SLA breach detection and mitigation mechanisms implemented as modular Cloud-native services.

Benzaid et al. [29], describe the concept of Zero Touch Networks (ZTNs) and how AI can be used to automate the service management of modern networks. However, beyond the advantages of AI-driven ZTNs, security and trust are considered open issues by the authors when AI is used. According to the authors, it has been proven that ML techniques are vulnerable to several attacks targeting both the training phase and the test phase. Since data are used by the AI mechanism, their integrity and provenance are important for the proper operation of the mechanism. Authors claim that blockchain technology can be the antidote to these security limitations, due to its immutability and distributed nature, without providing any architecture or details for the design of such a solution.

In [30], the authors discuss the considerations regarding trust in modern multi-stakeholder networks and propose the use of blockchain technology to deal with trust issues. Smart Contracts (SCs) deployed in blockchain networks are ideal to create Service Level Agreements (SLAs) among stakeholders and control SLA violations in a transparent and secure manner. Based on the table presented by the authors, blockchain can be combined with many other technologies to solve trust and security issues in modern networks. Some of these technologies are VNFs, AI and ML. Moreover, sensitive data in modern networks can be protected using blockchain technology in order to guarantee their integrity and provenance. The authors discuss a use case where data are used as fuel for AI and ML focusing on the importance of data security and highlighting that data security is extremely important in AI/ML-based solutions. Data must be untampered and protected in order to avoid dataset poisoning which may lead to wrong decisions taken by the AI and ML mechanisms. In this use case, the data can be relevant to the service deployment parameters and the measured quality while blockchain technology could solve the security and trust issues.

In their survey paper, Liyanage et al. [31] present the progress of ZSM standardization and highlight the main goals and challenges. The security threats highlighted in this work by the authors are ML/AI-based attacks, open API security threats, intent-based security threats, automated Closed-Loop network-based security threats, and threats due to programmable network technologies. Moreover, the multidomain and heterogeneous nature of modern networks labels trust among different entities as a major issue. According to the authors these open issues have not been sufficiently explored, although there are some published ideas where the use of blockchain is discussed as a solution.

Concluding, to the best of our knowledge none of the existing works clearly propose an architecture to answer how cross-domain resource management could be implemented in a secure manner using blockchain technology in ZSM-aligned networks. The current research aims to present a blockchain-enabled ZSM architecture for the secure E2E service deployment where blockchain and ZSM are combined. Combining several studies together to address all the issues would not necessarily result in solving all of them as unintended interactions/interplays may be revealed. Additionally, adopting the definition of complexity of [32], the combination of technologies and the introduction of large numbers of components would result in a significant complexity increase, while relying on a single technology (blockchain technology) introduces less complexity.

3. The ZSM Framework Overview: Architecture and Open Security Issues

ZSM is expected to become one of the dominating frameworks of NGNs according to ETSI and many articles in the literature such as [31]. The goal is the development of a framework that will include solutions and management services to achieve orchestration and automation of the emerging end-to-end network slicing technology, as well as of the end-to-end, cross-domain service orchestration and automation. Additionally, the ETSI standardization team works on generic enablers and solutions for closed-loop as well as on advanced topics for next-generation closed-loop operations. In this course of action, ZSM has highlighted the use of Artificial Intelligence (AI) and Machine Learning (ML) technologies in NGNs aiming to leverage the benefits they provide towards the automation and optimization of the service management process [33].

The reference architecture of the framework is presented in [34] and enables the definition of the functionality and the requirements that should be met in any ZSM implementation. For example, in cross-domain services, QoS requirements and service interoperability should be guaranteed across different management domains. In a multi-stakeholder scenario, a Management Domain (MD) is usually the administrative area of an NP that is responsible for the proper functioning of services running in this area. When E2E cross-domain services are deployed, the ZSM framework should guarantee the proper collaboration of MDs in order to support the E2E service with appropriate resources.

One of the main factors that affect the performance of the service is the time needed for the management tasks to be completed. The management tasks related to the deployment of E2E services should be executed within the limited processing time according to the ZSM reference architecture requirements. Functional and non-functional requirements defined by the ETSI standardization team in their manuscript determine the successful operation of the framework. The satisfaction of those requirements ensures the efficient operation of the network.

3.1. Description of ZSM Architecture and Main Elements

The main blocks in the ZSM architecture are illustrated in Figure 1 and are the following: the management services, the management functions, the management domains (MD), the end-to-end (E2E) service management domain, the cross-domain integration fabric and the data services. Management services are the core component as they can be offered and consumed by other services and ZSM participants to support network services and applications. The management services consumption and/or offering is conducted using management functions as presented in Figure 1. A management function can either be a "management service producer", a management "service consumer", or both at the same time. Moreover, management domains are used to define different areas of responsibility that belong to different ZSM participants. Each management domain can use its own management services or services offered for consumption by other management domains using the ZSM framework. The E2E service management domain depicted in the upper part of Figure 1, is a special management domain that provides end-to-end management of customer-facing services, composed of the customer-facing or resource-facing services provided by one or more management domains. The "cross-domain integration fabric" located at the center of Figure 1 is responsible for the interoperation and communication between the management functions within or across different domains. The registration, discovery and invocation of management services and the communication between management functions are implemented by the integration fabric. Finally, data services enable consistent means of shared management data access and persistence by authorized consumers across management services within or across management domains.

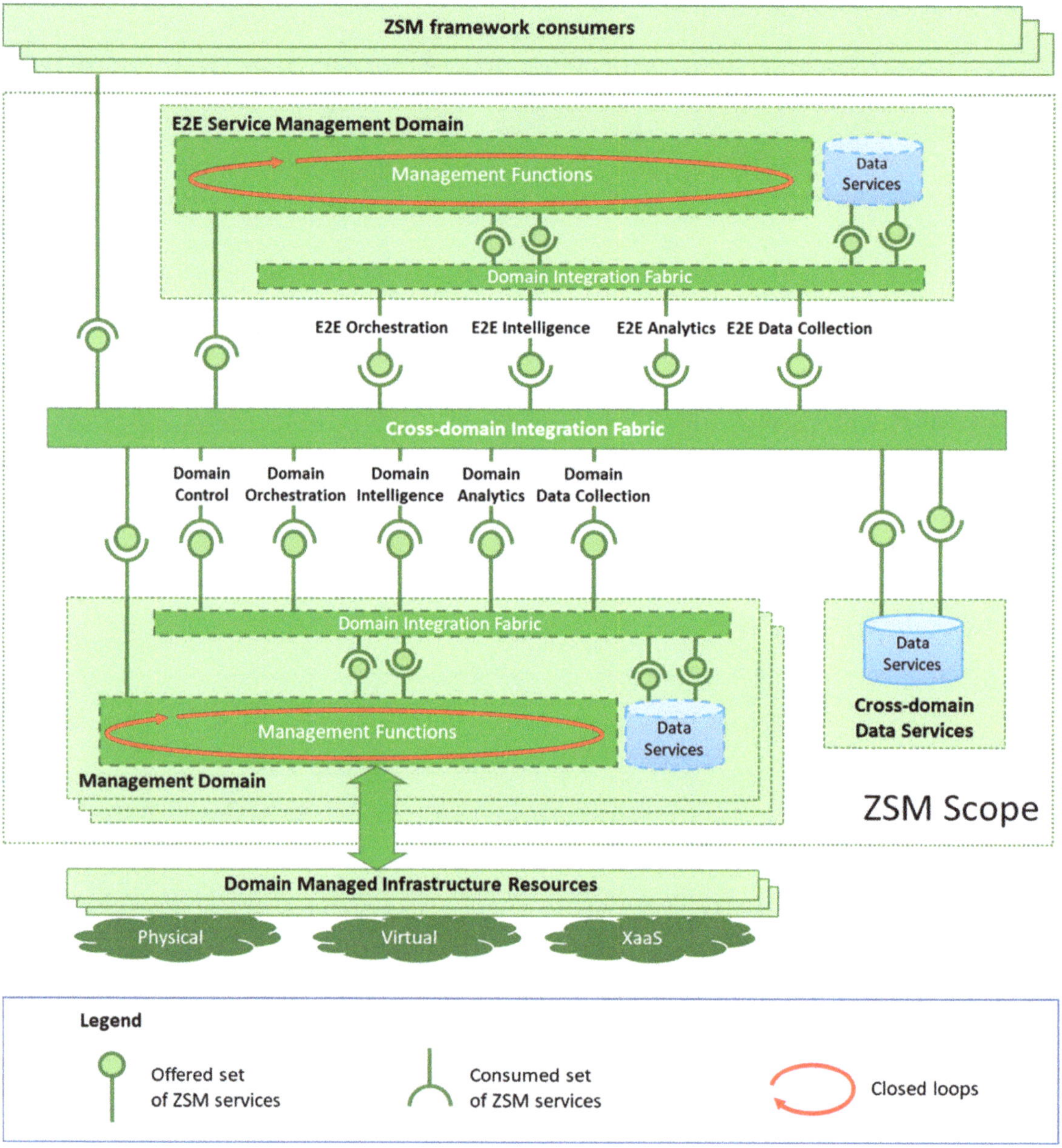

Figure 1. The ZSM architecture [34].

Considering the role of the aforementioned blocks, an example of ZSM operation is examined: supposing there is a multi-domain network where a network provider (say NP1) is responsible to support a demanding network service with characteristics that require a specific set of resources to provide the necessary QoS level. In the presented scenario, assume that NP1's resources are unavailable in the area where the service must be deployed. According to the standard's functionality, NP1 exploits the ZSM elements, such as management functions and cross-domain integration fabric, to find and consume a management service offered by another network provider (say NP2), which implements the necessary actions to cover the needs that NP1 has defined.

3.2. Open Security Issues

Among the requirements defined in [34], there are certain open security issues that are extremely vital for the ZSM's operation. The main issues and security risks of the ZSM framework as defined in [21], are as follows (and are also tabulated in Table 1):

- The trust relationship between multiple management domains: As new NPs form their management domain and embrace the ZSM concept, the collaboration among different domains in an automated manner requires a level of trust. Their proper operation is based on a service level agreement (SLA) signed among NPs (NP1 and NP2 in the example provided above), which must facilitate the proper network conditions for the desired operation of E2E service.
- Security risks introduced by the vulnerability of management function and security assurance of ZSM management function: Since the core functionality of ZSM is based on management services, the possibility of a security threat breach in the operation of those functions would be catastrophic. Therefore, the immutability and the high-security level of management functions are extremely important in ZSM networks.
- Security isolation and security requirement fulfillment in a multi-tenancy environment of ZSM framework: The multitenant nature of ZSM networks should not affect the security of services supported by virtualized resources. The isolation feature inherited by the virtualization technology that is used in modern networks increases the security level which should be high in every tenant of the network.
- Access control for management service provided by multiple domain service producers of ZSM framework: taking into consideration that numerous NPs provide a management service, the access over this service should be controlled and supervised in order to identify any malicious activity and avoid service malfunction. The normal functionality of these services should be safeguarded since they are the heart of the ZSM framework.
- Leverage existing security specifications to identify security risks of AI/ML models and protect AI/ML models in the ZSM framework: Although AI/ML are key technologies of the ZSM and increase the automation level of modern networks by introducing characteristics such as self-adaptation and self-optimization, their susceptibility in malicious attacks is a major issue. Models used in these technologies are trained using data sets that might be tampered. This type of attack is called dataset poisoning and may lead to wrong AI/ML decisions which are major threats to the framework's proper functionality [35].

Table 1. Solutions to security issues: Current ZSM vs. Blockchain-based ZSM.

Security issues and complexity	Trust relationship between multiple management domains	Security risks due to the vulnerability of management functions	Access control for management services in a multi domain scenario	Security risk of AI/ML model and protection of AI/ML models in ZSM	Complexity level
Current ZSM approach—Solutions	Reflective and adaptive trust model	GSMA Network Equipment Security Assurance Scheme (NESAS)	Authentication and authorization mechanisms, which check the trust relationship among entities	Risk assessment based on the Adversarial ML Threat Matrix	High—Use a bunch of technologies to increase ZSM security
Blockchain-based ZSM approach—Solutions	Achieve trust in trustless environment using blockchain	Use of SCs to eliminate the vulnerabilities which are automatically and securely executed	Use SCs for management services and apply visibility rules to achieve access control	Store the AI/ML decisions in the ledger and guarantee dataset integrity using both blockchain and IPFS	Low—Use blockchain technology and take advantage of its inherent characteristics

In [21], the standardization team proposes countermeasures to overcome the security issues mentioned above. Regarding the trust relationship among entities, they propose a reflective and adaptive trust model to build mutual trust among entities in the ZSM framework. The goal of this process is to ensure the confidentiality, integrity, availability, and regulatory compliance of every MD. To accomplish this, each entity that owns an MD needs to evaluate the trustworthiness of the other entity also owning an MD, based on threat and risk analysis and by examining the security policies applied in the entity. The outcome of this process leads to the building of a trust relationship among entities, followed by authentication procedures between parties and the formation of a secure channel where the behavior of each entity is tracked. Although this solution seems to tackle the trust problem, other approaches can also be investigated.

The safeguarding of management functions which are crucial for the operation of ZSM is addressed using the GSMA Network Equipment Security Assurance Scheme (NESAS). This methodology defines security requirements and performs an assessment for secure product development and product lifecycle processes, using 3GPP's defined security processes for the evaluation of network equipment. Although this solution is tested for the security assurance of network equipment following the 3GPP's Security Assurance Methodology (SECAM), other technologies could be studied to protect management functions. Additionally, the multitenancy issue is answered using policies applied to each tenant that uses the ZSM framework. The policy mechanism aims to provide a sufficient security layer for the users of the ZSM framework to avoid the exploitation of multi-tenancy which may lead to the loss of sensitive data of E2E services and the loss of the frameworks' reputation. However, this solution is based on security requirements defined by authors and cannot by itself be considered as a high-security level solution. Moreover, the access control of management services (MnS) is another major issue, as the exhaustive usage of management resources by a malicious entity may cause the mis-operation of these critical services. Robust access control mechanism, including identification processes, authentication, authorization and audit of MnS usage, should be applied to prevent MnSs and other management resources of the ZSM framework from being misused by MnS consumers, according to the authors. Considering that ZSM is implemented in a multi-domain environment, the standardization team proposes techniques to enhance the security of MnSs by introducing authentication and authorization mechanisms, which check the trust relationship among entities in ZSM.

AI and ML are the main technologies used in the ZSM framework and their reliability and robustness should not be left open to dispute. The decisions of AI/ML affect the ZSM network operation as they perform closed-loop operations for the efficient deployment of

E2E cross-domain services. A comprehensive risk assessment for the AI/ML vulnerability issue based on the Adversarial ML A Threat Matrix is presented in [21] and possible countermeasures are proposed at a high level only.

4. The Architecture of the Blockchain-Enabled ZSM Approach

The proposed blockchain approach aims at increasing the security of the ZSM framework by addressing the aforementioned security issues and at the same time maintaining the complexity level low (as will be explained) using the blockchain technology and the benefits it inherently provides. Focusing on the scenario of E2E service deployment in a multi-domain environment and having in mind the architecture described in [34], we propose the introduction of blockchain technology in the cross-domain integration fabric component as it is depicted in Figure 2.

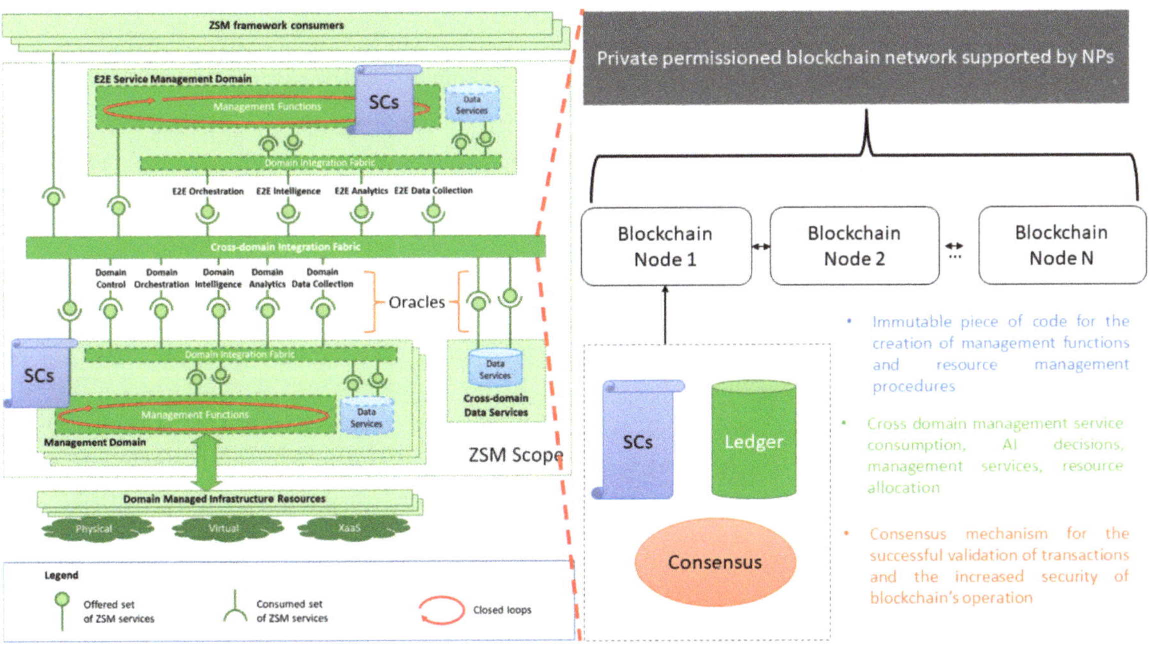

Figure 2. The architecture of blockchain-based ZSM.

In the presented approach, each NP is part of the ZSM framework and hosts a blockchain node that belongs to a *private permissioned blockchain network* as depicted in the figure above. The private and permissioned characteristic of the network increases the security of this approach as we are able to control which NPs participate in the network and at the same time minimize the possibility of a malicious participant. This is also the approach adopted in [36] for Secure Routing for Multidomain SDN-Enabled IoT Network. The NPs are registered in the blockchain and obtain a unique address (IDs) used for their identification in the network. In addition, the ledger of the blockchain includes not only the IDs of the NPs but also the addresses of the Smart Contracts (SCs) deployed in the network. Both the E2E service management domain and the management domain of the ZSM participant create and execute *management functions, which are deployed in the form of SC in the blockchain network*. According to the ZSM standard, the development and execution of management functions are implemented using closed loops. Closed loops are based on AI/ML technology which uses mathematical models trained by secure datasets of the framework. It is worth mentioning that no AI/ML code runs inside the blockchain. It is *the decisions* generated by the closed-loop mechanisms that *are registered in the blockchain* so as to ensure the immutability and traceability of the decisions. Additionally, every change in the ZSM network in a cross-domain scenario, which in our case can be *the consumption* of a management function, is considered a transaction and *is stored in the ledger*. The registration of an NP, the creation of an SC that utilizes a management function and the outcome of an SC are considered blockchain transactions and are permanently written in the ledger of the blockchain. Moreover, the blockchain interacts with other ZSM components using *oracle mechanisms to ensure that valid information is exchanged from and towards the blockchain*. Oracles in our case are software mechanisms developed to provide a secure interface between the blockchain network (including the SCs deployed in it) and ZSM services. To accomplish that, oracles use cryptography or/and consensus techniques applied on-chain or off-chain, to establish a secure connection between blockchain and other services outside of it. In the current research, oracles are used by the cross-domain integration fabric component as presented in Figure 2.

Let us examine how the multi-domain scenario described in the previous section changes with the integration of blockchain technology. Assuming that NP1 has a request to support a demanding streaming application based on predefined network services. In this scenario, NP1 cannot support the application using its own resources and uses the ZSM framework to complete the request. Using the management services, NP1 finds another management service in a different domain that can fulfill the request. NP1 decides to consume the management service of the other provider (i.e., NP2) by executing a management function in the form of SC. An example of this SC is described in an abstract manner by the following pseudocode.

```
NP1.ZSM(search_MnS[network_resources, MD]) = true;
ZSM.returns true; "This is the result of ZSM that triggers the SC via the oracle mechanism"
function requestResources(NP1, NP2, MnS) public returns{
    if NP2 is true:
            NP2[id].lease[MnS] to NP1
        transfer_resources(NP2, NP1);  "This is a transaction stored in the ledger"
        oracle.transaction(NP2,NP1, MnS); "Triggers ZSM to implement the agreed NSs"
  }
```

Pseudocode 1. Example of a Management Function in the form of SC.

The consumption of NP2's service by NP1 is registered as a transaction in the blockchain and the details of this transaction are defined by the SC. Finally, since the network is zero-touch, the decision regarding the consumption of a management service by the NP1 provider can be made by an AI/ML mechanism. Blockchain stores the decisions of AI in the form of transactions. (It could additionally be used to check the validity of datasets stored in an Inter Planetary File System (IPFS) structure which were used to train the ML model). Figure 3 illustrates the complete lifecycle of the described resource management operation that follows the blockchain-enabled ZSM approach.

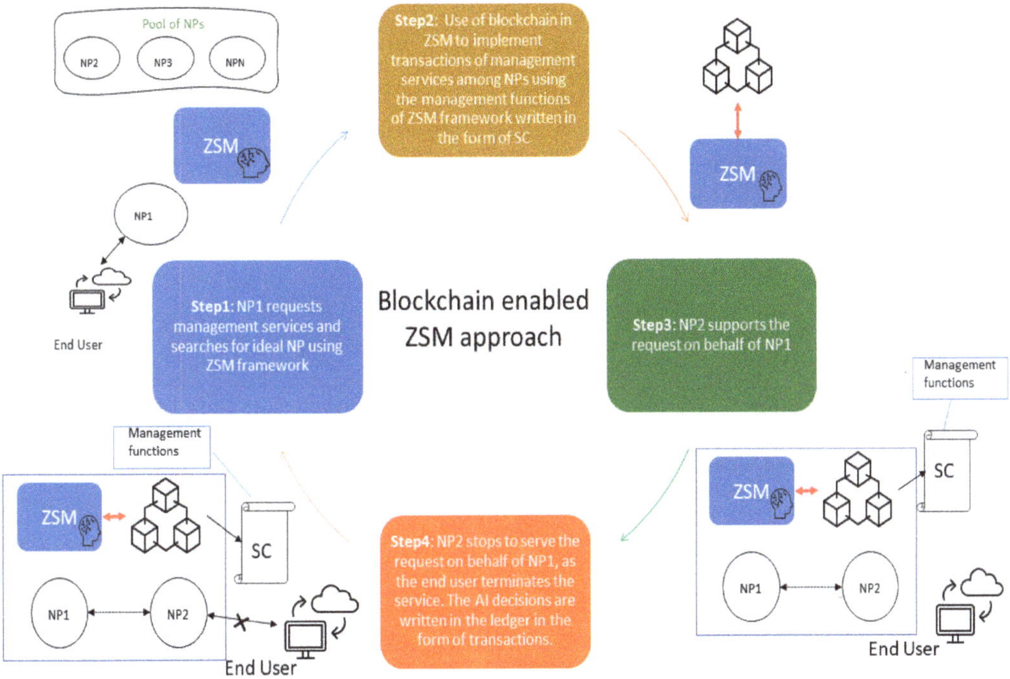

Figure 3. Lifecycle of a blockchain-enabled ZSM scenario.

5. Criteria for the Selection of the Blockchain/DLT Platform to Build the ZSM-Tailored Solution

The selection of the most suitable DLT solution for the development of a blockchain-based ZSM scenario should be based on specific criteria which should be clearly defined. The following section of this paper aims to present the characteristics that a DLT solution should present in order to satisfy the needs of the ZSM use case. Furthermore, we proceed to the identification of the most suitable blockchain and DAG solutions, and we analyze their main functionalities and characteristics. Then we evaluate their suitability for the particular use case and we propose modifications to fulfill the requirements of our scenario.

The performance of the blockchain network directly affects the performance of the overall system which means that characteristics of the network such as latency and throughput affect the end user's QoE. Furthermore, other characteristics such as the accessibility of the network, the resiliency, the scalability and the network's ability to easily accept new features, are vital for the systems' successful operation and future growth. To this end, we define certain characteristics that a DLT solution should present in order to build a secure, scalable and high-performance environment ready to be integrated with the ZSM framework. The DLT solution integrated with the ZSM framework must be:

- ✔ Access controlled: the access and the ability to perform actions in the network should be allowed only to permitted members and should be restricted to anyone else.
- ✔ Scalable: the nodes of the network are hosted by the NPs only. Every provider who wishes to join the network should be able to easily deploy a node and become an active member of the solution.
- ✔ Resilient: the proper functionality of the network should not be affected if a node or a number of nodes become unavailable for a period of time. The network should be resilient to network failures in order to present high availability which is one of the main requirements of NGNs.

- ✔ Very fast: the network should be able to perform multiple actions in a short period of time (in the unit of seconds or even milliseconds), to meet the dynamicity and resource management agility target by NGNs.
- ✔ Programmable: the DLT adopted in this particular use case should be able to support the development of code to implement the necessary functions which proliferate over time.
- ✔ Extensible: the DLT chosen for the implementation of this scenario should be able to accept new features that will upgrade its functionality and cover needs that may occur in the future.
- ✔ Interoperable: the DLT should be able to support interaction with services outside the network in order to successfully communicate with ZSM services. This characteristic increases the ability of the network to interact with the outside world and use services that may leverage the functionality of the whole solution.

The above characteristics are translated in blockchain terms as follows:

- ✔ Permissioned network: only the permitted entities can access the network and perform actions in the form of transactions. This feature increases the security of the system since every entity is known to others, and therefore, it is less possible to act maliciously. It should be noted that the government entity does not control the operation of the network and its role is restricted only to the authorization of the NPs.
- ✔ Private/Consortium network: the network is supported by the NPs only, which means that the nodes of the network are created, managed and maintained by the participants. Although this feature seems to question the sentiment of decentralization in the network, it increases the security of the system since only the NPs are responsible for the proper operation of the network. In this particular use case, an NP who wishes to become an active member of the network should be able to easily deploy a node that will automatically become part of the network. This results in the growth of the network as new NPs with new nodes can easily become part of the solution. As a result, the scalability feature of the entire solution is highlighted, which is a factor that attracts new members.
- ✔ Crash fault tolerance: the network should support fault-tolerant mechanisms in order to ensure that its operation is not affected if a number of nodes become unavailable. Considering that the ZSM framework uses the DLT network to perform crucial tasks, the resiliency of the network is vital in our scenario. The main element that is responsible for the network's proper operation is the consensus. Therefore, the network should be able to use consensus mechanisms that will increase its fault tolerance and guarantee the functionality of the system in hazardous situations.
- ✔ Low consensus convergence time: the proposed solution is expected to receive a large number of transactions which must be validated/executed with the minimum possible delay. As a result, the selection of a consensus algorithm with a low convergence time that is able to cope with a high number of transactions is crucial. This characteristic is related to the previous one, as the consensus mechanism used is able to increase the resiliency of the network and the speed of transaction validation. Therefore, the DLT solution should be able to support a consensus mechanism that can validate transactions fast, and at the same time tolerate failures.
- ✔ Support of SCs: the development of code in DLTs can be accomplished by the creation of SCs that can implement various functionalities of the network in a secure manner. Since the network should interact with ZSM services and provide solutions to many different problems regarding the management of modern networks, the capability of the DLT to support SCs is extremely important.
- ✔ Support of tokens: this feature is related to the previous one and highlights the need to support tokens implemented in the form of special SC functions, which improve the functionality of the solution. Tokens can be used to represent assets, currency and rights (i.e., use a token to access a website). The development of tokens is based on standards that ensure their smooth integration into the network. There are two main well-known token categories, Fungible Tokens (ERC 20) and Non-Fungible Tokens

(ERC 721). In our use case, a Fungible token could be used to create a currency used for the transactions among NPs in order to build a modern marketplace. Additionally, Non-Fungible Tokens (NFTs) could be developed to represent the reputation of an NP which could be related to the successful completion of a number of requests. This feature could be used by NPs as a criterion during the process of NP selection to support their request. It should be mentioned that these scenarios are only examples of how tokens could be used and have not yet been designed for our use case. However, the ability of a network to accept and use new programming features to enhance its functionality and solve any future issues is a very important characteristic.

✔ Interaction with oracle mechanisms: the DLT solution should be able to interact with entities outside the network in a secure manner to leverage the functionality of the whole system and support the ZSM processes. To this end, oracle mechanisms must be supported by the network, while the selection of the most suitable solution should be made based on the security level and performance. On the one hand, the information from and towards the network should be well protected while on the other hand, the latency introduced by the oracle should be the minimum. Having in mind that some oracle solutions use consensus, which automatically introduces extra latency to the system, the selection of other oracles that use different tools seems to be preferred. There are many oracle mechanisms available that use encryption to protect the content of their data and guarantee the origin of the information. The adoption of such a solution may not affect the overall system's performance dramatically.

6. Assessment of the Architecture

We examine how the proposed solution tackles the security issues mentioned in Section 3.2 one-by-one in the sequel and also tabulate them in Table 1.

As new NPs join the ZSM framework, the number of blockchain nodes increases and the network grows, assuming that each NP hosts/deploys at least one blockchain node. The private and permissioned characteristics of the network minimize the possibility of the existence of a malicious player which is also tackled by the applied consensus mechanism. Since the blockchain network is private and permissioned, we assume that a trusted governance entity is responsible for registering the NPs to the network. Automatically, a trust layer among competitive NPs is created and the *trust issue among multiple management domains* (security issue 1) highlighted by the ETSI team is addressed.

To reduce the *vulnerabilities of management functions* (security issue 2) we take advantage of the immutability feature of Smart Contracts (SCs). We propose the use of SCs for the implementation of the management functions defined in [34]. The rationale behind this is the following: an SC is an immutable deterministic piece of code stored and used in the blockchain network. An SC's functionality cannot be undermined, and its content cannot be tampered with as it is stored in the ledger. When an SC is created, it is related to a unique blockchain address used by other entities in the network in order to execute its functions. Additionally, an SC is a set of promises that are executed when predefined conditions are met. This feature allows SCs to execute functions automatically without human intervention. Given the *security concerns regarding the vulnerability of management functions* in ZSM, the use of SCs for their implementation is ideal.

Moreover, the ability to control an SC's visibility to other blockchain participants is supported in various blockchain solutions and can be used to increase the confidentiality of a transaction or the non-disclosure of SC's information in multitenant environments, if this is required. As a result, we can achieve *access control to sensitive information*, such as management functions, stored in the network.

Blockchain can also be used to monitor the behavior of AI/ML by storing the decisions in the form of transactions. The traceability feature of blockchain allows us to examine the decisions of AI/ML components during their operation and identify any suspicious activity. At the same time, the credibility of the decisions' history cannot be questioned since it is a valid blockchain transaction registered in the ledger. Furthermore, the training of mathematical models used in these technologies is based on datasets that should be safeguarded. Although the first thought would be to store datasets in the blockchain, the scalability issue of this technology forces us to design an alternative solution. Datasets which usually include enormous amounts of information can be stored in Inter Planetary File System (IPFS) distributed structures and the link that points to the data location can be stored in the form of a hash as a valid transaction. As a result, the origin and quality of data are guaranteed, and an *extra layer of security is added to the AI/ML components of ZSM.*

Finally, with respect to the complexity, we adopt the definition presented in [32] where the complexity level of the system is defined based on the number of parts comprising the system. According to the definition, a system that uses a smaller number of parts is less complex than a system that uses a higher number of parts. In this paper, the solutions proposed by the standardization team use many different technologies and techniques (e.g., NESAS, Adversarial ML Threat Matrix) to tackle security issues. Our approach aims to eliminate these issues by using only blockchain technology.

It is worth stressing that based on our experience from the implementation of a solution that supports resource trading among NPs (which is presented in [37,38]), the transaction speed using the Ethereum-Quorum platform as the basis for the solution is in the order of a few seconds and highly depends on the adopted consensus mechanisms while the number of nodes in the network does not have a significant impact on this performance aspect. Furthermore, the infrastructural resources required for the implementation of this functionality are definitely affordable by a network provider. In our experiments presented in [37,38], we implemented a blockchain network supported by blockchain nodes in the form of VMs with the following characteristics: four CPU cores, 8GB memory (RAM) and 30GB storage. Additional insights on the parameters that affect the performance of blockchain-based solutions are provided in [39,40].

7. Assessment of Blockchain and DAG Platforms to Be Used for the ZSM-Solution Implementation

Having defined the main characteristics that a DLT solution should present to become the ideal choice for our scenario, we proceed to the assessment of several well-known DLT solutions. In this section of the paper, we identify the main characteristics of Hyperledger Fabric (HLF) [41], Ethereum Quorum [42] and R3 Corda [43] which belong to the blockchain family and then we focus on two DAG solutions, the IOTA [44] and the Hedera Hashgraph [45]. The reason for their selection is that they present characteristics that are more likely to fulfill the requirements of our scenario. At the end of our assessment, we tabulate our findings to guide prospective implementers.

7.1. Candidate Blockchain and DAG Solutions for Our Approach

7.1.1. Hyperledger Fabric (HLF)

This blockchain solution is an open-source permitted platform that is established and maintained under the umbrella of the Linux Foundation. HLF's architecture is modular and configurable in order to easily adapt to a wide spectrum of industry use cases. The versatility of this platform makes it ideal for several sectors such as healthcare, supply chain and others, while its ability to support SCs written in general-purpose programming languages (i.e., Java, Go and Node.js) makes it very attractive to organizations. In addition, this platform is permission; therefore, the participants are known to each other which automatically grows a sentiment of security. This means that while the participants may not fully trust one another (they may, for example, be competitors in the same industry), a network can be operated under a governance model [41].

Another important characteristic of this platform is its ability to support pluggable consensus mechanisms. This feature allows HLF to be effectively customized in order to fit various use cases. For instance, in the ZSM scenario where only known NPs are members of the network, a fully byzantine fault tolerant mechanism might be considered unnecessary and an excessive drag on performance and throughput. In order to maintain high-performance standards and increase the availability of the solution, a crash fault-tolerant consensus might be the preferred option. This modular architecture allows the platform to rely on well-established toolkits for crash fault-tolerant or byzantine fault-tolerant ordering. Fabric currently offers a crash fault-tolerant ordering service implementation based on the etcd library of the Raft protocol. Moreover, Fabric can leverage consensus protocols that do not require a native cryptocurrency to incent costly mining or to fuel smart contract execution.

The aforementioned design features make HLF one of the better-performing platforms both in terms of transaction processing and transaction confirmation latency, and it enables privacy and confidentiality of transactions and the smart contracts that implement them. It should be mentioned that many research papers have been published where the performance metrics of the HLF are studied and tested using the Hyperledger Caliper. Authors in [46] scaled HLF to 20,000 transactions per second [41]. Concluding the presentation of HLF, this solution supports the creation and management of tokens and is able to use several oracle mechanisms in order to become suitable for several use cases that demand the interaction of blockchain with the outside world.

7.1.2. Ethereum Quorum

Quorum is a permitted implementation of Ethereum and it was initially developed by JP Morgan. The goal of this blockchain is to cover the needs of scenarios designed to operate in a controlled network where the identity of the members is known and access to the public is restricted. Therefore, it is considered an ideal solution for the implementation of private and consortium networks.

In contrast to the traditional Ethereum network, Quorum supports two different types of consensus mechanisms: the Raft and the IBFT. This feature allows developers to use a mechanism that suits better to their use case. For example, the Raft which belongs to the Crash Fault Tolerance (CFT) consensus family is preferred in cases where the existence of a malicious participant is unlikely and the need for fault tolerance is high. Nevertheless, in cases where many different entities are participating in the Quorum network and the likelihood of a malicious member is high, the IBFT mechanism is preferred since it introduces byzantine fault tolerance.

Similar to Ethereum, the network of Quorum supports the use of tokens and SCs that allow the creation of distributed secure applications. This feature broadens the application area of Quorum as many DApps can be developed to implement various scenarios. However, a major difference between these two blockchains is that Quorum supports privacy which was one of its main design goals. More specifically, it allows subsets of parties in a consortium to transact with one another without making the transactions public to members of the larger consortium. Quorum practically splits the ledger into a public and a private ledger. All nodes of the network can observe the public ledger, while the private ledger is visible only to the transacting parties. Only a hash of the private transaction appears on the public ledger and is visible to other nodes that are not counterparties to the transactions. This process can be conducted also for the deployment of private smart contracts which would be visible only to the transacting parties [47].

Moreover, another significant difference between Quorum and Ethereum is the fact that Quorum does not adopt the concept of adding cost to a transaction using gas. Although it is a fork of Ethereum and supports the use of gas, it sets this value to zero to run transactions without gas fees. Since Quorum is usually deployed in a consortium or private blockchain, the use of gas in Ethereum terms is not mandatory [47]. Additionally, the Quorum platform can be combined with the oracle mechanism in order to become part of a solution that is not limited only to the blockchain world.

7.1.3. R3 Corda

Another popular blockchain platform is the R3 Corda, which allows the implementation of private permitted networks ready to support various use cases. The consensus process can be implemented using either a crash or byzantine fault-tolerant algorithm. The selection of the desired algorithm depends on the use case scenario as it is stated earlier. Similar to the previous platforms, Corda supports both BFT and Raft mechanisms. Considering our use case presented previously, a notary [48] that uses Raft between nodes that form a network of NPs will present extremely good performance in terms of throughput and latency, at the cost of being more vulnerable to malicious attack by whichever node has been elected as a leader.

Moreover, the Corda platform supports SCs for the development of several solutions called Cor-Dapps. SCs are defined using a restricted form of Java Virtual Machine (JVM) bytecode, which automatically allows developers to implement the logic of their solution by writing code in a variety of programming languages. Developers are able to use well-developed toolchains and reuse code written in Java or other JVM-compatible languages, which is a fact that widens the application area of this platform. Additionally, Corda supports the development and use of tokens which can be used according to our use case to represent resources, such as CPU, memory and others [48].

Additionally, the privacy feature is supported in this blockchain as Corda uses several techniques to achieve this functionality. This means that the implementation of private transactions or the execution of private SCs is feasible. At the same time, Corda supports the use of oracles, defined as a network service that is trusted to sign transactions containing statements about the world outside the ledger only if the statements are true. This characteristic allows the secure communication of blockchain with entities outside the network such as the ZSM framework and its services. Additionally, Corda presents high-performance values as it can reach up to 20,000 transactions per second according to their benchmarking results illustrated on the platform's website [43,48].

7.1.4. IOTA

IOTA is a very popular DLT solution, which is based on the Tangle DAG. It is supported by the IOTA Foundation which aims at the development of new DLT-based solutions. On July 2016 the IOTA main net was activated and it is considered a public permissionless network. Some of IOTA's main characteristics are the increased scalability, the increased sentiment of decentralization and the zero transaction fees. In contrast to typical blockchain networks which present scalability issues as the transaction number increases, IOTA becomes more efficient and more powerful when the transaction number grows. Since IOTA does not use miners in the network, a node is at the same time the creator and the validator of a transaction. This means that everyone in the network contributes to the consensus process, which is a fact that highlights the decentralized nature of this particular solution. Moreover, the consensus mechanism implemented in the latest IOTA version is a probabilistic leaderless binary voting protocol called fast probabilistic consensus (FPC). This mechanism is responsible for Tangle's validity by addressing issues such as the double spending problem [44].

IOTA supports the creation of smart contracts called ISC and hence the development of several applications for various use case scenarios. ISC is agnostic regarding the virtual machine which executes the SC code. IOTA currently supports two types of SCs: the Rust/Wasm-based and Solidity/EVM-based [49]. Furthermore, IOTA allows the use of tokens that can be exchanged among entities in this DAG-based network. A well-known token used as a cryptocurrency in this solution is the MIOTA which can be purchased by a user in order to buy assets in the network. MIOTA is an example of a fungible token; however, the use of NFTs is also supported. Moreover, the use of oracles is supported in this DLT [50]. Oracles bring off-chain data to decentralized applications and smart contracts on the IOTA network. These mechanisms provide blockchains with outside information, typically for use in smart contracts, or provide interoperability between different distributed ledgers.

7.1.5. Hedera Hashgraph

Another DAG-based solution is presented by Hedera, named Hedera Hashgraph, which uses a distributed network, cryptographic tools and timestamps to store data in the form of transactions. This platform is considered a public permitted network, although it is currently governed by the Hedera Council which deploys and supports the network nodes. In the future, anyone will be able to host and operate a Hedera Hashgraph node.

The consensus mechanism used in this DLT is called Hashgraph and is based on the gossip protocol [45]. Every node that transacts with another one sends information regarding the current state of the network which is based on information previously received by other nodes. As a result, the information regarding the current state spreads like gossip among the nodes of the network. Therefore, every node in the network contributes to the consensus process and every node is aware of the current state. Hashgraph achieves a high-throughput with 10,000+ transactions per second today and low-latency finality in seconds from its innovative gossip about gossip protocol and virtual voting. Once consensus is reached, the transaction is immutable and available on the public ledger for everyone to transparently see. It should be mentioned that the nodes store only the latest state of the network in their ledger, which automatically increases the scalability of the network.

Additionally, Hedera Hashgraph offers a set of so-called Hedera Services that allow users to perform various tasks such as the creation of SCs and tokens. The Smart Contract service of Hedera allows the creation of contracts using the Solidity language similar to Ethereum-based networks, while Hedera promises fast SC execution with lower cost than blockchain alternatives. Moreover, tokens are supported by the Hedera Token service which allows the configuration and management of native fungible and non-fungible tokens. Having in mind our use case, an NP could receive a payment in the form of token for lending its resources to another NP. In addition, NFTs could be used as a reputation badge to highlight the reliability of an NP in our scenario. The support of SCs and tokens widens the application area of Hedera Hashgraph which can be used in various use cases.

Nevertheless, an additional factor that adds extra flexibility to this platform is its ability to cooperate with oracle mechanisms. Chainlink and Hedera Hashgraph announced in 2019 their collaboration to integrate Chainlink's decentralized oracle solution with Hedera's network. Chainlink is a well-known oracle mechanism that allows smart contracts to securely access and retrieve off-chain information when needed. It uses a similar model to a blockchain, as it implements a decentralized network of independent entities, called oracles, that collectively retrieve data from multiple sources, aggregate it, and deliver a validated single data point to the smart contract to trigger its execution, removing any centralized point of failure [51].

7.2. Assessment of Blockchain/DAG Platforms

Having examined the blockchain and DAG DLTs in the previous subsection, we proceed to the identification of their main characteristics in Table 2, which directly affect their suitability for our use case scenario. The columns of the table present the main attributes that describe the DLT's functionality while at the bottom of the table, the row with the ideal solution defines the properties of the most suitable solution for our scenario.

Table 2. Suitability of the examined blockchain and DAG solutions.

Solution	Public—Private/Consortium	Permissioned—Permissionless	Consensus Type	Support of SCs	Support of Tokens	Support of Oracles
HLF	Private/Consortium	Permissioned	CFT (Raft) or BFT (pBFT)	Yes	Yes	Yes
Quorum	Private/Consortium	Permissioned	CFT (Raft) or BFT (IBFT)	Yes (private SCs)	Yes	Yes
R3 Corda	Private/Consortium	Permissioned	CFT (Raft) or BFT (pBFT)	Yes (private SCs)	Yes	Yes
IOTA	Public	Permissionless	FPC	Yes	Yes	Yes
Hedera Hashgraph	Public	Permissioned	Hashgraph	Yes	Yes	Yes
Ideal solution	Private/Consortium	Permissioned	CFT—rapid convergence	Yes	Yes	Yes

Comparing each of the discussed solutions with the ideal one, we observe that every solution supports SCs, tokens and oracle mechanisms. However, in the Quorum and R3 Corda, we are able to use private SCs and implement private transactions if necessary. This is an extra feature that can be used to add extra functionalities to our solution in the future. For instance, some NPs in the network may sign an agreement of cooperation and fulfill requests with special terms which they might not want to unveil to other NPs in the network.

Furthermore, the blockchain solutions implement private/consortium and permitted blockchains which are the ideal characteristics of the network to be adopted in our scenario. On the contrary, IOTA and Hedera Hashgraph support only public implementations while the IOTA network allows access to anyone as it is a permissionless network. These characteristics decrease the suitability level of those solutions for our use case and should be modified.

In addition, the ideal solution should use a crash fault tolerant consensus mechanism with high convergence time in order to ensure the high availability of the system and achieve high transaction validation numbers, considering that the likelihood of a malicious participant is low. As illustrated in Table 2 the blockchain-based solutions can use consensus mechanisms with these features by implementing the Raft algorithm. However, DAG-based solutions use the FPC and Hashgraph mechanisms which present higher transaction validation speed which seems ideal for our use case. In terms of performance, the DAG solutions present higher numbers (transaction throughput) than the blockchain ones, and therefore, they are considered more suitable for our use case where the transaction number is expected to be extremely high.

The scalability of the DLT solution is also a significant factor that plays a crucial role in the selection of the ideal solution. As was mentioned in previous subsections, the blockchain solutions present scalability issues as every node holds a full copy of the ledger which increases as new transactions are validated. Nevertheless, DAG nodes store only parts of the graphs, and therefore, they scale well when the transaction number increases. Hence, they are considered more scalable than blockchain solutions.

To conclude, the qualitative assessment of blockchain/DAG platforms on which the prospective developer could build the ZSM-oriented solution, currently, none of the examined networks fulfill the criteria to become the ideal solution for our use case. The development of a private/consortium and permissionless DAG network could be the most suitable solution, bearing in mind the main characteristics of the presented DAG-based solutions.

8. Conclusions

ZSM networks are expected to lead the way toward the development of self-managed and self-optimized networks to form NGNs. The increased automation, in combination with the minimum human intervention, increases the performance of the network and at the same time exposes several security concerns. The current research examined the use of blockchain to tackle the security issues of the ZSM framework. The proposed blockchain architecture was found to address these security issues which further encouraged our interest in selecting the appropriate blockchain/platform on which such a solution would be implemented (to guide prospective implementers). In this course, we defined the selection criteria and examined the representative set of platforms. The conclusion was that for all of them, a modification will be needed as none meeting all the criteria was found. Although blockchain introduces valuable characteristics such as the ability to form a network of trust in a trustless environment without the existence of a trusted third party, the immutability of the data recorded as validated transactions in the ledger and the traceability feature, no technology is weakness-free. There are some drawbacks in this technology, which should be carefully considered during the design of the blockchain-based solution and deserve further research: (a) scalability is one of the main drawbacks of this technology since transactions written in the ledger cannot be deleted afterward; (b) the transaction validation time affects the dynamicity the network can support and mostly depends on the consensus mechanism used in the blockchain software [37,38]; (c) the support of tokens is also anticipated to add value in this environment for trading the resources. Moreover, the use of cryptocurrency could lead to the development of a modern marketplace where NPs could rent, buy or lease resources using the ZSM framework in a secure manner.

Author Contributions: Conceptualization, M.X. and H.C.L.; methodology, H.C.L.; investigation, M.X.; resources, M.X., D.G.K. and P.A.K.; writing—original draft preparation, M.X.; writing—review and editing, D.G.K., P.A.K. and H.C.L.; visualization, M.X.; supervision, H.C.L. All authors have read and agreed to the published version of the manuscript.

Funding: This research received no external funding.

Data Availability Statement: Data are available upon request. Please contact the authors for further information.

Conflicts of Interest: The authors declare no conflict of interest.

References

1. Saadon, G.; Haddad, Y.; Simoni, N. A survey of application orchestration and OSS in next-generation network management. *Comput. Stand. Interfaces* **2019**, *62*, 17–31. [CrossRef]
2. Medhat, A.M.; Taleb, T.; Elmangoush, A.; Carella, G.A.; Covaci, S.; Magedanz, T. Service function chaining in next generation networks: State of the art and research challenges. *IEEE Commun. Mag.* **2016**, *55*, 216–223. [CrossRef]
3. García-Otero, M.; Zahariadis, T.; Alvarez, F.; Leligou, H.C.; Población-Hernández, A.; Karkazis, P.; Casajús-Quirós, F.J. Secure geographic routing in ad hoc and wireless sensor networks. *EURASIP J. Wirel. Commun. Netw.* **2010**, *2010*, 1–12. [CrossRef]
4. Unleashing a New Breed of 5G Services: A 2021 Ecosystem Makeover. Available online: https://www.forbes.com/sites/forbestechcouncil/2021/03/01/unleashing-a-new-breed-of-5g-services-a-2021-ecosystem-makeover/?sh=643358bc4c83 (accessed on 20 September 2022).
5. How the Cloud Telecommunications Revolution Changes Business. Available online: https://www.forbes.com/sites/googlecloud/2021/06/21/how-the-cloud-telecommunications-revolution-changes-business/?sh=713312351ecb (accessed on 25 September 2022).
6. Cloud Players Reshape Telecom's Landscape—Industry Voices-Walker. Available online: https://www.fiercetelecom.com/telecom/cloud-players-reshape-telecom-s-landscape-industry-voices-walker (accessed on 27 September 2022).
7. Drampalou, S.F.; Miridakis, N.I.; Leligou, H.C.; Karkazis, P.A. A Survey on Optimal Channel Estimation Methods for RIS-Aided Communication Systems. *Signals* **2023**, *4*, 208–234. [CrossRef]
8. Prekas, S.; Karkazis, P.; Nikolakakis, V.; Trakadas, P. Comprehensive Comparison of VNE Solutions Based on Different Coordination Approaches. *Telecom* **2021**, *2*, 390–412. [CrossRef]
9. Barakabitze, A.A.; Barman, N.; Ahmad, A.; Zadtootaghaj, S.; Sun, L.; Martini, M.G.; Atzori, L. QoE management of multimedia streaming services in future networks: A tutorial and survey. *IEEE Commun. Surv. Tutor.* **2019**, *22*, 526–565. [CrossRef]
10. Chen, X.; Li, Z.; Zhang, Y.; Long, R.; Yu, H.; Du, X.; Guizani, M. Reinforcement learning–based QoS/QoE-aware service function chaining in software-driven 5G slices. *Trans. Emerg. Telecommun. Technol.* **2018**, *29*, e3477. [CrossRef]
11. Mahmoud HH, H.; Amer, A.A.; Ismail, T. 6G: A comprehensive survey on technologies, applications, challenges, and research problems. *Trans. Emerg. Telecommun. Technol.* **2021**, *32*, e4233. [CrossRef]
12. Shahraki, A.; Abbasi, M.; Piran, M.; Taherkordi, A. A comprehensive survey on 6G networks: Applications, core services, enabling technologies, and future challenges. *arXiv* **2021**, arXiv:2101.12475.
13. Yang, G.; Shin, C.; Yoo, Y.; Yoo, C. A Case for SDN-based Network Virtualization. In Proceedings of the 2021 29th International Symposium on Modeling, Analysis, and Simulation of Computer and Telecommu-nication Systems (MASCOTS), Houston, TX, USA, 3–5 November 2021; pp. 1–8. [CrossRef]
14. Alam, I.; Sharif, K.; Li, F.; Latif, Z.; Karim, M.M.; Biswas, S.; Nour, B.; Wang, Y. A survey of network virtualization techniques for Internet of Things using SDN and NFV. *ACM Comput. Surv.* **2020**, *53*, 1–40. [CrossRef]
15. Industry Specification Group (ISG); Zero Touch Network and Service Management (ZSM). Available online: https://www.etsi.org/committee/zsm (accessed on 10 October 2022).
16. Xevgenis, M.G.; Kogias, D.; Leligou, H.C.; Chatzigeorgiou, C.; Feidakis, M.; Patrikakis, C.Z. A Survey on the Available Blockchain Platforms and Protocols for Supply Chain Management. In Proceedings of the IOT4SAFE@ ESWC, Herakleion, Greece, 2 June 2020.
17. State of Dapps. Available online: https://www.stateofthedapps.com/ (accessed on 20 October 2022).
18. IBM Food Trust: A New Era in the World's Food Supply. Available online: https://www.ibm.com/blockchain/solutions/food-trust (accessed on 25 October 2022).
19. CargoX. Available online: https://cargox.io/ (accessed on 5 November 2022).
20. Farmers World. Available online: https://farmersworld.io/?utm_source=DappRadar&utm_medium=deeplink&utm_campaign=visit-website (accessed on 5 November 2022).

21. ETSI GR ZSM. General Security Aspects. In *Zero-Touch Network and Service Management (ZSM)*; Technical Report; Zero-touch network and Service Management (ZSM); ETSI Industry Specification Group (ISG); Sophia Antipolis Cedex: Valbonne, France, 2021.
22. Dalgkitsis, A.; Mekikis, P.V.; Antonopoulos, A.; Kormentzas, G.; Verikoukis, C. Dynamic Resource Aware VNF Placement with Deep Reinforcement Learning for 5G Networks. In Proceedings of the GLOBECOM 2020 IEEE Global Communications Conference, Taipei, Taiwan, 7–11 December 2020; pp. 1–6.
23. Uzunidis, D.; Karkazis, P.; Roussou, C.; Patrikakis, C.; Leligou, H.C. Intelligent Performance Prediction: The Use Case of a Hadoop Cluster. *Electronics* **2021**, *10*, 2690. [CrossRef]
24. Subramanya, T.; Riggio, R. Centralized and federated learning for predictive VNF autoscaling in multi-domain 5G networks and beyond. *IEEE Trans. Netw. Serv. Manag.* **2021**, *18*, 63–78. [CrossRef]
25. Boudi, A.; Bagaa, M.; Pöyhönen, P.; Taleb, T.; Flinck, H. AI-based resource management in beyond 5G cloud native environment. *IEEE Netw.* **2021**, *35*, 128–135. [CrossRef]
26. Short, A.; Leligou HCTheocharis, E.; Papoutsidakis, M. Using blockchain technologies to improve security in Federated Learning Systems. In Proceedings of the IEEE COMPSAC (Conference on Computers, Software and Applications), Madrid, Spain, 13–17 July 2020. [CrossRef]
27. Carrozzo, G.; Siddiqui, M.S.; Betzler, A.; Bonnet, J.; Perez, G.M.; Ramos, A.; Subramanya, T. AI-driven ze-ro-touch operations, security and trust in multi-operator 5G networks: A conceptual architecture. In Proceedings of the 2020 European Conference on Networks and Communications (EuCNC), Dubrovnik, Croatia, 15–18 June 2020; pp. 254–258.
28. Theodorou, V.; Lekidis, A.; Bozios, T.; Meth, K.; Fernández-Fernández, A.; Tavlor, J.; Diogo, P.; Martins, P.; Behravesh, R. Blockchain-based Zero Touch Service Assurance in Cross-domain Network Slicing. In Proceedings of the 2021 Joint European Conference on Networks and Communications & 6G Summit (EuCNC/6G Summit), Porto, Portugal, 8–11 June 2021; pp. 395–400.
29. Benzaid, C.; Taleb, T. AI-driven zero touch network and service management in 5G and beyond: Challenges and research directions. *IEEE Netw.* **2020**, *34*, 186–194. [CrossRef]
30. Benzaïd, C.; Taleb, T.; Farooqi, M.Z. Trust in 5G and beyond networks. *IEEE Netw.* **2021**, *35*, 212–222. [CrossRef]
31. Liyanage, M.; Pham, Q.V.; Dev, K.; Bhattacharya, S.; Maddikunta, P.K.R.; Gadekallu, T.R.; Yenduri, G. A survey on Zero touch network and Service (ZSM) Management for 5G and beyond networks. *J. Netw. Comput. Appl.* **2022**, *203*, 103362. [CrossRef]
32. Standish, R.K. Concept and definition of complexity. In *Intelligent Complex Adaptive Systems*; IGI Global: Hershey, PA, USA, 2008; pp. 105–124.
33. Gallego-Madrid, J.; Sanchez-Iborra, R.; Ruiz, P.M.; Skarmeta, A.F. Machine learning-based zero-touch network and service management: A survey. *Digit. Commun. Netw.* **2021**, *8*, 105–123. [CrossRef]
34. ETSI GS ZSM. Reference Architecture. In *Zero-Touch Network and Service Management (ZSM)*; Techical Report; ETSI Industry Specification Group (ISG); Sophia Antipolis Cedex: Valbonne, France, 2019.
35. Siriwardhana, Y.; Porambage, P.; Liyanage, M.; Ylianttila, M. AI and 6G security: Opportunities and challenges. In Proceedings of the 2021 Joint European Conference on Networks and Communications & 6G Summit (EuCNC/6G Summit), Porto, Portugal, 8–11 June 2021; pp. 616–621.
36. Zeng, Z.; Zhang, X.; Xia, Z. Intelligent Blockchain-Based Secure Routing for Multidomain SDN-Enabled IoT Networks. *Wirel. Commun. Mob. Comput.* **2022**, *2022*, 5693962. [CrossRef]
37. Xevgenis, M.; Kogias, D.G.; Karkazis, P.; Leligou, H.C.; Patrikakis, C. Application of Blockchain Technology in Dynamic Resource Management of Next Generation Networks. *Information* **2020**, *11*, 570. [CrossRef]
38. Xevgenis, M.; Kogias, D.G.; Christidis, I.; Patrikakis, C.; Leligou, H.C. Evaluation of a Blockchain-Enabled Resource Management Mechanism for NGNs. *Int. J. Netw. Secur. Appl.* **2021**, *13*, 1–16. [CrossRef]
39. Yang, G.; Lee, K.; Lee, K.; Yoo, Y.; Lee, H.; Yoo, C. Resource Analysis of Blockchain Consensus Algorithms in Hyperledger Fabric. *IEEE Access* **2022**, *10*, 74902–74920. [CrossRef]
40. Thakkar, P.; Nathan, S.; Viswanathan, B. Performance Benchmarking and Optimizing Hyperledger Fabric Blockchain Platform. In Proceedings of the 2018 IEEE 26th International Symposium on Modeling, Analysis, and Simulation of Computer and Tele-communication Systems (MASCOTS), Milwaukee, WI, USA, 25–28 September 2018; pp. 264–276. [CrossRef]
41. Hyperledger Fabric Documentation: Latest Release 25 January 2023. Available online: https://hyperledger-fabric.readthedocs.io/_/downloads/vi/latest/pdf/ (accessed on 28 January 2023).
42. Consensys Quorum. Available online: https://consensys.net/quorum/ (accessed on 10 March 2023).
43. R3 Corda. Available online: https://www.r3.com/products/corda/ (accessed on 1 February 2023).
44. Sealey, N.; Aijaz, A.; Holden, B. IOTA Tangle 2.0: Toward a Scalable, Decentralized, Smart, and Autonomous IoT Ecosystem. In Proceedings of the 2022 International Conference on Smart Applications, Communications and Networking (SmartNets), Palapye, Botswana, 29 November–1 December 2022; pp. 1–8.
45. Hedera How It Works. Available online: https://hedera.com/how-it-works (accessed on 3 February 2023).
46. Gorenflo, C.; Lee, S.; Golab, L.; Keshav, S. FastFabric: Scaling hyperledger fabric to 20,000 transactions per second. *Int. J. Netw. Manag.* **2020**, *30*, e2099. [CrossRef]
47. Baliga, A.; Subhod, I.; Kamat, P.; Chatterjee, S. Performance evaluation of the quorum blockchain platform. *arXiv* **2018**, arXiv:1809.03421.
48. Corda: A Distributed Ledger. Available online: https://www.r3.com/blog/corda-technical-whitepaper/ (accessed on 28 January 2023).
49. IOTA Smart Contracts. Available online: https://wiki.iota.org/shimmer/smart-contracts/overview/ (accessed on 2 February 2023).

50. IOTA Oracles. Available online: https://blog.iota.org/introducing-iota-oracles/ (accessed on 2 February 2023).
51. What Is Chainlink: A Beginner's Guide. Available online: https://blog.chain.link/what-is-chainlink/?_ga=2.209069778.1120344513.1675428709-842934195.1673628852 (accessed on 3 February 2023).

future internet

MDPI

Review

Securing Wireless Sensor Networks Using Machine Learning and Blockchain: A Review

Shereen Ismail [1,*], Diana W. Dawoud [2] and Hassan Reza [1]

1 School of Electrical Engineering and Computer Science, University of North Dakota, Grand Forks, ND 58202, USA; hassan.reza@ndus.edu
2 College of Engineering and Information Technology, University of Dubai, Dubai 14143, United Arab Emirates; ddawoud@ud.ac.ae
* Correspondence: shereen.ismail@ndus.edu

Abstract: As an Internet of Things (IoT) technological key enabler, Wireless Sensor Networks (WSNs) are prone to different kinds of cyberattacks. WSNs have unique characteristics, and have several limitations which complicate the design of effective attack prevention and detection techniques. This paper aims to provide a comprehensive understanding of the fundamental principles underlying cybersecurity in WSNs. In addition to current and envisioned solutions that have been studied in detail, this review primarily focuses on state-of-the-art Machine Learning (ML) and Blockchain (BC) security techniques by studying and analyzing 164 up-to-date publications highlighting security aspect in WSNs. Then, the paper discusses integrating BC and ML towards developing a lightweight security framework that consists of two lines of defence, i.e, cyberattack detection and cyberattack prevention in WSNs, emphasizing the relevant design insights and challenges. The paper concludes by presenting a proposed integrated BC and ML solution highlighting potential BC and ML algorithms underpinning a less computationally demanding solution.

Keywords: Internet of Things; wireless sensor networks; security; machine learning; blockchain; detection; prevention; cyberattacks; integration; review

Citation: Ismail, S.; Dawoud, D.W.; Reza, H. Securing Wireless Sensor Networks Using Machine Learning and Blockchain: A Review. *Future Internet* **2023**, *15*, 200. https://doi.org/10.3390/fi15060200

Academic Editors: Christoph Stach and Clémentine Gritti

Received: 27 April 2023
Revised: 15 May 2023
Accepted: 22 May 2023
Published: 30 May 2023

1. Introduction

Wireless Sensor Networks are the backbone that enables Internet of Things (IoT) at low cost and low power [1]. These networks have been considered for a wide range of applications, such as military, environmental, healthcare, and civilian, despite being vulnerable to attacks [2]. Indeed, Wireless Sensor Networks (WSNs) result in major concerns in terms of security. Concerns include the use of devices which have resource constraints in terms of energy, the adopted wireless broadcasting channels, the involvement of multi-hop relays, the dynamic network topology, variable medium-to-large network scales, heterogeneous sensor node fabrication, and most importantly, the different routing protocols employed. Securing WSNs is relevant to securing IoT [3], as the latter comprises one or more WSNs, which implies that developing prevention, detection, and mitigation security solutions for WSNs are essential for establishing secure and reliable IoT systems.

Classical WSN security techniques, such as spread spectrum, cryptography, and key management [4,5], may not efficiently detect attacks, and can demand sophisticated software and hardware changes, rendering these solutions insufficient to address WSN security concerns, as WSN devices constrain the network's power, storage, computational, and communication capabilities [6]. There has been growing interest in novel security paradigms, with cybersecurity companies investing as much as USD 119 billion to solve these problems [7]. This has led to newly evolved means aimed at strengthening WSN security against possible cyberattacks via Machine Learning (ML) and Blockchain (BC) [8].

Compared to classical techniques, ML techniques are particularly useful in WSNs and IoT applications, as computational complexity and communication overhead can be significantly decreased, no human intervention is required, and they perform better in dynamic environments. On the other hand, BC allows highly secure data transactions within any network similar to WSNs [9]. The fact that ML and BC can potentially provide promising solutions and effective mechanisms to protect and secure WSNs against cyberattacks has motivated several recent research works focused on evaluating the performance of BC and ML to secure WSNs. The performance of ML and BC is affected by the challenging characteristics of WSNs, such as its large generated data volume, which are extremely hard to manage, especially when considering highly dense networks. To this end, this paper attempts to answer the following overarching research questions. How is ML used to detect WSN cyberattacks? How is BC used to prevent WSN cyberattacks? How can the integration of ML and BC provide an effective framework to protect and secure WSNs against cyberattacks? Finally, What are the key technical challenges related to this integration? Thus, the main contributions of this survey are: (a) classification of WSN cyberattacks and the unique characteristics that complicate the design of effective detection and prevention mechanisms against cyberattacks; (b) a literature review of the existing Intrusion Detection System (IDS) architectures in the context of WSNs; (c) a comprehensive taxonomy of ML and BC along with an evaluation of relevant existing security techniques and challenges; (d) discussion of an integrated solution incorporating both technologies towards development of a WSN that is significantly immune against attacks; and (e) an ultimate overview of our approach to providing a lightweight and integrated ML and BC framework towards enhanced protection against cyberattacks in WSN contexts.

The rest of this paper is organized as follows: Section 2 reviews existing surveys on ML and BC solutions in the context of securing WSNs; Section 3 outlines the unique WSN characteristics that present network security challenges when developing such techniques; Section 4 illustrates the security requirements for designing a secure WSN; Section 5 classifies and defines cyberattacks that target WSNs; Section 6 discusses the underlying IDS architectures considered in conjunction with different WSN architectures; Section 7 extends the discussion to include different types of IDSs used for intrusion detection; Sections 8 and 9 focus on the respective taxonomies of ML and BC techniques used to detecting cyberattacks, along with related aspects; Section 10 explores the integration of BC and ML towards developing a lightweight security framework for WSNs and presents our approach to developing such a framework for cyberattack prevention and detection in WSN contexts; finally, Section 11 concludes this review.

2. Existing Surveys on ML and BC in WSN

This paper discusses ML and BC protection mechanisms in a comprehensive manner [10–15]; however, the emphasis is on securing WSNs. In this regard, a few previous surveys have focused on presenting state-of-the-art ML and BC techniques for WSN cybersecurity. Key surveys tackling WSN security are tabulated in Table 1, which highlights the different subtopics covered, including ML, BC, attack taxonomy, and ML–BC integration, among others. The surveyed sources were collected from popular academic databases, such as IEEE Xplore, Elsevier, and Scopus, as per the most recent citation provided by Google Scholar, and are detailed in Figure 1.

Table 1 reveals that research work on ML techniques is the primary subject of existing survey papers in the literature. A number of surveys that were published between 2012 and 2017, such as [16], did not examine WSN-related ML techniques, instead jointly discussing methods adopted in both IoT and WSN. On the other hand, surveys similar to [16–19] focused primarily on WSN. The authors of [19] considered only Denial of Service (DoS) attacks over the five TCP/IP layers. The authors of [17] provided a generalized and comprehensive review of ML techniques adopted to support WSNs against their inherent limitations, including security. The paper specifically focused on ML methods used to detect outlying and misleading measurements. The authors of [14] discussed the different types of attacks targeting WSNs and associated proposed ML solutions. Protecting WSNs using several ML methods was discussed in [16]. The authors of [20] explored using ML techniques with WSNs, including anomaly detection, with a focus on Deep learning (DL) techniques. A different research direction was analyzed in [20,21], where the authors focused on a specific type of WSN. The authors of [20] presented ML learning techniques targeting advanced WSN systems, and [21] reviewed ML techniques to secure industrial WSN systems. The authors of [22] reviewed ML algorithms and considered using software-defined networking (SDN) as a solution that can help enhance the node efficiency, creating a new foundation for using ML schemes to secure WSNs.

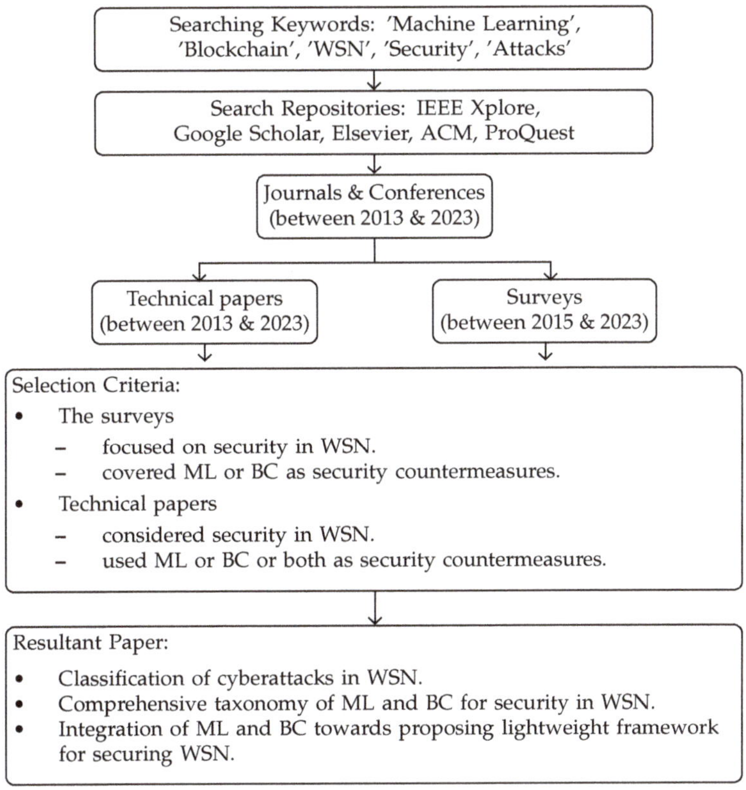

Figure 1. Paper collection criteria flowchart.

Considering BC techniques, several reviews have been conducted on securing IoT through the use of BC, such as [7,23–30]; however, Ref. [31] is the only article addressing BC for mitigating cyberattacks in WSNs. The study concluded that integrating BC techniques within WSNs has limitations, as BC is demanding in terms of both energy and computational complexity and is not expandable. Thus, to the best of our knowledge, our paper is the first work to review the integration of both technologies to improve WSN security, which is confirmed by Table 1.

Table 1. Existing works on applicability of ML and BC for WSN security.

Year	Ref.	Direction				
		ML	**BC**	**Attacks**	**IDS**	**Integration**
2015	[9]	✓		✓	✓	
2015	[19]	✓		DoS	✓	
2017	[32]	✓				
2018	[21]	✓		✓	✓	
2018	[16]	✓		✓		
2020	[17]	✓				
2020	[18]	✓			✓	
2021	[20]	✓			✓	
2022	[22]	✓		✓	✓	
2021	[31]		✓			
2023	our work	✓	✓	✓	✓	✓

3. WSN Security Requirements

The most important WSN security requirements include integrity, availability, scalability, non-repudiation, mutual authentication, confidentiality, and data freshness, defined in turn as follows:

1. Integrity: transmitted messages cannot be tampered with due to illegal actions when moving from one node to the other.
2. Availability: legitimate (and authenticated) nodes can effectively access the network/provided services.
3. Scalability: the network should be able to cope with increases in size and to adapt to the dynamic addition and removal of various nodes, and node functionalities must be incorporated with sensor nodes for every service without affecting the network's security level.
4. Mutual Authentication: the identities of any pair of nodes engaged in communication must be recognized before they interact.
5. Non-repudiation: the nodes cannot deny the implemented operations or alter the messages they send.
6. Confidentiality: the privacy of sensitive data transmitted over the network medium must be preserved by ensuring that any intruder or other neighboring network intercepting the communication channels cannot obtain any confidential information.
7. Data Freshness: the data must be recent in order to ensure that no old messages have been replayed and that attackers cannot confuse the network by replaying captured messages [33,34].

4. WSN Design Challenges and Unique Characteristics

WSN security solution design is highly affected by the unique features of these networks that make them more susceptible to cyberattacks than other technologies. This is primarily due to their challenging underlying infrastructure, which consists of a collection of sensor nodes utilizing scarce resources. The basic building blocks of a sensor node consist of four main units, namely, processing, sensing, communication, and power [31], as shown in Figure 2. The processing unit is the central unit, containing a processor or microcontroller that controls the sensor's activities and executes the communication protocols; however, it has limited storage memory. The processing unit is connected to the sensing unit by an Analog to Digital Converter (ADC). The sensing device captures surrounding data, which are then converted into an electrical signal by the ADC. The communication unit typically supports data exchange between the sensor and the other network elements using a transceiver. Finally, the power unit provides the electrical energy required by the other units using limited-lifespan batteries. Optional sensor hardware additions include power generation and mobilization units [35]. Certain nodes may include a location-finding unit for positional localization in reference to the node's neighbors. These special characteristics must be identified before they can be used in the design and development of more secure networks. The following points describe the dominant design considerations in WSNs in detail, which are further highlighted in Figure 3.

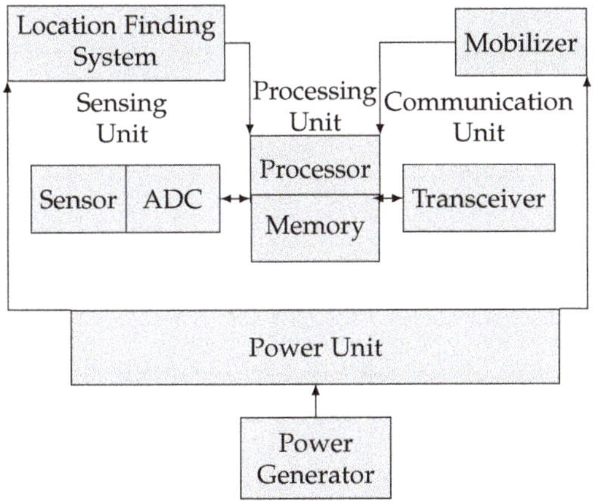

Figure 2. Illustration of sensor node building blocks.

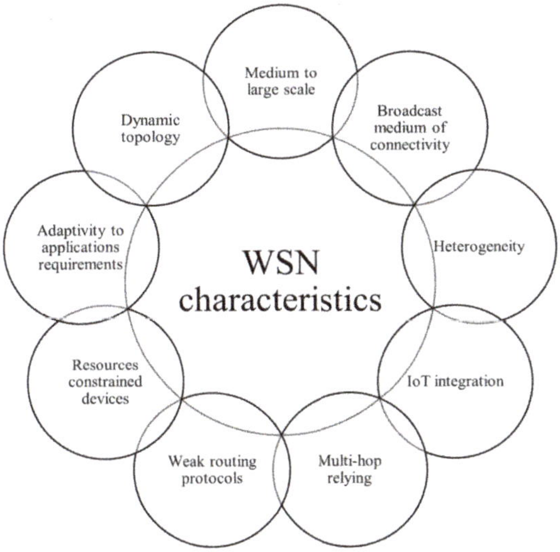

Figure 3. Unique characteristics of WSNs.

- WSNs can be used in a wide range of applications with different security requirements; however, they must be able to ensure privacy, confidentiality, integrity, freshness, and authentication.
- Sensor nodes must be heterogeneous in terms of fabrication and energy-saving strategies, such as sleep, idle, and wake-up modes, which dictates the need to provide different underlying network architectures for the different heterogeneous applications.

- WSNs have many appealing applications, creating a need for different levels of secure functionalities and service requirements, such as secure node selection, data aggregation [36], localization, and routing.
- Resource-constrained devices have limited memory, power, and transmitting bandwidth. For example, TelosB [37] is an ultra-low-power sensor with a a 16 bit processor and 8 MHz RISC microcontroller with only 10 Kb RAM, 48 Kb program memory, and 1024 Kb flash storage. The required total space for a typical code, such as TinyOS, which is the de facto standard operating system for wireless sensors, is approximately 4 Kb [38]. Therefore, any implemented security algorithm within the network must not be computationally demanding beyond these limitations.
- Security algorithms must be able to manage unsupervised sensors, which could be exposed to physical attack by demolishing the hardware or to attackers equipping sensors with extra hardware to perform hidden or malicious functions prior to their being deployed in the network area.
- Determining the adopted broadcast dynamic channel used as a wireless communication medium is challenging, as it is unattended and might be affected by collision and interference issues. WSN communication links are usually based on the 802.15.4 standard, and can be implemented via the use of other technologies as well, such as Bluetooth, ZigBee, PLC, WiFi, 4G, and 5G.
- The lack of fixed physical infrastructure is a significant design challenge due to the rapidly changing connectivity between nodes.
- A dynamic underlying network topology results from node failure, deployment of new nodes, possible variations in node position (which is especially the case under harsh environmental conditions), node mobility. The resulting flexibility in terms of link connectivity presents a design challenge for security algorithms, which must be able to adapt to network node variations in order to obtain the extra measure of protection provided by monitoring of corrupted nodes.
- WSN routing protocols have weaknesses, including malicious routing information injection, alteration, or spoofing, which might lead to network disruptions such as creation of routing loops, broadcasting of fake error messages to partition the network, attracting or repelling network traffic from particular nodes, extending or shortening route paths, and increasing end-to-end latency. These issues are likely to complicate the design of security routing techniques [39].
- Medium- to large-scale networks of hundreds or thousands of nodes deployed randomly or uniformly throughout the network field presents a challenge when designing security algorithms that are sufficiently flexible to support different security-level requirements.
- The scalability of WSNs implies handling large amounts of data that may have inconsistent, noisy, erroneous, redundant, and missing values, which requires designing intelligent security approaches that can correctly interpret data to drive intelligent decision-making.
- Data transmission over multi-hop relaying creates a significant threat, as relays could be eavesdroppers [40], and communicated data may be breached, tampered with, or forged.
- Time synchronization is an issue, as nodes are independently controlled in the field. Local clocks should be coordinated to avoid synchronization uncertainties, which could cause sensed data to become ambiguous and unreliable.
- Unexpected and unusual sensor behavior patterns may arise during WSN deployment in unpredictable and hazardous environments, potentially changing the entire historical pattern of the sensed data.

These characteristics render a completely secure WSN system almost impossible to establish, unlike its counterpart networks. The characteristics of WSN systems limit the available security options, including those similar to heavyweight classical security approaches such as spread spectrum, cryptography, and key management at either the device level or the overall network level. These options are demanding in terms of the resources required to protect the network. As existing security solutions for WSNs are insufficient due to these networks' unique characteristics, it is difficult to create lightweight and effective security mechanisms that can enable optimization of node resource usage while supporting network scalability and without compromising security, allow for a dynamic network topology with different possible configurations and node localization, and integrate heterogeneous hardware and software platforms for sensors to allow them to detect malfunctioning or faulty nodes.

5. Cyberattacks in WSN Contexts

Cyberattacks are the greatest challenge facing communication networks worldwide. The threat of cyberattacks affects any network's connectivity, availability, reliability, and confidentiality, limiting its efficient use. Mitigating this challenge is essential, especially because the frequency and the nature of attacks have increased tremendously over time [41]. For this reason, cyberattacks targeting WSNs have been the focus of several recent studies in the literature [4,42–46]. Cyberattacks occur when good nodes are communicating over a communication link and intruder or eavesdropper nodes interfere with or disturb that link. This malicious activity usually aims to obtain, alter, or prevent the flow of data within the network using different means; therefore, this activity should be prevented, detected, and mitigated in order to maintain a reliable communication channel [47]. Malicious acts targeting WSNs have been classified in the literature in different ways: the first classification divides attacks into active or passive attacks; the second classification is based on the physical location of the attack relative to the network's physical position, using this distinction to divide attacks into inside or outside attacks; and the third classification is based on the disrupted stack Open Systems Interconnection (OSI) layer, dividing attacks into physical layer, data link layer, and network layer attacks [42,48]. Table 2 classifies a selection of classical attacks targeting WSNs and provides their definitions.

Table 2. Classification of cyberattacks on WSNs.

Attack Type	Affected Stack Layer	Attack Name	Definition
Active	Multi-layer	Man-in-the-Middle	A malicious node intercepts a message passing between two sensor nodes with the aim of modifying, injecting, or deleting content before relaying the message again.
		Denial-of-Service	An attacker performs malicious activities to prevent original users from accessing system resources.
		Distributed Denial-of-Service	A more powerful version of DoS attack that overwhelms the targeted nodes with excessive messages to exhaust their resources, leading to a system overload that prevents it from answering some or all legitimate messages.
	Application	Deluge	An attacker tries to remotely reprogram a sensor node.
		Misdirection	An attacker forwards packets to the wrong destinations or paths by misdirecting packets or altering routes towards a malicious node.

Table 2. *Cont.*

Attack Type	Affected Stack Layer	Attack Name	Definition
		Clock skewing	Disrupts sensors that requir synchronization for successful communication; an attacker desynchronizes sensor clocks by generating false timing information, leading to desynchronization of the victim nodes.
		Selective Forwarding	Malicious nodes drop a portion of a received message while forwarding most of the message, impacting data integrity.
	Transport	Flooding	An attacker sends a large number of useless packets to a legitimate node, preventing it from communicating normally and consuming its resources.
		Session Hijacking	An attacker exploits a valid session, pretends to be a victim node, and obtains fake access to the session.
		De-synchronization	An attacker intercepts sequence numbers or controls flag packets that it attempts to forge; if the attacker can desynchronize two communicating nodes, the receiver node must request retransmission from the sender for the lost packet. Frequent retransmission consumes network resources and increases traffic over the network.
	Network	Reply	An attacker records the messages sent between nodes and re-transmits them later to waste the target node's resources.
		Selective Forwarding or Grayhole	A malicious node selectively, constantly, or randomly drops packets while forwarding the remaining packets to a particular destination, which happens when relay nodes do not forward messages they receive.
		Neglect and Greed	A special case of selective forwarding attack in which the attacker arbitrarily drops some of the received packets while acknowledging the source node (neglect attack) or sends its own packets with higher priority to other nodes (greed attack) [49].

Table 2. *Cont.*

Attack Type	Affected Stack Layer	Attack Name	Definition
		Homing	An attacker analyzes traffic using a traffic pattern analysis algorithm to recognize the nodes with special responsibilities, such as cluster heads (CHs) or base station (BS), which are the attack targets. Afterwards, additional DoS attacks may be launched toward these nodes to jam or destroy them.
		Spoofing	An attacker forges its identity by impersonating another node and falsifying the identity field in routing messages to launch DoS attacks by injecting fraudulent data packets, such as falsely advertising services to other nodes or providing incorrect routing and control information to compromise network operation [50].
		Blackhole	A malicious node, usually located in the center, does not forward traffic and drops the packets completely.
		Wormhole	A collusion-based attack in which two or more malicious nodes create a low-latency data delivery tunnel between two or more malicious nodes to perform other attacks, such as a blackhole attack. For instance, the nodes may establish a low-latency tunnel by which one malicious node misroutes the packets to be forwarded and sends them to its partner using a faked routing path to disrupt routing operations in the network.
		Sybil	A single attacker node assumes several identities or steals them from other authorized nodes to create several sybil nodes that can be virtually present in different neighborhoods, then attack the network to cause problems with multipath routing, network topology, storage access, and detection [50].
		Sinkhole	A malicious node identifies itself as a blackhole to attract network traffic. The attacker observes path requests and falsely offers the shortest or most power-efficient paths to the BS. As the attacker is in the relay path between the communicating nodes, it is able to change or alter the packets passing between them [44].

Table 2. *Cont.*

Attack Type	Affected Stack Layer	Attack Name	Definition
		Hello Flooding	An attacker broadcasts advertisement 'Hello' messages with high power, asking network nodes to join an existing WSN and tricking the nodes into believing that it is located in their neighborhood. The nodes choose to route their packets through the attacker, which has a longer transmission range than normal nodes, leading to additional delays and energy waste.
	Data link	Collision	An attacker sends signals while another node is transmitting a message, causing interference that alters data packets or causes them to be considered invalid. Collision usually occurs when multiple nodes transmit data at the same frequency and data rate.
		Denial of Sleep (Sleep Deprivation)	A Malicious node prevents legitimate nodes from entering low-power sleep mode, causing them to keep wasting their energy [51].
		Power Exhaustion	In order to drain the victim node's power, an attacker sends packets over the channel continually by requesting calculations or the receipt or transmission of unnecessary data, which leads to starvation. The source of the attack can be a PC or laptop.
		Unfairness	A malicious node continuously sends packets without waiting a reasonable time to let other nodes use the channel. This is a kind of exhaustion-based attack which disrupts equal load sharing in the WSN.
	Physical	Jamming	An attacker sends a radio signal that interferes with the sensor network's use of certain radio frequencies.
		Physical or Node Tampering	An attacker physically accesses a compromised node and takes over the control, for example, to obtain sensitive information such as transmission keys [52,53].
		Node Replication or Clone	An attacker captures a compromised node, obtains access to the stored credentials, purposefully duplicates the node's identity, and then deploys clones in key positions of the WSN [54] to initiate different internal attacks.

Table 2. *Cont.*

Attack Type	Affected Stack Layer	Attack Name	Definition
Passive		Camouflage Adversaries	A camouflaged node deceives the other nodes and attract packets from them in order to either misroute the packets or eventually drop the packets.
		Eavesdropping and Traffic Analysis	The most common attack on privacy, also called sniffing or snooping, where an attacker simply discovers the content of communications.
		Passive Information Gathering	If the content of messages from network communication media, such as message identification numbers (IDs), nodes locations, and timestamps, is not encrypted then an attacker with the appropriate receiver can collect and observe the information.
		Replay or Duplication	An attacker copies a stream of messages between communicating nodes, then replays the stream to one or more of the nodes [55]

Active attacks threaten network integrity and reduce availability by continuously attempting to modify the content of the network packets or flooding the victim nodes with surplus packets. The different types of active attacks are based on the underlying stack layer disrupted by the attack, as shown in Figure 4 [42,48]. Attacks such as link jamming, physical tampering, or node replication are hardware-oriented attacks that affect the node's physical layer. These attacks are more likely to occur when the sensor is exposed to a harsh environment or open to an adversary; therefore, they are unlikely to occur when the sensor node is placed in a secure indoor location. Other attacks, such as collision, exhaustion, and unfairness, are executed against the Media Access Control (MAC) protocol at the data link layer. These attacks cause collisions that result in packet re-transmission; therefore, copies of the same packets must traverse the network, overwhelming the communication channel and wasting limited sensors energy. The most common attacks, such as sinkhole, wormhole, blackhole, selective forwarding (grayhole), 'Hello' flooding, sybil, spoofing, and altered or replayed routing information attacks, all interrupt the network layer. These attacks prevent proper packet delivery to the destination through methods such as taking advantage of the multi-hop routing protocol, in which any node routes passing through malicious nodes are unable to deliver packets or are intentionally redirected to incorrect nodes. Examples of attacks impacting the functionality of the transport layer include session hijacking, flooding, and de-synchronization attacks. For example, flooding results in node failure, as the attacker consumes node resources by sending multiple connection requests. Attacks that target the application layer include selective forwarding, deluge, and clock skewing. The most difficult to detect among these attacks is selective forwarding, as the attacker does not block packet forwarding entirely, and only drops or alters some of the received packets from selected nodes. Deluge allows the sensor nodes to be reprogrammed remotely, and clock skewing disrupts those sensors that require synchronization for successful communication. Unlike active attacks, passive attacks do not affect network integrity, instead compromising network confidentiality. These attacks sniff and read unauthorized messages through the communication channels between nodes without disrupting their communication or interrupting network processes. Passive attacks may make the network more vulnerable to other kinds of attacks, such as camouflaged adversaries, physical tampering, eavesdropping, and traffic analysis.

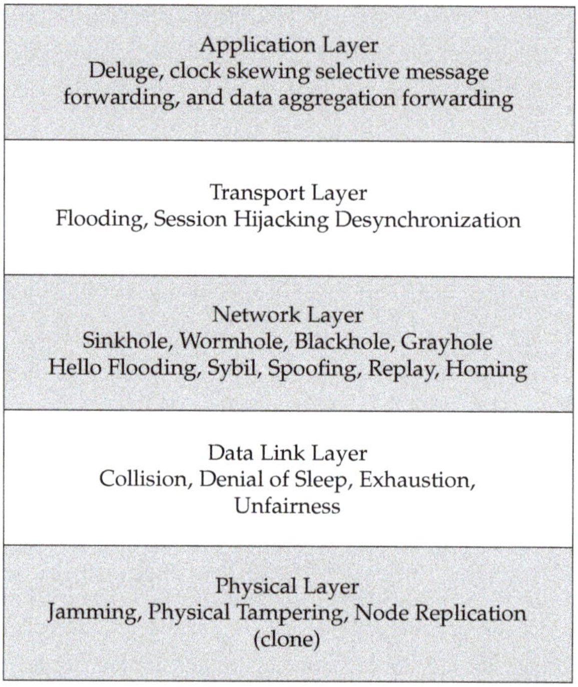

Figure 4. Active attack classification according to OSI stack layer.

Internal attacks are initiated from within the network's physical boundaries. These attacks control and utilize other network nodes to execute malicious acts. An inside attack can obtain the network transmission key or other network information from the transmitted packets within the network, then use this information to attack the entire network. A typical example of an internal attack is when an attacker takes advantage of a dump security implementation at an unsecured sensor node or a non-updated device's firmware, which allows the attacker to turn sensing devices into malicious nodes. The attacker then utilizes the node's network connectivity with other nodes to extract network data using eavesdropping, interfering, or misrouting. External attacks are initiated from beyond the network boundaries; therefore, they cannot obtain network information, such as node identification numbers or transmission keys, making attack recognition easier [48]. In addition, external attackers require powerful wireless transceivers to listen to data packets inside the network in order to accomplish malicious activities such as eavesdropping, replay, injection, and interference. Figure 5 depicts scenarios for external and internal attacks targeting a WSN.

In terms of OSI layers, the physical and network layers experience the most threatening attacks. The physical layer possesses a broadcasting channel and a dynamic topology, which allows attackers to easily listen to or sniff the communication channel and establish attacks. While, the network layer has a weak routing protocol that attackers can exploit to execute malicious acts. Another form of attack can be initiated over several WSN stack layers; such attacks across multiple layers include DoS and Man-in-the-middle (MITM) attacks [38]. DoS attacks are numerous, and include jamming and node tampering at the physical layer, collision, exhaustion, denial of sleep, unfairness at the data link layer, homing, blackhole, grayhole, wormhole, sinkhole, spoofing, 'Hello' flooding, TDMA scheduling, sybil, and replay attacks at the network layer, as well as flooding and desynchronization at the transport layer [56]. MITM attacks work as a relay between two victims [6]; this type of attack can be passive, where the attacker eavesdrops or intercepts the data traveling on the network between two legitimate nodes without altering the data, such as eavesdropping at the physical layer, or it can be active, where the attacker can delay, drop, or modify the content of a packet, such as a replay attack at the network layer [57].

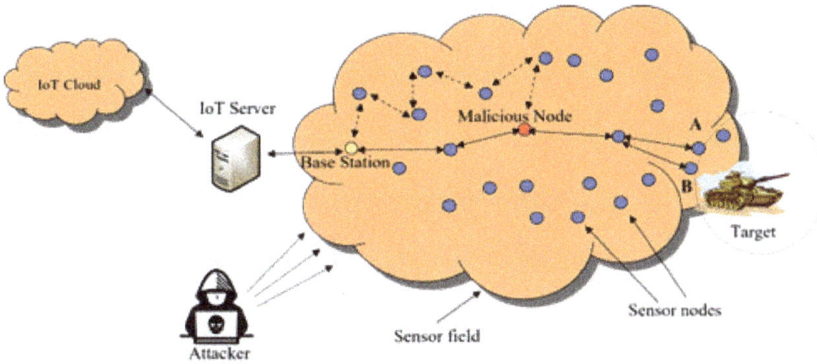

Figure 5. Internal and external cyberattack scenarios in WSN contexts.

6. Architecture of WSN vs. Architecture of IDS

Intrusions are similar to attacks in that they aim to disturb the network's normal operation or obtain access to the network's information. The IDS is the network's line of defense, designed to detect violations and tell the controller, or BS, to react appropriately.

6.1. Naive or Flat-Based WSN Architecture for Centralized IDS

In a centralized architecture, better known as a Naive WSN architecture, a central BS collects all the information sensed by all network nodes and forwards the collected information to the cloud IoT server. Similarly, in a centralized IDSs, the BS acts as a global reference that performs computationally demanding tasks to monitor and filter data traffic to facilitate attack detection. Several studies have considered executing the IDS at both the BS level and at the remote server level connected to the IoT cloud, which is called a multi-layer IDS scenario. This approach has multiple limitations, including attack detection latency, considerable communication overhead, and high energy consumption. Latency occurs when data traffic analysis is delayed until the information reaches the BS. Communication overhead is caused by the need to transmit all sensed information to the central BS over the communication link, increasing energy consumption as the node's distance relative to the central unit increases [58]. Due to these limitations, centralized IDS architectures are typically used only in very small networks.

6.2. Naive or Flat-Based WSN Architecture for Stand-Alone IDS

The opposite philosophy to centralized IDS is stand-alone IDS, which is a node-centric architecture. Each node individually uses an IDS detection model to detect any possible attack locally without needing to exchange any information with the adjacent nodes or a central BS unit. This approach does not exhibit latency when detecting node attacks or introducing communication overhead; however, energy consumption at the node level is higher than in a centralized IDS, and the nodes have lower battery life.

6.3. Naive or Flat-Based WSN Architecture for Distributed or Cooperative IDS

This approach assumes that each node has its own local IDS model to monitor the data traffic, then involves all network nodes in deciding whether an intrusion is present in the network based on the detected indicators. If a locally measured indicator is weak or inconclusive, the involved node can initiate a cooperative global intrusion detection procedure in which all network nodes cooperatively participate in reaching a global decision. Otherwise, if an intrusion is locally detected with sufficient evidence, the involved node can independently alert the rest of the nodes to the presence of a violation in the network. This approach reduces false attack stimulus events, which relate to scenarios in which a violation alarm is triggered even though no real threat is in progress within the network. In this approach, node power consumption is higher and node battery life is lower than the stand-alone IDS due to the an additional optional cooperative procedure.

6.4. Naive or Flat WSN Architecture for Agent-Based IDS

Agent-based IDS involves installing the detection model in a selected subset of sensor nodes, which are called Monitor Nodes (MNs), to reduce the detection overhead faced by the stand-alone and distributed approaches. In tis approach, selected nodes perform detection in addition to their normal sensing, communication, and routing activities in the case of flat WSN architecture. Agents' tasks are relocated to another predefined subset of nodes after a certain period of time or when performing a specific mission, which improves IDS detection efficiency and increases network lifetime. Agent-based IDS is especially suitable for WSNs, as nodes near the BS can be excluded from communicating all of their samples when developing the reference ML model at the BS because they do not contribute much to the determination of hypersphere of the developed ML model. Agent-based IDS is typically preferred over centralized IDS architecture, especially for networks with geographically dispersed nodes, as in a centralized approach the nodes consume more power when transmitting their data to the central location.

6.5. Hierarchical WSN Architecture for Distributed or Cooperative IDS

A WSN's hierarchical architecture is a variation of centralized architecture, which can be implemented as cluster-based or tree-based. In a cluster-based architecture all sensor nodes are partitioned into clusters, whereas in a tree-based architecture the nodes are partitioned into trees according to their topographical area. The nodes in a tree-based architecture are organized into a routing tree rooted at the BS. Cluster-based architectures can be static or dynamic. In static clustering the sensors are divided proactively into several clusters at the time of network deployment, while in dynamic clustering the formation of clusters is triggered reactively by detecting the event of interest. In a distributed IDS, the detection model is placed in every sensor network node, allowing nodes to collaborate in order to detect possible intrusions. The clear advantage of implementing a combined hierarchical and distributed architecture is that the communication overhead is significantly lower than in other approaches, as both hierarchical and distributed architectures involve less communication exchange between nodes [58]. A disadvantage of this approach is the need for each network node to have sufficient energy, processing, and storage capacity. Studies have considered using multi-layer instead of distributed IDS, with heterogeneous detection models placed only at the BS and CHs.

7. Types of IDS

IDS-based mechanisms are effective and lightweight solutions for detecting abnormal behavior in WSN sensor nodes. An IDS requires an IDS agent or detector node that analyses the network traffic to detect a abnormal behavior. Intrusion detection at the IDS agent level involves three phases: collection, processing, and action. Network data traffic is collected during a specific time period, then this collected information is processed according to a particular detection mechanism. Detection approaches can be classified as misuse-based, anomaly-based, and specification-based detection. In misuse-based or signature-based detection, the system searches for specific patterns or signatures to identify and detect an intrusion. This approach easily detects known attacks, but cannot detect new or unknown attacks. In specification-based detection, a set of rules or specifications have been set as a reference for normal system operation; any deviation from these specifications triggers an abnormal behavior alert, allowing the system to take proper preventive actions accordingly.

This approach has a low false positive rate; however, developing the required specifications is very time-consuming. Anomaly-based detection systems learn the normal behavior profile from normal network traffic and create a reference model accordingly. This model is then used to detect any deviation from the learned pattern or behavior exceeding a certain estimated threshold for use in identifying intrusions [32].

Anomaly-based detection is adaptive, and can detect new and unknown attacks efficiently; however, it has a higher false positive rate compared to previous approaches, as any deviation from the normal behavior profile is considered an attack even though it might be due to normal activity of an unlearned profile or a faulty node producing abnormal activities [15]. Especially in critical infrastructure applications, these types of anomalies are just as harmful as those caused by intruders, and should be identified by the developed reference model [32]. Anomaly-based detection is practical, flexible, computationally feasible, bandwidth (BW) and both spectrally and memory efficient [21]; therefore, it is widely used to secure WSNs. For this reason, the focus of this survey is on anomaly-based detection.

Anomaly-based detection techniques are classified into statistical and ML approaches. The stochastic network behavior in normal conditions is measured during a specific time window and is used to establish a baseline for future detection of patterns that are different from normal traffic [58]. However, the approach continuously generates other reference profiles with a given score for comparison to the reference profile during traffic monitoring. In this approach, the IDS is able to detect an anomalous occurrence if the score is above a certain threshold. On the other hand, ML approaches use classification algorithms to detect intrusions and malicious activities. ML classification algorithms build models capable of classifying packets to distinguish between normal and abnormal packets through training. The model is installed at the sensor level, and can classify upcoming packets after training. The advantage of ML is the ability of models to learn from experience without being explicitly reprogrammed, allowing them to be improved automatically [15,59].

8. ML and Cyberattack Detection

ML algorithms are used to build self-learning classifiers consisting of behaviors, which are able to act without human intervention by using mathematical techniques based on specialized datasets. These algorithms enhance the network nodes' ability to learn without being explicitly programmed. Such models are used to make future predictions based on new input data. ML algorithms are currently used in various applications, such as smart cities, energy, agriculture, intelligent transportation systems, industry and manufacturing, search engines, social media, cyberattack detection, spam email filtering, and recommendation systems. Different ML techniques are used to improve the functionalities of WSNs, such as data sensing, CH selection, routing and optimal path determination, data aggregation, minimizing packet delivery latency, duty cycle management, quality of service (QoS) provisioning, resource management, and to increase network lifetime. ML algorithms have been used to design lightweight detection and mitigation systems to secure WSNs against cyberattacks. They allow sensor nodes to detect possible attacks and immediately take appropriate actions to mitigate the impact of an attack by triggering an alarm, determining the degree of the risk, and isolating the attacker node from the next round of network progress [60]. The ML pipeline spans data collection and pre-processing feature selection, model training using proper ML algorithms, hyperparameter tuning, model testing, validation, and deployment.

8.1. ML Methodology

Several studies have developed and investigated effective ML techniques for cyberattack detection and mitigation. Figure 6 illustrates the generalized methodology of an ML algorithm applied to ML-based IDS. The workflow includes several phases corresponding to dataset collection, data preprocessing, features selection and extraction, ML model training, hyperparameter tuning, and model testing and validation. The first step is the availability of a dataset, which can be balanced (using of an equal number of samples for each attack type in addition to normal class samples) or imbalanced (consisting of an unequal distribution of the classes in the dataset). The next step is data preprocessing, which involves several stages: class rebalancing and sample size reduction, missing value imputations, cleaning or feature removal, data normalization, and transformation (i.e., encoding labeled data). The advantage of balancing the dataset before using it in training is to avoid bias towards the majority class. This is followed by feature selection, which involves determining an optimal set of features to help reduce dataset dimensionality, especially when considering a large dataset that may have irrelevant, redundant, erroneous, and correlated features. A lower number of dataset dimensions lead to less computational and training time being required. The reduced dataset is then utilized to train the ML model. Optimal hyperparameter values can be obtained by applying efficient tuning techniques. The final step is testing and validation, which entails using several evaluation metrics to assess ML model performance, such as the probability of detection P_d, probability of false alarm P_{fa}, probability of misdetection P_{md}, positive prediction value PPV, accuracy (ACC), F1-score, root mean square error RMSE, and receiver operating characteristics (ROC).

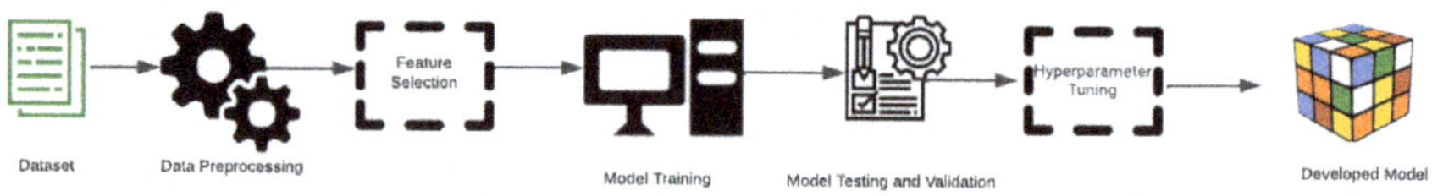

Figure 6. Illustration of generalized ML conceptual methodology.

8.2. Existing ML-Based approaches

8.2.1. Classical Machine Learning

ML algorithms are typically categorized as supervised, unsupervised, semi-supervised, and reinforcement learning. Supervised ML algorithms learn the inputs and their corresponding outputs to perform the learning process. Supervised algorithms are subdivided into regression and classification; well-known models include Logistic Regression (LR), K-Nearest Neighbors (K-NN), Support Vector Machine (SVM), Decision Trees (DT), Gaussian Naive Bayes (NB), Artificial Neural Network (ANN), and Random Forests (RF). Unsupervised learning ML algorithms only use the inputs while learning, as the associated outputs are not provided; the learning process is performed by classifying the provided input data into groups called clusters, and any new input is classified within its corresponding group. Clustering and dimensionality reduction are the two main categories of unsupervised learning. Semi-supervised learning works by combining a small amount of labeled data with a large amount of unlabeled data. In reinforcement learning, neither the inputs nor their corresponding outputs are provided, and the relationship between the input and the output is learned by interaction with the surrounding environment and a reward scheme. The reward scheme depends on the learning algorithm's performance when achieving a certain task such that a reward is provided if it achieves high performance. A popular example of reinforcement learning is Q-learning.

Several studies have discussed the efficiency of using classical ML techniques to tackle different cyberattack types [47,60–71]. For instance, research has examined well-known network layer DoS attacks (blackhole, grayhole, flooding, and TDMA scheduling). Another study used ANN and SVM to target common MAC layer attacks (collision, unfairness, and exhaustion) [72]. Other research work has considered the efficiency of RF techniques for the detection of physical layer clone attack [73]. The authors of [74] used a reinforcement learning (RL)-based IDS to detect DoS, remote-to-local, user-to-root, and probe attacks.

8.2.2. Deep Learning

DL requires a larger amount of data samples; therefore, more processing time and power are required than with classical ML techniques, which is not favorable in resource-constrained WSN contexts. DL models are more suitable for classification and prediction tasks in IoT applications that generate unstructured data, such as images, audio, and video.

DL techniques such as recurrent neural networks (RNN), deep belief networks, and Convolutional Neural Network (CNN) are largely used for security preservation and attack detection, due to their fundamental constraints when applied to WSNs. The computational complexity associated with their training, inference, and adaptation makes their use in sensor devices impractical. Several studies have been conducted on using DL techniques, such as [72], where the authors used autoencoder neural networks with a single hidden layer of neurons for lower complexity, which suits resource-constrained WSN contexts. The authors of [75] proposed a hybrid DL model using CNN and long short-term memory (LSTM) for blackhole and grayhole attack detection. The same techniques, CNN and LSTM, were used by the authors of [76] to detect DoS attacks. The authors of [77] investigated the performance of different DL techniques, including Deep Neural Network (DNN), CNN, RNN, and CNN, in combination with RNN for a single detection layer against DoS attacks. The authors of [78] proposed a DL model using a restricted Boltzmann machine with different numbers of hidden layers. The authors of [79] proposed a DL using CNN for the detection of DoS, UR2, R2L, and probe attacks. They proposed a hybrid algorithm consisting of the whale optimization technique and artificial bee colony optimization technique. Overall, both ML and DL techniques are promising for efficient IDSs in WSNs and IoT thanks to their ability to process high–dimensional data, extract useful features from network traffic payloads, and determine complex nonlinear relationships between inputs and outputs to enable informed and intelligent decisions on the part of networks.

8.2.3. Deep Reinforcement Learning

Adapting to new or constantly evolving attacks is a major drawback in classical ML and DL algorithms due to their dependence on the fixed features of existing attacks provided by the dataset for the learning process, which limits the implementation of algorithms in applications that are vulnerable to dynamic intrusions [80]. Research activities have searched for a more efficient solution by integrating DL methods with RL, which has proved effective in various IDS applications for detecting sophisticated types of cyberattacks, especially in real-time and adversarial environments [80]. For instance, attacks that affect both the physical and MAC layers were effectively detected using a proposed deep reinforcement learning (DRL) model that relied only on partial observations. In [81], a new DRL-based IDS for WSNs was designed considering link invulnerability and node importance.

8.2.4. Federated Learning

Federated Learning (FL) supports a distributed approach to perform model training at the sensor node, unlike ML or DL. WSN nodes sense and collect the data readings, then use the locally collected data for model training [82]. Afterwards, the full locally obtained model parameters in the network are shared with a powerful node, referred to as an aggregator, usually the IoT cloud server. The aggregator then merges the received trained model parameters and generates a global model that is deployed to all WSN nodes. A system based on FL structures is more robust and privacy-preserving than a traditional ML- or DL-based IDSs, because the sensor nodes collaboratively build a global learning model while safely preserving all training data locally at the sensor storage location. In a traditional ML- or DL-based IDSs, large volumes of raw data are continuously transmitted from sensors to the BS, which involves significant channel interference and energy consumption, keeping in mind that only a small fraction of the data readings are anomalous.

Recent studies have addressed the challenges of applying FL in the context of WSNs, as FL requires additional overhead and complexity, which may affect detection accuracy and convergence speed. Anomalous samples represent a very small fraction of the local data, meaning the accuracy of the training process is reduced because only the node's locally collected data is used for training, and the local dataset may lack enough training data for certain types of attacks. The node's resource heterogeneity and dynamic physical topology could lead to unexpected inconsistencies during the training process. Different nodes collect different numbers of data samples for training, meaning that attacks might only appear in very few nodes, and the same type of attack may have diverse distribution patterns at different nodes. This imbalanced distribution of data can slow down the training process at the aggregator and reduce performance due to diverging weights; thus, reducing the number of rounds required or the learning process to reach convergence is necessary in the context of a WSN in order to reduce power consumption.

Fast iteration convergence is a challenge when considering that training data samples at the local nodes are not independent and identically distributed (iid) in FL, as is the case with other ML techniques. This challenge is caused by issues such as non-uniform placement of sensors in space, faulty sensors, and high packet loss rates. Despite this, several studies support the assumption that the data samples collected at the sensor nodes are iid, as training on iid data is likely to converge faster than training on non-iid data. However, this assumption is not applicable in FL.

A promising clustering FL approach has recently been examined in the literature to solve these challenges. WSN nodes in a clustering architecture, known as MNs, send their observations to their current CHs, which performs the learning process on the aggregated data at the local cluster level. Each CH then uploads its model parameters to the FL cloud server through the BS, where they are combined into a global model with the minimum possible frequency to reach convergence [83]. This clustering approach can help to reduce overall network energy consumption, as one aggregated transmission is much more energy-efficient than multiple separate transmissions, especially when the data size is large [15]. In addition, it can help reduce communication overhead, as data compression is possible in this approach [82]. A clear challenge with the clustering FL approach is the need to optimize the number of the CHs and the number of cluster members (CMs) per cluster, in addition to the possibility that a CH may fail to train or send its local model to the server.

The different approaches mentioned above share a common challenge related to the high number of transmissions required for the BS or the aggregator to broadcast the parameters of the developed model with the rest of the nodes in the WSN, which introduces a different communication overhead that requires high energy consumption [84,85].

8.3. *ML Challenges in WSN*

This section discusses challenges introduced by network resources, application and routing algorithms, the classical ML framework and, cross-layer attack detection when implementing ML techniques for the detection of cyberattacks targeting WSNs.

8.3.1. Challenges Related to Constrained Resources

ML algorithm selection should include consideration of the computational complexity, memory usage, and balance between the quality of learning and the associated energy budget, as the developed models are intended for deployment on resource-scarce devices. Continuous or periodic collection of network traffic results in big data issues, leading to a prominent challenge for the ML framework [86]. Moreover, the frequency of uploading data samples is different from one network scenario to another. For instance, certain networks are configured to put sensor nodes into deep sleep mode in order to conserve network energy; however, important readings may be missed in these scenarios, and a body of knowledge may be lost. Another example involves the possibility that network resource consumption may not be relative to the frequency of global aggregation and model training accuracy [87].

8.3.2. Challenges Related to Applications and Routing Algorithms

Developing a suitable ML security model to detect attacks for diverse WSN applications is challenging, especially for mission-critical, highly sensitive, real-time, and adversarial environment applications. It is preferable that anomaly detection be performed locally at the local sensor nodes to avoid any communication with other nodes, the BS, or the IoT cloud due to high security requirements, which are not feasible for resource-scarce nodes. Another cyberattack detection challenge is the attacker's ability to exploit the routing algorithm and compromise its individual forwarding steps to attack the network. The purpose of these actions is to disrupt the routing and communication process by misdirecting or alternating the routing information or broadcasting fake information. The IDSs, on the other hand, can take advantage of the known behavior associated with the routing algorithm to build models of legitimate operation and compare them to the real exchange of routing messages between the nodes. Routing attacks belong to the network layer, and include sybil, 'Hello' flooding, sinkhole, blackhole, grayhole, and wormhole attacks [88,89]. Using secure routing protocols as a prevention technique and deploying proper ML-based IDS should be considered when securing these networks.

8.3.3. Challenges Related to the ML Framework

Pre-processing, feature selection and extraction, and hyperparameter tuning are essential for the success of any ML model learning process; however, collecting labeled data is not always possible in WSNs, as certain attacks may only appear in very few nodes and with low frequency. Thus, selecting an algorithm that can use minimally labeled data in a way that is sufficient for the learning process is crucial. Data reduction is required to reduce the processing time of the learning proccess on large datasets, especially for large-scale WSNs. ML preprocessing includes the process of adjusting the raw data to a format that can be used to train an ML model, such as removing features, sample size reduction, class rebalancing, missing data imputation, data normalization, encoding labeled data, and changing the data type of certain features. The process of reducing redundancy and correlation by selecting the most informative features during feature extraction while dropping irrelevant or partially relevant features from the dataset can be classified as follows: filter-based, wrapper-based, or embedded-based. Filter-based methods filter out irrelevant features independently of the learning algorithm, making it much faster and computationally effective than other methods and more suitable for WSNs [90]. Stacked-based feature extraction has been used as well; it combines several feature selection algorithms ordered as a stack and executed one after another, then applied to the dataset [91].

Hyperparameters are the set of parameters or arguments that are set manually before training and optimizing the ML model structure for better classification. These parameter value ranges are different for each ML algorithm. Hyperparameter selection significantly affects prediction results. Default parameters are the initial values that are pre-established when no values are explicitly provided. Optimized hyperparameters can be determined manually or automatically. Manual hyperparameter tuning is time-consuming, especially with a high number of possible combinations. Optimization algorithms can automate the process of finding the hyperparameters' this is called hyperparameter optimization. Different approaches include Bayesian optimization, grid search, random search, genetic algorithms, and particle swarm optimization. Several parameter combinations can be identified via search to determine the set of parameters that provide better detection results. Hyperparameter tuning is time-consuming when additional hyperparameters are added, as the number of parameter search combinations increases.

8.3.4. Challenges Related to Cross-Layer Attack Detection

Most of the existing techniques only mitigate specific types of attacks belonging to a single stack layer, excluding attacks on other layers. For instance, the network layer IDS can only detect routing attacks, and cannot recognize attacks belonging to the MAC, physical, or transport layers. It is essential to develop a cross-layer IDS that can detect different possible attacks that may occur at different WSN layers. Attacks can be identified by exploiting the information across the different layers to correlate the cross-layer features among them, such as between the MAC and network layers [86,88,89].

8.4. Datasets

Datasets are needed during the learning process to train and test ML models; therefore, dataset reliability and size are crucial to obtaining accurate results [92]. For instance, datasets that are large enough to have samples of normal traffic allow ML algorithms to learn normal network behavior, enabling the detection of unknown attacks by considering any deviation beyond the known usual behavior as unusual. It is challenging for the system to differentiate between the characteristics or signatures of a specific intrusion and a malfunction, which may cause false positives. Overall, any dataset collection for a specific network scenario should be performed over a sufficient time period to collect a sufficient volume of samples for each data class. Dataset parameters such as whether the dataset is balanced or imbalanced, the number of samples per class, dataset size, and dataset dimensionality can influence the selection of a proper ML classification technique and affect the behavior of the ML classifier.

Dataset quality affect the performance of ML models. First, ML algorithms used with a certain dataset may not be applicable for other datasets, as they may have differences in the number of classes to be distinguished, number of instances or samples for each class, and number of attributes that differentiate each class. Second, dataset characteristics such as being labeled or not (i.e., balanced or imbalanced), the number of features (i.e., dataset dimensionality), and feature importance can affect model quality. Feature extraction methods are usually used to filter out potential and relevant features. Third, the criticality and real-time nature of the WSN application at the time of data collection may result in noisy samples and irrelevant features, which can affect the final classification results and the ability of the trained model to differentiate between normal and abnormal behavior. For instance, increasing attack timespan traces and capture size can be used to control the imbalance within the dataset, thereby enhancing the learning process and allowing the algorithm to learn more differences between normal and attack samples. In addition, retraining ML models is possible, and can take place periodically during network progress as new or unknown attacks occur, allowing an ML model to modify its behavior and improve its detection accuracy.

Specialized datasets that consider the long list of cyberattacks targeting WSNs, whether collected using real-time experiments or computer-simulated, are limited. WSN-DS [92], NSL-KDD and its predecessor KDD Cup 1999 [93], CICIDS2017 [94], and UNSW-NB15 [95] are the most commonly used datasets been utilized for training and testing ML-based detection models in the context of securing WSNs. It is worth mentioning that none of these datasets except for WSN-DS are tailored to the need of developing ML models for WSNs, which motivates a need to generate new datasets or collect actual logs of real normal network data and simulated attacks.

8.5. Evaluation Metrics

Two types of cyberattack classifications are present in the literature, based on the number of classes (i.e., attacks): binary classification, in which there are only two classes, attack or normal; and multi-class classification, in which the number of considered classes is greater than two if more than one attack has been detected and sampled in the dataset. In both cases, the testing phase in the process of developing a ML model involves different evaluation metrics, which can include P_d, P_{fa}, P_{md}, positive prediction value (PPV), ACC, Error rate (ERR), geometric Mean (GM), root mean square error (RMSE), normalized RMSE, normalized RMSE (NRMSE), receiver operating characteristics (ROC), and $F1 - score$, which can be expressed as follows:

$$P_d = \frac{T_P}{T_P + F_N}$$
$$P_{fa} = \frac{F_P}{T_F + F_N}$$
$$P_{md} = \frac{F_N}{T_N + F_P}$$
$$PPV = \frac{T_P}{T_P + F_P}$$
$$ACC = \frac{T_P + T_N}{T_P + T_N + F_P + F_N}$$
$$GM = \sqrt{(P_d * P_{md})} = \sqrt{(T_p/(T_p + F_P) * T_N/(T_N + F_N))}$$
$$ERR = (1 - ACC) = \frac{F_P + F_N}{T_P + T_N + F_P + F_N}$$
$$F1 - score = \frac{2(P_d * PPV)}{(P_d + PPV)}$$

where T_P, T_N, F_P, and F_N are the number of true positives, true negatives, false positives, and false negatives, respectively, as per the confusion matrix illustrated in Figure 7.

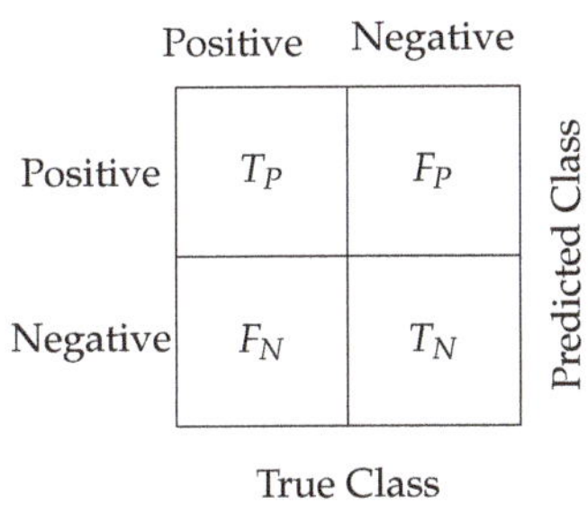

Figure 7. Confusion matrix.

These attributes are estimated after dataset testing and are calculated from the generated confusion matrix.

P_d, called the sensitivity, recall, and detection rate or true positive rate, corresponds to the number of correctly detected attacks vs. the total number of attacks. P_{fa}, or the false alarm rate, corresponds to the number of incorrectly detected attacks vs. the total number of normal traffic instances. P_{md}, or the false negative rate, is the number of undetected attacks vs. the total number of normal traffic instances. ACC is the measure of correctly detected traffic instances, whether normal or attack, vs. the total number of detected samples. ERR is the complement of ACC; it is the misclassification rate, which provides a measure of incorrectly detected traffic instances vs. the total number of detected samples. PPV represents the total number of correctly detected attacks vs. the total number of correctly and incorrectly detected attacks [96]. The F1-score or F-measure represents the harmonic mean of precision and recall; it uses F_N and F_P to efficiently classify noisy or imbalanced data [97]. High ACC, P_d, PPV, F1-score, and GM and low P_{fa} and P_{md} values generally indicate that an ML model has the potential to accurately detect attacks while ensuring that a low number of attacks go undetected. In addition, RMSE and NRMSE are used to evaluate different cyberattack detection methods numerically, and can be expressed mathematically as

$$\text{RMSE} = \sqrt{\frac{\sum_{i=1}^{N}(A_i - \hat{A}_i)}{N}} * 100\%$$

where i is the index of the evaluated sample, A_i is the actual value, $\hat{A}_i$ is the estimated or predicted value, and N is the total number of tested samples. NRMSE is defined as a measure of a model's predictive power against simple prediction using the mean of the observed data, and facilitates comparison between models with different scales; it is calculated as follows:

$$\text{NRMSE} = \frac{\text{RMSE}}{A}$$

where A represents the mean of the observed data values, which can be replaced with a range defined as the difference between the maximum and the minimum values of the observed samples. The ROC curve plot indicates the tradeoff between P_d (the Y-axis) and P_{md} (the X-axis). Preferably, the area under the ROC curve should be close to unity; low values are an indication of weak model performance in terms of detection [98]. NRMSE can be interpreted as a fraction of the overall range that is typically resolved by the model. A lower RMSE is preferable. This value is minimized when the predicted value matches the true observed value from the environment.

Evaluation metrics such as PPV, ACC, ERR, and F1-score are computed using values from the confusion matrix in both columns, and as such are sensitive to any change, especially with an imbalanced dataset. These metrics change as the distribution of data changes, even if the classifier's performance does not [96]. However, GM can be used with both balanced and imbalanced data, even if its calculation involves values from both columns of the confusion matrix, because the changes in the class distribution cancel each other out.

Other evaluation metrics commonly used to assess the ability of ML models to detect cyberattacks targeting WSNs are related to the required memory usage, buffer size, computational complexity, processing time, and prediction time, which are other elementary evaluation metrics. On-chip memory usage considers the random access memory (RAM) and Flash memory in the microcontroller unit, usually measured in kilobytes (KB). The amount of RAM directly affects processing speed. A larger amount can handle more data; however, WSN nodes have relatively low on-chip memory, which means that ML models must require low amounts of on-chip memory and be optimized for efficiency. Buffer size affects the rate of false alarms, as the node buffer usually stores certain fields of monitored traffic data that can be used as input for the detection model running within the nodes. In certain scenarios, a specific MN is responsible for monitoring its neighbors, listening to messages within radio range, and continuously examining traffic to look for intruders.

9. BC and Cyberattack Prevention

One of the earliest data security techniques of major significance is digital timestamping, which was proposed by the authors of [99] in 1991 and has drawn the attention of industry and academia ever since. The work in [99] proposed using a family of cryptographically secure collision-free hash functions, digital signatures, and linking schemes to preserve the sequential occurrence of the client's requests in the network. Digital timestamping is the precursor of the well-known BC technique [100], which is discussed in the following subsections.

9.1. *BC Background*

A BC is a set of blocks, with each block being a combination of an individual set of transactions. The number of transactions in each block depends on the block size and the transaction size. The blocks are linked using cryptographical sequential digital signatures [101,102]. These signatures are chained utilizing a hash value that involves data from the previous and current blocks to preserve the authenticity of the block's content against any data tampering [100,103]. The chain starts with a genesis block, which is the first block in the chain [104], and each subsequent block is added based on a distributed consensus with a hash value and timestamp. The shared ledger in each node connected to the BC network is updated through a consensus algorithm with each added block. The consensus mechanism ensures a common ledger database that is difficult to tamper with and has unified content on all nodes.

Figure 8 depicts the general structure of the block. Each block consists of a block header and block body. The block header contains that block's metadata, including its version number, previous block hash, nonce, Merkle root, timestamp, and nBits, while the block body consists of the transactions embedded in the block [105]. An explanation of each component is provided below.

- Version number: indicates validation rules that the BC must follow.
- Previous hash: a 256-bit value that points to the previous block (sometimes called a parent block) and affects the current block's hash to ensure the chain structure's uniqueness.
- Timestamp: the block's approximate creation time, which is required for traceability.
- Nonce: a one-time use number in the block header that is required in order to state the number of leading zeros for the hash value. This number can then be used to determine the level of difficulty when calculating the hash of a block and for verification to ensure consensus.
- Merkle root: sometimes called the 'hash of all hashes', this uniquely identifies the block; its calculation depends on the block's transactions [106]. The Merkle tree is used to verify the validity of the transactions instead of downloading the entire chain. Figure 8 illustrates the Merkle tree's structure, which is represented through the individual hashes of the transactions or leaf nodes; each set of child hashes is combined and hashed again up the tree until the root is reached [102]. Changing one transaction causes a change in the whole chain of hashes up to the Merkle root value [102].
- Hash target or nBits: a threshold value that the block header hash must not exceed in order for the block to be valid; the nBit value is usually continuously adjustable and increases with the number of leading zeros.

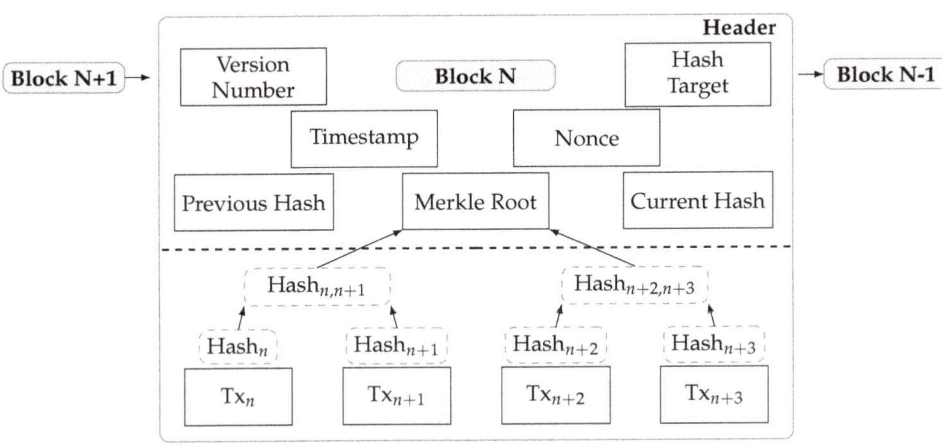

Figure 8. Block structure.

BC eliminates the need for a third-party central authority, as it is distributed or decentralized and comprises all committed transactions in the network; this makes it useful for securing cryptocurrency systems. In addition, BC has an ideal architecture for many applications that require ensuring distributed transactions between nodes and decentralizing computation and management in a trustless environment [107]. Using BC in IoT systems can reduce security risks by safely storing data, routing, accessing resources, and authenticating identities [108]. As discussed in [109], BC is a promising approach for securing data and authenticating identities in IoT because of its peer-to-peer (P2P) distributed ledger, which supports scalability and faster settlement for coordinating and securing joining nodes. However, the challenge of applying BC in WSN is its high demand in terms of storage and computational complexity, which causes additional delays and reduces network throughput. BC is often costly in terms of communication, memory usage, and power consumption, while sensor devices are typically designed to be low-cost with restricted resources; however, the cost of setting up and maintaining a centralized database can be reduced with BC. A node's idle state can be fully utilized in terms of the device's computational, storage, and bandwidth capabilities, thereby lowering overall network calculation and storage costs.

Overall, using BC to secure a WSN has many advantages; however, it is difficult to develop lightweight BC security mechanisms that carefully consider the tradeoffs between BC security and WSN design factors in terms of power and latency [110]. This work aims to investigate how BC can effectively protect sensor nodes from possible cyberattacks and determine its appropriateness for WSN applications.

9.2. BC Features

The main keywords or features that describe BC are illustrated below.

- Data immutability: data are protected using cryptographic hashes unique to each block, disallowing manipulation or alteration after registration in the BC network.
- Decentralization: the absence of a trusted supervised centralized authority; decentralization ensures a lower failure rate, makes the network less prone to malicious attacks, and reduces reliance on a third party.
- Transparency: every involved node in the network is aware of the updated stored data.
- Security and Resilience: any data manipulation requires the approval of more than half of the miner nodes, which is extremely difficult to obtain practically.
- Data Encryption: the provision of public and private keys for data encryption and decryption, respectively, via the use of an asymmetric encryption algorithm for every two communicating nodes; the public key is shared between all nodes in the network to encrypt the data, and the targeted receiver can decrypt the data using its own private key.
- Digital Signatures: digital signing of transactions using a digital signature algorithm, such as the elliptic-curve digital signature algorithm (ECDSA), to approve transaction content and originate node identities.
- Consensus: every node in the network should agree on the current state of the distributed ledger, which is made possible using one of several popular consensus mechanisms, such as Proof-of-Work (PoW), Proof of Authority (PoA), Proof of Capacity (PoC),Proof of Share Stake (PoS), Delegated Proof of Stake (DPoS), Raft, Proof of Elapsed Time (PoET), and practical Byzantine fault tolerance (PBFT) [111].
- Smart contract: a piece of code that adds customizability to a BC. It represents an arrangement and executes itself automatically under a predetermined set of rules and conditions without a third party. Smart contracts can be used for node verification and authentication. The input of the smart contract is the transaction, which is executed with a corresponding code that consists of the value, address, functions, and state to generate the output events (see Figure 9) [112].

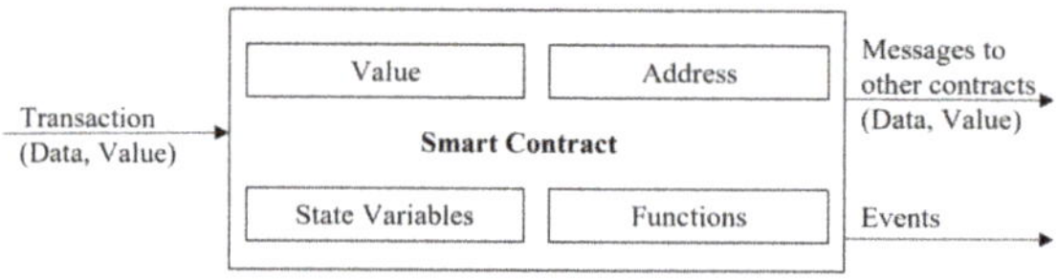

Figure 9. Smart contract structure.

9.3. Types of BC

There are four primary types of BC platforms: public (or permissionless), private (or permissioned), consortium (or federated), and hybrid (Figure 10). A BC is a fully or partially decentralized architecture that authenticates sensor devices joining the network and accepts or rejects transactions. A public BC is completely distributed; it allows any node to join the BC with similar access rights, generate new blocks, and validate data blocks. Public access to the BC provides data availability, transparency, and confidentiality. Examples of public BC platforms include Ethereum and Kadena [113]. A private BC has a central authority (or network manager) that determines which nodes may join, and does not provide each node with equal rights to perform tasks [114]. It differs from a public BC in that it restricts node participation and access to the BC depending on the authorization provided by the network [115]. Examples of private BC platform include Hyperledger Fabric, Hyperledger Burrow, IOTA, Quorum, Corda, Tendermint, Symbiont, HydraChain, Exonum, and Multichain. Both types, private and public, have disadvantages; for instance, public BCs tend to have a longer validation period for new data than private BCs, while private BCs are more exposed to certain types of cyberattacks. Compared to public BCs, several research works have considered private BCs to be advantageous when used in IoT systems, particularly in terms of network latency, due to the additional time required by public BCs to obtain consensus between all peers. PA private BC is fully controlled by one organization, and trust comes from preselecting which nodes are authorized to use the shared ledger and to verify transactions; as there are fewer trust difficulties, fewer security measures are necessary between nodes, creating a more responsive network, which is needed in IoT deployments in terms of scalability. Another advantage of private BC implementation is higher data privacy, as network data are limited to the private network and are entirely controlled by the network manager. Changes can only be made by certain nodes within the network, though all network nodes can read the data within the private BC. The role of network miners, called voters, validators, or peers, is to approve transactions and maintain copies of the BC ledger, which helps to secure and stabilize the private BC network. Because only a few nodes are delegated to publish blocks within the network in a private BC, they are more vulnerable to certain attacks types, as the authority may modify or tamper with rules or even data, and the organization may choose to revoke their BC to a previous time instant [116].

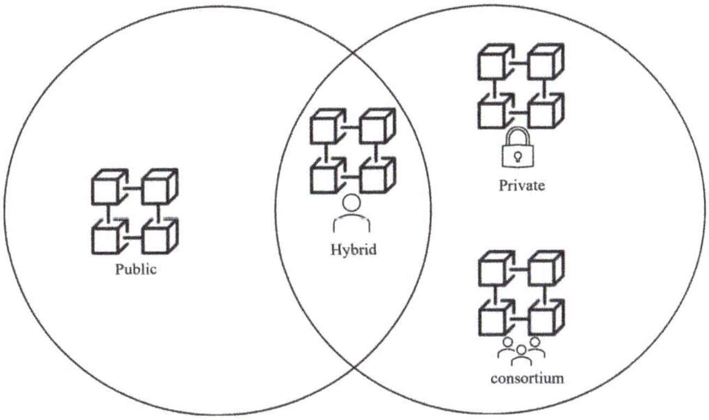

Figure 10. Types of BC.

When using a private BC for an IoT application, all nodes are identified before deployment using a Certificate Authority (CA) or Membership Services Provider, which releases identities, or key pairs, for IoT nodes. Each IoT node's registration is performed on the BC using its cryptographic hardware identity hash. Node registration is performed by mapping an IoT device's public keys and their identities, which must be stored on the BC [24]. This stops the BC from receiving information from unauthorized IoT nodes, securing it against potential attacks. A consortium BC is a type of permissionless BC; it is partially decentralized, as it is governed by a group of preselected nodes that directly participate in the consensus mechanism, instead of a single central entity as in a private

BC. A consortium BC is more decentralized than a private BC, and provides better security; however, establishing a consortium requires cooperation between a number of key nodes (sometimes called organizations), which presents logistical challenges and increases potential risk in cases where a majority of the consortium wants to tamper with the BC. A hybrid BC refers to a customizable BC architecture that combines features of both private and public BCs. Hybrid BCs are best suited for systems that cannot be fully private or public and involve a lack of trust, such as IoT, supply chains, finance, and banking.

9.4. Performance Evaluation Metrics

The performance of a BC security system used in a WSN depends on the effectiveness of its peer trust, node authentication, access control, smart contracts, consensus mechanisms, resources management, and big data processing and storage. There are multiple performance criteria of interest, including transaction throughput, response time, latency, storage overhead, and energy consumption, which are the most commonly used metrics for security analysis of BC-based WSNs, and can be defined as follows [117]:

- Transaction Throughput: the maximum amount of transactions that are processed and committed by the BC in a specific period of time, usually represented by transactions per second (tps).
- Response time: the time required to handle and verify the transactions processed by peers. The response time increases with increasing batch size, such as when the number of transactions in the queue grows; this can result in system congestion, as peers are required to handle more transactions.
- Latency: the period of time between when the transaction is invoked by a node and the time the transaction is added to the ledger.
- Transaction size: the amount of data in the transaction to be added to the next block.
- Block size: the size of the block, that is, the number of transactions included in the block.
- Storage Overhead: the storage capacity required for BC operations, which may exceed the node's storage capabilities due to the large amount of data accumulated by security tasks [118].
- Residual energy: the remaining energy in the sensor nodes; this metric is important to consider for energy-related attacks which shorten the network lifetime by wasting nodes' energy by launching malicious activities [119].

Other metrics include processing frequency, percentage of central processing unit (CPU) usage, computational complexity of encryption [120], and processing time of trust evaluation. In addition, the authors of [121] used the probability of attack success and strength of attack detection to evaluate secure mechanisms using BC. The probability of attack detection identifies how efficiently a secured mechanism can distinguish between legitimate and malicious entities targeting an IoT network, while attack strength is determined by an attacker's ability to compromise a certain node and force the network to behave maliciously.

9.5. Securing WSNs Using BC

A typical BC procedure in a P2P WSN begins when a transaction is launched between two sensor nodes. This transaction is hashed and broadcast to the P2P network. The nodes involved in the interaction sign the transaction using their public keys, as several nodes are involved in the forwarding path in a multi-path forwarding scenario. The transactions are verified and validated based on the consensus mechanism in terms of data and identity by miners or voters, then disseminated, stored, and grouped into a block. The new block is sent to the BC P2P network to be added to the chain when complete. The chain is shared, immutable, and tamper-proof across the participant nodes (Figure 11).

Two architectures are most common to build BC-based security systems in WSN contexts, namely, centralized (Figure 12) and cluster-based (Figure 13) [122]. In addition, there are two types of nodes, full and lightweight [123]. Full or Aggregator nodes store the complete ledger locally; therefore, they have access to the complete transaction history in the chain. In a WSN, a full node is usually a BS or CH. Lightweight nodes do not store a complete ledger; they only store the BC transactions with high importance and what is relevant to their operation, placing their "trust" in their associated full nodes. These nodes are usually the terminal nodes or CM. In this way, the download and storage requirements of these nodes are reduced. These architectures align with private BCs; however, it is not recommended to have a single central node similar to a BC or a CH as a master authority in charge of authentication and trust management in order to avoid any critical points of vulnerability in the network.

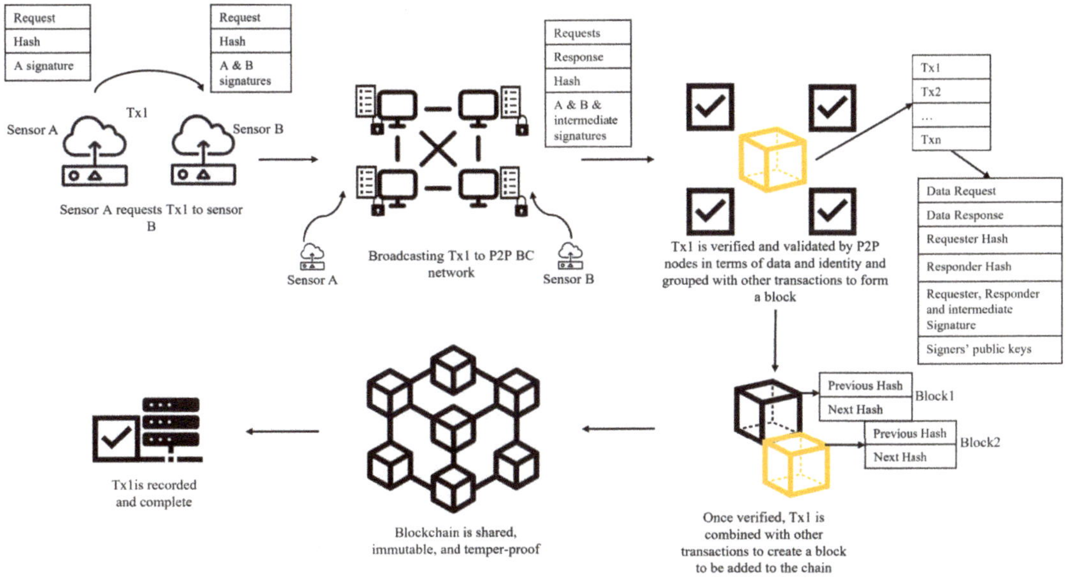

Figure 11. BC workflow in WSN.

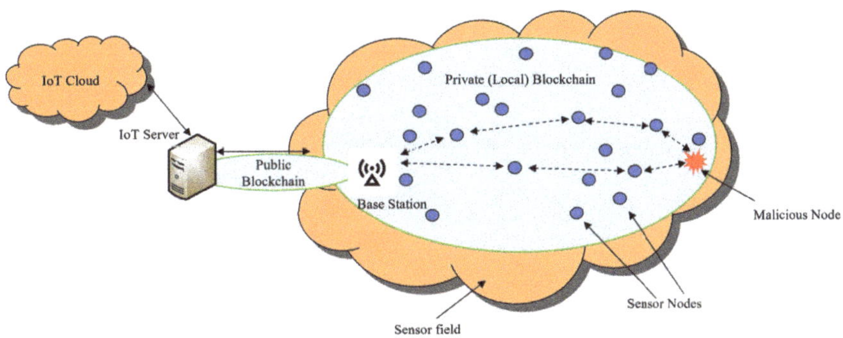

Figure 12. Centralized architecture in a BC-based WSN.

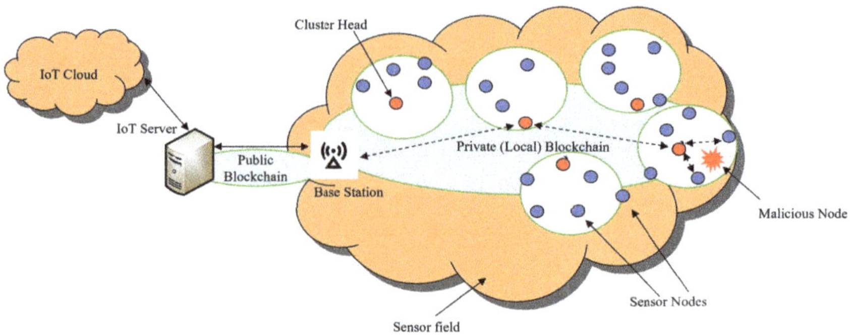

Figure 13. Cluster-based architecture in a BC-based WSN.

9.6. BC Challenges in WSNs

- Scalability and Storage: the amount of generated data grows exponentially with an increase in the number of devices deployed in different IoT applications, which leads to an increase in the rate of transaction execution and in the storage capacity needed to keep the ledger up to date [124]. Scalability becomes a severe bottleneck with an increasing number of transactions, and limits the practical development of BC in WSN contexts. With BC technology, blocks are not stored in a central server; however, a subset of the nodes need to keep a copy of the entire ledger in their own limited storage, which means that maintaining enough storage space for the ledger may not be feasible. Moreover, the size of the ledger increases over time, while most of nodes have low storage capacity of 10 KB to 100 KB memory at the most.
The ledger storage requirement remains an open research issue. According to [125], certain IoT devices are limited to up to 8 MB of memory, most of which is used for storing the software that manages the device; therefore, lightweight mechanisms that limit the ledger size and allow it to be stored by each node are highly recommended. The authors of [118] defined three strategies for data storage by IoT sensors: full storage, in which all nodes store the full current data; partial storage, in which each node stores only part of the data, allowing it to be restored when combined with data from other nodes; and persistent storage, in which low-priority or old data can be stored in a remote centralized database. Similar criteria can be suggested to reduce storage overhead.
Efficient consensus protocols, optimizing block size, sharding, pruning, lighting protocols, and off-chain storage have been proposed in the literature to address scalability issues. For instance, PBFT is considered a suitable protocol for fixed and small-size networks, although it is not scalable for larger numbers of IoT devices [126]. Sharding is one of the newer mechanisms to support scalability; it aims to split the overhead of processing transactions between multiple 'shards', or subgroups, of consensus nodes. These groups work in parallel to maximize performance while significantly avoiding the overhead due to duplication of communication, computation, and storage per full node, allowing the system to scale to larger networks [127]. Scalability can be increased by pruning the size of blocks on the BC, which includes removing older transactions to control memory usage [128]. Lighting protocols aim to lower the verification process period by only allowing full nodes to store the complete ledger, with lightweight nodes only keeping a portion of it. In off-chain storage, only hashes are stored in the ledger, whereas actual data are stored off-chain i.e., in the cloud, to support the scalability in dense WSNs.
- Consensus mechanisms: common consensus mechanisms such as PoA, PoS, and PoC are primarily designed to work for monetary transactions, and are not suitable for adoption in WSNs and constrained-resource IoT devices [129]. PoW is not common in IoT and WSN applications, as it is demanding in terms of computational power;

while PoC is energy efficient, it depends on a node's storage capacity and monetary stake, and monetary stakes do not exist in IoT and WSNs [130]. In addition, the most commonly known consensus mechanisms do not perform as desired in their raw mode because of massive requirements and scalability issues [131].

- Communication Overhead and Synchronization: a significant amount of communication overhead is required to synchronize the BC copies, as there is a need to forward every verified transaction to all peers. Establishing keys and authenticating nodes with cryptography, which is determined by the encryption type (either asymmetric or symmetric), causes high communication overhead and key storage [118]. Time consistency during time synchronization between sensor nodes requires exchanging a number of messages depending on synchronization frequency.
- Computation Overhead: heterogeneous IoT devices have different processing capabilities for running encryption and decryption, which leads to variations in processing time. Integrating a BC into the sensor network enables a logical peer-to-peer network to validate and store transactions locally, which is straightforward for personal computers or workstations; however, it might be difficult for tiny sensors with limited computational resources.
- Complexity and Energy Wastage: most widely employed BCs use PoW as a consensus mechanism, where the network participants must solve a mathematical problem or cryptographic puzzle in order to validate and authenticate transactions. PoW uses a significant amount of computational resources, causing energy losses; therefore, it is not practically suitable for IoT networks [30,111]. In addition, severe latency affects WSN stability in delay-sensitive applications [132].
- Guaranteeing Security: many malicious activities target IoT and WSNs. A single attack can harm a large number of devices or be used to destroy another system, as monitor nodes can be turned into malicious nodes to launch further attacks. The network's ability to manage advanced cyberattacks is degraded due to the constrained resources of IoT devices. However, a BC relies on sophisticated hash functions, which require heavy computation and consensus mechanisms that consume network bandwidth.
- Compatibility and Standardization: standardization for BC security applications is needed in to ensure that devices meet a reasonable set of standards and have fundamental security and privacy capabilities and to diminish risks associated with cyberattacks against IoT devices [133].

BC in WSN and IoT may never become a reality unless the storage, battery life, computation power, and bandwidth availability of sensor devices are improved [134]. Securing a network using a BC with resource-constrained devices remains challenging [111], and researchers are currently seeking lightweight mechanisms that can solve problem of excessive resources consumption by sensor networks when using BC to secure WSN and IoT networks against possible attacks such as data manipulation and tampering.

9.7. BC in the Literature

BC is a dynamic technology that has spurred tremendous technological advances in many fields in the last few years. BC has been recently proposed as a method to secure the systems applications associated with IoT, such as smart homes, supply chains, smart agriculture, and smart grids [135]. The recent focus in the literature has shifted to securing WSNs using BC, despite the challenging characteristics and performance limits discussed in Section 9.6. BC characteristics, which drive many WSN design challenges include ledger size, block size, number of transactions per block, smart contracts, miner selection criteria, number of miners, selected BC type, and hash function specifications. Using BC to secure WSNs includes proposing mechanisms to protect data sharing, establish trust, authenticate identities, secure routing tables, and secure localization against dangerous cyberattacks such as sybil, spoofing, DoS, message substitution, and replay attacks (see Table 3). The rest of this section examines BC techniques presented in the literature while highlighting their advantages and drawbacks.

Malicious node detection is one of the main applications of BC in the context of WSNs [31], and is the focus of [136], where the authors studied the use of BC in conjunction with smart contracts to identify malicious nodes. The authors proposed a trust model using smart contracts based on the processing delay, forwarding rate, and response time as evaluation metrics to distinguish malicious and benign nodes. The result of the detection process was then recorded in the BC. The work in [136] established that the proposed method is effective in terms of detecting malicious nodes and allows detection process traceability; however, the adopted traditional consensus method, PoW, is computationally demanding and is unsuitable for resource-limited WSN nodes.

Another major application is maintaining data authenticity, which is ensured using node authentication and trust management [52,106,109,136–143]. Trust management is tied to authentication mechanisms, which identify end-communicating nodes and ensure data validity and confidentiality. The authors of [138] proposed a BC-based trust model and node authentication using a smart contract at gateway nodes such as CHs and BS to reduce energy consumption, claiming that the model takes 0.000250 s, unlike Ethereum BC, which requires 14 s to achieve the same results. The benefits of this type of model were further discussed in [140] through a test-bed experiment, which was conducted to determine whether a BC-based data-driven trust mechanism could reduce network transaction throughput in the presence of grayhole and blackhole attacks. The authors of [143] examined ways of reducing the communication overhead associated with BC by reducing the size of public and private keys. This reduction was achieved in [139] by employing Hyperel-liptic Curve Cryptography (HECC), which can potentially provide a similar security level to other key generators along with a lower key size. Another practical implementation of integrating BC into a WSN to protect data against tampering was discussed and evaluated in [52,106,141,142]. The performance evaluation in [106] indicated that the computational complexity associated with evaluating the hash function increased as the amount of sensed data increased, as the ledger size increased accordingly, resulting in reduced data transfer efficiency in the network. The work of [52] proposed limiting the size of each BC and setting a time window with a circular buffer mechanism to minimize BC length as a solution to this limitation. The authors of [52] considered a hash and an additional time interval measure to determine nodes' reliability levels, which is equivalent to the dynamic value of the accumulated trust points of a certain node. This reliability level was controlled (increased or decreased) by tracking a ledger-checking message. The experimental results of [52] indicated that the proposed trust mechanism could reduce both communication overhead and memory requirements. A different approach to protecting data integrity was proposed in [120], where the authors focused on using a cryptographic algorithm in conjunction with a private BC to protect data during transmission between nodes. The authors proposed a methodology integrating BC and Advanced Encryption Standard (AES) symmetric encryption in a WSN system, with the hash value used to encrypt data during the transaction and AES used in the data transport layer as an encryption/decryption process carried out between any two communicating nodes. This methodology reduces resources consumption while protecting the network against linking, MITM, and Distributed denial of service (DDoS) attacks. The method proposed in [120] is not scalable, however, as it is limited to the use of private BCs. The authors of [129] proposed and simulated a new model for a consensus algorithm that can reduce the time required for mining. The results presented in [129] confirmed that the proposed model can potentially protect the network against spoofing and injecting phantom devices; however, the software associated with the model cannot be updated.

Protecting the network routing process using a BC was the focus of the work in [141,144,145]. The authors of [144] proposed a BC-based routing protocol that uses a BC to store the network's activities and broadcast the status of the nodes. In [144], the aim of the authors was to secure the route determination process by avoiding untrusted nodes and to resolve the load-balancing issues associated with routing. The authors of [141] proposed using the a BC-block approach with flow routing tables instead of converting the entire SDN-enabled

WSN network to a BC network, thereby preventing tampering with flow entries. This approach reduces the energy consumption associated with traditional BC algorithms; the results obtained in [141] using a simulated Riverbed model indicate that the proposed scheme can provide security against MITM, replay, and blackhole attacks, though with increased energy consumption and end-to-end delay. A different distributed ledger-based technology was considered in [142] as an effective lightweight network to authenticate and protect routing tables against sybil attacks in addition to protecting the network against fake identities more broadly. However, the proposed network in [142] is time-consuming, not scalable, and has a centralized architecture. The authors of [145] proposed a trust model for a decentralized architecture to secure WSN routing through a dual BC model. The first model is public and implements a PoW consensus to authenticate aggregating nodes (ANs), while the second model is private and uses PoA for authenticated sensor nodes. The PoW mechanism enhances the security level of the unauthenticated public BC, which includes the BSs, though at the cost of high computational complexity. On the other hand, PoA is less computationally complex, helping to reduce the overhead of the resource-limited AN, which is included in the authenticated private BC. Node integrity is evaluated using a trust evaluation metric to determine the legitimate nodes that take part in the routing process; however, the security analysis in [145] indicated that the proposed approach could be vulnerable to smart contract-based attacks resulting from possible bugs in smart contract code, such as integer underflow, overflow, parity multisig, timestamp dependency, transaction ordering dependence, call stack depth attack, and re-entrancy.

The authors of [109,139,146] proposed BC-based identity authentication mechanisms for sensor nodes joining the network. The authors of [109] proposed a secure identity authentication mechanism in a hierarchical architecture for a multi-WSN environment using public and private BC, where the former includes the BS and terminal users as miners, while the latter is composed of all authenticated CHs. This technique minimizes communication overhead, as sensing nodes are not connected directly to the unauthenticated public BC; therefore, frequent node authentication is not required. The focus of [146] was on securing an identity authentication scheme against worm attacks by using the IOTA Tangle BC to store the authentication data safely; however, the network proposed in [146] relies on a single point, namely, sink nodes, to authenticate other nodes, which means that the network has a centralized architecture. Another approach was proposed in [147], where the authors considered a sequential detection scheme that starts by validating the hash value of the node's ID, followed by validation of the node signature by each node, and ends with a voting technique that determines whether the node is malicious or benign. The results from the different stages are then used to decide whether a suspect node is kept or eliminated. The authors of [147] revealed that the security level of the proposed BC method is improved compared to other classical approaches in the literature; however, the latency introduced by the three combined phases could potentially be higher than classical approaches. The authors of [148] proposed another decentralized authentication and trust model, which stores the authentication and trust information in the BC and uses a subjective probability as a reputation level. The technique is limited by the origin block problem, which causes the system to misbehave in cases where malicious values are included in the first block in the chain.

Localizing WSN nodes accurately is another application, and was addressed in [149] by investigating a decentralized BC-based trust management model. Their model relied on a trust value consisting of both behavior and data trust values evaluated by a selected number of trusted nodes, such as the number of successful and unsuccessful interactions between sensor nodes and feedback metrics related to the integrity of each beacon node. Though the simulated results in [149] indicate that the proposed scheme outperforms other current techniques in several aspects, it requires an additional number of transactions associated with the evaluation processes, and lacks a complexity analysis of the proposed technique.

Table 3. Existing research on BC-based WSN security (N/A means not available).

BC Type	Ref.	Security Threat	Study Highlights
Private	[120]	DDOS and Linking attacks	Relies only on AES symmetric encryption for data integrity
	[138]	Data Tampering	BC-based trust model and node authentication using smart contracts to reduce latency.
	[106]	Data Tampering	Performance evaluation of the computational complexity associated with the ledger.
	[147]	Internal attacks	Three-phase sequential detection using sensor node hash values, node signatures, and voting degree
	[146]	Worm attack	Relies on IOTA Tangle
	[142]	Sybil attack	Relies on IOTA Tangle
Consortium	[136]	Internal attacks	Utilizes a smart contract in conjunction with BC
Hybrid	[109]	Sybil, MITM, DoS, Message Substitution, and Replay attacks	Identity management and secure authentication mechanism
	[139]	Internal attacks	Employs HECC to generate public and private keys
	[145]	DoS and Sybil attacks	Employs dual public and private BCs which implement a PoW and PoA consensus, respectively, for authentication of each BC.
N/A	[140]	Greyhole and blackhole internal attacks	Test-bed experiments using a data-driven trust mechanism to reduce network transaction throughput.
	[52]	Physical or logical data tampering	Limits the size of the BC and uses a time window with a circular buffer mechanism to reduce the BC's length.
	[149]	Attacks on the localization process	Uses both behaviour and data trust values to determine the reliability level.

Table 3. *Cont.*

BC Type	Ref.	Security Threat	Study Highlights
	[129]	Injection, data tampering, firmware modification, listening to traffic	Proposes a new consensus algorithm model to reduce mining time.
	[144]	Routing attacks	Utilizes BC as a shared memory for route determination to avoid untrusted nodes
	[141]	Routing attacks (blackhole, replay, and MITM)	Relies on BC block technology to protect the flow routing tables at the nodes against routing attacks in an SDN-enabled WSN
	[148]	Internal attacks	Uses a subjective probability measure as a reputation level model for peer trust

10. BC–ML Integration

It is evident from Sections 8 and 9 that BC technology has been found in the literature to be a useful framework for securely recording data transactions in a tamper-proof ledger with the help of embedded mechanisms such as consensus and smart contracts, whereas ML provides efficient classification models to identify attacks. Therefore, when considering integrated BC and ML approaches, BC technology can help to securely store data generated by WSN devices. This generates huge amounts of data, which can then be modified and organized to safely train ML classifiers, potentially achieving high detection accuracy. It is notable that the output of the ML detection process can be securely stored on the BC network to preserve the integrity of the detection process results. Figure 14 depicts key features of the integrated BC–ML security approach. Despite the potential benefits and the fact that their integration is possible, inevitable, and beneficial, integrating these two technologies simultaneously poses new challenges when adopted in any WSN application. Most of the existing literature has studied ML and BC separately when considering securing WSNs, unlike other IoT applications such as smart grids and supply chains. However, the gains to be achieved and the challenges faced when seeking to combine these two technologies for securing WSNs have not yet been extensively explored in the literature due to its being a relatively new research direction.

In this regard, our approach is to have two lines of defense utilizing the integration of BC and ML. The first line of defense is attack prevention using BC, while the second line of defense is attack detection using ML. In case the first line of defense fails to prevent an attack, the second should verify and examine the incoming traffic for any sign of vulnerability, alerting the network to the presence of a malicious attack [150].

The emphasis of this section is on those research works that consider BC–ML integration to secure WSNs, as discussed in Section 10.1. Following a discussion of the important open issues and research challenges involved in the interrelationship and integration of both technologies to protect WSNs against cyberattacks, which is detailed in Section 10.2, this section is closed by detailing our proposed approach to an integrated BC–ML solution.

10.1. Related Work

Integrating the technologies of ML and BC to secure WSNs has been considered in the literature in several different directions, namely, secure routing, secure authentication, malicious nodes detection, and trust mechanism, as outlined in Table 4.

Table 4. Existing research works on BC and ML integration for securing WSNs (N/A means not available).

BC	ML	Ref.	Attack	Study Highlights
Public	Deep CNN	[151]	Internal	Trusted distributed routing using BC while avoiding routing paths with congestion and malicious nodes using DL-CNN
	Hidden Markov Model (HMM)	[137]	Sybil	Trust model for identifying Sybil nodes; trust value is evaluated via HMM, and the trust values are then added to the BC
	Histogram Gradient Boost (HGB)	[152]	DoS	BC-based authentication mechanism in which IPFS is integrated with BC for data storage, and HGB detection module to mitigate DoS attacks
Private	Isolated forest algorithm	[153]	Internal	Isolated forest algorithm for anomaly detection in the BC
	Genetic Algorithm (GA)-based SVM and GA-based DT	[154]	Grayhole, mistreatment, and MITM	ML models for malicious node detection and registration along with authentication mechanism and data storage routing using BC
Consortium	RL	[155]	Blackhole	Trusted routing with the use of BC and reinforcement learning
	DNN	[156]	Routing, specifically Blackhole	Trusted routing with the use of BC and DNN
Hybrid	Gaussian NB	[122]	Internal	BC-based identity management and secure authentication mechanism with Gaussian NB detection module to mitigate DoS attacks
N/A	Generative Adversarial Networks (GAN)	[157]	Network layer	Authentication and validation of current routing data using a Generative Adversarial Network-based BC-enabled secured routing protocol

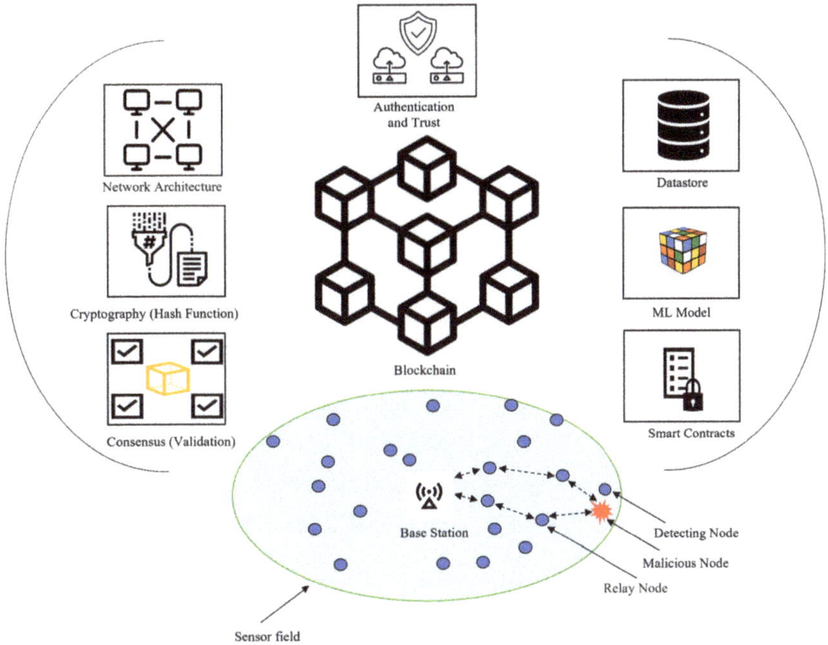

Figure 14. Key features of BC–ML integration for WSNs.

Securing WSN routing protocols using an integrated BC–ML approach has been considered in [151,155–157]. The proposed framework in [155] relies on the BC network to securely record the routing information via the use of a registration contract, token contract, and token transactions, as well as to preserve data integrity by the use of PoA due to its high processing efficiency. The routing protocol of the proposed framework in [155] exploits a reinforcement learning algorithm to dynamically provide trusted routes. The results in [155] confirm that the average packet delay is reduced by 81% compared to state of the art techniques thanks to the trusted queue length information released in the proposed framework [155], while the use of PoA helps to reduce token transaction latency. A PoA-based BC was considered in the proposed framework of [156] as well, which utilized a deep learning selection model through CNN to provide the validators required for the PoA smart contract instead of randomly selecting them. The proposed PoA–DL consensus mechanism was shown to require a steady latency that is less than the average transaction delays of the state-of-the-art techniques, and enhances the transaction processing capacity due to the preselected and limited number of validators. Another deep learning method, referred to as Fully Decentralized Generative Adversarial Network (FDGAN), was proposed in [157] in conjunction with GAN, IDS, and BC to design a new routing protocol named Block Chain enabled secured Routing Protocol (GBCRP).

Malicious node detection techniques using integrated ML and BC methods were the focus in [122,152–154]. The isolated forest algorithm anomaly detection model was studied in [153]; this model it is not computationally demanding and can deliver good detection performance, especially in the case of high-volume and high-dimensional processed data. The BC helps to ensure safe storage and adequate updating of the isolated forest global detection model by providing the required trusted blocks (isolated trees) to form the model. The results reported by [153] indicate that the proposed anomaly detection model integrating BC and the isolated forest algorithm can achieve a high detection level and accuracy rate for all types of attacks while requiring less communication and storage overhead compared to other similar BC-based anomaly detection models, as it only stores the detection model and not the detection results. A joint identity management and secure routing model was proposed in [154], in which the GA-SVM and GA-DT ML techniques were examined for the detection of malicious nodes. It was shown that GA-SVM is better than GA-DT in terms of detection accuracy; the outcome of the GA-DT process determines whether the node continues to be involved in the routing process or whether its registration in the BC network is revoked, and the safety of the routing transactions is secured using the PoA consensus mechanism. It was shown in [154] that when removing MNs, the packet delivery rate increased to 99.72%. Another consensus mechanism known as Verifiable Byzantine Fault Tolerance (VBFT) was used to validate transactions in [152], while the use of the HGB–ML classifier was proposed for detecting MN. Furthermore, [152] proposed storing data associated with normal nodes in an Interplanetary File System (IFS) to generate hashed chunks that can be then stored in the BC. Extensive comparisons were performed in [152], showing high precision of at least 98% obtained using HGB, which is more than could be achieved by its counterparts, and further demonstrating the lower transaction costs of VBFT compared to PoW. Our prior work in [122] proposed a BC-based identity management and secure authentication mechanism using a Gaussian NB detection module to mitigate possible internal DoS attacks targeting CH nodes.

10.2. Research Challenges

Developing a lightweight integrated framework that combines BC and ML while being WSN-compatible is a research area that remains in its infancy, and many open issues and challenges must be carefully addressed. The challenges associated with such systems combine the challenges related to each individual technology. Key technical challenges can be segmented into integration performance, scalability, lightweight architectures and schemes, managing network resources, legal issues, and vulnerabilities.

- Integration performance: BC and ML integration performance depends on each technology's performance; however, having both technologies operating within the same system rsises the idea of using each technology to improve the functional performance of the other. For example, ML model detection performance can be degraded by data tampering. In this regard, BC can protect the data transactions used to train the ML models along with the recorded decisions (i.e., output) of attack classification with confidence, disallowing tampering. These records can be reviewed and audited at any time by authorized nodes, and can be used to improve future ML detection decisions. In this way, incremental ML models can improve their future decision-making to detect novel attacks and handle drift in networks that change dynamically over time [65].
- Scalability: a measure of how well systems are used in conjunction with WSNs, scalability is related to network capacity in terms of the number of nodes that can join and the transaction volume that can be generated and processed over the network. The selection of the BC type and consensus mechanism highly affect scalability. For instance, the PBFT and PoA consensus mechanisms can improve transaction throughput compared to PoW, which usually supports only a few dozen transactions per second. Frequent authentication and peer trust requirements coupled with increased ledger size as the number of nodes and data increase present a challenge when aiming for a scalable ML–BC integrated security framework; however, many solutions have been presented in the literature that support scalability when employing BC technology. Among these solutions is the use of a hybrid BC, which utilizes a public BC connected to multiple private BCs wherein each private BC operates with one WSN. This structure limits the transaction volume and size of the ledger, ensuring better scalability. Among the known consensus mechanisms, voting or multiparty consensus works better with private BCs, and their combination is a candidate for use in cooperative WSNs. Another consensus mechanism is Proof-of-Authentication (PoAh), proposed in [158] for resource-scarce networks, adn which could be tested for WSNs. ML algorithms, on the other hand, can be used to code smart contracts for a more scalable approach to effective detection of malicious nodes.

- Lightweight schemes: to reduce overhead, the development and refinement of lightweight BC–ML integrated schemes while maintaining the same desired security level is essential. Deploying BC involves many elements, such as trust, authentication, access control, smart contracts, and consensus mechanisms, and each element can be implemented using a variety of options. The complexity of a BC can be refined by considering lightweight schemes in terms of storage, processing, and communication for each element involved in the deployed BC. For instance, in [152] the authors suggested Interplanetary File System (IPFS) to record the detection process, with the aim of reducing the cost of data storage in WSN; however, they did not consider the communication overhead required to upload and download data between IPFS and BS. In terms of consensus, PBFT and PoA are preferable, as they offer reduced computation and delay compared to PoW.
- Vulnerability: the ultimate goal of combining ML and BC into one system is the potential increase in security level; however, this integration does not completely eliminate threats. The root of these possible threats can be understood by considering that even though data my be safely protected by BC, it could be susceptible to tampering before it is securely recorded in the ledger.
  Considering the two approaches for BC implementation, namely, public and private, a public BC is open and accessible to all nodes, whereas a private BC is not. Therefore, a is preferable when higher levels of security are desired [159]. However, private BCs limit access to the large amount of data required to develop an efficient ML model, especially with the amount of continuously developed attack types, which makes an ML–BC integrated system vulnerable to newly developed attacks. Other possible threats might be due to malfunctioning or faulty sensors, or even sensors equipped with extra hardware allowing them to be operated maliciously, and which cannot be detected unless physically tested. These challenges add up when considering that nodes can become malicious and threaten the network security after joining the network. In addition, smart contracts can be vulnerable to possible smart contract-based attacks due to bugs in the smart contract code. ML can be used for smart contract verification and vulnerability detection [160].
- Managing network resources: limited-resource sensor nodes represent a key technical challenge when developing an ML–BC integrated solution considering encryption, trust and authentication, and validation of transactions through consensus. The ledger grows exponentially over time, and eventually may not fit within a node's memory. These technical challenges in terms of storage and processing translate into high power consumption, extending across all aspects of system design. The authors of [161] suggested a solution to this problem by switching to symmetric instead of asymmetric BC encryption in order to simplify the system's computational complexity. The computational complexity can be reduced using a simplified method for hash function calculation, such as SHA-256 [161]. Another proposed direction is dedicating specific nodes with high capabilities, such as CHs and BS nodes, for ledger storage, with other nodes only keeping the constant-length hash value of the data in the ledger to be referenced when needed. In addition, old data can be migrated from the CHs and BS toward the IoT cloud or external storage (i.e., IPFS).
- Legal issues: proliferation of different standards or a lack of security regulation can represent a challenge when designing systems involving two different technologies. Setting standards for such integrated solutions can potentially be done at the level of manufacturing and fabrication, that is, at the sensor stage.

Overall, a substantial amount of future research needs to be directed toward designing a robust ML–BC integrated solution to secure WSNs before they can be expected to work smoothly. A lightweight framework must be designed that considers sensor resource constraints and is able to effectively secure WSNs in terms of establishing trust in a trustless environment. Specially-developed consensus mechanisms, application-specific smart contracts, simple transaction verification, an alternative to block mining, and optimized architectures that balance computation and communication consumption between nodes are vital to promoting such integration in WSN applications. In this regard, we propose the BC–ML integrated system depicted in Figure 15. The proposed system includes an ML detection model that detects the malicious behaviour of nodes using neighboring information. The ML model can identify unknown types of attacks by recognizing any deviation from the normal operation of a system as malicious [162], which allows it to use transfer learning to detect new and unknown attacks by transferring its knowledge of known attacks [163]. Concurrently, a BC-based prevention model avoids possible attempts by malicious nodes to modify their data. The BC records the ML detection process securely on the BC ledger in order to maintain its integrity. Furthermore, an smart contract is used for identity management to prevent malicious nodes from becoming authorized to access the BC network (if they are newly deployed) or to revoked their access (should malicious behaviour be detected). In addition, a trust smart contract ensures end-to-end trustworthiness between communicating nodes and limits the negative impact caused by attacks to only the affected part of the network, specifically when a cluster-based architecture is employed [134]. Smart contracts can host ML models to establish trust between nodes, making smart contracts more effective. It has been proposed to use ML models to detect smart contract-based attacks or vulnerable smart contracts deployed by malicious nodes; however, studies have revealed that smart contracts may not be able to process ML tasks with high computational needs [164]. The proposed overall BC structure is a multi-layer or hybrid one, with private BCs deployed for internal authentication in the network and a public BC deployed between the BS and IoT cloud.

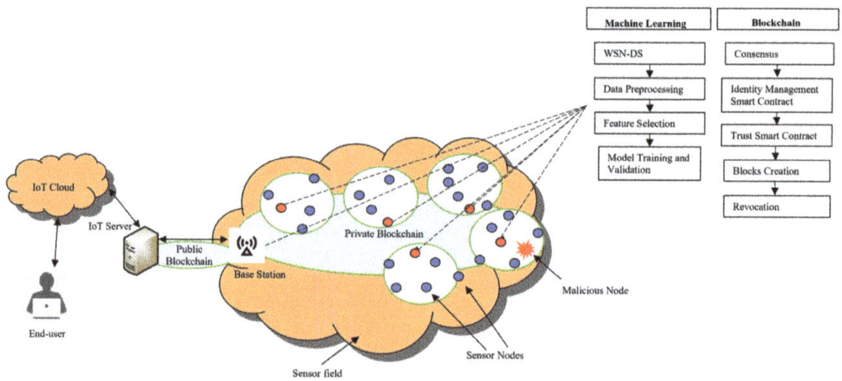

Figure 15. Proposed integrated BC–ML system for use in WSNs.

11. Conclusions

Several countermeasures to secure WSNs have been considered in this review, and extensive research efforts have been made to address the related security threats. However, at present these networks cannot manage the computational overhead necessary to implement many of the proposed defensive strategies. ML and BC are two promising technologies that we have focused on in this study for ensuring secure WSNs. In this paper, we have aimed to investigate the integration of both technologies towards a lightweight security framework for WSNs. Our review began by discussing existing surveys on ML and BC in WSN contexts, then provided a taxonomy of ML and BC approaches for WSN-related cyberattack detection and prevention. We next discussed related work and open issues for future research associated with both technologies. Finally, we illustrated the integration of ML and BC to secure WSNs, surveyed related work, and discussed the associated challenges. Finally, we ended our review by proposing the use of an integrated ML and BC system in two lines of defense to enhance the security of WSNs. In our future work, we will consider the implementation of the proposed framework and examine the performance of the integrated system with the goal of enhancing the security of WSNs.

Author Contributions: Conceptualization, S.I. and D.W.D.; Methodology, S.I. and D.W.D.; visualization, S.I. and D.W.D.; Supervision, H.R.; Writing—original draft, S.I. and D.W.D.; Writing—review, S.I., D.W.D. and H.R.; Writing—editing, H.R. All authors have read and agreed to the published version of the manuscript.

Funding: This research received no external funding.

Data Availability Statement: Not applicable.

Conflicts of Interest: The authors declare no conflict of interest.

References

1. Abbas, G.; Mehmood, A.; Carsten, M.; Epiphaniou, G.; Lloret, J. Safety, Security and Privacy in Machine Learning Based Internet of Things. *J. Sens. Actuator Netw.* **2022**, *11*, 38. [CrossRef]
2. Bajaj, K.; Sharma, B.; Singh, R. Integration of WSN with IoT applications: A vision, architecture, and future challenges. In *Integration of WSN and IoT for Smart Cities*; Springer: Berlin/Heidelberg, Germany, 2020; pp. 79–102.
3. Pavithran, D.; Shaalan, K.; Al-Karaki, J.N.; Gawanmeh, A. Towards building a blockchain framework for IoT. *Clust. Comput.* **2020**, *23*, 2089–2103. [CrossRef]
4. Sinha, P.; Jha, V.K.; Rai, A.K.; Bhushan, B. Security vulnerabilities, attacks and countermeasures in wireless sensor networks at various layers of OSI reference model: A survey. In Proceedings of the 2017 International Conference on Signal Processing and Communication (ICSPC), Coimbatore, India, 28–29 July 2017; pp. 288–293.
5. Panda, M. Security in wireless sensor networks using cryptographic techniques. *Am. J. Eng. Res. (AJER)* **2014**, *3*, 50–56.
6. Hussain, F.; Hussain, R.; Hassan, S.A.; Hossain, E. Machine learning in IoT security: Current solutions and future challenges. *IEEE Commun. Surv. Tutor.* **2020**, *22*, 1686–1721. [CrossRef]
7. Xu, L.D.; Lu, Y.; Li, L. Embedding Blockchain Technology Into IoT for Security: A Survey. *IEEE Internet Things J.* **2021**, *8*, 10452–10473. [CrossRef]
8. Kumar, D.P.; Amgoth, T.; Annavarapu, C.S.R. Machine learning algorithms for wireless sensor networks: A survey. *Inf. Fusion* **2019**, *49*, 1–25. [CrossRef]
9. Alsheikh, M.A.; Lin, S.; Niyato, D.; Tan, H.P. Machine Learning in Wireless Sensor Networks: Algorithms, Strategies, and Applications. *IEEE Commun. Surv. Tutor.* **2014**, *16*, 1996–2018. [CrossRef]
10. Bout, E.; Loscri, V.; Gallais, A. How Machine Learning Changes the Nature of Cyberattacks on IoT Networks: A Survey. *IEEE Commun. Surv. Tutor.* **2022**, *24*, 248–279. [CrossRef]
11. Tahsien, S.M.; Karimipour, H.; Spachos, P. Machine learning based solutions for security of Internet of Things (IoT): A survey. *J. Netw. Comput. Appl.* **2020**, *161*. [CrossRef]
12. da Costa, K.A.P.; Papa, J.P.; Lisboa, C.O.; Munoz, R.; de Albuquerque, V.H.C. Internet of Things: A survey on machine learning-based intrusion detection approaches. *Comput. Netw.* **2019**, *151*, 147–157. [CrossRef]
13. Ahmad, R.; Alsmadi, I. Machine learning approaches to IoT security: A systematic literature review. *Internet Things* **2021**, *14*, 100365. [CrossRef]
14. Haji, S.H.; Ameen, S.Y. Attack and Anomaly Detection in IoT Networks using Machine Learning Techniques: A Review. *Asian J. Res. Comput. Sci.* **2021**, *9*, 30–46. [CrossRef]
15. Faraj, O.; Megias, D.; Ahmad, A.M.; Garcia-Alfaro, J. Taxonomy and challenges in machine learning-based approaches to detect attacks in the internet of things. In Proceedings of the 15th International Conference on Availability, Reliability and Security, Virtual, 25–28 August 2020; pp. 1–10.
16. Mamdouh, M.; Elrukhsi, M.A.I.; Khattab, A. Securing the internet of things and wireless sensor networks via machine learning: A survey. In Proceedings of the 2018 International Conference on Computer and Applications (ICCA), Beirut, Lebanon, 25–26 August 2018; pp. 215–218.
17. Mehta, A.; Sandhu, J.K.; Sapra, L. Machine Learning in Wireless Sensor Networks: A Retrospective. In Proceedings of the 2020 Sixth International Conference on Parallel, Distributed and Grid Computing (PDGC), Solan, India, 6–8 November 2020; pp. 328–331. [CrossRef]
18. Baraneetharan, E. Role of machine learning algorithms intrusion detection in WSNs: A survey. *J. Inf. Technol.* **2020**, *2*, 161–173.
19. Gunduz, S.; Arslan, B.; Demirci, M. A Review of Machine Learning Solutions to Denial-of- Services Attacks in Wireless Sensor Networks. In Proceedings of the 2015 IEEE 14th International Conference on Machine Learning and Applications (ICMLA), Miami, FL, USA, 9–11 December 2015. [CrossRef]
20. Kim, T.; Vecchietti, L.F.; Choi, K.; Lee, S.; Har, D. Machine Learning for Advanced Wireless Sensor Networks: A Review. *IEEE Sens. J.* **2021**, *21*, 12379–12397. [CrossRef]
21. Ramotsoela, D.; Abu-Mahfouz, A.; Hancke, G. A survey of anomaly detection in industrial wireless sensor networks with critical water system infrastructure as a case study. *Sensors* **2018**, *18*, 2491. [CrossRef] [PubMed]

22. Ahmad, R.; Wazirali, R.; Abu-Ain, T. Machine Learning for Wireless Sensor Networks Security: An Overview of Challenges and Issues. *Sensors* **2022**, *22*, 4730. [CrossRef]
23. Jesus, E.F.; Chicarino, V.R.; De Albuquerque, C.V.; Rocha, A.A.A. A Survey of How to Use Blockchain to Secure Internet of Things and the Stalker Attack. *Secur. Commun. Netw.* **2018**, *2018*, 9675050. [CrossRef]
24. Liao, Z.; Pang, X.; Zhang, J.; Xiong, B.; Wang, J. Blockchain on Security and Forensics Management in Edge Computing for IoT: A Comprehensive Survey. *IEEE Trans. Netw. Serv. Manag.* **2021**, *19*, 1159–1175. [CrossRef]
25. Sengupta, J.; Ruj, S.; Das Bit, S. A Comprehensive Survey on Attacks, Security Issues and Blockchain Solutions for IoT and IIoT. *J. Netw. Comput. Appl.* **2020**, *149*, 102481. [CrossRef]
26. Khan, M.A.; Salah, K. IoT security: Review, blockchain solutions, and open challenges. *Future Gener. Comput. Syst.* **2018**, *82*, 395–411. [CrossRef]
27. Darla, S.; Naveena, C. Survey on Securing Internet of Things through Block chain Technology. In Proceedings of the 2022 International Conference on Electronics and Renewable Systems (ICEARS), Tuticorin, India, 16–18 March 2022; pp. 836–844. [CrossRef]
28. Uddin, M.A.; Stranieri, A.; Gondal, I.; Balasubramanian, V. A survey on the adoption of blockchain in IoT: Challenges and solutions. *Blockchain Res. Appl.* **2021**, *2*, 100006. [CrossRef]
29. Pohrmen, F.H.; Das, R.K.; Saha, G. Blockchain-based security aspects in heterogeneous Internet-of-Things networks: a survey. *Trans. Emerg. Telecommun. Technol.* **2019**, *30*, e3741. [CrossRef]
30. Miglani, A.; Kumar, N. Blockchain management and machine learning adaptation for IoT environment in 5G and beyond networks: A systematic review. *Comput. Commun.* **2021**, *178*, 37–63. [CrossRef]
31. Matin, M.A.; Islam, M.M. Overview of wireless sensor network. *Wirel. Sens.-Netw.-Technol. Protoc.* **2012**, *1*, 3.
32. Khan, Z.A.; Samad, A. A study of machine learning in wireless sensor network. *Int. J. Comput. Netw. Appl.* **2017**, *4*, 105–112. [CrossRef]
33. Rehana, J. Security of wireless sensor network. In Proceedings of the Seminar on Internetworking, Helsinki University of Technology, Glasgow, UK, 24–28 August 2009.
34. Sora, D. Security Issues in Wireless Sensor Networks. *Int. J. Online Biomed. Eng. (IJOE)* **2010**, *6*, 26–30. [CrossRef]
35. Patel, N.R.; Kumar, S. Wireless Sensor Networks' Challenges and Future Prospects. In Proceedings of the 2018 International Conference on System Modeling & Advancement in Research Trends (SMART), Moradabad, India, 23–24 November 2018; pp. 60–65. [CrossRef]
36. de Farias, C.M.; Pirmez, L.; Delicato, F.C.; Pires, P.F.; Guerrieri, A.; Fortino, G.; Cauteruccio, F.; Terracina, G. A multisensor data fusion algorithm using the hidden correlations in Multiapplication Wireless Sensor data streams. In Proceedings of the 2017 IEEE 14th International Conference on Networking, Sensing and Control (ICNSC), Calabria, Italy, 16–18 May 2017; pp. 96–102. [CrossRef]
37. Karray, F.; Jmal, M.W.; Garcia-Ortiz, A.; Abid, M.; Obeid, A.M. A comprehensive survey on wireless sensor node hardware platforms. *Comput. Netw.* **2018**, *144*, 89–110. [CrossRef]
38. Xie, H.; Yan, Z.; Member, S.; Yao, Z. Data Collection for Security Measurement in Wireless Sensor Networks: A Survey. *IEEE Internet Things J.* **2019**. *6*, 2205–2224. [CrossRef]
39. Alam, S.; De, D. Analysis of security threats in wireless sensor network. *arXiv* **2014**, arXiv:1406.0298.
40. Walters, J.P.; Liang, Z.; Shi, W.; Chaudhary, V. Wireless sensor network security: A survey. In *Security in distributed, Grid, and Pervasive Computing*; Auerbach Publications: Boca Raton, FL, USA, 2006; pp. 208–222.
41. Chapter 16—Wireless Sensor Network Security. In *Computer and Information Security Handbook*, 2nd ed.; Vacca, J.R., Ed.; Morgan Kaufmann: Boston, MA, USA, 2013; pp. 301–322. . [CrossRef]
42. Elhoseny, M.; Hassanien, A.E. Secure data transmission in WSN: An overview. In *Dynamic Wireless Sensor Networks*; Springer: Cham, Switzerland, 2019; pp. 115–143.
43. Shahzad, F.; Pasha, M.; Ahmad, A. A survey of active attacks on wireless sensor networks and their countermeasures. *arXiv* **2017**, arXiv:1702.07136.
44. Mathew, A.; Terence, J.S. A survey on various detection techniques of sinkhole attacks in WSN. In Proceedings of the 2017 International Conference on Communication and Signal Processing (ICCSP), Chennai, India, 6–8 April 2017; pp. 1115–1119.
45. Dewal, P.; Narula, G.S.; Jain, V.; Baliyan, A. Security attacks in Wireless sensor networks: A survey. In *Cyber Security*; Springer: Berlin/Heidelberg, Germany, 2018; pp. 47–58.
46. Kaur, R.; Kaur Sandhu, J. A Study on Security Attacks in Wireless Sensor Network. In Proceedings of the 2021 International Conference on Advance Computing and Innovative Technologies in Engineering (ICACITE), Greater Noida, India, 4–5 March 2021; pp. 850–855. [CrossRef]
47. Ismail, S.; Khoei, T.T.; Marsh, R.; Kaabouch, N. A Comparative Study of Machine Learning Models for Cyber-attacks Detection in Wireless Sensor Networks. In Proceedings of the 2021 IEEE 12th Annual Ubiquitous Computing, Electronics & Mobile Communication Conference (UEMCON), New York, NY, USA, 1–4 December 2021; pp. 1–5.
48. Pruthi, V.; Mittal, K.; Sharma, N.; Kaushik, I. Network layers threats & its countermeasures in WSNs. In Proceedings of the 2019 International Conference on Computing, Communication, and Intelligent Systems (ICCCIS), Greater Noida, India, 18–19 October 2019; pp. 156–163.

49. Yang, G.; Dai, L.; Wei, Z. Challenges, threats, security issues and new trends of underwater wireless sensor networks. *Sensors* **2018**, *18*, 3907. [CrossRef] [PubMed]
50. de Lima Pinto, E.M.; Lachowski, R.; Pellenz, M.E.; Penna, M.C.; Souza, R.D. A machine learning approach for detecting spoofing attacks in wireless sensor networks. In Proceedings of the 2018 IEEE 32nd International Conference on Advanced Information Networking and Applications (AINA), Krakow, Poland, 16–18 May 2018; pp. 752–758.
51. Bhattasali, T.; Chaki, R. A Survey of Recent Intrusion Detection Systems for wireless sensor network. In *Advances in Network Security and Applications, Proceedings of the 4th International Conference, CNSA 2011, Chennai, India, 15–17 July 2011*; Springer: Berlin/Heidelberg, Germany, 2011; pp. 268–269.
52. Tiberti, W.; Carmenini, A.; Pomante, L.; Cassioli, D. A Lightweight Blockchain-based Technique for Anti-Tampering in Wireless Sensor Networks. In Proceedings of the 2020 23rd Euromicro Conference on Digital System Design (DSD), Kranj, Slovenia, 26–28 August 2020; pp. 577–582.
53. Periyanayagi, S.; Sumathy, V. Swarm-based defense technique for tampering and cheating attack in WSN using CPHS. *Pers. Ubiquitous Comput.* **2018**, *22*, 1165–1179. [CrossRef]
54. Numan, M.; Subhan, F.; Khan, W.Z.; Hakak, S.; Haider, S.; Reddy, G.T.; Jolfaei, A.; Alazab, M. A Systematic Review on Clone Node Detection in Static Wireless Sensor Networks. *IEEE Access* **2020**, *8*, 65450–65461. [CrossRef]
55. Gupta, S.; Verma, H.K.; Sangal, A.L. Security attacks & prerequisite for wireless sensor networks. *Int. J. Eng. Adv. Technol. (IJEAT)* **2013**, *2*, 558–566.
56. Premkumar, M.; Sundararajan, T.V. DLDM: Deep learning-based defense mechanism for denial of service attacks in wireless sensor networks. *Microprocess. Microsyst.* **2020**, *79*, 103278. [CrossRef]
57. Mohapatra, H. Handling of Man-In-The-Middle Attack in WSN Through Intrusion Detection System. *Int. J. Emerg. Trends Eng. Res.* **2020**, *8*, 1503–1510. [CrossRef]
58. Yahyaoui, A.; Abdellatif, T.; Attia, R. Hierarchical anomaly based intrusion detection and localization in IoT. In Proceedings of the 2019 15th International Wireless Communications & Mobile Computing Conference (IWCMC), Tangier, Morocco, 24–28 June 2019; pp. 108–113.
59. Somaya, H.; Tomader, M. Build a malware detection software for IOT network Using Machine learning. In Proceedings of the 4th International Conference on Networking, Information Systems & Security, Kenitra, Morocco, 1–2 April 2021; pp. 1–8.
60. Dener, M.; Al, S.; Orman, A. STLGBM-DDS: An Efficient Data Balanced DoS Detection System for Wireless Sensor Networks on Big Data Environment. *IEEE Access* **2022**, *10*, 92931–92945. [CrossRef]
61. Park, T.; Cho, D.; Kim, H. An effective classification for DoS attacks in wireless sensor networks. In Proceedings of the 2018 Tenth International Conference on Ubiquitous and Future Networks (ICUFN), Prague, Czech Republic, 3–6 July 2018; pp. 689–692.
62. Quincozes, S.E.; Kazienko, J.F. Machine learning methods assessment for denial of service detection in wireless sensor networks. In Proceedings of the 2020 IEEE 6th World Forum on Internet of Things (WF-IoT), New Orleans, LA, USA, 2–16 June 2020; pp. 1–6.
63. Alsubaie, F.; Al-Akhras, M.; Alzahrani, H.A. Using machine learning for intrusion detection system in wireless body area network. In Proceedings of the 2020 First International Conference of Smart Systems and Emerging Technologies (SMARTTECH), Riyadh, Saudi Arabia, 3–5 November 2020; pp. 100–104.
64. Alsulaiman, L.; Al-Ahmadi, S. Performance evaluation of machine learning techniques for DOS detection in wireless sensor network. *arXiv* **2021**, arXiv:2104.01963.
65. Ifzarne, S.; Tabbaa, H.; Hafidi, I.; Lamghari, N. Anomaly detection using machine learning techniques in wireless sensor networks. *J. Phys. Conf. Ser.* **2021**, *1743*, 012021. [CrossRef]
66. Batiha, T.; Krömer, P. Design and analysis of efficient neural intrusion detection for wireless sensor networks. *Concurr. Comput. Pract. Exp.* **2021**, *33*, e6152. [CrossRef]
67. Al-Akhras, M.; Al-Issa, A.I.; Alsahli, M.S.; Alawairdhi, M. POSTER: Feature Selection to Optimize DoS Detection in Wireless Sensor Networks. In Proceedings of the 2020 First International Conference of Smart Systems and Emerging Technologies (SMARTTECH), Riyadh, Saudi Arabia, 3–5 November 2020; pp. 263–265.
68. Batiha, T.; Prauzek, M.; Krömer, P. *Intrusion Detection in Wireless Sensor Networks by an Ensemble of Artificial Neural Networks*; Springer: Singapore, 2019; Volume 142, pp. 323–333. [CrossRef]
69. Ismail, S.; Dawoud, D.; Reza, H. A Lightweight Multilayer Machine Learning Detection System for Cyber-attacks in WSN. In Proceedings of the 2022 IEEE 12th Annual Computing and Communication Workshop and Conference (CCWC), Virtual, 26–29 January 2022; pp. 481–486. [CrossRef]
70. Ismail, S.; Reza, H. Evaluation of Naïve Bayesian Algorithms for Cyber-Attacks Detection in Wireless Sensor Networks. In Proceedings of the 2022 IEEE World AI IoT Congress (AIIoT), Seattle, WA, USA, 6–9 June 2022; pp. 283–289.
71. Meng, D.; Dai, H.; Sun, Q.; Xu, Y.; Shi, T. Novel Wireless Sensor Network Intrusion Detection Method Based on LightGBM Model. *IAENG Int. J. Appl. Math.* **2022**, *52*, 1–7.
72. Luo, T.; Nagarajan, S.G. Distributed anomaly detection using autoencoder neural networks in WSN for IoT. In Proceedings of the 2018 IEEE International Conference on Communications (ICC), Kansas City, MO, USA, 20–24 May 2018; pp. 1–6.
73. Sherubha, P.; Amudhavalli, P.; Sasirekha, S. Clone attack detection using random forest and multi objective cuckoo search classification. In Proceedings of the 2019 International Conference on Communication and Signal Processing (ICCSP), Melmaruvathur, India, 4–6 April 2019; pp. 0450–0454.

74. Otoum, S.; Kantarci, B.; Mouftah, H. Empowering reinforcement learning on big sensed data for intrusion detection. In Proceedings of the ICC 2019-2019 IEEE international conference on communications (ICC), Shanghai, China, 20–24 May 2019; pp. 1–7.
75. Subasini, C.; Karuppiah, S.; Sheeba, A.; Padmakala, S. Developing an attack detection framework for wireless sensor network-based healthcare applications using hybrid convolutional neural network. *Trans. Emerg. Telecommun. Technol.* **2021**, *32*, e4336. [CrossRef]
76. Salmi, S.; Oughdir, L. CNN-LSTM Based Approach for Dos Attacks Detection in Wireless Sensor Networks. *Int. J. Adv. Comput. Sci. Appl.* **2022**, *13*. . [CrossRef]
77. Salmi, S.; Oughdir, L. Performance evaluation of deep learning techniques for DoS attacks detection in wireless sensor network. *J. Big Data* **2023**, *10*, 1–25. [CrossRef]
78. Otoum, S.; Kantarci, B.; Mouftah, H.T. On the feasibility of deep learning in sensor network intrusion detection. *IEEE Netw. Lett.* **2019**, *1*, 68–71. [CrossRef]
79. Hussain, K.; Xia, Y.; Onaizah, A.N.; Manzoor, T.; Jalil, K. Hybrid of WOA-ABC and Proposed CNN for Intrusion Detection System in wireless sensor networks. *Optik* **2022**, 170145. [CrossRef]
80. Nguyen, T.T.; Reddi, V.J. Deep reinforcement learning for cyber security. *IEEE Trans. Neural Netw. Learn. Syst.* **2019** . [CrossRef] [PubMed]
81. Benaddi, H.; Ibrahimi, K.; Benslimane, A.; Qadir, J. A deep reinforcement learning based intrusion detection system (drl-ids) for securing wireless sensor networks and internet of things. In Proceedings of the International Wireless Internet Conference, Taichung, Taiwan, 26–27 November 2019; pp. 73–87.
82. Niknam, S.; Dhillon, H.S.; Reed, J.H. Federated learning for wireless communications: Motivation, opportunities, and challenges. *IEEE Commun. Mag.* **2020**, *58*, 46–51. [CrossRef]
83. Kamel, R.M.; El Mougy, A. Retrospective sensing based on federated learning in the IoT. In Proceedings of the 2020 IEEE 45th LCN Symposium on Emerging Topics in Networking (LCN Symposium), Sydney, Australia, 16–19 November 2020; pp. 150–161.
84. Kim, S.; Cai, H.; Hua, C.; Gu, P.; Xu, W.; Park, J. Collaborative anomaly detection for internet of things based on federated learning. In Proceedings of the 2020 IEEE/CIC International Conference on Communications in China (ICCC), Chongqing, China, 9–11 August 2020; pp. 623–628.
85. Mertens, J.; Galluccio, L.; Morabito, G. Federated learning through model gossiping in wireless sensor networks. In Proceedings of the 2021 IEEE International Black Sea Conference on Communications and Networking (BlackSeaCom), Bucharest, Romania, 24–28 May 2021; pp. 1–6.
86. Banerjee, J.; Maiti, S.; Chakraborty, S.; Dutta, S.; Chakraborty, A.; Banerjee, J.S. Impact of Machine Learning in Various Network Security Applications. In Proceedings of the 2019 3rd International Conference on Computing Methodologies and Communication (ICCMC), Erode, India, 27–29 March 2019; pp. 276–281. [CrossRef]
87. Wang, S.; Tuor, T.; Salonidis, T.; Leung, K.K.; Makaya, C.; He, T.; Chan, K. Adaptive federated learning in resource constrained edge computing systems. *IEEE J. Sel. Areas Commun.* **2019**, *37*, 1205–1221. [CrossRef]
88. Zahariadis, T.; Trakadas, P.; Maniatis, S.; Karkazis, P.; Leligou, H.C.; Voliotis, S. Efficient detection of routing attacks in wireless sensor networks. In Proceedings of the 2009 16th International Conference on Systems, Signals and Image Processing, Chalkida, Greece, 18–20 June 2009; pp. 1–4.
89. Loo, C.E.; Ng, M.Y.; Leckie, C.; Palaniswami, M. Intrusion detection for routing attacks in sensor networks. *Int. J. Distrib. Sens. Netw.* **2006**, *2*, 313–332. [CrossRef]
90. Amouri, A.; Alaparthy, V.T.; Morgera, S.D. Cross layer-based intrusion detection based on network behavior for IoT. In Proceedings of the 2018 IEEE 19th Wireless and Microwave Technology Conference (WAMICON), Sand Key, FL, USA, 9–10 April 2018; pp. 1–4. [CrossRef]
91. Pande, S.; Khamparia, A.; Gupta, D. Feature selection and comparison of classification algorithms for wireless sensor networks. *J. Ambient. Intell. Humaniz. Comput.* **2021**, 1–13. [CrossRef]
92. Almomani, I.; Al-Kasasbeh, B.; Al-Akhras, M. WSN-DS: A Dataset for Intrusion Detection Systems in Wireless Sensor Networks. *J. Sens.* **2016**, *2016*. [CrossRef]
93. Tavallaee, M.; Bagheri, E.; Lu, W.; Ghorbani, A.A. A detailed analysis of the KDD CUP 99 data set. In Proceedings of the 2009 IEEE Symposium on Computational Intelligence for Security and Defense Applications, Ottawa, ON, Canada, 8–10 July 2009; pp. 1–6.
94. Sharafaldin, I.; Lashkari, A.H.; Ghorbani, A.A. Toward generating a new intrusion detection dataset and intrusion traffic characterization. *ICISSp* **2018**, *1*, 108–116.
95. Moustafa, N.; Slay, J. UNSW-NB15: A comprehensive data set for network intrusion detection systems (UNSW-NB15 network data set). In Proceedings of the 2015 Military Communications and Information Systems Conference (MilCIS), Canberra, Australia, 10–12 November 2015; pp. 1–6.
96. Tharwat, A. Classification assessment methods. *Appl. Comput. Inform.* **2018**, *17*, 168–192. [CrossRef]
97. Panda, M.; Abd Allah, A.M.; Hassanien, A.E. Developing an Efficient Feature Engineering and Machine Learning Model for Detecting IoT-Botnet Cyber Attacks. *IEEE Access* **2021**, *9*, 91038–91052. [CrossRef]
98. Alsahli, M.S.; Almasri, M.M.; Al-Akhras, M.; Al-Issa, A.I.; Alawairdhi, M. Evaluation of Machine Learning Algorithms for Intrusion Detection System in WSN. *Int. J. Adv. Comput. Sci. Appl.* **2021**, *12*, 617–626. [CrossRef]

99. Haber, S.; Stornetta, W.S. How to Time-Stamp a Digital Document. In *Proceedings of the Advances in Cryptology-CRYPTO' 90; Menezes*; Menezes, A.J., Vanstone, S.A., Eds.; Springer: Berlin/Heidelberg, Germany, 1991; pp. 437–455.
100. Nakamoto, S. Bitcoin: A Peer-to-Peer Electronic Cash System. *Decentralized Bus. Rev.* **2008**, 21260. Available online: https://bitcoin.org/bitcoin.pdf (accessed on 21 August 2018).
101. Wang, X.; Zha, X.; Ni, W.; Liu, R.P.; Guo, Y.J.; Niu, X.; Zheng, K. Survey on blockchain for Internet of Things. *Comput. Commun.* **2019**, *136*, 10–29. [CrossRef]
102. Gao, W.; Hatcher, W.G.; Yu, W. A Survey of Blockchain: Techniques, Applications, and Challenges. In Proceedings of the 2018 27th International Conference on Computer Communication and Networks (ICCCN), Hangzhou, China, 30 July–2 August 2018; pp. 1–11. [CrossRef]
103. Zhang, S.; Lee, J.H. Analysis of the main consensus protocols of blockchain. *ICT Express* **2020**, *6*, 93–97. [CrossRef]
104. Gagneja, K.; Gagneja, K.; Kiefer, R. Security Protocol for Internet of Things (IoT): Blockchain-based Implementation and Analysis. In Proceedings of the 2020 Sixth International Conference on Mobile And Secure Services (MobiSecServ), Miami, FL, USA, 22–23 February 2020. [CrossRef]
105. Monrat, A.A.; Schelén, O.; Andersson, K. A survey of blockchain from the perspectives of applications, challenges, and opportunities. *IEEE Access* **2019**, *7*, 117134–117151. [CrossRef]
106. Hsiao, S.J. Employing Blockchain Technology to Strengthen Security of Wireless Sensor Networks. *IEEE Access* **2021**, *9*, 72326–72341. [CrossRef]
107. Xu, R.; Chen, Y.; Blasch, E.; Chen, G. Blendcac: A blockchain-enabled decentralized capability-based access control for iots. In Proceedings of the IEEE 2018 International Congress on Cybermatics: 2018 IEEE Conferences on Internet of Things, Green Computing and Communications, Cyber, Physical and Social Computing, Smart Data, Blockchain, Computer and Information Technology, iThings/Gree Halifax, Canada, 30 July–3 August 2018; pp. 1027–1034. [CrossRef]
108. Khalil, A.A.; Franco, J.; Parvez, I.; Uluagac, S.; Rahman, M.A. A Literature Review on Blockchain-enabled Security and Operation of Cyber-Physical Systems. In Proceedings of the 2022 IEEE 46th Annual Computers, Software, and Applications Conference (COMPSAC), Los Alamitos, CA, USA, 27 June–1 July 2022.
109. Cui, Z.; Xue, F.; Zhang, S.; Cai, X.; Cao, Y.; Zhang, W.; Chen, J. A Hybrid BlockChain-Based Identity Authentication Scheme for Multi-WSN. *IEEE Trans. Serv. Comput.* **2020**, *13*, 241–251. [CrossRef]
110. Mamdouh, M.; Awad, A.I.; Khalaf, A.A.; Hamed, H.F. Authentication and Identity Management of IoHT Devices: Achievements, Challenges, and Future Directions. *Comput. Secur.* **2021**, *111*, 102491. [CrossRef]
111. Salimitari, M.; Chatterjee, M. A Survey on Consensus Protocols in Blockchain for IoT Networks. *arXiv* **2018**, 1–15. arXiv:1809.05613v4.
112. Mohanta, B.K.; Panda, S.S.; Jena, D. An overview of smart contract and use cases in blockchain technology. In Proceedings of the 2018 9th International Conference on Computing, Communication and Networking Technologies (ICCCNT), Bengaluru, India, 10–12 July 2018; pp. 1–4.
113. Teslya, N.; Ryabchikov, I. Blockchain platforms overview for industrial IoT purposes. In Proceedings of the Conference of Open Innovation Association, FRUCT, Jyvaskyla, Finland, 15–18 May 2018; pp. 250–256. [CrossRef]
114. Ismail, S.; Reza, H.; Zadeh, H.K.; Vasefi, F. A Blockchain-based IoT Security Solution Using Multichain. In Proceedings of the 2023 IEEE 13th Annual Computing and Communication Workshop and Conference (CCWC), Las Vegas, NV, USA, 8–11 March 2023; pp. 1105–1111. [CrossRef]
115. Alkurdi, F.; Elgendi, I.; Munasinghe, K.S.; Sharma, D.; Jamalipour, A. Blockchain in IoT Security: A Survey. In Proceedings of the 2018 28th International Telecommunication Networks and Applications Conference, ITNAC 2018, Sydney, Australia, 21–23 November 2018; pp. 1–4. [CrossRef]
116. Liu, Y.; Yu, F.R.; Li, X.; Ji, H.; Leung, V.C. Blockchain and Machine Learning for Communications and Networking Systems. *IEEE Commun. Surv. Tutor.* **2020**, *22*, 1392–1431. [CrossRef]
117. Honar Pajooh, H.; Rashid, M.; Alam, F.; Demidenko, S. Hyperledger fabric blockchain for securing the edge internet of things. *Sensors* **2021**, *21*, 359. [CrossRef]
118. Tian, Y.; Wang, Z.; Xiong, J.; Ma, J. A Blockchain-Based Secure Key Management Scheme With Trustworthiness in DWSNs. *IEEE Trans. Ind. Inform.* **2020**, *16*, 6193–6202. [CrossRef]
119. Goyat, R.; Kumar, G.; Alazab, M.; Saha, R.; Thomas, R.; Rai, M.K. A secure localization scheme based on trust assessment for WSNs using blockchain technology. *Future Gener. Comput. Syst.* **2021**, *125*, 221–231. [CrossRef]
120. Guerrero-Sanchez, A.E.; Rivas-Araiza, E.A.; Gonzalez-Cordoba, J.L.; Toledano-Ayala, M.; Takacs, A. Blockchain mechanism and symmetric encryption in a wireless sensor network. *Sensors* **2020**, *20*, 2798. [CrossRef]
121. Rathee, G.; Balasaraswathi, M.; Chandran, K.P.; Gupta, S.D.; Boopathi, C. A secure IoT sensors communication in industry 4.0 using blockchain technology. *J. Ambient. Intell. Humaniz. Comput.* **2021**, *12*, 533–545. [CrossRef]
122. Ismail, S.; Dawoud, D.; Reza, H. Towards A Lightweight Identity Management and Secure Authentication for IoT Using Blockchain. In Proceedings of the 2022 IEEE World AI IoT Congress (AIIoT), Seattle, WA, USA, 6–9 June 2022; pp. 77–83. [CrossRef]
123. Miraz, M.H. Blockchain of things (BCoT): The fusion of blockchain and IoT technologies. In *Advanced Applications of Blockchain Technology*; Springer: Berlin/Heidelberg, Germany, 2020; pp. 141–159.
124. Biswas, S.; Sharif, K.; Li, F.; Nour, B.; Wang, Y. A scalable blockchain framework for secure transactions in IoT. *IEEE Internet Things J.* **2019**, *6*, 4650–4659. [CrossRef]

125. Kushch, S.; Prieto-Castrillo, F. A rolling blockchain for a dynamic WSNs in a smart city. *arXiv* **2018**, *1*, 1–8. arXiv:1806.11399.
126. Lao, L.; Li, Z.; Hou, S.; Xiao, B.; Guo, S.; Yang, Y. A survey of IoT applications in blockchain systems: Architecture, consensus, and traffic modeling. *ACM Comput. Surv. (CSUR)* **2020**, *53*, 1–32. [CrossRef]
127. Zamani, M.; Movahedi, M.; Raykova, M. RapidChain: Scaling blockchain via full sharding. In Proceedings of the ACM Conference on Computer and Communications Security, Toronto, ON, Canada, 15–19 October 2018. [CrossRef]
128. Cherupally, S.R.; Boga, S.; Podili, P.; Kataoka, K. Lightweight and Scalable DAG based distributed ledger for verifying IoT data integrity. *Int. Conf. Inf. Netw.* **2021**, *2021*, 267–272. [CrossRef]
129. Buldin, I.D.; Gorodnichev, M.G.; Makhrov, S.S.; Denisova, E.N. Next Generation Industrial Blockchain-Based Wireless Sensor Networks. In Proceedings of the 2018 Wave Electronics and its Application in Information and Telecommunication Systems (WECONF), Saint Petersburg, Russia, 3–7 June 2018; pp. 1–5. [CrossRef]
130. Baig, M.A.; Ali Sunny, D.; Alqahtani, A.; Alsubai, S.; Binbusayyis, A.; Muzammal, M. A Study on the Adoption of Blockchain for IoT Devices in Supply Chain. *Comput. Intell. Neurosci.* **2022**, *2022*, 9228982. [CrossRef]
131. Goyal, H.; Saha, S. Reli: Real-time lightweight byzantine consensus in low-power iot-systems. In Proceedings of the 2022 18th International Conference on Network and Service Management (CNSM), Thessaloniki, Greece, 31 October–4 November 2022; pp. 275–281.
132. Gopalakrishnan, K. Security vulnerabilities and issues of traditional wireless sensors networks in IoT. In *Principles of Internet of Things (IoT) Ecosystem: Insight Paradigm*; Springer: Berlin/Heidelberg, Germany, 2020; pp. 519–549.
133. Waheed, N.; He, X.; Ikram, M.; Usman, M.; Hashmi, S.S.; Usman, M. Security and Privacy in IoT Using Machine Learning and Blockchain: Threats and Countermeasures. *ACM Comput. Surv.* **2021**, *53*, 1–37.
134. Marchang, J.; Ibbotson, G.; Wheway, P. Will blockchain technology become a reality in sensor networks? In Proceedings of the 2019 Wireless Days (WD), Manchester, UK, 24–26 April 2019, pp. 1–4.
135. Shammar, E.A.; Zahary, A.T.; Al-Shargabi, A.A. A survey of IoT and blockchain integration: Security perspective. *IEEE Access* **2021**, *9*, 156114–156150. [CrossRef]
136. She, W.; Liu, Q.; Tian, Z.; Chen, J.S.; Wang, B.; Liu, W. Blockchain trust model for malicious node detection in wireless sensor networks. *IEEE Access* **2019**, *7*, 38947–38956. [CrossRef]
137. Arifeen, M.M.; Al Mamun, A.; Ahmed, T.; Kaiser, M.S.; Mahmud, M. A Blockchain-Based Scheme for Sybil Attack Detection in Underwater Wireless Sensor Networks. In *Proceedings of International Conference on Trends in Computational and Cognitive Engineering*; Kaiser, M.S., Bandyopadhyay, A., Mahmud, M., Ray, K., Eds.; Springer: Singapore, 2021; pp. 467–476.
138. Chanana, R.; Singh, A.K.; Killa, R.; Agarwal, S.; Mehra, P.S. Blockchain Based Secure Model for Sensor Data in Wireless Sensor Network. In Proceedings of the 2020 6th International Conference on Signal Processing and Communication (ICSC), Noida, India, 5–7 March 2020; pp. 288–293. [CrossRef]
139. Mubarakali, A. An efficient authentication scheme using blockchain technology for wireless sensor networks. *Wirel. Pers. Commun.* **2021**, *1*, 1–15. [CrossRef]
140. Sivaganesan, D. A data driven trust mechanism based on blockchain in IoT sensor networks for detection and mitigation of attacks. *J. Trends Comput. Sci. Smart Technol. (TCSST)* **2021**, *3*, 59–69.
141. Karakoç, E.; Çeken, C. Black hole attack prevention scheme using a blockchain-block approach in SDN-enabled WSN. *Int. J. Hoc Ubiquitous Comput.* **2021**, *37*, 37–49. [CrossRef]
142. Soltani, R.; Saxena, L.; Joshi, R.; Sampalli, S. Protecting Routing Data in WSNs with use of IOTA Tangle. *Procedia Comput. Sci.* **2022**, *203*, 197–204. [CrossRef]
143. Javed, S.; Khan, M.A.; Abdullah, A.M.; Alsirhani, A.; Alomari, A.; Noor, F.; Ullah, I. An Efficient Authentication Scheme Using Blockchain as a Certificate Authority for the Internet of Drones. *Drones* **2022**, *6*, 264. [CrossRef]
144. Lazrag, H.; Chehri, A.; Saadane, R.; Rahmani, M.D. Efficient and secure routing protocol based on Blockchain approach for wireless sensor networks. *Concurr. Comput. Pract. Exp.* **2021**, 33, e6144. [CrossRef]
145. Awan, S.; Javaid, N.; Ullah, S.; Khan, A.U.; Qamar, A.M.; Choi, J.G. Blockchain Based Secure Routing and Trust Management in Wireless Sensor Networks. *Sensors* **2022**, *22*, 411. [CrossRef]
146. Chen, Y.; Yang, X.; Li, T.; Ren, Y.; Long, Y. A blockchain-empowered authentication scheme for worm detection in wireless sensor network. *Digit. Commun. Netw.* **2022** . [CrossRef]
147. Almaiah, M.A., A New Scheme for Detecting Malicious Attacks in Wireless Sensor Networks Based on Blockchain Technology. In *Artificial Intelligence and Blockchain for Future Cybersecurity Applications*; Maleh, Y., Baddi, Y., Alazab, M., Tawalbeh, L., Romdhani, I., Eds.; Springer International Publishing: Cham, Switzerland, 2021; pp. 217–234. [CrossRef]
148. Moinet, A.; Darties, B.; Baril, J.L. Blockchain based trust & authentication for decentralized sensor networks. *arXiv* **2017**, arXiv:1706.01730.
149. Kim, T.H.; Goyat, R.; Rai, M.K.; Kumar, G.; Buchanan, W.J.; Saha, R.; Thomas, R. A novel trust evaluation process for secure localization using a decentralized blockchain in wireless sensor networks. *IEEE Access* **2019**, *7*, 184133–184144. [CrossRef]
150. Pundir, S.; Wazid, M.; Singh, D.P.; Das, A.K.; Rodrigues, J.J.P.C.; Park, Y. Intrusion detection protocols in wireless sensor networks integrated to Internet of Things deployment: Survey and future challenges. *IEEE Access* **2019**, *8*, 3343–3363. [CrossRef]
151. Revanesh, M.; Sridhar, V. A trusted distributed routing scheme for wireless sensor networks using blockchain and meta-heuristics-based deep learning technique. *Trans. Emerg. Telecommun. Technol.* **2021**, *32*, e4259. [CrossRef]
152. Nouman, M.; Qasim, U.; Nasir, H.; Almasoud, A.; Imran, M.; Javaid, N. Malicious Node Detection using Machine Learning and Distributed Data Storage using Blockchain in WSNs. *IEEE Access*, 2023, *11*, 6106–6121. [CrossRef]

153. Yang, X.; Chen, Y.; Qian, X.; Li, T.; Lv, X. BCEAD: A blockchain-empowered ensemble anomaly detection for wireless sensor network via isolation forest. *Secur. Commun. Netw.* **2021**, *2021*, 9430132. [CrossRef]
154. Sajid, M.B.E.; Ullah, S.; Javaid, N.; Ullah, I.; Qamar, A.M.; Zaman, F. Exploiting Machine Learning to Detect Malicious Nodes in Intelligent Sensor-Based Systems Using Blockchain. *Wirel. Commun. Mob. Comput.* **2022**, *9*, 24695–24707. [CrossRef]
155. Yang, J.; He, S.; Xu, Y.; Chen, L.; Ren, J. A trusted routing scheme using blockchain and reinforcement learning for wireless sensor networks. *Sensors* **2019**, *19*, 970. [CrossRef]
156. Abd El-Moghith, I.A.; Darwish, S.M. Towards designing a trusted routing scheme in wireless sensor networks: A new deep blockchain approach. *IEEE Access* **2021**, *9*, 103822–103834. [CrossRef]
157. Rajasoundaran, S.; Kumar, S.; Selvi, M.; Ganapathy, S.; Rakesh, R.; Kannan, A. Machine learning based volatile block chain construction for secure routing in decentralized military sensor networks. *Wirel. Netw.* **2021**, *27*, 4513–4534. [CrossRef]
158. Puthal, D.; Mohanty, S.P.; Nanda, P.; Kougianos, E.; Das, G. Proof-of-Authentication for Scalable Blockchain in Resource-Constrained Distributed Systems. In Proceedings of the 2019 IEEE International Conference on Consumer Electronics (ICCE), Berlin, Germany, 8–11 September 2019; pp. 1–5. [CrossRef]
159. Mingxiao, D.; Xiaofeng, M.; Zhe, Z.; Xiangwei, W.; Qijun, C. A review on consensus algorithm of blockchain. In Proceedings of the 2017 IEEE International Conference on Systems, Man, and Cybernetics (SMC), Banff, AB, Canada, 5–8 October 2017; pp. 2567–2572. [CrossRef]
160. Zhou, Q.; Zheng, K.; Zhang, K.; Hou, L.; Wang, X. Vulnerability Analysis of Smart Contract for Blockchain-Based IoT Applications: A Machine Learning Approach. *IEEE Internet Things J.* **2022**, *9*, 24695–24707. [CrossRef]
161. Wang, S.Y.; Hsu, Y.J.; Hsiao, S.J. Integrating blockchain technology for data collection and analysis in wireless sensor networks with an innovative implementation. In Proceedings of the 2018 International Symposium on Computer, Consumer and Control (IS3C), Taichung, Taiwan, 6–8 December 2018; pp. 149–152.
162. Omar, S.; Ngadi, A.; Jebur, H.H. Machine learning techniques for anomaly detection: An overview. *Int. J. Comput. Appl.* **2013**, *79*, 33–41. [CrossRef]
163. Zhao, J.; Shetty, S.; Pan, J.W.; Kamhoua, C.; Kwiat, K. Transfer learning for detecting unknown network attacks. *Eurasip J. Inf. Secur.* **2019**, *2019*, 1. [CrossRef]
164. Lu, Y.; Tang, Q.; Wang, G. On enabling machine learning tasks atop public blockchains: A crowdsourcing approach. In Proceedings of the 2018 IEEE International Conference on Data Mining Workshops (ICDMW), Singapore, 17–20 November 2018; pp. 81–88.

future internet

MDPI

Review

Approaches and Challenges in Internet of Robotic Things

Aqsa Sayeed [1], Chaman Verma [2,*], Neerendra Kumar [1,*], Neha Koul [1] and Zoltán Illés [2]

[1] Department of Computer Science & IT, Central University of Jammu, Jammu 181143, India
[2] Department of Media and Educational Informatics, Faculty of Informatics, Eötvös Loránd University, 1053 Budapest, Hungary
* Correspondence: chaman@inf.elte.hu (C.V.); neerendra.csit@cujammu.ac.in (N.K.)

Abstract: The Internet of robotic things (IoRT) is the combination of different technologies including cloud computing, robots, Internet of things (IoT), artificial intelligence (AI), and machine learning (ML). IoRT plays a major role in manufacturing, healthcare, security, and transport. IoRT can speed up human development by a very significant percentage. IoRT allows robots to transmit and receive data to and from other devices and users. In this paper, IoRT is reviewed in terms of the related techniques, architectures, and abilities. Consequently, the related research challenges are presented. IoRT architectures are vital in the design of robotic systems and robotic things. The existing 3–7-tier IoRT architectures are studied. Subsequently, a detailed IoRT architecture is proposed. Robotic technologies provide the means to increase the performance and capabilities of the user, product, or process. However, robotic technologies are vulnerable to attacks on data security. Trust-based and encryption-based mechanisms can be used for secure communication among robotic things. A security method is recommended to provide a secure and trustworthy data-sharing mechanism in IoRT. Significant security challenges are also discussed. Several known attacks on ad hoc networks are illustrated. Threat models ensure integrity confidentiality and availability of the data. In a network, trust models are used to boost a system's security. Trust models and IoRT networks play a key role in obtaining a steady and nonvulnerable configuration in the network. In IoRT, remote server access results in remote software updates of robotic things. To study navigation strategies, navigation using fuzzy logic, probabilistic roadmap algorithms, laser scan matching algorithms, heuristic functions, bumper events, and vision-based navigation techniques are considered. Using the given research challenges, future researchers can get contemporary ideas of IoRT implementation in the real world.

Keywords: IoRT; robotics; sensors; augmented reality and virtual reality; robot navigation techniques; heuristic functions; bumper event; fuzzy logic; trust-based mechanism; IoRT security framework; threat model; trust model; machine learning; IoRT remote server access; IoRT energy efficiency

Citation: Sayeed, A.; Verma, C.; Kumar, N.; Koul, N.; Illés, Z. Approaches and Challenges in Internet of Robotic Things. *Future Internet* **2022**, *14*, 265. https://doi.org/10.3390/fi14090265

Academic Editors: Christoph Stach and Clémentine Gritti

Received: 9 August 2022
Accepted: 8 September 2022
Published: 14 September 2022

1. Introduction

Robotic systems have aided various technological developments during the previous decade. During the 1990s, robotic and network technologies were combined to expand the range of functional values of the robots [1]. IoRT was formulated to determine the structure in which sensor data from various sources are incorporated, and then explicated using local and distributed information. Thereafter, the data are used to monitor and verify things in the physical world [2,3]. According to the IEEE Society of Robotics and Automation, a networked robot is described as "a robotic device associated with a communication network through the internet or local area network (LAN) using standard network protocols such as TCP, UDP, or 802.11". Robotic engineering systems are used widely in the industry today. Robotic systems are seen as critical components for humanity's growth in the new digital era. The robotic systems were turned into industrial IoRT applications when technologies of IIoT, AI, robots, intelligent networking, and electric mobility emerged [4]. Robotic things can now be connected to anything and everyone at any time, at any location, via various paths/networks and services. Due to new advancements in intelligent networking.

Edge nodes, which are formed by networked robotic devices, might act as the pillar for IoRT applications in the future [4,5]. The IoT and robotic technologies focus on two goals: (1) to provide information services for detection, sensing, and tracking, and (2) to create movement and interaction behavior. The development of IoRT has been improved due to the combination of the above two goals. According to Vermesan [4], IoRT is defined as an active global network framework with self-adapting and self-configuring characteristics. The characteristics are based upon the standard communication protocol (rules for data access over the network) and the interoperability protocol (multiple system data exchanges). In this technology, to make decisions and act on various sets of rules, virtual and physical robotic things with varying degrees of mobility and autonomy use intelligent interfaces, cognition, and connectivity. IoRT enables the collaboration of people, devices, processes, and technology with actuators and sensors [1]. IoRT performs various functions including human–robot interactions (HRIs) and robotic interaction services (ROIs). A robotic system requires necessary equipment, commonly a microphone, camera, LIDAR, RADAR, and even sensors for performing interactions and reactions [6]. HRI is built into a robot for assisted living facilities, hotels, etc. Due to IoRT and HRI, various robots are deployed to monitor the work continuously. During IoRT communications, the data leakage problem is a big issue for data exchange. Data leakage affects the privacy of customers. For example, a stage subjected to IoRT security risk is associated with the transmission of data to IoRT systems by sensing units and sensors. Sensing units transmit data to the IoRT system to detect physical environments, while sensors give information to the device [7,8].

IoRT devices suffer from heterogeneity, interoperability, time variance, network inactivity, security, multirobot systems, quality of services, precise navigation, and standardization. This article discusses secure communication for IoRT devices to overcome leakage problems. This manuscript provides a review of the IoRT definition and technologies used in the functionalities of IoRT. The abilities of robotic components are very essential for the autonomous behavior of robotic things; various characteristics are illustrated in this article. Various organizations use architectures as per their requirements, and there are various architectures for IoRT devices. In this review, we discuss many IoRT architectures, among which five-tier architectures are the most advanced and feasible for intelligent IoRT devices. This article describes the IoRT key concept, abilities, evolution, applications, latest architectural designs, robotic navigation techniques for obstacle-free navigation, IoRT security, and technical challenges.

The primary findings of this work are as follows:

a. We present a novel taxonomy for Internet of robotic things strategies.
b. We provide an in-depth study and analysis of several IoRT literature approaches and techniques.
c. We briefly illustrate the security methods for IoRT.
d. We highlight some open research problems, as well as futuristic scope, in this active field of research.

Organization of Paper

The organization of the remainder of this paper is depicted in Figure 1. Section 2 gives an overview of IoRT techniques, architecture, and abilities. Section 3 delivers a summary of the recent literature survey. A focus on security and the taxonomy of security threats is presented in Section 4. Section 5 highlights some open research challenges in this active field of research, and Section 6 concludes the paper, along with the future scope.

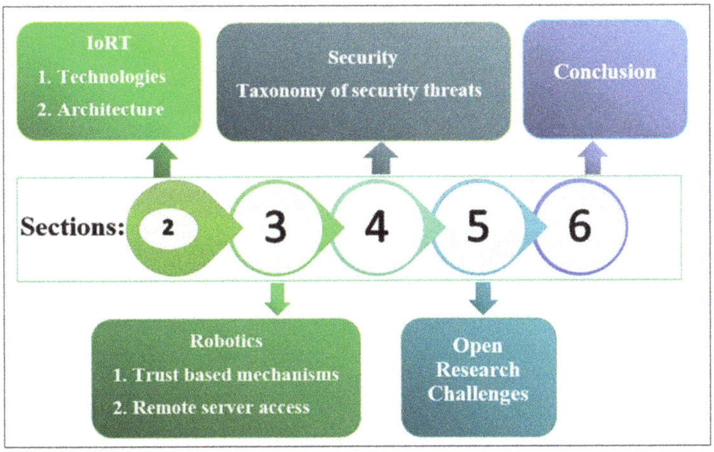

Figure 1. Graphical layout of article.

2. IoRT: An Overview

2.1. Definitions and Concept of IoRT

According to Ray [1], IoRT is described as a global framework for the information sector. IoRT facilitates the improvement of services by robots by affiliating robotic things on the basis of known and emerging compatible information and communication technologies. As per ABI Research [6], IoRT is an intelligent device that monitors procedures and merges sensor data from diverse sources. Robotic devices practice local and distributed intellect to conclude the best way of action.

IoRT is an explicit and dynamic internet framework. The association of IoRT and cloud results in the collection of data from all devices and brings out a report after examining and scrutinizing the data. IoRT allows a large number of distinguishable "things" to share and transfer information with other things over the available Internet or the compatible protocols of the network. Using basic protocols (TCP/IP), IoRT provides a powerful platform for connecting things to assist M2M and M2H data transmission [3,9]. Mark Weiser was the first to mention the idea of IoT in his Scientific American article "The Computer for the 21st Century", based on ubiquitous computing. After that, in 1999, the director of the Auto-ID Center (Kevin Ashton) coined the IoT term. Scientific efforts have enabled the IoRT to pursue real-time decisions by integrating robots and IoT technologies. No study has yet provided a proper and complete definition of IoRT. IoRT is usually proposed as a merger of IoT and robotics (cloud robotics) [3]. IoRT has boosted the IoT application market, as well as advanced the technology, by providing important features such as AI, robotics, and swarm technologies. Earlier robotic technologies relied on computer programs, while more recent robotic technologies rely on AI and ML algorithms, resulting in very effective IoRT technology [6]. Different types of technologies use different types of robots as per their needs. A wired robot is linked to a network (Internet or LAN), and the network (wired or wireless) uses many protocols such as TCP, UDP, and IEEE 802.11 for data transmission among multiple robots. IoRT is a new field, and many more new technologies are currently being developed. The sensing efficiency of robots is enhanced by a network of sensors (installed, repaired, and maintained by robots to increase their reliability and availability). The network sensors result in long-distance robot communications and activity maintenance [3]. A robot is a large-capacity closed system. A cloud robotic system is utilized to overcome the noise, congestion, and time-delay limitations of network robots. In addition to networked and cloud robots, IoRT employs more advanced IoT technologies and robotic devices for expanded capabilities. In addition, depending on the functionality and complexity based on the operability and sophistication of the robot, each robot has a network interface card (NIC) card with a unique NIC address, as well as the remaining hardware identifiers [10]. IoRT connects a variety of smart devices to a sophisticated IoRT infrastructure that includes cloud and edge technologies. For IoRT computation and control in the cloud, the robotic systems are connected to the cloud via a primary medium known as the "Internet". Cloud robotics is a new branch of robotics based on cloud storage, cloud computing, and other Internet technologies [11–15]. Figure 2 represents the basic ideas of IoRT, IoT, and cloud robotics and mentions their functionalities. Currently, the robotic operating system (ROS) is fully advanced in all aspects.

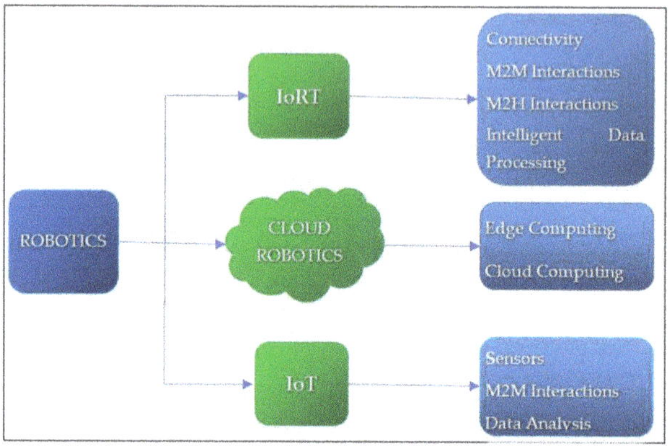

Figure 2. Diagrammatic representation robotics and robotic functionalities.

Hence, there is no threat of complexity in IoRT communication, and a simple API is required for all communication [7,16]. Figure 3, a block diagram of IoRT, mentions the functionalities of robotic things, the latest enabling technologies in robotics, and the application area of IoRT.

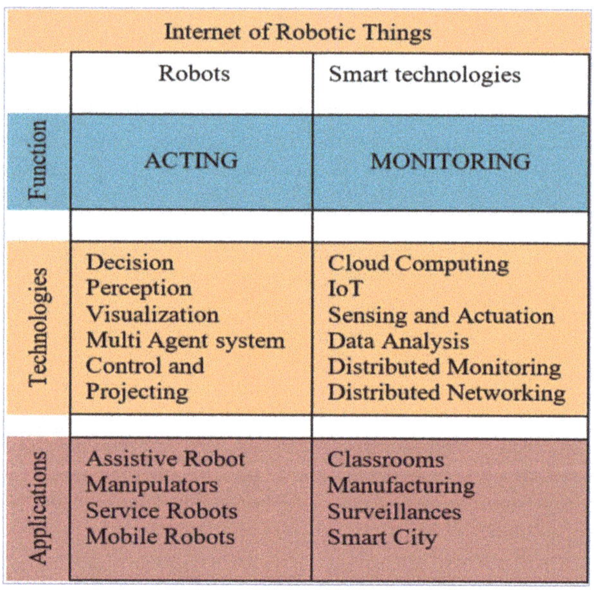

Figure 3. Block diagram of IoRT, including functionalities, technologies, and applications.

2.1.1. How Does Communication Take Place in IoRT?

In H2M interactions, humans provide input to IoT devices in the form of speech, text, and images, among other things. The IoT device, including sensors and actuators, then interprets the input, analyzes it, and reacts to the user via text or a visual display, such as facial recognition or speech recognition. By automating programs, machines may communicate with one another. M2M communication needs machine-level instructions. Communication can happen without human assistance. A point-to-point connection between two network devices is known as an M2M connection, e.g., alerts from a smart washing machine and smart meters. M2H communication is the most prevalent sort of communication utilized when robots assist humans in their regular activities. It is a type of interaction in which humans collaborate with smart systems and other machines to complete a task by using tools or gadgets, e.g., fire alarms and traffic lights [8,17,18]. The IoRT platform maintains the robotic thing's functionalities and technologies. The platform's major capabilities enable robotic things to achieve their main goals, such as communication among robotic things, data flow, IoRT device organization for accessing and maintaining devices, and IoRT device cooperation inside and between the platforms. This is all done to form IoRT applications via the IoRT platform infrastructure. IoRT platform technologies enable elasticity, usability, and productivity [4,9,19,20]. Sharing of data between robots is the responsibility of IoRT platforms to connect data (in the cloud and at control centers) to robotic objects, devices, and people (IoRT environment) [21,22].

2.1.2. How Does Robot-to-Human Communication Take Place?

The digital twin technique is used for robotic virtual commissioning over the lifespan of robotic things. This may be accomplished by combining data from physical IoRT devices with other inputs. All of this leads to real-time optimization, application scenarios, throughput, and possible issues. As a result, the system's virtual representation invokes and strengthens its ability to serve as a real and physical robotic device, as well as an HRI. Enhanced intelligent cognition at the control of IoRT applications enables the combination of AR and VR into human–robot interconnection [3,4].

2.1.3. Security Importance in Robot Communication

Robots are often wirelessly connected to a file server. The network associations create a subnet with the router's static IP address exposed globally, and this is the main reason for robotic data attacks. The server and robots create a subnet of local IP addresses. On the other hand, each robot possesses a static IP address. Distributed ledger technologies (DLTs) are linked with IoRT frameworks and provide systematic data management concerning security, privacy, and safety [10,23–25] The reliability of the IoRT system is increased by hardening end-to-end security, digital identities, services, and mobile data security. This is prompted by robotic cognition from new AI algorithms [4].

2.2. Abilities of IoRT

IoRT depends on the robot functionalities, which are categorized into basic-level abilities, higher-level abilities, and system-level abilities, as given in Figure 4. Some of the characteristics of IoRT are mentioned below, along with the taxonomy of abilities.

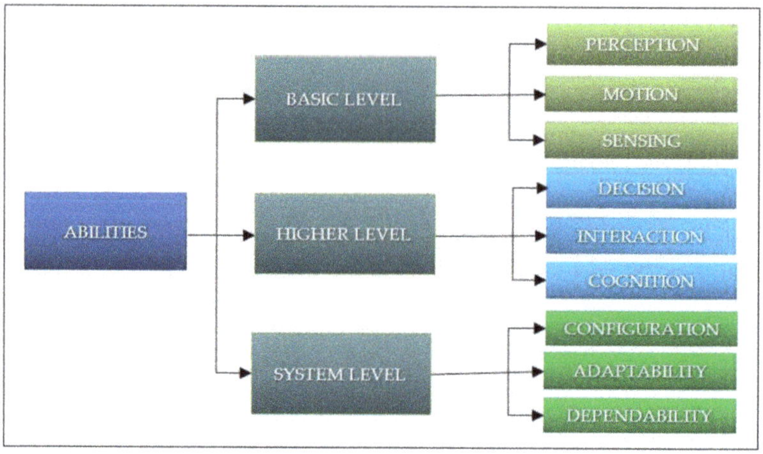

Figure 4. An evaluation of the characteristics of robotic things.

Perception: The performance of the robotic system relies on IoRT sensor information and data analytics technologies. Perception interprets vision, sound, smell, and touch using sensors. Perception is carried out through the utilization of technologies such as software engineering, cloud computing, and big data to accomplish M2M interaction, sensor communications, and AI. IoRT has become more sophisticated as a result of IoT. As a result, robots can sense the real-time world to perform complex tasks [6].

Motion: The important ability of IoRT in all technologies is the ability to travel. The important factor that plays a role in determining the locomotion of devices is mechanical architecture. For navigating the robots, IoT networking also plays an important role [26–28]. A robotic equation of motion defines its motion as a function of time and optional control inputs [29]. Equation (1) is written as

$$F(\boldsymbol{q}(t),\ \boldsymbol{q}'(t), \boldsymbol{q}''(t),\ \boldsymbol{k}(t), t) = 0, \qquad (1)$$

where t is the time variable, $\boldsymbol{q}$ is the vector of ordered coordinates, e.g., the vector of combined angles for a manipulator, $\boldsymbol{q}'$ is the first time derivative (velocity) of $\boldsymbol{q}$, $\boldsymbol{q}''$ is the second time derivative (acceleration) of $\boldsymbol{q}$, and k is the vector of control inputs.

Manipulation and sensing: Sensing as a service can be implemented for IoRT and robotic system interactions with IoT devices and people. The responsibility of the IoT is to sense the surroundings. The responsibility for catching, shifting, and directing the shape is taken by robots [13,30,31].

Decisional autonomy (DA): Choosing the best plan for completing a task by a system is called DA. IoT middleware neglects this characteristic and uses API execution (smarts) in its applications, which hides the intrinsic complexity [3,22,30].

Interaction: This is the ability of robots to communicate systematically and cognitively with other systems in an environment. In the industrial context, the interaction potential highlights how IoT technology may boost HRI. For manufacturers, IoT devices can enhance the robustness of HRI [3,22,32,33].

Cognition: This is the ability of IoRT to comprehend a robotic system by sensing the sensor data. IoRT can examine the data from varied systems in the surroundings and take the obligatory way of action. Through this, the intelligence of robots is leveraged [34,35].

Control: Control loops in IoRT can be simply mapped to nearly anything, from virtual things to physical items, from the cloud to multiple networks, granting IoRT autonomy [9,36].

Configurability: Robotic systems are modified for particular tasks or reconfigured for various tasks. IoT is useful in the manufacturing context for software configurability and the interactive configuration of several computers that contribute different functionality and collaborate to execute complex tasks or jobs. Let f_i be the degrees of freedom of a robot spatial procedure supplied by joint i, and let c_i be the number of constraints given by joint i; it follows that $f_i + c_i = m_i$ for all i [37,38]. Then, Grubler's formulas (Equations (2)–(4)) for the degrees of freedom (*dof*) of the robot are as follows:

$$dof = m(N-1) - \sum_{i=1}^{j} c_i. \tag{2}$$

$$dof = m(N-1) - \sum_{i=1}^{j} (m - f_i). \tag{3}$$

$$dof = m(N-1-j) + \sum_{i=1}^{j} f_i. \tag{4}$$

The formulas are only retained if all joint constraints are autonomous. If they are not, then the formulas give a lower bound on the number of degrees of freedom. In the above equations, the robot has N links and $N-1$ is the total number of degrees of freedom of the bodies if they are not contrived by joints [39].

Adaptability: The ability of a system to respond to a variety of problems, conditions, etc. is called adaptability. Adaptability adjusts robots in the environment to respond to unexpected circumstances and uncertain human behavior. Adaptability is possible through perception, decision planning, and the configuration of a robot [6].

2.3. Evolution in IoRT

In 1961, robots were first used in the industrial sector to unload parts in a die-casting factory. After 20 years, Japanese manufacturers developed new designs to incorporate robotic manufacturing lines. Robotics and artificial intelligence have advanced rapidly in recent years. Automated machines are now widely utilized in industry, marine exploration, space exploration, the military, and commercialized agriculture to undertake repetitive activities [4]. The IoRT evolution requires many robotic thing activities. The main robotic thing activities for IoRT evolution are secure data, robotic thing cognition, robotic thing collective and collaborative actions, real-time actions, authentic low-latency communication, and energy efficiency. The latest IoRT applications expedite the merging of IoT and autonomous intelligent systems. As a result, collaborative robotic objects may pass on to others, learn autonomously, and have more secure relationships with the environment (people and other things). To improve robotic technologies, future independent IoRT systems may consist of the following qualities: think, learn, sense, act, connect, collaborate, and locate [4,7]. Table 1 illustrates the evolution of IoRT.

Table 1. Evolution of IoRT.

Multidisciplinary Attributes	Evolution In Multidisciplinary Nature of IoRT
Think	Computing, cognition, connectivity, and control
Connect	Connectivity in robotic things and the environment
Locate	High-definition dynamic maps, GPS, GNSS, and location of networks
Learn	AI algorithms are used for learning robotic things
Sense	Collection and processing of data streams from the perception domain radars, LIDARs, cameras, and ultrasound sensors
Collaborate	Activities with their robotic things, autonomous vehicles, edge cloud, etc.
Act	Acting, speed, and stopping

2.4. Applications of IoRT

For the past few years, IoRT has been a rapidly growing field. IoRT applications interlinked with the Internet are found in every field. Examples include transferring resource-intensive activities to the cloud, accessing huge quantities of data, and exchanging data with other robots [3]. Some application fields are manufacturing, agriculture, healthcare, education, and surveillance [24]. The electronics industry is using the IoRT widely. In the modern era, robots do the work of humans in every sector, such as healthcare robots, agricultural robots, and home and hotel robots [23]. IoRT is significantly developing in terms of the revolution of numerous application fields. Hence, new techniques are emerging and required [2]. The human standard of living has been affected by the Internet of robotic technologies in numerous ways. Several manufacturers use robotics to do sophisticated, critical, and difficult jobs, including welding, product assemblage, product testing, packing, and quality control. Preprogrammed robotics has aided and improved industries to never-before-seen levels of precision and 24/7 operational capability. Robotics became more efficient as network technologies were merged, allowing them to perform in unstructured situations [2,40]. Figure 5 describes the overall percentage of IoRT in different fields such as the health sector, agriculture, manufacturing, and surveillance, giving us a brief idea of the latest use of IoRT in all sectors. Figure 6 classifies the robots on the basis of application areas, requirements, and features [4]. The IoRT physical operation classifications used by IoRT include ground and underground, space and planetary exploration, marine and underwater, hybrid location operations, and aerial. Each class has its own set of capabilities [22,41].

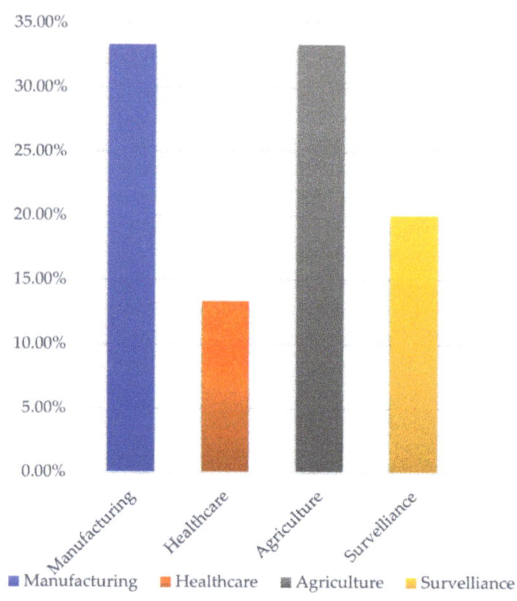

Figure 5. IoRT market usage [1,3].

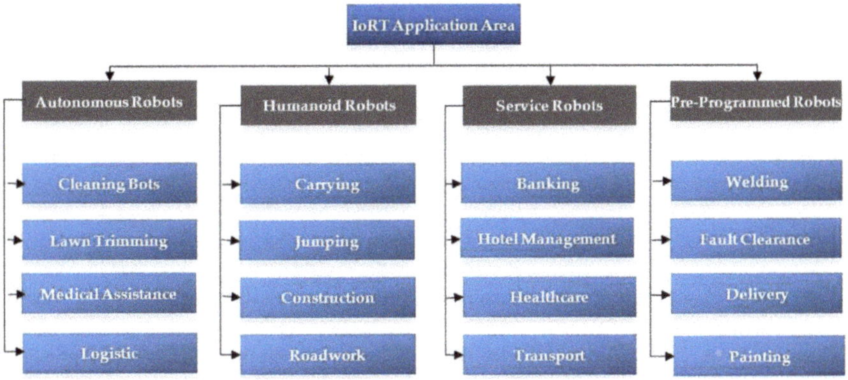

Figure 6. Summary of IoRT application areas.

2.5. *Robotic and IoRT Enabling Technologies*

Robotic and IoRT technologies are discussed in this section.

2.5.1. Robotic Technologies

The purpose of robotics is to create machines that can support and benefit people. Robotics is the study of creating machines that can replace people and perform human-like tasks. Robotic applications vary according to the environment. In this section, we discuss robots as per IoRT requirements, such as cloud robotics, collaborative robotics, cognitive robotics, fog robotics, network robotics, smart robotics, and swarm robotics [2]. The categorization of different robotic technologies is based on robot functions and numerous interconnected technologies. Cloud robotics uses cloud technology such as processing, storage, and data analysis. HRI is a difficult aspect of robotics, and collaborative robotics aids in the interaction between humans and robots. For intelligent decisions, cognitive behavior is a distinctive trait in robotics, and smart robotics and cognitive robotics play a key part. In communication and computing activities, network robotics and fog robotics are required [1]. Table 2 represents the functions of various robots according to the robotic technologies and purpose.

Table 2. Types of robotics.

Type	Description
Cloud robotics	Robots + cloud infrastructure
Collaborative robotics	Robot–human collaboration
Cognitive robotics	Robots use AI algorithms to learn and respond the complex tasks
Fog robotics	Robots use fog computing to process data and services
Network robotics	To complete a task, multiple robots collaborate and coordinate through networked communication
Smart robotics	AI + robots + ML + DL + cloud computing
Swarm robotics	Multiple robotic systems with physical robots

2.5.2. IoRT Enabling Technologies

IoRT requires many technologies such as sensors and actuators, communication technologies, processing, data fusion techniques, environments, objects, virtual and augmented reality, VR, VC, orchestration, decentralized cloud, adaptation, ML, end-to-end operation, Internet technologies, safety and security frameworks, blockchain, etc. All of these technologies work together to complete various tasks collaboratively. The major IoRT enabling technologies are defined below, and Table 3 provides a survey on existing robotic technologies.

Actuators and sensors: IoRT and IoT technologies obtain precise and accurate real-time data identification from sensors and actuators. The sensors and actuators are the fundamental gadgets that set the groundwork for the improvement of IoT and robotic systems. The present sensor industry focuses on 2D sensing information. However, with the upcoming IoRT boom, 2D sensing information might change to 4D [1,33].

AR and VR (digital twins): Augmented and virtual reality are counter-reflections of each other. VR provides digital leisure in a real-life scenario. AR provides virtual objects as a cover for the real world. The latest example is Meta's "meta-verse", which merges virtual reality with physical reality and blurs the gap between our interactions online and in real life [4,42].

Voice recognition and control system: For better HRIs, voice control and recognition systems play an important role. For HRI, the IoRT system must be able to communicate between humans and robotic things. Due to the critical nature of VR and VC, such technologies should be versatile and modular to remove the noise using information gathered from the robot's motions and expressions. In addition, the quality of the microphone and speech recognition procedures ought to be able to minimize noise. Multichannel systems with progressive methods such as side-lobe cancellers and feature-space noise clampdown should be included in IoRT systems [4,43].

AI and ML: IoRT technology combines IoT, AI, cloud computing, and other techniques. Due to this, IoRT systems become highly competent in real time and improve the learning experience. These techniques are used in the various layers of the IoRT frameworks to give data and perceptions, as well as maximize the functionality of individual robotic things. Adapting ML and DL techniques and algorithms to IoT-enabled devices enhances the intelligence in IoRT. The primary topics of ML are computational learning and pattern recognition. This provides systems with the capability to acquire data by researching the construction of models to predict and assimilate datasets. In the next few years, ML may be able to replace human learning for data analysis and prediction [4,5].

Connectivity and communication: Communication is the most necessary functionality of the IoRT system. Communication protocols are required to provide layer-by-layer information transmission. IoRT connectivity is preferred over wireless access methods. The new IoRT connectivity strategy permits pooled real-time computing and data stream exchange [19,42,44,45].

Table 3. A summary of enabling technologies in IoRT.

Technologies	Author	Domain	Findings
IoT/IIoT, autonomous robotic system, intelligent connectivity, AI, DL, ML, swarm technology, and VR and AR	Versemen et al. [4]	IoRT—intelligent connectivity and frameworks	• This paper mentions the merging of ML algorithms (CNN and RNN) with IoT and networks for combining the IoRT architecture with edge and fog computing • Role of digital twins, VR and AR, in HRI; collective tasks and efficient data management by swarm technologies and DLT

Table 3. *Cont.*

Technologies	Author	Domain	Findings
Voice recognition and voice control, ML, and security framework	Khalid et al. [3]	IoRT—detailed review	• The author explains how the sensors in different fields are used and how they work • The actuating of sensor data • The improvement in HRI is due to VR and AR; the way in which security attacks occur in networks.
Architecture and network framework, multi-robotic system, computing (edge, fog, cloud), and security	Ilya et al. [46]	IoRT—analysis	• A detailed summary of network layers and their functionality in communication and connectivity • The author mentions the protocols used in different scenarios, as well as the efficiency of multi-robotics

Swarm technology: Swarm robotics may be defined as the integration of multiple robots into a system. Multirobot systems consist of many simple physical robots to perform collective tasks. Combining the swarm robots with IoRT results in scalability, flexibility, and robustness for multirobot systems [1].

2.6. *IoRT Architectures*

There is no single architectural design that is agreed upon universally because each organization, company, or each user, for that matter, has different requirements. Moreover, the hierarchy of architectures includes three-tier architecture, four-tier architecture, five-tier architecture, and seven-layer architecture. IoRT is an interaction between the physical and digital worlds using sensors, actuators, and robots. In a few years, IoRT has framed so many novel designs, criteria, and platforms. Different architectures of IoRT were illustrated in [1,3,42,47].

2.6.1. Three-Tier Architecture

According to [3], IoRT has a three-tier architecture. The three-tier architecture of IoRT is illustrated in Table 4, featuring the hardware/physical/perception layer, network layer, and application layer, as discussed below.

Hardware layer/physical layer: The physical layer or robotic layer comprises actual IoRT devices. IoRT devices may vary from small sensors to a varied range of robotic devices to produce data [1]. This bottom-most layer comprises various robotic things such as sensors, vehicles, smartphones, home equipment, and actuators. The intelligent IoRT develops a multi-robotic system and delivers innovative features through distributed activities by contacting and integrating them. This layer is in charge of operating in the environment, sensing the data, acquiring information, and transmitting it to the higher layer. Above the robotic layer lies the network layer [48,49].

Table 4. An illustration of various components in a three-tier IoRT architecture.

Layers	Domain
Services and application layer	Smart environments Installation and execution of programs are carried out here by interconnected IoRT
Network and control layer	Routers, switches, local and cloud servers, and network and management protocols
Physical/hardware layer	Sensors, robots, actuators, robot-to-robot communication, and multi-robotic systems

Network layer: The network layer transfers the sensor data between different layers using networks of type 3G, 4G, 5G, RFID, LAN, Bluetooth, and NFC. The network layer contains components that communicate and control operations entailing several robotic things using several protocols. To offer the required connectivity, this layer can comprise routers, controllers, and gateways. Sensor and robot connectivity was explained in [50,51].

Application layer: The application layer is the uppermost layer in the IoRT architecture and defines all applications that use IoRT technology. The application layer interprets and monitors data using various application software. Records are prepared on the basis of data analysis [26]. The physical layer aims to distribute the client experience by investigating the offered sample of robotics-based applications. IoT-connected robots can actively participate in solving a variety of problems in fields [52,53].

2.6.2. Four-Tier Architecture

According to [47], IoRT has a four-tier architecture, divided into four layers for reliable data communication: (i) hardware layer, (ii) support layer, (iii) network layer, and (iv) application layer. The roles of three of the layers were discussed above; the fourth support layer is described below.

Support layer: The support layer provides security in the architecture of IoRT. In a three-tier architecture, data are directly communicated to the network layer, which is susceptible to attacks. The support layer consists of antiviruses and secure computing, overcoming the flaws of the three-layer architecture. Information obtained from the perception layer is sent to the support layer, which provides authenticity to the user. Then, the support layer sends information to the network layer.

2.6.3. Five-Tier Architecture

According to [1], IoRT has a five-tier architecture, which can be further subdivided for a better understanding of IoRT functionalities, thereby minimizing modification requirements to the underlying hardware and software logic: (i) hardware/robotic things layer, (ii) network layer, (iii) Internet layer, (iv) infrastructure layer, and (v) application layer. The five-tier architecture layers are summarized below.

Network layer: The network layer and transport layer are in charge of transmitting data from one end of a network to the other. Both layers are closely linked and are commonly mentioned collectively. Figure 7 depicts a five-tier design, with the network layer referred to as the transport layer.

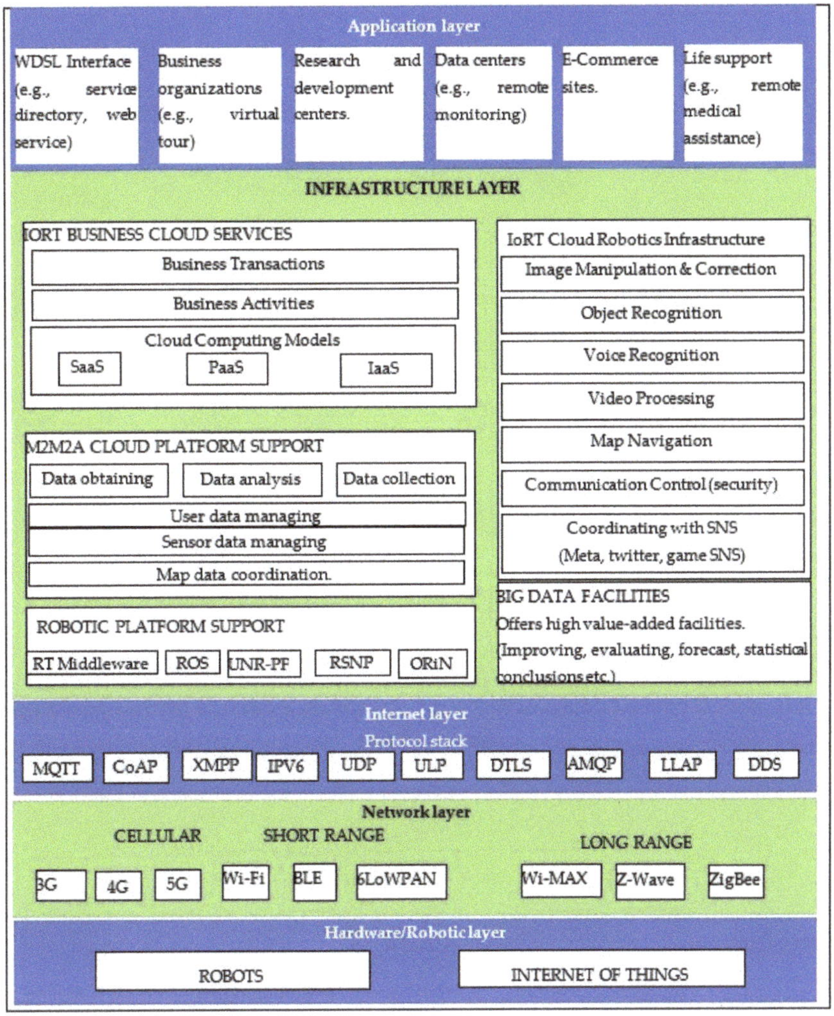

Figure 7. Five-layered IoRT infrastructural architecture.

Internet layer: Network connectivity is an option for facilitating device connectivity and the right to use information from wherever in the world. Internet connectivity provides connectivity for systems and access to data anywhere and anytime. Internet connectedness is regarded as the core part of communication in the IoRT architecture. As the IoRT is constructed on the basis of robotic things, it uses a variety of IoT-defined communication protocols to enable M2M and M2H communication, as well as lightweight processing of information in robotic systems [1,49].

Infrastructure layer: The robotic cloud stack transforms this portion of the architecture into the maximum managed service-centric methods for the cloud, middleware, business processes, and big data. The infrastructure layer is made up of five different but connected modules, including robotic cloud infrastructure, M2M2A cloud infrastructure support, IoT business cloud facilities, big data facilities, and IoT cloud robotics structure. All of these layers are well outlined in the architecture diagram (Figure 8) of IoRT [1,51].

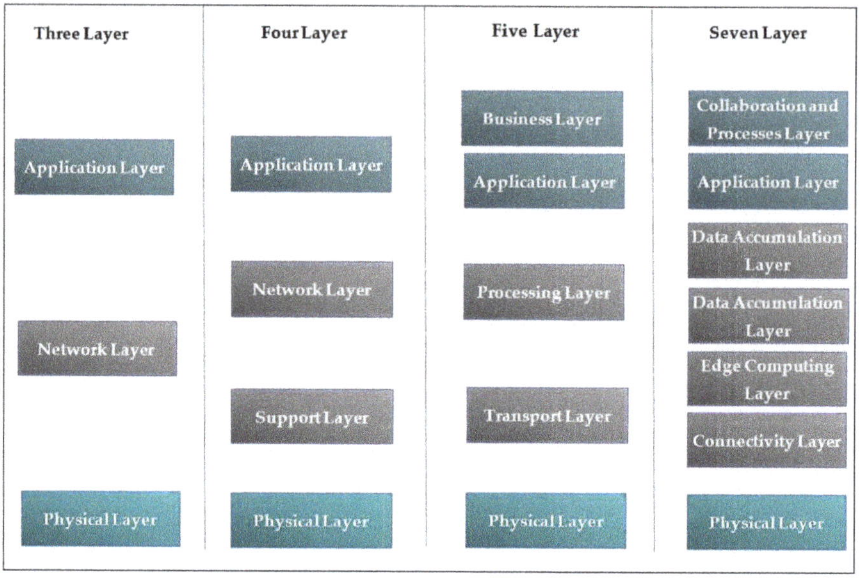

Figure 8. Evolution of IoRT architectures: three-, four-, five-, and seven-tier architectures.

Application layer: The application layer is the uppermost layer in the IoRT architecture. The physical layer aims to distribute the client experience by investigating the offered sample of robotics-based applications. IoT-connected robots can actively participate in solving a variety of problems in fields [53].

A conceptual diagram of the detailed architecture of robotic things and cloud computing is given in Figure 8. This architecture gives an overview of how the robotic platform support gives robot-specific service technologies such as middleware, robotic operating systems, service network protocol, and network interfaces. The M2M2A exchanges the data to the network with resource utilization in the Internet of things business cloud services [1].

2.6.4. Seven-Layer Architecture

According to [47], IoRT has a seven-tier architecture, which breaks down the intricate problem into manageable parts to acquire a complete sense. This IoRT architecture is more realistic rather than just conceptual. Moreover, the data control layer in the IoRT architecture grips data at the edge, fog, and cloud. The seven-layer architecture is summarized below.

Network connectivity layer: The connectivity layer, also called the network layer, performs packet forwarding, requiring virtual connections obtained from infrastructure suppliers to operate virtualization with the required environmental outline, trustworthiness, and efficiency for telecom operators. Studies have illustrated how low-cost IPTV distribution may be achieved via wide-area IP multicast, which tracks on the maximum of a trustworthy virtual network. This layer ensures accurate and consistent data transmission by implementing numerous protocols, switching and routing protocol interpretation, and networking inquiry [16].

Edge computing layer: This layer emphasizes the analysis, processing, and transformation of data.

Data accumulation layer: This layer interprets mobile data as fixed data [54].

Data abstraction layer: This layer is aware of the many languages used to express data where the information is stored. As a result, the layer is able to handle the communication needs of the appropriate information sources. This layer allows multiagent systems entities to access information via Java calls, regardless of the true data representation language. Different application programming interfaces (APIs) plus a new component called the data access layer make up the DAL. The APIs are a set of Java functions that serve as a link between data stored in one location and the remains of the network. The data abstraction layer uses data stored in various formats to create easy and more performant applications [54].

Collaboration and processes layer: This layer of architecture utilizes and distributes the application information with business processes and people [1].

Each organization requires a specific architecture for the development of a particular product, which means that architectures are used as per the requirement. A detailed architectural diagram is shown in Figure 7, which represents the evolution of architecture. IoRT is made up of several components, such as temperature, motion, light, gas, accelerometer, and pressure sensors. Gateways in IoRT are devices that connect to any network and store data in cloud centers. Analytics or mobile applications analyze the data according to the needs. Several IoRT structures such as three-layer, four-layer, five-layer, and seven-layer architectures are provided to thoroughly analyze these components. Data are acquired from sensors and actuators in the perception layer of a three-layer design. Data are collected and sent to cloud servers for storage and analysis. The application layer is in charge of providing services to users. Layers are further subdivided into a five-layer design for a better understanding of IoRT features. Data are transferred from the physical layer to the network layer in this architecture. A vast volume of data is stored and subsequently analyzed by the processing layer or middleware. Data processed in the application layer are used by users in a human-readable format. The business layer is at the head of IoRT technology, managing the whole system, user policy, profit model, and applications. For a better understanding of IoRT technology, the five layers are divided into seven layers, each of which has been addressed previously. As a result, an evolution of layers occurs as each tiered design is required by the organization. In the evolution diagram, architectural layers are segregated into the next layers for a better understanding of technology [47,48,55,56]. Table 5 mentions the various existing layered architectures. Figure 7 illustrates four-tier, three-tier, five-tier, and seven-tier architectures.

Table 5. A survey of layered IoRT architectures.

Author	IoRT Domain	Architecture
Ray et al. [1]	IoRT—infrastructure	Five-layered
Khalid et al. [3]	IoRT—applications	Three-layered
Anand et al. [6]	Intelligent robotics	Five-layered
Ilya et al. [46]	IoRT—architecture and components	Three-layered
Rana et al. [47]	IoT—energy efficiency and interoperability	Three-, four-, five-, and seven-layered
Sathish et al. [48]	IoRT—security and privacy	Three-layered

3. Related Work

3.1. IoRT: An Outline

IoRT is a fusion of several disciplines such as robotics, cloud computing, AI, and the IoT [2]. IoRT allows robotic objects to participate actively in diverse environments. In diverse surroundings, the robotic objects share data with other robotic devices, IoT devices, and people [3]. IoRT's most recent concepts, technologies, and challenges are useful for the future progress of robotic systems. The application of the IoRT system can be further intensified in industrial production and development, agriculture, and other areas of human importance [6]. Khalid et al. [3] showed a three-layer architecture of IoRT including related technologies such as actuators and sensors, and the Bricks View-RoIS. In addition, the authors of [3] presented HRI challenges and related charts for service robots. In [1], an IoRT architecture was recognized and understood considering five layers: robotic, network, Internet, infrastructure, and application layers. The infrastructure layer includes the robotic and M2M2A cloud platforms, IoT commercial cloud facilities, and IoT cloud robotics setup. The main capabilities of the five-layered architecture are awareness, interoperability, extensibility, virtualized diversity, dynamic, and self-adaptive behavior. The authors of [22] described the enabling technologies of the robotic system such as robots, AI, ANN, ML, fuzzy logic, and swarm technology, along with their application. The future difficulty of IoRT is data connectivity and security, which require a great deal of attention [7]. Gaze tracking, speech recognition, and biological recognition are HRI issues encountered by IoRT. HRI problems have not yet been put to the test. Instead, HRI issues are mostly being investigated. Computational issues, optimization, security concerns, and ethical concerns are among the IoRT challenges [3,5,11]. In our study, a taxonomy of IoRT, including IoRT technologies and capabilities, is proposed in Figure 9. IoRT, as we know, is a mix of different technologies, such as AI, which aids in intelligent decision making, and cloud robotics, which aids robots by employing cloud infrastructures such as cloud computing, cloud storage, and connectivity technologies. The Internet of things focuses on sensing, monitoring, and tracking, whereas robotics focuses on interactions, navigation, etc. [4]. As seen in the taxonomy graphic, all of these technologies interact with one another.

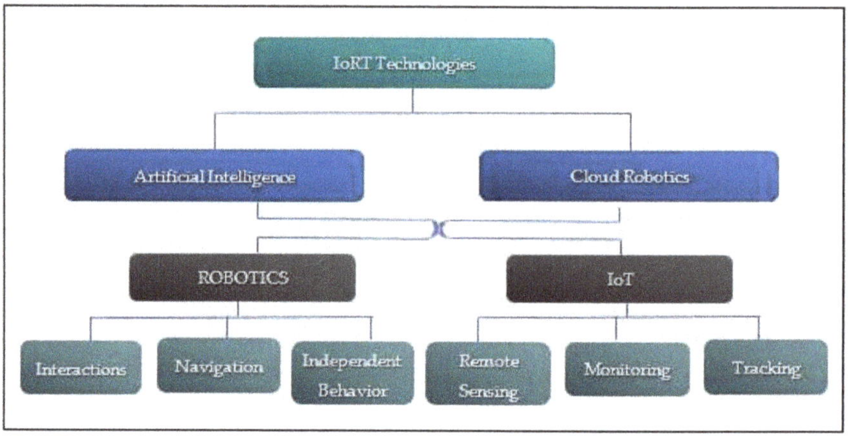

Figure 9. A taxonomy of IoRT technology.

3.2. *Secure Communication Mechanisms for IoRT*

Secure communication among IoRT devices is one of the primary concerns for industries, as well as society. Commonly used secure communication mechanisms for IoRT are as follows:

- Trust-based mechanisms,
- Encryption-based mechanisms.

3.2.1. Trust-Based Mechanisms for Robotic Devices

A technique for trust-based IoT VANET reveals security issues to make the system secure and trustworthy. The trustworthy cluster is identified as the "cluster head". The cluster head employs statistical models. Trust metrics are calculated by statistical models to identify maliciously infected nodes. RSU is in charge of calculating the clusters in the process. In the process, previous trust values surrounding the nodes are saved in special fixed storage with unique vehicle identification. For analysis of performance, the OMNet++ Simulator is employed. In this mechanism, a Sybil attack is detected by trust-based criteria to provide security. A malicious code is identified as one not being used to earn greater trust levels. It is possible to upgrade the technology to establish a bidirectional clustering technique for VANETs [11]. An enriched, reliable execution environment is employed for IIoT edge devices. The described environment focuses on the real-time and safety features of edge devices. The security features are represented by three CIA elements.

The model demonstrates that security is the most important aspect of most protected systems [9,12–15,40,45].

3.2.2. Encryption-Based Mechanism

Encryption intercepts data using computer algorithms and decodes it using a key provided by the sender. Encryption ensures that confidential information remains confidential, whether it is saved or in transit. Any illegitimate access to the data may produce a jumbled array of bytes. Data security is an issue in cloud computing, and it includes many aspects such as the CIA, surveillance, reliability, and telecommunications. The cloud introduces various types of data security solutions using encryption techniques [17,18,20].

3.3. Robot Navigation Techniques

The ability of a robot to establish its location and orientation within a frame of reference is referred to as robot navigation. Robots use sensors to extract information from their surroundings [57]. A robot navigating in unidentified terrain may encounter an impediment that must be avoided. Probabilistic roadmaps, bumper events, and some algorithms are used for clearing impediments in navigation [58,59]. The robot follows a path with a specific goal, avoiding obstacles along the way. For implementation, a real TurtleBot robot with sensors is used [60]. The navigation model just requires prior information for navigation at the beginning and places the goal. The navigation methods allow the avoidance of both static and dynamic obstacles [61]. A few navigation techniques are mentioned below.

- Robot navigation using fuzzy logic

The robot localization model uses two kinds of controllers, namely, fuzzy logic and pure pursuit. The controllers use labeled data input and output mapping FIS algorithms. The two algorithms control navigation and obstacle avoidance. The former determines the direct path without considering obstacles, while the latter does [62,63]. Using fuzzy logic, the unidentified territory is guessed. The fuzzy logic design, membership functions, and fuzzy rule base are all used in the fuzzy controller. For receiving inputs (minimum range, corresponding angle), the MATLAB-Simulink model is utilized, as well as the gazebo simulator. For pre-navigation, the system does not require data for the obstacles. As a result, a model for navigating robots in an unknown environment is worthy of consideration. With future improvements, the left or right turn can be eliminated [64–66].

- Robot navigation using probabilistic roadmap algorithm

For robot path pursuit, the probabilistic roadmap is implemented; a path is obtained from the beginning to the end of navigation. The phases of the navigation process are as follows:

a. Creating a map of the neighboring world,
b. Storing the map in an intelligible form,
c. Selecting a suitable path from start to finish on the preserved map,
d. Ultimately navigating the robot on the detected path.

The code is written in the MATLAB programming language. To achieve experimental findings, probabilistic roadmaps and path pursuit are employed. In the future, dynamic environments with moving obstacles can be built [58].

- Robot navigation using laser scan matching algorithm

A laser scan is executed for concurrent positioning and mapping in robot steering. The method is fulfilled by using two normal distribution transform algorithms [67]. The laser scan data from the robot are collected and kept using one algorithm. The other algorithm scans the matching and mapped buildings. To avoid obstacles, the laser sensor receives input that is converted into angular velocity. Neural network training parameters are required for scanning acceptable data quality. Laser scan measurements acquired at two places during navigation can be positioned using the rotation and translation of the robot's two coordinate frames [32,68,69]. Equation (5) displays the 3D plotting (f) mid (x1, y1) and (x2, y2) coordinate frames of the robot.

$$f: \begin{matrix} x2 \\ y2 \end{matrix} = \begin{bmatrix} \cos\phi & -\sin\phi \\ \sin\phi & \cos\phi \end{bmatrix} \begin{matrix} x1 \\ y1 \end{matrix} + \begin{matrix} \delta x \\ \delta y' \end{matrix} \tag{5}$$

where ϕ is the rotation between the two frames [δx, δy]; T is the transformation between *(x1, y1)* and *(x2, y2)* [67].

- Robot navigation using heuristic functions

Three heuristic functions are used to independently navigate a robot. A navigational map is obtained. Among the three functions, Euclidean distance yields the most nonuniform global path planner time. The octile distance yields the most uniform time throughout the navigation procedure [70,71]. The Manhattan distance between two points {$p_1(x_1, y_1)$ and $p_2(x_2, y_2)$} is given in Equation (6), while the Euclidean distance between two points and octile distances are represented in Equations (7) and (8), respectively.

$$h_1(p_1, p_2) = \mathbb{C}(|x_1 - x_2|, |y_1 - y_2|), \tag{6}$$

$$h_2(p_1, p_2) = \mathbb{C}\sqrt{(x1 - x2)^2 + (y1 - y2)^2}, \tag{7}$$

$$h_3(p_1, p_2) = \mathbb{C}(|x_1 - x_2| + |y_1 - y_2|) + (\mathcal{D} - 2\mathbb{C})Min(|x_1 - x_2|, |y_1 - y_2|), \tag{8}$$

where C and D are constants.

- Robot navigation using bumper event

The bumper event is used to remove obstacles from the robot's navigation. The bumper event algorithm is applied to the TurtleBot in the gazebo simulator. Bumper and state fields comprise the robot. The bumper sensor coupled with a TurtleBot is used to manage the hurdles. The robot is moved and turned using two different ROS velocities (linear and angular). C++ code is used to implement the algorithm. Because it reduces complexity, this approach is very beneficial in unfamiliar contexts. As a drawback, the algorithm does not give collision-free navigation; hence, the camera gets priority over the bumper sensor for collision-free navigation [23,60,72].

- Vision-based navigation

The robot's gaze direction can be chosen from a variety of directions according to the inclination of angles. More gaze directions necessitate more computational time. For vision-based navigation, an assessment function M is used to calculate the corresponding connection of feature lines between two images as defined in Equation (9) [70,72–75].

$$\mathrm{M} = \alpha \sum_{i=1}^{\mathrm{N1}} \mathrm{Di} + \beta \sum_{i=1}^{\mathrm{N1}} \mathrm{Li} + \gamma \sum_{j=0}^{\mathrm{N_2}} \mathrm{Pj}, \tag{9}$$

where L_i is the absolute variance between two location intervals. D_i is the absolute horizontal variance value of feature lines i in the first image and the parallel candidate image. α, β, and γ are the weights for each term; α + β + γ = 1. P_j is the penalty value when a feature line does not have a communicator in the second image. N1 is the number of feature lines, with contenders in the second image. N2 is the number of feature lines that do not have a communicator in the second image.

3.4. Remote Server Access in IoRT

Computer servers contain important data and software. The servers can be accessed by IoRT devices remotely. However, data exchange between the server and IoRT devices should be secure enough. Local ORM vehicle work is performed on VEC servers. This type of model aids in the execution of tasks such as distributed and trustworthy reputation maintenance, precision reputation updating, and accessible reputation usage [75].

A software update makes use of the MEC for high processing capability in the access network, despite the limited resources of IoT devices. IoT devices can use MEC's software functionalities [44,69]. Remote software update performed over trusted connections is done in five steps [44]:

1 Record the service profile on a cloud server,
2 Request to ASP server for the service package,
3 Send service package using ASP server,
4 Control function codes using the data core network,
5 Update function codes.

To achieve higher energy efficiency in IoT, fog data analytics of data has been stressed more than cloud to minimize latency [11,25,45]. The energy consumption model is concerned with calculating the total quantity of energy utilized by all nodes throughout the transmission. The major reasons for energy usage in an IoRT network are receiving and sending a packet on a trustworthy channel. The IEEE 802.15.4 communication standard accomplishes the entire process. The energy usage at the node p on link e [47], with E for packet rectifying, is given by Equation (10).

$$E^p{}_c = E_l{}^p + E^p{}_{tx} + E^p{}_{rx} + E^p{}_{sl} = \left(t_l{}^p I_l + \frac{(I_{tx} + I_{rx})L}{R} + t^p{}_{sl} I_{sl}\right)V, \tag{10}$$

where V is the node voltage, L is the packet size, and R is the data packet rate. I_l and $E_l{}^P$ are the current drawn and energy consumption during listening. I_{tx} and $E^P{}_{tx}$ are the current drawn and energy consumption during transmitting. I_{rx} and $E^P{}_{rx}$ are the current drawn and energy consumption during receiving. I_{sl} and $E^P{}_{sl}$ are the current drawn and energy consumption during sleeping.

Assumptions:

$$\begin{cases} E^p{}_{tx} = 0;\ \text{if}\ p\ \text{is}\ a\ \text{transmitter} \\ E^p{}_{rx} = 0;\text{if}\ p\ \text{is}\ a\ \text{receiver} \end{cases}.$$

Then,

$$E^p{}_c = \begin{cases} \left(t_l{}^p I_l + \frac{(I_{tx})L}{R} + t^p{}_{sl} I_{sl}\right)V, & \text{if}\ p\ \text{is}\ a\ \text{transmitter} \\ \left(t_l{}^p I_l + \frac{(I_{rx})L}{R} + t^p{}_{sl} I_{sl}\right)V, & \text{if}\ p\ \text{is}\ a\ \text{receiver} \end{cases}. \tag{11}$$

4. IoRT Security

Security is a major concern in the connectivity of robotic things [46]. IoRT has significant challenges in terms of security and protection to enable effective collaboration with networks, sensors, and robots. Companies that collect data from robotic systems face the biggest risk from IoRT. Because IoRT networks are still connected to the Internet, new sorts of data breach attacks can be launched against them [3]. There is always a need for communication protocols for data transmission and processing. Therefore, the communication between robotic things must be encrypted, which often does not occur. The Dieffie–Hellman concept is also used for data encryption for the security of a system's communication [24]. To address security concerns, a secure method has been developed that includes a requirement to register IoRT devices using a digital certificate, as well as a user to the cloud server. For a cloud-based IoRT network, we need a three-way (CIA triad) security architecture [76]. To convey information in a secure approach accumulated by robots, secure frameworks are needed with respect to integrity and confidentiality. The IoRT system should be encompassed with physical access security frameworks for verifying data, maintaining trust and privacy, and keeping the data confidential [6,19]. A security taxonomy is given in Figure 10, which describes the generic security threats and threats at the architecture level of IoRT. In addition, the Internet is the basic source of threats and vulnerabilities to robotic things because it is the basic building block of the IoRT device's communication and connection. Non-standardization of IoT technologies has increased the frequency of security breaches daily, which has increased the vulnerabilities. Some machinery or physical and boot process vulnerabilities are generic issues that apply to the whole IoRT system. Security assaults are also a result of the HRI. IoRT companies supply some security and data protection mechanisms for the safety of user data. However, the effectiveness of protection against vulnerabilities is uncertain and may or may not be guaranteed. Phishing and security breaches are also caused by users' and employees' lack of awareness. IoRT devices are also responsible for a large percentage of denial-of-service assaults (96%) [77]. Threats to the IoRT architectural layers exist as well. Eavesdropping, battery exhaustion, hardware crashes, data breaches, and unauthorized access to IoRT systems are all possible threats to the physical layer. Spoofing, node replication, and fraudulent message bombardment to gateways for denial-of-service assaults are all threats at the network layer. Because this layer connects numerous private LANs, the MAC or network layer is extremely vulnerable to attacks. The risks of brute-force attacks on encrypted data and malicious code at the application layer are also risks to IoRT devices. Thus, there is a necessity for a dependable data transfer service for IoRT [76,78–80]. There are several well-known attacks on ad hoc wireless networks, as listed below, including network attacks (a to d) and the trust model itself (e to g) [33].

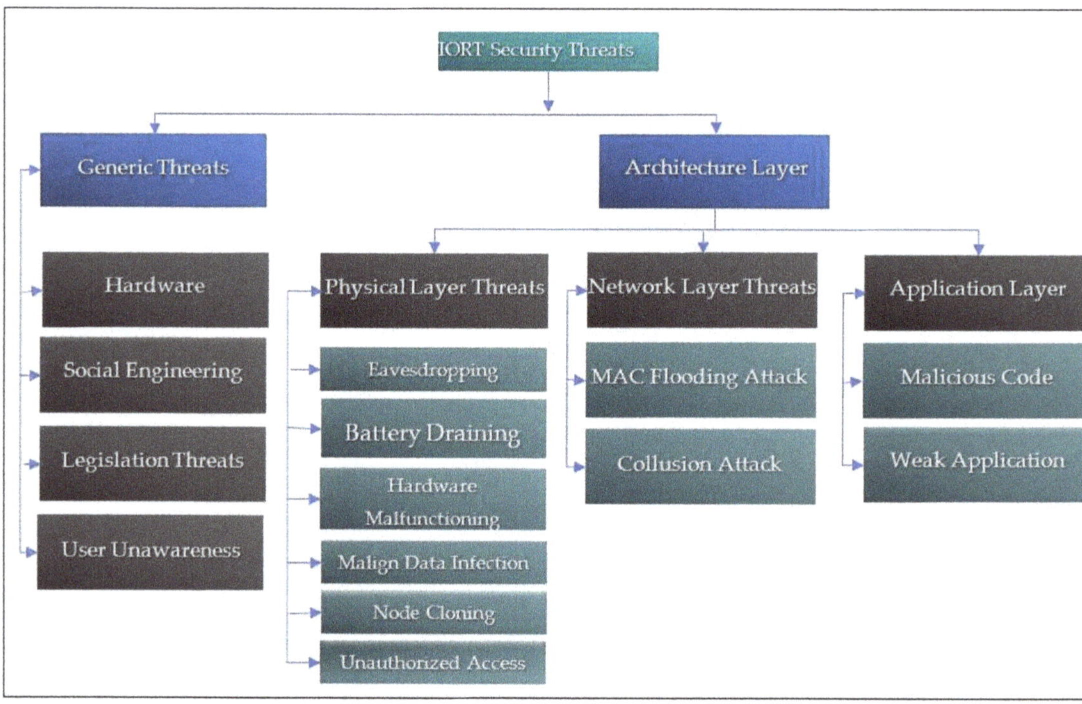

Figure 10. A taxonomy of security threats in data exchange.

a. Packet dropping or modification attacks—black hole and gray hole,
b. Wormhole attack,
c. Sybil attacks,
d. Newcomer attacks,
e. Badmouthing attacks,
f. On–off attacks,
g. Collusion attacks.

IoRT makes use of trust-based techniques to protect the system from vulnerabilities and threats. One of the trust-based mechanisms is threat modeling, which is used to launch data-sharing security attacks. A threat model is similar to assumptions about an intruder. The threat model's mechanism ensures that the data policy (data should have confidentiality, integrity, and availability) is followed as long as the intruder follows the threat model, which means that, if the threat model is right, it should be able to follow the policy. When security fails, the threat model mechanism is usually to blame. Furthermore, various approaches are used for the screw-up system's policy, such as "recovery questions" [80]. For example, when the threat model goes wrong, it is upgraded over time to ensure its effectiveness. In the 1980s, Kerberos was based on cryptography 56 keys; however, in this cypher-DES, the plausible size is less secure and not reasonable. Later on, it was advanced by applying 256 keys that are more secure [20]. Figure 11 describes the threat model security method in communication to depict secure data flow between the two nodes. The threat model ensures that this communication channel is secure since hackers are always attempting it [81,82]. A threat model is a logical representation of all the data that influence an application's security. Threat modeling (system or data) is the understanding of how a threat actor (external or internal, hostile or abusive) might target a certain asset. Threat modeling differs from application testing [33,58]. The threat model examines the ecosystem, processes, and the circumvention of ecosystem safeguards. If applied effectively, it is one of the finest prospects in solutions, systems, and data security [83]. In successful threat modeling, the following steps are implemented:

a. Uncovering the illegitimate mastermind in the organization,
b. Figuring out the breaking-in method,
c. Choosing the priority method,
d. Portraying the countermeasures,
e. Implementing the solution and testing it.

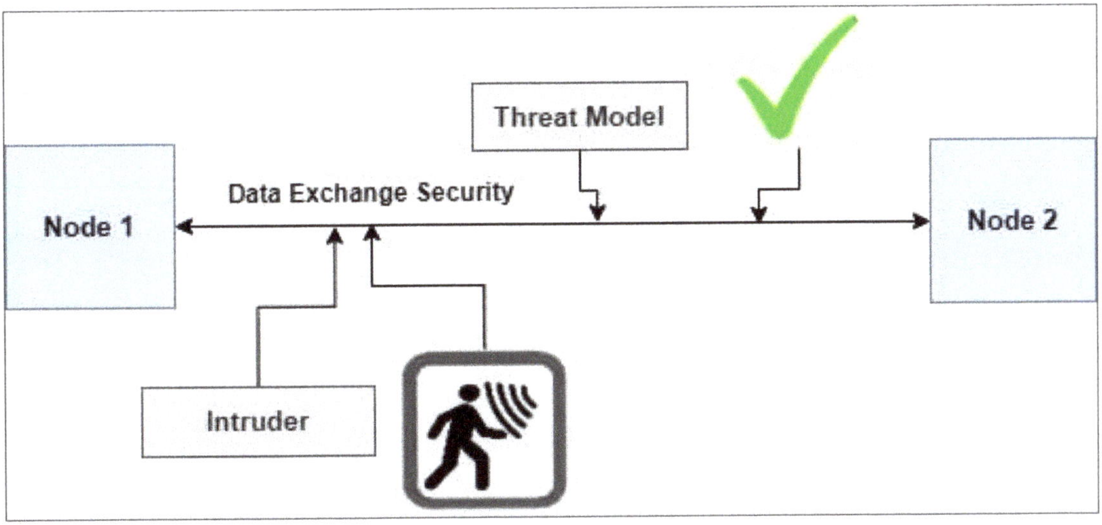

Figure 11. Threat model in secure data exchange.

The term "trust" refers to a set of relationships between the parties involved in a particular protocol. Trust is a belief or trust in other nodes or objects that are based on a defined protocol. Trust is dynamic and is not necessarily transitive. Trust is asymmetrical and dependent on the situation [33]. The computation of trust may be achieved in two ways: distributed or centralized. In distributed systems, we have direct trust, indirect trust, and hybrid trust. Trust agents can be found from the local standpoint in the network, just as in the centralized system [33,84]. A trusted operator can be denoted by the following formula:

$$T^{m}_{i,k}=M_i\left(T^{n}_{i,k},T^{n}_{j,k}\right). \quad (12)$$

Prefix M models the trust of the i-th agent. $T^{n}_{i,k}$ and $T^{n}_{j,k}$ are the trust values of i-th and j-th agents toward the k-th agent. Similarly, n is the pre-operation value, and m is the resulting value of the operation [85].

Threats in an application can be mitigated by using countermeasures in threat modeling. Table 6 illustrates various security techniques used in various application domains, with a description. Table 7 illustrates various mitigations (countermeasures) corresponding to various security services. Service-level agreement is a way of transferring risk to another company, such as hosting data in a third-party data center to prevent the risk within the facility. The Internet of things infrastructure, operations, cloud computing, and business technologies all work together and require end-to-end communication mechanisms to assure the security. IoRT is a growing technology, and it is necessary to use security evaluations on Internet-connected platforms, devices, and protocols on a regular basis. Currently, the security measures of products throughout the world contain security-related patterns. The security posture of the IoRT product can be evaluated by using the Common Vulnerability Scoring System (CVSS) standard. CVSS is used to rate the severity of each IoRT product's security vulnerabilities [21]. Some popular secure service products are the following:

i. ARMbed for ARM to develop IoT products,
ii. Brillo and Weave connectivity for IoT/IoRT devices by Google,
iii. Homekit by Apple,
iv. Kura Eclipse offersing application program interface access to hardware interfaces of IoT/IoRT ports,
v. Secure operations for robotic automation by BILA.

Table 6. An illustration of IoRT security techniques.

Security Techniques	Author	Domain	Description
Secure IoRT network for data transmission	Khalid et al. [3]	IoRT—analysis	The paper mentions the security challenges and the reasons for data breaches
Integrity, trust, and confidentiality of secure data.	Ray et al. [1]	IoRT—architecture, technologies	The author discusses the security issues, the trustworthy IoRT VM, and the idea of the protection of secure data.
IoT protocols	Neerendra et al. [59]	Modern communication protocols for IoT	On the basis of six key factors of protocols, IoT protocols are analyzed and compared for optimal communication
Automated key update mechanism for M2M communication, preshared key	Tsai et al. [53]	IoT security enhancement	This paper focuses on a technique for increasing security performance for IoT devices in M2M communication
Privacy filter framework, probabilistic model	Zahir et al. [7]	IoRT—applications	A privacy filter framework is designed for attacks in IoRT-HRI applications
Mobile phone security	Liao et al. [86]	Mobile computing used to evaluate IoT device security	The author discusses the security, accuracy, and limitations of IoT devices and mobile phones
Software-defined network	Waseem et al. [77]	IoT security requirements, challenges	This paper mentions the security challenges, the threats of various layers of the IoT architecture, and approaches to network security

Table 6. *Cont.*

Security Techniques	Author	Domain	Description
Three-way system authentication	Nida et al. [76]	Three-way security structure for cloud-based IoT network	This framework can offer the ability to register IoT devices using digital certificates and users on cloud servers
Cyber-security, encryption	Ilya et al. [46]	IoRT architecture analysis	The author draws attention to the authentication mechanism of data.
Blockchain, software-defined networking	Djamel et al. [87]	IoRT survey—securities, privacy, the blockchain	The effective mechanisms in IoT and the security issues surrounding the safety of systems
UML extension for IoT system security modeling	David et al. [88]	IoT security	According to the author, IoT security is a UML extension; to describe IoT systems, the extension attempts to encapsulate security knowledge
AI, DL algorithms, security	Hui-WU et al. [89]	IoT security—using AI	Different algorithms are employed in this study to improve secure networking
Intelligent community security system (ICSS)	Sathish et al. [90]	IoRT—security and privacy issues	The author discusses various ICSS and their subsystems

Table 7. An illustration of security services and their mitigations.

Security Services	Countermeasures
Authentication	Encryption, trusted server authentication
Authorization	Access controls are required
Data validation	Output encoding
User session management	Encrypted authentic cookies, secure sessions

5. Open Research Challenges

IoRT is a new research field and is in the early stages of development, with many obstacles to overcome. This in-depth and critical investigation of the state of the art in IoRT led to various open research challenges that may be carried out further by researchers in the field of IoRT. The major challenges or gaps that emerged from our study are listed in this section, as well as in Table 8, along with future tasks.

Table 8. A description of IoRT limitations and future tasks.

Author	Paper Focus	Limitations	Future Task
Burghart et al. [35]	Cognitive framework for an intelligent humanoid robotic system	A multimodal fusion of speech and motions	Access to active models through tight integration
Nagarajan et al. [91]	Physical HRI mechanism	One-wheeled, continuous position displacements of ballbot	Laser range finders and stereo cameras are needed for accurate localization
Yoo et al. [51]	Gaze control-based localization for mobile robots	The main issue is how to transmit and display various types of data at the same time	The presented design can be expanded to deal with arbitrarily formed and equally sized objects traveling in peculiar ways
Ariffin et al. [32]	ACI used to build a humanoid-led navigation mobile platform within an obstacle in the surroundings by integrating exterior laser sensing with a humanoid	Security concerns	Path planning and trust-based mechanisms can be involved to overcome navigation and security issues

Computational problems: Due to the competence of IoRT, the transfer of resource-intensive computational tasks for execution to the IoT cloud is possible. However, this process requires a more rigid and merged architectural framework and can handle several complex issues. To solve the above problem, the system's global area (shared pool) can be supported. The novel shared offloading policy can examine so many factors, such as vast data exchange by several robotic things and real-time retard limits, to conclude the specific task in a fixed order. Moreover, the IoRT should be able to determine the competence of performing tasks within the IoRT or not [1,78].

Data security: The most considerable challenges in IoRT are data processing and security [1]. The IoRT-VM environment must be reliable. Without the assistance of a real robot, a malicious IoRT-VM can effortlessly erode a critical mission. For example, in military exercises, IoRT-approved robotic objects must be able to distinguish between trustworthy IoRT-VM infrastructure and harmful IoRT-VM infrastructure to connect to respectable infrastructure. Robotic objects should avoid the dangerous IoRT-VM infrastructure. To address this issue, three approaches can be used: trust establishment, trust measurement, and reputation-based trust. Future robotic systems must have the confidence to commence computing tasks on IoRT-based clouds. In such a manner, the robotic system's owner or controller may perform verification. It must be ensured that no harmful code is operating in the background of these outsourced activities. Simultaneously, secret data can be permanently kept on IoT-enabled cloud servers with reasonable data being cloned to private cloud servers. To safeguard IoRT data, stringent approaches are required to preserve integrity, trust, and confidentiality [23,40].

Ethical issues: Robotics has been working on resolving this critical problem. Sir Isaac Asimov's three renowned laws should be followed in robotics. A robot may not harm a human person or cause injury to a human being through its actions. Except where such directives clash with the first law, a robot must obey directions given by humans. As long as this shielding does not clash with the first or second laws, a robot must defend its own existence [1,3].

Human–robot interactions: According to recent research, HRI is facing a variety of problems in gaze tracking, voice interactions, and biological recognition, but these problems have not yet been tested and are mostly being studied by researchers. HRI-defined human movements must be adapted by intelligent robots [3,42,92–94].

Emotional robots: Emotional robots, bring their emotional relationships to reality. Recent advances in the field of emotional computing involve intervening in the design and development of "emotional robots" to create an emotional attachment between humans and robots. Nevertheless, there are huge gaps that need to be corrected in the future. The artificial software agents (bots) of "Pepper" are paving the way for emotional interactions to become a reality [36,95,96]

Remote computation problems: Remote working has provided enormous advantages in recent years, especially in the COVID pandemic, as it helps to increase productivity through the best work/life balance. Remote education is also an important advantage of using remote education robots. The educational relationship between people and robots has to be further developed. For a better means of managing industrial operations, additional improvement in such IoRT technology is required [44].

Energy consumption by devices: Industrial technologies are facing a problem of energy demand. In smart environments, the assessment and optimization of energy quality lack a detailed understanding of energy consumption. To address this issue, smart sensor energy utilization should be prioritized [14,47,91].

Data processing: Robotic things are facing enormous IoRT security threats in data exchange. The security of the IoT and the safety of robots are big issues. Large amounts of data are processed in IoRT systems, causing cybersecurity issues. To overcome this, we need an advanced network for IoRT communication to avoid insecure communication between robots and users. The security issue needs to be further investigated [17].

Authorization to industrial IoT: In industrial IoT, data must be shared using the same encrypted protocol with any other compatible system anywhere in the world. There should be proper authorization and privacy for industrial output and management applications and the internal information of the company. Authorization plays an important role in data security. Authorization is required for sensitive data, as many IoRT programs usually gather data from both labs and engaged clients. This matter should be investigated [11,13,46].

Localization problems: The navigational duties performed by robots remain restricted to motion modeling and position analysis, with little discussion of trajectory planning [27].

Noise problems: Noise is a serious problem in robotic movement, depending on the surface resistance and pushback in the joints.

Accurate localization: The measurements produced by the small sensors that are frequently used with humanoid robots are noisy and inconsistent. As a result, precise navigation, which is thought to be mostly addressed for wheeled robots, remains a difficult challenge for humanoid robots.

6. Conclusions

IoRT technology is relatively a new research area. IoRT has boomed in the market due to its rapid growth and demand in the e-commerce manifesto, the education section, consumer arcade, and research areas in just a few years. The IoRT industry is expected to be worth 21.44 billion USD by 2022, with a compound annual growth rate of 29.7% between 2016 and 2022.

This review focuse on IoRT abilities, evolution, applications, enabling technologies, and IoRT architectures. It was found that collaboration between robots and IoT sensors results in a more advanced IoRT technology. Furthermore, collaboration assists in sensitive data transmission and connectivity. A detailed review of the architectures of IoRT was presented. The study provided an outline of the latest enabling technology of IoRT infrastructure based on M2M2A cloud platforms, IoT business cloud services, and big data analysis. Various methods for navigation of robotic things were reviewed. For robot navigation, different algorithms were studied to overcome the impediment of the robotic surroundings. A security method was presented for secure data transmission between robotic devices. IoRT systems require enormous quantities of data to be transmitted among robots, cloud storage, and other devices. The transmission can lead to data leaks and cyberattacks. Security issues are becoming more serious. For secure data transmission, secure and trusted data-sharing mechanisms were proposed to eliminate existing research gaps. To handle security threats, future systems can be prepared by considering the proposed security methods and trusted data sharing mechanisms.

Author Contributions: Conceptualization, A.S.; methodology, A.S. and N.K. (Neha Koul); formal analysis and investigation, A.S., N.K. (Neerendra Kumar) and N.K. (Neha Koul); writing—original draft preparation, A.S. and N.K. (Neha Koul); writing—review and editing, A.S., N.K. (Neha Koul), N.K. (Neerendra Kumar), C.V. and Z.I.; resources, N.K. (Neerendra Kumar), C.V. and Z.I. All authors have read and agreed to the published version of the manuscript.

Funding: The work of Chaman Verma, and Zoltán Illés was financially supported by Faculty of Informatics, Eötvös Loránd University (ELTE), Budapest, Hungary.

Data Availability Statement: Not applicable.

Acknowledgments: The work of Chaman Verma was supported under ÚNKP, MIT (Ministry of Innovation and Technology) and the National Research, Development, and Innovation (NRDI) Fund, Hungarian Government. Furthermore, the work of Chaman Verma and Zoltán Illés was supported by the Faculty of Informatics, Eötvös Loránd University (ELTE), Budapest, Hungary.

Conflicts of Interest: The authors declare no conflict of interest.

Abbreviations

Acronym	Description	Acronym	Description
IoRT	Internet of robotic things	**AR**	Augmented reality
IoT	Internet of things	**VR**	Virtual reality
AI	Artificial intelligence	**BLE**	Bluetooth Low Energy
ML	Machine learning	**BGAN**	Broadband global area network
VR	Voice recognition	**6LowPAN**	Low-power wireless area network
DT	Distributed technologies	**ROS**	Robotic operating system
DLTs	Distributed ledger technologies	**VC**	Voice control
TCP	Transmission control protocol	**LORA**	Long-range transmission with low power
IP	Internet protocol	**MQTT**	Message Queueing Telemetry Transport
M2H	Machine to human	**CoAP**	Constrained Application Protocol
LAN	Local area network	**XMPP**	Extensible Messaging and Presence Protocol
M2M	Machine to machine	**IPV6**	IP Version 6
UDP	User datagram protocol	**DTLS**	Datagram Transport Layer Security
HRI	Human–robot interfaces	**AMQP**	Advanced Message Queuing Protocol
RoIS	Robotic interface services	**LLAP**	Live Long and Process
M2M2A	Machine to machine to actuator	**DDS**	Data Distribution Service
VANET	Vehicular ad hoc network	**WSDL**	Web Services Description Language
ORM	Online reputation management	**ULP**	Upper Layer Protocol
CIA	Confidentiality, integrity, availability	**SNS**	Simple Notification Service
API	Application programming interface	**UNR-PF**	Open Source of Cloud Robotics
ANN	Artificial neural networks	**RSNP**	Robot Service Network Protocol
VEC	Vehicular edge computing	**ORiN**	Standard Network Interface for Factor Automation
MEC	Mobile edge computing	**RPL**	Robot Programming Language
ASP	Active server pages	**CORPL**	Cobalt-RPL

References

1. Ray, P.P. Internet of Robotic Things: Concept, Technologies, and Challenges. *IEEE Access* **2017**, *4*, 9489–9500. [CrossRef]
2. Simoens, P.; Dragone, M.; Saffiotti, A. The Internet of Robotic Things: A review of concept, added value and applications. *Int. J. Adv. Robot. Syst.* **2018**, *15*, 1729881418759424. [CrossRef]
3. Khalid, S. Internet of Robotic Things: A Review. *J. Appl. Sci. Technol. Trends* **2021**, *2*, 78–90. [CrossRef]
4. Vermesan, O.; Bahr, R.; Ottella, M.; Serrano, M.; Karlsen, T.; Wahlstrøm, T.; Sand, H.E.; Ashwathnarayan, M. Internet of Robotic Things Intelligent Connectivity and Platforms. *Front. Robot. AI* **2020**, *7*, 104. [CrossRef]
5. Vandewinckele, L.; Claessens, M.; Dinkla, A.; Brouwer, C.; Crijns, W.; Verellen, D.; Elmpt, W. Van Overview of artificial intelligence-based applications in radiotherapy: Recommendations for implementation and quality assurance. *Radiother. Oncol.* **2020**, *153*, 55–66. [CrossRef]
6. Nayyar, A. Internet of Robotic Things: Driving Intelligent Robotics of Future—Concept, Architecture, Applications and Technologies. In Proceedings of the 2018 4th International Conference on Computing Sciences (ICCS), Jalandhar, India, 30–31 August 2018; 2020; pp. 151–160. [CrossRef]
7. Alsulaimawi, Z. A Privacy Filter Framework for Internet of Robotic Things Applications. In Proceedings of the 2020 IEEE Security and Privacy Workshops (SPW), San Francisco, CA, USA, 21–21 May 2020; pp. 262–267, ISBN 9781479966646. [CrossRef]
8. Yuan, B.; Lin, C.; Zhao, H.; Zou, D.; Yang, L.T. Secure Data Transportation with Software-defined Networking and k-n Secret Sharing for High-confidence IoT Services. *IEEE Internet Things J.* **2020**, *7*, 7967–7981. [CrossRef]
9. Cao, Q.H.; Khan, I.; Farahbakhsh, R.; Madhusudan, G.; Lee, G.M.; Crespi, N. A Trust Model for Data Sharing in Smart Cities. In Proceedings of the 2016 IEEE International Conference on Communications (ICC), Kuala Lumpur, Malaysia, 22–27 May 2016; ISBN 9781479966646.
10. Yfantis, E.A.; Fayed, A. Authentication and secure robot communication. *Int. J. Adv. Robot. Syst.* **2014**, *11*, 10. [CrossRef]
11. Romeo, L.; Petitti, A.; Marani, R.; Milella, A. Internet of Robotic Things in Smart Domains: Applications and Challenges. *Sensors* **2020**, *20*, 3355. [CrossRef]
12. Goh, S. Three architectures for trusted data dissemination in edge computing. *Data Knowl. Eng.* **2006**, *58*, 381–409. [CrossRef]
13. Pinto, S.; Pereira, J.; Cabral, J. IIoTEED: An Enhanced, Trusted Execution Environment for Industrial IoT Edge Devices. *IEEE Internet Comput.* **2017**, *21*, 40–47. [CrossRef]
14. Khan, Z.A.; Herrmann, P.; Ullrich, J.; Voyiatzis, A.G. A trust-based resilient routing mechanism for the internet of things. In Proceedings of the 12th International Conference on Availability, Reliability and Security, Reggio Calabria, Italy, 29 August–1 September 2017; Volume Part F1305. [CrossRef]
15. Chahal, R.K.; Kumar, N.; Batra, S. Trust management in social Internet of Things: A taxonomy, open issues, and challenges. *Comput. Commun.* **2019**, *150*, 13–46. [CrossRef]

16. Zhu, Y.; Sampath, R.Z.; Jennifer, R. Cabernet: Connectivity architecture for better network services Cabernet: Connectivity Architecture for Better Network Services. In Proceedings of the 2008 ACM CoNEXT Conference, Madrid, Spain, 9–12 December 2008. [CrossRef]
17. Zhao, F.; Li, C.; Liu, C.F. A cloud computing security solution based on fully homomorphic encryption. In Proceedings of the 16th International Conference on Advanced Communication Technology, Pyeongchang, Korea, 16–19 February 2014; pp. 485–488. [CrossRef]
18. Hemalatha, N.; Jenis, A.; Cecil Donald, A.; Arockiam, L. A Comparative Analysis of Encryption Techniques and Data Security Issues in Cloud Computing. *Int. J. Comput. Appl.* **2014**, *96*, 1–6. [CrossRef]
19. Gulzar, M.; Abbas, G. Internet of Things Security: A Survey and Taxonomy. In Proceedings of the 2019 International Conference on Engineering and Emerging Technologies (ICEET), Lahore, Pakistan, 21–22 February 2019; pp. 1–6.
20. Kumar, N.; Chaudhary, P. Performance evaluation of encryption/decryption mechanisms to enhance data security. *Indian J. Sci. Technol.* **2016**, *9*, 1–10. [CrossRef]
21. Alqahtani, A.; Li, Y.; Patel, P.; Solaiman, E.; Ranjan, R. End-to-End Service level Agreement Specification for IoT Applications. In Proceedings of the 2018 International Conference on High Performance Computing & Simulation (HPCS), Orleans, France, 16–20 July 2018. [CrossRef]
22. Wu, H.; Han, H.; Wang, X.; Sun, S. Research on Artificial Intelligence Enhancing Internet of Things Security: A Survey. *IEEE Access* **2020**, *8*, 153826–153848. [CrossRef]
23. Abbasi, M.H. Deep Visual Privacy Preserving for Internet of Robotic Things. In Proceedings of the 2019 5th Conference on Knowledge Based Engineering and Innovation (KBEI), Tehran, Iran, 28 February–1 March 2019; pp. 292–296.
24. Su, J. Authentication and Encryption for a Robotic Ad Hoc Network using Identity-Based Cryptography. In Proceedings of the 2018 4th International Conference on Big Data Innovations and Applications (Innovate-Data), Barcelona, Spain, 6–8 August 2018. [CrossRef]
25. Kumar, N.; Jamwal, P. Analysis of Modern Communication Protocols for IoT applications. *Karbala International Journal of Modern Science* **2021**, *7*, 390–404. [CrossRef]
26. Zhong, C.; Zhu, Z.; Huang, R. Study on the IOT Architecture and Access Technology. In Proceedings of the 2017 16th International Symposium on Distributed Computing and Applications to Business, Engineering and Science (DCABES), Anyang, China, 13–16 October 2017. [CrossRef]
27. Li, Y.; Tan, D.; Wu, Z.; Zhong, H.; Zu, D. Dynamic stability analyses based on ZMP of a wheel-based humanoid robot. In Proceedings of the 2006 IEEE International Conference on Robotics and Biomimetics, Kunming, China, 17–20 December 2006; pp. 1565–1570. [CrossRef]
28. Yanjie, L.; Zhenwei, W.; Hua, Z. The dynamic stability criterion of the wheel-based humanoid robot based on ZMP modeling. In Proceedings of the 2009 Chinese Control and Decision Conference, Guilin, China, 17–19 June 2009; pp. 2349–2352. [CrossRef]
29. URL-Robot Motion. Available online: https://scaron.info/robot-locomotion/equations-of-motion.html (accessed on 2 May 2022).
30. Rostami, M.; Koushanfar, F.; Karri, R. A primer on hardware security: Models, methods, and metrics. *Proc. IEEE* **2014**, *102*, 1283–1295. [CrossRef]
31. Ariffin, I.M.; Rasidi, A.I.H.M.; Yussof, H.; Mohamed, Z.; Miskam, M.A.; Amin, A.T.M.; Omar, A.R. Sensor Based Mobile Navigation Using Humanoid Robot Nao. *Procedia Comput. Sci.* **2015**, *76*, 474–479. [CrossRef]
32. Ariffin, I.M.; Baharuddin, A.; Atien, A.C.; Yussof, H. Real-Time Obstacle Avoidance for Humanoid-Controlled Mobile Platform Navigation. *Procedia Comput. Sci.* **2017**, *105*, 34–39. [CrossRef]
33. Muzammal, S.M.; Murugesan, R.K.; Jhanjhi, N.Z. A Comprehensive Review on Secure Routing in Internet of Things: Mitigation Methods and Trust-Based Approaches. *IEEE Internet Things J.* **2021**, *8*, 4186–4210. [CrossRef]
34. Munawar, A.; De Magistris, G.; Pham, T.H.; Kimura, D.; Tatsubori, M.; Moriyama, T.; Tachibana, R.; Booch, G. MaestROB: A Robotics Framework for Integrated Orchestration of Low-Level Control and High-Level Reasoning. In Proceedings of the 2018 IEEE International Conference on Robotics and Automation (ICRA), Brisbane, QLD, Australia, 21–25 May 2018; pp. 527–534. [CrossRef]
35. Burghart, C.; Mikut, R.; Stiefelhagen, R.; Asfour, T.; Holzapfel, H.; Steinhaus, P.; Dillmann, R. A cognitive architecture for a humanoid robot: A first approach. In Proceedings of the 5th IEEE-RAS International Conference on Humanoid Robots, Tsukuba, Japan, 5 December 2005; pp. 357–362. [CrossRef]
36. Pessoa, L. Do Intelligent Robots Need Emotion? *Trends Cogn. Sci.* **2017**, *21*, 817–819. [CrossRef]
37. Zaraki, A.; Pieroni, M.; De Rossi, D.; Mazzei, D.; Garofalo, R.; Cominelli, L.; Dehkordi, M.B. Design and evaluation of a unique social perception system for human-robot interaction. *IEEE Trans. Cogn. Dev. Syst.* **2017**, *9*, 341–355. [CrossRef]
38. Chen, F.; Cao, L.; Tian, M.; Du, G. Research and Improvement of Competitive Double Arm Wheeled Humanoid Robot. In Proceedings of the 2020 IEEE 3rd International Conference on Information Systems and Computer Aided Education (ICISCAE), Dalian, China, 27–29 September 2020; pp. 599–601. [CrossRef]
39. Mordern Robotics. Available online: https://modernrobotics.northwestern.edu/nu-gm-book-resource/2-2-degrees-of-freedom-of-a-robot/ (accessed on 2 May 2022).
40. Mohammadi, V.; Rahmani, A.M.; Darwesh, A.M.; Sahafi, A. Trust-based recommendation systems in Internet of Things: A systematic literature review. *Hum. Cent. Comput. Inf. Sci.* **2019**, *9*, 21. [CrossRef]

41. Liu, R.; Yu, G.; Qu, F.; Zhang, Z. Device-to-Device Communications in Unlicensed Spectrum: Mode Selection and Resource Allocation. *IEEE Access* **2016**, *4*, 4720–4729. [CrossRef]
42. Razafimandimby, C.; Loscri, V.; Vegni, A.M. A neural network and IoT based scheme for performance assessment in Internet of Robotic Things. In Proceedings of the 2016 IEEE first international conference on internet-of-things design and implementation (IoTDI), Berlin, Germany, 4–8 April 2016. [CrossRef]
43. Möller, R.; Furnari, A.; Battiato, S.; Härmä, A.; Farinella, G.M. A survey on human-aware robot navigation. *Rob. Auton. Syst.* **2021**, *145*, 103837. [CrossRef]
44. Kim, D.; Kim, S.; Park, J.H. Remote Software Update in Trusted Connection of Long Range IoT Networking Integrated with Mobile Edge Cloud. *IEEE Access* **2018**, *6*, 66831–66840. [CrossRef]
45. Zhu, C.; Rodrigues, J.J.P.C.; Leung, V.C.M.; Shu, L.; Yang, L.T. Trust-based communication for the industrial internet of things. *IEEE Commun. Mag.* **2018**, *56*, 16–22. [CrossRef]
46. Afanasyev, I.; Mazzara, M.; Chakraborty, S.; Zhuchkov, N.; Maksatbek, A.; Yesildirek, A.; Kassab, M.; Distefano, S. Towards the Internet of Robotic Things: Analysis, Architecture, Components and Challenges. In Proceedings of the 2019 12th International Conference on Developments in eSystems Engineering (DeSE), Kazan, Russia, 7–10 October 2019; pp. 3–8. [CrossRef]
47. Rana, B. A systematic survey on internet of things: Energy efficiency and interoperability perspective. *Trans. Emerg. Telecommun. Technol.* **2020**, *32*, e4166. [CrossRef]
48. Kumar, J.S. A Survey on Internet of Things: Security and Privacy Issues. *Int. J. Comput. Appl.* **2014**, *90*, 20–26.
49. Mukherjee, A. Physical-Layer Security in the Internet of Things: Sensing and Communication Confidentiality Under Resource Constraints. *Proc. IEEE* **2015**, *103*, 1747–1761. [CrossRef]
50. Dizdarević, J.; Carpio, F.; Jukan, A.; Masip-Bruin, X. A survey of communication protocols for internet of things and related challenges of fog and cloud computing integration. *ACM Comput. Surv.* **2019**, *51*, 1–29. [CrossRef]
51. Jeong, J.; Yang, J.; Baltes, J. Robot magic show as testbed for humanoid robot interaction. *Entertain. Comput.* **2022**, *40*, 100456. [CrossRef]
52. Althumali, H.; Othman, M.; Member, S. A Survey of Random Access Control Techniques for Machine-to-Machine Communications in LTE/LTE-A Networks. *IEEE Access* **2018**, *6*, 74961–74983. [CrossRef]
53. Tsai, W.; Wang, T. An Automatic Key-update Mechanism for M2M Communication and IoT Security Enhancement. In Proceedings of the 2020 IEEE International Conference on Smart Internet of Things (SmartIoT), Beijing, China, 14–16 August 2020; pp. 354–355. [CrossRef]
54. Batet, M.; Gibert, K.; Valls, A. The data abstraction layer as knowledge provider for a medical multi-agent system. In *AIME Workshop on Knowledge Management for Health Care Procedures*; Springer: Berlin/Heidelberg, Germany, 2007; pp. 87–100. [CrossRef]
55. Hendrich, N.; Bistry, H.; Zhang, J. Architecture and Software Design for a Service Robot in an Elderly-Care Scenario. *Engineering* **2015**, *1*, 27–35. [CrossRef]
56. Ankele, R.; Marksteiner, S.; Nahrgang, K. Requirements and Recommendations for IoT/IIoT Models to automate Security Assurance through Threat Modelling, Security Analysis and Penetration Testing. In Proceedings of the 14th International Conference on Availability, Reliability and Security, Canterbury, UK, 26–29 August 2019.
57. Kumar, N.; Vámossy, Z.; Szabó-Resch, Z.M. Robot Path Pursuit Using Probabilistic Roadmap. In Proceedings of the 2016 IEEE 17th International Symposium on Computational Intelligence and Informatics (CINTI), Budapest, Hungary, 17–19 November 2016; pp. 139–144.
58. Kumar, N.; Vámossy, Z.; Szabó-Resch, Z.M. Robot Obstacle Avoidance Using Bumper Event. In Proceedings of the 2016 IEEE 11th International Symposium on Applied Computational Intelligence and Informatics (SACI), Timisoara, Romania, 12–14 May 2016; pp. 485–490.
59. Kishi, T.; Shimomura, S.; Futaki, H.; Yanagino, H.; Yahara, M.; Cosentino, S.; Nozawa, T.; Hashimoto, K.; Takanishi, A. Development of a Humorous Humanoid Robot Capable of Quick-and-Wide Arm Motion. *IEEE Robot. Autom. Lett.* **2016**, *1*, 1081–1088. [CrossRef]
60. Ravankar, A.; Ravankar, A.A.; Kobayashi, Y.; Hoshino, Y.; Peng, C.C. Path smoothing techniques in robot navigation: State-of-the-art, current and future challenges. *Sensors* **2018**, *18*, 3170. [CrossRef]
61. Kumar, N.; Takács, M.; Vámossy, Z. Robot Navigation in Unknown Environment using Fuzzy Logic. In Proceedings of the 2017 IEEE 15th International Symposium on Applied Machine Intelligence and Informatics (SAMI), Herl'any, Slovakia, 26–28 January 2017; pp. 279–284.
62. Rath, A.K.; Parhi, D.R.; Das, H.C.; Muni, M.K.; Kumar, P.B. Analysis and use of fuzzy intelligent technique for navigation of humanoid robot in obstacle prone zone. *Def. Technol.* **2018**, *14*, 677–682. [CrossRef]
63. Muni, M.K.; Parhi, D.R.; Kumar, P.B.; Sahu, C.; Kumar, S. Towards motion planning of humanoids using a fuzzy embedded neural network approach. *Appl. Soft Comput.* **2022**, *119*, 108588. [CrossRef]
64. Kashyap, A.K.; Parhi, D.R.; Pandey, A. Multi-objective optimization technique for trajectory planning of multi-humanoid robots in cluttered terrain. *ISA Trans.* **2021**, *125*, 591–613. [CrossRef]
65. Kumar, N.; Vámossy, Z. Laser Scan Matching in Robot Navigation. In Proceedings of the 2018 IEEE 12th International Symposium on Applied Computational Intelligence and Informatics (SACI), Timisoara, Romania, 17–19 May 2018; pp. 241–246.
66. Kastner, L.; Lambrecht, J. Augmented-Reality-Based Visualization of Navigation Data of Mobile Robots on the Microsoft Hololens—Possibilities and Limitations. In Proceedings of the 2019 IEEE International Conference on Cybernetics and Intelligent

Systems (CIS) and IEEE Conference on Robotics, Automation and Mechatronics (RAM), Bangkok, Thailand, 18–20 November 2019; pp. 344–349. [CrossRef]
67. Oh, H.S.; Lee, C.W.; Mitsuru, I. Navigation control of a mobile robot based on active vision. In Proceedings of the IECON'91: 1991 International Conference on Industrial Electronics, Control and Instrumentation, Kobe, Japan, 28 October–1 November 1991; Volume 2, pp. 1122–1126. [CrossRef]
68. Kumar, N.; Vámossy, Z.; Szabó-resch, Z.M. Heuristic Approaches in Robot Navigation. In Proceedings of the 2016 IEEE 20th Jubilee International Conference on Intelligent Engineering Systems (INES), Budapest, Hungary, 30 June–2 July 2016; pp. 219–222.
69. Tang, L. Vision Based Navigation for Mobile Robots in Indoor Environment bly Teaching and Playing-back Scheme. In Proceedings of the 2001 ICRA, IEEE International Conference on Robotics and Automation, Seoul, Korea, 21–26 May 2001; pp. 3072–3077.
70. Al-Mutib, K. Smart stereovision based gaze control for navigation in low-feature unknown indoor environments. In Proceedings of the 2014 5th International Conference on Intelligent Systems, Modelling and Simulation, Langkawi, Malaysia, 27–29 January 2014; 2015; Volume 2015, pp. 121–126. [CrossRef]
71. Yoo, J.K.; Kim, J.H. Gaze Control-Based Navigation Architecture with a Situation-Specific Preference Approach for Humanoid Robots. *IEEE/ASME Trans. Mechatron.* **2015**, *20*, 2425–2436. [CrossRef]
72. Adachi, Y.; Tsunenari, H.; Matsumoto, Y.; Ogasawara, T. Guide robot's navigation based on attention estimation using gaze information. In Proceedings of the 2004 IEEE/RSJ International Conference on Intelligent Robots and Systems (IROS), Sendai, Japan, 28 September–2 October 2004; Volume 1, pp. 540–545. [CrossRef]
73. Awan, K.A.; Ud, I.; Senior, D.I.N.; Almogren, A.; Member, S.; Fellow, M.G.; Khan, S. StabTrust—A Stable and Centralized Trust-based Clustering Mechanism for IoT enabled Vehicular Ad-hoc Networks. *IEEE Access* **2020**, *8*, 21159–21177. [CrossRef]
74. Zeeshan, N.; Member, M.R. Three-way Security Framework for Cloud based IoT Network. In Proceedings of the 2019 International Conference on Computing, Electronics & Communications Engineering (iCCECE), London, UK, 22–23 August 2019; pp. 183–186.
75. Iqbal, W.; Abbas, H.; Daneshmand, M.; Rauf, B.; Abbas, Y. An In-Depth Analysis of IoT Security Requirements, Challenges and their Countermeasures via Software Defined Security. *IEEE Internet Things J.* **2020**, *7*, 10250–10276. [CrossRef]
76. Wu, A.; Guo, J.; Yang, P. Research on Data Sharing Architecture for Ecological Monitoring Using Iot Streaming Data. *IEEE Access* **2020**, *8*, 195385–195397. [CrossRef]
77. Rostami, M.; Koushanfar, F.; Rajendran, J.; Karri, R. Hardware security: Threat models and metrics. In Proceedings of the 2013 IEEE/ACM International Conference on Computer-Aided Design (ICCAD), San Jose, CA, USA, 18–21 November 2013; pp. 819–823. [CrossRef]
78. Hussain, S.; Erwin, H.; Dunne, P. Threat modeling using Formal Methods: A New Approach to Develop Secure Web Applications. In Proceedings of the 2011 7th International Conference on Emerging Technologies, Islamabad, Pakistan, 5–6 September 2011.
79. Bradbury, M.; Jhumka, A.; Watson, T.I.M.; Burton, J.; Butler, M.; Data, M.M. Threat-modeling-guided Trust-based Task Offloading for Resource-constrained Internet of Things. *ACM Trans. Sens. Netw.* **2022**, *18*, 1–41. [CrossRef]
80. Maciel, R.; Araujo, J.; Dantas, J.; Melo, C.; Guedes, E.; Maciel, P. Impact of a DDoS Attack on Computer Systems: An Approach Based on an Attack Tree Model. In Proceedings of the 2018 Annual IEEE International Systems Conference (SysCon), Vancouver, BC, Canada, 23–26 April 2018.
81. Fei, Y.; Ning, J.; Jiang, W. A quantifiable Attack-Defense Trees model for APT attack. In Proceedings of the 2018 IEEE 3rd Advanced Information Technology, Electronic and Automation Control Conference (IAEAC), Chongqing, China, 12–14 October 2018; pp. 2303–2306. [CrossRef]
82. Trček, D. A formal apparatus for modeling trust in computing environments. *Math. Comput. Model.* **2009**, *49*, 226–233. [CrossRef]
83. Liao, B.I.N.; Ali, Y.; Nazir, S. Security Analysis of IoT Devices by Using Mobile Computing: A Systematic Literature Review. *IEEE Access* **2020**, *8*, 120331–120350. [CrossRef]
84. Kouicem, D.E.; Bouabdallah, A.; Lakhlef, H.P.T. Internet of things security: A top-down survey. *Comput. Netw.* **2018**, *141*, 199–221. [CrossRef]
85. Robles-Ramirez, D.A.; Escamilla, P.J.; Tryfonas, T. IoTsec: UML extension for Internet of things systems security modelling. In Proceedings of the 2017 International Conference on Mechatronics, Electronics and Automotive Engineering (ICMEAE), Cuernavaca, Mexico, 21–24 November 2017. [CrossRef]
86. Huang, X.; Yu, R.; Kang, J. Distributed Reputation Management for Secure and Efficient Vehicular Edge Computing and Networks. *IEEE Access* **2017**, *5*, 25408–25420. [CrossRef]
87. Tandon, A.; Srivastava, P. Trust-based Enhanced Secure Routing against Rank and Sybil Attacks in IoT. In Proceedings of the 2019 Twelfth International Conference on Contemporary Computing (IC3), Noida, India, 8–10 August 2019; pp. 1–7. [CrossRef]
88. Nagarajan, U.; Kantor, G.; Hollis, R. The ballbot: An omnidirectional balancing mobile robot. *Int. J. Rob. Res.* **2014**, *33*, 917–930. [CrossRef]
89. Gurunath, R.; Agarwal, M.; Nandi, A.; Samanta, D. An Overview: Security Issue in IoT Network. In Proceedings of the 2018 2nd International Conference on I-SMAC (IoT in Social, Mobile, Analytics and Cloud) (I-SMAC), Palladam, India, 30–31 August 2018; pp. 1–4.
90. Shiomi, M.; Shatani, K.; Minato, T.; Ishiguro, H. How Should a Robot React before People's Touch?: Modeling a Pre-Touch Reaction Distance for a Robot's Face. *IEEE Robot. Autom. Lett.* **2018**, *3*, 3773–3780. [CrossRef]
91. Aucouturier, J.-J. Cheek to Chip: Dancing Robots and AI's Future. *IEEE Intell. Syst.* **2008**, *23*, 74–84. [CrossRef]

92. Velásquez, J.D. When robots weep: Emotional memories and decision-making. In *Proceedings of the Fifteenth National Conference on Artificial Intelligence (AAAI-98)*; The AAAI Press: Menlo Park, CA, USA, 1998; pp. 70–75. Available online: https://www.aaai.org/Papers/AAAI/1998/AAAI98-010.pdf (accessed on 2 May 2022).
93. Dorner, D.; Hille, K. Artificial Souls: & Motivated Emotional Robots. In Proceedings of the 1995 IEEE International Conference on Systems, Man and Cybernetics. Intelligent Systems for the 21st Century, Vancouver, BC, Canada, 22–25 October 1995; pp. 3828–3832.
94. Hornung, A.; Wurm, K.M.; Bennewitz, M. Humanoid robot localization in complex indoor environments. In Proceedings of the 2010 IEEE/RSJ International Conference on Intelligent Robots and Systems, Taipei, Taiwan, 18–22 October 2010; pp. 1690–1695. [CrossRef]
95. Taylor, C.; Ward, C.; Sofge, D.; Lofaro, D.M. LPS: A Local Positioning System for Homogeneous and Heterogeneous Robot-Robot Teams, Robot-Human Teams, and Swarms. In Proceedings of the LPS: A Local Positioning System for Homogeneous and Heterogeneous Robot-Robot Teams, Robot-Human Teams, and Swarms, Jeju, Korea, 24–27 June 2019; pp. 200–207. [CrossRef]
96. Huang, K.; Xian, Y.; Zhen, S.; Sun, H. Robust control design for a planar humanoid robot arm with high strength composite gear and experimental validation. *Mech. Syst. Signal Process.* **2021**, *155*, 107442. [CrossRef]

future internet

Review

Redactable Blockchain: Comprehensive Review, Mechanisms, Challenges, Open Issues and Future Research Directions

Shams Mhmood Abd Ali [1,2], Mohd Najwadi Yusoff [1,*] and Hasan Falah Hasan [1,3]

1 School of Computer Science, Universiti Sains Malaysia, Gelugor 11800, Penang, Malaysia
2 College of Literature, Aliraqia University, Hayba Katoon, Street 22, Avenue 308, Adhamyh, Baghdad 7366, Iraq
3 College of Engineering, Aliraqia University, Hayba Katoon, Street 22, Avenue 308, Adhamyh, Baghdad 7366, Iraq
* Correspondence: najwadi@usm.my

Abstract: The continuous advancements of blockchain applications impose constant improvements on their technical features. Particularly immutability, a highly secure blockchain attribute forbidding unauthorized or illicit data editing or deletion, which functions as crucial blockchain security. Nonetheless, the security function is currently being challenged due to improper data stored, such as child pornography, copyright violation, and lately the enaction of the "Right to be Forgotten (RtbF)" principle disseminated by the General Data Protection Regulation (GDPR), where it requires blockchain data to be redacted to suit current applications' urgent demands, and even compliance with the regulation is a challenge and an unfeasible practice for various blockchain technology providers owing to the immutability characteristic. To overcome this challenge, mutable blockchain is highly demanded to solve previously mentioned issues, where controlled and supervised amendments to certain content within constrained privileges granted are suggested by several researchers through numerous blockchain redaction mechanisms using chameleon and non-chameleon hashing function approaches, and methods were proposed to achieve reasonable policies while ensuring high blockchain security levels. Accordingly, the current study seeks to thoroughly define redaction implementation challenges and security properties criteria. The analysis performed has mapped these criteria with chameleon-based research methodologies, technical approaches, and the latest cryptographic techniques implemented to resolve the challenge posed by the policy in which comparisons paved current open issues, leading to shaping future research directions in the scoped field.

Keywords: blockchain; security; distributed ledger; immutability; redaction mechanism; chameleon hash

Citation: Abd Ali, S.M.; Yusoff, M.N.; Hasan, H.F. Redactable Blockchain: Comprehensive Review, Mechanisms, Challenges, Open Issues and Future Research Directions. *Future Internet* **2023**, *15*, 35. https://doi.org/10.3390/fi15010035

Academic Editors: Christoph Stach and Clémentine Gritti

Received: 27 November 2022
Revised: 27 December 2022
Accepted: 31 December 2022
Published: 12 January 2023

1. Introduction

Blockchain technology has recently acquired significant attention from academicians and the industry as evidenced by multiple applications, including copyright dispute resolution [1], product traceability [2], electronic voting [3], storage services [4], healthcare services [5], product tracking throughout the entire supply chain [6], and data management [7] in the internet-of-things (IoT) [8]. Prior researchers [9] predicted global expenditures on blockchain solutions to expand from 1.5 billion dollars in 2018 to 15.9 billion dollars annually by 2023. Blockchain technology provides pertinent data solutions through decentralization, collection, storage, and processing by recording transactions into respective blocks, which are appended to one another through cryptology to provide high security and validity levels. Specifically, each block is equipped with a reference number, known as a hash, to be attached to the subsequent block [10,11] to maintain technical immutability, which requires all attached blocks to be edited if a specific block is to be amended [12]. Blockchain functions as a distributed database, which is regulated by a peer-to-peer network wherein all stakeholders (nodes) comply with a stipulated software protocol (consensus) to communicate and validate existing records or blocks [13,14]. With every node possessing all copies of other blocks without centralized and official replicas, blockchain serves as a 'trustless' system without a third party to validate every performed transaction and facilitate interactions between corporations and clients [15,16]. Immutability is a fundamental attribute of conventional blockchain technology to maintain high data integrity, which prevents transaction data modification while expediting the data auditing process [17,18]. Notwithstanding, establishing procedures to authorize blockchain data redaction is also essential owing to three factors. Correspondingly, immutability might be abused for malevolent motives, such as illegal information storage and dissemination [19]. In [20], eight records related to sexual content were discovered in the Bitcoin blockchain, of which two comprised 274 child pornography website links, with 142 associated with darknet services. Therefore, the available arbitrary data storage is exploited due to child pornography, improper content, and illicit material violating intellectual rights [21]. With the misusage, users might be reluctant to adopt blockchain technology to avoid unintentional possession of illicit or inappropriate digital content and law violation. If the negative situation persists, thousands of blockchain nodes might be affected, consequently diminishing the Bitcoin ecosystem and functionality. Hence, continuously appending the latest digital information would not be feasible without illegal content being removed from the blockchain, which poses an essential prerequisite to further adoption by being obligatory for law enforcement agencies, including Interpol [22]. Another rationale for requiring the implementation of a redactable blockchain is triggered by imposed laws and regulations. According to the European General Data Protection Regulation (GDPR) [23,24], users' data contain the "right to be forgotten", which allows all users to remove personal details and relevant duplicates while simultaneously amending personal data anonymously attached to former blocks [25]. Furthermore, blockchain immutability might not be pertinent to emerging blockchain-based applications, which adamantly request a certain flexibility level for data redaction. Examples include, but are not restricted to, the data stored on the blockchain, that may concern users' confidential and sensitive information, such as healthcare and insurance records [26,27]. The users might prefer removing sensitive details from the platform while updating the information if required, including the contract amendment service, and deleting redundant data to free more space in the IoT-based blockchain systems [28,29].The current study would allow blockchain developers and researchers to capture a holistic view and ease establishing future blockchain redaction. Summarily, this study contributed to the existing knowledge corpus by reviewing present redaction techniques utilizing the chameleon and non-chameleon hashing function, analyzing blockchain security and performance features thoroughly, outlining blockchain information redaction challenges, and differentiating redactable blockchain systems comprehensively before providing open issues and future directions. The following content of this paper is organized as follows: the present study scrutinizes in Section 2 the blockchain technology research model with the respective characteristics elucidated. Section 3 represents blockchain construct features discussed in depth. Section 4 illustrates blockchain types. Section 5 thoroughly discusses the security properties of blockchain technology before examining multiple design challenges for redactable blockchain in Section 6. In Section 7, approaches and mechanisms are primarily illustrated, analyzed, and heavily discussed. Section 8 represents comparisons to the state of the art discussed in Section 7. Open issues and future research directions are explained in Section 9, which draws the path for researchers to further investigations. Section 10 contains conclusions and, finally, Appendix A contains Table A1 that contains a list of acronyms.

2. Blockchain Technology

Satoshi Nakamoto introduced Bitcoin as an innovative, decentralized, and peer-to-peer cryptocurrency system founded in 2008 [30], which seeks to develop a trustless yet credible financial platform to facilitate transactions through Bitcoins without existing financial institutions or third parties as intermediaries. The fundamental technology is blockchain, which harnesses the strength of peer-to-peer networks, Merkle trees, consensus, and public-key cryptology to maintain high assurance and validity degrees of every transaction and resolve multiple issues [31]. Essentially, a blockchain could be defined as a decentralized, transparent, and transactional database shared among all network nodes (computers) to smoothen the exchange process of various values and media [32,33]. A conventional database, either relational or object-oriented, employs referential integrity methods to maintain satisfactory data validity and synchronization across numerous copies. Conversely, a blockchain is equipped with a consensus algorithm that determines the block orders and relevant transactions before being included in the ledger. The algorithm automates conflict resolutions, such as double spending, by accepting solely one value as a valid transaction [34]. The latest transaction present in the blockchain repository would be authenticated by the entire node before including the record in a pool of pending transactions. All pending transactions would subsequently be categorized into respective blocks by miners, which are computers that classify and publish the transactions on the blockchain by competing with peers to complete a cryptographic puzzle in acquiring the publication priority. As the puzzle difficulty elevates gradually, feigning transactions on the public ledger would be significantly unlucrative with all nodes ably revalidating the transactions upon receiving the latest blocks. Hence, the blockchain validation feature ensures the mining process is highly secure and fraud-resistant, which contributes high integrity to the public ledger. Each block is timestamped and connected to the previous blocks through hashing to create a sequence of blocks attached to the genesis block (the first block in the network). Hashing is a cryptographic technique that converts all information with various file sizes to a fixed output, namely the hash [35–37]. Correspondingly, alterations to the original blocks, regardless of minimal or significant, would result in a different hash [38,39]. Specifically, all stored blockchain data would be subject to several verification stages, which render the information virtually irrevocable and immutable, thus forbidding data tampering even by the initial network entities which publish, process, and store the data. Summarily, the process is performed by miners employing software to decode a relevant cryptogram to publish blocks on the blockchain in guaranteeing network governance [40]. A block is constituted of two main parts, header and body, in which the header contains a previous block hash, timestamp, nonce and Merkle root while the body contains a list of transactions stored in the block. The genesis block is the father of the following blocks in the chain in which it does not contain any previous block hash value nonetheless, genesis block hash is generated by SHA256 hash function, and all subsequent blocks can be traced through it. The previous block's header is hashed via a hashing function in order to be stored in the current block and, thus, blocks are linked, chain growth is increased and simultaneously integrity is preserved against tampering. A timestamp indicates the block creation time, whereas nonce is utilized in the block's production and verification. Merkle tree is a binary tree that contains leaf nodes represented by hashed transactions wherein non-leaf nodes result from the concatenation of its hashed child transactions leading to the production of a Merkle root. Merkle root represents a unique single hash value attained from hashing Merkle tree as the advantage gained to ease transaction verification in the block; however, any amendments performed in any transaction will invalidate its hash and hence, a mismatched Merkle root hash value is denoted [41,42], as portrayed in Figure 1.

An eclectic range of consensus protocols is available contingent on respective network sources, including proof-of-work (PoW), proof-of-stake (PoS), proof-of-activity (PoA), Byzantine fault tolerance (BFT), and hybrid BFT algorithms. Particularly, the PoW protocols function by each node independently decoding the PoW cryptogram to create blocks with authentic transactions, wherein the blocks could be validated in the main chain. Furthermore, honest nodes would frequently mine on the main chain, which was selected based on the protocol rules. For instance, the longest sequence is chosen as the main chain in the Bitcoin blockchain protocol, although the PoW protocol and relevant variants pose exorbitant computation costs and low throughput [43,44]. Contrarily, the PoS blocks are produced by stakeholders [45,46], in which a PoS miner's probability of proposing a block is corresponding to the stake value. The PoA protocol [47] is a hybrid approach composed of both PoW and PoS protocols, in which blocks are created by stakeholders associated with a pseudo-random series, with their probabilities in the series equivalent to the possessed stake volumes. Meanwhile, each consensual participant in a BFT protocol could suggest an alternative block, such as a common set of transactions, to be consented to by a group of participants [48,49]. The protocol could accomplish a larger transaction throughput, although a communication overhead explosion might be induced. Summarily, every consensus protocol aims to ensure that the most honest nodes could concur on a unified blockchain history.

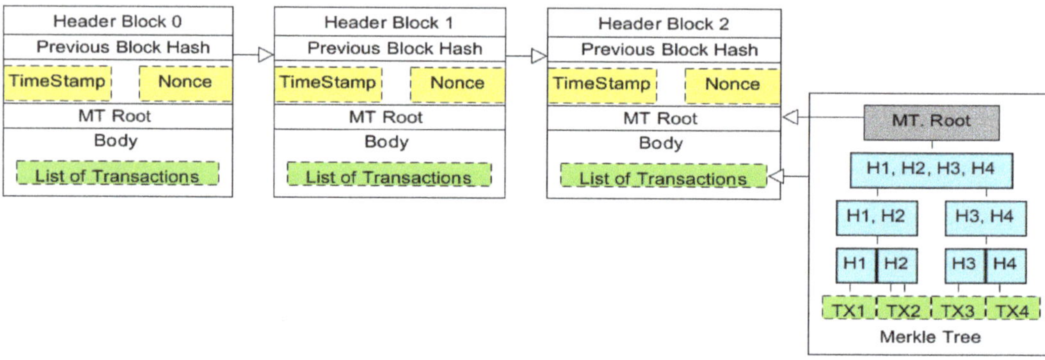

Figure 1. Blockchain construction.

3. Blockchain Construction Features

Different lower-level blockchain technology construct features. In [50,51], decentralization allows the blockchain to serve as a decentralized data repository (ledger) duplicated and disseminated across all users of a specific network. Moreover, neutrality facilitates continuous valid transaction registration in a blockchain regardless of the origination, wherein all individuals with adequate credibility in terms of payments or trust levels could include the transaction records in the public ledger. Concurrently, the replication feature would authorize the consensus models to replicate all valid transactions across all network nodes, while audibility ensures all transactions are publicly displayed and audited due to every performed transaction being appended to the genesis block. Furthermore, integrity is defined as transactions being verified through a hashing algorithm before being published to the distributed ledger. Specifically, the SHA256 algorithm is frequently employed to provide a digital fingerprint which cannot be reverse-engineered [52]. Thus, every amendment could not be executed without invalidating the signature before eventually invalidating the transaction. Anonymity allows users to conduct transactions through pseudonyms generated by respective secret keys rather than actual identities or real-life addresses to prevent exposing personal particulars [53]. Traceability provides detailed records between transactions to trace all transactions in the blockchain and reveal the entire flow of transactions. Simultaneously, immutability issues signatures to guarantee transaction authenticity and integrity, and the Merkel tree (MT) structure maintains an efficient integrity check process, prohibiting malicious block tampering with the hash [54]. As displayed in Figure 2, the transaction TX3 highlighted in red was modified on the nth block, which caused the hash values of the current branch from TX3 to the MT root to become divergent from other nodes' duplicates. Particularly, the header hash of the altered block was also inconsistent with the copy in the (n + 1)th block. As such, the modification was invalid and difficult to be approved without other nodes possessing over 51% hashing power to accept the modification and regenerate an alternative chain from the (n + 1)th block. Resultantly, data immutability is ensured.

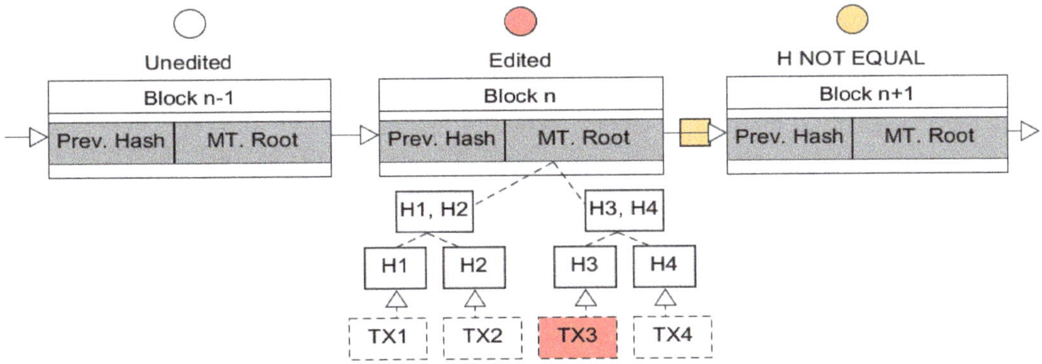

Figure 2. An instance of blockchain immutability.

4. Blockchain Types

Blockchains are categorized into numerous types depending on their respective purposes and distinctive features as illustrated in Table 1. The permissionless or public blockchain does not necessitate platform users to participate in the network [55], which is fully decentralized as the participation is based on the process of consensus and perusal before sharing the transaction record to maintain the distributed ledger [56]. Additional blocks are posted, examined, and verified by all users by possessing a copy of the entire blockchain, which ensures high security in establishment and operation as the blocks are corroborated by computationally challenging consensus procedures, such as decoding cryptograms or investing more personal cryptocurrency. Therefore, any form of data tampering in the blockchain content is prevented by hashes and decentralized concord with other benefits of anonymity and privacy [57]. Nonetheless, permissionless blockchains pose multiple research issues. For example, efficiency is negatively impacted by numerous participants and computationally expensive consensus procedures. Another type is permissioned or private blockchains developed for a sole organization, wherein participants join the network through invitation and are obliged to sustain the blockchain decentralization characteristic [55]. Permissioned blockchains contrast with permissionless blockchains in that only approved entities are permitted to participate in the network and maintain the blocks, which provides higher security and efficiency levels by preventing data tampering through hashes and consensus. Nevertheless, private blockchain nodes are not anonymous [56] as the networks could be breached by internally authorized users. Meanwhile, consortium blockchains serve as private blockchains for various organizations, wherein merely invited and approved users are permitted to participate and support the network. The consensus process is comparatively more time-consuming than in permissioned blockchains, although swifter than in permissionless platforms. In terms of security, consortium blockchains could process data in a more protected manner for averting alterations and hacking activities than permissioned blockchains, owing to the monitoring by different organizations [55].

Table 1. Blockchain types.

Type	Public (Permissionless)	Private (Permissioned)	Consortium
Network	Decentralized	Decentralized Partially	Hybrid Among Public and Private
Access	Any participant	Predefined Participant	Multiple Predefined Participants
Concept	1. Read and write transactions. 2. Vote to add pooled transactions. 3. Validate transactions and consequently they are secured.	1. Conditional read and write operations. 2. Conditional verification. 3. Public read might be allowed.	1. Permission read/write by multi controlling nodes. 2. Controlling nodes selection differ among participated entities. 3. Public read might be allowed.
Consensus	PoW/PoS	Multi party	Multi party
Approval Time	10 Minutes	100 ms	100 ms
Scalability	Slow	Fast, Light	Fast, Light
Security, Privacy	Lack of privacy and anonymity.	High Privacy	High Privacy
Cost	Costive.	Costive.	Costive.
Energy	High Consumption	Low Consumption	Low Consumption
Efficiency	Non-Efficient	Efficient	Efficient
Immutability	Non-tempered	Can Be Tampered	Can Be Tampered
Centralization	No	Yes	Partially
Use Cases	Cryptocurrency	Supply Chain	Banking, Insurance
Application	Bitcoin	Ethereum	Edexa

5. Security Properties

A mature blockchain system should fulfill three security attributes established [58,59]. The security properties guarantee blockchain accuracy, consensus, and validity with a high probability, whereas redaction designs must not impact them. They are classified as follows:

- **Chain Quality:**

  It refers to the ratio of the blocks controlled by an adversary in a chunk of an honest party's chain that cannot exceed a certain fraction X where it represents the total amount of adversary-dominated resources.

- **Common Prefix:**

  Formally common prefix property represents two honest node chains S1 and S2, whereas the shortest chain is a common prefix for the longest chain at different time segments T1 and T2. The common prefix $K \in \mathbb{Z}$ indicates the number of blocks required to be removed from a timely older chain. If $S1 \leq S2$ where $T1 \leq T2$, it means removing k blocks at the end of S1 and becoming a common prefix to S2.

- **Chain Growth:**

  The growth of an honest chain must be compatible with the number of blocks produced regardless of adversarial employed methodology.

The security properties guarantee blockchain accuracy, consensus, and validity with a high probability. Specifically, accuracy maintains the "healthy" degree of a redactable blockchain, while consensus necessitates all honest nodes to concur on a particular sequence. Meanwhile, validity specifies that most blocks emanate from honest nodes [59]. A redaction technique is required to not impact the aforementioned blockchain security properties.

6. Redactable Blockchain Implementation Challenges

Redaction contradicts the principle of blockchain immutability, thus requiring the process to be conducted under stringent rules. Accordingly, the redaction procedure is restricted to authorized personnel and contingent on specific situations without compromising system reliability and constancy. Contemporarily, multitudinous blockchain data redaction approaches and models are developed by suggesting different pertinent methods. Similarly, every relevant redaction approach should prioritize validating all redacted transactions or blocks. Figure 3 presents several redaction challenges to thoroughly investigate and differentiate every proposed model and they are as follows.

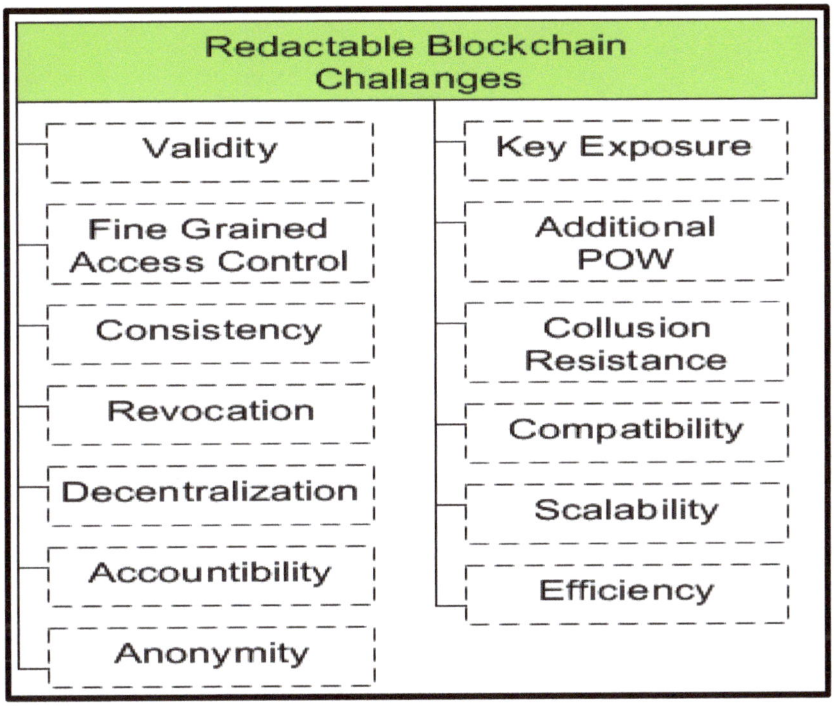

Figure 3. Redactable blockchain implementation challenges.

6.1. Validity

Consensus protocol approved policy must be met and applied by redacted transaction/block to acquire validity. Once redaction is verified, then it is considered honest by other nodes.

6.2. Consistency

Authentic nodes observe the chain consistently. Connected transactions/block validity must not be affected by redaction at any period. Thus, avoiding consistency and traceability damages at some points caused by honest redaction.

6.3. Fine Grained

Who has the right to modify it remains an unsolved question. Fine-grained access control authorizes various rights to diverse users to modify contents, however coarse-grained access control cannot identify who made the certain modification.

6.4. *Decentralization*

Centrality remains a bottleneck, which has been solved by decentralization through transferring power and control to users; consequently, system security is improved as consistency and corruption are fairly reduced.

6.5. *Accountability*

Redacting transactions by authorized users requires certain accountable mechanisms to observe their activities.

6.6. *Revocation*

Access control methods play a vital role in controlling users' privileges where they must act effectively once the user's rights are revoked, left, or degraded without affecting system efficiency in terms of cost and computation overhead.

6.7. *Anonymity*

Anonymity aims to achieve privacy to protect transaction contents via authentication and preserve personal identifications of modifiers from any adversarial activities.

6.8. *Collusion Resistance*

Collusion is the illegal accumulation of rights of cooperated parties who aim to gain unlawful access to data and the system must prohibit these actions.

6.9. *Scalability*

User amount growth must not affect system performance.

6.10. *Efficiency*

Time is the efficiency factor in admitting/ratifying certain redactions.

6.11. *Computation Overhead*

Low cost must be achieved while applying all the above measures.

6.12. *Additional PoW*

Redaction mechanisms at the block level might be required to resolve PoW puzzles again, and hence it leads to being undesirable to miners due to its high hashing power consumption.

6.13. *Key Exposure*

It is an obstacle that prevents CH from being used. Trapdoor keys can be disclosed by finding several collisions for a certain function.

7. Redaction Mechanisms in Blockchain

This section represents comprehensively the taxonomy of two main different categories, namely chameleon and non-chameleon-based redaction mechanisms in blockchain, which involves analyzing emerging challenge-related issues in order to produce feasible perception. Figure 4 shows a detailed discussion in the following subsections.

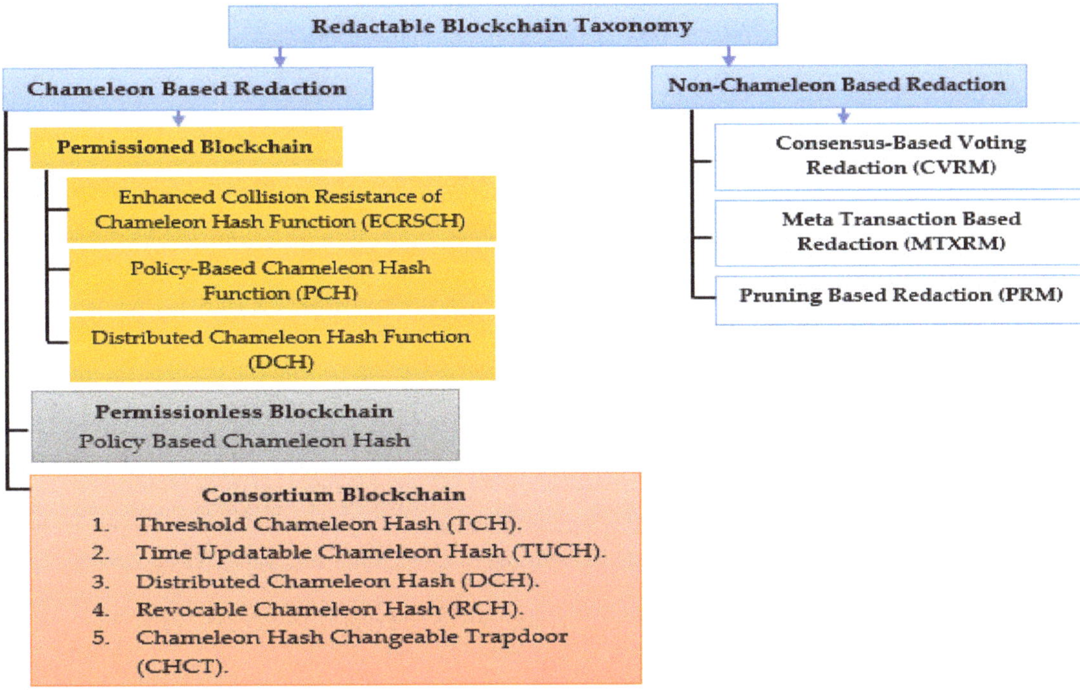

Figure 4. Paper taxonomy analyzing redactable blockchain state of art.

7.1. Redactable Blockchain-Based Chameleon

Redactable blockchains leveraging chameleon hash functions, which contain a trapdoor key, function as a standard cryptographic hash under conventional circumstances. A chameleon hash-based redactable blockchain is collision-resistant, which allows the trapdoor key holder to effortlessly perform data amendments by seeking the same hash as the original version. The holder could be either a centralized entity or belong to a defined set of independent entities. Summarily, a chameleon hash function allows the following procedures to be executed: (i) each function contains a couple of trapdoors and hashing keys, (ii) individuals with hash keys could create the hash function, (iii) trapdoor key holders could locate collisions in the function domain, (iv) the function is collision-resistant for individuals without trapdoor keys [60]. Table 2 provides an overview of all proposed methods discussed below.

Table 2. Chameleon redaction methods outlines.

Schema	Methods	Security Assumption	Setting (Blockchain Type)	Redaction Type
[61], (2017)	CH and PKE	DLP	Private	BL
[62], (2020)	CH and ZR	SXDH	Private	BL
[63], (2020)	CH and ZR	SXDH	Private	BL
[64], (2019)	CH and ABE	DLP	Private	TL
[65], (2020)	CH and ABE and DS	DLP	Private	TL
[66], (2021)	CH and DS	DDH	Private	TL
[67], (2022)	CH and ABE and DS	DLP	Private	TL\BL
[68], (2021)	CH and MA-ABE and DGS	DLIN	Private	TL
[69], (2021)	CH	DLP	Private\Public	TFL
[70], (2021)	CH and ABE and DS	DLIN	Private	TL
[71], (2021)	CH and MA_ABE	DLP	Private	TL
[72], (2022)	CH and ABE and DS	DLP	Private	TL
[73], (2021)	CH and ABE	DBDH	Private	TL
[74], (2021)	CH and ABE and DS	DLP	Private	TL
[75], (2021)	CH and TEE	DLP	Private	BL
[76], (2021)	CH and Lattice	SIS	Private	BL
[77], (2022)	CH	DLP	Private\Public	TL
[78], (2021)	CH and MA_ABE and DS	DLP	Public	TL
[79], (2019)	TCH and DS	CDHP	Consortium	TL
[80], (2019)	RCH	CDHP	Consortium	TL
[81], (2020)	TTCH	CDHP	Consortium	TL
[82], (2020)	CH	DLP	Consortium	TL
[83], (2021)	TCH and DS	CDHP	Consortium	TL
[84], (2021)	TCH and DS	CDHP	Consortium	TL
[85], (2021)	CH	DLP	Consortium	TL
[86], (2022)	CH	CDHP	Consortium	TL

7.1.1. Redactable Permissioned Blockchain

According to the taxonomy, permissioned blockchain redaction mechanisms have been divided into three main sets in which are analyzed and discussed thoroughly as follows:

- Enhanced Collision-Resistance of Chameleon Hash Function (ECRSCH)

Ateniese et al. [61] have proposed the first redactable blockchain with an enhanced standard chameleon hash function (ECRSCH) by incorporating chameleon hash function (CH), public-key cryptography (PKE), and non-zero knowledge proof (NIZKP), which is adopted by Accenture. Every block structure in the blockchain will extend to include an additional field to efficiently record the randomness (r), also termed as a chameleon hash check value, whereas solely trapdoor key holders consequently are able to compute (r′), as the computation of additional redacted block (block n) randomness (r) is highly challenging without a trapdoor key. The randomness maintains the identicality of redacted blocks and header hashes (prior hashes) as before reduction to ensure the blockchain is intact. Figure 5 represents Ateniese's proposal. The approach is limited as the procedure is constrained to rewriting the blockchain based on the block level where it impacts system security and efficiency regardless of mechanism employed, whereas chameleon hash function is utmost risky where the whole block privacy depends on a single trapdoor key for all the transactions maintained; once a modifier granted permission to alter any transaction, the rest can be modified too. Simultaneously, the procedure diminishes transaction validity in the blocks, including other relevant transactions. In addition, coarse-grained decision-making in the function does not define individuals entitled to compute the collision when the hash is frequently produced by the public key.

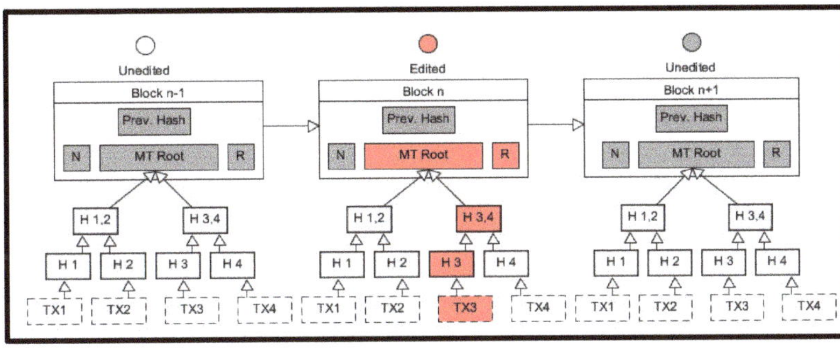

Figure 5. Ateniese's proposal: randomness update in redacted block (edit).

Conversely, secret keys are employed to search for collisions without hash makers being aware of the individual performing the collision. Meanwhile, block traceability is not supported, and accountability is excluded from the chameleon hash function, thus posing the possibility of malicious modifications. With the presence of the trapdoor key and centralized storage, distributed management is not feasible due to the missing decentralization characteristic. Subsequently, Khalili et al. [62] and Derler et al. [63] enhanced the collision resistance feature of the CH corresponding to the Ateniese proposal [61]. However, better efficiency is achieved while similar construction is utilized. Both proposals sought trapdoor exposure prevention; meanwhile, they still suffer from similar Ateniese's proposal drawbacks.

- Policy-based Chameleon Hash Function (PCH)

Authority regulation to redact data in a fine-grained model, Derler et al. [64] proposed the policy-based CH function (PCH) as shown in Figure 6.

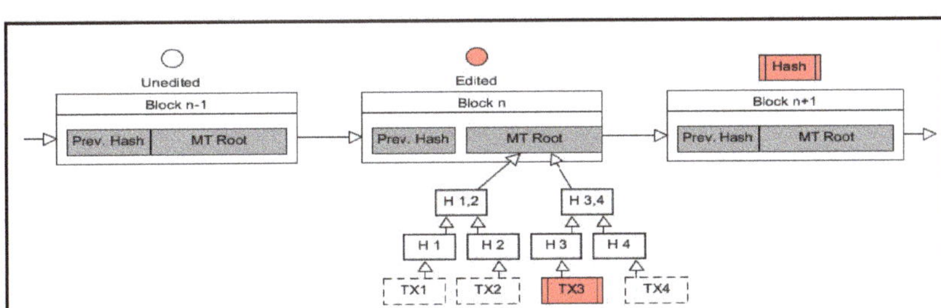

Figure 6. Derler proposal.

It is inspired by the CH with an ephemeral trapdoors (CHET) function [87], and cyphertexts policy attribute-based encryption (CP-ABE) [88] to serve as a hybridization function. The PCH function is associated with the user's attributes to control redaction privileges. The present study fulfilled two main questions by determining the individual authorized to perform a redaction and the specific information that could be redacted.

Specifically, users of private blockchains are permitted to perform redaction activities as their participation is authorized before becoming involved in the network. The CP-ABE provides the PCH function with high indistinguishability, collision, and resistance levels due to the security measures from indistinguishability under the non-adaptive and adaptive chosen cyphertexts attack (IND-CCA2), which prevents attackers from recovering trapdoor keys in conducting double-spending and tampering attacks. Meanwhile, the PCH chain allows transaction-level (TX-level) redaction without negatively impacting the system security, although collusion attacks are highly possible between participants by unifying personal attributes and updating or sharing access to an MT to be consistent with a specific strategy. Nonetheless, non-traceable evidence modifications are a security issue as the changes allowed by the PCH function would not be visible to other users. ABE is not guaranteed to encrypt valid one-time trapdoor keys without NIZK inclusion. The comprises of CHET base RSA in PCH is ineffective in terms of cost and time overhead. Other limitations of Ateniese's proposal [61], including non-accountability, distributed key management, trapdoor centralized storage, and central control of secret keys wherein decentralization remain unresolved, as in Figure 6 showing a delineated Derler's proposal [64]. Tian et al. [65] propounded accountability to be the main contributing solution by proposing the policy-based CH function with black-box accountability (PCHBA) by combining the CHET, the CP-ABE, and the digital signature. The PCHBA is advantageous in prohibiting a transaction modifier to perform malicious rewriting with the accountability property to halt any exploitation form of rewriting privilege. Nonetheless, the PCHBA violates user anonymity as the function supports key generation by central authority, while privilege revocation and decentralization are not considered. PCHBA deprives an effective penalty mechanism to stop malicious redaction activities. As such, Xu et al. [66] proposed the k-time modifiable and epoch-based redactable blockchain (KERB), which consists of a monetary penalty to regulate rewriting privileges and punish malevolent demeanors. The proposal improved the [64] algorithm by ensuring sufficient accountability and traceability to account for the lack of punishment mechanisms and negligence of collusion attack possibility. The KERB comprises the central authority (CA), the miner, and the modifier. The modifier sends a key creation request, which includes the transaction to be modified, the required modification, and the key expiry time. The CA approves both the modifier's identity and request by signing the request with a conditional deposit to produce a token key before being sent to the modifier. A violated modifier triggers punishment by withdrawing agreement on the deposit. Meanwhile, it is committed to any conducted modification approvals wherein modifications are not stored immediately to facilitate checking by the miner. Only approved transactions will be stored in the blockchain public ledger. Nevertheless, security weakening, scalability and flexibility are less common due to CA centralization. Concurrently, computation overhead and malicious modifiers' privileges are not minimized as the deposit amount might be smaller than the damage cost. The authors in [67] have followed the previous suggestion by proposing a blockchain rewriting approach (CDEdit) that allows modifiers to acquire controllable redaction privileges and diversified editing types based on [66] frameworks. The CDEdit model, which is identified as a privilege token service (PTS), authorizes multilevel editing in on-chain permissioned blockchains, such as Hyperledger, to manage and disseminate redaction prerogatives. The entire process is triggered by modifiers upon sending a rewriting request. The modifiers are divided into four main categories, including one-time modifiers on the transaction and block level, and multi-time modifiers on the transaction and block level, without producing conflicts. Specifically, the process commences with requesting a token before receiving a token key (TK) which comprises the request of required rewriting times, expiry times, object indexes, and modifier public key (MPK) issued by the PTS. Before sending the TK to the modifier, a deposit is necessary to account for potential malicious actions after the PTS validates and signs the modifier's request. The PTS remains vulnerable due to centralization, which would negatively affect scalability with the presence of latency and computation overhead, and the absence of an effective punishment mechanism. Panwar et al. [68] adapted the [64] model by developing a revocable and traceable blockchain rewriting framework (ReTRACe) from the revocable CHET (RCHET) and revocable attribute-based encryption (RABE). Two primary dissimilarities exist wherein the RCHET hashing algorithm is a private procedure permitting only trapdoor holders to generate the CHs and subsequently conduct relevant adaptation, hence being inconsistent with the CH definition. Meanwhile, the hashing algorithm is intended to be a public procedure in the RABE to allow attribute revocation. Decentralization is attained via replacing ABE with MA-ABE; nonetheless, a public RABE is unrealistic as a user is required to acquire another decryption key through a secure medium when the user's attribute is repealed by the attribute authority (AA), although the group manager's (GM) attributes could not be revoked. Subsequently, Jia et al. [69] developed a fine-grained blockchain rewriting mechanism to support hierarchical revocation via two stages. The first two data types in a mutable transaction, which were personal information (required to be recorded in a chain) and security-related details (senders, receivers, and transaction amounts), were considered. Users would be authorized by a semi-trusted regulator to amend existing personal records and modify security-related data. As the entire process concentrated on certain mutable transaction fields, the procedures would be more precise compared to the block-level and transaction-level blockchain redaction. Simultaneously, the regulator possessing a master key could repeal a user's rewriting privilege by generating an alternative subkey to disable the original key. Notwithstanding, the model did not establish a decentralization concept or indistinguishability. Xu et al. [70] established a revocable policy CH function (RPCH) adapted from the PCH [64] to aid in fine-grained and revocable blockchain rewriting approaches. The RPCH development referred to CHET based on the Rivest, Shamir, and Adleman (RSA) [89] proposal with the RABE method [90], and the fast attribute-based message encryption (FAME) process [91]. The RPCH also did not include a blockchain decentralization feature in its design via permitting a trusted party to practically repeal a chameleon trapdoor possessor's redaction prerogative, outsource the decoding process, and transmit confidentiality without another semi-trusted entity however CHET performs based on RSA, which affects scalability issues. Meanwhile, Zhang et al. [71] and Ma et al. [72] proposed a multi-authority policy-based CH (MAPCH) function, wherein the CA in the [64] model is replaced with multiple authorities to manage the attributes applied to a permissioned blockchain. The performance in both proposals has been improved due to the decentralized technique proposed, however the reliance on RSA remains a scalability drawback and the absence of revocation requires further investigation. Guo et al. [73] suggested a hybrid blockchain rewriting process both online and offline via an auditable outsource computation (OO-RB-AOC) scheme, which was adapted from [64]. The hybrid model is an enhancement to produce two PCH types, namely online and offline. Particularly, high-value transaction operations could be performed offline to provide adequate credibility by incorporating multiple authorities to generate rewriting keys. Beneficiaries who acquired the keys would be verified on whether the keys were issued by legitimate authorities. The previous model has been built based on ring signature [92] and is employed to create a rewriting key in which external resources could be continuously applied due to the limited device capacity. Nevertheless, the limitations include the lack of a pertinent mechanism for revocation and misbehavior observation. Meanwhile, Hou et al. [74] established a fine-grained and governable redactable blockchain model to allow destructive information forced erasure. Specifically, the transaction creator possesses a delicate regulation over selective individuals to conduct data redaction by specifying the components to be edited, after receiving sufficient miners' votes. All miners could also disclose the block index with detrimental data generated by malevolent users to be subsequently recompensed after authoritatively erasing the deleterious information based on the index. Malevolent users would be prohibited from performing further transactions if the penalty was not paid as miners' compensation within a stipulated period. Moreover, the fine-grained framework facilitates additional information and unexpended transaction output (UTXO) redaction concurrently. Nonetheless, revocation is unavailable for any misdemeanors and decentralization is not recognized with the presence of a trusted authority, significantly influencing the model's performance.

- Distributed Chameleon Hash

Liu et al. [75] have proposed the inclusion of the distributed key management function by using distributed CH function and trusted execution environment (TEE) [93]. Distributed key management is employed to guarantee high trapdoor security levels, which could be strengthened by the TEE to elevate asset confidentiality and integrity degree in a secure TEE enclave. Nevertheless, as the rewriting procedure is performed throughout the blockchain level, weak security is contributed by the exclusion of collusion, accountability, and tractability. Wu et al. [76] further improved the existing CH function. In the present study, the development was based on enhancing the collision resistance characteristic which was perceived to be highly vital to managing redactable blockchain. Particularly, quantum-resistant key-exposure-free CH functions via the concrete single trapdoor were the foundation to optimize the existing CH function based on the proposed framework. Two operations were included, namely distributing trapdoor keys on a semi-honest secure part and voting on the redaction to provide public accountability. However, collusion is not prevented. Matzutt et al. [77] proposed Redact Chain, which comprises several parallel local jury committees that are rotated periodically to perform redactions instead of utilizing external tools. The committee includes solely authorized members who can conduct redactions jointly by employing the CH function and threshold cryptographic method. Each committee will be located at a randomly selected node to act immediately when certain content is reported. Disputed transaction cases are solved by the jury committees voting in an off-chain mode to approve the modifications. The CH function is applied to grant redaction authority while limiting the possessed authority from being abused. Keys will be distributed amongst the jury members respectively to equalize the trapdoor key control. All modifications could only be conducted once before the committee is replaced by another batch selected from recently successful miners. As such, Redact Chain supports global redaction to organize respective mining activities. Nonetheless, attackers could dominate the jury committee as no accountability mechanism is enforced on the committee.

7.1.2. Redactable Permissionless Blockchain

Developing a redactable public blockchain protocol is challenging yet engaging as the accomplishment requires highly sophisticated solutions. Correspondingly, researchers in [65,74] stated that their proposed redactable mechanisms would be realized in the permissionless blockchain as their future work. In 2021, Tian et al. [78] created an enhancement protocol by expanding the current blockchain rewriting ability with a fine-grained access control policy in permissionless blockchains, including Bitcoin and Ethereum, which required no trusted parties to provide access privileges. To achieve strong security, dynamic proactive secrete sharing (DPSS) as a decentralized set of procedures were implemented to remove all forms of trusted parties, while a committee of several users was established with each user holding a segment of trust. All users are authorized to join the committee whenever available to utilize the KP-ABE to safeguard fine-grained access control. The previous author proposes another algorithm that contains two major phases of fine-grained and public traceability. Subkeys are divided from the master key and distributed amongst committee users, which ensures all committee users provide unanimous agreement before granting access privileges requested by modifiers through master key recovery. Any shift in the committee structure would not affect the master key as the algorithm remains intact. Subsequently, public traceability, digital signature, and the KP-ABE, which constitute the second phase, necessitate the modifier to sign a modified transaction with the private key to guarantee the modification is publicly verifiable. The public traceability function is enabled by the attribute-based encryption traceability (ABET) through a set of rewriting privileges produced from the interaction with the access black box to preclude unauthorized access. Simultaneously, the presence of the public key proves that the modifier was authorized to rewrite a block by the committee. Nonetheless, committee members' collusion behavior could not be effectively averted. Similarly, security issues remain present from the high flexibility provided to the modifier and the lack of effective misbehavior punishment mechanisms and revocation methods.

7.1.3. Redactable Consortium Blockchain

Huang et al. [79] demonstrated that [61] blockchain protocol was not optimally applicable to the conventional industrial IoT (IIOT) scenario and, therefore, a redactable consortium blockchain (RCB) was established to increase redacted transaction validity. The RCB concentrates on forfeiting transaction redactions when modifiers are offline while upgrading the CH function as threshold CH (TCH) through linkage with participants' identities (authorized sensors). Authorized sensors with genuine identities could effectively calculate additional randomness assisted by holders with trapdoor keys. By adhering to the hash-and-sign model, the TCH function hashes.

The transactions before being endorsed by an accountable-and-sanitizable chameleon signature (ASCS) algorithm [94] as portrayed in Figure 7. Participants with authentic identities could also assess the same signature to authenticate redacted transactions. Nevertheless, the scalability issue exists owing to the high cost of generating trapdoor key holders' pertinent signatures, which increases proportionally to the number of nodes. Redaction power exploitation could also occur concurrently. Huang et al. [80] developed smart-chained redactions supported by the self-redactable blockchain proposal (SRB), which required no cooperation to perform transaction redaction in the blockchain. The revocable hash function (RCH) optimizes the CH function by linking identities within a cyclic time and being relied on by the revocable chameleon signature (RCS) to generate transactions. Participants are only allowed to perform periodic redactions based on the defined cyclic time in the blockchain. Specifically, private redactions are conducted with trapdoor keys contained, and (X-1) public redaction only if any X transactions are attached beyond the performed redaction. Meanwhile, public reaction is conducted through the signature to facilitate every individual performing transaction authentication. When a trapdoor expiration period is stipulated, the trapdoor will be updated periodically. Nevertheless, due to high cryptographic complexity, the SRB is rarely adopted in IoT devices. In addition, low adoption is due to the SRB and the RCB sharing a fixed expiration period, which renders the trapdoor key management inflexible while allowing the old trapdoor to be continuously utilized. Hence, key exposure would pose security issues. Zhang et al. [81] proposed the reusable and redactable blockchain (Re-chain) for the RCB, which was a redactable and reusable blockchain established by the threshold trapdoor CH (TTCH) function to address blockchain storage issues in accounting for information explosion. The proof-of-concept consensus was also realized when Re-chain achieved maximum storage size to rewrite previous blocks with security and authority, while maintaining block connection safety. Consensus guarantees robustness even with the presence of N-node acceptance (N/2-1) faults in the edge nodes. Furthermore, Re-chain is beneficial for the IIOT which frequently encounters serious storage limitations. Past experiments revealed that the rewriting process was efficient when performance achieved a medium scale within the limit of 30 nodes. Nonetheless, the deficiency of scalability due to high computation overhead would not support user accountability and anonymity on the platform. Time could be defined as a physical entity in terms of hours and days, or as a virtual being in terms of the number of blocks in the blockchain. As such, time is the main factor in generating an efficient key to redact and generate transactions. Correspondingly, LV et al. [82] suggested a decentralized CH function to gain enhanced security, time efficiency, and reduced space overhead in offering higher RCB flexibility. Modifiers would be necessitated to sign redaction requests to be validated by authorized nodes on personal identities by comparing the signatures and redaction requests. The requests are stored in modified transactions while the signature key is stored in the local redaction record. Thus, accountability is maintained as redaction records could be compared with the stored signatures. Nonetheless, an additional storage cost and low anonymity would be the issues. Huang et al. [83] improved the previous work of [79] by elevating the anonymity and scalability degrees with the time-updatable CH (TUCH) and linkable-and-redactable ring signature (LRRS) as theoretical fundamentals to develop a decentralized, scalable and redactable algorithm with sufficient anonymity similar to the SRB. The TUCH enables automatic ring configuration without interacting with other entities when producing an access key. As the chameleon randomness in holding a hash collision is subject to cyclical expiration, redaction could fulfill verification criteria only within a pre-determined period. Meanwhile, the LRRS ensures modifiers anonymously provide personal signatures on a message without trusted parties. The process is completed by forming a ring of users automatically, without the users being informed that their public keys are employed in producing the signature. Nevertheless, increasing scalability would lower the anonymity level owing to the requirement of the signature, which simultaneously renders the process to be costly. Conversely, Gao et al. [84] criticized the [79] TCH proposal after manifesting that the TCH was vulnerable to malicious redactions. Correspondingly, an enhanced TCH was established to elevate the collision n-resistance degree, which was recognized as one-more preimage resistance. The enhanced protocol ensured sufficient resistance even when preimages with the same hash values and identities were exposed. Nonetheless, similar to the RCB, the protocol was seldom adopted in IoT devices due to high cryptographic complexity. Security issues, such as collusion and scalability, were not adequately considered. Zhang et al. [85] discovered that credible industrial blockchain data management could optimize blockchain reducibility by enhancing public key generation through the CH function in three phases, namely off-chain setup, on-chain setup, and data management. During the setup different data types are included in the blockchain where trapdoor partial authenticity verification is conducted and aided by smart contracts to determine accountability levels. Particularly, the off-chain setup deploys blockchain parameters to create trapdoor segments and perform trapdoor distribution to support trapdoor holders. The on-chain setup is executed by employing the redactable blockchain to collect various data before gauging the authenticity degree. Subsequently, data management separation from other transactions is performed by two different blockchain approaches to realize management supervision on transaction redaction, while reducing coupling in the blockchain-based trapdoor recovery accountability mechanism. Smart contracts would verify the validity of trapdoor segments published by the holders, which are concurrently monitored by supervision nodes before jointly approving editing requests. Trapdoor performers would redact information upon receiving approval. Nonetheless, the entire process is costly, which is unsuitable for the IIoT environment. Accordingly, Wei et al. [86] proposed the federal learning model to be applied in the IIOT circumstances, such as the medical blockchain (RMB), by incorporating the CH changeable trapdoor (CHCT) function. When collisions are identified, the CHCT would be updated to govern the process without revealing the trapdoor. The CHCT could be canceled in in all circumstances to avoid key exposure owing to the highly restricted trapdoor contributed by the owner's autonomy to define the expiration time.

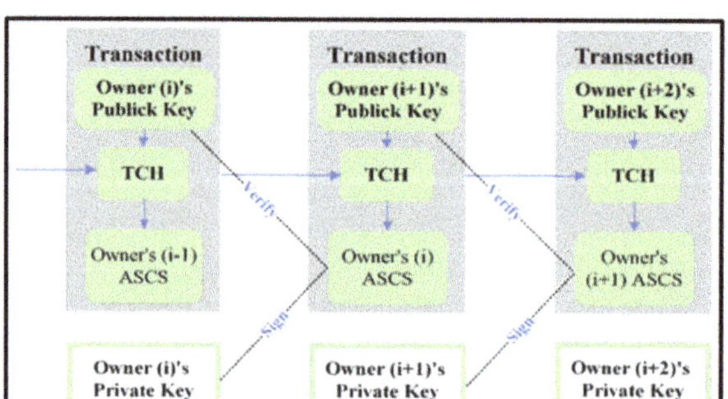

Figure 7. The Huang et al. (2019) RCB signature chain.

7.2. Non-Chameleon Hash-Based Redactable Blockchain

In this section, non-chameleon hash-based mechanisms have been thoroughly discussed and further analyzed as illustrated in Table 3 in terms of methods, security assumption, blockchain type, and finally redaction level type.

Table 3. Non-chameleon redaction approaches outlines.

Schema	Methods	Security Assumption	Setting (Blockchain Type)	Redaction Type
[95], (2019)	CVRA	N/A	Public	TL
[96], (2020)	CVRA	N/A	Private/Public	TL
[97], (2017)	MTRA	N/A	Private/Public	TL
[98], (2020)	MTRA	N/A	Private	TL
[99], (2019)	PRA	N/A	Public	BL
[100], (2020)	PRA	N/A	Public	BL

7.2.1. Consensus-Based Voting Redaction Approach (CVRA)

The CVRA leverages standard regulations to concur on the eventual condition of redacted transactions or blocks via on-chain voting. A redaction would only be approved when all honest nodes consent to the altered state. Past studies [101] employed the hard fork to amend the information history, resolve crucial security risks in codes, include additional functionality, restore previous transactions, and overcome negative hacking consequences. For instance, Ethereum applied the hard fork to restore past transactions in recovering substantial funds lost to "The DAO" attack. When a fork simultaneously appears in at least two block versions of the same height, an alternative chain will be produced. Although accidental forks could be resolved via identical consensus principles, attackers' chains are more difficult to be removed due to the longer series than the honest version (51% attack). Consequently, the hard fork adhering to the latest consensus principles would generate an irreversible sequence independent of the original chain, which is typically employed to divide blockchain cryptocurrency. For example, Bitcoin Gold and Bitcoin Cash were separated from Bitcoin [102]. Nonetheless, several limitations exist as the hard fork would not erase exact data history as both original and alternative chains would be maintained by respective nodes. Moreover, a high cost is incurred as the procedure consumes significant computation bandwidth.

Deuber et al. [95] suggested authorizing data redaction in the public blockchain with the CV (consensus voting) sequence expanding the block header structure to provide additional fields encompassing previous states or hash values, which enabled connectivity between redacted and subsequent blocks. All blocks are connected by two hash chains to represent the latest prevHash and prior hashes, with the following block constantly being attached to the previous block after referring to the prior condition. For multiple redactions, the old state consists of all states of all revisions. Moreover, the CV series adequately sustains consistency through honest nodes, holding a unified monitor of redaction activities by publicly validating before voting. The redaction request on a specific transaction would be announced in the peer-to-peer (P2P) network, where miners securing a subsequent block could vote for approving a valid revision while recording the redacted block hash in personal blocks. The revision would be approved when the editing request on a particular block received sufficient votes. Resultantly, honest nodes would update personal copies with the performed redaction. Scalability and efficiency degrees are low in the process owing to the significant amount of time required for approval of valid redactions. Marsalek et al. [103] formulated an amendable blockchain protocol in two hash chains, wherein the standard chain stored the original data, whereas the correction chain comprised numerous blocks to store correction data. The CV approach was also applied to perform a decentralized decision on requested corrections, in which a correction block consisting of the actual correction data replaced the redacted block with the same height in the standard chain after the voting session. Nevertheless, several challenges persist, including a long voting period and poor redaction efficiency. Meanwhile, Thyagarajan et al. [96] proposed the Reparo approach to conduct redactions on a repair layer in both public and private blockchains. Particularly, redaction would be sought through a repair message to specify the block hash, the object, and the chain state to be amended in an off-chain for validation. A repair witness would be voted on-chain by miners. Similar to the procedures in the CV series, the witness obtaining adequate votes would be approved to maintain transaction consistency by Reparo recording the original version and the approved repair message in data repositories. The repositories would also be announced and managed by the P2P network nodes. Furthermore, the repair message requires monetary expenses to be stored in the blockchain. Limited storage size, extended voting time, large bandwidth requirement, and computation overhead contributed to low process efficiency.

7.2.2. Meta-Transaction-Based Redaction Approach

The meta-transaction-based redaction approach (MTRA) is regulated by fiat and mate-transaction (mate-T), which is an alternative type to authorize redaction activities while preventing heavy cryptographic primitives and an extended voting period. Puddu et al. [97] recommended extracting coins from performed transactions, namely μchain, to enable blockchain data redaction by providing additional meta-T as mutants and extending transactions. The process commences with μchain separating the entire transactions, such as currency and smart contract deployment transactions, into mutable and immutable versions. The mutable version would be identified as active in each stage while the redaction would be subject to the stipulated rules delineating the redacted objects, the redactor, and the time window. The redactor would produce the meta-Tx to activate redaction procedures. When the extending transaction is permitted in an efficient off-chain voting method, the mutant transaction would substitute the active transaction. Subsequently, μchain encodes the mutable transaction to conceal the different transaction history. The secret key is distributed via the DPSS protocol [104] to fulfill the security threshold. Nevertheless, μchain violates consistency owing to high computation overhead and bandwidth in the enciphering procedure. Blockchain transparency is also diminished as limited verification operations would engender low audibility. Dorri et al. [98] established flexible memory management in the IoT, namely the memory-optimized and flexible blockchain (MOF-BC), to allow certain blockchain transaction deletion or summarization within a period in extensive networks. Concurrently, the deleted and previous transaction hashes would remain to guarantee record consistency through the redactor's signature to corroborate that personal editing legitimacy MOF-BC records essential data about the summarized transactions, including the MT root, timestamp, and order. Nevertheless, the CA verification is not robust owing to key management ineffectively preventing denial-of-service (DOS) and double spending attacks. Florian et al. [99] proposed the functionality preserving local erasure (FPLE) to delete improper information of transaction outputs in the local node storage as transaction outputs, including arbitrary information, would be improbably further employed, which enables rewriting in the local UTXO database. To cautiously delete the data, the FPLE adjusts the transaction output, which was the script public key (ScriptPubKey), to develop a removal database to store relevant data before performing redaction, while inspecting unedited inputs. The database could also validate the following transactions which refer to the amended records. After inspecting and validating, the original record would be substituted with the amended block before being completely removed. Meanwhile, drawbacks are present due to the unguaranteed credibility of the removal database source and a trusted entity is necessitated to manage the database, which violates decentralization and security principles.

7.2.3. Pruning-Based Redaction Approach (PRA)

The PRA is devised to be applied in specific situations in circumventing massive cryptographic primitives. Pyoung et al. [105] established two different PRAs to allow more edge node storage space in the IoT ecosystem. As edge nodes possessed finite amounts of storage and computing power in the IoT ecosystem, memory would be swiftly depleted and complete blockchain generation would be impeded. Accordingly, the blockchain Life-Time (LiTiChain) resolves the limitations by removing expired blocks, in which edge nodes achieve a distributed concord based on the BFT algorithm [106] through the corruption tolerance threshold (t) fulfills $n \geq 3t + 1$ for n nodes. Each LiTiChain block persists from the creation time until the stipulated end time and would be removed from the chain when the block lifetime expires. When an obsolete block consists of unexpired transactions, LiTichain restores and attaches the unexpired blocks to the latest block. Subsequently, LiTichain formulates a graph congruent with the block end-time order, wherein the child block possesses a shorter lifespan than that of the parent block. To maintain block consistency, every block header comprises two reference hashes, namely the prevHash of the former block and the Parent BlockHash of the parent block, in the graph. Although certain blocks would be erased, the remaining blocks remain in connection with one another. Removing expired blocks would minimize the percentage of honest blocks in the chain, while compromising the chain quality and common prefixes. Moreover, LiTiChain could not effectively sustain block consistency as the deletion process would negatively impact previous and upcoming transactions, which might pose security vulnerabilities. For instance, malevolent nodes could effortlessly perform double spending and transaction forgery on blocks legally removed from the series. LiTiChain is also vulnerable to the DoS attack as the removed block hashes would not be stored, therefore leading to low chain traceability and integrity. Matzutt et al. [100] suggested a snapshot method, which was CoinPrune, to minimize blockchain size in optimizing the storage and bandwidth criteria. CoinPrune generates a snapshot for constant intervals, such as every 10,000 blocks, which contains block headers and serialized UTXOs in a group of pruned blocks. The snapshot is openly broadcasted to be authenticated within a period. If the authentication process exceeds the stipulated threshold, the snapshot is regarded as genuine and would be subsequently accepted. Contrarily, the snapshot is considered invalid, and the pruning process would be postponed when the threshold was not satisfied. The latest joining nodes necessitate only the validated snapshot and several complete blocks to be concurrent with the system. CoinPrune could decrease the required device storage and synchronization period, although continuous pruning could render a costly overhead. As data are not entirely removed from the blockchain, CoinPrune enables pertinent resolutions in preventing certain nodes from storing improper information in local devices. The CoinPrune nodes generating the blockchain via the snapshot technique are comparable to the lightweight counterparts in requiring complete nodes to store the entire blockchain. As such, blockchain consistency and security are ensured when decentralization is facilitated with all nodes composed of a low network percentage. Resultantly, pruning is a feasible approach to performing redaction, which serves as a high-interest research topic.

8. Comparisons and Discussion

Blockchain data redaction studies have been started by using the [61] proposal, which has been the core for the aftercoming suggestions. This section is dedicated to comparing applied implementation challenges and security properties for previously analyzed literature respectively for chameleon based, as illustrated in Table 4, and non-chameleon redaction mechanisms in Table 5. The discussion below illustrates the advantages and disadvantages of the proposed methods in both the chameleon and non-chameleon approaches analyzed above.

Table 4. Chameleon redaction mechanism challenge comparisons.

Schema (Year)	Security Properties						Challenges										
	Chain Growth	**Chain Quality**	**Common Prefix**	**Validity**	**Block Consistency**	**Transaction Consistency**	**Scalability**	**Additional POW**	**Compatibility**	**Fined Grained**	**Secret Sharing**	**Accountably**	**Anonymity**	**Decentralization**	**Revocation**	**Anti-Collusion**	**Efficiency (High (H)/Medium (M), Poor (P))**
[61], **(2017)**							√				√					√	H
[62], **(2020)**					√		√				√					√	H
[63], **(2020)**					√		√				√					√	H
[64], **(2019)**	√	√	√		√		√		√	√						√	H
[65], **(2020)**	√	√	√	√	√	√	√		√	√		√				√	H
[66], **(2021)**	√	√	√	√	√	√	√		√	√		√				√	M
[67], **(2022)**	√	√	√	√	√	√	√	√	√	√		√		√		√	H
[68], **(2021)**	√	√	√	√	√	√	√		√	√		√	√	√	√		M
[69], **(2021)**	√	√	√	√	√	√			√		√	√	√		√	√	M
[70], **(2021)**	√	√	√	√	√	√	√	√		√		√			√	√	H
[71], **(2021)**	√	√	√	√	√	√	√		√	√				√		√	H
[72], **(2022)**	√	√	√	√	√	√	√		√	√		√		√		√	H
[73], **(2021)**	√	√	√	√	√	√	√		√	√		√					H
[74], **(2021)**	√	√	√	√	√	√	√		√	√		√				√	H
[75], **(2021)**					√		√		√		√						H
[76], **(2021)**	√	√	√	√	√			√	√		√			√		√	H
[77], **(2022)**	√	√	√	√	√	√	√	√	√		√	√					M
[78], **(2021)**	√	√	√	√	√	√	√		√	√		√		√		√	H
[79], **(2019)**	√	√	√	√	√	√					√	√		√			M
[80], **(2019)**	√	√	√	√	√	√					√						M
[81], **(2020)**	√	√	√	√	√						√						M
[82], **(2020)**	√	√	√	√	√	√					√	√		√			L
[83], **(2021)**	√	√	√	√	√	√					√	√	√	√			M
[84], **(2021)**	√	√	√	√	√	√					√	√		√		√	M
[85], **(2021)**	√	√	√	√	√	√	√	√	√		√						M
[86], **(2022)**	√	√	√	√	√	√					√	√			√	√	M

Table 5. Non-chameleon redaction approaches challenges comparisons.

	Security Properties						Challenges										
Schema (Year)	**Chain Growth**	**Chain Quality**	**Common Prefix**	**Validity**	**Block Consistency**	**Transaction Consistency**	**Scalability**	**Additional POW**	**Compatibility**	**Fined Grained**	**Secret Sharing**	**Accountably**	**Anonymity**	**Decentralization**	**Revocation**	**Anti-Collusion**	**Efficiency (High (H)/Medium (M), Poor (P))**
[95], (2019)	√	√	√	√	√	√		√						√			P
[96], (2020)	√	√	√	√	√	√		√	√					√			P
[97], (2017)	√	√	√	√				√						√			M
[98], (2020)				√		√		√									M
[99], (2019)								√	√								P
[100], (2020)	√	√	√	√	√			√	√								P

9. Open Challenges and Future Research Direction

Recent advancements in designing mutable blockchains are still immature, and consequent modeling redaction techniques require further investigations. This section is dedicated to enriching interested researchers' knowledge and shaping future research direction through open issues listed below:

- **Permissionless settings.** Permissioned and consortium blockchain concepts have adopted most of the proposed solutions where permissionless settings remain a challenge due to their openness and unrestricted characteristics and, hence, it is still unclear how to solve chameleon-based redaction in public blockchain. However, there are several suggestions for a non-chameleon method which proved to be inefficient due to time overhead. Non-chameleon is out of the scope of this research.
- **Redaction exceptional circumstances.** Redaction in total is a delicate case due to violating major blockchain immutability features. The balance requires special supplements, and careful procedures in order to succeed in redaction without any contradiction.
- **Revocation scalability consequences.** Centralization revoking mechanisms lack efficiency and effectiveness. Consequently, it overwhelmingly poses cost, time, communication overhead and even other equipment that might be required.
- **Trapdoor key exposure.** The trapdoor key is the main weakness of the chameleon hashing function; meanwhile, the entire collision being indistinguishable relies on its secrecy.
- **Revocation centralization authorities.** Studies in the current domain have employed central authorities to perform redaction; however, centralization is considered a drawback and decentralization is highly recommended.
- **Punishment methodologies.** Punishment has been replaced by accountability where violators remain safe without any further actions taken against them.
- **IOT-based redaction in blockchain performance.** Scalability is an ineffective factor due to a lack of blockchain redaction in the IoT domain wherein performance criteria are still low due to edge device limitations not being considered. According to authentic requirements, blockchain-based IOT/IIOT designs cannot provide networking resources. Current designs mainly suffer from a dispute among network resources, security, and redaction.

- **The balance between accountability and anonymity.** Privacy is a legal right to preserve identity concealment. Proposals mainly prioritize accountability over personal privacy. However, GDPR legislation strictly states that hidden identities must be maintained which is acutely confronted by the recent redaction mechanisms concepts.
- **Rewriting flexibility limiting.** Absolute power offered to the rewriting modifiers sabotaged data integrity and confined rewriting abilities to themselves.
- **Redactor granting privileges agreement.** The cooperation between owners and modifiers demands prearranged agreements.
- **Collusion resistance:** Colluding must be prevented due to illegal privilege accumulation among different colluded, revoked users who are willing to either access authentic data or try to falsely redact data.
- **Consistency preservation:** Redacted chain stability remains a major obstacle in designing any redaction mechanism. Removal operation performed after storing transaction/block state prior redaction in the current proposal's methods is as yet vulnerable due to verifications and transaction chain failure.

10. Conclusions

The current study aimed to investigate blockchain redaction concepts which emerge as an urgent need to legally violate immutability features due to blockchain technology employment in the wide business sector. Immutability ensures data persistence once approved and published. Current scenarios include adding illicit content that is essentially permitted according to the nature of public blockchain, where immutability becomes an issue that cannot be removed from the blockchain conceptual architecture. Mutability was suggested as an alternative solution to immutability to be applied under special circumstances and highly monitored atmosphere to remove malicious added content; however, it remains difficult to be achieved. The latter has attracted both academia and industry to invest in this research direction to increase blockchain growth effectiveness, wherein several suggested hypotheses were deployed in order to solve previously mentioned drawbacks without effecting either security or scalability. The state of the art is thoroughly analyzed in this paper as it has been classified based on chameleon-based and non-chameleon-based redactable blockchain mechanisms. Non-chameleon suffers mainly from poor efficiency, while chameleon based keeps better efficiency records but at certain domains, such as IoT sectors, its efficiency still needs more enhancements. Future directions must focus on chameleon hash based on its simplicity and it does not add any complication to the blockchain architecture; rather, no certain requirements need to be amended in blockchain infrastructure. However, key exposure, blockchain type and cryptographic methods used must be noted extensively in order to establish balanced solutions among traditionally conflicted security and efficiency.

Author Contributions: Visualization, S.M.A.A.; supervision, S.M.A.A. and H.F.H.; project administration, M.N.Y.; funding acquisition, M.N.Y.; writing—original draft, CS/USM.; writing review and editing, H.F.H.; all authors equally contributed to this work. All authors have read and agreed to the published version of the manuscript.

Funding: Fundamental Research Grant Scheme (FRGS) of grant no. FRGS/1/2019/ICT03/USM/02/1. Postgraduate funding scheme (PFS) of grant no. PFS/9/2019/PhD 19166/01/Art College/ Aliraqia University.

Informed Consent Statement: Not applicable.

Acknowledgments: The authors sincerely acknowledge and dedicate this paper to the support of the Ministry of Higher Education, Malaysia, and Universiti Sains Malaysia (USM). We would also like to express our deepest gratitude to the college of Art, Aliraqia University, Baghdad, Iraq, for their highly appreciated sponsorship.

Conflicts of Interest: The authors declare no conflict of interest.

Appendix A

Table A1. List of acronyms.

Abbreviation	Description
CH	Chameleon Hash Function
PKE	Public Key Encryption
ABE	Attribute Base Encryption
TCH	Threshold Chameleon Hash
DS	Digital Signature
ZR	Zero Knowledge Proof
TTCH	Threshold Trapdoor Chameleon Hash
MA_ABE	Multi Authority Attribute Base Encryption
TEE	Trusted Execution Environment
DGS	Digital Group Signature
TUCH	Time Update Chameleon Hash
CHCT	Chameleon Hash Changeable Trapdoor
MTRA	Meta Transaction Base Redactable Approach
CVRA	Consensus Voting Base Redaction Approach
PRA	Pruning Base redaction Approach
P	Permissioned Blockchain (Privat Blockchain)
PL	Permissionless Blockchain (Public Blockchain)
CP	Consortium Blockchain
BL	Block Level
TL	Transaction Level
TFL	Transaction Field Level
DLP	Discrete Logarithm Problem
CDHP	Computational Diffie Hillman Problem
SXDH	Standard Symmetric External Diffie Hillman
SIS	Small Integer Solution
DLIN	Decision Linear Assumption
DPDH	Decisional Bilinear Diffie Hillman

References

1. Ma, Z.; Jiang, M.; Gao, H.; Wang, Z. Blockchain for Digital Rights Management. *Future Gener. Comput. Syst.* **2018**, *89*, 746–764. [CrossRef]
2. Aitzhan, N.Z.; Svetinovic, D. Security and Privacy in Decentralized Energy Trading Through Multi-Signatures, Blockchain and Anonymous Messaging Streams. *IEEE Trans. Dependable Secur. Comput.* **2018**, *15*, 840–852. [CrossRef]
3. Khan, K.M.; Arshad, J.; Khan, M.M. Investigating Performance Constraints for Blockchain Based Secure E-Voting System. *Future Gener. Comput. Syst.* **2020**, *105*, 13–26. [CrossRef]
4. Khan, N.; Aljoaey, H.; Tabassum, M.; Farzamnia, A.; Sharma, T.; Tung, Y.H. Proposed Model for Secured Data Storage in Decentralized Cloud by Blockchain Ethereum. *Electronics* **2022**, *11*, 3686. [CrossRef]
5. Wang, B.; Li, Z. Healthchain: A Privacy Protection System for Medical Data Based on Blockchain. *Future Internet* **2021**, *13*, 247. [CrossRef]
6. Yiu, N.C.K. Toward Blockchain-Enabled Supply Chain Anti-Counterfeiting and Traceability. *Future Internet* **2021**, *13*, 86. [CrossRef]
7. Abidi, M.H.; Alkhalefah, H.; Umer, U.; Mohammed, M.K. Blockchain-Based Secure Information Sharing for Supply Chain Management: Optimization Assisted Data Sanitization Process. *Int. J. Intell. Syst.* **2021**, *36*, 260–290. [CrossRef]
8. Kapassa, E.; Themistocleous, M.; Christodoulou, K.; Iosif, E. Blockchain Application in Internet of Vehicles: Challenges, Contributions and Current Limitations. *Future Internet* **2021**, *13*, 313. [CrossRef]

9. Petroc Taylor. Worldwide Spending on Blockchain Solutions from 2017 to 2024. Available online: https://www.statista.com/statistics/800426/w.orldwide-blockchain-solutions-spending (accessed on 23 May 2022).
10. Hyla, T.; Pejaś, J. EHealth Integrity Model Based on Permissioned Blockchain. *Future Internet* **2019**, *11*, 76. [CrossRef]
11. Sanka, A.I.; Irfan, M.; Huang, I.; Cheung, R.C.C. A Survey of Breakthrough in Blockchain Technology: Adoptions, Applications, Challenges and Future Research. *Comput. Commun.* **2021**, *169*, 179–201. [CrossRef]
12. Narayanan, A.; Bonneau, J.; Felten, E.; Miller, A.; Goldfeder, S.; Clark, J. *Bitcoin and Cryptocurrency Technologies: A Comprehensive Introduction*; Princeton University Press: Princeton, NJ, USA, 2016; ISBN 978-0-691-17169-2.
13. Voulgaris, S.; Fotiou, N.; Siris, V.A.; Polyzos, G.C.; Jaatinen, M.; Oikonomidis, Y. Blockchain Technology for Intelligent Environments. *Future Internet* **2019**, *11*, 213. [CrossRef]
14. Przytarski, D.; Stach, C.; Gritti, C.; Mitschang, B. Query Processing in Blockchain Systems: Current State and Future Challenges. *Future Internet* **2022**, *14*, 1. [CrossRef]
15. Al-Abdullah, M.; Alsmadi, I.; AlAbdullah, R.; Farkas, B. Designing Privacy-Friendly Data Repositories: A Framework for a Blockchain That Follows the GDPR. *Digit. Policy Regul. Gov.* **2020**, *22*, 389–411. [CrossRef]
16. Zhang, D.; Le, J.; Mu, N.; Liao, X. An Anonymous Off-Blockchain Micropayments Scheme for Cryptocurrencies in the Real World. *IEEE Trans. Syst. Man Cybern. Syst.* **2020**, *50*, 32–42. [CrossRef]
17. Zheng, X.; Zhu, Y.; Si, X. A Survey on Challenges and Progresses in Blockchain Technologies: A Performance and Security Perspective. *Appl. Sci.* **2019**, *9*, 4731. [CrossRef]
18. El Ioini, N.; Pahl, C. A Review of Distributed Ledger Technologies. In Proceedings of the Lecture Notes in Computer Science (Including Subseries Lecture Notes in Artificial Intelligence and Lecture Notes in Bioinformatics), Valletta, Malta, 22–26 October 2018; Volume 11230.
19. Casino, F.; Politou, E.; Alepis, E.; Patsakis, C. Immutability and Decentralized Storage: An Analysis of Emerging Threats. *IEEE Access* **2020**, *8*, 4737–4744. [CrossRef]
20. Matzutt, R.; Hiller, J.; Henze, M.; Ziegeldorf, J.H.; Müllmann, D.; Hohlfeld, O.; Wehrle, K. A Quantitative Analysis of the Impact of Arbitrary Blockchain Content on Bitcoin. In Proceedings of the Lecture Notes in Computer Science (Including Subseries Lecture Notes in Artificial Intelligence and Lecture Notes in Bioinformatics), Nieuwpoort, The Netherlands, 26 February–2 March 2018; Volume 10957.
21. Jordan Pearson The Bitcoin Blockchain Could Be Used to Spread Malware, INTERPOL Says: (27 March 2015). Available online: https://www.vice.com/en/article/ezv8jn/the-bitcoin-blockchain-could-be-used-to-spread-malware-interpol-says (accessed on 23 May 2022).
22. Tziakouris, G. Cryptocurrencies—A Forensic Challenge or Opportunity for Law Enforcement? An INTERPOL Perspective. *IEEE Secur. Priv.* **2018**, *16*, 92–94. [CrossRef]
23. Schellinger, B.; Völter, F.; Urbach, N.; Sedlmeir, J. Yes, I Do: Marrying Blockchain Applications with GDPR. In Proceedings of the 55th Hawaii International Conference on System Sciences, Maui, HI, USA, 4–7 January 2022.
24. Bai, P.; Kumar, S.; Kumar, K.; Kaiwartya, O.; Mahmud, M.; Lloret, J. GDPR Compliant Data Storage and Sharing in Smart Healthcare System: A Blockchain-Based Solution. *Electronics* **2022**, *11*, 3311. [CrossRef]
25. Campanile, L.; Iacono, M.; Marulli, F.; Mastroianni, M. Designing a GDPR Compliant Blockchain-Based IoV Distributed Information Tracking System. *Inf. Process. Manag.* **2021**, *58*, 102511. [CrossRef]
26. Bigini, G.; Freschi, V.; Lattanzi, E. A Review on Blockchain for the Internet of Medical Things: Definitions, Challenges, Applications, and Vision. *Future Internet* **2020**, *12*, 208. [CrossRef]
27. Gatteschi, V.; Lamberti, F.; Demartini, C.; Pranteda, C.; Santamaría, V. Blockchain and Smart Contracts for Insurance: Is the Technology Mature Enough? *Future Internet* **2018**, *10*, 20. [CrossRef]
28. Zheng, J.; Dike, C.; Pancari, S.; Wang, Y.; Giakos, G.C.; Elmannai, W.; Wei, B. An In-Depth Review on Blockchain Simulators for IoT Environments. *Future Internet* **2022**, *14*, 182. [CrossRef]
29. Politou, E.; Alepis, E.; Patsakis, C. Forgetting Personal Data and Revoking Consent under the GDPR: Challenges and Proposed Solutions. *J. Cybersecur.* **2018**, *4*, tyy001. [CrossRef]
30. Nakamoto, S. Bitcoin: A Peer-to-Peer Electronic Cash System, October 2008. *Cited* 2008. Available online: https://bitcoin.org/bitcoin.pdf (accessed on 23 May 2022).
31. Swan, M. *Blockchain: Blueprint for a New Economy*; O'Reilly Media, Inc.: Sebastopol, CA, USA, 2015.
32. Koblitz, N.; Menezes, A.J. Cryptocash, Cryptocurrencies, and Cryptocontracts. *Des. Codes Cryptogr.* **2016**, *78*, 87–102. [CrossRef]
33. Workie, H. Distributed Ledger Technology: Implications of Blockchain for the Securities Industry. *J. Secur. Oper. Custody* **2017**, *9*, 347–355.
34. Chaudhary, K.; Fehnker, A.; Van De Pol, J.; Stoelinga, M. Modeling and Verification of the Bitcoin Protocol. In Proceedings of the Electronic Proceedings in Theoretical Computer Science, EPTCS, Suva, Fiji, 23 November 2015; Volume 196.
35. Tschorsch, F.; Scheuermann, B. Bitcoin and beyond: A Technical Survey on Decentralized Digital Currencies. *IEEE Commun. Surv. Tutor.* **2016**, *18*, 2084–2123. [CrossRef]
36. Christodoulou, K.; Iosif, E.; Inglezakis, A.; Themistocleous, M. Consensus Crash Testing: Exploring Ripple's Decentralization Degree in Adversarial Environments. *Future Internet* **2020**, *12*, 53. [CrossRef]
37. Chen, Y.; Guo, J.; Li, C.; Ren, W. FaDe: A Blockchain-Based Fair Data Exchange Scheme for Big Data Sharing. *Future Internet* **2019**, *11*, 225. [CrossRef]

38. Conte de Leon, D.; Stalick, A.Q.; Jillepalli, A.A.; Haney, M.A.; Sheldon, F.T. Blockchain: Properties and Misconceptions. *Asia Pac. J. Innov. Entrep.* **2017**, *11*, 286–300. [CrossRef]
39. Khanal, Y.P.; Alsadoon, A.; Shahzad, K.; Al-Khalil, A.B.; Prasad, P.W.C.; Rehman, S.U.; Islam, R. Utilizing Blockchain for IoT Privacy through Enhanced ECIES with Secure Hash Function. *Future Internet* **2022**, *14*, 77. [CrossRef]
40. Buccafurri, F.; De Angelis, V.; Lazzaro, S. A Blockchain-Based Framework to Enhance Anonymous Services with Accountability Guarantees. *Future Internet* **2022**, *14*, 243. [CrossRef]
41. Chen, Y.-C.; Chou, Y.-P.; Chou, Y.-C. An Image Authentication Scheme Using Merkle Tree Mechanisms. *Future Internet* **2019**, *11*, 149. [CrossRef]
42. Feng, H.; Wang, J.; Li, Y. An Efficient Blockchain Transaction Retrieval System. *Future Internet* **2022**, *14*, 267. [CrossRef]
43. Eyal, I.; Gencer, A.E.; Sirer, E.G.; Van Renesse, R. Bitcoin-NG: A Scalable Blockchain Protocol. In Proceedings of the 13th USENIX Symposium on Networked Systems Design and Implementation, NSDI 2016, Santa Clara, CA, USA, 16–18 March 2016.
44. Qu, Q.; Xu, R.; Chen, Y.; Blasch, E.; Aved, A. Enable Fair Proof-of-Work (PoW) Consensus for Blockchains in IoT by Miner Twins (MinT). *Future Internet* **2021**, *13*, 291. [CrossRef]
45. Bentov, I.; Gabizon, A.; Mizrahi, A. Cryptocurrencies without Proof of Work. In Proceedings of the Lecture Notes in Computer Science (Including Subseries Lecture Notes in Artificial Intelligence and Lecture Notes in Bioinformatics), Christ Church, Barbados, 26 February 2016; Volume 9604.
46. Gazi, P.; Kiayias, A.; Zindros, D. Proof-of-Stake Sidechains. In Proceedings of the IEEE Symposium on Security and Privacy, San Francisco, CA, USA, 19–23 May 2019; Volume 2019.
47. Bentov, I.; Lee, C.; Mizrahi, A.; Rosenfeld, M. Proof of Activity: Extending Bitcoin's Proof of Work via Proof of Stake. *Cryptol. Eprint Arch.* **2014**, *452*, 34–37.
48. Duan, S.; Reiter, M.K.; Zhang, H. BEAT: Asynchronous BFT Made Practical. In Proceedings of the ACM Conference on Computer and Communications Security, Toronto, ON, Canada, 15–19 October 2018.
49. Xiao, Y.; Zhang, N.; Lou, W.; Hou, Y.T. A Survey of Distributed Consensus Protocols for Blockchain Networks. *IEEE Commun. Surv. Tutor.* **2020**, *22*, 1432–1465. [CrossRef]
50. Antonopoulo, A.M. *Mastering Bitcoin Unlocking Digital Cryptocurrencies*; O'Reilly Media, Inc.: Sebastopol, CA, USA, 2014; Volume 9.
51. Habib, G.; Sharma, S.; Ibrahim, S.; Ahmad, I.; Qureshi, S.; Ishfaq, M. Blockchain Technology: Benefits, Challenges, Applications, and Integration of Blockchain Technology with Cloud Computing. *Future Internet* **2022**, *14*, 341. [CrossRef]
52. Makridakis, S.; Christodoulou, K. Blockchain: Current Challenges and Future Prospects/Applications. *Future Internet* **2019**, *11*, 258. [CrossRef]
53. Gong, Y.; van Engelenburg, S.; Janssen, M. A Reference Architecture for Blockchain-Based Crowdsourcing Platforms. *J. Theor. Appl. Electron. Commer. Res.* **2021**, *16*, 937–958. [CrossRef]
54. Yu, H.; Yang, Z.; Sinnott, R.O. Decentralized Big Data Auditing for Smart City Environments Leveraging Blockchain Technology. *IEEE Access* **2019**, *7*, 6288–6296. [CrossRef]
55. Cai, W.; Wang, Z.; Ernst, J.B.; Hong, Z.; Feng, C.; Leung, V.C.M. Decentralized Applications: The Blockchain-Empowered Software System. *IEEE Access* **2018**, *6*, 53019–53033. [CrossRef]
56. Xie, J.; Tang, H.; Huang, T.; Yu, F.R.; Xie, R.; Liu, J.; Liu, Y. A Survey of Blockchain Technology Applied to Smart Cities: Research Issues and Challenges. *IEEE Commun. Surv. Tutor.* **2019**, *21*, 2794–2830. [CrossRef]
57. Bodziony, N.; Jemioło, P.; Kluza, K.; Ogiela, M.R. Blockchain-Based Address Alias System. *J. Theor. Appl. Electron. Commer. Res.* **2021**, *16*, 1280–1296. [CrossRef]
58. Garay, J.; Kiayias, A.; Leonardos, N. The Bitcoin Backbone Protocol: Analysis and Applications. In Proceedings of the Lecture Notes in Computer Science (Including Subseries Lecture Notes in Artificial Intelligence and Lecture Notes in Bioinformatics), Sofia, Bulgaria, 26–30 April 2015; Volume 9057.
59. Kiayias, A.; Russell, A.; David, B.; Oliynykov, R. Ouroboros: A Provably Secure Proof-of-Stake Blockchain Protocol. In Proceedings of the Lecture Notes in Computer Science (Including Subseries Lecture Notes in Artificial Intelligence and Lecture Notes in Bioinformatics), Santa Barbara, CA, USA, 20–24 August 2017; Volume 10401.
60. Krawczyk, H.; Rabin, T. Chameleon Hashing and Signatures. *IACR Cryptol. Eprint Arch.* **1998**, *1998*. Available online: https://eprint.iacr.org/1998/010 (accessed on 23 May 2022).
61. Ateniese, G.; Magri, B.; Venturi, D.; Andrade, E.R. Redactable Blockchain—Or—Rewriting History in Bitcoin and Friends. In Proceedings of the 2nd IEEE European Symposium on Security and Privacy, EuroS and P 2017, Paris, France, 26–28 April 2017.
62. Khalili, M.; Dakhilalian, M.; Susilo, W. Efficient Chameleon Hash Functions in the Enhanced Collision Resistant Model. *Inf. Sci.* **2020**, *510*, 155–164. [CrossRef]
63. Derler, D.; Samelin, K.; Slamanig, D. Bringing Order to Chaos: The Case of Collision-Resistant Chameleon-Hashes. In Proceedings of the Lecture Notes in Computer Science (Including Subseries Lecture Notes in Artificial Intelligence and Lecture Notes in Bioinformatics), Edinburgh, UK, 4–7 May 2020; Volume 12110.
64. Derler, D.; Samelin, K.; Slamanig, D.; Striecks, C. Fine-Grained and Controlled Rewriting in Blockchains: Chameleon-Hashing Gone Attribute-Based. IACR Crypto ePrint Archive. 2019. 406p. Available online: https://eprint.iacr.org/2019/406 (accessed on 23 May 2022).
65. Tian, Y.; Li, N.; Li, Y.; Szalachowski, P.; Zhou, J. Policy-Based Chameleon Hash for Blockchain Rewriting with Black-Box Accountability. In Proceedings of the ACM International Conference Proceeding Series, Austin, TX, USA, 7–11 December 2020.

66. Xu, S.; Ning, J.; Ma, J.; Huang, X.; Deng, R.H. K-Time Modifiable and Epoch-Based Redactable Blockchain. *IEEE Trans. Inf. Forensics Secur.* **2021**, *16*, 4507–4520. [CrossRef]
67. Chen, X.; Gao, Y. CDEdit: A Highly Applicable Redactable Blockchain with Controllable Editing Privilege and Diversified Editing Types. *arXiv* **2022**, arXiv:2205.07054.
68. Panwar, G.; Vishwanathan, R.; Misra, S. ReTRACe: Revocable and Traceable Blockchain Rewrites Using Attribute-Based Cryptosystems. In Proceedings of the ACM Symposium on Access Control Models and Technologies, SACMAT, Virtual, 16–18 June 2021.
69. Jia, Y.; Sun, S.F.; Zhang, Y.; Liu, Z.; Gu, D. Redactable Blockchain Supporting Supervision and Self-Management. In Proceedings of the ASIA CCS 2021—Proceedings of the 2021 ACM Asia Conference on Computer and Communications Security, Hong Kong, China, 7–11 June 2021.
70. Xu, S.; Ning, J.; Ma, J.; Xu, G.; Yuan, J.; Deng, R.H. Revocable Policy-Based Chameleon Hash. In Proceedings of the Lecture Notes in Computer Science (Including Subseries Lecture Notes in Artificial Intelligence and Lecture Notes in Bioinformatics), Darmstadt, Germany, 4–8 October 2021; Volume 12972.
71. Zhang, Z.; Li, T.; Wang, Z.; Liu, J. Redactable Transactions in Consortium Blockchain: Controlled by Multi-Authority CP-ABE. In Proceedings of the Lecture Notes in Computer Science (Including Subseries Lecture Notes in Artificial Intelligence and Lecture Notes in Bioinformatics), Virtual Event, 1–3 December 2021; Volume 13083.
72. Ma, J.; Xu, S.; Ning, J.; Huang, X.; Deng, R.H. Redactable Blockchain in Decentralized Setting. *IEEE Trans. Inf. Forensics Secur.* **2022**, *17*, 1227–1242. [CrossRef]
73. Guo, L.; Wang, Q.; Yau, W.C. Online/Offline Rewritable Blockchain with Auditable Outsourced Computation. *IEEE Trans. Cloud Comput.* **2021**, *14*, 1–16. [CrossRef]
74. Hou, H.; Hao, S.; Yuan, J.; Xu, S.; Zhao, Y. Fine-Grained and Controllably Redactable Blockchain with Harmful Data Forced Removal. *Secur. Commun. Netw.* **2021**, *2021*, 3680359. [CrossRef]
75. Liu, L.; Tan, L.; Liu, J.; Xiao, J.; Yin, H.; Tan, S. Redactable Blockchain Technology Based on Distributed Key Management and Trusted Execution Environment. In Proceedings of the Communications in Computer and Information Science, Guangzhou, China, 5–6 August 2021; Volume 1490.
76. Wu, C.; Ke, L.; Du, Y. Quantum Resistant Key-Exposure Free Chameleon Hash and Applications in Redactable Blockchain. *Inf. Sci.* **2021**, *548*, 438–449. [CrossRef]
77. Matzutt, R.; Ahlrichs, V.; Pennekamp, J.; Karwacik, R.; Wehrle, K. A Moderation Framework for the Swift and Transparent Removal of Illicit Blockchain Content. In Proceedings of the IEEE International Conference on Blockchain and Cryptocurrency, ICBC 2022, Shanghai, China, 2–5 May 2022; Institute of Electrical and Electronics Engineers Inc.: New York, NY, USA, 2022.
78. Tian, Y.; Liu, B.; Li, Y.; Szalachowski, P.; Zhou, J. Accountable Fine-Grained Blockchain Rewriting in the Permissionless Setting. *arXiv* **2021**, arXiv:2104.13543.
79. Huang, K.; Zhang, X.; Mu, Y.; Wang, X.; Yang, G.; Du, X.; Rezaeibagha, F.; Xia, Q.; Guizani, M. Building Redactable Consortium Blockchain for Industrial Internet-of-Things. *IEEE Trans. Ind. Inform.* **2019**, *15*, 3670–3679. [CrossRef]
80. Huang, K.; Zhang, X.; Mu, Y.; Rezaeibagha, F.; Du, X.; Guizani, N. Achieving Intelligent Trust-Layer for Internet-of-Things via Self-Redactable Blockchain. *IEEE Trans. Ind. Inform.* **2020**, *16*, 2677–2686. [CrossRef]
81. Zhang, J.; Lu, Y.; Liu, Y.; Yang, X.; Qi, Y.; Dong, X.; Wang, H. Serving at the Edge: A Redactable Blockchain with Fixed Storage. In Proceedings of the Lecture Notes in Computer Science (Including Subseries Lecture Notes in Artificial Intelligence and Lecture Notes in Bioinformatics), Guangzhou, China, 23–25 September 2020; Volume 12432.
82. Lv, W.; Wei, S.; Li, S.; Yu, M. Verifiable Blockchain Redacting Method for a Trusted Consortium with Distributed Chameleon Hash Authority. In Proceedings of the Lecture Notes in Computer Science (Including Subseries Lecture Notes in Artificial Intelligence and Lecture Notes in Bioinformatics), Dallas, TX, USA, 11–13 December 2020; Volume 12575.
83. Huang, K.; Zhang, X.; Mu, Y.; Rezaeibagha, F.; Du, X. Scalable and Redactable Blockchain with Update and Anonymity. *Inf. Sci.* **2021**, *546*, 25–41. [CrossRef]
84. Gao, W.; Chen, L.; Rong, C.; Liang, K.; Zheng, X.; Yu, J. Security Analysis and Improvement of a Redactable Consortium Blockchain for Industrial Internet-of-Things. *Comput. J.* **2022**, *65*, 2430–2438. [CrossRef]
85. Zhang, C.; Ni, Z.; Xu, Y.; Luo, E.; Chen, L.; Zhang, Y. A Trustworthy Industrial Data Management Scheme Based on Redactable Blockchain. *J. Parallel Distrib. Comput.* **2021**, *152*, 167–176. [CrossRef]
86. Wei, J.; Zhu, Q.; Li, Q.; Nie, L.; Shen, Z.; Choo, K.K.R.; Yu, K. A Redactable Blockchain Framework for Secure Federated Learning in Industrial Internet of Things. *IEEE Internet Things J.* **2022**, *9*, 17901–17911. [CrossRef]
87. Camenisch, J.; Derler, D.; Krenn, S.; Pöhls, H.C.; Samelin, K.; Slamanig, D. Chameleon-Hashes with Ephemeral Trapdoors and Applications to Invisible Sanitizable Signatures. In Proceedings of the Lecture Notes in Computer Science (Including Subseries Lecture Notes in Artificial Intelligence and Lecture Notes in Bioinformatics), Amsterdam, The Netherlands, 28–31 March 2017; Volume 10175.
88. Goyal, V.; Pandey, O.; Sahai, A.; Waters, B. Attribute-Based Encryption for Fine-Grained Access Control of Encrypted Data. In Proceedings of the ACM Conference on Computer and Communications Security, Taipei, Taiwan, 21–24 March 2006.
89. Rivest, R.L.; Shamir, A.; Adleman, L. A Method for Obtaining Digital Signatures and Public-Key Cryptosystems. *Commun. ACM* **1978**, *21*, 120–126. [CrossRef]

90. Boldyreva, A.; Goyal, V.; Kumart, V. Identity-Based Encryption with Efficient Revocation. In Proceedings of the ACM Conference on Computer and Communications Security, Alexandria, VA, USA, 27–31 October 2008.
91. Agrawal, S.; Chase, M. FAME: Fast Attribute-Based Message Encryption. In Proceedings of the ACM Conference on Computer and Communications Security, Dallas, TX, USA, 30 October–3 November 2017.
92. Liu, J.K.; Wei, V.K.; Wong, D.S. Linkable Spontaneous Anonymous Group Signature for Ad Hoc Groups (Extended Abstract). *Lect. Notes Comput. Sci. (Incl. Subser. Lect. Notes Artif. Intell. Lect. Notes Bioinform.)* **2004**, *3108*, 325–335. [CrossRef]
93. Sabt, M.; Achemlal, M.; Bouabdallah, A. Trusted Execution Environment: What It Is, and What It Is Not. In Proceedings of the 14th IEEE International Conference on Trust, Security and Privacy in Computing and Communications, TrustCom 2015, Helsinki, Finland, 20–22 August 2015; Volume 1.
94. Ateniese, G.; Chou, D.H.; Medeiros, B.D.; Tsudik, G. Sanitizable Signatures. In Proceedings of the n European Symposium on Research in Computer Security, Milan, Italy, 12–14 September 2005; Springer: Berlin/Heidelberg, Germany, 2005; pp. 159–177.
95. Deuber, D.; Magri, B.; Thyagarajan, S.A.K. Redactable Blockchain in the Permissionless Setting. In Proceedings of the Proceedings—IEEE Symposium on Security and Privacy, San Francisco, CA, USA, 19–23 May 2019.
96. Thyagarajan, S.A.K.; Bhat, A.; Magri, B.; Tschudi, D.; Kate, A. Reparo: Publicly Verifiable Layer to Repair Blockchains. In Proceedings of the Lecture Notes in Computer Science (Including Subseries Lecture Notes in Artificial Intelligence and Lecture Notes in Bioinformatics), Virtual Event, 1–5 March 2021; Volume 12675.
97. Puddu, I.; Zurich, E.; Dmitrienko, A.; Capkun, S. Mchain: How to Forget without Hard Forks. *arXiv* **2017**. Available online: https://eprint.iacr.org/2017/106.pdf (accessed on 23 May 2022).
98. Dorri, A.; Kanhere, S.S.; Jurdak, R. MOF-BC: A Memory Optimized and Flexible Blockchain for Large Scale Networks. *Future Gener. Comput. Syst.* **2019**, *92*, 357–373. [CrossRef]
99. Florian, M.; Henningsen, S.; Beaucamp, S.; Scheuermann, B. Erasing Data from Blockchain Nodes. In Proceedings of the 4th IEEE European Symposium on Security and Privacy Workshops, EUROS and PW 2019, Stockholm, Sweden, 17–19 June 2019.
100. Matzutt, R.; Kalde, B.; Pennekamp, J.; Drichel, A.; Henze, M.; Wehrle, K. How to Securely Prune Bitcoin's Blockchain. In Proceedings of the IFIP Networking 2020 Conference and Workshops, Networking 2020, Paris, France, 22–26 June 2020.
101. Yiu, N.C.K. An Overview of Forks and Coordination in Blockchain Development. *arXiv* **2021**, arXiv:2102.10006.
102. Webb, N. A Fork in the Blockchain: Income Tax and the Bitcoin/Bitcoin Cash Hard Fork. *North Carol. J. Law Technol.* **2018**, *19*, 283.
103. Marsalek, A.; Zefferer, T. A Correctable Public Blockchain. In Proceedings of the 2019 18th IEEE International Conference on Trust, Security and Privacy in Computing and Communications/13th IEEE International Conference on Big Data Science and Engineering, TrustCom/BigDataSE 2019, Rotorua, New Zealand, 5–8 August 2019.
104. Ostrovsky, R.; Yung, M. How to Withstand Mobile Virus Attacks (Extended Abstract). In Proceedings of the Tenth Annual ACM Symposium on Principles of Distributed Computing—PODC '91, Montreal, QC, Canada, 19–21 August 1991.
105. Pyoung, C.K.; Baek, S.J. Blockchain of Finite-Lifetime Blocks with Applications to Edge-Based IoT. *IEEE Internet Things J.* **2020**, *7*, 2102–2116. [CrossRef]
106. Xu, X.; Zhu, D.; Yang, X.; Wang, S.; Qi, L.; Dou, W. Concurrent Practical Byzantine Fault Tolerance for Integration of Blockchain and Supply Chain. *ACM Trans. Internet Technol.* **2021**, *21*, 1–17. [CrossRef]

MDPI
St. Alban-Anlage 66
4052 Basel
Switzerland
www.mdpi.com

Future Internet Editorial Office
E-mail: futureinternet@mdpi.com
www.mdpi.com/journal/futureinternet

www.ingramcontent.com/pod-product-compliance
Lightning Source LLC
Chambersburg PA
CBHW061506220926
51850CB00004B/184
9783036587721